UNITEXT for Physics

UNITEXT for Physics series, formerly UNITEXT Collana di Fisica e Astronomia, publishes textbooks and monographs in Physics and Astronomy, mainly in English language, characterized of a didactic style and comprehensiveness. The books published in UNITEXT for Physics series are addressed to graduate and advanced graduate students, but also to scientists and researchers as important resources for their education, knowledge and teaching.

More information about this series at http://www.springer.com/series/13351

Paul R. Berman

Introductory Quantum Mechanics

A Traditional Approach Emphasizing Connections with Classical Physics

Paul R. Berman
University of Michigan
Ann Arbor, MI, USA

ISSN 2198-7882 ISSN 2198-7890 (electronic)
UNITEXT for Physics
ISBN 978-3-319-88628-2 ISBN 978-3-319-68598-4 (eBook)
https://doi.org/10.1007/978-3-319-68598-4

Printed on acid-free paper

This Springer imprint is published by Springer Nature
The registered company is Springer International Publishing AG
The registered company address is: Gewerbestrasse 11, 6330 Cham, Switzerland

I would like to dedicate this book to
Debra Berman, my wife of 30+ years,
for her support and positive outlook on life.

Preface

This book is based on junior and senior level undergraduate courses that I have given at both New York University and the University of Michigan. You might ask, in heavens name, why anyone would want to write yet another introductory text on quantum mechanics. And you would not be far off base with this assessment. There are many excellent introductory quantum mechanics texts. Moreover, with the material available on the internet, you can access almost any topic of your choosing. Therefore, I must agree that there are probably no compelling reasons to publish this text. I have undertaken this task mainly at the urging of my students, who felt that it would be helpful to students studying quantum mechanics.

For the most part, the approach taken is a traditional one. I have tried to emphasize the relationship of the quantum results with those of classical mechanics and classical electromagnetism. In this manner, I hope that students will be able to gain physical insight into the nature of the quantum results. For example, in the study of angular momentum, you will see that the absolute squares of the spherical harmonics can be given a relatively simple physical interpretation. Moreover, by using the effective potential in solving problems with spherical symmetry, I am able to provide a physical interpretation of the probability distributions associated with the eigenfunctions of such problems and to interpret the structures seen in scattering cross sections. I also try to stress the time-dependent aspects of problems in quantum mechanics, rather than focus simply on the calculation of eigenvalues and eigenfunctions.

The book is intended to be used in a year-long introductory course. Chapters 1–13 or 1–14 can serve as the basis for a one-semester course. I do not introduce Dirac notation until Chap. 11. I do this so students can try to master the wave function approach and its implications before engaging in the more abstract Dirac formalism. Dirac notation is developed in the context of a more general approach in which different representations, such as the position and momentum representations, appear on an equal footing. Most topics are treated at a level appropriate to an undergraduate course. Some topics, however, such as the hyperfine interactions described in the appendix of Chap. 21, are at a more advanced level. These are included for reference purposes, since they are not typically included in undergraduate (or graduate) texts. There is a web site for this book, http://www-personal.umich.edu/~pberman/qmbook.html, that contains an Errata, Mathematica subroutines, and some additional material.

The problems form an integral part of the book. Many are standard problems, but there are a few that might be unique to this text. Quantum mechanics is a difficult subject for beginning students. I often tell them that falling behind in a course such as this is a disease from which it is difficult to recover. In writing this book, my foremost task has been to keep the students in mind. On the other hand, I know that no textbook is a substitute for a dedicated instructor who guides, excites, and motivates students to understand the material.

I would like to thank Bill Ford, Aaron Leanhardt, Peter Milonni, Michael Revzen, Alberto Rojo, and Robin Shakeshaft for their insightful comments. I would also like to acknowledge the many discussions I had with Duncan Steel on topics contained in this book. Finally, I am indebted to my students for their encouragement and positive (as well as negative) feedback over the years. I am especially grateful to the Fulbright foundation for having provided the support that allowed me to offer a course in quantum mechanics to students at the College of Science and Technology at the University of Rwanda. My interactions with these students will always remain an indelible chapter of my life.

Ann Arbor, MI, USA Paul R. Berman

Contents

Chapter 1
Introduction

As a science or engineering major, you are about to embark on what may be the most important course of your undergraduate career. Quantum mechanics is the foundation on which our current picture of the structure of matter is built. For students, an introductory course in quantum mechanics can be difficult and frustrating. When you studied Newtonian physics, it was easy to envision experiments involving the motion of particles moving under the influence of forces. Electromagnetism and optics were a bit more abstract once the concept of fields was introduced, but you are familiar with many optical effects such as colors of thin films, the rainbow, and diffraction from single or double slits. In quantum mechanics, it is more difficult conceptually to understand what is going on since you have very little day to day experience with the wave nature of matter. However, you can exploit your knowledge of both classical mechanics and optics to get a better feel for quantum mechanics.

I often begin a course in quantum mechanics asking students "Why and when is quantum mechanics needed to explain the dynamics of a particle moving in a potential. In other words, when does a classical particle description fail?" To understand and appreciate many aspects of quantum mechanics, you should always have this and a few other questions in the back of your mind. Question two might be "How is the quantum mechanical solution of a given problem related to the corresponding problem in classical mechanics?" Question three could be "Is there a problem in optics with which I am familiar that can shed light on a given problem in quantum mechanics?" You can add other questions to this list, but these are a good start.

Another key point in the study of quantum mechanics is not to lose track of the physics. There are many special functions, such as Hermite and Laguerre polynomials, that emerge from solutions of the Schrödinger equation, and the algebra can get a little complicated. As long as you remind yourself of the physical nature of the solution rather than the mathematical details, you will be in great

P.R. Berman, *Introductory Quantum Mechanics*, UNITEXT for Physics,
https://doi.org/10.1007/978-3-319-68598-4_1

shape. The reason for this is that most of the solutions share many common features. The specific form of the solutions may change, but the overall qualitative nature of the solutions is remarkably similar for many different problems. Moreover, with the availability of symbolic manipulation programs such as Mathematica, Maple, or Matlab, it is now easy to plot and evaluate any of these functions using a few keystrokes.

To help introduce the subject matter, I will present a very broad, qualitative overview of the way in which quantum mechanics was born, a birth that took about 25–30 years. I will not worry about historical accuracy here, but simply try to give you a reasonable picture of the manner in which it became appreciated that a wave description of matter was needed in certain limits.

1.1 Electromagnetic Waves

Quantum mechanics is a wave theory. Since the wave properties of matter have many similarities to the wave properties of electromagnetic radiation, it won't hurt to review some of the fundamental properties of the electromagnetic field. The possible existence of electromagnetic waves followed from Maxwell's equations. By combining the equations of electromagnetism, Maxwell arrived at a wave equation, in which the wave propagation speed in vacuum was equal to $v = 1/\sqrt{\epsilon_0\mu_0}$, where ϵ_0 is the permittivity and μ_0 the permeability of free space. Since these were known quantities in the nineteenth century, it was a simple matter to calculate v, which turned out to equal the speed of light. This result led Maxwell to conjecture that light was an electromagnetic phenomenon.

All electromagnetic waves travel in vacuum with speed $c = 2.99792458 \times 10^8$ m/s, now *defined* as the speed of light. What distinguishes one type of electromagnetic wave from another is its wavelength λ or frequency f, which are related by

$$c = f\lambda. \tag{1.1}$$

Instead of characterizing a wave by its frequency (which has units of cycles per second or Hz) and wavelength, we can equally well specify the *angular frequency*, $\omega = 2\pi f$, (which has units of radians per second or s^{-1}) and the magnitude of the *propagation vector* (or *wave vector*), defined by $k = 2\pi/\lambda$. With these definitions, Eq. (1.1) can be replaced by

$$\omega = ck, \tag{1.2}$$

which is known as a *dispersion relation*, relating frequency to wave vector. For electromagnetic radiation in vacuum the dispersion relation is linear.

The *source* of electromagnetic waves is oscillating or accelerating charges, which give rise to propagating electric and magnetic fields. The wave equation for the electric field vector $\mathbf{E}(\mathbf{R}, t)$ at position $\mathbf{R}$ at time t in vacuum is

$$\nabla^2\mathbf{E}(\mathbf{R}, t) = \frac{1}{c^2}\frac{\partial^2\mathbf{E}(\mathbf{R}, t)}{\partial t^2}. \tag{1.3}$$

The simplest solution of this equation is also the most important, since it is a building block solution from which all other solutions can be constructed. The building block solution of the wave equation is the infinite, monochromatic, plane wave solution, having an electric field vector given by

$$\mathbf{E}(\mathbf{R}, t) = \boldsymbol{\epsilon} E_0 \cos(\mathbf{k}\cdot\mathbf{R} - \omega t), \tag{1.4}$$

where E_0 is the field amplitude and the *polarization* of the field is specified by a unit vector $\boldsymbol{\epsilon}$ that is perpendicular to the propagation vector $\mathbf{k}$ of the field. There are two independent field polarizations possible for each propagation vector. The fact that $\boldsymbol{\epsilon}\cdot\mathbf{k} = 0$ follows from the requirement that $\nabla\cdot\mathbf{E}(\mathbf{R},t) = 0$ in vacuum. The field (1.4) corresponds to a wave that is infinite in extent and propagates in the $\mathbf{k}$ direction. Of course, no such wave exists in nature since it would uniformly fill all space.

As a consequence of the linearity of the wave equation, the sum of any two solutions of the wave equation is also a solution. This is known as a *superposition principle*. It is not difficult to visualize how waves add together. If, at the same time you are putting your finger in and out of water in a lake, someone else is also putting their finger in and out of the water, the resulting wave results from the actions of *both* your fingers. The important thing to remember is that it is the *displacements* or *amplitudes* of the waves that add, not their intensities. The *intensity* of a wave is proportional to the square of the wave amplitude.

For the time being, let us consider two waves having the same frequency and the same amplitude. If the two waves propagate in opposite directions, the total electric field vector is

$$\begin{aligned}\mathbf{E}(\mathbf{R}, t) &= \boldsymbol{\epsilon} E_0 [\cos(\mathbf{k}\cdot\mathbf{R} - \omega t) + \cos(-\mathbf{k}\cdot\mathbf{R} - \omega t)] \\ &= 2\boldsymbol{\epsilon} E_0 \cos(\omega t)\cos(\mathbf{k}\cdot\mathbf{R}),\end{aligned} \tag{1.5}$$

a *standing wave* pattern. If you plot the wave amplitude at several different times as a function of position, you will find that there is an envelope for the wave that is *fixed* in space—the wave "stands" there and oscillates within the envelope. The points of zero amplitude are called *nodes* of the field and the maxima (or minima) are called *antinodes* of the field. For a standing wave field in the x direction that is confined between two parallel mirrors separated by a distance L, the standing wave pattern will "fit in" for wavelengths equal to $2L$, L, $2L/3$, $L/2$, etc., as shown in

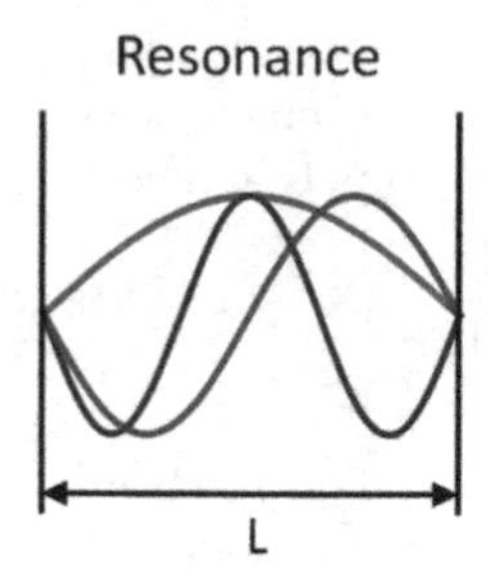

Fig. 1.1 Resonance involving standing waves with clamped endpoints

Fig. 1.1. In this limit, the tangential component of the electric field vanishes at the mirrors, as required by the boundary conditions on the field. The condition

$$\lambda_n = 2L/n; \qquad n = 1, 2, 3, \ldots. \tag{1.6}$$

is known as a *resonance* condition.

1.1.1 Radiation Pulses

To get a *pulse* of radiation, it is necessary to add together monochromatic waves having a *continuous* distribution of frequencies. It is *easy* to create a pulse of radiation. Simply turn a laser or other light source on and off and you have created a pulse. Why worry about the frequencies contained in the pulse? It turns out that it is important, even central, to understand this concept if you are going to have some idea of what quantum mechanics is about.

Let us assume the pulse is propagating in the x direction and has a duration Δt corresponding to a spatial width $\Delta x = c\Delta t$. Clearly this cannot be a monochromatic wave since a monochromatic wave is not localized. Instead, the pulse must be a superposition of waves having a range of frequencies Δf centered around some average frequency f_0. Using the theory of Fourier analysis it is possible to show that Δf and Δt are related by

$$\Delta f \Delta t \approx \frac{1}{2\pi}. \tag{1.7}$$

The quantity Δf is known as the *spectral width* of the pulse. If the frequency of the pulse is known *precisely*, as in a monochromatic wave, there is no *range* of frequencies and $\Delta f = 0$. In general there is a central frequency f_0 in a pulse and a range of frequencies Δf about that central frequency as shown in Fig. 1.2. If $\Delta f \ll f_0$, then the frequency is pretty well-defined and the field is said to be *quasi-monochromatic*. The field associated with the light of a green laser pointer

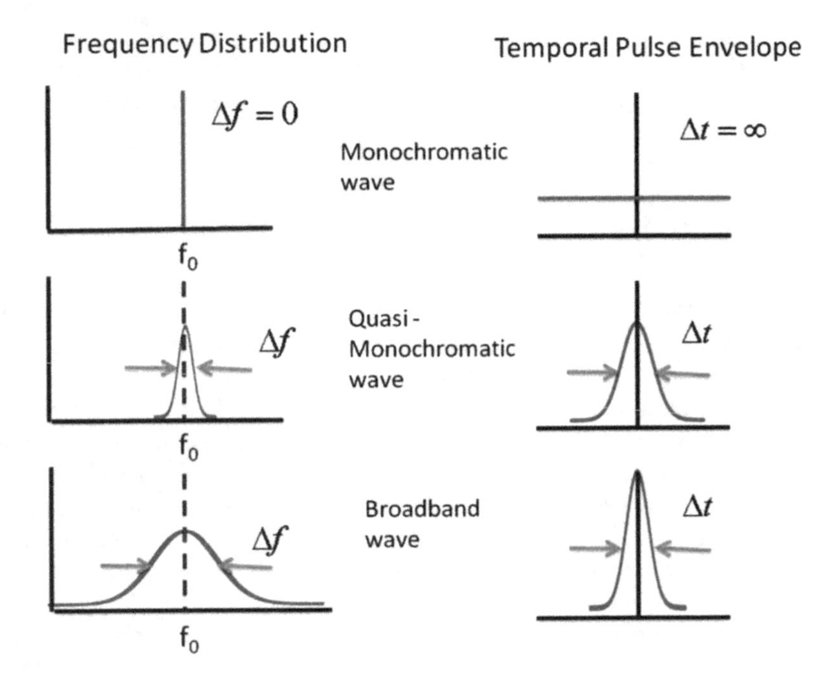

Fig. 1.2 Illustration of the equation $\Delta f \Delta t \approx 1/2\pi$. As the spectral frequency distribution of a pulse broadens, the temporal pulse width narrows

and most other laser fields are quasi-monochromatic, as are the fields from neon discharge tubes. Radio waves are also quasi-monochromatic, the central frequency is the frequency of the station (about 1000 kHz for AM and 100 MHz for FM) broadcasting the signal. On the other hand, if Δf is comparable with f_0, as in a light bulb, the source is said to be incoherent or broadband.

For example, consider a pulse of green laser light that has an ns (10^{-9} s) duration, $\Delta t = 1$ ns. The spatial extent of this pulse is 3×10^8 m/s · 10^{-9} s = 30 cm. The frequency range in the pulse is given by

$$\Delta f = 1/\left(2\pi 10^{-9}\ \text{s}\right) = 1.6 \times 10^8\ \text{Hz}. \tag{1.8}$$

Since the light from the green laser pointer has a central wavelength of 532 nm, it has a central frequency of about $f_0 = 5.6 \times 10^{14}$ Hz. As a consequence, $\Delta f \ll f_0$ and this pulse is quasi-monochromatic. That is, a one nanosecond pulse of this light appears to have a single frequency or color if you look at it. On the other hand, if you try to make a one femtosecond (fs) pulse with this light (1 fs = 10^{-15} s) which has a spatial extent of 300 nm, then

$$\Delta f = 1/\left(2\pi 10^{-15}\ \text{s}\right) = 1.6 \times 10^{14}\ \text{Hz}, \tag{1.9}$$

which is comparable to the central frequency. As a result this is a broadband pulse which no longer appears green, but closer to white. Note that the spatial extent of the pulse, 300 nm is comparable to the central wavelength. This is an equivalent test for broadband radiation. If the spatial extent is comparable to the central wavelength, the radiation is broadband; if it is much *larger* than the central wavelength, it is quasi-monochromatic.

1.1.2 Wave Diffraction

The final topic that we will need to know something about is wave diffraction. Diffraction is a purely wave phenomenon, but sometimes waves don't appear to diffract at all. If you shine a laser beam at a tree or a pencil, it will reflect off of these and not bend around them. Moreover, if you shine a laser beam through an open door, it will not bend around into the hallway. *Diffraction is only important when waves meet obstacles (including apertures or openings) that are comparable to or smaller than the central wavelength of the waves.* This is actually true for the laser beam itself!—it will stay a beam of approximately constant diameter only if its diameter is much greater than its wavelength, otherwise, it will spread significantly.

I can make this somewhat quantitative. Imagine there is a circular opening in a screen having diameter a through which a laser beam passes, or that it passes through a slit having width a, or that a laser beam having diameter a propagates in vacuum with no apertures present. In each of these cases, there is diffraction that leads to a spreading of the beam by an amount

$$\sin\theta \approx \frac{\lambda}{a}, \tag{1.10}$$

where θ is the angle with which the beam spreads. If $\lambda/a \ll 1$, the spreading angle is very small and diffraction is relatively unimportant. *This is the limit where the wave acts as a particle, moving on straight lines.* On the other hand, when a gets comparable to a wavelength, the spreading of the beam becomes significant. If $\lambda/a > 1$, the spreading is over all angles. For example, if a pinhole is illuminated with light, you can see the diffracted light at any angle on the other side of the pinhole.

Even if the diffraction angle is small, the effects can get large over long distances. If a laser beam having diameter 1.0 cm and central wavelength 600 nm is sent to the moon, the diffraction angle is

$$\theta \approx \frac{\lambda}{a} = \frac{6\times10^{-7}\,\text{m}}{10^{-2}\,\text{m}} = 6\times10^{-5}. \tag{1.11}$$

By the time it gets to the moon the spot size diameter d is

$$d \approx R_{\text{EM}}\theta = 3.8\times10^{8}\,\text{m}\times6\times10^{-5} = 2.3\times10^{4}\,\text{m} = 23\,\text{km}, \tag{1.12}$$

a lot larger than 1 cm! In this equation R_{EM} is the Earth-moon distance.

The bottom line on diffraction. If you try to confine a wave to a distance less than or comparable to its wavelength, it will diffract significantly.

1.2 BlackBody Spectrum: Origin of the Quantum Theory

Thermodynamics was developed in the nineteenth century and involves the study of the properties of vapors, liquids, and solids in terms of such parameters as temperature, pressure, volume, density, conductivity, etc. At the end of the nineteenth century, the theory of statistical mechanics was formulated in which the properties of systems of particles are explained in terms of their statistical properties. It was shown that the two theories of thermodynamics and statistical mechanics were consistent—one could explain macroscopic properties of vapors, liquids, and solids by considering them to be made up of a large number of particles following Newton's laws. One result of this theory is that for a system of particles (or waves) in thermodynamic equilibrium at a temperature T, each particle or wave has $(1/2)k_BT$ of energy for each "degree of freedom." This is known as the *equipartition theorem*.

The quantity $k_B = 1.38 \times 10^{-23}$ J/°K is Boltzmann's constant and T is the absolute temperature in degrees Kelvin. A "degree of freedom" is related to an independent motion a particle can have (translation, rotation, vibration, etc.). For a free particle (no forces acting on it), there are three degrees of freedom, one for each independent direction of motion. For a transverse wave, there are two degrees of freedom, corresponding to the two possible independent directions for the polarization of the wave.

At the beginning of the twentieth century there was a problem in trying to formulate the theory of a *blackbody*. A blackbody is an object that, in equilibrium, absorbs and emits radiation at the same rate. At a given temperature, the radiation emitted by a blackbody is spread over a wide range of frequencies, but the peak intensity occurs at a wavelength $\lambda_{\max}$ governed by Wien's law,

$$\lambda_{\max} T = 2.90 \times 10^{-3} \text{ m} \cdot^{\circ} \text{K}, \tag{1.13}$$

where T is the temperature in degrees Kelvin. For example, the surface of the sun is about 5000 °K. If the sun is approximated as a blackbody, then $\lambda_{\max} \approx 580$ nm is in the yellow part of the spectrum. As an object is heated it emits radiation at higher frequencies; an object that is "blue" hot is hotter than an object that is "red" hot.

The experimental curve for emission from a blackbody as a function of wavelength is shown in Fig. 1.3. To try to explain this result, Rayleigh and Jeans considered a model for a blackbody consisting of a cavity with a small hole in it. Standing wave patterns of radiation fill the cavity and each standing wave has k_BT of energy associated with it ($k_BT/2$ for each independent polarization of the field). A standing wave pattern or *mode* is characterized by the number of (half) wavelengths in the x, y, and z directions. There is a maximum wavelength $\lambda = 2L$ in each direction, but there is *no* limit on the minimum wavelength. Since the number

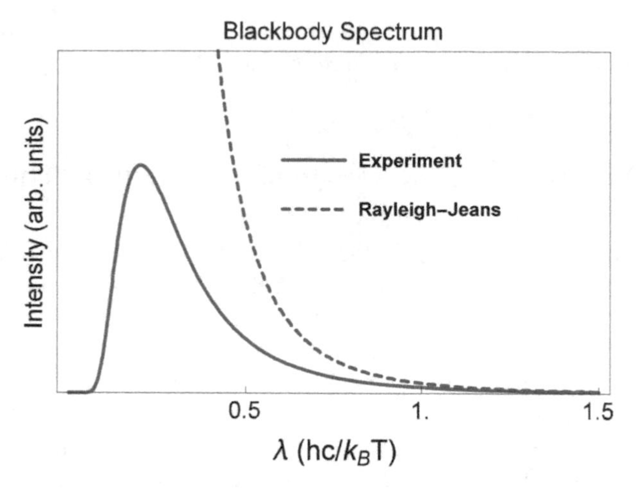

Fig. 1.3 Intensity per unit wavelength (in arbitrary units) as a function of wavelength [in units of (hc/k_BT)] for a blackbody. The maximum occurs at $\lambda_{\text{max}} \approx hc/(5k_BT)$

of small wavelength (high frequency) modes that can fit is arbitrarily large and since each mode has k_BT of energy, the amount of energy needed to reach equilibrium becomes infinite. Another way of saying this is that the energy density (energy per unit frequency interval) approaches infinity as $\lambda \sim 0$ or $f \sim \infty$. The fact that it does not take an infinite amount of energy to bring a blackbody into thermal equilibrium or that the energy density does not approach infinity at high frequencies is known as the "ultraviolet catastrophe"; theory and experiment were in disagreement for high or "ultraviolet" frequencies (smaller wavelengths) (see Fig. 1.3).

Planck tried to overcome this difference by making an *ad hoc* hypothesis.[1] In effect, he said, that to excite a mode of frequency f, a specific *minimum* amount of energy hf (h is defined below) is needed that must be provided in an *all or nothing* fashion by electrons oscillating in the cavity walls. This went against classical ideas that energy can be fed continuously to build up oscillation in a given mode. By adopting this theory, he found that he could explain the experimental data if he chose

$$h = 6.63 \times 10^{-34}\ \text{J} \cdot \text{s} = 4.14 \times 10^{-15}\ \text{eV} \cdot \text{s}, \tag{1.14}$$

which is now known as *Planck's constant* [recall that one eV (electron-volt) is equal to 1.60×10^{-19} J (Joules)]. The quantum theory was born. Some details of the Planck solution are given in the Appendix.

In thermal equilibrium at temperature T, electrons in the walls of the cavity have a maximum energy of order of a few k_BT, so that they are capable of exciting only

[1] *On the law of the distribution of energy in the normal distribution*, Annalen der Physik **4**, 553–563 (1901).

those radiation modes having $hf \lesssim k_B T$. Quantitatively, it can be shown that the maximum in the blackbody spectrum as a function of frequency occurs at

$$f_{\max} = \frac{2.82 k_B T}{h} = 5.88 \times 10^{10} T \text{ Hz/}^\circ\text{K}. \tag{1.15}$$

As a function of wavelength the maximum in the blackbody spectrum occurs at

$$\lambda_{\max} = \frac{hc}{4.965 k_B T} = \frac{2.90 \times 10^{-3}}{T} \text{ m} \cdot^\circ \text{K}. \tag{1.16}$$

The numerical factors appearing in these equations and the fact that

$$f_{\max} \lambda_{\max} \neq c \tag{1.17}$$

follow from the details of the Planck distribution law (see Appendix). Visible radiation having $\lambda_{\max} = 500$ nm corresponds to a blackbody temperature of $T \approx 6,000\,^\circ$K, about the temperature of the surface of the sun. In the visible part of the spectrum, hf is of order of a few eV.

In 1907, Einstein also used the Planck hypothesis to explain a feature that had been observed in the specific heat of solids as a function of temperature.[2] Many solids have a specific heat at constant volume equal to $3R$ at room temperature, where $R = 8.31$ J/mole/°K is the gas constant; however, some substances such as diamond have a much lower specific heat. Based on a model of solids composed of harmonic oscillators, Einstein used Planck's radiation law to show that the contributions to the specific heat of diamond from the oscillations are "frozen out" at room temperature. To fit the data on diamond, Einstein used a value of 2.73×10^{13} Hz for the frequency of oscillations, which would imply that the vibrational degrees of freedom begin to diminish for temperatures below $T = hf/k_B \approx 1300$ °K and are effectively frozen out for temperatures below $T = hf/\left(2.82 k_B\right) \approx 460$ °K.

1.3 Photoelectric Effect

The *all or nothing* idea surfaced again in Einstein's explanation of the photoelectric effect in Section 8 of his 1905 paper, *On a heuristic point of view concerning the production and transformation of light.*[3] It had been noticed by Hertz and J. J. Thomson in 1897 and 1899, respectively, that electrons could be ejected from metal surfaces when light was shined on the surfaces. Einstein gave his explanation of the effect in 1905 (this explanation and *not* relativity theory was noted in his Nobel Prize

[2] *Planck theory of radiation and the theory of specific heats,* Annalen der Physik **22**, 180–190 (1907).

[3] Annalen der Physik **17**, 132–148 (1905).

citation) and further experiments by Millikan confirmed Einstein's explanation. It turns out that Einstein's explanation is not really evidence for *photons*, even though the authors of many textbooks assert this to be the case. To understand the photoelectric effect, it is necessary to know that different metals have different *work functions*. The work function (typically on the order of several eV) is the minimum energy needed to extract an electron from a metal surface. In some sense it is a measure of the height of a hill that the electron must climb to get out of the metal. From a physical perspective, when an electron leaves the surface, it creates an image charge inside the surface that attracts the charge back to the surface; the work function is the energy needed to escape from this attractive force.

The experimental observations for the photoelectric effect can be summarized as follows:

1. When yellow light is shined on a specific metal, no electrons are ejected. Increasing the intensity of the light still does not lead to electrons being ejected.
2. When ultraviolet light is shined on the same metal, electrons are ejected. When the intensity of the ultraviolet radiation is increased, the number of electrons ejected increases, but the *maximum kinetic energy of the emitted electrons does not change.*

To explain these phenomena, Einstein stated that radiation consists of particles (subsequently given the name *photons* by the chemist Gilbert Lewis in 1926). For radiation of frequency f, the photons have energy hf. When such particles are incident on a metal surface, they excite the electrons by giving up their energy in an all or nothing fashion. Thus, if the work function is denoted by W, the frequency of the radiation must satisfy $hf > W$ for the radiation to cause electrons to be ejected. If $hf < W$, the photons cannot excite the electrons, no matter what the intensity of the field—this explains why increasing the intensity of the yellow light does not lead to electrons being ejected. On the other hand, if $hf > W$, the photon can excite the electron with any extra energy going into kinetic energy KE of the electrons (some of which may be lost on collisions). Thus the maximum energy of the emitted electrons is

$$KE_{\text{max}} = hf - W. \tag{1.18}$$

Increasing the intensity of the radiation does not change this maximum kinetic energy since it results from single photon events, it affects only the intensity of the emitted electrons.

1.4 Bohr Theory

By 1910, it was known that there existed negatively charged particles having charge $-e$ equal to -1.6×10^{-19} C and mass m equal to 9.1×10^{-31} kg. Moreover, the size of these particles could be estimated using theoretical arguments related to their energy content to be about 10^{-15} m. From atomic densities and measurements in gases, atoms were known to have a size of about 10^{-10} m. Since matter is neutral, there must be positively charged particles in atoms that cancel the negative charge.

But how were the charges arranged? Was the positive charge spread out over a large sphere and the negative charge embedded in it or was there something like a planetary model for atoms? This question was put to rest in 1909 by Geiger and Madsen who collided alpha particles (helium nuclei that are produced in radioactivity) on thin metal foils. They found "back-scattering" that indicated the positive charges were small. In 1911 Rutherford analyzed the data and estimated the positive charges to have a size on the order of 10^{-15} m. He then proposed a planetary model of the atom in which the electron orbited the positive nucleus.

However, there were problems in planetary—model—ville. It is easy to calculate that an accelerating electron in orbit around the nucleus should radiate its energy in about a nanosecond, yet atoms were stable. Moreover, the spectrum of radiation emitted by the hydrogen atom consisted of a number of discrete (quasi-monochromatic) lines, but accelerating electrons in different orbits should produce a broadband source of radiation. How could this be explained?

To explain the experimental data within the context of the planetary model, Bohr in 1913 came up with the following postulates:

1. The electron orbits the nucleus in *circular orbits* (back to Ptolemy and Copernicus) having discrete values of *angular momentum* given by

$$L = mvr = n\hbar, \tag{1.19}$$

where $n = 1, 2, 3, 4, \ldots$ (actually he formulated the law in terms of energy rather than angular momentum, but the two methods yield the same results). The quantity r is the radius of the orbit, v is the electron's speed, and L is the magnitude of the angular momentum. The mass in this equation should actually be the reduced mass of the electron which is smaller than the electron mass by a factor of $1/1.00054$.

2. The electrons radiate energy only when they "jump" from a larger to a smaller orbit. The frequency of the radiation emitted is given by

$$f_{n_2,n_1} = \frac{E_{n_2} - E_{n_1}}{h}, \tag{1.20}$$

where E_{n_i} is the energy associated with an orbit having angular momentum $L = n_i\hbar$.

Although these are seemingly benign postulates, they have extraordinary consequences. To begin with, I combine the first postulate with the laws of electrical attraction and Newton's second law. The magnitude of the electrostatic force on the electron produced by the proton is

$$F = \frac{1}{4\pi\epsilon_0}\frac{e^2}{r^2}. \tag{1.21}$$

Combing this with Newton's second law, I find

$$\frac{1}{4\pi\epsilon_0}\frac{e^2}{r^2} = ma = \frac{mv^2}{r}, \tag{1.22}$$

or

$$mvr = \frac{e^2}{4\pi\epsilon_0}. \tag{1.23}$$

Equations (1.19) and (1.23) can be solved for the allowed radii and speeds of the electron. The possible speeds are

$$v_n = \frac{\alpha_{FS} c}{n}, \tag{1.24}$$

where

$$\alpha_{FS} = \frac{1}{4\pi\epsilon_0}\frac{e^2}{\hbar c} \approx \frac{1}{137} \tag{1.25}$$

is known as the *fine-structure constant*. The allowed radii are

$$r_n = n^2 a_0, \tag{1.26}$$

where

$$a_0 = \frac{4\pi\epsilon_0\hbar^2}{me^2} = \frac{\bar{\lambda}_c}{\alpha_{FS}} \approx 0.0529\,\text{nm} \tag{1.27}$$

is the *Bohr radius*, $\bar{\lambda}_c = \lambda_c/2\pi$, and

$$\lambda_c = h/mc = 2.43 \times 10^{-12}\ \text{m} \tag{1.28}$$

is the *Compton wavelength* of the electron. The allowed energies are given by

$$E_n = \frac{mv_n^2}{2} - \frac{1}{4\pi\epsilon_0}\frac{e^2}{r_n} = -\frac{E_R}{n^2}, \tag{1.29}$$

where the *Rydberg energy* E_R is defined as

$$E_R = \frac{1}{2}mc^2\alpha_{FS}^2 \approx 13.6\ \text{eV}\ = 2.18 \times 10^{-18}\ \text{J}. \tag{1.30}$$

Equations (1.19), (1.24), (1.26), and (1.29) are amazing in that they predict that only orbits having *quantized* values of angular momentum, speed, radius, and energy are permitted, totally in contradiction to any classical dynamic models. Moreover the smallest allowed radius $r_1 = a_0$ gives the correct order of magnitude for the size of atoms. In addition, the energies could be used to explain the discrete nature of the spectrum of the hydrogen atom.

Bohr's second postulate addresses precisely this point. Since the energies are quantized, the frequencies emitted by an excited hydrogen atom are also quantized, having possible values

$$f_{nq} = \frac{E_n - E_q}{h} = \frac{E_R}{h}\left(\frac{1}{q^2} - \frac{1}{n^2}\right) = 3.2880 \times 10^{15}\left(\frac{1}{q^2} - \frac{1}{n^2}\right) \text{ Hz,} \tag{1.31}$$

with $n > q$. The corresponding wavelengths are

$$\lambda_{nq} = 91.18\frac{n^2q^2}{n^2 - q^2} \text{ nm.} \tag{1.32}$$

The different frequencies of the radiation emitted on transitions ending on the $n = 1$ level are referred to as the *Lyman* series, those ending on $n = 2$ as the *Balmer* series, and those ending on $n = 3$ as the *Paschen* series. All Lyman transitions are in the ultraviolet or soft X-ray part of the spectrum. The three lowest frequency transitions in the Balmer series are in the visible (red—$\lambda_{32} = 656$ nm, blue-green—$\lambda_{42} = 486$ nm, and violet—$\lambda_{52} = 434$ nm), with the remaining Balmer transitions in the ultraviolet. All Paschen transitions and those terminating on levels having $n > 3$ are in the infrared or lower frequency spectral range. The second postulate also "explained" the stability of the hydrogen atom, since an electron in the $n = 1$ orbital had nowhere to go.

Thus, Bohr's theory explained in quantitative terms the spectrum of hydrogen. That is, the Bohr theory gave very good agreement with the experimental values for the frequencies of the emitted lines. Bohr theory leads to correct predictions for the emitted frequencies because Bohr got the energies right. The Bohr theory also predicts the correct characteristic values for the radius and speed of the electron in the various orbits. As we shall see, however, the radius and speed are not quantized in a correct theory of the hydrogen atom. Moreover, while angular momentum is quantized in the quantum theory, the allowed quantized values for the magnitude of the angular momentum differ somewhat from those given by the Bohr theory.

1.5 De Broglie Waves

It turns out that the photoelectric effect and quantization conditions of the Bohr theory of the hydrogen atom can be explained if one allows for the possibility that matter is described by a wave equation. The first suggestion of this type originates with the work of Louis de Broglie in 1923. The basic idea is to explain the stable orbits of the electron in the hydrogen atom as a repeating wave pattern. Just what type of wave remains to be seen. De Broglie actually arrived at a wavelength for matter using two different approaches.

First let us consider the circular orbits of the electron in hydrogen, for which $L_n = mv_n r_n = n\hbar$. If we imagine that an orbit fits in n wavelengths of matter, then

$$n\left(\lambda_{dB}\right)_n = 2\pi r_n = \frac{nh}{mv_n}, \tag{1.33}$$

where λ_{dB} is the *de Broglie wavelength*. Solving for λ_{dB}, I find

$$\left(\lambda_{dB}\right)_n = \frac{h}{mv_n}. \tag{1.34}$$

Thus, if we assign a wavelength to matter equal to Planck's constant divided by the momentum of the particle, then the wave will "repeat" in its circular orbit. If Planck's constant were equal to zero, there would be no wave-like properties to matter.

De Broglie reached this result in another manner starting with Einstein's theory of relativity, combined with aspects of the photoelectric effect. In special relativity the equation for the length of the momentum-energy 4-vector is

$$p^2c^2 - E^2 = m^2c^4, \tag{1.35}$$

where p is the momentum, m the mass, and E the energy of the particle. If I apply this equation to "particles" of light having zero mass, I find $p = E/c$, which relates the momentum and energy of light. Considering light of frequency f to be composed of photons having energy $E = hf$, we are led to the conclusion that the momentum of a photon is $p = E/c = hf/c = h/\lambda$, where λ is the wavelength of the light. Thus, for photons, $\lambda = h/p$. De Broglie then suggested that this is the correct prescription for assigning a wavelength to matter as well.

You can calculate the de Broglie wavelength of any macroscopic object and will find that it is extremely small compared with atomic dimensions. Since the de Broglie wavelength of macroscopic matter is much smaller than the size of an atom, macroscopic matter could exhibit its wave-like properties only if it encounters interaction potentials that vary on such a scale, an unlikely scenario. On the other hand, if you calculate the de Broglie wavelength of the electron in the ground state of hydrogen, you will find a value close to the Bohr radius. Since the electron is confined to a distance on the order of its wavelength, it exhibits wave-like properties.

In 1927, Davisson and Germer observed the diffraction and interference of electron waves interacting with a single crystal of nickel, providing confirmation of the de Broglie hypothesis. About the same time Schrödinger was formulating a wave theory of quantum mechanics. The wave theory of matter was born!

1.6 The Schrödinger Equation and Probability Waves

With de Broglie's hypothesis, it is not surprising that scientists tried to develop theories in which matter was described by a wave equation. In 1927, Schrödinger developed such a theory at the same time that Heisenberg was developing a theory based on matrices. Eventually the two theories were shown to be equivalent. I will focus on the Schrödinger approach in this discussion; the matrix approach is similar to Dirac's formalism, which is discussed in Chap. 11.

The Schrödinger equation is a partial differential equation for a function $\psi(\mathbf{r}, t)$ which is called the *wave function*. Before thinking about how to interpret the wave function, it is useful to describe how to solve the Schrödinger equation. The Schrödinger equation is different than the wave equation for electromagnetic waves, although there are some similarities. To solve the Schrödinger equation, one must find the "building block" solutions, that is, those solutions analogous to the infinite plane monochromatic waves that served as the building block solutions of the wave equation in electromagnetism. The building block solutions of the Schrödinger equation are called *eigenfunctions*. Eigenfunctions or *eigenstates* are solutions of the Schrödinger equation for which $|\psi(\mathbf{r}, t)|^2$ is constant in time. It is not obvious that such solutions exist, but it can be shown that they always can be found. The eigenfunctions are labeled by *eigenvalues* which correspond to dynamic constants of the motion such as energy, momentum, angular momentum, etc. Let me give you some examples.

A free particle is a particle on which no forces act. For such a particle both momentum $\mathbf{p} = m\mathbf{v}$ and energy $E = mv^2/2$ are constant. Thus momentum and energy can be used as eigenvalue labels for a free particle. But which is a more encompassing label? I can write the energy of the free particle as $E = p^2/2m$. If I give you the momentum, you can tell me the energy. On the other hand, if I give you the energy you can tell me the *magnitude* of the momentum, but not its direction. Thus momentum uniquely determines the eigenfunction since, to each momentum, there is associated exactly one eigenstate. On the other hand, for a given energy there is an infinite number of eigenstates since the momentum can be in any direction. In this case the eigenfunctions or eigenstates are said to be infinitely *degenerate*. We will see that whenever there is a degeneracy of this type, there is an underlying symmetry of nature. In this case there is translational symmetry, since the particle can move in any direction without experiencing a change the force acting on it (since there is no force). The eigenfunctions for the free particle are our old friends, infinite plane waves, but they differ from electromagnetic waves. The wavelength of the free particle waves is just the de Broglie wavelength $\lambda = h/p$—one momentum implies one wavelength. Thus it is the momentum that determines the wavelength of the free particle eigenfunctions.

For the bound states of the hydrogen atom, the electron is subjected to a force so that momentum is not constant and cannot be used as a label for the hydrogen atom eigenfunctions. On the other hand, angular momentum is conserved for any central force. For reasons to be discussed in Chap. 9, the three components of the angular momentum vector cannot be used to label the eigenfunctions; instead, the magnitude of the angular momentum and one of its components are used.

Let me return to the free particle. What is the significance of the wave function $|\psi(\mathbf{r},t)|^2$? The interpretation given is that $|\psi(\mathbf{r},t)|^2$ is the probability density (probability per unit volume) to find the particle at position $\mathbf{r}$ at time t. To give you some idea of what this means, imagine throwing darts at a wall. You throw one dart and it hits somewhere on the wall. Now you throw a second and a third dart. Now you throw a million darts. After a million darts, you will have a pretty good idea where the next dart will land—in other words, you will know the probability distribution for where the darts will land. This will not tell you where the next dart will hit, only the probability. The probability of hitting a *specific* point is essentially equal to zero—you must talk about the probability density (in this case, probability per unit area). In quantum mechanics we have something similar. You prepare a quantum system in some initial state. At some later time you measure the particle somewhere. Now you start with an *identical* initial state, wait the same amount of time and measure the particle again. You repeat this a very large number of times and you will have the probability distribution to find the particle at a time t after it was prepared in this state. This will not tell you where the particle will be the next time you do the experiment—only the probability. If you send single particles, acting as waves, through the two-slit apparatus, each particles will set off only one detector on the screen. If you repeat this many times however, eventually you will build up the same type of interference pattern that occurs in the double slit experiment for light.[4]

Since the eigenfunctions for a free particle are infinite mono-energetic (single momentum or velocity) waves, the probability density associated with the free particle eigenfunction is constant over the entire universe! In other words, the eigenfunctions of the free particle are spread out over all space. Clearly no such state exists. Any free particles that we observe in nature are in a superposition state that is referred to as a *wave packet*, which is the matter analogue of a radiation pulse. As with radiation, we make up a superposition state by adding several waves or eigenfunctions together. But radiation pulses and wave packets differ in a fundamental way. All frequency components of a radiation pulse propagate at the same speed in vacuum, the speed of light. For a free particle wave packet, the different eigenfunctions that compose the wave packet correspond to different wavelengths, which, in turn, implies that they correspond to different momenta, since wavelength for matter waves is related to momentum. Thus different parts of a free particle wave packet move at different rates—the wave packet's shape changes in time! In the case of matter waves, Eq. (1.7) is replaced by

$$\Delta x \Delta p \geq \hbar/2, \tag{1.36}$$

[4]A beautiful video of an experiment demonstrating the buildup of a two-slit interference pattern for electrons can be found at http://www.hitachi.com/rd/portal/highlight/quantum/.

where Δx is the spread of positions and Δp the spread of momenta in the wave packet. Equation (1.36) is a mathematical expression of the famous *Heisenberg Uncertainty Principle.*

Matter can be considered to be "particle-like" as long as it is not confined it to a distance less than its de Broglie wavelength. If a particle is confined to a distance on the order of its de Broglie wavelength or encounters changes in potential energy that vary significantly in distances of order of its de Broglie wavelength, the particle exhibits wave-like properties.

1.7 Measurement and Superposition States

Measurement is one of the most difficult and frustrating features of quantum mechanics. The problem is that the measuring apparatus itself must be classical and not described by quantum mechanics. You will hear about "wave-function collapse" and the like, but quantum mechanics does not describe the dynamics in which a quantum state is measured by a classical apparatus. As long as you ask questions related to the probability of one or a succession of measurements, you need not run into any problems.

1.7.1 What Is Truly Strange About Quantum Mechanics: Superposition States

When I discussed electromagnetic waves, I showed that a superposition of plane wave states can result in a radiation pulse. There is nothing unusual about a radiation pulse. However, for quantum-mechanical waves, the superposition of eigenfunctions can sometimes lead to what at first appears to be rather strange results. The strangeness is most readily apparent if we look at the electrons in atoms or other quantum particles that are in a superposition of such *bound* states. Consider the electron in the hydrogen atom. It is possible for the electron to be in a superposition of two or more of its energy states (just as a free particle can be in a superposition of momentum eigenstates). Imagine that you prepare the electron in a superposition of its $n = 1$ and $n = 2$ energy states. What does this mean? If you measure the energy of the electron you will get *either* -13.6 eV *or* -3.4 eV. If you prepare the electron in the *same* manner and again measure the energy, you will get either -13.6 eV or -3.4 eV. After many measurements on identically prepared electrons, you will know the relative probability that the electron will be measured in the $n = 1$ state or the $n = 2$ state. You might think that sometimes the electron was prepared in the $n = 1$ state and sometimes in the $n = 2$ state, but this is not the case since it is assumed that the electron is always prepared in an *identical* manner which can yield either measurement—that is, it is in a superposition state of the $n = 1$ and $n = 2$ states. When you measure the energy, some people (myself excluded) like to say that the wave function *collapses* into the state corresponding to the energy you measure. I would say that quantum mechanics just allows you to predict the probabilities for various measurements and cannot provide answers to questions regarding how the wave function evolves when a measurement is made.

The idea that the observer *forces* a quantum system into a given state has led to some interesting, but what I consider misleading, ideas. Perhaps the most famous is the question of "Who killed Schrödinger's cat?" In the "cat" scenario, a cat is put into a box with a radioactive nucleus. If the nucleus decays, it emits a particle that activates a mechanism to release a poisonous gas. Since the nucleus is said to be in a superposition state of having decayed and not having decayed, the cat is also said to be in a superposition state of being dead and alive! Only by looking into the box does the observer know if the nucleus has decayed. In other words, the observation forces the nucleus into either its decayed or undecayed state. Thus it appears that looking into the box can result in the death of the cat.

My feeling is that this is all a lot of nonsense. The reason that it is nonsense is somewhat technical, however. When the nucleus decays, it emits a gamma ray, so the appropriate superposition state must include this radiated field as well. The transition from initial state to the decayed state plus gamma ray essentially occurs *instantaneously*, even though the *time* at which the decay occurs follows statistical laws. Thus when you look into the box, the cat is either alive or it is dead, but not in a superposition state. In other words, the health of the cat is a measure of whether or not the nucleus has decayed, but the cat itself is not in a superposition state. On many similarly prepared systems, you will simply find that the cat dies at different times according to some statistical law. You needn't worry about being convicted of "catacide" if you open the box.

Note that if it were true that you could force the nucleus into a given state by observing it, you could prevent the nucleus from decaying simply by looking at it continuously! Since it starts in its initial state and you continuously force it to stay in its initial state by looking at it, it never decays. This is known as the Zeno effect, in analogue with Zeno's paradoxes. It should be pointed out that it *is* possible to keep certain quantum systems in a given state by continuous observation. As long as you make measurements on the system on a time scale that is short compared with the time scale with which the system will evolve into a superposition state, you can keep the system in its initial state. For the decay of particles, however, the transition from initial to final states occurs essentially instantaneously and it is impossible to use continuous "measurements" to keep the particle in its undecayed state. I return to a discussion of the Zeno effect in Chap. 24.

1.7.2 The EPR Paradox and Bell's Theorem

Although a single quantum system in a superposition state has no classical analogue and already represents a strange animal, things get *really* strange when we consider two, interacting quantum systems that are prepared in a specific manner.

1.7.2.1 The EPR Paradox

In 1935, Einstein, Podolsky, and Rosen (EPR) published a paper entitled *Can Quantum-Mechanical Description of Physical Reality Be Considered Complete,* which has played an important role in the development of quantum mechanics.[5] For an excellent account of this paper and others related to it, see *Quantum Paradoxes and Physical Reality* by Franco Selleri.[6] In their paper, EPR first pose questions as to what constitutes a satisfactory theory: "*Is the theory correct? Is the description given by the theory complete?* They then go on to define what they mean by an element of physical reality: *If, without in any way disturbing a system, we can predict with certainty (i.e. with probability equal to unity) the value of a physical quantity, then there exists an element of physical reality corresponding to this physical quantity.*" Based on this definition and giving an example in which the wave function corresponds to an eigenfunction of one of two non-commuting operators,[7] they conclude that "*either (1) the quantum-mechanical description of reality given by the wave function is not complete or (2) when the operators corresponding to two physical quantities do not commute the two quantities cannot have simultaneous reality.*"

To illustrate the EPR paradox, one can consider the decay of a particle into two identical particles. The initial particle has no intrinsic spin angular momentum, whereas each of the emitted particles has a spin angular momentum of $1/2$ (spin angular momentum is discussed in Chap. 12). The spin of the emitted particles can be either "up" or "down" relative to some quantization axis, but the sum of the components of spin relative to this axis must equal zero to conserve angular momentum (the spin of the original particle is zero so the total spin of the composite particles must be zero). The particles are emitted in opposite directions to conserve linear momentum and in a superposition of two states, a state in which one particle (particle A) has spin up and the other (particle B) has spin down plus a state in which one particle (particle A) has spin down and the other (particle B) has spin up, written symbolically as

$$|\psi\rangle = \frac{1}{\sqrt{2}}\left(|\uparrow\downarrow\rangle - |\downarrow\uparrow\rangle\right), \tag{1.37}$$

where $\uparrow$ refers to spin up and $\downarrow$ to spin down. The spin states of the two particles are said to be *correlated* or *entangled.*

Consider the situation in which the particles are emitted in this correlated state and fly off so they are light years apart. According to quantum mechanics, the spin of each of the particles (by "spin," I now mean either up or down) is not fixed until

[5] A. Einstein, B. Podolsky, and N. Rosen, Physical Review **47**, 777–780 (1935).

[6] Franco Selleri, *Quantum Paradoxes and Physical Reality* (Kluwer Academic Publishers, Dordrecht, The Netherlands, 1990).

[7] The commutator of two operators is defined in Chap. 5.

it is measured. In other words, the spin of each of the particles can be either up or down. If you don't measure the spin of particle A, the spin of particle B will be either up or down when measured relative to the z-axis. However, if you measure the spin of particle (A) to be up relative to the z-axis, you are guaranteed that particle (B) will have its spin down relative to the z-axis.

Of course you are free to measure the spin component of particle A along *any* axis. If you measure the spin of particle A as up relative to the z-axis, the spin of particle B will have physical reality along the $-z$ direction. On the other hand, if you measure the spin of particle A as up relative to the x-axis, the spin of particle B will have physical reality along the $-x$ direction. This is a situation in which both states of particle B correspond to the *same* physical reality and these states correspond to eigenfunctions of non-commuting spin operators, which is not allowed in quantum mechanics. Since this outcome violates condition (2) stated above, the conclusion in EPR is that quantum mechanics is not a complete theory. This is known as the EPR paradox.

In the conventional description of quantum mechanics, there is no paradox. For these correlated states, the probability that you measure a spin down for the second particle after you have measured a spin up for the first particle is unity when both spins are measured relative to the same axis. Quantum mechanics does not tell you *why* this is the case or *how* it occurs, it just gives the probability for the outcome. The idea of one measurement *influencing* the other is not particularly useful or meaningful.

According to EPR, the state represented by Eq. (1.37) cannot represent the complete state of the system. Although it is nowhere mentioned in their paper, one often interprets this to imply that there must be additional labels or *hidden variables* needed in Eq. (1.37) In other words, there are some variables encoded in the quantum system that determine the outcome of the measurements. Quantum mechanics was inconsistent with EPR's element of reality, but a hidden variables theory might be. Which is right?

1.7.2.2 Bell's Theorem

The EPR paradox was and continues to be disturbing to some physicists. The idea that a measurement on one physical system can influence the outcome of a measurement on another system that is not causally connected with it can be somewhat unnerving. Motivated by the issues raised by the EPR paradox, John Bell in 1964 published a paper in which he discussed the idea of elements of reality.[8] Without any reference to quantum mechanics, he was able to prove that certain inequalities must be obeyed on measurements of correlated systems to be

[8]John Bell, *On the Einstein Podolsky Rosen Paradox*, Physics **1,** 195–200 (1964). See also, *Hidden variables and the two theorems of John Bell* by N. David Mermin in the *Review of Modern Physics*, **65**, 803–815 (1993).

consistent with the EPR idea of physical reality. In other words, if Bell's inequalities are violated, then physical observables may not have physical reality of the type described by EPR, independent of whether or not quantum mechanics is a valid theory. It turns out that there are now experiments in which Bell's inequalities are violated. Moreover, the results of these experiments are correctly predicted by quantum mechanics. I will give a proof of Bell's theorem and discuss examples in Chap. 13. Although Bell's theorem is generally accepted, there are some who question its validity.

Measurement in quantum mechanics is a subject that continues to attract a great deal of attention. At this level, maybe it's better not to worry about it too much and instead concentrate on mastering the basic elements of the quantum theory. In the words of Richard Feynman (or not),[9] *Shut up and calculate!*

1.8 Summary

In this chapter, I gave a qualitative introduction to several aspects of the quantum theory that we will encounter in the following chapters. As I stated, quantum mechanics is challenging the first go-around for many students. To help master the material it is important to try as many of the problems as possible. It can also help to consult other texts that may treat the material in a different fashion that you find more accessible. To begin, it may prove useful to review some mathematical concepts. You can skip Chap. 2 if you are familiar with these concepts.

1.9 Appendix: Blackbody Spectrum

Blackbody radiation is discussed in almost every quantum mechanics textbook, but the manner in which a blackbody achieves thermal equilibrium is almost never discussed. I will return to this question after discussing the equilibrium blackbody spectrum. There are essentially two ways to derive the blackbody spectrum. One employs the quantum statistics of a Bose gas plus some additional assumptions (radiation is described by Bose statistics and the energy of a photon having frequency ω is $\hbar\omega$). Since you will see such a treatment in your course on statistical mechanics, I give an alternative derivation, more along the lines given by Planck.

The first step is to get the distribution of radiation field modes in a box, each of whose sides has length L. There are two ways of doing this, both giving the same final *density of states* of the radiation field.

[9] N. D. Mermin, *Could Feynman have said this?*, Physics Today, May, 2004, page 10.

1.9.1 Box Normalization with Field Nodes on the Walls

In this case I assume that either the electric or magnetic field has a node on the perfectly conducting walls of a metal cavity which is bounded by $0 \leq x \leq L$, $0 \leq y \leq L, 0 \leq z \leq L$. For the appropriate field to vanish at the boundaries, the field mode function must be of the form $\sin(k_x x)$, $\sin\left(k_y y\right)$, or $\sin(k_z z)$, with

$$k_x = \frac{\pi n_x}{L}; \tag{1.38a}$$

$$k_y = \frac{\pi n_y}{L}; \tag{1.38b}$$

$$k_x = \frac{\pi n_z}{L}, \tag{1.38c}$$

where n_x, n_y, n_z, are *positive* integers [note that $\sin\left(\frac{\pi x}{L}\right)$ and $\sin\left(-\frac{\pi x}{L}\right)$ correspond to the same mode, differing only by a phase factor]. The field propagation vector is given by

$$\mathbf{k} = k_x \mathbf{u}_x + k_x \mathbf{u}_y + k_x \mathbf{u}_z, \tag{1.39}$$

where $\mathbf{u}_j$ $(j = x, y, z)$ is a unit vector in the j-direction. The angular frequency ω of a field mode is equal to kc, where $k = \sqrt{k_x^2 + k_y^2 + k_z^2}$.

Thus, it follows from $\omega = kc$ that

$$\omega_n = k_n c = \frac{\pi c n}{L}, \tag{1.40}$$

where

$$n = \sqrt{n_x^2 + n_y^2 + n_z^2}. \tag{1.41}$$

Each set of values (n_x, n_y, n_z) corresponds to a given mode of the field. If there are many modes, I can replace the discrete modes by a continuum, essentially making n a continuous variable with $\omega = kc = \pi cn/L$. In other words, the number of modes in a shell having inner radius n and outer radius $n + dn$ is simply $2\frac{1}{8}\left(4\pi n^2\right) dn$. The factor of 1/8 is present since only the positive quadrant for (n_x, n_x, n_z) is allowed, the factor of 2 is added to allow for two polarization components for each spatial mode, $\left(4\pi n^2\right)$ is the surface area of the shell, and dn is the width of the shell. It then follows from Eq. (1.40) that the number of modes having frequency between ω and $\omega + d\omega$ is

$$\mathcal{N}(\omega)d\omega = 2\frac{1}{8}\left(4\pi n^2\right) dn = \pi n^2 \frac{dn}{d\omega} d\omega = \frac{L^3 \omega^2 d\omega}{\pi^2 c^3}, \tag{1.42}$$

and the number of modes per unit volume between ω and $\omega + d\omega$, denoted by $n(\omega)d\omega$, is

$$n(\omega)d\omega = \frac{\mathcal{N}(\omega)d\omega}{L^3} = \frac{\omega^2 d\omega}{\pi^2 c^3}. \tag{1.43}$$

1.9.2 Periodic Boundary Conditions

In this case I assume that either the electric or magnetic field is subject to *periodic boundary conditions* in which the field values repeat at spatial intervals L in each direction. Since the field phase of a monochromatic wave varies as $e^{i\mathbf{k}\cdot\mathbf{r}}$ this implies that

$$k_x = \frac{2\pi n_x}{L}; \tag{1.44a}$$

$$k_y = \frac{2\pi n_y}{L}; \tag{1.44b}$$

$$k_x = \frac{2\pi n_z}{L}, \tag{1.44c}$$

where n_x, n_y, n_z, are integers, positive, negative or zero [now $e^{i\frac{2\pi x}{L}}$ and $e^{-i\frac{2\pi x}{L}}$ correspond to *different* modes]. The calculation proceeds as in box normalization, except that all quadrants are allowed. Thus,

$$\omega = kc = \frac{2\pi cn}{L}, \tag{1.45}$$

the number of modes in a shell having inner radius n and outer radius $n + dn$ is $2\left(4\pi n^2\right) dn$ (there is no factor of 1/8 is present since all quadrants for (n_x, n_x, n_z) are allowed). The number of modes having frequency between ω and $\omega + d\omega$ is

$$\mathcal{N}(\omega)d\omega = 2\left(4\pi n^2\right) dn = 8\pi n^2 dn = 8\pi n^2 \frac{dn}{d\omega} d\omega = \frac{L^3\omega^2 d\omega}{\pi^2 c^3}, \tag{1.46}$$

such that the number of modes per unit volume, $n(\omega)d\omega$, between ω and $\omega + d\omega$ is

$$n(\omega)d\omega = \frac{\mathcal{N}(\omega)d\omega}{L^3} = \frac{\omega^2 d\omega}{\pi^2 c^3}, \tag{1.47}$$

in agreement with Eq. (1.43).

1.9.3 Rayleigh-Jeans Law

To get the Rayleigh-Jeans law one simply assigns an energy of $k_B T$ to each mode, resulting in an energy density $u(\omega)$ (energy per unit volume per unit frequency ω) given by

$$u(\omega) = k_B T \frac{\omega^2}{\pi^2 c^3}. \tag{1.48}$$

Clearly, the energy density diverges at large frequency; this is known as the *ultraviolet catastrophe*.

1.9.4 Planck's Solution

Implicit in the equipartition theorem is the assumption that each mode or degree of freedom can have a *continuous* energy distribution. In other words for the radiation modes in a cavity the energy distribution at temperature T, assumed to be a Boltzmann distribution, is

$$W(E) = \frac{1}{k_B T} e^{-E/k_B T}, \tag{1.49}$$

which has been normalized such that

$$\int_0^\infty W(E)dE = 1. \tag{1.50}$$

You can verify that

$$\bar{E} = \int_0^\infty EW(E)dE = k_B T, \tag{1.51}$$

is *independent of frequency*, for these modes.

To explain the blackbody spectrum, Planck conjectured that the energies of the modes were discrete rather than continuous. A mode having frequency ω could have only those energies that are an integral multiple of a constant $\hbar$ times ω. In other words, there is now a *separate* energy distribution for *each* mode having frequency ω. The probability of having energy $n\hbar\omega$ in a mode having frequency ω is given by

$$W(n, \omega) = Ae^{-n\hbar\omega/k_B T}, \tag{1.52}$$

where A is a normalization constant that I do not have to specify. The average energy for a mode having frequency ω is then

$$\langle W(\omega)\rangle = \frac{A\sum_{n=0}^{\infty} n\hbar\omega e^{-n\hbar\omega/k_BT}}{A\sum_{n=0}^{\infty} e^{-n\hbar\omega/k_BT}} = \hbar\omega\frac{\sum_{n=0}^{\infty} nx^n}{\sum_{n=0}^{\infty} x^n}, \tag{1.53}$$

where

$$x = \exp\left(-\frac{\hbar\omega}{k_BT}\right). \tag{1.54}$$

Using the fact that

$$\sum_{n=0}^{\infty} x^n = \frac{1}{1-x}; \tag{1.55a}$$

$$\sum_{n=0}^{\infty} nx^n = x\frac{d}{dx}\sum_{n=0}^{\infty} x^n = \frac{x}{(1-x)^2}, \tag{1.55b}$$

you can show that

$$\langle W(\omega)\rangle = \frac{x\hbar\omega}{(1-x)} = \frac{\hbar\omega}{\left(\frac{1}{x}-1\right)} = \frac{\hbar\omega}{\left(e^{\hbar\omega/k_BT}-1\right)}. \tag{1.56}$$

It follows that $\langle W(\omega)\rangle \sim k_BT$ for $\hbar\omega/k_BT \ll 1$, reproducing the equipartition result, but $\langle W(\omega)\rangle \sim \hbar\omega e^{-\hbar\omega/k_BT} \ll 1$ for $\hbar\omega/k_BT \gg 1$. In essence, Planck ruled out the possibility of exciting high frequency modes. The energy density per unit frequency is obtained by combining Eqs. (1.43) and (1.56),

$$u(\omega) = n(\omega)\langle W(\omega)\rangle = \frac{\hbar\omega^3}{\pi^2c^3}\frac{1}{\left(e^{\hbar\omega/k_BT}-1\right)}, \tag{1.57}$$

which is the Planck distribution. If $\hbar$ is taken to be

$$\hbar = 1.055\times 10^{-34}\ \mathrm{J\cdot s}, \tag{1.58}$$

Planck's distribution agrees with experiment.

The total energy density is given by

$$\begin{aligned} u &= \int_0^{\infty}\frac{\hbar\omega^3}{\pi^2c^3}\frac{d\omega}{\left(e^{\hbar\omega/k_BT}-1\right)} \\ &= \frac{\hbar}{\pi^2c^3}\left(\frac{k_BT}{\hbar}\right)^4\int_0^{\infty}\frac{x^3dx}{(e^x-1)} = \frac{4\sigma}{c}T^4, \end{aligned} \tag{1.59}$$

which is the *Stefan-Boltzmann Law* and

$$\sigma = \frac{k_B^4}{4\pi^2\hbar^3c^2}\int_0^\infty \frac{x^3dx}{(e^x - 1)} \tag{1.60}$$

is Stefan's constant. The integral is tabulated

$$\int_0^\infty \frac{x^3dx}{(e^x - 1)} = \frac{\pi^4}{15}, \tag{1.61}$$

such that

$$\sigma = \frac{\pi^2 k_B^4}{60\hbar^3c^2} = 5.67 \times 10^{-8}\text{W/m}^2/\left(^\circ\text{K}\right)^4. \tag{1.62}$$

Equation (1.57) must also give Wien's displacement law; the wavelength corresponding to maximum emission multiplied by the temperature is a constant. The frequency that gives rise to the maximum in the energy density is obtained by setting $du/d\omega = 0$, yielding

$$\frac{3\omega_{\max}^2}{\left(e^{\hbar\omega_{\max}/k_BT} - 1\right)} - \frac{\hbar}{k_BT}e^{\hbar\omega_{\max}/k_BT}\frac{\omega_{\max}^3}{\left(e^{\hbar\omega_{\max}/k_BT} - 1\right)^2} = 0, \tag{1.63}$$

or

$$e^{\hbar\omega_{\max}/k_BT}\omega_{\max} = 3\frac{k_BT}{\hbar}\left(e^{\hbar\omega_{\max}/k_BT} - 1\right); \tag{1.64a}$$

$$\omega_{\max} = 3\frac{k_BT}{\hbar}\left(1 - e^{-\hbar\omega_{\max}/k_BT}\right). \tag{1.64b}$$

Setting

$$y = \frac{\hbar\omega_{\max}}{k_BT}, \tag{1.65}$$

I find that the maximum occurs for

$$y = 3\left(1 - e^{-y}\right), \tag{1.66}$$

which can be solved graphically to obtain $y = 2.82$, leading to

$$\hbar\omega_{\max} = 2.82k_BT; \quad f_{\max} = \frac{2.82k_BT}{h} = 5.88 \times 10^{10}T\ \text{Hz}/^\circ\text{K}. \tag{1.67}$$

If I set $\lambda_{\max} = c/f_{\max}$, I find

$$\lambda_{\max} T = \frac{cT}{f_{\max}} = 5.10 \times 10^{-3} \text{ m} \cdot^{\circ} \text{K}, \tag{1.68}$$

which is not quite Wien's law,

$$\lambda_{\max} T = 2.9 \times 10^{-3} \text{ m} \cdot^{\circ} \text{K}. \tag{1.69}$$

The reason for the difference is that Wien's law is derived from the energy density *per unit wavelength* $w(\lambda)$ rather than energy density per unit frequency and the two methods give different maxima. That is, if you measure the energy density per unit frequency as a function of frequency and the energy density per unit wavelength as a function of wavelength, $f_{\max}\lambda_{\max} \neq c$. To see this I set

$$u(\omega)d\omega = -w(\lambda)d\lambda. \tag{1.70}$$

Since $d\omega = -2\pi c d\lambda/\lambda^2$, I find

$$\begin{aligned} w(\lambda) &= \frac{2\pi c}{\lambda^2} u\left(\frac{2\pi c}{\lambda}\right) \\ &= \frac{2\pi c}{\lambda^2} \frac{\hbar \left(\frac{2\pi c}{\lambda}\right)^3}{\pi^2 c^3} \frac{1}{\left(e^{hc/\lambda k_B T} - 1\right)} \\ &= \frac{8\pi hc}{\lambda^5} \frac{1}{\left(e^{hc/\lambda k_B T} - 1\right)}. \end{aligned} \tag{1.71}$$

Now, instead of Eq. (1.66), I must solve

$$y = 5\left(1 - e^{-y}\right). \tag{1.72}$$

The solution is $y = hc/\lambda_{\max} k_B T = 4.965$, such that

$$\lambda_{\max} T = \frac{hc}{4.965 k_B} = 2.90 \times 10^{-3} \text{ m} \cdot \text{K}, \tag{1.73}$$

which is Wien's law.

1.9.5 Approach to Equilibrium

Let me now return to the question of the approach to equilibrium. Consider first a box with two point particles, one at rest somewhere in the box and one having speed v_0 and energy $E_0 = mv_0^2/2$ located at some other position in the box. The collisions between particles and between the particles and the walls are assumed to be perfectly elastic. You might think that, on average, each particle will have energy $E_0/2$ if you wait long enough, but this is not true. There are many initial conditions that will not result in collisions between the particles. For a given set of initial conditions, the dynamics is perfectly determined.

As the number of particles in the box increases to a very large number, it becomes more and more likely that all the atoms undergo collisions at a rapid rate. In this case, the equilibrium distribution is Maxwellian. Returning to the modes of the radiation field in the cavity, there is no obvious way they can exchange energy to achieve equilibrium. They must exchange energy with the charges in the cavity walls. Thus one is faced with modeling this interaction and then having some model for the energy distribution of the charges in the cavity. As far as I know, no one has ever solved this approach to equilibrium in a satisfactory manner.

1.10 Problems

1. What is the ultraviolet catastrophe and how did it lead to Planck's quantum hypothesis? Specifically how did theory and experiment differ in describing the spectrum of a blackbody? What hypothesis did Planck make to minimize the contribution of the high frequency modes?

2. Describe the photoelectric effect experiments and Einstein's explanation for both the number and energy of the emitted electrons as a function of the frequency of the incident light.

3. What were Bohr's postulates in his theory of the hydrogen atom? How do these postulates explain the spectrum and stability of the hydrogen atom?

4. Draw an energy level diagram for the hydrogen atom and indicate on it the energy of the four lowest energy states. Also indicate the radius, electron velocity, and angular momentum of these states as given by the Bohr theory.

5. Calculate the wavelength of the $n = 7$ to $n = 2$ and of the $n = 3$ to $n = 2$ transitions in hydrogen. What frequency of radiation is needed to excite a hydrogen atom from its $n = 1$ to $n = 3$ state? How much energy is required to ionize a hydrogen atom from its $n = 2$ state?

6. Radiation from the $n = 2$ to $n = 1$ state of hydrogen is incident on a metal having a work function of 2.4 eV. What is the maximum energy of the electrons emitted from the metal?

7. What is the significance of the equation $\lambda_{dB} = h/p$? What is meant by the wave particle duality of matter? What determines when matter acts as a particle and when it acts as a wave?

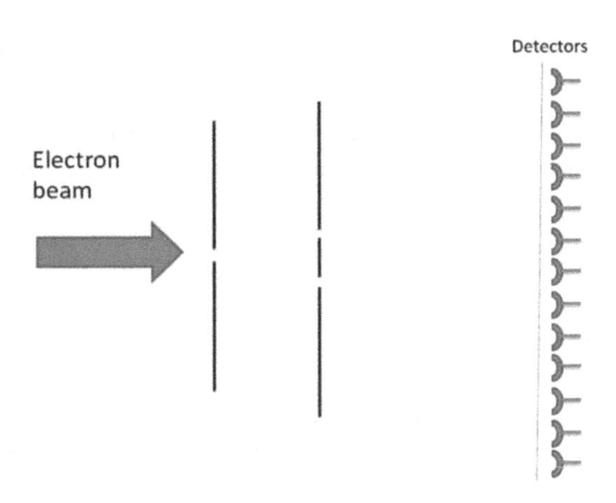

Fig. 1.4 Problem 1.10

8. Calculate the de Broglie wavelength for a particle of mass 1.0 g moving with a speed of 1.0 cm/yr. Calculate the de Broglie wavelength for the electron in the $n = 1$ state of the Bohr atom.

9. In general terms, discuss the measurement process in quantum mechanics. Why is it necessary to make measurements on a large number of identically prepared systems to obtain $|\psi(\mathbf{r}, t)|^2$? How does this differ from Newtonian mechanics? Why is a single particle in a superposition state an intrinsically quantum object?

10. In a two-slit experiment, particles are sent into the apparatus one particle at a time (see Fig. 1.4). How many detectors does a single particle trigger? If the experiment is repeated many times, what will a graph of the number of counts in a detector *vs* detector position look like? Explain.

11. When light is incident on a glass slab, some of the light is reflected. This is a wave-like phenomenon (if a classical particle encounters a change in potential, it simply slows down or speeds up with no reflection), even though this corresponds to a geometrical optics limit (neglect of diffraction). Why does a wave-like effect occur in this case? Is there any connection of this result with the rainbow?

12. The blackbody spectrum as a function of frequency $u_f(f)$ can be obtained using $u_f(f)df = u(\omega)d\omega = 2\pi u(2\pi f)df$. Plot $\left[cu_f(f)/4\pi\right] \times 10^{18}$ as a function of frequency for $T = 2.73\,^\circ\mathrm{K}$ and find the maximum frequency [$cu_f(f)/4\pi$ is the power per unit area per unit frequency per unit solid angle—this corresponds to the flux incident on a detector per unit frequency per steradian (sr)]. This Planck distribution corresponds to the *cosmic microwave background*. What is the energy per unit volume of the cosmic microwave background?

Extra Reading

There are many excellent texts on quantum mechanics. My suggestion is to go to your library and try to find texts that you find especially helpful. Some possible recommendations are:

Undergraduate Texts

David Bohm, *Quantum Theory* (Prentiss Hall, New York., 1951).

Ashok Das and Adrian Melissinos, *Quantum Mechanics – A Modern Introduction*, (Gordan and Breach, New York, 1986).

Peter Fong, *Elementary Quantum Mechanics*, Expanded Edition (World Scientific, Hackensack, N.J., 2005).

Stephen Gasiorowicz, *Quantum Physics,* Second Edition (John Wiley and Sons, New York, 1996).

David Griffiths, *Introduction to Quantum Mechanics*, Second Edition (Pearson Prentiss Hall, Upper Saddle River, N.J., 2005).

Richard Liboff, *Introductory Quantum Mechanics* 4th Edition (Addison Wesley, San Francisco, 2003).

David Park: *Introduction to the Quantum Theory*, Second Edition (McGraw Hill, New York, 1974).

John Powell and Bernd Crasemann, *Quantum Mechanics* (Addison Wesley, Reading, MA, 1961).

David Saxon, *Elementary Quantum Mechanics*, (Holden-Day, San Francisco, 1968).

George Trigg, *Quantum Mechanics*, (Van Nostarnd, Princeton, N.J., 1968).

Graduate Texts

Ernst Abers, *Quantum Mechanics* (Pearson Education, Upper Saddle River, N.J., 2004).

Claude Cohen-Tannoudji, Bernard Liu, and Franck Laloë, *Quantum Mechanics* (Wiley Interscience, Paris, 1977).

Robert Dicke and James Wittke, *Introduction to Quantum Mechanics*, (Addison Wesley, Reading, MA, 1960).

P. A. M. Dirac, *The Principles of Quantum Mechanics*, Fourth Edition (Oxford University Press, London, 1958).

Eugen Merzbacher, *Quantum Mechanics,* Third Edition (John Wiley and Sons, New York, 1998).

Albert Messiah, Quantum Mechanics, Vols. 1 and 2, (Dover Publications, New York, 2014).

Leonard Schiff, *Quantum Mechanics,* Third Edition (McGraw Hill, New York, 1968).

Chapter 2
Mathematical Preliminaries

Anyone taking a course in quantum mechanics usually has had several semesters of calculus and some additional advanced courses in mathematics and mathematical physics. In this chapter I review briefly some of the mathematical concepts we will need in our study of quantum mechanics.

2.1 Complex Function of a Real Variable

Since the wave function in quantum mechanics is complex, I will often be dealing with complex functions. If $\psi(x)$ is a complex function of a real variable x, then it can be written as

$$\psi(x) = u(x) + iv(x) \tag{2.1a}$$

$$= |\psi(x)| e^{i\phi(x)}, \tag{2.1b}$$

where the real functions,

$$u(x) = \mathrm{Re}[\psi(x)]; \tag{2.2a}$$

$$v(x) = \mathrm{Im}[\psi(x)], \tag{2.2b}$$

are related to the *magnitude* $|\psi(x)|$ and *argument* $\phi(x)$ of $\psi(x)$ by

$$|\psi(x)| = \sqrt{[u(x)]^2 + [v(x)]^2}, \tag{2.3a}$$

$$\phi(x) = \tan^{-1}[v(x)/u(x)], \tag{2.3b}$$

P.R. Berman, *Introductory Quantum Mechanics*, UNITEXT for Physics,
https://doi.org/10.1007/978-3-319-68598-4_2

with $-\pi/2 \le \phi(x) \le \pi/2$.[1] The *complex conjugate* of $\psi(x)$, denoted by $\psi^*(x)$, is defined as

$$\psi^*(x) = u(x) - iv(x). \tag{2.4}$$

Some equations that you will find useful are

$$|\psi(x)|^2 = \psi(x)\psi^*(x) = [u(x)]^2 + [v(x)]^2, \tag{2.5a}$$

$$e^{i\theta} = \cos\theta + i\sin\theta, \tag{2.5b}$$

$$\left|e^{i\theta}\right| = 1, \qquad \theta \text{ real}; \tag{2.5c}$$

$$e^{2\pi in} = 1, \qquad n \text{ integer}; \tag{2.5d}$$

$$e^{\pi in} = (-1)^n, \qquad n \text{ integer}. \tag{2.5e}$$

I assume that you are familiar with complex functions of a real variable, but include some review problems at the end of the chapter.

2.2 Functions and Taylor Series

In physics, we are always making approximations. Most potentials can be approximated as quadratic in a region near a potential minimum. Often we want to know the value of a function in the region of a particular point. To be able to get this information, we must know the value of the function at the point and the values of the derivatives of the function at the point. The more derivatives we know, the better we can approximate the function.

2.2.1 *Functions of One Variable*

This is the simplest case and the one with which you are most familiar. Consider the function shown in Fig. 2.1.

Suppose we know the value $f(x')$ at $x' = b$ and want to approximate the function at $x' = x$, when $x \approx b$. If the function were a straight line between the two points, then a knowledge of the slope of the line would be sufficient to calculate $f(x)$. One can get a better and better approximation to $f(x)$ by approximating the function between the points as a polynomial—the higher the order of the polynomial, the

[1]The restriction of $\phi(x)$ to values $-\pi/2 \le \phi(x) \le \pi/2$ corresponds to what is known as the *principal value* of $\tan^{-1}$. In some problems, such as those to be encountered in scattering theory, $\phi(x)$ can correspond to a physical quantity that should not be restricted to these limits.

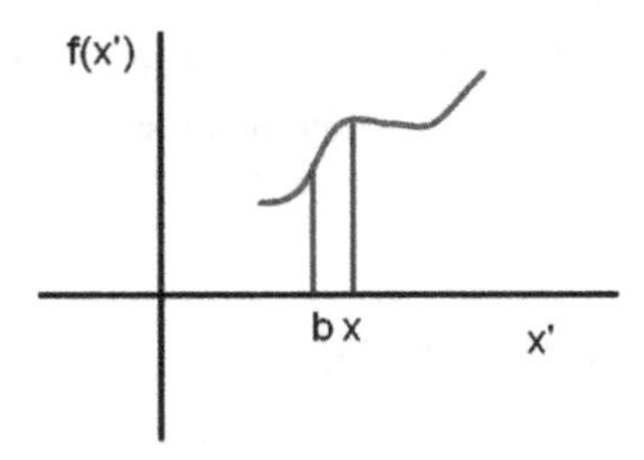

Fig. 2.1 Approximating a function $f(x')$ near $x' = b$

better the approximation. In effect, each time we add a power to the polynomial fit, we need to know one higher derivative of the function in the vicinity of $x' = b$. To proceed formally one writes

$$f^{(n-1)}(x) - f^{(n-1)}(b) = \int_b^x f^{(n)}(x'')\,dx'', \tag{2.6}$$

where $f^{(n)}(x)$ is the nth derivative of $f(x)$ and $f^{(n)}(b)$ is a shorthand notation for $f^{(n)}(x)$ evaluated at $x = b$. If you change the x to x' in this equation and integrate both sides with respect to x' from b to x, you can obtain

$$f^{(n-2)}(x) - f^{(n-2)}(b) - (x-b)f^{(n-1)}(b) = \int_b^x dx' \int_b^{x'} f^{(n)}(x'')\,dx''. \tag{2.7}$$

This procedure can be repeated to arrive at

$$f^{(n-3)}(x) - f^{(n-3)}(b) - (x-b)f^{(n-2)}(b) - \frac{(x-b)^2}{2!}f^{(n-1)}(b)$$
$$= \int_b^x dx_3 \int_b^{x_3} dx_2 \int_b^{x_2} f^{(n)}(x_1)\,dx_1. \tag{2.8}$$

Continuing up to n integrations, I find

$$f(x) - f(b) - (x-b)f^{(1)}(b)$$
$$-\frac{(x-b)^2}{2!}f^{(2)}(b) - \cdots - \frac{(x-b)^{n-1}}{(n-1)!}f^{(n-1)}(b)$$
$$= \int_b^x dx_n \ldots \int_b^{x_4} dx_3 \int_b^{x_3} dx_2 \int_b^{x_2} f^{(n)}(x_1)\,dx_1. \tag{2.9}$$

Solving Eq. (2.9) for $f(x)$, I obtain

$$f(x) = f(b) + (x-b)f^{(1)}(b) + \frac{(x-b)^2}{2!}f^{(2)}(b) + \cdots + \frac{(x-b)^{n-1}}{(n-1)!}f^{(n-1)}(b), \tag{2.10}$$

with a remainder that is of order $\frac{(x-b)^n}{(n)!}f^{(n)}(b)$. Equation (2.10) is known as Taylor's theorem and the sum on the right-hand side of the equation is a Taylor series of the function $f(x)$ about $x = b$. One can set $b \to y$ and $x \to y + a$ in Eq. (2.10) to write Taylor's theorem in the form

$$f(y+a) = f(y) + af^{(1)}(y) + \frac{a^2}{2!}f^{(2)}(y) + \cdots \frac{a^{n-1}}{(n-1)!}f^{(n-1)}(y), \tag{2.11}$$

with a remainder that is of order $\frac{a^n}{(n)!}f^{(n)}(y)$. In general, Eq. (2.10) converges only for $|x - b| < r_c(b)$ and Eq. (2.11) only for $|a| < r_c(y)$, where r_c is the *radius of convergence*.

As an example, I can approximate $\sqrt{27}$ by taking $b = 25$, $x = 27$, and $f(x) = \sqrt{x}$ in Eq. (2.10). If I keep only three terms in the series, I find

$$\sqrt{27} \approx \sqrt{25} + \frac{2}{2\sqrt{25}} + \frac{4}{2!}\left(-\frac{1}{2}\right)\frac{1}{2}\frac{1}{25^{3/2}} = 5.196, \tag{2.12}$$

with an expected error of order $\frac{8}{3!}\frac{1}{2}\frac{1}{2}\frac{3}{2}\frac{1}{25^{5/2}} = 0.00016$. The exact value is 5.19615 so the error is 0.00015, in agreement with the estimate of the error. The Taylor series of $\sqrt{25+x}$ converges for $|x - 25| < 25$.

2.2.2 *Scalar Functions of Three Variables*

Next I consider scalar functions of the form $f(\mathbf{r})$. What does this mean? In this expression, $\mathbf{r} = x\mathbf{u}_x + y\mathbf{u}_y + z\mathbf{u}_z$ is the coordinate vector (I use a notation in which $\mathbf{u}_j$ is a unit vector in the j direction). If one specifies (x, y, z), then $f(\mathbf{r}) \equiv f(x, y, z)$ gives a prescription for evaluating the value of the function at that point. For example, the scalar potential associated with a point charge q located at position $\mathbf{a} = a_x\mathbf{u}_x + a_y\mathbf{u}_y + a_z\mathbf{u}_z$ is

$$V(\mathbf{r}) = V(x, y, z) = \frac{q}{4\pi\epsilon_0}\frac{1}{\sqrt{(x-a_x)^2 + (y-a_y)^2 + (z-a_z)^2}}. \tag{2.13}$$

Taylor's theorem can be extended to functions of three variables, such as $V(\mathbf{r})$. To generalize the one variable result, I use a "trick." The trick is to express the function at the new point in terms of a *single dependent variable*, as one does when writing a parametric equation for a line in terms of a single variable [e.g., $x = at, y = bt^2$]. Thus I write

$$f(\mathbf{r} + \mathbf{a}) = f(x + a_x, y + a_y, z + a_z) = f(x + \alpha t, y + \beta t, z + \gamma t), \tag{2.14}$$

where

$$a_x = \alpha t, \quad a_y = \beta t, \quad a_z = \gamma t. \tag{2.15}$$

Now I can consider $f(\mathbf{r}+\mathbf{a})$ as a function of a single variable t and use the chain rule to obtain

$$f(\mathbf{r}+\mathbf{a}) = f(\mathbf{r}) + \left.\frac{df}{dt}\right|_{t=0} t + \frac{1}{2!}\left.\frac{d^2f}{dt^2}\right|_{t=0} t^2 + \cdots, \tag{2.16}$$

where

$$\begin{aligned}\left.\frac{df}{dt}\right|_{t=0} &= \left[\begin{array}{c}\frac{\partial f}{\partial(x+\alpha t)}\frac{d(x+\alpha t)}{dt} + \frac{\partial f}{\partial(y+\beta t)}\frac{d(y+\beta t)}{dt} \\ +\frac{\partial f}{\partial(z+\gamma t)}\frac{d(z+\gamma t)}{dt}\end{array}\right]_{t=0} \\ &= \frac{\partial f}{\partial x}\alpha + \frac{\partial f}{\partial y}\beta + \frac{\partial f}{\partial z}\gamma. \end{aligned} \tag{2.17}$$

Similarly,

$$\begin{aligned}\left.\frac{d^2f}{dt^2}\right|_{t=0} &= \left[\frac{\partial f'}{\partial(x+\alpha t)}\alpha + \frac{\partial f'}{\partial(y+\beta t)}\beta + \frac{\partial f'}{\partial(z+\gamma t)}\gamma\right]_{t=0} \\ &= \frac{\partial^2 f}{\partial x^2}\alpha^2 + \frac{\partial^2 f}{\partial x\partial y}\beta\alpha + \frac{\partial^2 f}{\partial x\partial z}\gamma\alpha + \frac{\partial^2 f}{\partial y\partial x}\alpha\beta + \frac{\partial^2 f}{\partial y^2}\beta^2 \\ &\quad + \frac{\partial^2 f}{\partial y\partial z}\gamma\beta + \frac{\partial^2 f}{\partial z\partial x}\alpha\gamma + \frac{\partial^2 f}{\partial z\partial y}\beta\gamma + \frac{\partial^2 f}{\partial z^2}\gamma^2. \end{aligned} \tag{2.18}$$

Combining all terms and using Eq. (2.15), I obtain

$$\begin{aligned} f(\mathbf{r}+\mathbf{a}) &= f(\mathbf{r}) + \frac{\partial f}{\partial x}a_x + \frac{\partial f}{\partial y}a_y + \frac{\partial f}{\partial z}a_z \\ &\quad + \frac{1}{2!}\left[\begin{array}{c}\frac{\partial^2 f}{\partial x^2}a_x^2 + \frac{\partial^2 f}{\partial y^2}a_y^2 + \frac{\partial^2 f}{\partial z^2}a_z^2 \\ +2\frac{\partial^2 f}{\partial x\partial y}a_xa_y + 2\frac{\partial^2 f}{\partial x\partial z}a_xa_z + 2\frac{\partial^2 f}{\partial z\partial y}a_za_y\end{array}\right] + \cdots. \end{aligned} \tag{2.19}$$

It is easy to generate higher order terms. Note that the first derivative terms can be written as $\nabla f \cdot \mathbf{a}$ and can be used to *define* the gradient in arbitrary coordinate systems.

2.2.3 Vector Functions of Three Variables

Quantities such as scalars, vectors, tensors are often defined in terms of their transformation properties under some symmetry operation. For example, a scalar function under rotation is one that is unchanged as the coordinate axes are rotated. A vector function such as the electric field $\mathbf{E}(\mathbf{r})$ consists of *three* scalar functions $[E_x(\mathbf{r}), E_y(\mathbf{r}), E_z(\mathbf{r})]$ whose components change in a prescribed manner under a rotation of the coordinate axes. If you want to make a Taylor series expansion of a vector function, you must expand *each* of the component functions in a Taylor series. Thus,

$$\begin{aligned}\mathbf{E}(\mathbf{r}+\mathbf{a}) \simeq \mathbf{E}(\mathbf{r}) &+ \left[\frac{\partial E_x}{\partial x}a_x + \frac{\partial E_x}{\partial y}a_y + \frac{\partial E_x}{\partial z}a_z\right]\mathbf{u}_x + \\ &+ \left[\frac{\partial E_y}{\partial x}a_x + \frac{\partial E_y}{\partial y}a_y + \frac{\partial E_y}{\partial z}a_z\right]\mathbf{u}_y \\ &+ \left[\frac{\partial E_z}{\partial x}a_x + \frac{\partial E_z}{\partial y}a_y + \frac{\partial E_z}{\partial z}a_z\right]\mathbf{u}_z \\ &= \mathbf{E}(\mathbf{r}) + (\mathbf{a}\cdot\nabla)\,\mathbf{E}(\mathbf{r}).\end{aligned} \tag{2.20}$$

The vector form in the last line is useful only in rectangular coordinates. Note that if you write $\mathbf{E}(\mathbf{r})$ using spherical or cylindrical coordinates and make a Taylor's expansion, you must make a Taylor's expansion of *both* the components *and* the unit vectors, since the unit vectors depend on the coordinates.

2.3 Vector Calculus

Most of you are familiar with the divergence or Gauss theorem and Stokes theorem. There are generalized versions of these theorems that I will need at some later time. The generalized Gauss and Stokes theorems can be stated as:

Generalized Gauss theorems:

$$\oint_S d\mathbf{a} = \int_V d\tau \boldsymbol{\nabla}; \tag{2.21a}$$

$$\oint_S d\mathbf{a} f(\mathbf{r}) = \int_V d\tau \boldsymbol{\nabla} f(\mathbf{r}); \tag{2.21b}$$

$$\oint_S d\mathbf{a}\cdot\mathbf{F}(\mathbf{r}) = \int_V d\tau \boldsymbol{\nabla}\cdot\mathbf{F}(\mathbf{r}); \tag{2.21c}$$

$$\oint_S d\mathbf{a}\times\mathbf{F}(\mathbf{r}) = \int_V d\tau \boldsymbol{\nabla}\times\mathbf{F}(\mathbf{r}), \tag{2.21d}$$

where $\oint_S$ implies a surface integral containing the volume V and $d\mathbf{a}$ is an outward normal to the surface. The differential da is an element of surface area and $d\tau$ is a volume element.

Generalized Stokes theorems:

$$\oint_C d\mathbf{l} = \int_S (d\mathbf{a} \times \nabla); \tag{2.22a}$$

$$\oint_C d\mathbf{l} f(\mathbf{r}) = \int_S (d\mathbf{a} \times \nabla) f(\mathbf{r}); \tag{2.22b}$$

$$\oint_C d\mathbf{l} \cdot \mathbf{F}(\mathbf{r}) = \int_S (d\mathbf{a} \times \nabla) \cdot \mathbf{F}(\mathbf{r}) = \int_S d\mathbf{a} \cdot [\nabla \times \mathbf{F}(\mathbf{r})]; \tag{2.22c}$$

$$\oint_C d\mathbf{l} \times \mathbf{F}(\mathbf{r}) = \int_S (d\mathbf{a} \times \nabla) \times \mathbf{F}(\mathbf{r}), \tag{2.22d}$$

where $\oint_C$ implies a line integral containing the surface area S and $d\mathbf{l}$ is a differential element tangent to the line. Note that $d\mathbf{a} \times \nabla$ takes on a simple form only in rectangular coordinates.

We will often encounter expressions that involve taking the gradient or Laplacian of exponential functions, namely

$$\nabla e^{iax} = iae^{iax}\mathbf{u}_x; \tag{2.23a}$$

$$\begin{aligned} \nabla e^{i\mathbf{a}\cdot\mathbf{r}} &= \nabla e^{i\left(a_x x + a_y y + a_z z\right)} \\ &= i\left(a_x \mathbf{u}_x + a_y \mathbf{u}_y + a_z \mathbf{u}_z\right) e^{i\mathbf{a}\cdot\mathbf{r}} = i\mathbf{a}e^{i\mathbf{a}\cdot\mathbf{r}}; \end{aligned} \tag{2.23b}$$

$$\nabla^2 e^{iax} = -a^2 e^{iax}; \tag{2.23c}$$

$$\nabla^2 e^{i\mathbf{a}\cdot\mathbf{r}} = -\left(a_x^2 + a_y^2 + a_z^2\right) e^{i\mathbf{a}\cdot\mathbf{r}} = -a^2 e^{i\mathbf{a}\cdot\mathbf{r}}, \tag{2.23d}$$

where

$$\nabla = \mathbf{u_x}\frac{\partial}{\partial x} + \mathbf{u}_y\frac{\partial}{\partial y} + \mathbf{u}_z\frac{\partial}{\partial z} \tag{2.24}$$

is the gradient operator,

$$\nabla^2 = \frac{\partial^2}{\partial x^2} + \frac{\partial^2}{\partial y^2} + \frac{\partial^2}{\partial z^2} \tag{2.25}$$

is the Laplacian operator, and $\mathbf{u}_j$ is a unit vector in the j direction.

2.4 Probability Distributions

You are all familiar with elementary concepts of probability theory. If you throw one die, there is a probability of $1/6$ that any number comes up. If you increase the number of sides of the die to 3 million, then the probability for any side to come up is $1/\left(3\times 10^{6}\right)$ (welcome to the lottery). As the number of sides increases without limit, the probability for any *specific* event to occur goes to zero, even if we know one event must occur on each trial. In the limit of the number of sides going to infinity, the probability of individual events is replaced by what is called the *probability density* $P(x)$ for an event to occur.

To illustrate the concept of a probability density, consider the probability that a point chosen at random on a line having length L is at the midpoint of the line. Of course, this probability is zero, as it is to obtain any single point in a single measurement. On the other hand, we can ask for the probability that a point chosen at random lies between x and $x+dx$. This probability is no longer equal to zero, but is given by

$$P(x)dx = dx/L, \tag{2.26}$$

since

$$P(x) = \begin{cases} 1/L & 0 \leq x \leq L \\ 0 & \text{otherwise} \end{cases} \tag{2.27}$$

is the probability density for a uniform distribution of points on the line.

I limit the discussion to one dimension, but extensions to higher dimensions are obvious. The probability distribution is *normalized* such that

$$\int_{-\infty}^{\infty} P(x)dx = 1. \tag{2.28}$$

The nth moment of the distribution is defined as

$$\langle x^n \rangle = \int_{-\infty}^{\infty} P(x)x^n dx. \tag{2.29}$$

The *average value* of x is the first moment,

$$\langle x \rangle = \bar{x} = \int_{-\infty}^{\infty} P(x)xdx, \tag{2.30}$$

while the *variance* of x is defined as

$$\Delta x^2 = \left\langle (x - \langle x \rangle)^2 \right\rangle = \int_{-\infty}^{\infty} P(x)\left(x - \langle x \rangle\right)^2 dx = \left\langle x^2 \right\rangle - \langle x \rangle^2 = \overline{x^2} - \bar{x}^2. \tag{2.31}$$

The *standard deviation* of x, denoted by Δx, is the square root of the variance. Often, but not always, the standard deviation is a measure of the width of the distribution about its mean value.

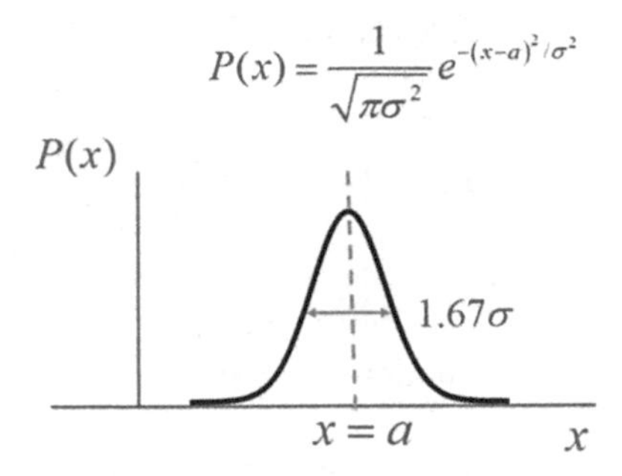

Fig. 2.2 Gaussian probability distribution

One of the most important probability distributions is the *Gaussian* or *normal* distribution defined by

$$P(x) = Ne^{-(x-a)^2/\sigma^2}, \tag{2.32}$$

where N is a normalization factor (see Fig. 2.2). To determine N, I require that

$$\int_{-\infty}^{\infty} Ne^{-(x-a)^2/\sigma^2} dx = 1. \tag{2.33}$$

The only way to evaluate *any* integral is to already know the answer—that is, you guess a solution and see if it works.[2] Fortunately, there are tables of integrals and built-in functions in computer programs that allow you to benefit from the collected guesses of many mathematicians and physicists. In this case, you will find that

$$\int_{-\infty}^{\infty} Ne^{-(x-a)^2/\sigma^2} dx = N\sigma\sqrt{\pi}, \tag{2.34}$$

which, when combined with Eqs. (2.32) and (2.33), leads to the normalized distribution

$$P(x) = \frac{1}{\sqrt{\pi}\sigma} e^{-(x-a)^2/\sigma^2}. \tag{2.35}$$

The mean or average value of x is

$$\bar{x} = \frac{1}{\sqrt{\pi}\sigma} \int_{-\infty}^{\infty} xe^{-(x-a)^2/\sigma^2} dx = a, \tag{2.36}$$

[2] Of course, you have undoubtedly learned many techniques for evaluating integrals, but these are all based on guesses that work.

as should be obvious from Fig. 2.2. The variance is

$$\Delta x^2 = \frac{1}{\sqrt{\pi} a} \int_{-\infty}^{\infty} e^{-(x-a)^2/\sigma^2} (x-a)^2 \, dx = \frac{\sigma^2}{2}, \tag{2.37}$$

so $\Delta x = \sigma/\sqrt{2}$.

I can calculate the half-width at half maximum (HWHM) or full-width at half maximum (FWHM) of this distribution by solving the equation

$$\left| f(x_{1/2}) \right|^2 = |f(\bar{x})|^2 / 2 = |f(a)|^2 / 2 \tag{2.38}$$

or

$$\frac{1}{\sqrt{\pi}\sigma} e^{-\left(x_{1/2}-a\right)^2/\sigma^2} = \frac{1}{2\sqrt{\pi}\sigma}. \tag{2.39}$$

The solution of this equation is

$$\left| x_{1/2} - a \right| = \sigma \sqrt{\ln 2} \approx 0.8326\sigma. \tag{2.40}$$

The HWHM is equal to 0.8326σ and the FWHM is equal to 1.665σ.

There can be some confusion as to the meaning of the "width" of a probability distribution. For a smooth distribution such as a Gaussian, one can define the "width" as either the HWHM or FWHM. For other distributions, the HWHM and FWHM may have no significance at all. On the other hand, the standard deviation of a probability distribution is always defined as the square root of the variance of the distribution. For some distributions, such as a Gaussian, the standard deviation and HWHM are not all that different, but for other distributions, they can differ dramatically. Some examples will help to illustrate this point.

For a uniform probability distribution

$$P(x) = \begin{cases} 1 & |x| \leq 1/2 \\ 0 & \text{otherwise} \end{cases}, \tag{2.41}$$

(x is now taken as a dimensionless variable), the HWHM is equal to $1/2$, while $\Delta x = 1/\sqrt{12} = 0.29$. For a Gaussian probability distribution

$$P(x) = \frac{1}{\pi^{1/2}} e^{-x^2}, \tag{2.42}$$

the HWHM is equal to 0.83, while $\Delta x = 1/\sqrt{2} = 0.707$. For both the uniform and Gaussian distributions the HWHM's and standard deviations are comparable. However, for the *Lorentzian* probability distribution

$$P(x) = \frac{1}{\pi} \frac{1}{x^2 + 1}, \tag{2.43}$$

the HWHM is equal to 1, while $\Delta x = \infty$. Finally if I take $P(x)$ as the sum of two Gaussian distributions,

$$P(x) = \frac{1}{2\pi^{1/2}} \left(e^{-(x-10)^2} + e^{-(x+10)^2} \right), \tag{2.44}$$

the HWHM has no real meaning [although you could talk about the HWHM of *each* term in Eq. (2.44)], while $\Delta x = \sqrt{201/2}$. Thus, although I will often use the terms "width" and "standard deviation" interchangeably, they need not be the same at all.

2.5 Fourier Transforms

In quantum mechanics, we often need to use Fourier transforms. The wave functions in coordinate space and momentum space are Fourier transforms of one another. This topic is covered in most textbooks on mathematical physics, so I will just sketch a few of the results. The functions $f(x)$ and $a(k)$ are Fourier transforms of one another if

$$f(x) = \frac{1}{\sqrt{2\pi}} \int_{-\infty}^{\infty} a(k) e^{ikx} dk, \tag{2.45}$$

$$a(k) = \frac{1}{\sqrt{2\pi}} \int_{-\infty}^{\infty} f(x) e^{-ikx} dx, \tag{2.46}$$

assuming that these integrals exist. We can also talk about frequency-time Fourier transforms defined by

$$g(t) = \frac{1}{\sqrt{2\pi}} \int_{-\infty}^{\infty} b(\omega) e^{-i\omega t} d\omega, \tag{2.47a}$$

$$b(\omega) = \frac{1}{\sqrt{2\pi}} \int_{-\infty}^{\infty} g(t) e^{i\omega t} dt. \tag{2.47b}$$

In essence, the Fourier transform of a spatial function is an expansion in functions having different wavelengths or propagation constants k, while the Fourier transform of a time-dependent function is an expansion in terms of its frequency components (including negative frequencies).

There are many properties of Fourier transforms that are derived in standard textbooks. The most important property that we will encounter relates the variances Δx^2 and Δk^2. Let $f(x)$ and $a(k)$ be Fourier transforms of one another and define

$$\bar{x} = \int_{-\infty}^{\infty} x \, |f(x)|^2 \, dx, \tag{2.48a}$$

$$\Delta x^2 = \int_{-\infty}^{\infty} (x - \bar{x})^2 \, |f(x)|^2 \, dx, \tag{2.48b}$$

$$\bar{k} = \int_{-\infty}^{\infty} k \, |a(k)|^2 \, dk, \tag{2.48c}$$

$$\Delta k^2 = \int_{-\infty}^{\infty} \left(k - \bar{k}\right)^2 |a(k)|^2 \, dk. \tag{2.48d}$$

It then follows that

$$\Delta x \Delta k = \sqrt{\Delta x^2 \Delta k^2} \geq \frac{1}{2}, \tag{2.49}$$

provided all these quantities exist and the normalization is such that

$$\int_{-\infty}^{\infty} |f(x)|^2 \, dx = 1. \tag{2.50}$$

I will prove Eq. (2.49) in Chap. 5. In essence, $|f(x)|^2$ can be considered to be a probability distribution in coordinate space and $|a(k)|^2$ a probability distribution in k-space. The quantities $\bar{x}$ and $\bar{k}$ are the average values or mean of the functions $|f(x)|^2$ and $|a(k)|^2$, respectively, and Δx^2 and Δk^2 are the variances in coordinate and k-space, respectively, while

$$\Delta x = \sqrt{\Delta x^2}; \qquad \Delta k = \sqrt{\Delta k^2} \tag{2.51}$$

are the standard deviations in coordinate and k-space, respectively. Thus the narrower the distribution is in k space, the wider it is in coordinate space and visa versa.

As a simple example, consider

$$f(x) = \frac{1}{\left(\pi\sigma^2\right)^{1/4}} e^{-x^2/2\sigma^2} e^{ik_0 x}, \tag{2.52}$$

for which

$$|f(x)|^2 = \frac{1}{\left(\pi\sigma^2\right)^{1/2}} e^{-x^2/\sigma^2} \tag{2.53}$$

is a Gaussian having full-width at half maximum (FWHM) equal to 1.67σ. The Fourier transform of $f(x)$ is

$$
\begin{aligned}
a(k) &= \frac{1}{\sqrt{2\pi}} \int_{-\infty}^{\infty} dx f(x) \exp(-ikx) \\
&= \frac{1}{\sqrt{2\pi}} \frac{1}{(\pi\sigma^2)^{1/4}} \int_{-\infty}^{\infty} dx\, e^{-x^2/2\sigma^2} e^{-i(k-k_0)x}.
\end{aligned}
\tag{2.54}
$$

The integral is tabulated or can be evaluated using contour integration. Explicitly, one finds

$$a(k) = \left(\frac{\sigma^2}{\pi}\right)^{1/4} e^{-(k-k_0)^2\sigma^2/2}; \tag{2.55}$$

$$|a(k)|^2 = \left(\frac{\sigma^2}{\pi}\right)^{1/2} e^{-(k-k_0)^2\sigma^2}. \tag{2.56}$$

The k-space distribution, $|a(k)|^2$, is *also* a Gaussian, centered at $k = k_0$, having FWHM equal to $1.67/\sigma$.

The variance of x is

$$\Delta x^2 = \int_{-\infty}^{\infty} dx\, x^2 \, |\psi(x,0)|^2 = \frac{\sigma^2}{2} \tag{2.57}$$

and the variance of k is

$$\Delta k^2 = \left\langle (k-k_0)^2 \right\rangle = \int_{-\infty}^{\infty} dk\, (k-k_0)^2 \, |a(k)|^2 = \frac{1}{2\sigma^2}, \tag{2.58}$$

such that

$$\Delta x \Delta k = \frac{1}{2}, \tag{2.59}$$

the minimum possible value.

Similarly, in the time domain for the distributions $|g(t)|^2$ and $|b(\omega)|^2$, one finds

$$\Delta\omega \Delta t \geq \frac{1}{2}, \tag{2.60}$$

which is known as the frequency-time uncertainty relation. The narrower the bandwidth (frequency spread) of a pulse, the wider is its time extent. As we shall see, the coordinate—k-space uncertainty relation follows from the postulates of quantum mechanics, but this is *not* the case for the frequency-time uncertainty relation. In effect, there is no probability distribution that can be associated with the time in quantum mechanics.

2.6 Dirac Delta Function

The Dirac delta function appears in many branches of physics. It is not difficult to understand why this is the case. For example, what is the density of a point charge having total charge q? Clearly, the charge density $\rho(\mathbf{r})$ equals zero everywhere but at the location of the charge. Since the volume of a point equals zero, the density at the position of the charge must be infinite. This leads us to define a function in three dimensions, $\delta(\mathbf{r})$, such that for a point charge,

$$\rho(\mathbf{r}) = q\delta(\mathbf{r}), \tag{2.61}$$

where

$$\delta(\mathbf{r}) = 0 \text{ if } \mathbf{r} \neq 0; \tag{2.62a}$$

$$\delta(\mathbf{r}) = \infty \text{ if } \mathbf{r} = 0; \tag{2.62b}$$

$$\int \delta(\mathbf{r})d\mathbf{r} = 1 \text{ if the origin is inside the integration volume;} \tag{2.62c}$$

$$\int \delta(\mathbf{r})d\mathbf{r} = 0 \text{ if the origin is outside the integration volume.} \tag{2.62d}$$

With this definition

$$\int \rho(\mathbf{r})d\mathbf{r} = q \tag{2.63}$$

as required, provided the charge is in the integration volume. Note that $\delta(\mathbf{r})$ has units of inverse volume.

The Dirac delta function is usually defined by its *integral* properties. That is, for any analytic function $f(\mathbf{r})$, it is defined by

$$\int f(\mathbf{r})\delta(\mathbf{r})d\mathbf{r} = f(0). \tag{2.64}$$

In other words, the Dirac delta function picks out the value of the function at the point where the argument of the delta function is equal to zero. The definitions given in Eqs. (2.62) and (2.64) are consistent.

I will first consider the one-dimensional Dirac delta function and then briefly discuss the Dirac delta function in two and three dimensions. The Dirac delta function in one dimension can be defined by

$$\delta(x) = 0 \text{ if } x \neq 0; \tag{2.65a}$$

$$\delta(x) = \infty \text{ if } x = 0; \tag{2.65b}$$

$$\int \delta(x)dx = 1 \text{ if the origin is inside the integration volume;} \tag{2.65c}$$

$$\int \delta(x)dx = 0 \text{ if the origin is outside the integration volume;} \tag{2.65d}$$

or by

$$\int f(x)\delta(x)dx = f(0), \tag{2.66}$$

where $f(x)$ is an arbitrary analytic function. Note that $\delta(x)$ has units of inverse length.

The Dirac delta function $\delta(x)$ is an infinitely narrow function of x centered at $x = 0$. The area under the function is equal to unity. Perhaps the easiest way to envision the Dirac delta function is the limit of a Gaussian having unit area whose height goes to infinity and width goes to zero,

$$\delta(x) = \lim_{a\to 0} \frac{1}{\sqrt{\pi a^2}} e^{-x^2/a^2}. \tag{2.67}$$

This function has all the required properties. Other representations of the Dirac delta function are

$$\delta(x) = \frac{1}{\pi} \lim_{a\to 0} \frac{a}{x^2 + a^2}, \tag{2.68}$$

$$\delta(x) = \lim_{K\to\infty} \frac{\sin(Kx)}{\pi x}, \tag{2.69}$$

and

$$\delta(x) = \frac{1}{2\pi} \int_{-\infty}^{\infty} e^{ikx} dk. \tag{2.70}$$

Equations (2.67) and (2.68) are obvious representations of a narrow function having area equal to unity, but what about Eqs. (2.69) and (2.70)? These equations do not define "proper" mathematical functions and are meaningful only when integrated with functions that vanish as $|x| \sim \infty$. Equation (2.70) is interesting in that the Dirac delta function can be defined as the Fourier transform of a constant. One way of seeing that this works is to write

$$\delta(x) = \lim_{K\to\infty} \frac{1}{2\pi} \int_{-K}^{K} e^{ikx} dk = \lim_{K\to\infty} \frac{\sin(Kx)}{\pi x}. \tag{2.71}$$

As K gets larger, the width of the central peak gets narrower and its height grows as K/π. With increasing K, the function oscillates so rapidly outside the central peak that, when integrated with an analytic function $f(x)$, any contributions to the integral average to zero, except near the central peak. Equation (2.70) is one that you will encounter many times in this course and other physics courses. Don't forget it! For any variable $\varnothing$ (e.g. $\varnothing = x$, $\varnothing = x - x_0$, etc.),

$$\delta(\varnothing) = \frac{1}{2\pi} \int_{-\infty}^{\infty} e^{ik\varnothing} dk. \tag{2.72}$$

Some useful properties of the Dirac delta function are:

$$\delta(-x) = \delta(x); \tag{2.73a}$$

$$\int_{-\infty}^{\infty} dx' f(x') \delta(x - x') = f(x); \tag{2.73b}$$

$$\delta(ax) = \frac{\delta(x)}{|a|}; \tag{2.73c}$$

$$\delta[g(x)] = \sum_i \frac{\delta(x - x_i)}{|g'(x_i)|}; \tag{2.73d}$$

$$\delta\left(x^2 - a^2\right) = \frac{1}{2|a|}\left[\delta(x - a) + \delta(x + a)\right]; \tag{2.73e}$$

$$\int_{-\infty}^{\infty} dx' f(x') \frac{d\delta(x - x')}{dx'} = -f'(x). \tag{2.73f}$$

The sum over i is over all the roots of $g(x_i) = 0$. The proofs are left to the problems.

The Dirac delta function can be used to derive the inverse Fourier transform. Given

$$f(x) = \frac{1}{\sqrt{2\pi}} \int_{-\infty}^{\infty} a(k') e^{ik'x} dk', \tag{2.74}$$

I can multiply this equation by e^{-ikx} and integrate over x to arrive at

$$\begin{aligned} \frac{1}{\sqrt{2\pi}} \int_{-\infty}^{\infty} dx f(x) e^{-ikx} &= \frac{1}{2\pi} \int_{-\infty}^{\infty} dk'\, a(k') \int_{-\infty}^{\infty} dx\, e^{i(k'-k)x} \\ &= \int_{-\infty}^{\infty} dk'\, a(k') \delta\left(k' - k\right) = a(k), \end{aligned} \tag{2.75}$$

where I used

$$\frac{1}{2\pi} \int_{-\infty}^{\infty} dx\, e^{i(k'-k)x} = \delta\left(k' - k\right), \tag{2.76}$$

which follows from Eq. (2.72).

In fact, if $f(x)$ and $a(k)$ are Fourier transforms of one another, then

$$f(x) = \frac{1}{\sqrt{2\pi}} \int_{-\infty}^{\infty} a(k) e^{ikx} dk = \int_{-\infty}^{\infty} dx' \left[\frac{1}{2\pi} \int_{-\infty}^{\infty} dk e^{ik(x-x')} \right] f(x'), \tag{2.77}$$

from which it follows that

$$\frac{1}{2\pi} \int_{-\infty}^{\infty} dk e^{ik(x-x')} = \delta\left(x - x'\right). \tag{2.78}$$

In other words, the existence of Fourier transforms leads us to the representation of the Dirac delta function given by Eq. (2.70). You can also use Eq. (2.70) to show that

$$\int_{-\infty}^{\infty} dx \, |f(x)|^2 = \int_{-\infty}^{\infty} dk \, |a(k)|^2, \tag{2.79}$$

which is known as *Parseval's theorem.*

In two dimensions, the Dirac delta function is

$$\delta(\boldsymbol{\rho} - \boldsymbol{\rho}') = \delta\left(x - x'\right) \delta\left(y - y'\right) = \frac{1}{\rho} \delta\left(\rho - \rho'\right) \delta\left(\varphi - \varphi'\right) \tag{2.80}$$

[ρ and ϕ are cylindrical coordinates] and, in three dimensions, it is given by

$$\begin{aligned} \delta(\mathbf{r} - \mathbf{r}') &= \delta\left(x - x'\right) \delta\left(y - y'\right) \delta\left(z - z'\right) \\ &= \frac{1}{r^2} \delta\left(r - r'\right) \delta\left(\cos\theta - \cos\theta'\right) \delta\left(\varphi - \varphi'\right) \end{aligned} \tag{2.81}$$

[r, θ, and ϕ are spherical coordinates].

2.7 Problems

Note: Problems with two or more problem numbers are an indication that the problem might take longer to solve than an average problem.

1. Evaluate $e^{i\pi}$, $e^{i\pi/2}$, and $e^{2.3i}$. If

$$\psi(x) = \frac{e^{iax} e^{-gx^2/2}}{x + ib} = u + iv = re^{i\theta},$$

find u, v, r, θ, assuming that x, a, b, g, r, θ are real. Evaluate $|\psi(x)|^2$.

2. Given the function

$$f(x) = \frac{N}{b^2 + x^2},$$

find N such that $f^2(x)$ is normalized; that is, find N such that

$$\int_{-\infty}^{\infty} f^2(x)dx = 1,$$

assuming that b is real. Find the Fourier transform $a(k)$ of $f(x)$. Evaluate

$$\Delta x^2 = \int_{-\infty}^{\infty} (x - \bar{x})^2 f^2(x)dx$$
$$\Delta k^2 = \int_{-\infty}^{\infty} \left(k - \bar{k}\right)^2 |a(k)|^2 dk$$

and the product $\Delta x \Delta k$. Does $\Delta x \Delta k = 1/2$ for these functions? To evaluate the integrals, you can use integral tables or Mathematica, Maple, or Matlab.

3–4. Suppose that the k space amplitude for a free particle in quantum mechanics is given by

$$a(k) = \begin{cases} 1/\sqrt{2k_0} & -k_0 \le k \le k_0 \\ 0 & \text{otherwise} \end{cases}.$$

The wave function $\psi(x, 0)$ is the Fourier transform of $a(k)$. Plot both $k_0 |a(k)|^2$ as a function of k/k_0 and $|\psi(x, 0)|^2 / k_0$ as a function of $k_0 x$. By "eyeballing" the graphs, estimate Δx, Δk, and their product. Now calculate Δk analytically and show that Δx is infinite.

5. If the functions $f(x)$ and $a(k)$ are Fourier transforms of one another, prove Parseval's Theorem,

$$\int_{-\infty}^{\infty} |f(x)|^2 dx = \int_{-\infty}^{\infty} |a(k)|^2 dk.$$

6. Prove

$$\delta(ax) = \frac{\delta(x)}{|a|}$$
$$\delta\left(x^2 - a^2\right) = \frac{1}{2|a|}\left[\delta(x - a) + \delta(x + a)\right]$$
$$\int_{-\infty}^{\infty} dx' f(x') \frac{d\delta(x - x')}{dx'} = -f'(x).$$

7. Given the probability distribution

$$P(x) = \begin{cases} 1 & 0 \leq x \leq 1 \\ 0 & \text{otherwise} \end{cases} .$$

Calculate $\bar{x}$ and Δx. If the probability distribution is shifted so it is centered at $x = 0$, do $\bar{x}$ and Δx change?

8. Use Taylor's theorem to estimate $29^{1/3}$ correct to order 0.001. Compare your answer with the exact value.

9. Show that $\int_0^\infty dx \frac{d\delta(x)}{dx}$ is not well defined. To do this use the definition of the delta function as the limit of a Gaussian,

$$\delta(x) = \lim_{a \to 0} \frac{1}{\sqrt{\pi a^2}} e^{-x^2/a^2},$$

to show that the integral diverges as $a \to 0$.

Chapter 3
Free-Particle Schrödinger Equation: Wave Packets

The Schrödinger equation is the fundamental equation of non-relativistic quantum mechanics. As with any equation in physics, its validity relies on experimental verification of the predictions of the equation. So far, it appears that there are no experiments that are inconsistent with quantum mechanics. As you shall see, it is not always easy to test the predictions of the Schrödinger equation. In other words, mapping out the probability distribution associated with a quantum system can represent a formidable task. Moreover, quantum mechanics is far from a complete theory since it does not address the dynamic evolution of the wave function when a measurement is made. Nevertheless, the success of the Schrödinger equation in describing the wave nature of matter and the energy level structure of atoms, molecules, and solids is beyond question.

In this chapter, I discuss the Schrödinger equation for a free particle, a particle not subjected to forces. Even though I use the word "particle" throughout this book, it is a misnomer in many cases since the particle is actually acting as a wave. It is not possible to *derive* Schrödinger's equation; it is essentially a postulate of quantum mechanics. However, it is possible to use an analogy with the wave equation of electromagnetism, plus some additional ingredients, to cook up an equation that turns out to be the time-dependent Schrödinger equation for a free particle. This is the recipe I shall follow.

I want to remind you of an admonition given in Chap. 1. The following chapters contain many mathematical expressions. Rather than focus on the mathematical details, you should always try to have a general idea of where the calculations are going. In other words, what physical features of a specific problem are being analyzed? Try not to let the mathematics obscure the underlying physical processes under investigation.

P.R. Berman, *Introductory Quantum Mechanics*, UNITEXT for Physics,
https://doi.org/10.1007/978-3-319-68598-4_3

3.1 Electromagnetic Wave Equation: Pulses

Maxwell's equations imply that both the electric and magnetic fields obey wave equations. The wave equation for the electric field vector $\mathbf{E}(\mathbf{R}, t)$ of light in vacuum is

$$\nabla^2 \mathbf{E}(\mathbf{R}, t) = \frac{1}{c^2}\frac{\partial^2 \mathbf{E}(\mathbf{R}, t)}{\partial t^2}, \tag{3.1}$$

where

$$\nabla^2 = \frac{\partial^2}{\partial x^2} + \frac{\partial^2}{\partial y^2} + \frac{\partial^2}{\partial z^2} \tag{3.2}$$

is defined here in terms of its rectangular components and c is the speed of light. A solution of this equation that also satisfies Maxwell's equations for the electric field is

$$\mathbf{E}(\mathbf{r}, t) = \hat{\boldsymbol{\epsilon}} E e^{i(\mathbf{k}\cdot\mathbf{r} - \omega t)} + \text{c.c.}, \tag{3.3}$$

provided

$$\omega = kc \tag{3.4}$$

and

$$\mathbf{k} \cdot \hat{\boldsymbol{\epsilon}} = 0, \tag{3.5}$$

where $\hat{\boldsymbol{\epsilon}}$ is a unit polarization vector for the field. Equation (3.5) must be satisfied to insure that $\nabla\cdot\mathbf{E}(\mathbf{r}, \mathbf{t}) = 0$. There are two independent polarizations for each field frequency and the direction of polarization is perpendicular to the propagation vector $\mathbf{k}$ of the field (the field is *transverse*). The abbreviation c.c. in Eq. (3.3) stands for *complex conjugate*.

The field in Eq. (3.3) corresponds to an infinite, monochromatic, plane wave and is a basic building block solution of Maxwell's wave equation. It is possible to construct *any* field pulse using a superposition of such states. To simplify matters, I consider only one polarization component of the field and a plane wave field. For an arbitrary plane wave field propagating in the *positive* x direction and polarized in the z direction, the electric field can be written as

$$\mathbf{E}(\mathbf{r}, t) = \mathbf{u}_z E(x, t), \tag{3.6}$$

where $\mathbf{u}_z$ is a unit vector in the z direction and the amplitude $E(x, t)$ can be expanded as

$$E(x, t) = \frac{1}{\sqrt{2\pi}} \int_0^\infty dk\, A(k) e^{i(kx - \omega t)} + \text{c.c.}. \tag{3.7}$$

The integral over k has been restricted to positive values to ensure that each component of the wave propagates in the positive x direction. It is a simple matter to show that Eq. (3.6) with $E(x,t)$ given by Eq. (3.7) is also a solution of the wave equation, provided $\omega = kc > 0$. Moreover, since $\omega = kc$,

$$E(x,t) = \frac{1}{\sqrt{2\pi}} \int_0^\infty dk\, A(k) e^{ik(x-ct)} + \text{c.c.} = E(x-ct,0). \tag{3.8}$$

The pulse amplitude is simply the original pulse amplitude translated by ct; in other words, the pulse propagates *without distortion* at a speed equal to the speed of light.

3.2 Schrödinger's Equation

I want to use Einstein's concept of photons and de Broglie's concept of matter waves to make a plausible transition from the wave equation of electromagnetism to the Schrödinger equation of quantum mechanics. I start by using Einstein's expression for the energy E (not to be confused with the field amplitude) associated with "photons" having frequency $f = \omega/2\pi$,

$$E = hf = \hbar\omega, \tag{3.9}$$

to transform Eq. (3.8) into

$$E(x,t) = \frac{1}{\sqrt{2\pi}} \int_0^\infty dk\, A(k) e^{i(kx-Et/\hbar)} + \text{c.c.}. \tag{3.10}$$

I now look for a wave function for *matter waves* having a similar form, namely

$$\psi(x,t) = \frac{1}{\sqrt{2\pi}} \int_{-\infty}^\infty dk\, \Phi(k) e^{i(kx-Et/\hbar)}. \tag{3.11}$$

Although the electric field amplitude $E(x,t)$ is real, it is assumed that the *wave function* $\psi(x,t)$ can be complex. The final step of the "derivation" is to use de Broglie's relation

$$p = \hbar k, \tag{3.12}$$

to write the energy E for a free particle having mass m in terms of k as

$$E = \frac{p^2}{2m} = \frac{\hbar^2 k^2}{2m}. \tag{3.13}$$

With this assignment, Eq. (3.11) becomes

$$\psi(x,t) = \frac{1}{\sqrt{2\pi}} \int_{-\infty}^\infty dk\, \Phi(k) \exp[i(kx - \hbar k^2 t/2m)]. \tag{3.14}$$

It is not difficult to prove that $\psi(x,t)$ satisfies the partial differential equation

$$i\hbar\frac{\partial\psi(x,t)}{\partial t}=-\frac{\hbar^2}{2m}\frac{\partial^2\psi(x,t)}{\partial x^2}, \tag{3.15}$$

which can be generalized to three dimensions as

$$i\hbar\frac{\partial\psi(\mathbf{r},t)}{\partial t}=-\frac{\hbar^2}{2m}\nabla^2\psi(\mathbf{r},t). \tag{3.16}$$

Equation (3.16) is recognized as the time-dependent Schrödinger equation for a free particle. Of course, some physical interpretation must be given to $\psi(\mathbf{r},t)$. As an additional postulate, I assume that $|\psi(\mathbf{r},t)|^2$ is the probability density to find the particle at position $\mathbf{r}$ at time t.

Equation (3.14) differs from Eq. (3.7) in a fundamental way since the dispersion relation (relation between ω and k) in Eq. (3.14) is not linear,

$$\omega=\frac{E}{\hbar}=\frac{\hbar k^2}{2m}. \tag{3.17}$$

As a result, waves having different values of k or p propagate at different velocities. Since the matter wave pulse, referred to as a *wave packet*, has components that propagate with different velocities and since no forces act on the particle, the shape of the wave packet changes in time, unlike that for a plane wave optical field pulse in vacuum.

3.2.1 Wave Packets

Now that I have defined the wave function, I can try to construct something that looks like a "particle." The term "particle" can be somewhat confusing. First of all, I am considering a wave theory, so I have to define what I mean by a "particle." Moreover, particles in classical physics necessarily have some internal structure and finite spatial extent. For the most part, the particles to which I refer in quantum mechanics correspond to idealized *point particles* in classical physics.

Let me start from the wave function in three dimensions for a matter wave corresponding to a point particle having mass m, that is, a generalization of Eq. (3.14) to three dimensions. It is not difficult to show that the wave function

$$\psi(\mathbf{r},t)=\frac{1}{(2\pi)^{3/2}}\int d\mathbf{k}\,\Phi(\mathbf{k})e^{i\left(\mathbf{k}\cdot\mathbf{r}-\hbar k^2 t/2m\right)}. \tag{3.18}$$

is a solution of time-dependent Schrödinger equation in three dimensions, Eq. (3.16). To model a particle, the momentum, or equivalently, the propagation vector $\mathbf{k}$ (since $\mathbf{k} = \mathbf{p}/\hbar$) must be fairly well defined. In other words, I must choose $|\Phi(\mathbf{k})|^2$ to be non-vanishing only for those values of $\mathbf{k}$ satisfying $|\mathbf{k} - \mathbf{k}_0| \lesssim k_0$, where $\mathbf{k}_0 = \bar{\mathbf{k}}$ and Δk is the standard deviation of k for the distribution $|\Phi(\mathbf{k})|^2$. Moreover, if $|\Phi(\mathbf{k})|^2$ is meant to represent a particle, it should be a smooth function having a maximum at $\mathbf{k} = \mathbf{k}_0$ that falls monotonically to zero in all directions (in k space) for $|\mathbf{k} - \mathbf{k}_0| > \Delta k$. In other words, $|\Phi(\mathbf{k})|^2$ is a sharply peaked function centered at $\mathbf{k} = \mathbf{k}_0$ having width of order Δk. I need to approximate the exponential function appearing in Eq. (3.18) when $\mathbf{k} \approx \mathbf{k}_0$.

It helps to write

$$\begin{aligned} \mathbf{k} &= \mathbf{k}_0 + (\mathbf{k} - \mathbf{k}_0)\,, \\ k^2 &= [\mathbf{k}_0 + (\mathbf{k} - \mathbf{k}_0)]^2 = -k_0^2 + 2\mathbf{k}_0 \cdot \mathbf{k} + |\mathbf{k} - \mathbf{k}_0|^2\,, \end{aligned} \tag{3.19}$$

allowing me to transform Eq. (3.18) into

$$\begin{aligned} \psi(\mathbf{r}, t) = &\frac{e^{i\hbar k_0^2 t/2m}}{(2\pi)^{3/2}} \int_{-\infty}^{\infty} d\mathbf{k}\, \Phi(\mathbf{k}) \\ &\times \exp\left\{i\left[\mathbf{k}\cdot(\mathbf{r} - \hbar\mathbf{k}_0 t/m) - \hbar\,|\mathbf{k} - \mathbf{k}_0|^2\, t/2m\right]\right\}. \end{aligned} \tag{3.20}$$

For the moment, suppose that I can neglect the $\hbar\,|\mathbf{k} - \mathbf{k}_0|^2\, t/2m$ term in the exponent,

$$\hbar\,|\mathbf{k} - \mathbf{k}_0|^2\, t/2m \ll 1. \tag{3.21}$$

Then, for all values of $\mathbf{k}$ that contribute significantly to the integral,

$$\begin{aligned} \psi(\mathbf{r}, t) &\approx \frac{e^{i\hbar k_0^2 t/2m}}{(2\pi)^{3/2}} \int_{-\infty}^{\infty} d\mathbf{k}\, \Phi(\mathbf{k}) e^{i\mathbf{k}\cdot(\mathbf{r} - \hbar\mathbf{k}_0 t/m)} \\ &= e^{i\hbar k_0^2 t/2m} \psi(\mathbf{r} - \mathbf{v}_0 t, 0), \end{aligned} \tag{3.22}$$

where

$$\mathbf{v}_0 = \frac{\hbar\mathbf{k}_0}{m} = \frac{\mathbf{p}_0}{m} \tag{3.23}$$

is the average velocity of the particle.

Aside from a phase factor, $\psi(\mathbf{r}, t)$ propagates as an undistorted wave having momentum $\mathbf{p}_0$. If the initial distribution $|\psi(\mathbf{r}, 0)|^2$ is non-vanishing only in a small volume centered at $\mathbf{r} = \mathbf{0}$, then the distribution $|\psi(\mathbf{r}, t)|^2$ will be non-vanishing only in a small volume centered at $\mathbf{r} = \mathbf{v}_0 t$; in other words, the distribution function can mirror the behavior of "particle" that is moving with velocity $\mathbf{v}_0$. How localized can the particle be? When can condition (3.21) be satisfied? It certainly fails if I let

t get arbitrarily large. Many of these questions can be answered by considering a specific wave function and seeing how it propagates. I will calculate $|\psi(x,t)|^2$ for a one-dimensional Gaussian wave packet that is centered at $x = 0$ at $t = 0$.

3.2.1.1 Gaussian Wave Packet

At $t = 0$, I take the wave function for a particle having mass m to be

$$\psi(x,0) = \frac{1}{(\pi\sigma^2)^{1/4}} e^{-x^2/2\sigma^2} e^{ik_0x}, \tag{3.24}$$

such that

$$|\psi(x,0)|^2 = \frac{1}{(\pi\sigma^2)^{1/2}} e^{-x^2/\sigma^2} \tag{3.25}$$

is a Gaussian centered at the origin with a full-width at half maximum (FWHM) equal to 1.67σ. The normalization has been chosen such that

$$\int_{-\infty}^{\infty} dx\, |\psi(x,0)|^2 = 1. \tag{3.26}$$

I want to find $\psi(x,t)$. To do so, I must first find $\Phi(k)$ and then use Eq. (3.14) to get $\psi(x,t)$. The factor of e^{ik_0x} in Eq. (3.24) leads to an average velocity for the packet equal to $\hbar k_0/m$, as you shall see.

It follows from Eq. (3.14) that

$$\psi(x,0) = \frac{1}{\sqrt{2\pi}} \int_{-\infty}^{\infty} dk\, \Phi(k) e^{ikx}. \tag{3.27}$$

I take the inverse Fourier transform of this equation to obtain

$$\begin{aligned}\Phi(k) &= \frac{1}{\sqrt{2\pi}} \int_{-\infty}^{\infty} dx\, \psi(x,0) e^{-ikx} \\ &= \frac{1}{\sqrt{2\pi}} \frac{1}{(\pi\sigma^2)^{1/4}} \int_{-\infty}^{\infty} dx\, e^{-x^2/2\sigma^2} e^{ik_0x} e^{-ikx}.\end{aligned} \tag{3.28}$$

The integral is tabulated or can be evaluated using contour integration. In either case, one finds

$$\Phi(k) = \left(\frac{\sigma^2}{\pi}\right)^{1/4} e^{-(k-k_0)^2\sigma^2/2}; \tag{3.29}$$

$$|\Phi(k)|^2 = \left(\frac{\sigma^2}{\pi}\right)^{1/2} e^{-(k-k_0)^2\sigma^2}. \tag{3.30}$$

The k-space distribution is *also* a Gaussian, centered at $k = k_0$, having FWHM equal to $1.67/\sigma$.

The variance of x at $t = 0$ is

$$[\Delta x(t=0)]^2 = \langle x^2(t=0)\rangle = \int_{-\infty}^{\infty} dx\, x^2 \left|\psi(x,0)\right|^2 = \frac{\sigma^2}{2} \tag{3.31}$$

and the variances of k and p are

$$(\Delta k)^2 = \left\langle (k-k_0)^2 \right\rangle = \int_{-\infty}^{\infty} dk\; (k-k_0)^2 \left|\Phi(k)\right|^2 = \frac{1}{2\sigma^2}; \tag{3.32a}$$

$$(\Delta p)^2 = \frac{\hbar^2}{2\sigma^2}, \tag{3.32b}$$

such that

$$\Delta x(t=0)\Delta k = \frac{1}{2}; \tag{3.33a}$$

$$\Delta x(t=0)\Delta p = \frac{\hbar}{2}. \tag{3.33b}$$

As you shall see, this corresponds to what is called a *minimum-uncertainty wave packet*, having the minimum value of $\Delta x \Delta p$ allowed for solutions of Schrödinger's equation. The momentum distribution and Δp do not change in time since no forces act on the particle.

Owing to the spread of momenta in the wave packet, however, Δx *does* change as a function of time. The wave packet is no longer a minimum uncertainty packet for $t > 0$. Using Eqs. (3.14) and (3.29), I calculate

$$\begin{aligned}\psi(x,t) &= \frac{1}{\sqrt{2\pi}}\left(\frac{\sigma^2}{\pi}\right)^{1/4} \int_{-\infty}^{\infty} dk\, e^{-(k-k_0)^2\sigma^2/2} e^{i\left(kx - \hbar k^2 t/2m\right)} \\ &= \frac{e^{ik_0\left(x-\frac{v_0 t}{2}\right)}}{\sqrt{2\pi}}\left(\frac{\sigma^2}{\pi}\right)^{1/4} \int_{-\infty}^{\infty} dk'\, e^{-k'^2\left(\sigma^2 + i\hbar t/m\right)/2} e^{ik'(x-v_0 t)},\end{aligned} \tag{3.34}$$

where $k' = k - k_0$ and

$$v_0 = \hbar k_0/m = p_0/m. \tag{3.35}$$

The integral is tabulated or can be evaluated using contour integration and the result is

$$\psi(x,t) = \left(\frac{\sigma^2}{\pi}\right)^{1/4} \frac{e^{ik_0\left(x-\frac{v_0 t}{2}\right)}}{\left[\sigma^2 + \frac{i\hbar t}{m}\right]^{1/2}} \exp\left(\frac{-(x-v_0 t)^2}{2\left[\sigma^2 + \frac{i\hbar t}{m}\right]}\right). \tag{3.36}$$

As a consequence,

$$|\psi(x,t)|^2 = \left(\frac{1}{\pi\sigma(t)^2}\right)^{1/2} e^{-(x-v_0 t)^2/\sigma(t)^2}, \tag{3.37}$$

with

$$\sigma(t)^2 = \sigma^2 + \left(\frac{\hbar t}{m\sigma}\right)^2. \tag{3.38}$$

The FWHM at any time is $1.67\sigma(t)$ and $\Delta x(t) = \sigma(t)/\sqrt{2}$, such that

$$\Delta x(t)\Delta p = \frac{\hbar\sigma(t)}{2\sigma} = \frac{\hbar}{2}\left[1 + \left(\frac{\hbar t}{m\sigma^2}\right)^2\right]^{1/2} \geq \frac{\hbar}{2}. \tag{3.39}$$

The packet remains Gaussian but spreads owing to the spread of momenta in the original packet. To see how $\Delta x(t)$ depends on Δp, I use the relationships $\Delta x(0) = \sigma/\sqrt{2}$ and $\Delta v = \Delta p/m = \hbar/\left(\sqrt{2}m\sigma\right)$ to rewrite $\Delta x(t)^2$ as

$$\begin{aligned}\Delta x(t)^2 &= \frac{\sigma(t)^2}{2} = \frac{\sigma^2}{2} + \frac{1}{2}\left(\frac{\hbar t}{m\sigma}\right)^2 \\ &= \Delta x(0)^2 + (\Delta v)^2 t^2.\end{aligned} \tag{3.40}$$

Although I have chosen a Gaussian wave packet, Eq. (3.40) turns out to be *exact* for any square-integrable initial wave function of the form $\psi(x,0) = f(x)e^{ik_0 x}$, for real $f(x)$ (see Problem 5.14–15 in Chap. 5). The variance of the wave packet is its *initial* variance plus a contribution attributable to the variance of the velocity components contained in the packet. For sufficiently large times, $\Delta x(t) \sim \Delta v t$.

I am now in a position to see when the wave packet can correspond to a classical particle. Free particles have never heard about wave packets; wave packets are a construct of physicists. For the wave packet to correspond to a particle, however,the uncertainties in position and momentum must satisfy

$$\Delta x(t) \ll x_0; \quad \Delta p \ll p_0 \tag{3.41}$$

subject to the restriction

$$\Delta x(0)\Delta p \geq \frac{\hbar}{2}. \tag{3.42}$$

The quantities x_0 and p_0 are determined by the problem. You can think of them as the smallest possible resolution in position and momentum that can be detected in a given experiment. In bound state problems they could correspond to some typical bound state radius and magnitude of bound state momentum for the bound particle. For the free particle, let's take $x_0 = 10^{-8}$ m and $v_0 = 10^{-7}$ m/s, which locates the particle to better than an optical wavelength and fixes its velocity to about three meters per year. We might be able to accomplish this by using spatial filters (e.g., slits) to select both the position and range of velocities. For a one gram mass, suppose we take $\Delta x(0) = 10^{-11}$ m which implies that $\Delta v = \Delta p/m \approx \hbar/[m\Delta x(0)] \approx 10^{-20}$ m/s [admittedly, it would be difficult to create such a small wave packet]. This is a "classical" particle at $t = 0$ according to my definition since it obeys conditions (3.41). At what time t would the spreading be sufficient to render the particle "unclassical" ? Arbitrarily, let's say the particle is no longer classical if $\Delta x(t) = x_0/100 = 10^{-10}$ m, which occurs for $t = 10^{10}$ s, 300 years! The bottom line is that spreading is unimportant as long as the de Broglie wavelength is much smaller than any characteristic length in the problem, such as the width of the initial wave packet. On the other hand, if you confine a free particle wave packet to a distance equal to its de Broglie wavelength, the spread in momentum in the wave packet is of order of the average momentum in the packet; as such spreading is important and the particle can no longer be viewed as a classical particle.

A simple example that illustrates the necessity of using a quantum description of matter can be found in an experiment related to *atom optics*. Suppose a well-collimated, pulsed atomic beam having velocity v_0 in the z direction is incident on a circular aperture having diameter d that is located in the xy plane. Moreover, assume that the de Broglie wavelength of the atoms, $\lambda_{dB} \ll d$. The atoms are treated as point particles, so another implicit assumption is that the atomic size is much smaller than d as well. After traversing the aperture at $t = 0$, you can think of the initial wave packet as a short pulse in the z direction having a cross-sectional area equal to $\pi d^2/4$. The transverse uncertainty in the momentum of this beam is of order $\Delta p_\perp \approx \hbar/d$. Matter wave effects become important when the transverse spreading is of order d, that is, for times t greater than some critical time t_F defined by

$$\Delta p_\perp t_F/m \approx \hbar t_F/(dm) \approx d, \tag{3.43}$$

where m is the mass of an atom in the beam. Since $t \approx z/v_0$ where z is the distance from the screen containing the aperture, the distance z_F corresponding to the time t_F is

$$z_F = v_0 t_F \approx \frac{mv_0 d^2}{\hbar} \approx \frac{d^2}{\lambda_{dB}}. \tag{3.44}$$

For $z \ll z_F$, the scattering of the particles by the slit is in the *shadow region* and the atomic motion can be treated classically. However for $z \gtrsim z_F$, diffraction plays an important role and a wave theory is needed. The situation is analogous to the scattering of optical radiation having wavelength λ by an aperture having diameter d. For distances $z \ll z_F = d^2/\lambda$ from the diffracting screen, a geometrical picture of light rays can be used, but once $z \gtrsim z_F$, diffraction effects become important and a wave theory of light is needed. In the optical case the region with $z \approx z_F$ corresponds to *Fresnel diffraction*.

Perhaps the best way to create a well-defined wave packet is to trap and cool an atom in the potential well of an *optical lattice*. An optical lattice is formed by using pairs of counter-propagating laser beams. These pairs of fields form standing wave patterns that can be used to trap neutral atoms owing to a spatially varying potential that is experienced by the atoms in the fields. Moreover the atoms can be cooled to the point where they are in the ground state of the potential. As such, if you trap one atom in one well, you have a pretty good idea of its wave function. If you suddenly remove the potential by turning off the fields, you have an initial condition in which the atom is in its ground state and has a center-of-mass wave function given by the ground state of the potential. You could let this wave packet propagate for some time and then restore the lattice and determine how far the packet has moved by seeing which well it is in. Although this experiment has yet to be carried out, the technology is now at the point where it is feasible.

3.2.2 Free-Particle Propagator

Instead of calculating $\Phi(k)$ from $\psi(x,0)$ for each wave packet, it is possible to relate $\psi(x,t)$ *directly* to an integral of $\psi(x,0)$. To do so I first calculate

$$\Phi(k) = \frac{1}{\sqrt{2\pi}} \int_{-\infty}^{\infty} dx' \, \psi(x',0) e^{-ikx'} \tag{3.45}$$

and substitute the result into Eq. (3.14) to obtain

$$\psi(x,t) = \frac{1}{2\pi} \int_{-\infty}^{\infty} dx' \, \psi(x',0) \int_{-\infty}^{\infty} dk \, \exp\left\{ i \left[k \left(x - x' \right) - \hbar k^2 t/2m \right] \right\}. \tag{3.46}$$

The integral over k is tabulated and I can write the final result as

$$\psi(x,t) = \int_{-\infty}^{\infty} dx' \, K\left(x - x', t\right) \psi(x',0), \tag{3.47}$$

where the *free-particle propagator* $K(x - x', t)$ is given by

$$K\left(x - x', t\right) = \frac{1}{2\sqrt{\pi b i}} e^{i(x-x')^2/4b} \tag{3.48}$$

and

$$b = \hbar t/2m. \tag{3.49}$$

As an example of the use of the propagator, I consider an initial state wave function

$$\psi(x,0) = \begin{cases} \frac{1}{\sqrt{a}} & |x| \le a/2 \\ 0 & \text{otherwise} \end{cases}. \tag{3.50}$$

Since the probability density $|\psi(x,0)|^2$ has sharp boundaries, we should expect these sharp boundaries to give rise to diffraction. From Eqs. (3.47) to (3.50), I calculate

$$\psi(x,t) = \frac{1}{2\sqrt{\pi abi}} \int_{-a/2}^{a/2} dx' \, e^{i(x-x')^2/4b}, \tag{3.51}$$

where b is given in Eq. (3.49). The limits on the integral have been set equal to $\pm a/2$ since $\psi(x',0) = 0$ for $|x'| > a/2$. The integral is tabulated in terms of error functions, but I present the results rather than give formal expressions for $|\psi(x,t)|^2$.

It is usually best to give plots in terms of dimensionless variables. In this case, it is clear that x/a is an appropriate dimensionless coordinate. It would make sense to choose a dimensionless time as

$$\tau = \frac{\Delta v t}{a} = \frac{\Delta p}{ma} t, \tag{3.52}$$

where $\Delta p = m\Delta v$ is the momentum uncertainty. Unfortunately, $\Delta p = \infty$ for this wave packet, since the envelope of the absolute square of the Fourier transform of the packet,

$$|\Phi(p)|^2 = \frac{1}{2\pi\hbar a} \left| \int_{-a/2}^{a/2} dx \, e^{-ipx/\hbar} \right|^2 = \frac{a}{2\pi\hbar} \frac{\sin^2(pa/2\hbar)}{(pa/2\hbar)^2}, \tag{3.53}$$

falls off as p^{-2} for large p, resulting in $\langle p^2 \rangle = \infty$. However, the *central* lobe of $|\Phi(p)|^2$ has a HWHM of order $\delta p \equiv \hbar/\Delta x(0)$, where $\Delta x(0) = a/\sqrt{12}$ is the standard deviation of the initial wave packet in coordinate space.[1]

If I set $\Delta p = \delta p \equiv \hbar/\Delta x(0) = 2\sqrt{3}\hbar/a$ in Eq. (3.52), then

$$\tau = \frac{2\sqrt{3}\hbar t}{ma^2} = \frac{4\sqrt{3}b}{a^2} \tag{3.54}$$

is an appropriate dimensionless time, and Eq. (3.51) can be written as

$$\psi(\xi,\tau) = \sqrt{\frac{\sqrt{3}}{\pi a\tau i}} \int_{-1/2}^{1/2} d\xi' \, e^{\sqrt{3}i(\xi-\xi')^2/\tau}, \tag{3.55}$$

with $\xi = x/a$ and $\xi' = x'/a$. The integral can be evaluated numerically or in terms of error functions. The dimensionless quantity $a|\psi(\xi,\tau)|^2$ is plotted in Fig. 3.1 for $\tau = 0, 0.25, 1$. The probability distribution $a|\psi(\xi,\tau)|^2$ evolves into a Fresnel-like diffraction pattern for $0 \lesssim \tau \lesssim 1$ and into a Fraunhofer diffraction pattern for $\tau \gtrsim 1$.

[1]The calculated value of the HWHM of the central lobe is $0.81\hbar/\Delta x(0)$.

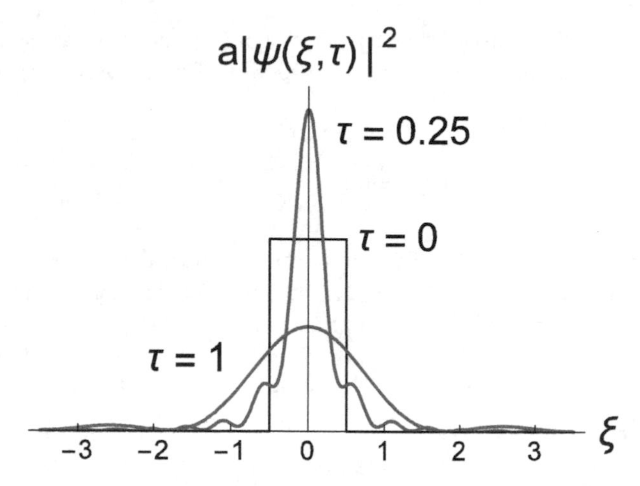

Fig. 3.1 Time evolution of the dimensionless probability distribution for a "square" wave packet. The original square packet ($\tau = 0$) undergoes Fresnel-like diffraction ($\tau = 0.25$) before assuming the Fraunhofer diffraction pattern of a single slit ($\tau = 1.0$). The dimensionless time τ is defined in Eq. (3.54)

Diffraction effects from the sharp edges of the wave packet are seen clearly in these diagrams. Although not evident from the figure, it turns out that $\Delta\xi(\tau) = \infty$ for any $\tau > 0$, since for fixed τ and $\xi \gg 1$ (see problems),

$$a\,|\psi(\xi,\tau)|^2 \sim \tau \frac{\sin^2\left(\sqrt{3}\xi/\tau\right)}{\sqrt{3}\pi\xi^2}. \tag{3.56}$$

This asymptotic form can be obtained by expressing the integral in Eq. (3.55) in terms of error functions and taking the asymptotic limit of the error functions. Equation (3.56) is also a good approximation to $a\,|\psi(\xi,\tau)|^2$ for fixed ξ and $\tau \gtrsim 1$. It represents the Fraunhofer diffraction pattern of a single slit.

The fact that $\langle\hat{p}^2\rangle = \infty$ and $\Delta x(\tau) = \infty$ for any $\tau > 0$ is linked to sharp edges of the initial coordinate space wave packet. In fact, it is possible to show that the same features occur for any initial wave function that possesses a point jump discontinuity. In practice it is impossible to create a wave packet having a point jump discontinuity. To do so would require an infinite amount of energy since the resulting packet has $\langle p^2\rangle = \infty$. Of course, the Schrödinger equation is a non-relativistic equation so that the momentum distribution is suspect for momenta $|p| \gtrsim mc$.

3.3 Summary

I have shown that it is possible to obtain Schrödinger's equation for a free particle, using an analogy with optical pulse propagation, along with de Broglie's definition of the wavelength of matter and Einstein's definition of the energy of a photon. The resultant free particle wave packet differs in a fundamental way from that of an optical pulse, since the dispersion relation relating energy to the momentum is quadratic for matter and linear for light. As a consequence, optical pulses in vacuum propagate without changing their shape, while free-particle, matter wave packets contain several momentum components and necessarily change their shape as a function of time.

3.4 Problems

1. Suppose that a smooth wave packet has $\Delta x(0) = a$ and $\Delta p = \alpha\hbar/a$, where $\alpha \geq 1/2$ is a constant. Explain why

$$\Delta x(t) = \sqrt{a^2 + \alpha^2 \left(\frac{\hbar t}{ma}\right)^2},$$

is not a bad guess for the width of the wave packet at any time. Using this guess with $\alpha = 1$, find the time it takes for a wave packet that is confined to its de Broglie wavelength to spread to twice its initial width in terms of the particle's energy and Planck's constant.

2. Using the result of Problem 3.1 and assuming that you are a point particle having mass 50 kg and are localized to 1.0×10^{-11} m, calculate how long it would take for you to spread by an amount equal to this initial localization distance.

3–4. Start with a one-dimensional wave packet in coordinate space. Take a wave function such as $\psi(x) = Ne^{-x^4}$ or $\psi(x) = Ne^{-x^6}$, for which no *simple* analytic solution exists for its Fourier transform. Normalize your wave function using numerical integration to find the value of N. Plot $|\psi(x)|^2$. Using Mathematica, Matlab, or Maple (or any other program you have), use numerical integration [NIntegrate in Mathematica] to obtain and plot the k-space distribution associated with the wave packet you chose. How does it differ from a Gaussian? Why? Calculate the value of $\Delta x \Delta k$ using numerical integration. In this problem, both x and k are dimensionless variables.

5. Sodium atoms moving at 1000 m/s are incident on a slit having a width of 100 nm. At 3 m from the slit, what is the approximate transverse width associated with the sodium atom's wave function? Is diffraction important in this case? These are typical values for experiments in atom optics.

6. Prove that

$$\psi(\mathbf{r},t) = e^{i\mathbf{p}\cdot\mathbf{r}/\hbar}e^{-ip^2t/2m\hbar}$$

is a solution of

$$i\hbar\frac{\partial\psi(\mathbf{r},t)}{\partial t} = -\frac{\hbar^2}{2m}\nabla^2\psi(\mathbf{r},t)$$

As a consequence, why is

$$\psi(\mathbf{r},t) = \frac{1}{(2\pi\hbar)^{3/2}}\int d\mathbf{p}e^{i\mathbf{p}\cdot\mathbf{r}/\hbar}e^{-ip^2t/2m\hbar}\Phi(\mathbf{p})$$

also a solution of the equation, where $\Phi(\mathbf{p})$ is some arbitrary function for which $|\psi(\mathbf{r},t)|^2$ is square integrable? Note that you can also write this equation as

$$\psi(\mathbf{r},t) = \frac{1}{(2\pi\hbar)^{3/2}}\int d\mathbf{p}e^{i\mathbf{p}\cdot\mathbf{r}/\hbar}\Phi(\mathbf{p},t),$$

where

$$\Phi(\mathbf{p},t) = e^{-ip^2t/2m\hbar}\Phi(\mathbf{p}).$$

In general it is assumed that $\psi(\mathbf{r},t)$ and $\Phi(\mathbf{p},t)$ are Fourier transforms of one another, even for cases when a potential is present. When a spatially varying potential is present, explain why $\Phi(\mathbf{p},t)$ can no longer be equal to $e^{-ip^2t/2m\hbar}\Phi(\mathbf{p})$.

7–8. Return to Problem 2.3–4 for an initial (normalized) wave function

$$\psi(x,0) = \frac{1}{2^{3/8}\sqrt{a}\sqrt{\Gamma(5/4)}}\exp\left(-x^4/a^4\right) = \frac{0.810}{\sqrt{a}}\exp\left(-x^4/a^4\right),$$

where a is a real constant and Γ is the gamma function. Calculate Δk^2 numerically and show that it is equal to $1.43/a^2$. Suppose you want to check the validity of Eq. (3.40) for this initial wave packet. Show that Eq. (3.40) can be written as

$$\Delta x^2(b) = \Delta x^2(0) + 4\Delta k^2 b^2,$$

where $b = \hbar t/2m$ and m is the mass of the particle. Evaluate $\Delta x^2(b)/a^2$ for $b/a^2 = 0, 0.1, 0.5, 0.75, 1, 5, 10$. Now use Eq. (3.47) to obtain an integral expression for $\psi(x,b)$, and numerically evaluate $\Delta x^2(b)/a^2$ for the same values of b/a^2 to see how well the equation $\Delta x^2(b) = \Delta x^2(0) + 4\Delta k^2 b^2$ agrees with the exact result. As noted in the text, the agreement should be exact in this case.

9. Use Mathematica or some other program to evaluate the integral in Eq. (3.55) in terms of error functions. Take the asymptotic limit of the result for $\xi \gg 1$ to derive Eq. (3.56) and show that $\Delta x(\tau) = \infty$ for any $\tau > 0$. Also, starting from Eq. (3.55), show that Eq. (3.56) is the correct asymptotic limit for $\tau \gg 1$ and any ξ.

10. Plot $a\left|\psi(\xi,\tau)\right|^2$ for $\psi(\xi,\tau)$ given in Eq. (3.55) as a function of ξ for $\tau = 0.001, 0.05, 0.1, 0.2$. This will show you the transition from the shadow region to that of Fresnel diffraction. Also plot $a\left|\psi(0,\tau)\right|^2$ for $0 \le \tau \le 0.5$ and find the maximum value it can have.

11. Assume that a wave function has the form

$$\psi(x,0) = f(x)\Theta(a-x)\Theta(b+x),$$

where a and b are positive and $f(x)$ is a real analytic function that is non-vanishing at $x = a, b$. The Heaviside function $\Theta(x)$ is equal to 0 for $x < 0$ and 1 for $x \ge 1$, so the Heaviside functions truncate the wave function and confine it to $-b < x < a$. By evaluating the Fourier transform of $\psi(x,0)$ for large momenta, prove that $\left\langle p^2\right\rangle = \infty$. [Hint: an integration by parts might help.]

12. For the initial wave packet of the previous problem, use Eqs. (3.47) and (3.48) to show that $\left\langle x^2\right\rangle = \infty$ for any $t > 0$.

Chapter 4
Schrödinger's Equation with Potential Energy: Introduction to Operators

The wave equations of electromagnetism are a natural consequence of Maxwell's equations. The time and spatial derivatives of the fields are found to be connected in a simple manner. But what *physical interpretation* can be given to the quantity $\nabla^2\psi(\mathbf{r},t)$ that appears in Schrödinger's equation? To understand the physical significance of this term, I have to introduce the concept of *operators*. Operators play a very important role in quantum mechanics but can be a bit confusing; specifically a distinction must be made between operators and functions. The plan is to look at the free particle Schrödinger equation and to show that the $-\left(\hbar^2/2m\right)\nabla^2$ term can be represented as $\hat{p}^2/2m = \hat{\mathbf{p}}\cdot\hat{\mathbf{p}}/2m$, where $\hat{\mathbf{p}}$ is an operator. I will then be able to generalize Schrödinger's equation to account for situations in which an external potential is present. To simplify matters, I work with the Schrödinger equation in one dimension; in the Appendix, an analogous development is given for the Schrödinger equation in three dimensions.

4.1 Hamiltonian Operator

So far I have been talking about free-particle solutions of Schrödinger's equation, which, in one dimension, is written as

$$i\hbar\frac{\partial\psi(x,t)}{\partial t} = -\frac{\hbar^2}{2m}\frac{\partial^2\psi(x,t)}{\partial x^2}. \tag{4.1}$$

The quantity $\partial^2/\partial x^2$ appearing in this equation is an *operator* that acts on functions in coordinate space. The operation simply involves second order differentiation of the function. Operators do just that, they *operate* on a function to produce a new function. For example, you could define the "Chicago operator" that translates the wave function to the top of the Sear's tower in Chicago.

P.R. Berman, *Introductory Quantum Mechanics*, UNITEXT for Physics,
https://doi.org/10.1007/978-3-319-68598-4_4

I've already looked at an integral solution of Eq. (4.1), namely

$$\psi(x,t) = \frac{1}{(2\pi)^{1/2}} \int_{-\infty}^{\infty} dk\, \Phi(k) \exp[i\left(kx - \hbar k^2 t/2m\right)]. \tag{4.2}$$

Actually I could equally well *guess* this as a solution of the free-particle Schrödinger equation. Substituting this guess into Eq. (4.1), I find

$$i\hbar \frac{\partial \psi(x,t)}{\partial t} = \frac{\hbar^2}{2m\,(2\pi)^{1/2}} \int_{-\infty}^{\infty} dk\, \Phi(k) k^2 \exp[i\left(kx - \hbar k^2 t/2m\right)] \tag{4.3}$$

and

$$-\frac{\hbar^2}{2m} \frac{\partial^2 \psi(x,t)}{\partial x^2} = \frac{\hbar^2}{2m\,(2\pi)^{1/2}} \int_{-\infty}^{\infty} dk\, \Phi(k) k^2 \exp[i\left(kx - \hbar k^2 t/2m\right)], \tag{4.4}$$

implying that the solution works. I shall assume that $\psi(x,t)$ is normalized,

$$\int_{-\infty}^{\infty} dx\, |\psi(x,t)|^2 = 1. \tag{4.5}$$

Instead of working in coordinate space, I could work equally well in *momentum space*; that is, I can try to get a differential equation for $\Phi(p,t)$, where $\Phi(p,t)$ is the Fourier transform of $\psi(x,t)$ defined by

$$\Phi(p,t) = \frac{1}{(2\pi\hbar)^{1/2}} \int_{-\infty}^{\infty} dx\, \psi(x,t) e^{-ipx/\hbar} \tag{4.6}$$

and

$$p = \hbar k \tag{4.7}$$

is given by the de Broglie relationship. It is *assumed* that $|\Phi(p,t)|^2$ is the probability density in momentum space, that is the probability density to find a momentum p for the particle at time t. Note that I need a factor of $\hbar^{-1/2}$ when momentum rather than k is used in defining the Fourier transform. This assures that $\Phi(p,t)$ has the correct units of $p^{-1/2}$ and that $\int_{-\infty}^{\infty} dp\, |\Phi(p,t)|^2 = 1$ if $|\psi(x,t)|^2$ is normalized.

Differentiating Eq. (4.6) with respect to time and using Eq. (4.1), I find

$$\begin{aligned} i\hbar \frac{\partial \Phi(p,t)}{\partial t} &= \frac{i\hbar}{(2\pi\hbar)^{1/2}} \int_{-\infty}^{\infty} dx\, \frac{\partial \psi(x,t)}{\partial t} e^{-ipx/\hbar} \\ &= -\frac{\hbar^2}{2m\,(2\pi\hbar)^{1/2}} \int_{-\infty}^{\infty} dx\, e^{-ipx/\hbar} \frac{\partial^2 \psi(x,t)}{\partial x^2}. \end{aligned} \tag{4.8}$$

I now integrate by parts two times and assume that the wave function and its derivative vanish as x goes to plus or minus infinity (the wave packet is not infinite in extent). In this manner I obtain

$$i\hbar\frac{\partial\Phi(p,t)}{\partial t} = \frac{p^2}{2m\left(2\pi\hbar\right)^{1/2}}\int_{-\infty}^{\infty} dx\, e^{-ipx/\hbar}\psi(x,t) = \frac{p^2}{2m}\Phi(p,t). \tag{4.9}$$

When this equation is generalized to three dimensions (see Appendix), it becomes

$$i\hbar\frac{\partial\Phi(\mathbf{p},t)}{\partial t} = \frac{p^2}{2m}\Phi(\mathbf{p},t) = \frac{\hat{p}^2}{2m}\Phi(\mathbf{p},t), \tag{4.10}$$

where the *operator* $\hat{p}^2$ is defined such that

$$\hat{p}^2 g(\mathbf{p}) = p^2 g(\mathbf{p}). \tag{4.11}$$

for any function $g(\mathbf{p})$. In other words, the operator $\hat{p}^2$ acting on a function $g(\mathbf{p})$ in momentum space simply *multiplies* $g(\mathbf{p})$ by p^2.

This is an important result. In coordinate space, the free particle Schrödinger equation is

$$i\hbar\frac{\partial\psi(\mathbf{r},t)}{\partial t} = -\frac{\hbar^2}{2m}\nabla^2\psi(\mathbf{r},t). \tag{4.12}$$

The right-hand side of the free particle Schrödinger equation in coordinate space is $-\left(\hbar^2/2m\right)\nabla^2$, while it is $\hat{p}^2/2m$ in momentum space. This implies that the operator $\hat{p}^2$ acting on a function $f(\mathbf{r})$ in *coordinate* space results simply in $-\hbar^2\nabla^2 f(\mathbf{r})$; that is,

$$\hat{p}^2 = -\hbar^2\nabla^2 \tag{4.13}$$

when acting in coordinate space. As a consequence, the Schrödinger equation can be written as

$$i\hbar\frac{\partial\psi(\mathbf{r},t)}{\partial t} = \frac{-\hbar^2\nabla^2\psi(\mathbf{r},t)}{2m} = \frac{\hat{p}^2}{2m}\psi(\mathbf{r},t). \tag{4.14}$$

Since $p^2/2m$ is equal to the energy of the free particle, I can rewrite Eq. (4.14) as

$$i\hbar\frac{\partial\psi(\mathbf{r},t)}{\partial t} = \hat{H}\psi(\mathbf{r},t), \tag{4.15}$$

where $\hat{H}$ is the *Hamiltonian* or energy operator.

I now conjecture that Eq. (4.15) remains valid even when there is a potential $V(\mathbf{r})$ present, with the Hamiltonian operator generalized as

$$\hat{H} = \frac{\hat{p}^2}{2m} + \hat{V}. \tag{4.16}$$

In coordinate space, the operator $\hat{V}$ is defined such that

$$\hat{V}f(\mathbf{r}) = V(\mathbf{r})f(\mathbf{r}) \tag{4.17}$$

for any function $f(\mathbf{r})$. In other words, operator $\hat{V}$ acting on a function $f(\mathbf{r})$ in coordinate space simply picks out the value of the potential energy at the position of the function and *multiplies* it by $f(\mathbf{r})$.

Since $\hat{p}^2$ is equal to $-\hbar^2\nabla^2$ in the coordinate representation, I have a complete description of the Hamiltonian operator in the coordinate representation, namely

$$\hat{H}\psi(\mathbf{r},t) = \left[\frac{\hat{p}^2}{2m} + \hat{V}\right]\psi(\mathbf{r},t) = \left[-\frac{\hbar^2\nabla^2}{2m} + V(\mathbf{r})\right]\psi(\mathbf{r},t). \tag{4.18}$$

In the momentum representation, however, things are not so clear, since I do not know the effect of the potential operator in momentum space. To do so, I must use my assumption that $\psi(\mathbf{r},t)$ and $\Phi(\mathbf{p},t)$ are Fourier transforms of one another and make the assumption that the average value of the operators is identical in the coordinate and momentum representations. A derivation of Schrödinger's equation in momentum space is given in Chap. 11.

4.2 Time-Independent Schrödinger Equation

In coordinate space, the time-dependent Schrödinger equation is

$$i\hbar\frac{\partial\psi(\mathbf{r},t)}{\partial t} = \hat{H}\psi(\mathbf{r},t) = \left(-\frac{\hbar^2}{2m}\nabla^2 + V(\mathbf{r})\right)\psi(\mathbf{r},t). \tag{4.19}$$

I guess a solution

$$\psi(\mathbf{r},t) = e^{-iEt/\hbar}\psi_E(\mathbf{r}), \tag{4.20}$$

substitute it into Eq. (4.19) and find that it works, *provided*

$$\hat{H}\psi_E(\mathbf{r}) = E\psi_E(\mathbf{r}), \tag{4.21}$$

an equation which is known as the *time-independent Schrödinger equation*. There may be many solutions of Eq. (4.21). Since Eq. (4.19) is linear, its general solution is found by forming a linear superposition of all the solutions of Eq. (4.21), namely

$$\psi(\mathbf{r},t) = \sum_E a_E e^{-iEt/\hbar}\psi_E(\mathbf{r}), \tag{4.22}$$

where the expansion coefficients a_E are determined by the initial conditions on $\psi(\mathbf{r},t)$.

Equation (4.22) is a *critically important* result. *If* we can solve the *time-independent* Schrödinger equation, we can build a solution to the time-dependent problem. This is why so much time is spent in quantum mechanics courses on the time-independent Schrödinger equation. However you should not forget that the interesting dynamics of a quantum system is obtained only through a solution of the *time-dependent* Schrödinger equation. In order to solve equations of the type given in Eq. (4.21), I will need the additional ammunition to be provided in Chap. 5.

4.3 Summary

The concept of an operator in quantum mechanics has been introduced. I found that the square of the momentum operator is proportional to the Laplacian operator in coordinate space. We have seen that it is possible to obtain a solution of the time-dependent Schrödinger equation if we are able to solve the time-independent Schrödinger equation. The formal method for solving the time-independent Schrödinger is given in the next chapter, where I discuss the properties of operators in a more systematic fashion.

4.4 Appendix: Schrödinger Equation in Three Dimensions

Schrödinger's equation in three dimensions is

$$i\hbar\frac{\partial\psi(\mathbf{r},t)}{\partial t} = -\frac{\hbar^2}{2m}\nabla^2\psi(\mathbf{r},t). \tag{4.23}$$

As in the one-dimensional case, I *guess* a solution of the form

$$\psi(\mathbf{r},t) = \frac{1}{(2\pi)^{3/2}}\int d\mathbf{k}\,\Phi(\mathbf{k})\exp[i\left(\mathbf{k}\cdot\mathbf{r} - \hbar k^2 t/2m\right)], \tag{4.24}$$

which is easily shown to be a solution of Eq. (4.23).

In three dimensions, $\Phi(\mathbf{p},t)$ is defined by

$$\Phi(\mathbf{p},t) = \frac{1}{(2\pi\hbar)^{3/2}}\int d\mathbf{r}\,\psi(\mathbf{r},t)e^{-i\mathbf{p}\cdot\mathbf{r}/\hbar}, \tag{4.25}$$

where

$$\mathbf{p} = \hbar \mathbf{k} \tag{4.26}$$

is given by the de Broglie relationship.

Differentiating Eq. (4.25) with respect to time and using Eq. (4.23), I find

$$\begin{aligned} i\hbar \frac{\partial \Phi(\mathbf{p}, t)}{\partial t} &= \frac{i\hbar}{(2\pi\hbar)^{3/2}} \int d\mathbf{r} \frac{\partial \psi(\mathbf{r}, t)}{\partial t} e^{-i\mathbf{p}\cdot\mathbf{r}/\hbar} \\ &= -\frac{\hbar^2}{2m(2\pi\hbar)^{3/2}} \int d\mathbf{r}\, e^{-i\mathbf{p}\cdot\mathbf{r}/\hbar} \nabla^2 \psi(\mathbf{r}, t). \end{aligned} \tag{4.27}$$

I now use the vector identity

$$\begin{aligned} \nabla^2 \left[\psi(\mathbf{r}, t) e^{-i\mathbf{p}\cdot\mathbf{r}/\hbar} \right] &= -\frac{p^2}{\hbar^2} \psi(\mathbf{r}, t) e^{-i\mathbf{p}\cdot\mathbf{r}/\hbar} - 2i e^{-i\mathbf{p}\cdot\mathbf{r}/\hbar} \frac{\mathbf{p}}{\hbar} \cdot \nabla \psi(\mathbf{r}, t) \\ &\quad + e^{-i\mathbf{p}\cdot\mathbf{r}/\hbar} \nabla^2 \psi(\mathbf{r}, t) \end{aligned} \tag{4.28}$$

to rewrite the integral appearing in Eq. (4.27) as

$$\begin{aligned} \int d\mathbf{r}\, e^{-i\mathbf{p}\cdot\mathbf{r}/\hbar} \nabla^2 \psi(\mathbf{r}, t) &= \int d\mathbf{r}\, \nabla^2 \left[\psi(\mathbf{r}, t) e^{-i\mathbf{p}\cdot\mathbf{r}/\hbar} \right] \\ &+ \frac{p^2}{\hbar^2} \int d\mathbf{r}\, \psi(\mathbf{r}, t) e^{-i\mathbf{p}\cdot\mathbf{r}/\hbar} + 2i \frac{\mathbf{p}}{\hbar} \cdot \int d\mathbf{r}\, e^{-i\mathbf{p}\cdot\mathbf{r}/\hbar} \nabla \psi(\mathbf{r}, t). \end{aligned} \tag{4.29}$$

The first integral on the right-hand side (rhs) of Eq. (4.29) can be evaluated using the divergence theorem as

$$\begin{aligned} \int d\mathbf{r}\, \nabla^2 \left[\psi(\mathbf{r}, t) e^{-i\mathbf{p}\cdot\mathbf{r}/\hbar} \right] &= \int d\mathbf{r}\, \nabla \cdot \nabla \left[\psi(\mathbf{r}, t) e^{-i\mathbf{p}\cdot\mathbf{r}/\hbar} \right] \\ &= \oint_S \nabla \left[\psi(\mathbf{r}, t) e^{-i\mathbf{p}\cdot\mathbf{r}/\hbar} \right] \cdot d\mathbf{a}. \end{aligned} \tag{4.30}$$

The surface integral is taken over a surface whose radius approaches infinity; this integral vanishes if the wave function falls off more rapidly than r^{-2} as $r \to \infty$, that is,

$$\oint_S \nabla \left[\psi(\mathbf{r}, t) e^{-i\mathbf{p}\cdot\mathbf{r}/\hbar} \right] \cdot d\mathbf{a} \sim 0. \tag{4.31}$$

The third integral on the rhs Eq. (4.29) can be written as

$$\begin{aligned}
&2i\frac{\mathbf{p}}{\hbar}\cdot\int d\mathbf{r}\, e^{-i\mathbf{p}\cdot\mathbf{r}/\hbar}\nabla\psi(\mathbf{r},t) \\
&= 2i\frac{\mathbf{p}}{\hbar}\cdot\int d\mathbf{r}\left\{\nabla\left[e^{-i\mathbf{p}\cdot\mathbf{r}/\hbar}\psi(\mathbf{r},t)\right]+\left(\frac{i\mathbf{p}}{\hbar}\right)e^{-i\mathbf{p}\cdot\mathbf{r}/\hbar}\psi(\mathbf{r},t)\right\} \\
&= 2i\frac{\mathbf{p}}{\hbar}\cdot\left\{\oint_S \psi(\mathbf{r},t)e^{-i\mathbf{p}\cdot\mathbf{r}/\hbar}d\mathbf{a}+\left(\frac{i\mathbf{p}}{\hbar}\right)\int d\mathbf{r}\, e^{-i\mathbf{p}\cdot\mathbf{r}/\hbar}\psi(\mathbf{r},t)\right\} \\
&= -2\frac{p^2}{\hbar^2}\int d\mathbf{r}\,\psi(\mathbf{r},t)e^{-i\mathbf{p}\cdot\mathbf{r}/\hbar},
\end{aligned} \tag{4.32}$$

where the generalized Gauss' theorem given in Eq. (2.21b) was used to convert the integral containing $\nabla\left[e^{-i\mathbf{p}\cdot\mathbf{r}/\hbar}\psi(\mathbf{r},t)\right]$ into a surface integral that vanishes as the radius of the surface approaches infinity. Combining Eqs. (4.29)–(4.32) and using Eq. (4.25), I find

$$\int d\mathbf{r}\, e^{-i\mathbf{p}\cdot\mathbf{r}/\hbar}\nabla^2\psi(\mathbf{r},t) = -\frac{p^2}{\hbar^2}\int d\mathbf{r}\,\psi(\mathbf{r},t)e^{-i\mathbf{p}\cdot\mathbf{r}/\hbar} = -\frac{p^2}{\hbar^2}(2\pi\hbar)^{3/2}\Phi(\mathbf{p},t). \tag{4.33}$$

Substituting this result into Eq. (4.27), I finally arrive at

$$i\hbar\frac{\partial\Phi(\mathbf{p},t)}{\partial t} = \frac{p^2}{2m}\Phi(\mathbf{p},t), \tag{4.34}$$

which is Eq. (4.10).

4.5 Problems

1. Given a Hamiltonian of the form

$$\hat{H} = -\frac{\hbar^2}{2m}\frac{d^2}{dx^2} + V(x)$$

in coordinate space, show that the most general solution of the time-dependent Schrödinger equation is

$$\psi(x,t) = \sum_E a_E\psi_E(x)e^{-iEt/\hbar} = \sum_E a_E(t)\psi_E(x)$$

provided $a_E(t) = a_E e^{-iEt/\hbar}$ and

$$\hat{H}\psi_E(x) = E\psi_E(x).$$

Prove that the probability to be in a given quantum state, given by $|a_E(t)|^2$, is constant in time, but that, in general, the probability $|\psi(x,t)|^2$ varies in time.

2. Why doesn't the momentum distribution change in time for a free particle, even though the coordinate space distribution changes? How would you expect the wave function in momentum space to change for a particle falling in a uniform gravitational field? Explain. What is the difference between the operator $\hat{p}^2$ in momentum space and in coordinate space?

3. Suppose you are given a Hamiltonian of the form

$$\hat{H} = -\frac{\hbar^2}{2m}\frac{d^2}{dx^2} + V(x),$$

with $\psi(x,0) = f(x)$. In terms of the eigenfunctions and eigenenergies of $\hat{H}$, derive an expression for $\psi(x,t)$.

4. Prove that

$$\int \psi^*(x,t)\hat{H}\psi(x,t)dx$$

is constant in time. Why must this be the case?

5. Consider a wave function for a free particle having mass m of the form

$$\psi(x) = Ne^{-x^2/a^2}\Theta(b-x)\Theta(b+x),$$

where $b \gg a > 0$, N is a normalization constant, and $\Theta(x)$ is a Heaviside function equal to 0 for $x < 0$ and 1 for $x \geq 1$. For $b \gg a$, you might think that it would be an excellent approximation to consider the wave function as a Gaussian packet having average energy $\langle E\rangle = \langle \hat{p}^2\rangle/2m = \hbar^2/\left(4ma^2\right)$. Use the result of Problem 2.9 to show that this is *not* the case—for any finite b, the average energy is infinite. This occurs because of the point jump discontinuity in the wave function.

Chapter 5
Postulates and Basic Elements of Quantum Mechanics: Properties of Operators

In this chapter I present a somewhat more formal introduction to the theory that underlies quantum mechanics. Although the discussion is limited mainly to single particles, many of the results apply equally well to many-particle systems. Some of the postulates of the theory depend on the properties of *Hermitian* operators, operators that play a central role in quantum mechanics. First I state the postulates, then discuss Hermitian operators, and finally explore some results that follow directly from Schrödinger's equation.

The postulates are:

1. The absolute square of the wave function $|\psi(\mathbf{r},t)|^2$ that characterizes a particle corresponds to the probability density of finding the particle at position $\mathbf{r}$ at time t.
2. To each dynamic physical observable in classical mechanics (such as position, momentum, energy), there corresponds a Hermitian operator in quantum mechanics.
3. The time dependence of $\psi(\mathbf{r},t)$ is governed by the time-dependent Schrödinger equation,

$$i\hbar\frac{\partial\psi(\mathbf{r},t)}{\partial t}=\hat{H}\psi(\mathbf{r},t) \tag{5.1}$$

   where $\hat{H}$ is the energy operator of the system.
4. The only possible outcome of a measurement on a *single* quantum system of a physical observable associated with a given Hermitian operator is one of the eigenvalues of the operator.
5. I must add one additional postulate. This postulate can take different forms. The one I use at this juncture is that the wave functions in coordinate space, $\psi(\mathbf{r},t)$, and in momentum space, $\Phi(\mathbf{p},t)$, are Fourier transforms of one another, namely

P.R. Berman, *Introductory Quantum Mechanics*, UNITEXT for Physics,
https://doi.org/10.1007/978-3-319-68598-4_5

$$\psi(\mathbf{r},t) = \frac{1}{(2\pi\hbar)^{3/2}} \int d\mathbf{p}\, \Phi(\mathbf{p},t) e^{i\mathbf{p}\cdot\mathbf{r}/\hbar}; \tag{5.2}$$

$$\Phi(\mathbf{p},t) = \frac{1}{(2\pi\hbar)^{3/2}} \int d\mathbf{r}\, \psi(\mathbf{r},t)\, e^{-i\mathbf{p}\cdot\mathbf{r}/\hbar}. \tag{5.3}$$

The function $|\Phi(\mathbf{p},t)|^2$ is the probability density in momentum space for the particle to have momentum $\mathbf{p}$ at time t.

You cannot measure $|\psi(\mathbf{r},t)|^2$ or $|\Phi(\mathbf{p},t)|^2$ directly. The only way you can build up information on the wave function is to make a series of measurements on *identically* prepared quantum systems, since each measurement generally modifies the system in some way. Quantum mechanics does not contain a prescription for carrying out the measurements. In fact, in the model I have adopted, the measuring apparatus is *external* to the quantum system. Moreover, it is often not trivial to construct experiments that *directly* measure physical observables such as energy, angular momentum, position, and momentum.

For the moment I associate quantum-mechanical operators with the coordinate, momentum, kinetic energy and potential energy functions of classical mechanics. For the coordinate and potential energy *operators*, $\hat{\mathbf{r}}$ and $\hat{V}$, corresponding to the classical variables $\mathbf{r}$ and $V(\mathbf{r})$, I define their operations in *coordinate space* by

$$\hat{\mathbf{r}}f(\mathbf{r}) = \mathbf{r}f(\mathbf{r}) \tag{5.4a}$$

$$\hat{V}f(\mathbf{r}) = V(\mathbf{r})f(\mathbf{r}) \tag{5.4b}$$

for any function $f(\mathbf{r})$. Similarly for the momentum and kinetic energy operators, $\hat{\mathbf{p}}$ and $\hat{p}^2/2m$, corresponding to the classical variables $\mathbf{p}$ and $p^2/2m$, I define their operations in *momentum space* by

$$\hat{\mathbf{p}}g(\mathbf{p}) = \mathbf{p}g(\mathbf{p}) \tag{5.5a}$$

$$\frac{\hat{p}^2}{2m}g(\mathbf{p}) = \frac{p^2}{2m}g(\mathbf{p}) \tag{5.5b}$$

for any function $g(\mathbf{p})$. It will turn out that postulate 5 will allow me to determine how $\hat{\mathbf{r}}$ and $\hat{V}$ act on functions of momentum, as well as how $\hat{\mathbf{p}}$ and $\hat{p}^2$ act on functions of position. We have already seen in Chap. 4 that $\hat{p}^2 f(\mathbf{r}) = -\hbar^2\nabla^2 f(\mathbf{r})$, but this relationship will be rederived using the postulates.

5.1 Hermitian Operators: Eigenvalues and Eigenfunctions

The time-independent Schrödinger equation,

$$\hat{H}\psi_E(\mathbf{r}) = E\psi_E(\mathbf{r}), \tag{5.6}$$

represents an *eigenvalue equation* in which E is the *eigenvalue* and $\psi_E(\mathbf{r})$ is the corresponding *eigenfunction*. It turns out that there are always solutions of equations of this type, provided the operators are Hermitian. To define what I mean by a Hermitian operator, I need to introduce the concept of the expectation value of an operator.

The *expectation value* of an operator $\hat{A}$ for a quantum system described by wave functions in coordinate and momentum space, $\psi(\mathbf{r},t)$ or $\Phi(\mathbf{p},t)$, respectively, is defined as

$$\begin{aligned}\left\langle\hat{A}\right\rangle &= \int d\mathbf{r}\,\psi^*(\mathbf{r},t)\hat{A}\psi(\mathbf{r},t)\\ &= \int d\mathbf{p}\,\Phi^*(\mathbf{p},t)\hat{A}\Phi(\mathbf{p},t).\end{aligned} \tag{5.7}$$

The symbols $d\mathbf{r}$ and $d\mathbf{p}$ correspond to volume elements in coordinate and momentum space, respectively, and the integrals in Eq. (5.7) are over all coordinate or momentum space. The expectation value is independent of whether the coordinate or momentum representation for the wave function is used. The operator $\hat{A}$ is assumed to be time-independent, but $\left\langle\hat{A}\right\rangle$ is a function of t, in general, owing to the time-dependence of the wave function. In writing Eq. (5.7), I assumed that the wave functions were normalized probability distributions; that is

$$\int d\mathbf{r}\,|\psi(\mathbf{r},t)|^2 = \int d\mathbf{p}\,|\Phi(\mathbf{p},t)|^2 = 1. \tag{5.8}$$

If the distributions are not normalized, then

$$\begin{aligned}\left\langle\hat{A}\right\rangle &= \frac{\int d\mathbf{r}\,\psi^*(\mathbf{r},t)\hat{A}\psi(\mathbf{r},t)}{\int d\mathbf{r}\,|\psi(\mathbf{r},t)|^2}\\ &= \frac{\int d\mathbf{p}\,\Phi^*(\mathbf{p},t)\hat{A}\Phi(\mathbf{p},t)}{\int d\mathbf{p}\,|\Phi(\mathbf{p},t)|^2}.\end{aligned} \tag{5.9}$$

Let us work in coordinate space and use a shorthand notation in which

$$\left\langle\hat{A}\right\rangle = \int d\mathbf{r}\,\psi^*(\mathbf{r},t)\hat{A}\psi(\mathbf{r},t) = \left(\psi,\hat{A}\psi\right). \tag{5.10}$$

Suppose that we demand that $\left\langle\hat{A}\right\rangle$ is *real* for an *arbitrary* $\psi(\mathbf{r},t)$. Then

$$\left\langle\hat{A}\right\rangle^* = \int d\mathbf{r}\,\psi(\mathbf{r},t)\left[\hat{A}\psi(\mathbf{r},t)\right]^* = \left(\hat{A}\psi,\psi\right) = \left\langle\hat{A}\right\rangle = \left(\psi,\hat{A}\psi\right) \tag{5.11}$$

or

$$\left(\hat{A}\psi,\psi\right)=\left(\psi,\hat{A}\psi\right). \tag{5.12}$$

Equation (5.12) can be used as the definition of a *Hermitian* operator. I will *assume* that all the operators we encounter in quantum mechanics are Hermitian operators. This is not unreasonable, since the expectation value of any physical dynamic variable must be real.

It turns out that the properties of Hermitian operators have been well-established by mathematicians. One of the most important properties is that there *exists* a *complete set* of *eigenfunctions* $\psi_a(\mathbf{r})$ associated with a time-independent Hermitian operator $\hat{A}$ for which

$$\hat{A}\psi_a(\mathbf{r})=a\psi_a(\mathbf{r}) \tag{5.13}$$

where a is the eigenvalue associated with the eigenfunction $\psi_a(\mathbf{r})$. The proof of existence and completeness is not trivial and is not given here. For the problems we encounter in quantum mechanics, there are certain boundary conditions that apply. The Hermitian nature of the operators and the completeness of the eigenfunctions depend implicitly on the appropriate boundary conditions. Examples are given as we go along.

At this point you might be getting frustrated and confused. What relationship do the eigenfunctions and eigenvalues have to physical systems? If we solve for these quantities, what have we gained? It turns out that the eigenfunctions and eigenvalues give us a *complete* picture of what is going on in quantum mechanics. The reason for this is that we can associate a Hermitian operator with every dynamic physical observable that can be measured. The outcome of a physical measurement of the operator associated with a *single* quantum system *must* be *one and only one* of the eigenvalues associated with that operator. Thus, if we measure the energy of a particle, we get one possible eigenvalue of the energy operator for that particle. If we measure the angular momentum of a particle, we get one eigenvalue of the angular momentum operator. Moreover, if we find the eigenfunctions and eigenvalues of the energy operator, we completely determine the wave function for the quantum system, given the initial conditions. The wave function allows you to calculate all properties of the quantum system. Many of these ideas will become clear with the examples that are given after I establish some properties of Hermitian operators.

5.1.1 Eigenvalues Real

I start from the eigenvalue equation (5.13),

$$\hat{A}\psi_a(\mathbf{r})=a\psi_a(\mathbf{r}), \tag{5.14}$$

and take the *inner product* (or integral) of both sides with $\psi_a^*(\mathbf{r})$,

$$\left(\psi_a, \hat{A}\psi_a\right) = (\psi_a, a\psi_a) = a\,(\psi_a, \psi_a), \tag{5.15}$$

where (ϕ, ψ) is a shorthand notation for

$$(\phi, \psi) = \int d\mathbf{r}\,\phi^*(\mathbf{r}, t)\psi(\mathbf{r}, t) = (\psi, \phi)^*. \tag{5.16}$$

Next, I take the complex conjugate of Eq. (5.13),

$$\left[\hat{A}\psi_a(\mathbf{r})\right]^* = [a\psi_a(\mathbf{r})]^*, \tag{5.17}$$

and take the *inner product* (or integral) of both sides with $\psi_a(\mathbf{r})$,

$$\left(\hat{A}\psi_a, \psi_a\right) = (a\psi_a, \psi_a) = a^*\,(\psi_a, \psi_a). \tag{5.18}$$

Subtracting Eq. (5.18) from Eq. (5.15), using Eq. (5.12), I find

$$\left(\psi_a, \hat{A}\psi_a\right) - \left(\hat{A}\psi_a, \psi_a\right) = 0 = \left(a - a^*\right)(\psi_a, \psi_a). \tag{5.19}$$

But since $(\psi_a, \psi_a) > 0$, it follows that $a = a^*$. The eigenvalues of a Hermitian operator are real.

5.1.2 Orthogonality

You are familiar with the concept of two vectors being orthogonal. The concept of orthogonality can be extended to *functions* by defining two functions $\phi_1(\mathbf{r})$ and $\phi_2(\mathbf{r})$ to be orthogonal if

$$(\phi_1, \phi_2) = \int d\mathbf{r}\phi_1^*(\mathbf{r})\,\phi_2(\mathbf{r}) = 0. \tag{5.20}$$

I will now show that the eigenfunctions of a Hermitian operator are either automatically orthogonal or can be chosen to be orthogonal. First, I prove a useful lemma (for some unknown reason, I love the word "lemma").

Lemma: If ϕ_1 and ϕ_2 are two *arbitrary* functions and if $\hat{A}$ is a Hermitian operator, then

$$\left(\phi_1, \hat{A}\phi_2\right) = \left(\hat{A}\phi_1, \phi_2\right). \tag{5.21}$$

The proof is straightforward. I use the definition given in Eq. (5.12), but replace ψ by

$$\psi(\mathbf{r}) = b_1\phi_1(\mathbf{r}) + b_2\phi_2(\mathbf{r}) \tag{5.22}$$

where b_1 and b_2 are *arbitrary* complex numbers. Then, for any b_1 and b_2

$$\left(b_1\phi_1 + b_2\phi_2, \hat{A}\left[b_1\phi_1 + b_2\phi_2\right]\right) = \left(\hat{A}\left[b_1\phi_1 + b_2\phi_2\right], b_1\phi_1 + b_2\phi_2\right) \tag{5.23}$$

or

$$\begin{aligned} &|b_1|^2\left(\phi_1, \hat{A}\phi_1\right) + |b_2|^2\left(\phi_2, \hat{A}\phi_2\right) + b_1^* b_2\left(\phi_1, \hat{A}\phi_2\right) + b_1 b_2^*\left(\phi_2, \hat{A}\phi_1\right) \\ &\quad = |b_1|^2\left(\hat{A}\phi_1, \phi_1\right) + |b_2|^2\left(\hat{A}\phi_2, \phi_2\right) + b_1^* b_2\left(\hat{A}\phi_1, \phi_2\right) + b_1 b_2^*\left(\hat{A}\phi_2, \phi_1\right). \end{aligned} \tag{5.24}$$

Since $\left(\phi_j, \hat{A}\phi_j\right) = \left(\hat{A}\phi_j, \phi_j\right)$ $(j = 1, 2)$ as a consequence of Eq. (5.12), Eq. (5.24) reduces to

$$b_1^* b_2\left(\phi_1, \hat{A}\phi_2\right) + b_1 b_2^*\left(\phi_2, \hat{A}\phi_1\right) = b_1^* b_2\left(\hat{A}\phi_1, \phi_2\right) + b_1 b_2^*\left(\hat{A}\phi_2, \phi_1\right). \tag{5.25}$$

The only way this can be satisfied for *arbitrary* complex b_1 and b_2 is if

$$\begin{aligned} \left(\phi_1, \hat{A}\phi_2\right) &= \left(\hat{A}\phi_1, \phi_2\right); \\ \left(\phi_2, \hat{A}\phi_1\right) &= \left(\hat{A}\phi_2, \phi_1\right). \end{aligned} \tag{5.26}$$

This is an alternative way to define a Hermitian operator.

5.1.2.1 Nondegenerate Eigenvalues

I first consider two eigenfunctions ψ_{a_1}, ψ_{a_2} whose eigenvalues are unequal or *nondegenerate*. I start from

$$\begin{aligned} \hat{A}\psi_{a_1}(\mathbf{r}) &= a_1\psi_{a_1}(\mathbf{r}); \\ \left[\hat{A}\psi_{a_2}(\mathbf{r})\right]^* &= \left[a_2\psi_{a_2}(\mathbf{r})\right]^* = a_2\left[\psi_{a_2}(\mathbf{r})\right]^*, \end{aligned} \tag{5.27}$$

having used the fact that a_2 is real, take the inner product of the first equation with ψ_{a_2} and the second equation with ψ_{a_1}, namely

$$\left(\psi_{a_2}, \hat{A}\psi_{a_1}\right) = a_1 \left(\psi_{a_2}, \psi_{a_1}\right);$$
$$\left(\hat{A}\psi_{a_2}, \psi_{a_1}\right) = a_2 \left(\psi_{a_2}, \psi_{a_1}\right), \tag{5.28}$$

subtract these equations, and use Eq. (5.21) to obtain

$$\left(\psi_{a_2}, \hat{A}\psi_{a_1}\right) - \left(\hat{A}\psi_{a_2}, \psi_{a_1}\right) = 0 = (a_1 - a_2)\left(\psi_{a_2}, \psi_{a_1}\right). \tag{5.29}$$

But since $a_1 \neq a_2$, it follows that

$$\left(\psi_{a_2}, \psi_{a_1}\right) = \int d\mathbf{r}\psi^*_{a_2}(\mathbf{r})\,\psi_{a_1}(\mathbf{r}) = 0; \tag{5.30}$$

eigenfunctions of a Hermitian operator corresponding to nondegenerate eigenvalues are *automatically* orthogonal. Moreover, I can *normalize* the eigenfunctions by setting

$$\left(\psi_{a_n}, \psi_{a_{n'}}\right) = \delta_{n,n'}, \tag{5.31}$$

where $\delta_{n,n'}$ is the Kronecker delta function that is equal to 1 if $n = n'$ and zero otherwise.

5.1.2.2 Degenerate Eigenvalues

The above proof fails if $a_1 = a_2$, but it still is possible to construct orthogonal eigenfunctions using a method called Schmidt orthogonalization. You know that in three-dimensional space, any three non-collinear unit vectors can serve as basis vectors. It is just convenient to choose orthogonal unit vectors. The same ideas apply here. Suppose that there are N eigenfunctions having the same eigenvalue a_n. I label these eigenfunctions by $\psi^{(m)}_{a_n}$ with m going from 1 to N. It is clear that any linear combination of these eigenfunctions also has eigenvalue a_n, since

$$\hat{A}\sum_{m=1}^{N} b_m \psi^{(m)}_{a_n} = \sum_{m=1}^{N} b_m \hat{A}\psi^{(m)}_{a_n} = a_n \sum_{m=1}^{N} b_m \psi^{(m)}_{a_n}. \tag{5.32}$$

The point is that it is always possible to choose N linear combinations of the eigenfunctions

$$\tilde{\psi}^{(q)}_{a_n} = \sum_{m=1}^{N} b_{qm}\psi^{(m)}_{a_n}; \qquad q = 1, 2, \ldots, N \tag{5.33}$$

which have eigenvalue a_n and are orthogonal to each other. Often the choice is easy to make by inspection; for example, symmetric and anti-symmetric combinations of two functions. In fact it is rare to actually have to use the Schmidt orthogonalization procedure.

If you *must* use it, you could proceed as follows: Suppose $\left(\psi_{a_n}^{(1)}, \psi_{a_n}^{(2)}\right) = c_{12} \neq 0$. If the $\psi_{a_n}^{(j)}$ are normalized, as is assumed, then $|c_{12}| < 1$.[1] Take $\tilde{\psi}_{a_n}^{(1)} = \psi_{a_n}^{(1)}$ and

$$\tilde{\psi}_{a_n}^{(2)} = b_{21}\psi_{a_n}^{(1)} + b_{22}\psi_{a_n}^{(2)}. \tag{5.34}$$

For $\tilde{\psi}_{a_n}^{(1)}$ and $\tilde{\psi}_{a_n}^{(2)}$ to be orthogonal, I must require that

$$\left(\tilde{\psi}_{a_n}^{(1)}, \tilde{\psi}_{a_n}^{(2)}\right) = b_{21} + b_{22}c_{12} = 0 \tag{5.35}$$

or

$$b_{21} = -c_{12}b_{22}. \tag{5.36}$$

The values of b_{21} and b_{22} can be determined if I normalize $\tilde{\psi}_{a_n}^{(2)}$,

$$\begin{aligned}
\left(\tilde{\psi}_{a_n}^{(2)}, \tilde{\psi}_{a_n}^{(2)}\right) &= \left(b_{21}\psi_{a_n}^{(1)} + b_{22}\psi_{a_n}^{(2)}, b_{21}\psi_{a_n}^{(1)} + b_{22}\psi_{a_n}^{(2)}\right) \\
&= |b_{22}|^2 \left(-c_{12}\psi_{a_n}^{(1)} + \psi_{a_n}^{(2)}, -c_{12}\psi_{a_n}^{(1)} + \psi_{a_n}^{(2)}\right) \\
&= |b_{22}|^2 \left(|c_{12}|^2 - |c_{12}|^2 - |c_{12}|^2 + 1\right) \\
&= |b_{22}|^2 \left(1 - |c_{12}|^2\right) = 1,
\end{aligned} \tag{5.37}$$

having used Eq. (5.36) and the fact that

$$\left(\psi_{a_n}^{(1)}, \psi_{a_n}^{(2)}\right) = \left(\psi_{a_n}^{(2)}, \psi_{a_n}^{(1)}\right)^* = c_{12}. \tag{5.38}$$

Thus, if

$$b_{22} = \left(1 - |c_{12}|^2\right)^{-1/2}; \quad b_{21} = -c_{12}b_{22} = -c_{12}\left(1 - |c_{12}|^2\right)^{-1/2}, \tag{5.39}$$

then $\tilde{\psi}_{a_n}^{(1)}$ and $\tilde{\psi}_{a_n}^{(2)}$ are orthonormal wave functions.

Now suppose $\left(\tilde{\psi}_{a_n}^{(1)}, \psi_{a_n}^{(3)}\right) = c_{13} \neq 0$ and $\left(\tilde{\psi}_{a_n}^{(2)}, \psi_{a_n}^{(3)}\right) = c_{23} \neq 0$. I set

$$\tilde{\psi}_{a_n}^{(3)} = g_{31}\tilde{\psi}_{a_n}^{(1)} + g_{32}\tilde{\psi}_{a_n}^{(2)} + g_{33}\psi_{a_n}^{(3)}, \tag{5.40}$$

---

[1]The fact that $|c_{12}| < 1$ follows from the Schwarz inequality, $\left|\left(\psi_{a_n}^{(1)}, \psi_{a_n}^{(2)}\right)\right|^2 \leq \left(\psi_{a_n}^{(1)}, \psi_{a_n}^{(1)}\right)\left(\psi_{a_n}^{(2)}, \psi_{a_n}^{(2)}\right) = 1$.

(the g's differ from the b's in Eq. (5.33) since I expand in terms of the *new* first two eigenfunctions) and demand that

$$\left(\tilde{\psi}_{a_n}^{(1)}, \tilde{\psi}_{a_n}^{(3)}\right) = 0, \tag{5.41a}$$

$$\left(\tilde{\psi}_{a_n}^{(2)}, \tilde{\psi}_{a_n}^{(3)}\right) = 0, \tag{5.41b}$$

$$\left(\tilde{\psi}_{a_n}^{(3)}, \tilde{\psi}_{a_n}^{(3)}\right) = 1. \tag{5.41c}$$

These constitute three complex equations that allow you to solve for the complex numbers g_{31}, g_{32}, g_{33}. And so on, for the remaining of the N degenerate eigenfunctions. In the end you have an orthonormal basis for these degenerate eigenfunctions. Thus the eigenfunctions corresponding to degenerate eigenvalues are not *automatically* orthogonal, but can be chosen to be so.

5.1.3 Completeness

Although the proof of the completeness of the eigenvalues is not trivial, a statement of the completeness *condition* is not difficult to obtain. If the eigenfunctions are complete, any function $\psi(\mathbf{r})$ can be expanded as

$$\psi(\mathbf{r}) = \sum_m b_m \psi_{a_m}(\mathbf{r}). \tag{5.42}$$

If I take the inner product of this equation with ψ_{a_n}, I find

$$\left(\psi_{a_n}, \psi\right) = \sum_m b_m \left(\psi_{a_n}, \psi_{a_m}\right) = \sum_m b_m \delta_{m,n} = b_n. \tag{5.43}$$

Note that Eq.(5.43) can be used to calculate the expansion coefficients b_n. Substituting Eq. (5.43) into Eq. (5.42), I obtain

$$\psi(\mathbf{r}) = \sum_m \left(\psi_{a_m}, \psi\right) \psi_{a_m}(\mathbf{r}) = \int_{-\infty}^{\infty} d\mathbf{r}' \left[\sum_m \psi_{a_m}^*(\mathbf{r}')\psi_{a_m}(\mathbf{r})\right] \psi(\mathbf{r}'). \tag{5.44}$$

For the equality to hold, the term in square brackets must equal $\delta(\mathbf{r} - \mathbf{r}')$, implying that

$$\sum_m \psi_{a_m}^*(\mathbf{r}')\psi_{a_m}(\mathbf{r}) = \delta\left(\mathbf{r} - \mathbf{r}'\right). \tag{5.45}$$

Equation (5.45) is a condition that must be satisfied if the eigenfunctions are complete.

Equation (5.42) is analogous to the expansion of a vector in terms of unit vectors, except that the unit vectors are now replaced by orthonormal *functions*. The dot product of vectors is replaced by an integral of the form given in Eq. (5.16). Moreover, Eq. (5.43) is equivalent to projecting out the component of a vector. The analogue with vector spaces will become exact when I consider Dirac notation in Chap. 11.

5.1.4 Continuous Eigenvalues

So far it has been assumed implicitly that the eigenvalues are discrete. As you shall see, this is always the case if the particles are confined to a finite volume. However unbound particles have *continuous* eigenvalues. For example, consider a free particle having mass m. The time-independent Schrödinger equation for this particle is

$$\hat{H}\psi_E(\mathbf{r}) = \frac{\hat{p}^2}{2m}\psi_E(\mathbf{r}) = -\frac{\hbar^2}{2m}\nabla^2\psi_E(\mathbf{r}). \tag{5.46}$$

Clearly the momentum of a free particle can take on any value. Moreover, for a given positive energy, there is an *infinite* number of momenta corresponding to a given energy—there is infinite degeneracy. To label each of the degenerate eigenfunctions, I can use the momentum and take as eigenfunctions

$$\psi_{\mathbf{p}}(\mathbf{r}) = e^{i\mathbf{p}\cdot\mathbf{r}/\hbar}, \tag{5.47}$$

which are solutions of Eq. (5.46), provided

$$E = \frac{p^2}{2m}. \tag{5.48}$$

The question arises as to how to normalize these eigenfunctions. Clearly $\int d\mathbf{r}\left|\psi_{\mathbf{p}}(\mathbf{r})\right|^2 = \infty$, so I cannot normalize as I did in the case of discrete eigenvalues. However, from the definition of the Dirac delta function, we know that

$$\int d\mathbf{r}\, e^{i(\mathbf{p}-\mathbf{p}')\cdot\mathbf{r}/\hbar} = (2\pi\hbar)^3\,\delta\left(\mathbf{p}-\mathbf{p}'\right). \tag{5.49}$$

By convention, I use this result and take as free-particle eigenfunctions,

$$\psi_{\mathbf{p}}(\mathbf{r}) = \frac{1}{(2\pi\hbar)^{3/2}}e^{i\mathbf{p}\cdot\mathbf{r}/\hbar} \tag{5.50}$$

or, in $\mathbf{k}$ space ($\mathbf{p} = \hbar\mathbf{k}$),

$$\psi_{\mathbf{k}}(\mathbf{r}) = \frac{1}{(2\pi)^{3/2}} e^{i\mathbf{k}\cdot\mathbf{r}}. \tag{5.51}$$

With this choice, the normalization conditions for the free particle eigenfunctions are

$$\int d\mathbf{r}\, \psi^*_{\mathbf{p}}(\mathbf{r})\psi_{\mathbf{p}'}(\mathbf{r}) = \delta\left(\mathbf{p} - \mathbf{p}'\right); \tag{5.52a}$$

$$\int d\mathbf{r}\, \psi^*_{\mathbf{k}}(\mathbf{r})\psi_{\mathbf{k}'}(\mathbf{r}) = \delta\left(\mathbf{k} - \mathbf{k}'\right). \tag{5.52b}$$

Equations (5.52) can still be used as the normalization condition in the more general case of unbound motion in the presence of a potential; however, the eigenfunctions are no longer given by Eqs. (5.50) and (5.51).

In going from discrete to continuous eigenvalues, the units for the wave functions and the expansion coefficients change. For discrete eigenenergies, the eigenfunctions in coordinate space have units of $1/\sqrt{\text{volume}}$. For continuous eigenvalues, $\psi_{\mathbf{p}}(\mathbf{r})$ has units of $(\hbar)^{-3/2}$ while $\psi_{\mathbf{k}}(\mathbf{r})$ is dimensionless. Similar differences arise for the expansion coefficients. In the case of continuous eigenvalues, an arbitrary function can be expanded as

$$\psi(\mathbf{r}) = \int d\mathbf{p}\, b(\mathbf{p})\psi_{\mathbf{p}}(\mathbf{r}); \tag{5.53a}$$

$$\psi(\mathbf{r}) = \int d\mathbf{k}\, \tilde{b}(\mathbf{k})\psi_{\mathbf{k}}(\mathbf{r}). \tag{5.53b}$$

In contrast to the case of discrete eigenenergies for which the expansion coefficients of the wave function are dimensionless, the expansion coefficients $b(\mathbf{p})$ have units of $(\text{momentum})^{-3/2}$,while the expansion coefficients $\tilde{b}(\mathbf{k})$ have units of $(\text{volume})^{3/2}$.

For continuous variables, the completeness conditions are

$$\int d\mathbf{p}\, \psi^*_{\mathbf{p}}(\mathbf{r})\psi_{\mathbf{p}}(\mathbf{r}') = \delta\left(\mathbf{r} - \mathbf{r}'\right); \tag{5.54}$$

$$\int d\mathbf{k}\, \psi^*_{\mathbf{k}}(\mathbf{r})\psi_{\mathbf{k}}(\mathbf{r}') = \delta\left(\mathbf{r} - \mathbf{r}'\right). \tag{5.55}$$

A formal method for going from discrete to continuous variables is given in the Appendix.

5.1.5 Relationship Between Operators

I have already derived an expression for the square of the momentum operator in coordinate space in Chap. 4. I now want to rederive this result from the postulates. To do so, I calculate the momentum operator in the coordinate representation. The expectation value of any operator is the same in both the coordinate and momentum representations. Let's see how this works for the momentum operator. I start from

$$\begin{aligned}\int d\mathbf{r}\,[\psi(\mathbf{r},t)]^*\,\hat{p}_x\psi(\mathbf{r},t) &= \int d\mathbf{p}\,[\Phi(\mathbf{p},t)]^*\,\hat{p}_x\Phi(\mathbf{p},t)\\ &= \int d\mathbf{p}\,[\Phi(\mathbf{p},t)]^*\,p_x\Phi(\mathbf{p},t),\end{aligned} \tag{5.56}$$

where I made use of Eq. (5.5a). I substitute the expression $\Phi(\mathbf{p},t)$ given in Eq. (5.3) into the right-hand side of this equation (being careful to use different dummy variables for the integrals) and obtain

$$\begin{aligned}\int d\mathbf{p}\,[\Phi(\mathbf{p},t)]^*\,p_x\Phi(\mathbf{p},t) &= \frac{1}{(2\pi\hbar)^3}\int d\mathbf{r}\int d\mathbf{r}'\,[\psi(\mathbf{r},t)]^*\,\psi(\mathbf{r}',t)\\ &\quad\times\int d\mathbf{p}\,p_x e^{i\mathbf{p}\cdot(\mathbf{r}-\mathbf{r}')/\hbar}.\end{aligned} \tag{5.57}$$

Since

$$\frac{\hbar}{i}\frac{\partial}{\partial x'}e^{i\mathbf{p}\cdot(\mathbf{r}-\mathbf{r}')/\hbar} = -p_x e^{i\mathbf{p}\cdot(\mathbf{r}-\mathbf{r}')/\hbar}, \tag{5.58}$$

I find that

$$\begin{aligned}\int d\mathbf{p}\,[\Phi(\mathbf{p},t)]^*\,p_x\Phi(\mathbf{p},t) &= -\frac{1}{(2\pi\hbar)^3}\int d\mathbf{r}\int d\mathbf{r}'\,[\psi(\mathbf{r},t)]^*\,\psi(\mathbf{r}',t)\\ &\quad\times\frac{\hbar}{i}\frac{\partial}{\partial x'}\int d\mathbf{p}\,e^{i\mathbf{p}\cdot(\mathbf{r}-\mathbf{r}')/\hbar}.\end{aligned} \tag{5.59}$$

The integral over $\mathbf{p}$ yields $(2\pi\hbar)^3\,\delta(\mathbf{r}-\mathbf{r}')$, such that, with the help of the chain rule,

$$\begin{aligned}&\int d\mathbf{p}\,[\Phi(\mathbf{p},t)]^*\,p_x\Phi(\mathbf{p},t)\\ &= -\int d\mathbf{r}\int d\mathbf{r}'\,[\psi(\mathbf{r},t)]^*\,\psi(\mathbf{r}',t)\frac{\hbar}{i}\frac{\partial}{\partial x'}\delta(\mathbf{r}-\mathbf{r}')\\ &= -\frac{\hbar}{i}\int d\mathbf{r}\,[\psi(\mathbf{r},t)]^*\int d\mathbf{r}'\frac{\partial}{\partial x'}\left[\psi(\mathbf{r}',t)\delta(\mathbf{r}-\mathbf{r}')\right]\\ &\quad+\frac{\hbar}{i}\int d\mathbf{r}\,[\psi(\mathbf{r},t)]^*\int d\mathbf{r}'\delta(\mathbf{r}-\mathbf{r}')\,\frac{\partial}{\partial x'}\psi(\mathbf{r}',t).\end{aligned} \tag{5.60}$$

The second integral in the first term involves an exact differential whose integral vanishes for a wave function having finite extent. I am left with

$$\int d\mathbf{p}\,[\Phi(\mathbf{p},t)]^* p_x\Phi(\mathbf{p},t) = \int d\mathbf{r}\,[\psi(\mathbf{r},t)]^* \frac{\hbar}{i}\frac{\partial}{\partial x}\psi(\mathbf{r},t), \tag{5.61}$$

which, according to Eq. (5.56), implies that

$$\hat{p}_x = \frac{\hbar}{i}\frac{\partial}{\partial x}. \tag{5.62}$$

In three dimensions the analogous equation is

$$\hat{\mathbf{p}} = \frac{\hbar}{i}\nabla_r, \tag{5.63}$$

which is consistent with $\hat{p}^2 = -\hbar^2\nabla^2$.

The same technique can be used for *any* operator, except that the functions corresponding to the operators must be Fourier transformed as well. For example, suppose that $B(\mathbf{p})$ is a function of momentum only. I associate a quantum-mechanical operator $\hat{B}(\hat{\mathbf{p}})$ with this function and *assume* that

$$\hat{B}(\hat{\mathbf{p}})\Phi(\mathbf{p}) = B(\mathbf{p})\Phi(\mathbf{p}) \tag{5.64}$$

for any function $\Phi(\mathbf{p})$ having a Fourier transform $\psi(\mathbf{r})$. Next I consider

$$\begin{aligned}
&\frac{1}{(2\pi\hbar)^{3/2}}\int d\mathbf{p}\, e^{i\mathbf{p}\cdot\mathbf{r}/\hbar}B(\mathbf{p})\Phi(\mathbf{p}) \\
&\quad= \frac{1}{(2\pi\hbar)^{3/2}}\frac{1}{(2\pi\hbar)^3}\int d\mathbf{r}\int d\mathbf{r}'\int d\mathbf{p}\, e^{i\mathbf{p}\cdot(\mathbf{r}-\mathbf{r}'-\mathbf{r}'')/\hbar}\tilde{B}(\mathbf{r}'')\psi(\mathbf{r}') \\
&\quad= \frac{1}{(2\pi\hbar)^{3/2}}\int d\mathbf{r}\int d\mathbf{r}'\,\delta(\mathbf{r}-\mathbf{r}'-\mathbf{r}'')\tilde{B}(\mathbf{r}'')\psi(\mathbf{r}') \\
&\quad= \frac{1}{(2\pi\hbar)^{3/2}}\int d\mathbf{r}'\,\tilde{B}(\mathbf{r}-\mathbf{r}')\psi(\mathbf{r}'),
\end{aligned} \tag{5.65}$$

where $\tilde{B}(\mathbf{r})$ is the Fourier transform of $B(\mathbf{p})$. I interpret this result to imply that an operator $\hat{B}(\hat{\mathbf{p}})$ acting on a function $\psi(\mathbf{r})$ produces the *integral* operation,

$$\hat{B}(\hat{\mathbf{p}})\psi(\mathbf{r}) = \frac{1}{(2\pi\hbar)^{3/2}}\int d\mathbf{r}'\,\tilde{B}(\mathbf{r}-\mathbf{r}')\psi(\mathbf{r}'). \tag{5.66}$$

Similarly, for an operator $\hat{C}(\hat{\mathbf{r}})$ that is associated with a function $C(\mathbf{r})$ that is a function of coordinates only and for which it is assumed that

$$\hat{C}(\hat{\mathbf{r}})\psi(\mathbf{r}) = C(\mathbf{r})\psi(\mathbf{r}), \tag{5.67}$$

the action of the operator $\hat{C}(\hat{\mathbf{r}})$ acting on a function $\Phi(\mathbf{p})$ produces the integral operation,

$$\hat{C}(\hat{\mathbf{r}})\Phi(\mathbf{p}) = \frac{1}{(2\pi\hbar)^{3/2}}\int d\mathbf{p}'\,\tilde{C}(\mathbf{p}-\mathbf{p}')\Phi(\mathbf{p}'), \tag{5.68}$$

where $\tilde{C}(\mathbf{p})$ is the Fourier transform of $C(\mathbf{r})$ and $\Phi(\mathbf{p})$ is the Fourier transform of $\psi(\mathbf{r})$. We now know how operators act on the wave function in both coordinate and momentum space. In fact, Eq. (5.68) implies that Schrödinger's equation in momentum space is

$$i\hbar\frac{\partial\Phi(\mathbf{p},t)}{\partial t} = \frac{p^2}{2m}\Phi(\mathbf{p},t) + \frac{1}{(2\pi\hbar)^{3/2}}\int_{-\infty}^{\infty} d\mathbf{p}'\,\tilde{V}(\mathbf{p}-\mathbf{p}')\Phi(\mathbf{p}',t), \tag{5.69}$$

where $\tilde{V}(\mathbf{p})$ is the Fourier transform of $V(\mathbf{r})$. I will derive the time-independent Schrödinger equation in momentum space in Chap. 11 using Dirac notation.

5.1.6 *Commutator of Operators*

I now look at some additional properties of operators. The *commutator* $\hat{C}$ of two operators $\hat{A}$ and $\hat{B}$ is defined as

$$\hat{C} = \left[\hat{A},\hat{B}\right] = \hat{A}\hat{B} - \hat{B}\hat{A} = -\left[\hat{B},\hat{A}\right]. \tag{5.70}$$

For example, the operators $\hat{x}$ and $\hat{p}_x = \frac{\hbar}{i}\frac{d}{dx}$ do not commute since

$$\begin{aligned}
[\hat{x},\hat{p}_x]\,\psi(x) &= \frac{\hbar}{i}\left(x\frac{d}{dx} - \frac{d}{dx}x\right)\psi \\
&= \frac{\hbar}{i}\left(x\frac{d\psi}{dx} - \frac{d}{dx}(x\psi)\right) = -\frac{\hbar}{i}\psi(x),
\end{aligned} \tag{5.71}$$

which can be satisfied for *arbitrary* $\psi(x)$ only if

$$[\hat{x},\hat{p}_x] = i\hbar. \tag{5.72}$$

Similarly,

$$\left[\hat{y},\hat{p}_y\right] = [\hat{z},\hat{p}_z] = i\hbar. \tag{5.73}$$

As you can see, the commutator of two operators can be calculated by looking at its action on functions. Using this method, it is easy to show that any two components of the position operator commute and any two components of the momentum operator commute,

$$[\hat{x}, \hat{y}] = [\hat{x}, \hat{z}] = [\hat{y}, \hat{z}] = [\hat{p}_x, \hat{p}_z] = \left[\hat{p}_x, \hat{p}_y\right] = \left[\hat{p}_y, \hat{p}_z\right] = 0. \tag{5.74}$$

Moreover, *different* components of the position and momentum operators commute as well,

$$\left[\hat{x}, \hat{p}_y\right] = [\hat{x}, \hat{p}_z] = [\hat{y}, \hat{p}_x] = [\hat{y}, \hat{p}_z] = [\hat{z}, \hat{p}_x] = \left[\hat{z}, \hat{p}_y\right] = 0. \tag{5.75}$$

Equations (5.72)–(5.75) are the fundamental commutator relations. Remember them at all times!

Commuting operators play a central role in quantum mechanics. I first prove that two Hermitian operators commute *if and only if* they possess simultaneous eigenfunctions. There are two parts to the proof. First suppose that Hermitian operators $\hat{A}$ and $\hat{B}$ possess simultaneous eigenfunctions ψ_{ab},

$$\hat{A}\psi_{ab} = a\psi_{ab}; \quad \hat{B}\psi_{ab} = b\psi_{ab}\,. \tag{5.76}$$

Then

$$\left[\hat{A}, \hat{B}\right]\psi_{ab} = \left(\hat{A}\hat{B} - \hat{B}\hat{A}\right)\psi_{ab} = (ba - ab)\,\psi_{ab} = 0 \tag{5.77}$$

and the operators commute. Conversely, suppose that ψ_a is an eigenfunction of $\hat{A}$ and that $\left[\hat{A}, \hat{B}\right]\psi_a = 0$. Then

$$\begin{aligned}\left[\hat{A}, \hat{B}\right]\psi_a &= \left(\hat{A}\hat{B} - \hat{B}\hat{A}\right)\psi_a = 0;\\ \hat{A}\left(\hat{B}\psi_a\right) &= a\left(\hat{B}\psi_a\right).\end{aligned} \tag{5.78}$$

Equation (5.78) is nothing but a statement of the fact that $\left(\hat{B}\psi_a\right)$ is an eigenfunction of $\hat{A}$ with eigenvalue a, having the most general form

$$\hat{B}\psi_a = b\psi_a, \tag{5.79}$$

where b is some constant. Therefore, ψ_a is a simultaneous eigenfunction of $\hat{B}$ with eigenvalue b. Actually the proof is valid only if the eigenfunctions ψ_a are nondegenerate. If there are N degenerate eigenfunctions $\psi_a^{(m)}$ associated with eigenvalue a, Eq. (5.78) implies only that

$$\hat{B}\psi_a^{(n)} = \sum_{m=1}^{N} b_{nm}\psi_a^{(m)}. \tag{5.80}$$

However, as in the Schmidt orthogonalization procedure, it is always possible to find linear combinations of the degenerate eigenfunctions that are simultaneous eigenfunctions of $\hat{B}$.

Thus if two Hermitian operators commute and one of these operators has nondegenerate eigenvalues, then its eigenfunctions are *automatically* eigenfunctions of the other operator. On the other hand, if two Hermitian operators commute and one of these operators has degenerate eigenvalues, then a degenerate eigenfunction of one of the operators is not *automatically* an eigenfunction of the other operator, but some linear combinations of the degenerate eigenfunctions can be chosen that *is* an eigenfunction of the other operator. Examples are given later in this chapter.

The central problem in quantum mechanics is to solve the time-independent Schrödinger equation. As long as there is no energy degeneracy, for discrete energy eigenvalues, you can always label the energy by a quantum number n with the lowest value of n corresponding to the lowest energy, the second value to the next highest energy, etc. For example, if $V(x) = ax^4 + bx^2 + cx$ with $a > 0$, it is not possible to find analytic expressions for the eigenenergies and eigenfunctions, but you can still label the lowest energy state and eigenfunction by $n = 0$, the next by $n = 1$, etc. When there is *energy degeneracy*, however, we need additional labels to distinguish states having the same energy; that is we need additional quantum numbers. Where can we get these quantum numbers? There may be a number of ways to specify the quantum numbers, but the most systematic way is to identify additional operators that commute with the Hamiltonian. You can then label the states by the eigenvalues of the simultaneous eigenfunctions of the commuting operators. It turns out, whenever there is energy degeneracy, it is usually possible to identify an operator that commutes with the Hamiltonian that is in some way associated with the degeneracy.

We have already seen one example of energy degeneracy and will see many more throughout this book. For the free particle in one dimension, the eigenfunctions are two-fold degenerate for each positive energy, $\psi_E(x) = e^{\pm i\sqrt{2mE}x/\hbar}$. That is, given the energy, you cannot uniquely label the eigenfunction. However the momentum and energy operators commute. Moreover, the momentum state eigenfunctions, given by $e^{ip_x x/\hbar}$, are *nondegenerate* (each momentum eigenfunction is associated with a different momentum). As a consequence the momentum state eigenfunctions *must* be simultaneous eigenfunctions of the energy operator. If we label the eigenfunctions by p_x alone, we completely specify the energy eigenfunctions as well. In other words, the eigenfunctions

$$\psi_{p_x}(x) = \frac{1}{\sqrt{2\pi\hbar}} e^{ip_x x/\hbar} \tag{5.81}$$

are simultaneous eigenfunctions of the energy operator if $E = p_x^2/2m$. Similarly in three dimensions, where there is infinite degeneracy for each positive energy,

$$\psi_{\mathbf{p}}(\mathbf{r}) = \frac{1}{(2\pi\hbar)^{3/2}} e^{i\mathbf{p}\cdot\mathbf{r}/\hbar} \tag{5.82}$$

uniquely labels the eigenfunctions, provided $E = p^2/2m$. We shall see later that the fact that the momentum operator commutes with the Hamiltonian is linked to the translational symmetry of the Hamiltonian.

5.1.6.1 Commutator Algebra Relationships

It is very easy to prove the following relationships for commutators:

$$\left[\hat{A}\hat{B}, \hat{C}\right] = \hat{A}\left[\hat{B}, \hat{C}\right] + \left[\hat{A}, \hat{C}\right]\hat{B}; \tag{5.83}$$

$$\left[\hat{A}, \hat{B}\hat{C}\right] = \hat{B}\left[\hat{A}, \hat{C}\right] + \left[\hat{A}, \hat{B}\right]\hat{C}. \tag{5.84}$$

Also, if $\hat{C} = \left[\hat{A}, \hat{B}\right]$, and if $\left[\hat{A}, \hat{C}\right] = 0$ and $\left[\hat{B}, \hat{C}\right] = 0$, then

$$e^{\hat{A}+\hat{B}} = e^{\hat{A}} e^{\hat{B}} e^{-\hat{C}/2}. \tag{5.85}$$

Another useful identity is

$$e^{\hat{A}}\hat{B}e^{-\hat{A}} = \hat{B} + \left[\hat{A}, \hat{B}\right] + \frac{1}{2!}\left[\hat{A}, \left[\hat{A}, \hat{B}\right]\right] + \cdots \tag{5.86}$$

which is known as the Baker-Campbell-Hausdorff theorem. Moreover, if $\hat{A}$ and $\hat{B}$ are Hermitian operators, then $\hat{A}\hat{B} + \hat{B}\hat{A}$ and $i\left[\hat{A}, \hat{B}\right]$ are Hermitian, but $\hat{A}\hat{B}$ is Hermitian only if $\left[\hat{A}, \hat{B}\right] = 0$ (see problems).

The basic commutation relations for coordinate and momentum space operators, obtained in a manner similar to the one used to arrive at Eq. (5.72) are

$$\left[\hat{\mathbf{r}}, \hat{B}(\hat{\mathbf{p}})\right] = i\hbar\nabla_p B(\mathbf{p}); \tag{5.87a}$$

$$\left[\hat{\mathbf{p}}, \hat{C}(\hat{\mathbf{r}})\right] = -i\hbar\nabla_r C(\mathbf{r}); \tag{5.87b}$$

$$\left[\hat{\mathbf{r}}, \hat{C}(\hat{\mathbf{r}})\right] = 0; \tag{5.87c}$$

$$\left[\hat{\mathbf{p}}, \hat{B}(\hat{\mathbf{p}})\right] = 0. \tag{5.87d}$$

For commutators involving higher powers of r and p, such as r^2 and p^2, one usually uses a combination of Eqs. (5.83), (5.84), and (5.87), rather than calculate the effect of the commutator on functions as I did in deriving Eq. (5.72). For example,

$$\left[\hat{x}, \hat{p}_x^2\right] = \left[\hat{x}, \hat{p}_x \hat{p}_x\right] = \hat{p}_x \left[\hat{x}, \hat{p}_x\right] + \left[\hat{x}, \hat{p}_x\right] \hat{p}_x = 2i\hbar \hat{p}_x. \tag{5.88}$$

5.1.7 Uncertainty Principle

Many people have heard about *Heisenberg's Uncertainty Relation*, even if they do not understand it. Actually, it is possible to derive an uncertainty relation for the product $\Delta A^2 \Delta B^2$ for any two non-commuting Hermitian operators $\hat{A}$ and $\hat{B}$. To do so I start from the inequality

$$\int d\mathbf{r} \left[\left(\hat{A} + i\lambda\hat{B}\right)\psi(\mathbf{r})\right]^* \left(\hat{A} + i\lambda\hat{B}\right)\psi(\mathbf{r}) \geq 0, \tag{5.89}$$

where λ is a constant taken to be real (a somewhat more general uncertainty relation can be derived if λ is taken to be complex). Without loss of generality, I take $\left\langle\hat{A}\right\rangle = 0$ and $\left\langle\hat{B}\right\rangle = 0$. If this were not the case, I would replace $\hat{A}$ by $\hat{A} - \left\langle\hat{A}\right\rangle$ and $\hat{B}$ by $\hat{B} - \left\langle\hat{B}\right\rangle$ in Eq. (5.89). Since $\hat{A}$ and $\hat{B}$ are Hermitian, Eq. (5.89) can be rewritten as

$$\begin{aligned} &\int d\mathbf{r} \left[\psi(\mathbf{r})\right]^* \left(\hat{A} - i\lambda\hat{B}\right)\left(\hat{A} + i\lambda\hat{B}\right)\psi(\mathbf{r}) \\ &= \int d\mathbf{r} \left[\psi(\mathbf{r})\right]^* \left(\hat{A}^2 + i\lambda\hat{A}\hat{B} - i\lambda\hat{B}\hat{A} + \lambda^2\hat{B}^2\right)\psi(\mathbf{r}) \geq 0, \end{aligned} \tag{5.90}$$

or

$$\lambda^2 \Delta B^2 + i\lambda\left\langle\hat{C}\right\rangle + \Delta A^2 \geq 0, \tag{5.91}$$

where

$$\hat{C} = \left[\hat{A}, \hat{B}\right]. \tag{5.92}$$

The minimum occurs for

$$\lambda = \frac{-i\left\langle\hat{C}\right\rangle}{2\Delta B^2} \tag{5.93}$$

and, when this result is substituted into Eq. (5.91), I find

$$\Delta A^2 \Delta B^2 \geq -\frac{\left\langle \hat{C} \right\rangle^2}{4}, \tag{5.94}$$

which implies that the expectation value of the commutator of two Hermitian operators either vanishes or is purely imaginary (this is not surprising since one prescription for quantization is to obtain the commutator as $i\hbar$ times the Poisson bracket of the classical variables associated with the operators). For $\hat{A} = \hat{x}$ and $\hat{B} = \hat{p}_x$, $\hat{C} = i\hbar$ and

$$\Delta x^2 \Delta p_x^2 \geq \frac{\hbar^2}{4}, \tag{5.95}$$

which is known as the Heisenberg Uncertainty Relation.

From Eq. (5.89), it follows that the equality in Eq. (5.94) holds only if

$$\left(\hat{A} + i\lambda \hat{B}\right) \psi\left(x\right) = 0. \tag{5.96}$$

For $\hat{A} = \hat{x}$ and $\hat{B} = \hat{p}_x$, $\lambda = \hbar/\left(2\Delta p_x^2\right)$ [see Eq. (5.93)] and the *minimum uncertainty wave function* $\psi_{\text{min}}\left(x\right)$ must satisfy the differential equation

$$\begin{aligned}\left(\hat{x} + i\lambda \hat{p}_x\right) \psi_{\text{min}}\left(x\right) &= \left[x + i\left(\frac{\hbar}{2\Delta p_x^2}\right)\frac{\hbar}{i}\frac{d}{dx}\right]\psi_{\text{min}}\left(x\right) \\ &= \frac{\hbar^2}{2\Delta p_x^2}\frac{d\psi_{\text{min}}\left(x\right)}{dx} + x\psi_{\text{min}}\left(x\right) = 0.\end{aligned} \tag{5.97}$$

The solution of this equation is

$$\psi_{\text{min}}\left(x\right) = N\exp\left(-\frac{x^2 \Delta p_x^2}{\hbar^2}\right) = N\exp\left(-\frac{x^2}{4\Delta x^2}\right), \tag{5.98}$$

where N is a normalization constant. The minimum uncertainty wave packet is a Gaussian and *only* a Gaussian! For example, we will see that the lowest energy state wave function for a particle confined to an infinite potential well is not a Gaussian—consequently $\Delta x^2 \Delta p_x^2$ *must* be greater than $\hbar^2/4$ for this wave function.

The uncertainty principle is often illustrated by examples in which you show that by measuring the position of a particle to a given precision, you necessarily introduce an uncertainty in the momentum that satisfies the uncertainty principle. You will note that my derivation has nothing to do with measurement, *per se*. In effect, the position-momentum uncertainty relation is related directly to the fact that the corresponding operators do not commute. Equivalently, it is linked to the fact that matter is described by a wave theory in which the wave functions in coordinate and momentum space are Fourier transforms of one another. Any measurements must be consistent with the theory, but the measurements themselves are not directly related to the uncertainty principle.

5.1.8 *Examples of Operators*

In most of the examples below, I consider one-dimensional motion only, with p standing for the x component of the momentum. Moreover, I work in the coordinate representation only. The generalization to two and three dimensions is often obvious. The Hamiltonian is assumed to be of the form

$$\hat{H} = \frac{\hat{p}^2}{2m} + \hat{V}. \tag{5.99}$$

5.1.8.1 Position Operator $\hat{x}$

The position operator is not often discussed in textbooks. I have assumed that, in coordinate space,

$$\hat{x}\psi(x') = x'\psi(x'). \tag{5.100}$$

Clearly $\hat{x}$ is Hermitian since

$$\begin{aligned}\int dx\, \psi^*(x)\,[\hat{x}\psi(x)] &= \int dx\, \psi^*(x) x\psi(x) \\ &= \int dx\, [x\psi(x)]^*\, \psi(x) = \int dx\, [\hat{x}\psi(x)]^*\, \psi(x)\end{aligned} \tag{5.101}$$

for real x.

The operator $\hat{x}$ does not commute with the momentum operator $\hat{p}$, nor with the Hamiltonian operator $\hat{H}$. As a consequence, it is impossible to find simultaneous eigenfunctions of $\hat{x}$ and $\hat{p}$ and it is also impossible to find simultaneous eigenfunctions of $\hat{x}$ and $\hat{H}$. But what *are* the eigenfunctions of $\hat{x}$? The eigenvalue equation in coordinate space is

$$\hat{x}\psi_a(x) = a\psi_a(x), \tag{5.102}$$

where the eigenvalue is designated by a to avoid confusion. From Eq. (5.100), we know that

$$\hat{x}\psi_a(x) = x\psi_a(x) \tag{5.103}$$

which, together with Eq. (5.102), implies that

$$x\psi_a(x) = a\psi_a(x) \tag{5.104}$$

for all x. The only way this can be true is if

$$\psi_a(x) = \delta(x-a). \tag{5.105}$$

This has the proper normalization for continuous eigenvalues,

$$\int_{-\infty}^{\infty} dx\, \psi_a^*(x)\psi_{a'}(x) = \int_{-\infty}^{\infty} dx\, \delta(x-a)\,\delta(x-a') = \delta(a-a'). \tag{5.106}$$

Thus the eigenfunctions of the position operator are Dirac delta functions.

5.1.8.2 Momentum Operator $\hat{p}$

The momentum operator in coordinate space is

$$\hat{p} = \frac{\hbar}{i}\frac{d}{dx}. \tag{5.107}$$

This operator is Hermitian since

$$\begin{aligned}
\int_{-\infty}^{\infty} dx\, \psi^*(x)\hat{p}\psi(x) &= \int_{-\infty}^{\infty} dx\, \psi^*(x)\frac{\hbar}{i}\frac{d}{dx}\psi(x) \\
&= \frac{\hbar}{i}\left[\left|\psi(x)\right|^2\Big|_{-\infty}^{\infty} - \int_{-\infty}^{\infty} dx\,\frac{d\psi^*(x)}{dx}\psi(x)\right] \\
&= -\frac{\hbar}{i}\int_{-\infty}^{\infty} dx\,\frac{d\psi^*(x)}{dx}\psi(x) = \int_{-\infty}^{\infty} dx\, \left[\hat{p}\psi(x)\right]^*\psi(x),
\end{aligned} \tag{5.108}$$

where it is assumed that the boundary conditions are such that the endpoint term, $\left|\psi(x)\right|^2\Big|_{-\infty}^{\infty}$, vanishes. This is true for any localized wave function. It would also be true for periodic boundary conditions in which $\psi(L/2) = \psi(-L/2)$. This example shows how the Hermiticity of an operator depends on boundary conditions.

The momentum operator does not commute with the Hamiltonian, except in the case where $V(x)$ is a constant C, independent of x, which can be taken equal to zero without loss of generality (all energies in the problem are simply shifted by C). Thus, it is only for the free particle that it is possible to find simultaneous eigenfunctions of $\hat{p}$ and $\hat{H}$, which are

$$\psi_{p,E}(x) = \frac{1}{\sqrt{2\pi\hbar}}e^{ipx/\hbar}, \tag{5.109}$$

with $E = p^2/2m$. In this equation, p can be positive or negative. The two independent, degenerate energy eigenfunctions are

$$\psi_E(x) = \frac{1}{\sqrt{2\pi\hbar}} e^{\pm i\sqrt{2mE}x/\hbar} \tag{5.110}$$

with $E > 0$.[2]

5.1.8.3 Parity Operator $\hat{P}$

An important operator in quantum mechanics is the *parity operator* that simply inverts the signs of coordinates. In other words

$$\hat{P}\psi(x) = \psi(-x). \tag{5.111}$$

You can prove easily that $\hat{P}$ is a Hermitian operator. Therefore, it has real eigenvalues. To find these eigenvalues, I note that

$$\hat{P}^2\psi(x) = \hat{P}\psi(-x) = \psi(x), \tag{5.112}$$

implying that the eigenvalue of $\hat{P}^2$ is one, which can be realized only if the (real) eigenvalues of $\hat{P}$ are ± 1. That is, there are only two eigenvalues. For the eigenvalue $+1$,

$$\hat{P}\psi_+(x) = \psi_+(-x) = \psi_+(x), \tag{5.113}$$

which implies that *any* even function is an eigenfunction of $\hat{P}$ having eigenvalue $+1$. Similarly,

$$\hat{P}\psi_-(x) = \psi_-(-x) = -\psi_-(x); \tag{5.114}$$

any odd function is an eigenfunction of $\hat{P}$ having eigenvalue -1. For example, $\cos(ax)$ has even parity and is an eigenfunction of $\hat{P}$ having eigenvalue $+1$, $\sin(ax)$ has odd parity and is an eigenfunction of $\hat{P}$ having eigenvalue -1, while $\exp(iax)$ does not have well-defined parity and is not an eigenfunction of the parity operator. The eigenfunctions of the parity operator are *infinitely* degenerate; any even function has eigenvalue $+1$ and any odd function has eigenvalue -1.

Since $\hat{P}$ does not commute with $\hat{x}$ and does not commute with $\hat{p} = \frac{\hbar}{i}\frac{d}{dx}$, it is not possible to find simultaneous eigenfunctions of $\hat{P}$ and $\hat{x}$ nor of $\hat{P}$ and $\hat{p}$. On

[2]Note that, in the *momentum* representation, the eigenfunctions of $\hat{p}$ are $\Phi_q(p) = \delta(p-q)$ and the eigenfunctions of $\hat{x}$ are $\Phi_x(p) = e^{-ipx/\hbar}/\sqrt{2\pi\hbar}$.

the other hand, $\left[\hat{P},\hat{p}^2\right] = 0$ and, as a consequence $\left[\hat{P},\hat{H}\right] = 0$ if $V(x)$ is an even function of x. This is an important result, that will become even more important when generalized to problems in three dimensions. For any Hamiltonian that is invariant under an inversion of coordinates, the eigenfunctions can be written as simultaneous eigenfunctions of the energy and the parity operator. That is, the eigenfunctions can be written as either even or odd functions of the coordinates. In an introductory quantum mechanics course, students often forget this very important result, nor do they appreciate the importance of the parity operator.

You might argue that the eigenfunctions for the free particle given in Eq. (5.109) are not eigenfunctions of the parity operator and you would be right. Those eigenfunctions are simultaneous eigenfunctions of the momentum and energy operators and cannot be simultaneous eigenfunctions of the parity operator since it is not possible to find simultaneous eigenfunctions of $\hat{P}$ and $\hat{p}$. On the other hand, we could have equally well taken our (unnormalized) *energy* eigenfunctions as

$$\psi_{p,E}(x) = \begin{cases} \cos\left(px/\hbar\right) \\ \sin\left(px/\hbar\right) \end{cases}, \tag{5.115}$$

which *are* simultaneous eigenfunctions of the parity operator, but no longer eigenfunctions of the momentum operator. This is an example where there is a two-fold energy degeneracy; for each value of the energy (other than zero), there are two independent eigenfunctions. These eigenfunctions can be taken as simultaneous eigenfunctions of the momentum operator *or* the parity operator, but not *both* (since the momentum and parity operators do not commute). With either choice, however, we have a unique way to label all the eigenfunctions. Recall that, if operators $\hat{A}$ and $\hat{B}$ commute, and if the eigenfunctions of operator $\hat{A}$ are degenerate, they are not *automatically* eigenfunctions of operator $\hat{B}$, but that some linear combination of the degenerate eigenfunctions of operator $\hat{A}$ can be chosen to be an eigenfunction of operator $\hat{B}$

5.2 Back to the Schrödinger Equation

5.2.1 How to Solve the Time-Dependent Schrödinger Equation

With our knowledge of the properties of Hermitian operators, it is a simple matter to construct a solution of the time-dependent Schrödinger equation if we know the eigenfunctions and eigenvalues of the Hamiltonian, as well as the initial condition for the wave function. The general solution of the time-dependent Schrödinger equation is

$$\psi(\mathbf{r},t) = \sum_E a_E e^{-iEt/\hbar}\psi_E(\mathbf{r}). \tag{5.116}$$

It is important to recognize that the summation index E appearing in this equation is a *dummy* index; any other letter works equally well. As a good general practice, when you have a product of two summations such as

$$\left(\sum_E a_E e^{-iEt/\hbar}\psi_E(\mathbf{r})\right)\left(\sum_E a_E e^{-iEt/\hbar}\psi_E(\mathbf{r})\right)^*,$$

you should use *different* summation indices to avoid getting into trouble. That is, write

$$\begin{aligned}
&\left(\sum_E a_E e^{-iEt/\hbar}\psi_E(\mathbf{r})\right)\left(\sum_E a_E e^{-iEt/\hbar}\psi_E(\mathbf{r})\right)^* \\
&= \left(\sum_E a_E e^{-iEt/\hbar}\psi_E(\mathbf{r})\right)\left(\sum_{E'} a_{E'} e^{-iE't/\hbar}\psi_{E'}(\mathbf{r})\right)^* \\
&= \sum_{E,E'} a_E a_{E'}^* e^{-i(E-E')t/\hbar}\psi_E(\mathbf{r})\psi_{E'}^*(\mathbf{r}).
\end{aligned} \tag{5.117}$$

If you use the same summation index, you cannot obtain the correct form in the double summation.

I assume that the initial condition is

$$\psi(\mathbf{r},0) = \psi_0(\mathbf{r}), \tag{5.118}$$

allowing me to solve for the expansion coefficients a_E by taking the inner product of Eq. (5.116) with $\psi_{E'}$, namely

$$(\psi_{E'},\psi_0) = \sum_E a_E\left(\psi_{E'},\psi_E\right) = \sum_E a_E\delta_{E,E'} = a_{E'}. \tag{5.119}$$

Therefore, $a_E = (\psi_E,\psi_0)$ and

$$\psi(\mathbf{r},t) = \sum_E \left(\psi_E,\psi_0\right)e^{-iEt/\hbar}\psi_E(\mathbf{r}). \tag{5.120}$$

For continuous variables in **k**-space for a particle having mass m moving in a potential for which the eigenfunctions are denoted by $\psi_{\mathbf{k}}(\mathbf{r})$ and the eigenenergies by E_k, the corresponding equation is

$$\psi(\mathbf{r},t) = \int d\mathbf{k}\,\left(\psi_{\mathbf{k}},\psi_0\right)e^{-iE_kt/\hbar}\psi_{\mathbf{k}}(\mathbf{r}), \tag{5.121}$$

where

$$(\psi_{\mathbf{k}}, \psi_0) = a(\mathbf{k}) = \int d\mathbf{r}\, \psi_0(\mathbf{r})\psi^*_{\mathbf{k}}(\mathbf{r}). \tag{5.122}$$

Although I concentrate on solutions of the time-independent Schrödinger equation in the next several chapters, you should not forget Eqs. (5.120) and (5.121). They are *central* to an understanding of quantum dynamics. In effect, there is a *three-step program* for solving any problem in quantum mechanics, given an initial condition for the wave function. The first step, and the most difficult, is to solve the time-independent Schrödinger equation for the eigenfunctions and eigenenergies. The second step is to obtain the expansion coefficients, a_E or $a(\mathbf{k})$ in terms of the initial wave function and the final step is to use Eq. (5.120) or (5.121) to obtain the time-dependent wave function. Often one is content just to find the eigenenergies and eigenfunctions.

Equation (5.116) can be written as

$$\psi(\mathbf{r}, t) = \sum_E a_E(t)\psi_E(\mathbf{r}), \tag{5.123}$$

where

$$a_E(t) = a_E e^{-iEt/\hbar} \tag{5.124}$$

is the *probability amplitude* for the particle to be in state E at time t. Although the probability to be in a specific state

$$|a_E(t)|^2 = \left|a_E e^{-iEt/\hbar}\right|^2 = |a_E|^2, \tag{5.125}$$

is *constant* in time, the probability density

$$\begin{aligned} |\psi(\mathbf{r}, t)|^2 &= \left|\sum_E a_E e^{-iEt/\hbar}\psi_E(\mathbf{r})\right|^2 \\ &= \sum_{E,E'} a_E a^*_{E'} e^{-i(E-E')t/\hbar}\psi_E(\mathbf{r})\psi^*_{E'}(\mathbf{r}) \end{aligned} \tag{5.126}$$

is a function of time if any two state amplitudes corresponding to nondegenerate eigenenergies are non-vanishing. Equation (5.126) contains all the dynamics.

One final point to engrave in your memory bank. *Each potential energy function gives rise to its own set of eigenfunctions and eigenenergies.* Some (or even all) of the eigenenergies may be the same for different potentials, but the eigenfunctions will *always* differ. For example, for the step potential to be considered in the next chapter, the eigenenergies take on all non-negative values, just as for a free particle, but the eigenfunctions differ.

5.2.2 *Quantum-Mechanical Probability Current Density*

The total probability is conserved for a single-particle quantum system. That is, if I consider a finite volume, the time rate of change of the probability to find the particle in the volume must equal the rate at which probability flows *into* the volume. In other words,

$$\frac{\partial}{\partial t}\int_V |\psi(\mathbf{r},t)|^2 d\mathbf{r} = -\oint_S \mathbf{J}(\mathbf{r},\mathbf{t})\cdot\mathbf{n}da, \tag{5.127}$$

where S is the surface enclosing the volume V and $\mathbf{n}$ is a unit vector pointing normally *outwards* from the volume. The quantity $\mathbf{J}$ is called the *probability current density*. By using the divergence theorem, I find

$$\int_V \left[\frac{\partial}{\partial t}\rho(\mathbf{r},t) + \nabla\cdot\mathbf{J}(\mathbf{r},t)\right] d\mathbf{r} = 0, \tag{5.128}$$

where

$$\rho(\mathbf{r},t) = |\psi(\mathbf{r},t)|^2 \tag{5.129}$$

is the *probability density*. Since Eq. (5.128) must hold for an arbitrary volume, it can be satisfied only if

$$\frac{\partial}{\partial t}\rho(\mathbf{r},t) + \nabla\cdot\mathbf{J}(\mathbf{r},t) = 0, \tag{5.130}$$

an equation known as the *equation of continuity*.

To get an expression for $\mathbf{J}$, I use Schrödinger's equation with

$$\hat{H}\psi(\mathbf{r},t) = \frac{\hbar^2}{2m}\nabla^2\psi(\mathbf{r},t) + V(\mathbf{r})\psi(\mathbf{r},t)$$

to write

$$\begin{aligned}
\frac{\partial}{\partial t}\rho(\mathbf{r},t) &= \frac{\partial}{\partial t}|\psi(\mathbf{r},t)|^2 = \frac{\partial}{\partial t}\left[\psi^*(\mathbf{r},t)\psi(\mathbf{r},t)\right] \\
&= -\frac{1}{i\hbar}\left[\hat{H}\psi(\mathbf{r},t)\right]^*\psi(\mathbf{r},t) + \left(\frac{1}{i\hbar}\right)\psi^*(\mathbf{r},t)\left[\hat{H}\psi(\mathbf{r},t)\right] \\
&= \frac{\hbar}{2mi}\psi(\mathbf{r},t)\nabla^2\psi^*(\mathbf{r},t) - \frac{\hbar}{2mi\hbar}\psi^*(\mathbf{r},t)\nabla^2\psi(\mathbf{r},t) \\
&= \frac{\hbar}{2mi}\nabla\cdot\left[\psi(\mathbf{r},t)\nabla\psi^*(\mathbf{r},t) - \psi^*(\mathbf{r},t)\nabla\psi(\mathbf{r},t)\right].
\end{aligned} \tag{5.131}$$

By comparing this equation with Eq. (5.130), I obtain the probability current density

$$\mathbf{J}(\mathbf{r},t) = \frac{i\hbar}{2m}\left[\psi(\mathbf{r},t)\,\nabla\psi^*(\mathbf{r},t) - \psi^*(\mathbf{r},t)\,\nabla\psi(\mathbf{r},t)\right]. \tag{5.132}$$

For a plane wave,

$$\psi(\mathbf{r},t) = \frac{1}{(2\pi\hbar)^{3/2}} e^{i\mathbf{p}\cdot\mathbf{r}/\hbar} e^{-ip^2 t/(2m)}, \tag{5.133}$$

$$\begin{aligned}\mathbf{J}(\mathbf{r},t) &= \frac{1}{(2\pi\hbar)^3}\frac{i\hbar}{2m}\left[-\frac{i\mathbf{p}}{\hbar} - \frac{i\mathbf{p}}{\hbar}\right] \\ &= \frac{1}{(2\pi\hbar)^3}\frac{\mathbf{p}}{m} = |\psi(\mathbf{r},t)|^2\frac{\mathbf{p}}{m} = \rho(\mathbf{r},t)\,\mathbf{v},\end{aligned} \tag{5.134}$$

where $\mathbf{v} = \mathbf{p}/m$. The probability current has the general form of a spatial density times a velocity, as expected. For real wave functions, the probability current density vanishes.

In general, it is easy to show that, for an arbitrary $\psi(\mathbf{r},t)$,

$$\langle\hat{\mathbf{p}}(t)\rangle/m = \int d\mathbf{r}\,\mathbf{J}(\mathbf{r},t). \tag{5.135}$$

Equation (5.135) follows from the fact that $\hat{\mathbf{p}}$ is Hermitian,

$$\begin{aligned}\langle\hat{\mathbf{p}}(t)\rangle &= \frac{\hbar}{i}\int d\mathbf{r}\,\psi^*(\mathbf{r},t)\,\nabla\psi(\mathbf{r},t) \\ &= \langle\hat{\mathbf{p}}(t)\rangle^* = -\frac{\hbar}{i}\int d\mathbf{r}\,\psi(\mathbf{r},\mathbf{t})\,\nabla\psi^*(\mathbf{r},t),\end{aligned} \tag{5.136}$$

since this equation implies that

$$\frac{\langle\hat{\mathbf{p}}(t)\rangle}{m} = \frac{i\hbar}{2m}\int d\mathbf{r}\left[\psi(\mathbf{r},t)\,\nabla\psi^*(\mathbf{r},t) - \psi^*(\mathbf{r},t)\,\nabla\psi(\mathbf{r},t)\right] = \int d\mathbf{r}\,\mathbf{J}(\mathbf{r},t). \tag{5.137}$$

In one dimension, the equation of continuity is

$$\frac{\partial}{\partial t}\rho(x,t) + \frac{\partial J_x(x,t)}{\partial x} = \frac{\partial}{\partial t}|\psi(x,t)|^2 + \frac{\partial J_x(x,t)}{\partial x} = 0, \tag{5.138}$$

where

$$J_x(x,t) = \frac{i\hbar}{2m}\left[\psi(x,t)\frac{\partial\psi^*(x,t)}{\partial x} - \psi^*(x,t)\frac{\partial\psi(x,t)}{\partial x}\right]. \tag{5.139}$$

The probability current must be conserved in problems where there is no loss. I will use the probability current density to get reflection and transmission coefficients in problems involving wells or barriers.

5.2.3 Operator Dynamics

The operators I consider are time-independent, but the expectation value of an operator for a quantum system characterized by the wave function $\psi(\mathbf{r}, t)$ is time-dependent, in general. To see this I write

$$\begin{aligned}
i\hbar \frac{d\left\langle \hat{A} \right\rangle}{dt} &= i\hbar \frac{d}{dt} \int d\mathbf{r}\, \psi^*(\mathbf{r}, t) \hat{A} \psi(\mathbf{r}, t) \\
&= i\hbar \int d\mathbf{r} \frac{\partial \psi^*(\mathbf{r}, t)}{\partial t} \hat{A} \psi(\mathbf{r}, t) + i\hbar \int d\mathbf{r} \psi^*(\mathbf{r}, t) \hat{A} \frac{\partial \psi(\mathbf{r}, t)}{\partial t} \\
&= -\int d\mathbf{r} \left[\hat{H} \psi(\mathbf{r}, t) \right]^* \hat{A} \psi(\mathbf{r}, t) + \int d\mathbf{r} \psi^*(\mathbf{r}, t) \hat{A} \hat{H} \psi(\mathbf{r}, t) \\
&= -\int d\mathbf{r}\, \psi^*(\mathbf{r}, t) \hat{H} \hat{A} \psi(\mathbf{r}, t) + \int d\mathbf{r} \psi^*(\mathbf{r}, t) \hat{A} \hat{H} \psi(\mathbf{r}, t);
\end{aligned}$$

$$i\hbar \frac{d\left\langle \hat{A} \right\rangle}{dt} = \left\langle \left[\hat{A}, \hat{H} \right] \right\rangle, \tag{5.140}$$

where the fact that $\hat{H}$ is Hermitian has been used. The expectation value of any operator that commutes with the Hamiltonian is constant in time. Another way of saying this is that the dynamic variable associated with any Hermitian operator that commutes with the Hamiltonian is a constant of the motion. For the free particle, momentum is conserved, consistent with the fact that the momentum operator commutes with the Hamiltonian. In problems with spherically symmetric potentials, the angular momentum is conserved classically, implying that the angular momentum operator commutes with the Hamiltonian. If you know that a classical variable is a constant of the motion for a given potential, then the eigenvalues of the associated quantum operator can sometimes be used to distinguish between degenerate eigenfunctions of the Hamiltonian.

Although Eq. (5.140) appears to provide a simple prescription for obtaining the expectation value of an operator, it is deceiving. In calculating the commutator $\left[\hat{A}, \hat{H} \right]$, one normally introduces new operators. As such one is often led to a never-ending set of coupled equations for the expectation values of different operators. It is only for potentials such as those for the free particle and simple harmonic oscillator that Eq. (5.140) can be used to obtain a *closed* set of equations for the expectation values of the position and momentum operators.

An alternative method for obtaining $\left\langle \hat{A} \right\rangle$ is to calculate the expectation value of any operator directly from the wave function. For example, if

$$\psi(\mathbf{r}, t) = \sum_E a_E e^{-iEt/\hbar} \psi_E(\mathbf{r}), \tag{5.141}$$

then

$$\begin{aligned} \left\langle \hat{A} \right\rangle &= \int d\mathbf{r}\, \psi^*(\mathbf{r}, t) \hat{A} \psi(\mathbf{r}, t) \\ &= \sum_{E,E'} a_E a_{E'}^* e^{-i(E-E')t/\hbar} \int d\mathbf{r}\, \psi_{E'}^*(\mathbf{r}) \hat{A} \psi_E(\mathbf{r}). \end{aligned} \tag{5.142}$$

In general there is time dependence in $\left\langle \hat{A} \right\rangle$ resulting from the exponential factors. For operators that commute with $\hat{H}$ the time dependence must disappear since $\left\langle \hat{A} \right\rangle$ is constant in this limit.

Since $\hat{H}$ commutes with itself, the average energy is time-independent. Explicitly,

$$\begin{aligned} \left\langle \hat{H} \right\rangle &= \sum_{E,E'} a_E a_{E'}^* e^{-i(E-E')t/\hbar} \int d\mathbf{r}\, \psi_{E'}^*(\mathbf{r}) E \psi_E(\mathbf{r}) \\ &= \sum_{E,E'} a_E a_{E'}^* e^{-i(E-E')t/\hbar} E \delta_{E,E'} \\ &= \sum_E |a_E|^2 E, \end{aligned} \tag{5.143}$$

simply a weighted sum of the energy E with the probability to be in the state corresponding to energy E. In fact, for any operator $\hat{G}(E)$ that correspond to a classical dynamic variable $G(E)$, it is not difficult to prove that

$$\left\langle \hat{G}(E) \right\rangle = \sum_E |a_E|^2 G(E). \tag{5.144}$$

5.2.4 *Sum of Two Independent Quantum Systems*

Finally I consider two *independent* quantum systems characterized by Hamiltonians $\hat{H}_1$ and $\hat{H}_2$, where

$$\begin{aligned} \hat{H}_1 \psi_{E_1} &= E_1 \psi_{E_1}, \\ \hat{H}_2 \psi_{E_2} &= E_2 \psi_{E_2}, \end{aligned} \tag{5.145}$$

and the total Hamiltonian is

$$\hat{H} = \hat{H}_1 + \hat{H}_2. \tag{5.146}$$

Since the Hamiltonians correspond to independent quantum systems, they must satisfy $\left[\hat{H}_1, \hat{H}_2\right] = 0$; consequently, the operators can possess simultaneous eigenfunctions. I can guess a solution in which the eigenfunctions for the composite system are simply the *product* of the eigenfunctions of the individual systems, while the eigenenergies are the *sums* of the individual energies. This guess works since

$$\begin{aligned}
\hat{H}\psi_{E_1}\psi_{E_2} &= \left(\hat{H}_1 + \hat{H}_2\right)\psi_{E_1}\psi_{E_2} = \hat{H}_1\psi_{E_1}\psi_{E_2} + \hat{H}_2\psi_{E_1}\psi_{E_2} \\
&= E_1\psi_{E_1}\psi_{E_2} + \psi_{E_1}\hat{H}_2\psi_{E_2} = E_1\psi_{E_1}\psi_{E_2} + E_2\psi_{E_1}\psi_{E_2} \\
&= (E_1 + E_2)\psi_{E_1}\psi_{E_2}.
\end{aligned} \tag{5.147}$$

Although this result is extremely simple, students often have trouble accepting or remembering it. For two independent systems, the eigenfunctions are products of the individual system eigenfunctions and the eigenenergies are the sum of the individual system eigenenergies.

5.3 Measurements in Quantum Mechanics: "Collapse" of the Wave Function

It has already been stated that a measurement on a single quantum system of a dynamic variable yields one and only one eigenvalue of the Hermitian operator associated with that dynamic variable. The wave function can be expanded in terms of the eigenfunctions of any Hermitian operator, provided the appropriate boundary conditions are met. For discrete eigenvalues, this implies that

$$\psi(\mathbf{r}, t) = \sum_n b_{a_n}(t)\psi_{a_n}(\mathbf{r}) \tag{5.148}$$

where a_n is an eigenvalue of some operator $\hat{A}$ and $\psi_{a_n}(\mathbf{r})$ is the corresponding eigenfunction. If this operator corresponds to a physical observable, then $|b_{a_n}(t)|^2$ corresponds to the probability that a measurement on a single quantum system at time t of the classical variable associated with $\hat{A}$ will yield the value a_n. For continuous eigenvalues a, this equation is replaced by

$$\psi(\mathbf{r}, t) = \int da\, b(a, t)\psi_a(\mathbf{r}), \tag{5.149}$$

where $|b(a,t)|^2\,da$ is the probability that a measurement with on a single quantum system at time t of the classical variable associated with $\hat{A}$ will yield a value between a and $a + da$.

Often it is stated in quantum mechanics texts that the wave function *collapses* into the eigenfunction associated with the eigenvalue that was measured. I have never been a big fan of this terminology. I have already noted that any *direct* measurement of a physical variable associated with a quantum system generally modifies the system, implying that the state of the system following the measurement is no longer the eigenfunction associated with that eigenvalue.

For example, in the two-slit experiment involving a single particle, the wave function before detection on a screen is spread out over an interference pattern associated with two-slit interference. One detector on the screen fires, localizing the particle, but the state of the particle is altered by the measurement. You can say that the measurement has collapsed the wave function, but I do not think this is a particularly useful image. Collapse tends to imply a physical collapse of the wave function, but quantum mechanics say nothing about the collapse process itself. As long as you deal with the probabilistic predictions of quantum mechanics you will not run into any problems. However if you try to associate a physical mechanism with the collapse process, you will be led down a path that seems to lead nowhere.

Although direct measurement of a physical variable that leaves a quantum system in an eigenfunction of the Hermitian operator associated with that variable is not possible, *indirect* measurements of the variable can be made that leave the quantum state of the system unchanged. Such measurements fall into two general classes: *quantum nondemolition measurements* on a single quantum system or measurements on a *correlated* or *entangled* state of a two-particle quantum system.

In a quantum nondemolition (QND) experiment,[3] one measures one of two non-commuting operators associated with a quantum system without introducing any noise (modification) of the dynamics associated with this operator. All the noise that is introduced goes into the dynamics associated with the other operator, but that operator is assumed *not* to appear in the Hamiltonian. In this *back-evading noise* scheme, the operator of interest can be measured with absolute accuracy, at least in principle. Such schemes were proposed as a means for measuring the small displacements produced on mechanical systems by gravitational waves. Note that such a scheme does not work if we measure the position of a free particle to arbitrary accuracy. Such a measurement would introduce an uncertainty in the momentum that acts back on the particle to modify its subsequent position. On the other hand, we could, in principle, measure the momentum of a free particle to arbitrary accuracy. Although the measurement would affect the position of the particle, the position does not appear in the Hamiltonian. To measure the momentum we would have to couple the particle of interest to another quantum system and make a measurement of the second quantum system that determines the momentum of the

[3]See, for example, M. O. Scully and M. S. Zubairy, *Quantum Optics* (Cambridge University Press, Cambridge, U.K., 1997), Chapter 19.

system of interest, using a method similar to that described in the next paragraph. To make QND work, the probe must provide a measure of the quantum variable being probed, without affecting its value. For example, when atoms are sent through a microwave cavity, they can acquire a phase shift that depends on the microwave intensity in the cavity; measuring this phase shift is an indirect way of measuring the microwave intensity, without altering its value.[4]

In an entangled state measurement, one measures properties of one particle or system and infers the properties of the other particle or system. The most famous example of this type of measurement is related to the so-called Einstein-Podolsky-Rosen (EPR) paradox. We shall see that a particle such as an electron has an intrinsic angular momentum and that a measure of the z-component of this intrinsic angular momentum can be either $\hbar/2$ (spin $\uparrow$) or $-\hbar/2$ (spin $\downarrow$). The electron is said to have a spin of $1/2$. If a spinless particle decays into two identical spin 1/2 particles, then the quantum state of the combined system is

$$\psi_{1,2} = \frac{1}{\sqrt{2}} \left[\psi_{1\uparrow}\psi_{2\downarrow} - \psi_{2\uparrow}\psi_{1\downarrow} \right], \tag{5.150}$$

since we don't know which particle is in either of the spin states. This is an *entangled* wave function since it cannot be written as the product of individual wave functions for each particle. However, if we measure the spin of one of the particles as "up" along some direction we are guaranteed that the other particle is in its spin "down" state along the same direction. In several quantum computation schemes one entangles the internal state of an atom with the polarization of the radiation emitted from the atom. In this way, a measurement of the polarization of the emitted radiation can be correlated with a given superposition of the internal states of the atom.

5.4 Summary

In this chapter, the basic postulates of quantum mechanics were stated. A detailed catalogue was constructed giving various properties of Hermitian operators and their eigenfunctions and eigenvalues. The central problem in quantum mechanics is reduced to finding the eigenfunctions and eigenvalues of the Hamiltonian operator, from which all properties of quantum systems can be derived.

---

[4]Experiments of this type were pioneered in the group of Serge Haroche, who was awarded the Nobel prize in recognition of these and other experiments.

5.5 Appendix: From Discrete to Continuous Eigenvalues

One way of going over to continuous from discrete eigenvalues is to use *periodic boundary conditions*. In one dimension a model to accomplish this goal involves a mapping of the one-dimensional problem onto a circular path having length L. The free-particle wave functions in this case are of the form

$$\psi_E(x) = \sqrt{1/L}e^{ikx}, \tag{5.151}$$

where $E = \hbar^2k^2/2m$. The boundary condition that must be imposed is $\psi_E(0) = \psi_E(L)$. As a consequence of this requirement, it is necessary that

$$\begin{aligned}\sqrt{1/L}e^0 &= \sqrt{1/L}e^{ik_nL};\\ e^{ik_nL} &= 1;\\ k_n &= \frac{2\pi n}{L},\end{aligned} \tag{5.152}$$

where n is an integer, positive, negative, or zero. The eigenfunctions are

$$\psi_n(x) = \sqrt{1/L}e^{2\pi inx/L} \tag{5.153}$$

and they form an orthonormal basis since

$$\int_0^L dx\psi_n^*(x)\psi_m(x) = \frac{1}{L}\int_0^L dxe^{2\pi i(m-n)x/L} = \delta_{m,n}. \tag{5.154}$$

An arbitrary wave function can be written as

$$\psi(x,t) = \sum_n a_{k_n}(t)\psi_{k_n}(x), \tag{5.155}$$

where the k_ns take on discrete values. To take the limit that $L \to \infty$ you must do two things. First, the sum over n must be converted to an integral over a continuous variable k. Second, $a_{k_n}(t)$ must be replaced by a continuous amplitude $a(k,t)$ that has units of $1/\sqrt{k}$. To go over to a continuum, I set

$$\begin{aligned}\psi(x,t) &= \sum_n a_{k_n}(t)\psi_{k_n}(x)\\ &= \sqrt{1/L}\sum_{k_n} a_{k_n}(t)e^{ik_nx}\\ &= \sqrt{1/L}\frac{1}{\Delta k}\sum_{k_n} a_{k_n}(t)e^{ik_nx}\Delta k_n.\end{aligned} \tag{5.156}$$

where $\Delta k_n = 2\pi\left[(n+1-n)\right]/L = 2\pi/L = \Delta k$. To achieve the final result I now make the replacements

$$\sum_{k_n} \Delta k_n \rightarrow \frac{L}{2\pi}\int_{-\infty}^{\infty} dk \tag{5.157}$$

and

$$a_{k_n}(t) \rightarrow \sqrt{2\pi/L}a(k,t). \tag{5.158}$$

to arrive at

$$\psi(x,t) = \sqrt{\frac{1}{2\pi}}\int_{-\infty}^{\infty} dk\, a(k,t)e^{ikx}. \tag{5.159}$$

This method can be generalized to three dimensions by adopting periodic boundary conditions in all three directions, namely

$$(k_x)_n = \frac{2\pi n_x}{L};\ \left(k_y\right)_n = \frac{2\pi n_y}{L}\ (k_z)_n = \frac{2\pi n_z}{L}. \tag{5.160}$$

The sum over n is converted to an integral over $\mathbf{k}$ using the prescription

$$\sum_{n_x,n_y,n_z} \rightarrow \left(\frac{L}{2\pi}\right)^3 \int d\mathbf{k} \tag{5.161}$$

and Eq. (5.158) is replaced by

$$a\left[(k_x)_{n_x}, \left(k_y\right)_{n_y}, (k_z)_{n_z}; t\right] \rightarrow \left(\frac{2\pi}{L}\right)^{3/2} a(\mathbf{k},t). \tag{5.162}$$

5.6 Problems

Note: You can always assume that the eigenfunctions have been chosen to be orthonormal unless specifically told otherwise.

1. Why are Hermitian operators important in quantum mechanics? What is the possible outcome of a single measurement on a single quantum system of the physical observable associated with a Hermitian operator? Why does the energy operator play such an important role in quantum mechanics? If the eigenvalues of the Hamiltonian are nondegenerate, what do you know about the eigenfunctions of that Hamiltonian? If the eigenvalues of the Hamiltonian are degenerate, are the eigenfunctions of that Hamiltonian necessarily orthogonal? Explain.

2. Suppose a Hermitian operator $\hat{A}$ has eigenfunctions $\psi_{a_n}(\mathbf{r})$ and eigenvalues a_n. At a given time, the (normalized) state of a single quantum system is equal to

$$\psi(\mathbf{r}) = \sum_n b_n \psi_{a_n}(\mathbf{r}).$$

Prove that

$$\left\langle \hat{A}^m \right\rangle = \sum_n (a_n)^m |b_n|^2.$$

If

$$\psi(\mathbf{r}) = \sum_E b_E \psi_E(\mathbf{r}),$$

where the $\psi_E(\mathbf{r})$ are (normalized) eigenfunctions of the Hamiltonian, derive an expression the variance of the energy (assume that $\psi(\mathbf{r})$ is normalized) in terms of the b_E's and the energy eigenvalues.

3. Consider a Hermitian operator $\hat{A}$ having eigenfunctions $\psi_{a_n}(\mathbf{r})$ and eigenvalues a_n. At a given time, the state of a single quantum system is equal to

$$\psi(\mathbf{r}) = N\left[\psi_{a_1}(\mathbf{r}) + 2\psi_{a_2}(\mathbf{r}) + 3\psi_{a_3}(\mathbf{r})\right]$$

with $a_1 = 1, a_2 = 3, a_3 = 5$ in some appropriate units.

(a) Find N such that $\psi(\mathbf{r})$ is normalized.
(b) For this state, what are the only possible values of the dynamic variable associated with the operator $\hat{A}$ that could be obtained in a single measurement?
(c) Find $\left\langle \hat{A} \right\rangle$ in this state.
(d) Find the variance of $\hat{A}$ in this state.

[Hint: This is an easy problem and doesn't require any complicated expressions. Use the results of Problem 5.2.]

4. Use the fact that

$$\langle \hat{x} \rangle = \int \psi^*(\mathbf{r},t)\, \hat{x} \psi(\mathbf{r},t)\, d\mathbf{r} = \int \Phi^*(\mathbf{p},t)\, \hat{x} \phi(\mathbf{p},t)\, d\mathbf{p}$$

to prove that

$$\hat{x} = i\hbar \frac{\partial}{\partial p_x}$$

when operating on functions of momentum.

5. (a) Prove

$$\left[\hat{A}\hat{B}, \hat{C}\right] = \hat{A}\left[\hat{B}, \hat{C}\right] + \left[\hat{A}, \hat{C}\right]\hat{B}$$

and

$$\left[\hat{A}, \hat{B}\hat{C}\right] = \hat{B}\left[\hat{A}, \hat{C}\right] + \left[\hat{A}, \hat{B}\right]\hat{C}.$$

(b) Prove that if $\hat{A}$ and $\hat{B}$ are Hermitian, then $\hat{A}\hat{B} + \hat{B}\hat{A}$ and $i\left[\hat{A}, \hat{B}\right]$ are Hermitian, but $\hat{A}\hat{B}$ is Hermitian only if $\left[\hat{A}, \hat{B}\right] = 0$.

6. Evaluate $[\hat{p}_x, V(x)]$, $\left[\hat{p}_x^2, x\right]$, and $\left[\hat{p}_x, \frac{\hat{p}_x^2}{2m} + \frac{kx^2}{2}\right]$. If $\hat{p}_x$ does not commute with the Hamiltonian, how do you know that $e^{ip_x/\hbar}$ is not an eigenfunction of the Hamiltonian?

7. Show that, to second order in the operators,

$$e^{\hat{A}+\hat{B}} = e^{\hat{A}}e^{\hat{B}}e^{-\hat{C}/2}$$

where $\hat{C} = \left[\hat{A}, \hat{B}\right]$. Note that this equation is true to higher order only if both $\hat{A}$ and $\hat{B}$ commute with $\hat{C}$. Also prove that to second order in the operator $\hat{A}$,

$$e^{\hat{A}}\hat{B}e^{-\hat{A}} = \hat{B} + \left[\hat{A}, \hat{B}\right] + \frac{1}{2!}\left[\hat{A}, \left[\hat{A}, \hat{B}\right]\right] + \cdots$$

8. (a) Prove by a counterexample that $\left[\hat{A}, \hat{B}\right] = 0$ and $\left[\hat{B}, \hat{C}\right] = 0$ does not imply $\left[\hat{A}, \hat{C}\right] = 0$.

(b) The *projection operator* $\hat{P}_a$ is defined by

$$\hat{P}_a\psi_{a'} = \delta_{a,a'}\psi_{a'}$$

where a is an eigenvalue of the Hermitian operator $\hat{A}$. Prove that $\hat{P}_a$ is Hermitian, that $\hat{P}_a^2 = \hat{P}_a$, $\sum_a \hat{P}_a = \hat{1}$, and that $\hat{A}\hat{P}_a = a\hat{P}_a$.

9. Suppose you are given a Hamiltonian

$$\hat{H} = \frac{\hat{p}^2}{2m} + \hat{V},$$

having eigenfunctions $\psi_E(x)$. At $t = 0$ the wave function can be written as

$$\psi(x, 0) = \sum_E a_E\psi_E(x).$$

(a) Write the solution valid for any $t > 0$.
(b) Obtain an integral expression for $\langle \hat{p} \rangle$ and show that, in general, $\langle \hat{p} \rangle$ is a function of time. Why is this so?

10. How do you know that eigenfunctions of the momentum operator must be eigenfunctions of the free particle Hamiltonian? Why is it that eigenfunctions of the free particle Hamiltonian are not necessarily eigenfunctions of the momentum operator even though the two operators commute? For problems in one dimension, why is the maximum energy degeneracy equal to 2?

11. Prove that $[\hat{P}, \hat{p}_x] \neq 0$, $[\hat{P}, \hat{p}_x^2] = 0$, and that $[\hat{P}, \hat{H}] = 0$ only if $V(x) = V(-x)$, where $\hat{H} = \frac{\hat{p}_x^2}{2m} + \hat{V}$ and $\hat{P}$ is the parity operator. Under what condition are you guaranteed that the energy eigenfunctions are either even or odd functions of x?

12. Consider that we have two *independent* quantum Hamiltonians, $\hat{H}_1$ and $\hat{H}_2$ and the total Hamiltonian is

$$\hat{H} = \hat{H}_1 + \hat{H}_2.$$

The eigenfunctions and eigenenergies of $\hat{H}_1$ and $\hat{H}_2$ are

$$\hat{H}_1 \psi_{E_{1n}}(\mathbf{r}_1) = E_{1n} \psi_{E_{1n}}(\mathbf{r}_1);$$

$$\hat{H}_2 \psi_{E_{2m}}(\mathbf{r}_2) = E_{2m} \psi_{E_{2m}}(\mathbf{r}_2).$$

Since the Hamiltonians correspond to independent quantum systems,

$$\left[\hat{H}_1, \hat{H}_2\right] = 0$$

and the operators can possess simultaneous eigenfunctions. Prove that the eigenfunctions for the composite system are simply the *product* of the eigenfunctions of the individual systems, while the eigenenergies are the *sums* of the individual energies.

13. (a) Prove that the current density $J_x(x)$ vanishes at all points in space only if $\psi(x)$ is purely real or purely imaginary.
(b) As a consequence show that $\langle \hat{p}_x \rangle = 0$ if $J_x(x) = 0$.
(c) Prove that the converse is not true. In other words, there are wave functions such as eigenfunctions of the parity operator for which $\langle \hat{p}_x \rangle = 0$, but for which $J_x(x)$ is not necessarily equal to zero.

14–15. For a particle of mass m having an initial normalized square integrable wave function of the form

$$\psi(x, 0) = f(x) e^{ikx},$$

where $f(x)$ is real, prove that

$$\langle \hat{x}\hat{p} + \hat{p}\hat{x} \rangle = 2\hbar k \int_{-\infty}^{\infty} dx f^2(x) x = 2\hbar k \langle x(0) \rangle .$$

Use this result to prove that, for such an initial wave function,

$$\Delta x(t)^2 = \Delta x(0)^2 + (\Delta v)^2 t^2,$$

where $\Delta v = \Delta p/m$.

16. For a free particle wave packet, prove that

$$\frac{d\left\langle \hat{x}^2(t) \right\rangle}{dt} = \langle \hat{x}\hat{p} + \hat{p}\hat{x} \rangle /m,$$

where m is the mass of the particle. Specifically, for the free particle wave packet whose wave function is given by Eq. (3.36) with $k_0 = 0$, prove *explicitly* that

$$\frac{d \int_{-\infty}^{\infty} dx \psi(x,t)^* \hat{x}^2 \psi(x,t)}{dt} = \frac{1}{m} \int_{-\infty}^{\infty} dx \psi(x,t)^* \left(\hat{x}\hat{p} + \hat{p}\hat{x} \right) \psi(x,t)$$

by evaluating both sides of the equation.

Chapter 6
Problems in One-Dimension: General Considerations, Infinite Well Potential, Piecewise Constant Potentials, and Delta Function Potentials

The simplest solutions of the Schrödinger equation are those involving one-dimensional problems. Of course, nature is three dimensional, but sometimes problems can be reduced to an effective one-dimensional problem. For example, if an optical field is incident normally on a dielectric slab, the problem is essentially a one-dimensional problem. Even more important, however, is that many features of quantum mechanics are illustrated using one-dimensional problems. In this chapter, I consider some general features of solutions of the Schrödinger equation in one dimension, discuss the infinite square well potential, look at other piecewise constant potentials, and examine the one-dimensional Dirac delta function potential. In the Appendix, I discuss periodic potentials and their relation to so-called *Bloch state* wave functions. The harmonic oscillator potential in one dimension is analyzed in Chap. 7.

6.1 General Considerations

Without specifying the exact form of the potential, I can characterize the types of solutions that can exist in one dimension. That is, I can determine if there is energy degeneracy, if bound states might exist, and if the eigenvalues are continuous or discrete. For example, we know already that for a free particle in one-dimension there is a two-fold energy degeneracy and that the energy eigenvalues are continuous. The degeneracy can be understood as arising from the fact that, for the same energy, a particle can be moving to the right or left. This can be viewed as a *left-right degeneracy*. In the examples given below, I always set the zero of energy such that the potential as $|x| \to \infty$ is positive or zero. In one-dimensional problems there can never be more than a two-fold degeneracy since the time-independent Schrödinger equation is a second order ordinary differential equation.

P.R. Berman, *Introductory Quantum Mechanics*, UNITEXT for Physics,
https://doi.org/10.1007/978-3-319-68598-4_6

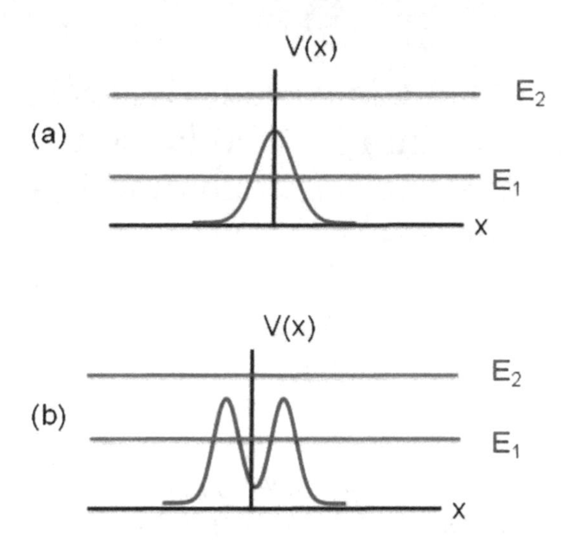

Fig. 6.1 Potential barriers

Motion in classical mechanics can be bounded or unbounded. Bounded, one-dimensional motion in classical mechanics is restricted to a finite region of space, while unbound motion can extend to either $x = \infty$ and/or $x = -\infty$. Bound states in quantum mechanics correspond to states that are, for the most part, localized to finite regions of space. In other words, a necessary condition for a bound state in quantum mechanics is that the eigenfunction associated with a bound state goes to zero as $|x| \to \infty$ (in three dimensions the eigenfunction must go to zero as $r \to \infty$). Unbound states in quantum mechanics correspond to states whose eigenfunctions do not vanish for either $x = \infty$ and/or $x = -\infty$. The eigenfunctions of the free particle do not correspond to a bound state. I now examine the classical motion and quantum-mechanical properties associated with several generic classes of one-dimensional potentials.

6.1.1 Potentials in which $V(x) > 0$ and $V(\pm\infty) \sim 0$

Potentials falling into this class are shown in Fig. 6.1. In both cases shown, there are continuous eigenenergies $E > 0$ and a two-fold degeneracy for each energy, since particles (waves) can be incident from the left or right. Were I to solve the time-independent Schrödinger equation for potentials of this type, I would find that the eigenenergies correspond to all positive energies. Classically, in case (a), a particle having energy E_1 incident from the left would be reflected by the potential. Quantum-mechanically there is some probability that the particle *tunnels* to the other side and is transmitted through the barrier. Tunneling is a wave-like phenomenon. A classical particle having energy E_2 incident from the left is always transmitted with no reflection, although its kinetic energy changes as it moves by the potential barrier. Quantum-mechanically there is some probability that the particle is reflected. For a potential that varies slowly over a de Broglie wavelength of the particle, this reflection is very small, but for a rectangular barrier it can be significant. As you shall see, the rectangular barrier case is analogous to one in which light is reflected by a dielectric slab. In case (b), classically, a particle having energy E_1 can be bound if it is located in the potential well. Quantum-mechanically there are no bound states, a particle prepared inside the well eventually tunnels out. A classical particle having energy E_1 incident from the left would always be reflected. Quantum-mechanically there is some probability that the particle is transmitted as a result of tunneling. A classical particle having energy E_2 incident from the left would always be transmitted with no reflection. Quantum-mechanically there is some probability that the particle is reflected.

6.1.2 Potentials in which $V(x) > 0$ and $V(-\infty) \sim 0$ while $V(\infty) \sim \infty$

For the potential of (Fig. 6.2) (a), a classical particle is reflected by the potential, as is a quantum wave packet incident from the left. Since a wave packet cannot be incident from the left, there is no degeneracy, although the eigenenergies are continuous. In case (b) there are no bound states quantum-mechanically, although there could be bound states classically for energy E_1, if the particle is located in the potential well.

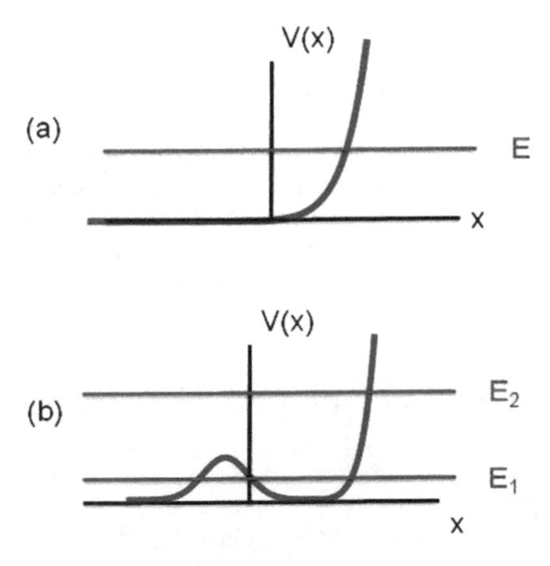

Fig. 6.2 Reflecting potentials

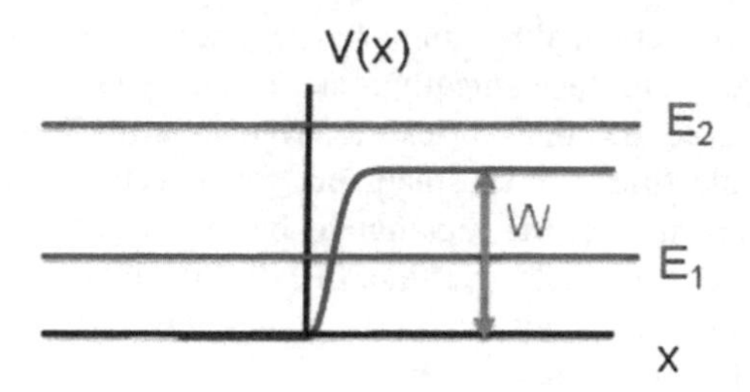

Fig. 6.3 Step potential

6.1.3 Potentials in which $V(x) > 0$ and $V(-\infty) \sim 0$ while $V(\infty) = W > 0$

In the case of a step potential (Fig. 6.3), a wave packet cannot be incident from the right if $E < W$. Therefore, for $E < W$, the eigenenergies are continuous, but there is no degeneracy. A wave packet incident from the left is totally reflected. On the other hand, for $E > W$, there are continuous eigenenergies and a two-fold degeneracy since wave packets can be incident from the left or right. Classically, a particle incident from the left is reflected if $E < W$ and transmitted with reduced kinetic energy (with no reflection) if $E > W$. Quantum mechanically there can be some reflection for a wave packet incident from the left having energy $E > W$.

6.1.4 Potentials in which $V(x) > 0$ and $V(\pm\infty) \sim \infty$

For the potentials of (Fig. 6.4), a wave packet can be incident neither from the right nor the left. The wave function must vanish as $|x| \to \infty$. In order to fit the waves in the potential and satisfy this boundary condition, the energy must take on discrete values. There is no degeneracy in this case, but there is an infinite number of discrete energies possible. Classically, any energy greater than the minimum value of the potential energy is allowed. A particle could be bound in one of the sub-wells in case (b) for energy E_1. Quantum-mechanically, there are no bound states that are localized *entirely* in only *one* sub-well.

6.1.5 Potentials in which $V(x) < 0$ and $V(\pm\infty) \sim 0$

In this case (Fig. 6.5) of a potential well, for $E > 0$, a wave packet can be incident from the left or right. Therefore, for $E > 0$, the eigenenergies are continuous, and there is a two-fold degeneracy. A wave packet incident from the left will be partially reflected and partially transmitted. There can also be *resonance* phenomena, as with

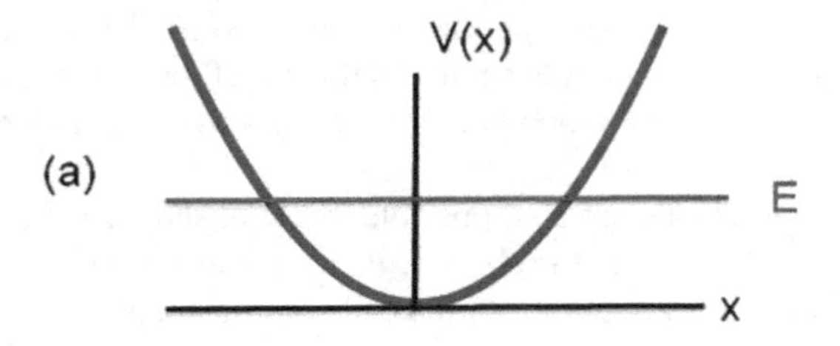

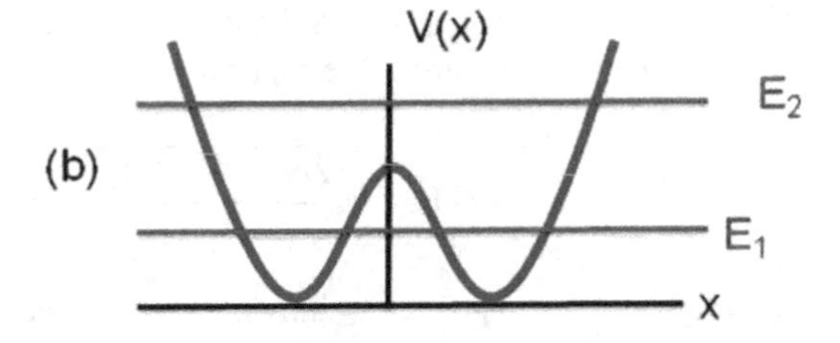

Fig. 6.4 Bound state potentials

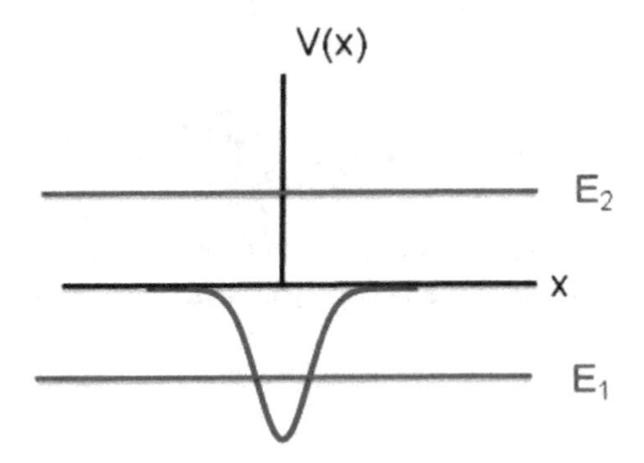

Fig. 6.5 Potential well

reflection of light from thin dielectric films. Classically there is no reflection for $E > 0$. For $E < 0$, quantum-mechanically there is no degeneracy and there is a finite number of bound states. It can be proven that there is always at least one bound state, regardless of the depth of the potential.[1] This might seem a little surprising, but, for low potential depths, the wave function extends significantly into the *classically*

[1] Somewhat more precise requirements that guarantee the existence of a bound state are $V(-\infty) = V(\infty) = V_0$ and

$$\int_{-\infty}^{\infty} [V(x) - V_0]\, dx < 0.$$

You are asked to prove this using the variational method in Problem 15.8.

forbidden regions (regions where the kinetic energy would be negative) giving rise to a large Δx and a correspondingly small Δp that is sufficiently small to prevent the particle from being freed from the well. Classically a particle having energy $E < 0$ is bound in the well.

Other types of potentials are also possible, but you should get the idea by now. In *bound state* one-dimensional problems, *there is never energy degeneracy*—the actual number of bound states (if any) depends on the details of the potential. For one-dimensional problems giving rise to continuous eigenenergies, there is two-fold degeneracy if the energy is greater than the potential as $|x| \to \infty$.

Values of x for which $E > V(x)$ correspond to the *classically allowed region* for the particle and to one for which Schrödinger's equation is

$$\frac{d^2\psi}{dx^2} = -k^2(x)\psi; \qquad k(x) = \sqrt{\frac{2m\left[E - V(x)\right]}{\hbar^2}} > 0. \tag{6.1}$$

If V is constant, the solutions are sines or cosines of kx. In general the solution is oscillatory in such classically allowed regions. On the other hand, values of x for which $E < V(x)$ corresponds to a *classically forbidden region* for the particle, since its kinetic energy would have to be *negative*. When $E < V(x)$, Schrödinger's equation is

$$\frac{d^2\psi}{dx^2} = \kappa^2(x)\psi; \qquad \kappa(x) = \sqrt{\frac{2m\left[V(x) - E\right]}{\hbar^2}} > 0. \tag{6.2}$$

For constant V, the solutions are real exponentials of $\pm\kappa x$. In general the solution is a smooth decaying function the deeper you penetrate into classically forbidden regions.

6.2 Infinite Well Potential

An important model problem in one dimension is the infinite square well potential, represented schematically in Fig. 6.6. The potential vanishes between the "walls" of

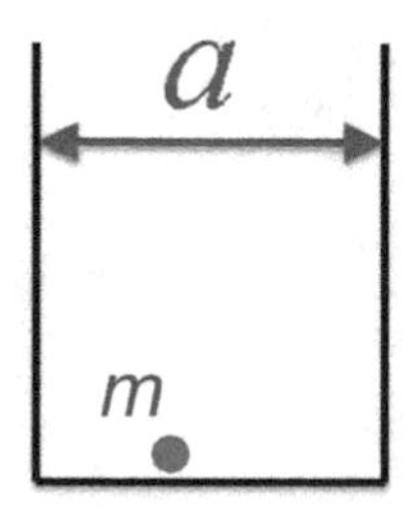

Fig. 6.6 Infinite square well potential

the potential and is infinite otherwise. This is a wonderful problem since it illustrates many of the features of bound state problems in quantum mechanics.

First, consider the classical problem. A particle having mass m is constrained to move between the walls of the potential having width a. Between the walls, the particle acts as a free particle having speed v, momentum $\mathbf{p} = \pm mv\mathbf{u}_x$ (+ if it moves to the right and − if it moves to the left) and energy $E = mv^2/2$. When the particle hits the wall, it undergoes an elastic collision in which the sign of its velocity (and momentum) is changed, but its energy remains unchanged. Since the velocity changes direction on collisions with the walls, momentum is not conserved. Also, since the velocity changes on collisions with the walls, the particle accelerates during each collision. The particle can have any kinetic energy whatsoever (the potential energy is zero inside the box), which remains constant during the particle's motion, and the position of the particle is determined precisely as it moves back and forth between the two walls.

What are the classical (time-averaged) distribution functions for this particle? The energy is fixed so the energy distribution is a Dirac delta function. For a given energy E, however, *two* possible momenta are possible,

$$\mathbf{p} = \pm\sqrt{2mE}\mathbf{u}_x \equiv p\mathbf{u}_x, \tag{6.3}$$

implying that the *time-averaged* momentum distribution in one dimension is

$$W_{\text{class}}(p) = \frac{1}{2}\left[\delta\left(p - \sqrt{2mE}\right) + \delta\left(p + \sqrt{2mE}\right)\right]. \tag{6.4}$$

On the other hand, *on average*, the particle is found with equal probability *anywhere* in the well, so the *time-averaged* spatial distribution is

$$P_{\text{class}}(x) = \frac{1}{a}. \tag{6.5}$$

Of course, the particle follows a classical trajectory given some initial condition; in other words, the probability *density* for the particle is always a Dirac delta function centered at the classical particle position. That is, the classical particle mass density is given by

$$\rho(x,t) = m\delta(x - x(t)), \tag{6.6}$$

where $x(t)$ is the position of the particle at time t.

Now let's turn to the quantum problem. The Hamiltonian for the particle when it is between the walls is just the free particle Hamiltonian,

$$\hat{H} = \frac{\hat{p}^2}{2m} = -\frac{\hbar^2}{2m}\frac{d^2}{dx^2}. \tag{6.7}$$

As a consequence, the time-independent Schrödinger equation is

$$-\frac{\hbar^2}{2m}\frac{d^2\psi_E(x)}{dx^2} = E\psi_E(x) \tag{6.8}$$

in this region. I have some freedom in choosing the origin of the coordinate system. If I take the well centered at $x = 0$, the potential is symmetric about the origin and the Hamiltonian commutes with the parity operator. With this choice the eigenfunctions *must* have definite parity, since there is no energy degeneracy in this problem. On the other hand, if I take the well located between 0 and a, the Hamiltonian does not commute with the parity operator and the eigenfunctions do not possess definite parity (this is clear since the wave function must vanish for $x < 0$ in this case). Of course the eigenenergies must be the same, since the particle's energy must be independent of the choice of origin. Let's solve for the eigenfunctions and eigenenergies using both coordinate systems.

6.2.1 *Well Located Between* $-a/2$ *and* $a/2$

In this case, the potential is given by

$$V(x) = \begin{cases} 0 & |x| < a/2 \\ \infty & |x| > a/2 \end{cases}. \tag{6.9}$$

The fact that the potential is *infinite* at the wall leads us to the assumption that the eigenfunctions must vanish for $|x| \geq a/2$; that is, the particle cannot penetrate into the walls. The *boundary condition*

$$\psi_E(\pm a/2) = 0 \tag{6.10}$$

can be obtained formally by taking a finite height for the potential in the regions $|x| > a/2$, and then letting this height approach infinity. To solve Eq. (6.8) in the region $-a/2 < x < a/2$, I guess a solution. It is not difficult to show that any of

$$\sin(kx), \ \cos(kx), \ \exp(ikx) \ \exp(-ikx) \tag{6.11}$$

are solutions *provided*

$$k = \frac{\sqrt{2mE}}{\hbar}. \tag{6.12}$$

Since a second order differential equation has two independent solutions, I can use any *two* of the linearly independent solutions given in Eq. (6.11) that are consistent with the boundary conditions. In this case, however, there is no degeneracy and the Hamiltonian commutes with the parity operator. Thus the eigenfunctions must also be eigenfunctions of the parity operator; that is, the eigenfunctions *must* be the trigonometric solutions. The eigenfunctions fall into two classes, those having even parity and those having odd parity, namely

$$\psi_k^+(x) = N_+ \cos\left(k^+ x/2\right) \tag{6.13}$$

and

$$\psi_k^-(x) = N_- \sin\left(k^- x/2\right), \tag{6.14}$$

where the plus and minus refer to even and odd parity solutions, respectively, and the $N_\pm$ are normalization constants.

The eigenfunctions satisfy the boundary condition given in Eq. (6.10) if

$$k^+ \to k_n^+ = \frac{n\pi}{a}; \quad n = 1, 3, 5, \ldots \tag{6.15a}$$

$$k^- \to k_n^- = \frac{n\pi}{a}; \quad n = 2, 4, 6, \ldots. \tag{6.15b}$$

The $k^\pm$ values are *quantized* and, as a consequence, so is the energy

$$E \to E_n = \frac{\hbar^2 k_n^2}{2m} = \frac{\hbar^2 \pi^2 n^2}{2ma^2}; \quad n = 1, 2, \ldots. \tag{6.16}$$

The normalized eigenfunctions (now labeled by n rather than k) are

$$\psi_n(x) = \begin{cases} \begin{cases} \sqrt{\frac{2}{a}} \cos[n\pi x/a] & n = 1, 3, 5, \ldots \\ \sqrt{\frac{2}{a}} \sin[n\pi x/a] & n = 2, 4, 6, \ldots \end{cases} & |x| \le a/2 \\ 0 & |x| > a/2 \end{cases}, \tag{6.17}$$

where the normalization constants were obtained by demanding that

$$\int_{-\infty}^{\infty} |\psi_n(x)|^2 \, dx = \int_{-a/2}^{a/2} |\psi_n(x)|^2 \, dx = 1. \tag{6.18}$$

The first few eigenfunctions are shown in Fig. 6.7, plotted as the dimensionless quantity $\sqrt{a}\psi_n(x)$. Many of the results for the infinite well potential are generic. For example, the lowest energy eigenfunction has no nodes in the classically allowed region and is symmetric about the origin. There is one additional node in the classically allowed region for each increase in n and the eigenfunctions alternate between symmetric and antisymmetric functions. As you shall see, these features are common to all bound state problems for potentials that are an even function of x. Even if the potential is not symmetric about the origin, the same nodal structure is to be expected, although the eigenfunctions no longer correspond to states of definite parity.

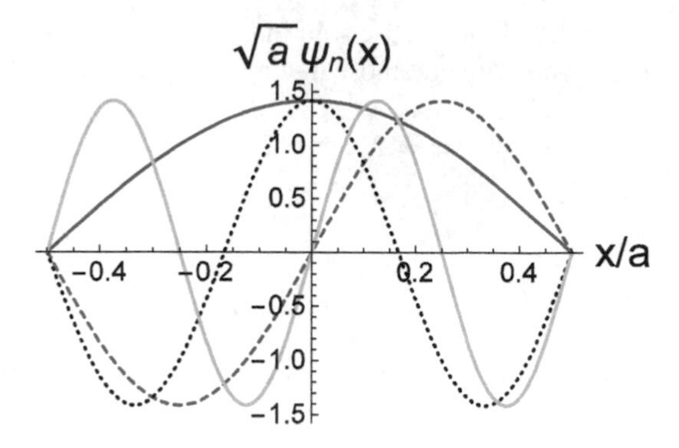

Fig. 6.7 Eigenfunctions in dimensionless units as a function of x/a for an infinite potential well centered at the origin: $n = 1$ (red, solid); $n = 2$ (blue, dashed); $n = 3$ (black, dotted); $n = 4$ (green, solid)

6.2.2 Well Located Between 0 and a

With this choice, the potential is

$$V(x) = \begin{cases} 0 & 0 < x < a \\ \infty & \text{otherwise} \end{cases} \tag{6.19}$$

and the boundary conditions are

$$\psi_E(0) = \psi_E(a) = 0. \tag{6.20}$$

The only solution of Eq. (6.8) in the region $0 < x < a$ that satisfies the boundary condition at $x = 0$ is of the form $\sin(kx)$ with $E = \hbar^2 k^2/2m$. To also satisfy the boundary condition that the wave function vanish at $x = a$, it is necessary that

$$k \to k_n = \frac{n\pi}{a}; \quad n = 1, 2, 3, 4, \ldots, \tag{6.21}$$

which leads to the quantized energy levels

$$E \to E_n = \frac{\hbar^2 \pi^2 n^2}{2ma^2}; \quad n = 1, 2, 3, 4, \ldots; \tag{6.22}$$

as was already mentioned, the energy cannot depend on the choice of origin. The normalized eigenfunctions are

$$\psi_n(x) = \begin{cases} \sqrt{2/a}\sin[n\pi x/a] & 0 \le x \le a \\ 0 & \text{otherwise} \end{cases}. \tag{6.23}$$

Whether it is convenient to use eigenfunctions in the form of Eq. (6.17) or Eq. (6.23) depends on what properties of the solution you are investigating.

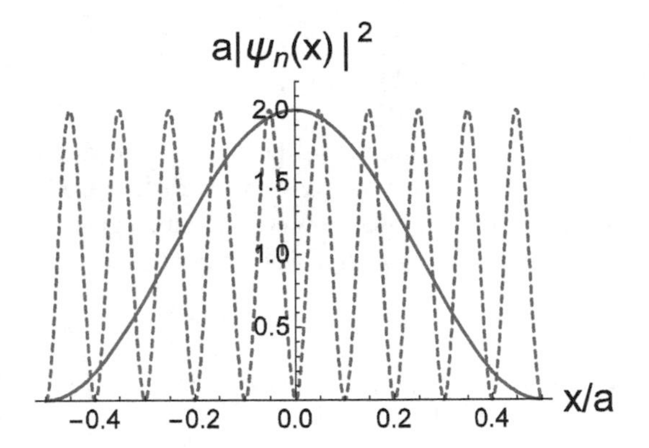

Fig. 6.8 Dimensionless probability distributions for the $n = 1$ and $n = 10$ eigenfunctions. The solid curve is for $n = 1$ and the dashed curve for $n = 10$

6.2.3 Position and Momentum Distributions

The dimensionless probability distribution $a\,|\psi_n(x)|^2$ is shown in Fig. 6.8 for $n = 1$ and $n = 10$ for the well located between $-a/2$ and $a/2$. You see that, for $n = 1$, the function is "bell-shaped," but for $n = 10$, there are many oscillations. If I average these oscillations for large n, I find that $\left\langle |\psi_n(x)|^2 \right\rangle = 1/a$, in agreement with the classical distribution given in Eq. (6.5), suggesting that $|\psi_n(x)|^2$ can be interpreted as a probability distribution. This is a bit of a swindle, however, since, in a state of given n, the quantum probability distribution is very different from the classical one. The classical distribution is a delta function centered at the classical particle position, while there are many places in the quantum distribution where the particle *cannot be found at all*. To get a *true* classical limit, you must take a superposition of a large number of quantum states to form a wave packet that will bounce back and forth between the walls with minimal spreading, simulating the classical particle motion. On the other hand, I have uncovered an important link between the classical and quantum problem. In the limit of large quantum numbers (high n), the quantum distribution, *averaged over oscillations in the classically allowed region* is approximately equal to the *time-averaged* classical particle density for a particle having an energy equal to that associated with the quantum state n.

The situation in momentum space is a bit closer to the classical picture. The eigenfunctions can be expanded as

$$\psi_n(x) = \frac{1}{(2\pi\hbar)^{1/2}} \int_{-\infty}^{\infty} dp \Phi_n(p) e^{ipx/\hbar}, \tag{6.24}$$

where the expansion coefficients $\Phi_n(p)$ are simply the Fourier transform of the spatial ones

$$\Phi_n(p) = \frac{1}{(2\pi\hbar)^{1/2}} \int_{-\infty}^{\infty} dx \psi_n(x) e^{-ipx/\hbar}. \tag{6.25}$$

I will assume that $|\Phi_n(p)|^2$ corresponds to the momentum distribution associated with the eigenfunction $\psi_n(x)$. Since the *momentum* distribution is independent of the choice of coordinates, I can choose the wave functions given by Eq. (6.23) since it allows me to get an expression for all n. Using Eq. (6.23) for $\psi_n(x)$, I calculate the momentum eigenfunctions as

$$\begin{aligned}\Phi_n(p) &= \frac{1}{(2\pi\hbar)^{1/2}} \sqrt{\frac{2}{a}} \int_0^a dx \sin(n\pi x/a)\, e^{-ipx/\hbar} \\ &= -\frac{n}{\pi\sqrt{p_c}} \frac{[1-(-1)^n \exp(-i\pi p/p_c)]}{(p/p_c)^2 - n^2},\end{aligned} \tag{6.26}$$

where

$$p_c = \pi\hbar/a. \tag{6.27}$$

The momentum distribution is then given by

$$|\Phi_n(p)|^2 = \frac{2n^2}{\pi^2 p_c} \frac{1-(-1)^n \cos(\pi p/p_c)}{\left[(p/p_c)^2 - n^2\right]^2}, \tag{6.28}$$

valid for any integer $n \geq 1$.

In Fig. 6.9, the dimensionless momentum distribution $p_c |\Phi_n(p)|^2$ is plotted as a function of p/p_c for $n = 1, 2, 10$. For $n = 1$, the distribution is a smooth curve having HWHM $\Delta p_{1/2}(n = 1)$ approximately equal to $1.19 p_c$. For $n \geq 2$, the distribution consists of two peaks whose centers are separated by

$$\delta p_n = 2np_c = 2n\pi\hbar/a. \tag{6.29}$$

In the problems you are asked to show that, for $n \gg 1$, the height of each peak, $p_c |\Phi_n(np_c)|^2$, approaches a value equal to $1/4$ and the HWHM of each peak approaches a value equal to $\Delta p_{1/2} \approx 2.79 p_c/\pi = 2.79\hbar/a$, *independent of n.* You can think of the width of each peak as being determined from the uncertainty principle. Since Δx equals $a/\sqrt{12}$ for large n (see below), the *magnitude* of the momentum cannot be determined to better than $\hbar/(2\Delta x) = \sqrt{3}\hbar/a = 1.73\hbar/a$.

For large n the distribution mirrors that of the classical distribution given in Eq. (6.4), since it consists of two peaks centered at

$$p_n = \pm(n\pi\hbar/a) = \pm\sqrt{2mE_n}, \tag{6.30}$$

as in the classical case. The peaks do *not* approach delta functions as in the classical case, but $\Delta p_{1/2}/\delta p_n \sim 0$ as $n \to \infty$.

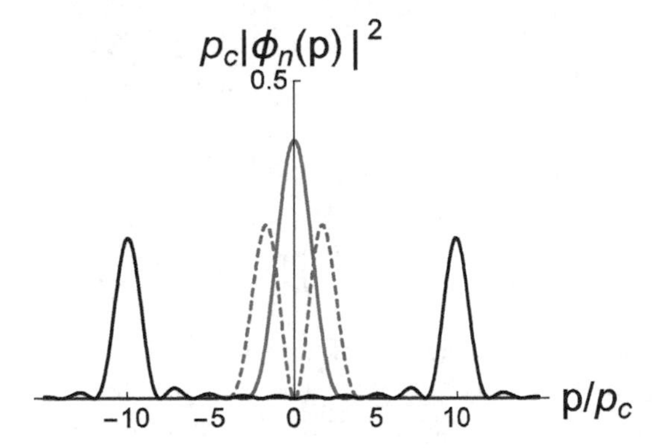

Fig. 6.9 Dimensionless eigenfunction momentum probability distributions as a function of p/p_c for $n = 1$ (solid single-peaked curve), $n = 2$ (dashed curve), and $n = 10$ (solid double-peaked curve)

It is not difficult to calculate the variance in position and momentum associated with each eigenfunction,

$$\Delta x_n^2 = \int_{-a/2}^{a/2} x^2 \left|\psi_n(x)\right|^2 dx = \left(\frac{1}{12} - \frac{1}{2n^2\pi^2}\right) a^2; \tag{6.31a}$$

$$\Delta p_n^2 = \int_{-\infty}^{\infty} p^2 \left|\Phi_n(p)\right|^2 dp = \hbar^2 n^2 \pi^2 / a^2, \tag{6.31b}$$

implying that

$$\Delta x_n \Delta p_n = \hbar \left(\frac{n^2\pi^2}{12} - \frac{1}{2}\right)^{1/2} \geq 0.568\hbar. \tag{6.32}$$

The uncertainty Δx_n grows with increasing n because the wave function becomes more spread out over the well. In the limit that $n \gg 1$, $\Delta x_n \sim a/\sqrt{12}$, the standard deviation of the classical probability distribution, $P_{\text{class}}(x) = 1/a$. Although the momentum distribution consists of two very sharp peaks for large values of n, Δp_n^2 grows with increasing n since the separation of the peaks is proportional to n for large n. A similar result holds for the classical momentum distribution with increasing energy.

6.2.4 Quantum Dynamics

To remind you that the solution of the time-independent Schrödinger equation allows you to calculate the quantum dynamics, I consider an initial state wave function (for a well centered at $x = 0$)

$$\psi(x,0) = \begin{cases} Ne^{-x^2/2b^2} & |x| \leq a/2 \\ 0 & \text{otherwise} \end{cases}, \tag{6.33}$$

where $b \ll a$ and N is a normalization factor given by[2]

$$N = \left(\int_{-a/2}^{a/2} dx e^{-x^2/b^2} \right)^{-1/2} \approx \frac{1}{\sqrt{b\pi^{1/2}}}. \tag{6.34}$$

The particle has an average momentum of zero and is localized at the center of the well with a position uncertainty of order b. The goal is to calculate $\psi(x,t)$.

Without going into the details of the exact solution, I can get a qualitative picture of what is going to happen. In other words, I can ask questions such as "How many eigenfunctions are needed in the expansion of the initial state wave function?", "When does the particle know that it was contained in a potential well?", "Does the particle ever return to its initial shape?"

The uncertainty in the momentum of the particle is $\Delta p \approx \hbar/\sqrt{2}b$ and the energy associated with this uncertainty is $\Delta E = \hbar^2/4mb^2$. As a consequence, I would expect to need to include energies at least equal to ΔE in the sum over eigenfunctions if I am to correctly approximate the initial state wave function. In other words, for some $n_{\max}$, I can approximate the dimensionless wave function as

$$\sqrt{a}\psi_{\text{app}}(x,0) = \sum_{n=1}^{n_{\max}} a_n \psi_n(x), \tag{6.35}$$

where

$$a_n = \begin{cases} \sqrt{2} \int_{-a/2}^{a/2} dx\, Ne^{-x^2/2b^2} \cos \frac{n\pi x}{a} & n \text{ odd} \\ \sqrt{2} \int_{-a/2}^{a/2} dx\, Ne^{-x^2/2b^2} \sin \frac{n\pi x}{a} = 0 & n \text{ even} \end{cases}. \tag{6.36}$$

I would expect the wave function (6.35) to be a good approximation to the exact initial wave function provided

$$n_{\max} \gg \sqrt{\frac{2ma^2\Delta E}{\pi^2\hbar^2}} = \frac{a}{\sqrt{2}\pi b}. \tag{6.37}$$

With $a/b = 10$ ($a/\sqrt{2}\pi b \approx 2.25$), $\sqrt{a}\psi_{\text{app}}(x,0)$ is shown in Fig. 6.10 for $n_{\max} = 1, 5, 9$, and compared with $\sqrt{a}\psi(x,0)$. For $n_{\max} = 9$, the two curves pretty much overlap.

As time progresses, the wave function spreads and $\Delta x(t)$, the standard deviation at time t, increases. For large times when $\Delta pt/m \gg \Delta x(0)$, $\Delta x(t) \approx \Delta pt/m$, so the "particle" spreads to the wall in a time of order

[2]In principle, the wave function in Eq. (6.33) should be multiplied by a factor such as $\cos(\pi x/a)$ to insure that $\psi(x,0)$ satisfies the correct boundary conditions at $x = \pm a/2$. However, if $b \ll a$, $e^{-x^2/2b^2}\cos(\pi x/a) \approx e^{-x^2/2b^2}$ for $-a/2 < x < a/2$. For this reason the $\cos(\pi x/a)$ factor has been omitted.

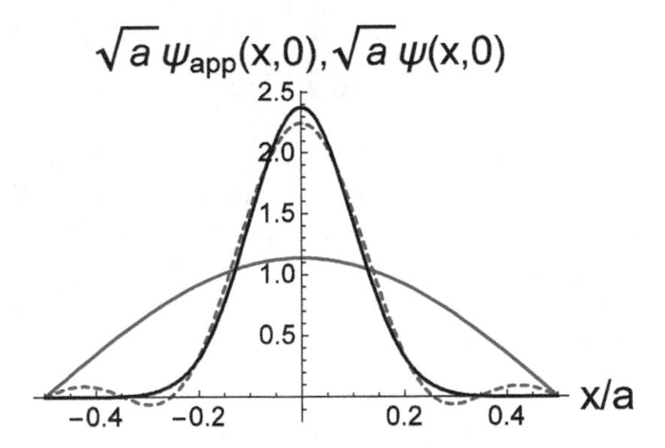

Fig. 6.10 Wave function $\psi(x,0)$ in dimensionless units as a function of x/a. The solid, black curve is the exact wave function given by Eq. (6.33) with $b/a = 0.1$. The other curves are approximations to the wave function calculated using Eq. (6.35) for different values of $n_{\max}$; the red, solid curve corresponds to $n_{\max} = 1$, the blue, dashed curve to $n_{\max} = 5$, and the green, solid curve to $n_{\max} = 9$. The $n_{\max} = 9$ curve is barely distinguishable from the original wave function

$$t_{sp} = \frac{a}{2\Delta v} = \frac{mab}{\sqrt{2}\hbar}, \tag{6.38}$$

where $\Delta v = \Delta p/m$ is the speed uncertainty in the initial packet. Note that t_{sp} depends inversely on $\hbar$ so the particle reaching the wall is a quantum effect related to wave-packet spreading (in other words, $t_{sp} \sim \infty$ as $\hbar \sim 0$). The time t_{sp} is the characteristic time it takes for the initial wave packet to acquire a width of order a as a result of spreading. It is the time need for the particle to "know" it was confined to the infinite potential well.

I can also simulate a classical particle moving back and forth in the well by taking as an initial state wave function

$$\psi(x,0) = \begin{cases} Ne^{-x^2/2b^2}e^{ik_0x} & -a/2 \leq x \leq a/2 \\ 0 & \text{otherwise} \end{cases}, \tag{6.39}$$

where $b \ll a$ and N is given by Eq. (6.34). The factor e^{ik_0x} leads to an initial average momentum of the packet equal to $p_0 = \hbar k_0 \mathbf{u}_x$. The momentum spread is still of order $\Delta p \approx \hbar/\sqrt{2}b$; however, now the particle, which might have thought it was free, is in for a rude awakening when it strikes the wall of the well in a time of order $a/2v_0$, where $v_0 = p_0/m$. The range Δn of energy eigenfunctions needed to construct this packet is still of order $a/2\pi b$, but the states are now centered about the integer closest to

$$n_0 = \frac{p_0 a}{\pi\hbar} = \frac{p_0}{p_c}, \tag{6.40}$$

obtained by setting $E_{n_0} = mv_0^2/2$. The particle makes of order

$$t_{sp}/\left(\frac{a}{v_0}\right) = t_{sp}/\left(\frac{ma}{p_0}\right) = \frac{bp_0}{\sqrt{2}\hbar} = \frac{\sqrt{2}\pi b}{\lambda_{dB}} \tag{6.41}$$

wall collisions before spreading of the packet is significant ($\lambda_{dB} = h/p_0$ is the average de Broglie wavelength of the initial packet). As long as $b/\lambda_{dB} \gg 1$, the quantum wave packet can be considered to represent a classical particle for times $t \ll t_{sp}$.

There is one additional interesting feature in this problem. If I write the general form for the wave function at time t as

$$\psi(x,t) = \sum_{n=1}^{\infty} a_n e^{-in^2\omega_1 t} \psi_n(x), \tag{6.42}$$

where

$$\omega_1 = \frac{E_1}{\hbar} = \frac{\hbar\pi^2}{2ma^2}, \tag{6.43}$$

it is clear that the initial wave packet is reproduced at integral multiples of the *revival time*

$$t_r = \frac{2\pi}{\omega_1} = \frac{4ma^2}{\hbar\pi}. \tag{6.44}$$

Such *quantum revivals* are a purely quantum effect since the revival time t_r goes to infinity as $\hbar$ goes to zero. Although a little harder to prove (see problems), quantum revivals $[|\psi(x,t)|^2 = |\psi(x,0)|^2]$ occur for times t that are integral multiples of $t_r/8$ if the initial wave function is symmetric about the origin, for times t that are integral multiples of $t_r/4$ if the initial wave function is antisymmetric about the origin. Moreover, regardless of the functional form of the initial wave function, $|\psi(-x,t)|^2 = |\psi(x,0)|^2$ for times t that are half-integral multiples of t_r.

6.3 Piecewise Constant Potentials

I now examine problems involving *piecewise constant potentials*. In other words, I look at problems in which the potential is constant in several regions, but undergoes point jump discontinuities between regions. An analogous problem in optics is transmission and reflection of light at a dielectric surface, in which the index of refraction is constant on either side of the dielectric interface, but undergoes a point jump discontinuity at the interface. Of course, no physical boundary can be infinitely sharp. In the optical case, the change in index is assumed to occur over a distance small compared with a wavelength, which can be satisfied quite easily using polished surfaces. In quantum mechanics, it is assumed that the change in the potential occurs over a distance that is small compared to a de Broglie

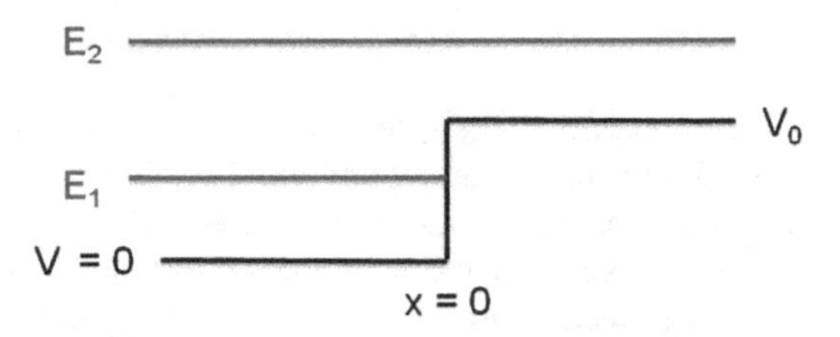

Fig. 6.11 Step potential

wavelength, a condition that is much harder to achieve experimentally. Let's forget about these complications for the moment and begin a systematic attack on this problem. I concentrate on solutions of the time-independent Schrödinger equation, but discuss wave packet dynamics as well. In problems involving reflection and transmission, I use the probability current density to obtain the reflection and transmission coefficients. I will not be concerned about normalizing the wave functions in problems involving reflection and transmission; I simply calculate the ratio of transmitted and reflected probability current densities to the incident probability current density.

6.3.1 Potential Step

Consider the potential step shown in Fig. 6.11,

$$V(x) = \begin{cases} 0 & x < 0 \\ V_0 & x > 0 \end{cases}. \tag{6.45}$$

There is a point jump discontinuity in the potential at $x = 0$, but, as long as the potential contains no singularities or infinities, both the wave function and its derivative are continuous at all points, since they are solutions of a well-behaved, second order, linear differential equation. To obtain the eigenfunctions, I solve Schrödinger's equation for $x < 0$ and for $x > 0$ and then equate the wave functions of the two solutions and their derivatives at $x = 0$. On physical grounds, the continuity of the wave function is consistent with the idea that the probability density must be a single valued function.

The procedure I follow for the step potential can be used in any problem involving piecewise constant potentials. That is, I solve the Schrödinger equation in each region of constant potential and use the continuity of the wave function and its derivative to connect the solutions. Additional boundary conditions are often needed for the solutions as $|x| \to \infty$. For the potential step, I consider $E < V_0$ and $E > V_0$ separately.

6.3.1.1 $E < V_0$

In the classical problem, a particle that approaches the barrier is reflected with the same speed. In the quantum-mechanical problem, I could simulate the classical problem by sending a wave packet towards the barrier. When I discuss scattering theory in Chap. 17, you will see that the wave packet actually penetrates into the barrier, but is then totally reflected with a *time delay* that depends on $(V_0 - E)/E$. In this chapter, I consider the time-independent problem only and calculate the reflection coefficient using the probability current density.

I need to solve the Schrödinger equation for $x < 0$ and for $x > 0$ and then match the wave functions of the two regions and their derivatives at $x = 0$. I know already that the eigenvalues are continuous and there is no degeneracy. For region I, in which $x < 0$ and $V(x) = 0$,

$$\frac{d^2\psi_{IE}(x)}{dx^2} = -k_E^2\psi_{IE}(x) \tag{6.46}$$

where

$$k_E = \frac{\sqrt{2mE}}{\hbar} > 0. \tag{6.47}$$

The general solution of this equation is exponentials or sines and cosines. It is better to choose exponentials since it will then be possible to interpret the results in terms of an incident and reflected probability current. Although there is no degeneracy I must take the most general possible solution of Eq. (6.46) or I will not be able to match the *two* boundary conditions at $x = 0$ for the continuity of the wave function and its derivative. Thus I take

$$\psi_{IE}(x) = Ae^{ik_E x} + Be^{-ik_E x}; \qquad x < 0. \tag{6.48}$$

For region II, in which $x > 0$ and $V(x) = V_0$, Schrödinger's equation is

$$\frac{d^2\psi_{IIE}(x)}{dx^2} = \kappa_E^2\psi_{IIE}(x), \tag{6.49}$$

where

$$\kappa_E = \frac{\sqrt{2m(V_0 - E)}}{\hbar} > 0. \tag{6.50}$$

The possible solutions are $e^{\pm\kappa_E x}$ but I must reject the $+$ exponential since it blows up for large x. That is, the boundary condition requiring the wave function to be finite at all points in space requires me to reject a solution of the form $e^{\kappa_E x}$ as $x \to \infty$. Thus, I take

$$\psi_{IIE}(x) = Ce^{-\kappa_E x}; \qquad x > 0. \tag{6.51}$$

Combining Eqs. (6.48) and (6.51), I find that the eigenfunction corresponding to an energy $0 < E < V_0$ is

$$\psi_E(x) = \begin{cases} Ae^{ik_E x} + Be^{-ik_E x} & x < 0 \\ Ce^{-\kappa_E x} & x > 0 \end{cases}. \tag{6.52}$$

It is possible to normalize this solution if some type of convergence factor is introduced (see problems), but the normalization is unimportant for our considerations.

Equating the wave function and its derivative at $x = 0$, I obtain the two equations

$$A + B = C; \tag{6.53a}$$

$$ik_E (A - B) = -\kappa_E C, \tag{6.53b}$$

from which I find

$$\frac{B}{A} = \frac{k_E - i\kappa_E}{k_E + i\kappa_E}; \tag{6.54a}$$

$$\frac{C}{A} = \frac{2k_E}{k_E + i\kappa_E}. \tag{6.54b}$$

What do these ratios mean?

To interpret them, I look at the probability current density associated with *each* component of the eigenfunction for $x < 0$. The probability current density [see Eq. (5.139)] associated with the $Ae^{ik_E x}$ part of the eigenfunction is $J_i = v_E |A|^2$ while that associated with the $Be^{-ik_E x}$ part is $J_r = -v_E |B|^2$, where $v_E = \hbar k_E/m$ and the i and r subscripts stand for "incident" and "reflected," respectively. I interpret

$$R = \frac{B}{A} \tag{6.55}$$

as an *amplitude reflection coefficient* and

$$\mathcal{R} = |R|^2 = -\frac{J_r}{J_i} = \frac{v_E |B|^2}{v_E |A|^2} = \left|\frac{B}{A}\right|^2 = 1 \tag{6.56}$$

as an *intensity reflection coefficient*. Since $\mathcal{R} = 1$, the wave is totally reflected. The probability current density *inside* the potential step vanishes since the wave function is real, but the quantity

$$\left|\frac{C}{A}\right|^2 = \frac{4k_E^2}{\kappa_E^2 + k_E^2} = \frac{4E}{V_0} \tag{6.57}$$

turns out to be a measure of the distance that a wave packet penetrates into the potential step before it is totally reflected (you have to solve the scattering problem to see this, as I do in Chap. 17). The problem is analogous to total reflection of light by a lossless plasma, when the frequency of the light is below the plasma frequency.

In the limit that $V_0 \to \infty$, $C/A \sim 0$, consistent with the assumption that the wave function vanishes in extended regions where there is an infinite potential. Moreover, as $V_0 \to \infty$, $B/A \sim -1$ in Eq. (6.54a), which implies a phase change of π on reflection. Therefore, as $V_0 \to \infty$, Eq. (6.52) reduces to

$$\psi_E(x) \sim \begin{cases} 2iA \sin(k_E x) & x < 0 \\ 0 & x > 0 \end{cases}. \tag{6.58}$$

The eigenfunctions are now *standing waves*, with a node at $x = 0$. The analogous situation for electromagnetic radiation is reflection at a perfect metal, where the tangential component of the electric field must vanish at the surface.

Another interesting limit occurs for $E \approx V_0$ ($E = V_0 - \epsilon$, with $0 < \epsilon \ll V_0$), for which $B/A \sim 1$, $C/A \sim 2$, and

$$\psi_E(x) \sim \begin{cases} 2A\cos(k_E x) & x < 0 \\ 2Ae^{-\kappa_E x}; \quad \kappa_E = \frac{\sqrt{2m\epsilon}}{\hbar^2}; & x > 0 \end{cases}. \tag{6.59}$$

In this limit the intensity reflection coefficient is still equal to unity, but there is no phase change on reflection. The wave function penetrates deeply into the potential step. In the classically allowed region the wave functions is a standing wave with an *antinode* at $x = 0$. If a wave packet having fairly well-defined energy $E \approx V_0$ is incident on the potential step, there is a long time delay before the packet is totally reflected.

6.3.1.2 $E > V_0$

In the classical problem, a particle approaches the potential step and is transmitted with a lower speed. There is no reflection in the classical problem. In the quantum-mechanical problem, a wave packet incident on the potential step is partially transmitted and partially reflected, with the reflection coefficient depending on $(E - V_0)/E$. The eigenenergies are continuous and there is a two-fold degeneracy.

For region I, in which $x < 0$,

$$\frac{d^2\psi_{IE}(x)}{dx^2} = -k_E^2\psi_{IE}(x) \tag{6.60}$$

as before, but for region II, in which $x > 0$, Eq. (6.49) is replaced by

$$\frac{d^2\psi_{IIE}(x)}{dx^2} = -k_E'^2\psi_{IIE}(x) \tag{6.61}$$

where

$$k_E' = \frac{\sqrt{2m(E - V_0)}}{\hbar} > 0. \tag{6.62}$$

You might think that I need to try a solution of the form

$$\psi_E(x) = \begin{cases} Ae^{ik_E x} + Be^{-ik_E x} & x < 0 \\ Ce^{ik'_E x} + De^{-ik'_E x} & x > 0 \end{cases}; \tag{6.63}$$

however, I will run into problems with such a solution. There are four unknowns in this equation. Using the boundary conditions at $x = 0$ gives two constraints (continuity of the wave function and its derivative) and normalization a third, but I am one short since I need four constraints. The reason for this dilemma is that there is a two-fold degeneracy in this problem. I must take two *independent* solutions for each energy. One way of doing this is to arbitrarily set one of the coefficients equal to zero in each of two separate solutions.

I do this in a manner that allows me simulate waves incident from the left or right; that is, I take

$$\psi_E^L(x) = \begin{cases} A_L e^{ik_E x} + B_L e^{-ik_E x} & x < 0 \\ C_L e^{ik'_E x} & x > 0 \end{cases}, \tag{6.64}$$

corresponding to a wave incident from the left and

$$\psi_E^R(x) = \begin{cases} C_R e^{-ik_E x} & x < 0 \\ A_R e^{-ik'_E x} + B_R e^{ik'_E x} & x > 0 \end{cases}, \tag{6.65}$$

corresponding to a wave incident from the right. These are two independent solutions for each energy $E > V_0$. Of course, any two, linearly independent combinations of Eqs. (6.64) and (6.65) could be used as well. I consider only the solution for the wave incident from the left [Eq. (6.64)] and drop the L superscript. The solutions (6.64) and (6.65) contain plane wave *components*, but these eigenfunctions are *not* plane waves, even if they extend over all space. As I have stressed, each potential has its own set of eigenfunctions; the only potential allowing for plane wave eigenfunctions is $V = 0$ (or a constant) in all space.

Matching the wave functions in the two regions and their derivatives at $x = 0$, I find

$$A + B = C; \tag{6.66a}$$

$$k_E (A - B) = k'_E C, \tag{6.66b}$$

from which I can obtain

$$R = \frac{B}{A} = \frac{k_E - k'_E}{k_E + k'_E}; \tag{6.67a}$$

$$T = \frac{C}{A} = \frac{2k_E}{k_E + k'_E}, \tag{6.67b}$$

as the amplitude reflection and transmission coefficients, respectively. The probability current density associated with Ae^{ik_Ex} is $J_i = v_E |A|^2$, that associated with Be^{-ik_Ex} is $J_r = -v_E |B|^2$, while that associated with $Ce^{ik'_Ex}$ is $J_t = v'_E |C|^2$, where $v_E = \hbar k_E/m$, $v'_E = \hbar k'_E/m$, and the t subscript stand for "transmitted." Thus

$$\mathcal{R} = |R|^2 = -\frac{J_r}{J_i} = \frac{v_E}{v_E}\left|\frac{B}{A}\right|^2 = \left(\frac{k_E - k'_E}{k_E + k'_E}\right)^2 = \left(\frac{\sqrt{E} - \sqrt{E - V_0}}{\sqrt{E} + \sqrt{E - V_0}}\right)^2 \tag{6.68}$$

is the (intensity) reflection coefficient and

$$\begin{aligned}\mathcal{T} = |T|^2 = \frac{J_t}{J_i} &= \frac{v'_E}{v_E}\left|\frac{C}{A}\right|^2 = \frac{k'_E}{k_E}\left(\frac{2k_E}{k_E + k'_E}\right)^2 \\ &= \frac{4k_E k'_E}{\left(k_E + k'_E\right)^2} = \frac{4\sqrt{E}\sqrt{E - V_0}}{\left(\sqrt{E} + \sqrt{E - V_0}\right)^2}\end{aligned} \tag{6.69}$$

is the (intensity) transmission coefficient. The fact that

$$\mathcal{R} + \mathcal{T} = 1, \tag{6.70}$$

is a statement of conservation of probability.

It is interesting to note that both $\mathcal{R}$ and $\mathcal{T}$ are independent of $\hbar$. That is, even if I take a classical limit in which $\hbar \to 0$, I do not recover the classical result of $\mathcal{R} \to 0$. The reason is simple. For the classical limit to hold, *changes* in the potential must occur on a length scale that is large compared with the de Broglie wavelength. Since the potential changes abruptly this is not possible. If the potential rose smoothly over a distance large compared with the de Broglie wavelength, there would be virtually no reflection as $\hbar \to 0$. In fact, for a smooth potential step of the form

$$V(x) = \frac{V_0}{1 + e^{-x/a}}, \tag{6.71}$$

where $a > 0$ is the length scale of the step, it is possible to solve Schrödinger's equation exactly in terms of hypergeometric functions and to show analytically that the reflection coefficient is[3]

$$\mathcal{R} = \left[\frac{\sinh\left[\pi\left(k_E - k'_E\right)a\right]}{\sinh\left[\pi\left(k_E + k'_E\right)a\right]}\right]^2, \tag{6.72}$$

which reduces to Eq. (6.68) when $k_E a = 2\pi a/\lambda_{dB} \ll 1$, but varies as $\exp\left(-4\pi k'_E a\right) \sim 0$ in the limit that a is finite and $\hbar \to 0$.

[3]See L. D Landau and E. M. Lifshitz, *Quantum Mechanics, Non-Relativistic Theory* (Pergamon Press, London, 1958), pp. 75–76.

The quantum step potential problem with $E > V_0$ is analogous to the reflection of light at a dielectric. As long as the interface is sharper than a wavelength, there is *always* a reflected wave. For normal incidence from vacuum to a medium having index of refraction n, the ratio of reflected to incident pulse *amplitudes* is

$$R = \left.\frac{B}{A}\right|_{\text{light}} = \frac{1-n}{1+n} \tag{6.73}$$

and the speed of light in the dielectric is c/n. This agrees with Eq. (6.67a) if I set

$$n \to n_{\text{eff}} = \sqrt{\frac{E-V_0}{E}} = \sqrt{1-\frac{V_0}{E}} < 1 \tag{6.74}$$

Thus, even though the *particle* speed *decreases*, the effective *index* is less than unity. The analogue with reflection at a dielectric is not exact, although the results take on the same form. Changes in the incident wavelength for light do not seriously affect the index of refraction, but the effective index depends in a significant way on the incident energy for matter waves.

In solving the Schrödinger equation for both $E < V_0$ and $E > V_0$, I automatically determined the eigenergies and the eigenfunctions. For $E < V_0$, I found that any energy in the range $0 < E < V_0$ gives rise to a solution and that the eigenfunctions are nondegenerate. For $E > V_0$, I found that any energy gives rise to a doubly-degenerate solution.

6.3.2 *Square Well Potential*

Now I turn my attention to the square well potential shown in Fig. 6.12 for which

$$V(x) = \begin{cases} -V_0 < 0 & |x| < a/2 \\ 0 & |x| > a/2 \end{cases}. \tag{6.75}$$

I consider $E < 0$ and $E > 0$ separately. I must solve the Schrödinger equation in three regions, $x < -a/2$, $-a/2 < x < a/2$, $x > a/2$, and equate the wave functions of the solutions and their derivatives at $x = -a/2$ and $x = a/2$.

6.3.2.1 $E < 0$

In the classical problem, a particle is always bound in the well for $E < 0$. In the quantum mechanics problem, you will see that there is always at least one bound state. However, the wave function penetrates into the classically forbidden regime.

You might think that there is no bound state for sufficiently small well depths based on the following argument. Since Δx is of order a, Δp is of order $\hbar/a$,

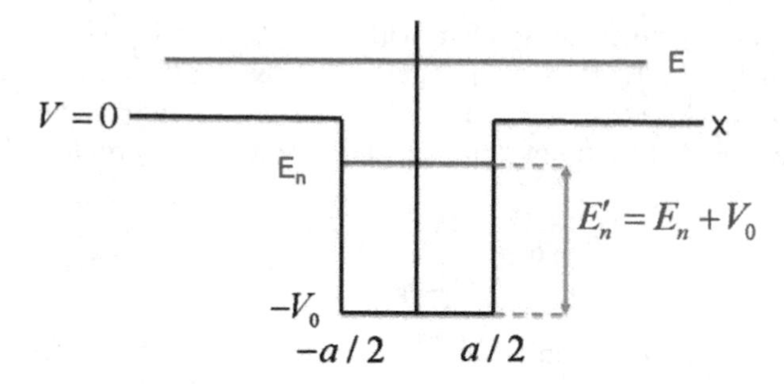

Fig. 6.12 Square well potential. There are unbound eigenfunctions for all positive energies and a finite number of bound states E_n for negative energies. The number of bound states is equal to the integer value of $(1 + \beta/\pi)$, where $\beta^2 = 2mV_0a^2/\hbar^2$

corresponding to an energy of $\hbar^2/2ma^2$. Therefore if $V_0 < \hbar^2/2ma^2$, the well is not deep enough to bind the particle. You will see what is wrong with this argument after I analyze the problem in detail.

For $E < 0$, the eigenenergies are discrete and there is no degeneracy. I can simplify the problem somewhat by noting that the Hamiltonian commutes with the parity operator. Therefore the energy eigenfunctions are guaranteed to be simultaneous eigenfunctions of the parity operator. The boundary conditions are such that the wave function must vanish as x approaches $\pm\infty$. The even parity solutions of Schrödinger's equation satisfying the boundary conditions at $x = \pm\infty$ are

$$\psi_E^+(x) = \begin{cases} B^+ e^{\kappa_E^+ x} & x < -a/2 \\ A^+ \cos\left(k_E'^+ x\right) & -a/2 < x < a/2 \\ B^+ e^{-\kappa_E^+ x} & x > a/2 \end{cases} \tag{6.76}$$

and the odd parity solutions are

$$\psi_E^-(x) = \begin{cases} B^- e^{\kappa_E^- x} & x < -a/2 \\ A^- \sin\left(k_E'^- x\right) & -a/2 < x < a/2 \;, \\ -B^- e^{-\kappa_E^- x} & x > a/2 \end{cases} \tag{6.77}$$

where

$$k_E'^\pm = \frac{\sqrt{2mE'^\pm}}{\hbar} > 0, \tag{6.78}$$

$$\kappa_E^\pm = \frac{\sqrt{-2mE^\pm}}{\hbar} > 0, \tag{6.79}$$

$$E'^\pm = \left(E^\pm + V_0\right), \tag{6.80}$$

m is the particle mass, and $+$ $(-)$ corresponds to even (odd) parity. The energy $E'^{\pm}$ is the *difference* between $E^{\pm}$ and the energy $-V_0$ at the bottom of the well (see Fig. 6.12).

By choosing the energy eigenfunctions to be simultaneous eigenfunctions of the parity operator, I guarantee that if I satisfy the boundary conditions on the wave function and its derivative at $x = a/2$, they are automatically satisfied at $x = -a/2$. Matching the wave functions their derivatives at $x = a/2$, I find

$$A^{+} \cos\left(\frac{k_E'^{+} a}{2}\right) = B^{+} \exp\left(-\frac{\kappa_E^{+} a}{2}\right) \tag{6.81a}$$

$$A^{+} k_E'^{+} \sin\left(\frac{k_E'^{+} a}{2}\right) = B^{+} \kappa_E^{+} \exp\left(-\frac{\kappa_E^{+} a}{2}\right) \tag{6.81b}$$

for the even parity solutions and

$$A^{-} \sin\left(\frac{k_E'^{-} a}{2}\right) = B^{-} \exp\left(-\frac{\kappa_E^{-} a}{2}\right) \tag{6.82a}$$

$$A^{-} k_E'^{-} \cos\left(\frac{k_E'^{-} a}{2}\right) = -B^{-} \kappa_E^{-} \exp\left(-\frac{\kappa_E^{-} a}{2}\right) \tag{6.82b}$$

for the odd parity solutions.

Equations (6.81) and (6.82) are typical of the type encountered in solving bound state problems for piecewise constant potentials. They are homogeneous equations with the same number of equations as unknowns. The only way to have non-trivial solutions of such equations is for the determinant of the coefficients to vanish. In solving the determinant equation, you find solutions for only *specific values of the energy.* This is why bound state motion leads to discrete or quantized eigenenergies.

Instead of setting the determinant of the coefficients in Eqs. (6.81) and (6.82) equal to zero, it is simpler to divide the equations to obtain

$$\tan\left(\frac{k_E'^{+} a}{2}\right) = \frac{\kappa_E^{+}}{k_E'^{+}} \tag{6.83}$$

for the even parity solutions and

$$\tan\left(\frac{k_E'^{-} a}{2}\right) = -\frac{k_E'^{-}}{\kappa_E^{-}} \tag{6.84}$$

for the odd parity solutions. Note that $k_E'^{-} = 0$ is not an acceptable odd parity solution [even though it is a solution of Eq. (6.84)] since it is not a solution of Eqs. (6.82). In other words, Eq. (6.84) gives the solutions to Eqs. (6.82) provided $k_E'^{-} \neq 0$.

I define dimensionless quantities

$$\beta^2 = \frac{2mV_0}{\hbar^2}a^2, \tag{6.85}$$

$$\left(y^{\pm}\right)^2 = \left(k_E'^{\pm}\right)^2 a^2, \tag{6.86}$$

such that

$$\kappa_E^{\pm} a = \sqrt{\frac{2m\left(V_0 - E'^{\pm}\right)}{\hbar^2}}a = \sqrt{\beta^2 - \left(y^{\pm}\right)^2}. \tag{6.87}$$

The quantity β^2 is a dimensionless measure of the strength of the well that we will encounter often. The condition determining the even parity eigenenergies is

$$\tan\left(\frac{y^+}{2}\right) = \frac{\kappa_E^+}{k_E'^+} = \sqrt{\frac{\beta^2}{\left(y^+\right)^2} - 1} > 0, \tag{6.88}$$

while the condition for the odd parity eigenfunctions is

$$\tan\left(\frac{y^-}{2}\right) = -\frac{1}{\sqrt{\frac{\beta^2}{\left(y^-\right)^2} - 1}} < 0. \tag{6.89}$$

Equations (6.88) and (6.89) can be solved graphically. The graphical solution for the even parity solution, Eq. (6.88), is shown in Figs. 6.13 and 6.14 for $\beta = 0.5$ and $\beta = 20$, respectively. As you can see there is always *at least one solution*, irrespective of the value of β. Why does the uncertainty principle argument given above fail? For $\beta \ll 1$, the value of $\kappa_E^+ a$ becomes small and the eigenfunction penetrates a long distance into the classically forbidden region. Thus the estimate that $\Delta x = a$ is wrong—I should use $\Delta x = \left(\kappa_E^+\right)^{-1} \approx a/\beta \gg a$, giving a corresponding ΔE which is less than V_0. The corresponding odd parity solutions are left to the problems. Using the graphical solutions, it is easy to show that the number of bound states in the well is equal to the integer value of $(1 + \beta/\pi)$.

I can estimate the energy E^+ of the bound state in the limit of a weakly binding well, $\beta \ll 1$. I define

$$z = \sqrt{-\frac{2mE^+}{\hbar^2}}a > 0, \tag{6.90}$$

such that

$$\beta^2 = \left(y^+\right)^2 + z^2. \tag{6.91}$$

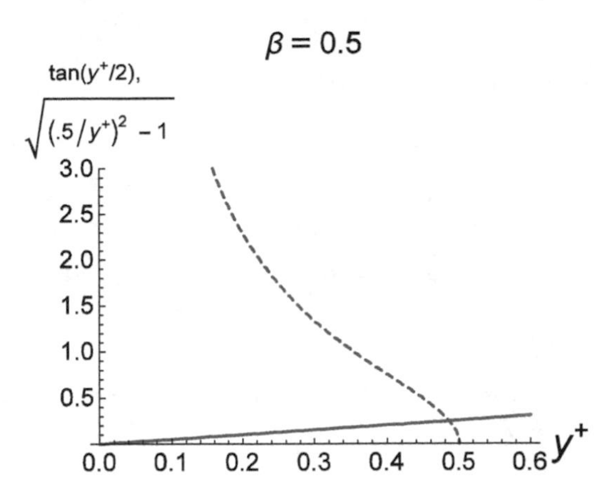

Fig. 6.13 Graphical solution of Eq. (6.88) for $\beta = 0.5$. The blue dashed curve is $\sqrt{(\beta/y^+)^2 - 1}$ and the red solid curve is $\tan\left(y^+/2\right)$

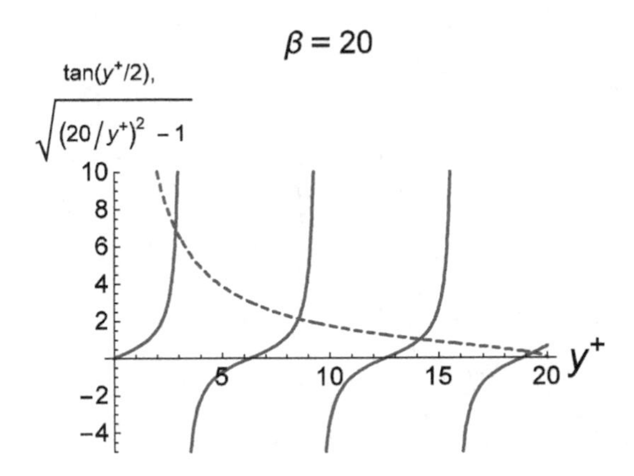

Fig. 6.14 Graphical solution of Eq. (6.88) for $\beta = 20$. The blue dashed curve is $\sqrt{(\beta/y^+)^2 - 1}$ and the red solid curve is $\tan\left(y^+/2\right)$

Setting

$$y^+ = \beta - \epsilon \tag{6.92}$$

and assuming that $\epsilon \ll \beta$ and $\beta \ll 1$, I can approximate Eq. (6.88) as

$$\tan\left(\frac{\beta}{2}\right) \approx \frac{\beta}{2} = \sqrt{\frac{\beta^2}{(\beta - \epsilon)^2} - 1} \approx \sqrt{\frac{2\epsilon}{\beta}} \tag{6.93}$$

or

$$\epsilon \approx \frac{\beta^3}{8}. \tag{6.94}$$

This result, in turn, implies that

$$z^2 = \beta^2 - \left(y^+\right)^2 = \beta^2 - (\beta - \epsilon)^2 \approx 2\epsilon\beta = \beta^4/4 \tag{6.95}$$

or

$$E = -\frac{\hbar^2 z^2}{2ma^2} = -\frac{\hbar^2 \beta^4}{8ma^2} = -\frac{\beta^2}{4} V_0. \tag{6.96}$$

Even if $\beta \ll 1$, there is always a bound state having an energy whose absolute value is much less than the well depth.[4]

In the opposite limit of a very deep well, that is, when $V_0 \to \infty$ and $E' = (E + V_0) \ll V_0$, I should recover the eigenfunctions and eigenenergies of the infinite potential well. In the limit that $V_0 \to \infty$ and $(E + V_0) \ll V_0$, Eqs. (6.88) and (6.89) reduce to

$$\tan\left(\frac{k_E'^+ a}{2}\right) = \infty, \tag{6.97a}$$

$$\tan\left(\frac{k_E'^- a}{2}\right) = 0. \tag{6.97b}$$

The first condition is satisfied if

$$k_E'^+ a = (2n + 1)\pi; \quad n = 0, 1, \ldots \tag{6.98}$$

and the second if

$$k_E'^- a = 2n\pi; \quad n = 1, 2, \ldots \tag{6.99}$$

[4]Equation (6.96) can be written as

$$E = -\left(m/2\hbar^2\right) V_0^2 a^2 = -\left(m/2\hbar^2\right)\left[\int_{-\infty}^{\infty} V(x)dx\right]^2.$$

This is a general result for "weak" potential wells having arbitrary shape—see L. D Landau and E. M. Lifshitz, *Quantum Mechanics, Non-Relativistic Theory* (Pergamon Press, London, 1958), pp. 155–156.

which, taken together, yield

$$k'_E a = n\pi; \qquad n = 1, 2, 3, \ldots, \tag{6.100}$$

where

$$k'_E = \frac{\sqrt{2m(E + V_0)}}{\hbar} = \frac{\sqrt{2mE'}}{\hbar}. \tag{6.101}$$

Equation (6.100) is recognized as the equation for the energy levels in an infinite square well. Similarly, the eigenfunctions go over to

$$\psi_{II}^{+}(x) = A^{+} \cos\left(k_E'^{+} x\right); \tag{6.102}$$

$$\psi_{II}^{-}(x) = A^{-} \sin\left(k_E'^{-} x\right), \tag{6.103}$$

which are the corresponding eigenfunctions.

6.3.2.2 $E > 0$

Since I am interested in transmission and reflection coefficients, I consider only the solution corresponding to a wave incident from the left. The mathematics can be simplified somewhat if I now take the well located between 0 and a. Although the potential no longer commutes with the parity operator with this choice of origin, the solution of interest does not have definite parity in any event since I am considering a wave incident from the left. The eigenfunctions are

$$\psi_E(x) = \begin{cases} Ae^{ik_E x} + Be^{-ik_E x} & x < 0 \\ Ce^{ik'_E x} + De^{-ik'_E x} & 0 < x < a \\ Fe^{ik_E x} & x > a \end{cases} \tag{6.104}$$

with

$$k_E = \frac{\sqrt{2mE}}{\hbar}; \tag{6.105a}$$

$$k'_E = \frac{\sqrt{2m(E + V_0)}}{\hbar}. \tag{6.105b}$$

I now equate the wave functions in the various regions and their derivatives at both $x = 0$ and $x = a$. The appropriate equations are

$$A + B = C + D; \tag{6.106a}$$

$$k_E(A - B) = k'_E(C - D); \tag{6.106b}$$

$$Ce^{ik'_E a} + De^{-ik'_E a} = Fe^{ik_E a}; \tag{6.106c}$$

$$k'_E \left(Ce^{ik'_E a} - De^{-ik'_E a} \right) = k_E Fe^{ik_E a}. \tag{6.106d}$$

From these four equations I can calculate B/A, C/A, D/A, and F/A by setting the determinant of the coefficients equal to zero. The algebra is a little complicated but the solution can be obtained easily using a symbolic program such as Mathematica. Explicitly, you can show that the amplitude reflection and transmission coefficients are equal to

$$R = \frac{B}{A} = \frac{i\left(k_E'^2 - k_E^2\right)\sin\left(k'_E a\right)}{2k_E k'_E \cos\left(k'_E a\right) - i\left(k_E'^2 + k_E^2\right)\sin\left(k'_E a\right)}; \tag{6.107a}$$

$$T = \frac{F}{A} = \frac{2k_E k'_E e^{-ik_E a}}{2k_E k'_E \cos\left(k'_E a\right) - i\left(k_E'^2 + k_E^2\right)\sin\left(k'_E a\right)}. \tag{6.107b}$$

The solution for the intensity transmission coefficient can be written as

$$\mathcal{T} = \frac{J_t}{J_i} = \frac{k_E}{k_E}\left|T\right|^2 = \frac{1}{\cos^2\left(k'_E a\right) + \frac{\epsilon'^2}{4}\sin^2\left(k'_E a\right)} \tag{6.108}$$

and for the intensity reflection coefficient as

$$\mathcal{R} = -\frac{J_r}{J_i} = \frac{k_E}{k_E}\left|R\right|^2 = 1 - \mathcal{T}, \tag{6.109}$$

where

$$\epsilon' = \frac{k_E}{k'_E} + \frac{k'_E}{k_E} = \frac{k_E^2 + k_E'^2}{k_E k'_E} = \frac{2E + V_0}{\sqrt{E}\sqrt{E + V_0}}. \tag{6.110}$$

The intensity transmission coefficient can be written in an alternative way as

$$\mathcal{T} = \frac{1}{1 + \frac{V_0^2}{4E(E+V_0)}\sin^2\left(k'_E a\right)}. \tag{6.111}$$

In the limit that $E \gg V_0$, $\mathcal{T} \sim 1$, as expected, since the energy is much higher than the well depth (recall that, classically, $\mathcal{T} = 1$ for *any* energy $E > 0$). On the other hand, it is not so clear as to what to expect when $E \to 0$, since this corresponds to the quantum regime (de Broglie wavelength greater than well size a). From Eq. (6.111), you see that the transmission goes to zero as $E \to 0$, *unless* $k'_E a = m\pi$, for integer m. That is, there is a *resonance* (sharp increase) in transmission for low energy scattering if $k'_E a \approx \beta = m\pi$. This corresponds approximately to the condition for having a bound state whose energy is very close to zero. For arbitrary energies, there are maxima in transmission whenever $k'_E a = m\pi$. At these points the transmission is equal to unity, but the resonances become broader with increasing energy.

This problem is somewhat analogous to light incident on a thin dielectric film from vacuum if the index of refraction of the film is replaced by

$$n \to n_{\text{eff}} = \sqrt{1 + V_0/E}. \tag{6.112}$$

In optics, when light is incident from vacuum normally on a thin dielectric slab having index of refraction n and thickness d, the reflection and transmission coefficients are

$$\mathcal{R} = 1 - \mathcal{T}; \tag{6.113a}$$

$$\mathcal{T} = \frac{1}{\cos^2(kd) + \frac{\epsilon_n^2}{4}\sin^2(kd)}, \tag{6.113b}$$

where

$$\epsilon_n = n + \frac{1}{n}, \tag{6.114}$$

$$k = nk_0, \tag{6.115}$$

and $k_0 = 2\pi/\lambda_0$ is the free-space propagation constant. The maxima in transmission occur when $kd = m\pi$ or $2d = m\lambda_n = m\lambda_0/n$; that is, when twice the thickness is an integral number of wavelengths in the medium. These are the same resonances that occur in the quantum problem; however, in the quantum problem, the effective index depends significantly on the energy while the index of refraction in the optical problem is approximately constant for a wide range of wavelengths. Details are left to the problems.

If I construct an initial wave packet and send it in from the left, the actual dynamics depends *critically* on the width of the packet. The reflection and transmission coefficients are derived for a *monoenergetic* wave. The range of energies ΔE in the packet must be sufficiently small to satisfy $\Delta E\tau/\hbar \ll 1$, where τ is the time it takes for the packet to be scattered (including bounces back and forth between $x = 0$ and $x = a$) if one is to find transmission and reflection coefficients given by Eqs. (6.108) and (6.109). For example, a wave packet having spatial width less than a could never have a transmission resonance at low energy—it would be totally reflected at the $x = 0$ discontinuity in the potential. Scattering of wave packets in one dimension is discussed in Chap. 17.

The analogy between the quantum and radiation problems is useful only when considering the reflection and transmission coefficients associated with nearly monoenergetic wave packets and nearly monochromatic radiation pulses. The analogy breaks down for wave packets or radiation pulses whose spatial extents are much smaller than the scattering region. For example, you can see from Eq. (6.112)

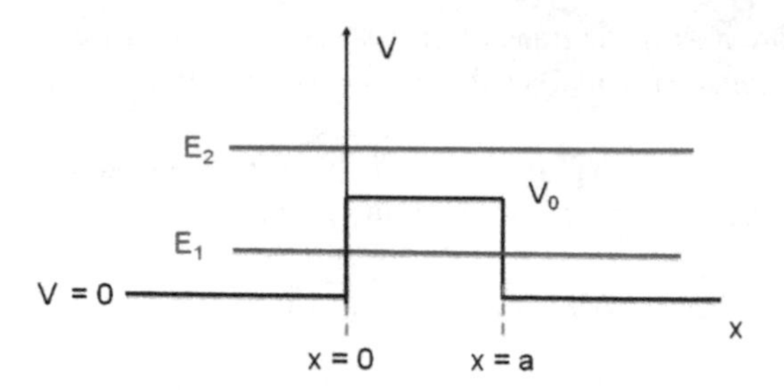

Fig. 6.15 Potential barrier

that a potential well corresponds to an index of refraction $n_{\text{eff}} > 1$. A narrow wave packet would speed up as it passes through the potential region, whereas an optical pulse would propagate at a slower speed in a dielectric corresponding to this potential well.

6.3.3 Potential Barrier

Now I turn my attention to the barrier potential shown in Fig. 6.15 for which

$$V(x) = \begin{cases} V_0 > 0 & 0 < x < a \\ 0 & \text{otherwise} \end{cases}. \tag{6.116}$$

I consider only $E < V_0$. For $E > V_0$, the results of the square well with $E > 0$ can be taken over directly by replacing $-V_0$ with V_0 in Eqs. (6.111) and (6.105b). In the classical problem, a particle is always reflected by the barrier when $E < V_0$. In the quantum-mechanical problem, you will see that the particle can *tunnel* through the barrier. The eigenenergies are continuous and there is a two-fold degeneracy.

As in the case of the potential well, I consider only the eigenfunction corresponding to a wave incident from the left, namely

$$\psi_E(x) = \begin{cases} Ae^{ik_E x} + Be^{-ik_E x} & x < 0 \\ Ce^{\kappa_E x} + De^{-\kappa_E x} & 0 < x < a \\ Fe^{ik_E x} & x > a \end{cases}, \tag{6.117}$$

where

$$k_E = \frac{\sqrt{2mE}}{\hbar} > 0; \tag{6.118}$$

$$\kappa_E = \frac{\sqrt{2m(V_0 - E)}}{\hbar} > 0. \tag{6.119}$$

I should now equate the wave functions in the various regions and their derivatives at both at $x = 0$ and $x = a$. It is not necessary to do so, however, since a comparison of Eqs. (6.104) and (6.117) shows they are identical if I replace ik'_E by κ_E or, equivalently, V_0 by $-V_0$ in Eqs. (6.104). With this replacement, $\sin^2\left(k'_E a\right) \rightarrow -\sinh^2\left(\kappa_E a\right)$ and Eq. (6.111) goes over into

$$\mathcal{T} = \frac{1}{1 + \frac{V_0^2}{4E(V_0 - E)} \sinh^2\left(\kappa_E a\right)}. \tag{6.120}$$

As $E \rightarrow V_0$ ($\kappa_E a \ll 1$), the energy approaches the barrier height and you might expect that significant transmission is possible. In this limit,

$$\mathcal{T} \sim \frac{1}{1 + \frac{V_0^2 \kappa_E^2 a^2}{4V_0(V_0 - E)}} = \frac{1}{1 + \frac{\beta^2}{4}}, \tag{6.121}$$

where β is defined as in Eq. (6.85). You see that $\mathcal{T} \sim 1$ only if the dimensionless barrier strength β is much less than unity. Note that Eq. (6.121) agrees with Eq. (6.111) when V_0 is replaced by $-V_0$ in that equation and the limit $E \rightarrow V_0$ is taken. In other words, the solutions for $E < V_0$ and $E > V_0$ match each other in the limit that $E \rightarrow V_0$, as you would expect.

On the other hand, for $\kappa_E a \gg 1$,

$$\mathcal{T} \sim \frac{16\left(V_0 - E\right) E}{V_0^2} e^{-2\kappa_E a}, \tag{6.122}$$

which represents *tunneling* through the barrier. [If you play tennis, you are familiar with tunneling—you swear you hit the ball, but it appears to have tunneled through your racket. Unfortunately this argument does not hold water since the tunneling probability is negligibly small)]. In the limit that $\hbar \rightarrow 0$, $\kappa_E \sim \infty$, the de Broglie wavelength goes to zero, and there is no tunneling. Tunneling can occur in optics if two prisms are separated by a small amount, as was originally discovered by Newton. Light that would normally be totally internally reflected by the first prism can tunnel into the second prism if the separation between the prisms is less than or on the order light's wavelength. Tunneling is a wave phenomenon.

6.4 Delta Function Potential Well and Barrier

The limit of a *delta function potential*,

$$V(x) = \pm V_0 a \delta(x), \tag{6.123}$$

can be approximated if I let the potential well or barrier width a go to zero while its amplitude V_0 goes to infinity, keeping the product $V_0 a$ constant. In Eq. (6.123) both V_0 and a are positive; the plus sign corresponds to a barrier and the minus sign to a potential well. From the nature of the solutions of the potential barrier or well problems, it follows that the wave function is continuous at the position of the delta function potential. On the other hand, the derivative of the wave function undergoes a jump. To see this, I start from the Schrödinger equation

$$\frac{d^2\psi(x)}{dx^2} = -\frac{2m}{\hbar^2}\left[E - V(x)\right]\psi(x) \tag{6.124}$$

and integrate about $x = 0$ to obtain

$$\begin{aligned}\left.\frac{d\psi(x)}{dx}\right|_{x=\epsilon} - \left.\frac{d\psi(x)}{dx}\right|_{x=-\epsilon} &= -\lim_{\epsilon\to 0}\frac{2m}{\hbar^2}\int_{-\epsilon}^{\epsilon}\left[E - V(x)\right]\psi(x)dx \\ &= -\lim_{\epsilon\to 0}\frac{2m\psi(0)}{\hbar^2}\int_{-\epsilon}^{\epsilon}\left[E - V(x)\right]dx \\ &= \pm\frac{2mV_0a\psi(0)}{\hbar^2}.\end{aligned} \tag{6.125}$$

The derivative of the wave function undergoes a point jump discontinuity at the position of the delta function potential.

6.4.1 Square Well with E < 0

As $a \to 0$ in the square well problem, the dimensionless strength parameter $\beta^2 = 2mV_0a^2/\hbar^2 \to 0$, since V_0a goes to a constant and $a \to 0$. If $\beta^2 \to 0$, the only bound state solution is the lowest energy, even parity solution. The energy is determined from Eq. (6.96),

$$E = -\frac{\hbar^2\beta^4}{8ma^2} = -\frac{mV_0^2a^2}{2\hbar^2}. \tag{6.126}$$

Now let's solve the problem directly. With $\kappa_E = \sqrt{-2mE/\hbar^2}$, the eigenfunction of the bound state can be taken as

$$\psi_E(x) = \begin{cases} Be^{\kappa_E x} & x < 0 \\ Be^{-\kappa_E x} & x > 0 \end{cases}, \tag{6.127}$$

which satisfies continuity of the wave function at $x = 0$. Using Eq. (6.125), I find

$$\begin{aligned}\left.\frac{d\psi(x)}{dx}\right|_{x=\epsilon} - \left.\frac{d\psi(x)}{dx}\right|_{x=-\epsilon} &= -2B\kappa_E = -2B\sqrt{-\frac{2mE}{\hbar^2}} \\ &= -\frac{2mV_0a\psi(0)}{\hbar^2} = -\frac{2mV_0aB}{\hbar^2}.\end{aligned} \tag{6.128}$$

Therefore,

$$-\frac{2mE}{\hbar^2} = \left(\frac{mV_0 a}{\hbar^2}\right)^2; \tag{6.129}$$

$$E = -\frac{mV_0^2 a^2}{2\hbar^2}, \tag{6.130}$$

in agreement with Eq. (6.126).

6.4.2 Barrier with $E > 0$

From Eq. (6.120), with $V_0 \gg E$ and $\kappa_E a \approx \beta \to 0$,

$$\mathcal{T} = \frac{1}{1 + \frac{V_0^2}{4E(V_0 - E)} \sinh^2(\kappa_E a)} \simeq \frac{1}{1 + \frac{V_0 \beta^2}{4E}} = \frac{1}{1 + \frac{mV_0^2 a^2}{2\hbar^2 E}}. \tag{6.131}$$

To solve the problem directly, I take

$$\psi_E(x) = \begin{cases} Ae^{ik_E x} + Be^{-ik_E x}; & x < 0 \\ Fe^{ik_E x} & x > 0 \end{cases}, \tag{6.132}$$

where $k_E = \sqrt{2mE/\hbar^2}$. The wave function is continuous at $x = 0$ and its derivative undergoes a jump discontinuity,

$$A + B = F; \tag{6.133}$$

$$\left.\frac{d\psi(x)}{dx}\right|_{x=\epsilon} - \left.\frac{d\psi(x)}{dx}\right|_{x=-\epsilon} = ik_E(F - A + B)$$
$$= \frac{2mV_0 a \psi(0)}{\hbar^2} = \frac{2mV_0 aF}{\hbar^2}, \tag{6.134}$$

which can be rewritten as

$$-\frac{B}{A} + \frac{F}{A} = 1; \tag{6.135}$$

$$\frac{B}{A} + \frac{F}{A}\left(1 + \frac{2imV_0 a}{k_E \hbar^2}\right) = 1. \tag{6.136}$$

Solving for the transmission coefficient, I find

$$\mathcal{T} = \left|\frac{F}{A}\right|^2 = \frac{4}{\left|2 + \frac{2imV_0 a}{k_E \hbar^2}\right|^2} = \frac{1}{1 + \frac{mV_0^2 a^2}{2\hbar^2 E}} = \frac{1}{1 + \left(\frac{\beta^2}{2k_E a}\right)^2}, \tag{6.137}$$

in agreement with Eq. (6.131). In the limit $\beta^2 V_0/4E = mV_0^2 a^2/2\hbar^2 E \gg 1$, $\mathcal{T} \sim 0$, which is the *strong barrier limit*. On the other hand, for $\beta^2 V_0/4E \ll 1$, the transmission goes to unity. If you consider the problem of transmission for a negative delta function potential, the transmission coefficient is unchanged, since $\mathcal{T}$ depends only on V_0^2.

6.5 Summary

I have examined a number of prototypical one-dimensional problems in quantum mechanics involving piecewise constant potentials. In all these problems, I was able to solve the Schrödinger equation in a number of distinct regions and piece together the solutions using the continuity of the wave function and its derivative. In considering the motion of particles in potentials that change abruptly at a given point, we always encounter wave-like properties of the particles since the potential changes in a distance small compared with the de Broglie wavelength of the particle. In the limit that $\hbar \to 0$ in such problems, we recover the geometrical or ray optics limit of optics for light incident on a dielectric interface. Processes such as transmission and reflection resonances, as well as tunneling, have optical analogues.

6.6 Appendix: Periodic Potentials

To arrive at the band structure of solids, one often models the problem of electrons interacting with atomic sites in a crystal by considering the electrons to move in a periodic potential having period d. If periodic boundary conditions are imposed, it is possible to find eigenfunctions $\psi_E(x)$ that satisfy Bloch's theorem,

$$\psi_E(x) = e^{i\alpha x/d} u_E(x), \tag{6.138}$$

where $u_E(x)$ is a periodic function having period d and α is the *Bloch phase.* Both α and $u_E(x)$ are functions of the energy E and the detailed nature of the potential. It is possible to relate the Bloch states to the transmission resonances that occur when matter waves are incident on an equally spaced array of identical potential barriers.[5] The problem can then be mapped onto one involving periodic boundary conditions by imposing the requirement that the wave function at the entrance of the array be equal to the wave function at the exit. I will consider only the problem of the transmission resonances in detail, but then make a connection with the Bloch states.

[5]For a more detailed discussion with references to earlier work, see P. R. Berman, *Transmission resonances and Bloch states from a periodic array of delta function potentials,* American Journal of Physics, **81**, 190–201 (2013).

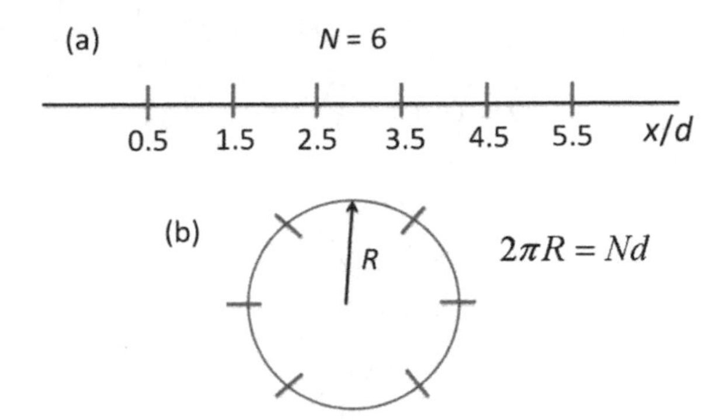

Fig. 6.16 (**a**) A finite array of delta function potentials on a line. (**b**) A periodic array of delta function potentials on a ring

To simplify the problem, I take the potential as

$$V(x) = V_0 a \sum_{j=1}^{N} \delta\left(x - \frac{2j-1}{2}d\right), \tag{6.139}$$

where V_0 and a are positive constants. The array has period d in the interval from $x = 0$ to $x = Nd$ with a delta function at the middle of each period [see Fig. 6.16a]. Periodic potentials with rectangular barriers constitute the so-called *Kronig-Penney model*. The wave function is taken as

$$\psi(x) = \begin{cases} A_0 e^{izx} + B_0 e^{-izx} = e^{izx} + R_N e^{-izx} & -\infty < x < 1/2 \\ A_n e^{iz(x-n)} + B_n e^{-iz(x-n)} & \left(n - \frac{1}{2}\right) < x < \left(n + \frac{1}{2}\right) \\ A_N e^{iz(x-N)} + B_N e^{-iz(x-N)} = T_N e^{iz(x-N)} & x > \left(N - \frac{1}{2}\right) \end{cases}, \tag{6.140}$$

where $z = kd$, $k = \sqrt{2mE}/\hbar$, and $1 \leq n \leq N - 1$. The quantity x has been redefined to be dimensionless, measured in units of the period d, and I have set $A_0 = 1$ and $B_0 = R_N$. With this choice, the quantities R_N and T_N are the reflection and transmission *amplitudes* for an N-period array, for a wave incident from the left.

By matching the wave function and its (discontinuous) derivative at $x = \left(n - \frac{1}{2}\right)$ using Eq. (6.140), I find

$$\begin{pmatrix} A_{n-1} \\ B_{n-1} \end{pmatrix} = \mathbf{M} \begin{pmatrix} A_n \\ B_n \end{pmatrix}, \tag{6.141}$$

where

$$\mathbf{M} = \begin{pmatrix} we^{-iz} & y \\ y^* & e^{iz}w^* \end{pmatrix}, \tag{6.142}$$

and

$$w = (1 + i\chi)\,; \quad y = i\chi, \tag{6.143}$$

with

$$\chi = z_0/z, \tag{6.144a}$$

$$z = kd, \tag{6.144b}$$

$$z_0 = \frac{mV_0 a}{\hbar^2} d. \tag{6.144c}$$

The matrix **M** is a *transfer matrix*. Had I considered potentials other than a delta function potential, the form of **M** would remain the same, but the values of w and y would change, subject to the constraint

$$|w|^2 - |y|^2 = 1, \tag{6.145}$$

provided by unitarity. Many of the equations are left in terms of w and y, but all calculations are performed for delta function potentials.

The transmission coefficient can now be calculated by writing

$$\begin{pmatrix} 1 \\ R_N \end{pmatrix} = \mathbf{M}_N \begin{pmatrix} T_N \\ 0 \end{pmatrix}, \tag{6.146}$$

with

$$\mathbf{M}_N = \mathbf{M}^N. \tag{6.147}$$

Thus, it is clear that

$$T_N = 1/\left(M_N\right)_{11}\,; \tag{6.148a}$$

$$R_N = \left(M_N\right)_{21} T_N = \left(M_N\right)_{21} / \left(M_N\right)_{11}. \tag{6.148b}$$

The amazing thing is that $\mathbf{M}_N$ can be calculated analytically for any N. It is possible to write the result for $\mathbf{M}_N$, T_N, and R_N in the compact form

$$\mathbf{M}_N = \frac{\mathbf{M}\sin(N\phi) - \mathbf{1}\sin[(N-1)\phi]}{\sin\phi}, \tag{6.149a}$$

$$\begin{aligned} \frac{1}{T_N} &= \frac{M_{11}\sin(N\phi) - \sin[(N-1)\phi]}{\sin\phi} \\ &= \frac{we^{-iz}\sin(N\phi) - \sin[(N-1)\phi]}{\sin\phi}, \end{aligned} \tag{6.149b}$$

$$R_N = \frac{M_{21}}{M_{11}} \left[1 + \frac{\sin [(N-1) \phi]}{\sin \phi} T_N \right]$$
$$= \frac{y^* e^{iz}}{w} \left[1 + \frac{\sin [(N-1) \phi]}{\sin \phi} T_N \right], \tag{6.149c}$$

where ϕ is defined by

$$\cos \phi = \mathrm{Re}\,(M_{11}) = \mathrm{Re}\,(1/T_1) = \mathrm{Re}\,(w \cos z) + \mathrm{Im}\,(w \sin z)\,, \tag{6.150}$$

and **1** is the 2×2 identity matrix. This completes the solution to the problem.

It is not difficult to calculate the energies for which the transmitted intensity is equal to unity using Eq. (6.149b). I need to find values $\phi = \phi_T$ for which

$$\frac{1}{T_N} = \frac{M_{11} \sin (N\phi_T) - \sin [(N-1) \phi_T]}{\sin \phi_T} = \pm 1. \tag{6.151}$$

For arbitrary M_{11} the solution is obtained by setting

$$\frac{\sin (N\phi_T)}{\sin \phi_T} = 0, \tag{6.152a}$$

$$-\frac{\sin [(N-1) \phi_T]}{\sin \phi_T} = \cos(N\phi_T) = \pm 1. \tag{6.152b}$$

These equations are satisfied if

$$\phi_T = q\pi/N; \qquad q = 1, 2, \ldots, N-1. \tag{6.153}$$

Note that q cannot be equal to zero or N since $\phi_T = 0$ and π are *not* solutions of Eq. (6.151). For each value of q, there is an infinite number of solutions of Eq. (6.150) for $z = kd$, with successive solutions corresponding to different *energy bands*. Since q can take on $N-1$ distinct values, there are $N-1$ transmission resonances in each band. The transmission amplitude at each *transmission resonance* is $T_N = (-1)^q$.

The transmission coefficient $\mathcal{T}_N = |T_N|^2$ is shown in Fig. 6.17 for $N = 10$ and $z_0 = 5$ as a function of $z/\pi = kd/\pi$. You can see that the transmission peaks are contained in bands which (almost) go over into the band structure of crystals. The upper band edge of the mth band is close, but not equal to $z = m\pi$. A blow-up of the first band is shown in Fig. 6.18 and the real part of the transmission amplitude for this band is shown in Fig. 6.19.

In general, there are $(N-1)$ transmission peaks contained in each band. Of the $N-1$ resonances, there are $(N-2)/2$ values where $T_N = 1$ when N is even and $(N-1)/2$ values where $T_N = 1$ when N is odd. The other resonances correspond to $T_N = -1$. I will return to this shortly.

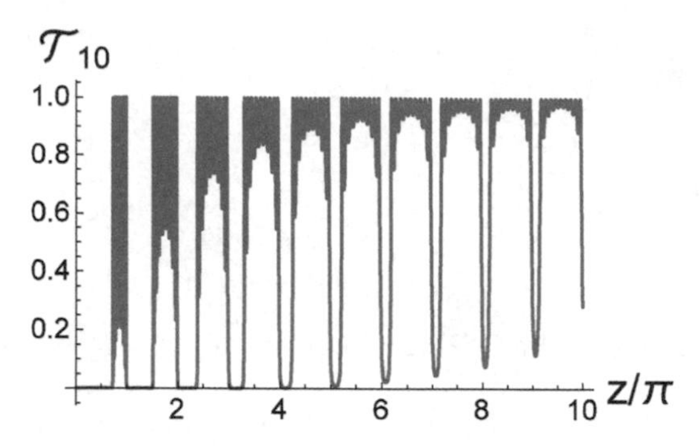

Fig. 6.17 Intensity transmission coefficient for a periodic array of 10 delta functions with $z_0 = 5$

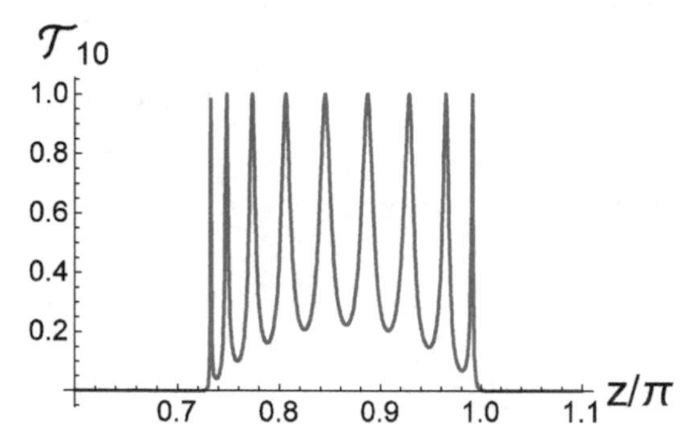

Fig. 6.18 Blow-up of the first "band" of the intensity transmission coefficient for a periodic array of 10 delta functions with $z_0 = 5$. There are $(N-1) = 9$ transmission resonances in each band

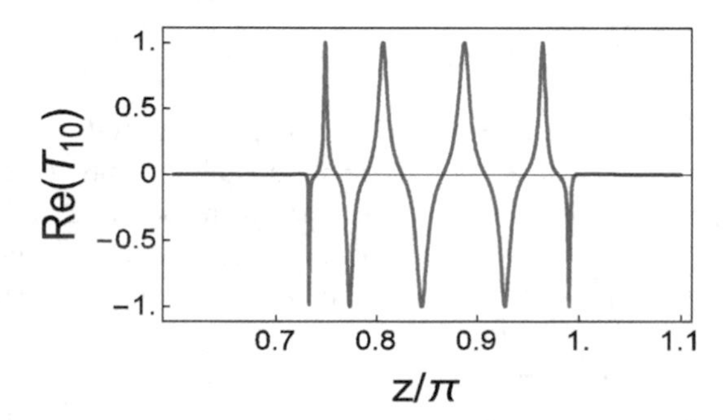

Fig. 6.19 Real part of the transmission *amplitude* in the first "band" of a periodic array of 10 delta functions with $z_0 = 5$. There are four resonances with $T = 1$ and five with $T = -1$

The net result is that, in the low energy bands where $\chi = z_0/m\pi \gg 1$, the transmission resonances are resolved and correspond to quasi-bound states of the "lattice." The quasi-resonances are confined to these energy bands and the width of the mth band, with upper band edge at $z \approx m\pi$, is of order $\Delta z = \Delta kd \sim 2m\pi/z_0$ for delta function potentials. On the other hand, for high energies, $\chi = z_0/m\pi \ll 1$, the transmission resonances are not resolved and form a series of bands with narrow gaps between the bands, having a gap width $\Delta z = \Delta kd \sim 2z_0/m\pi$ for delta function potentials. In these high energy bands (not yet seen in Fig. 6.17) $T_N \approx 1$. The high energy bands correspond to quasi-free particle states that can exist at these energies.

6.6.1 Bloch States

The Bloch theory of a crystal assumes a *periodic potential* in *all* space. Of course, no crystal exists in all space, so what is usually done is to imagine the periodic potential on a ring, that is, N identical delta function potentials, periodically placed on a ring [see Fig. 6.16b]. I can map the ring, having radius R, onto a line of length $L = 2\pi R = Nd$ with $\psi(x = 0) = \psi(x = L/d)$, and

$$\theta = 2\pi x/N. \tag{6.154}$$

The ring radius R grows with increasing N for fixed d.

It can be shown that the Bloch energies correspond to a *subset* of the solutions associated with Eq. (6.153), containing those values for which $T_N = 1$, but *not* those for $T_N = -1$. The wave functions with $T_N = -1$ are out of phase by π when they return to the same physical point. This implies that there are approximately *half* as many Bloch state energies as transmission energies, but each of these Bloch states is two-fold degenerate, since waves can move either clockwise or counterclockwise. Moreover, there are additional Bloch energies that occur that are absent in the transmission resonances, namely values of ϕ which are equal to zero or π.

As a consequence, the energy "bands," each contain $(N + 2)/2$ discrete energy levels for N even and $(N + 1)/2$ discrete energy levels for N odd. The discrete energies correspond to the transmission resonances for which $T_N = 1$, *plus* an energy eigenfunction for $\phi = 0$ and, if N is even, an additional one for $\phi = \pi$. Each state is two-fold degenerate, *except* for those corresponding to $\phi = 0$ or π. Thus there are $2[(N + 2)/2 - 2] + 2 = N$ states for N even and $2[(N + 1)/2 - 1] + 1 = N$ states for N odd. Each band contains exactly N states. Note that as $N \to \infty$, the number of states in each band goes to infinity but the states remain discrete. Despite the fact that these states remain discrete, the resulting structure is referred to as the (continuous) band structure of solids.

6.7 Problems

1. Show that solutions of

$$\frac{d^2f}{dx^2} = -k^2 f$$

are $e^{\pm ikx}$ and solutions of

$$\frac{d^2f}{dx^2} = k^2 f$$

are $e^{\pm kx}$.

2. Suppose you are given a constant potential V_0 in some region of space in a one-dimensional problem. For energies $E > V_0$ (classically allowed region), prove that possible solutions of the Schrödinger equation are $e^{ik'x}$, $e^{-ik'x}$, $\cos(k'x)$, $\sin(k'x)$, provided $k' = \sqrt{2m(E - V_0)/\hbar^2}$. For energies $E < V_0$ (classically forbidden region), prove that possible solutions of the Schrödinger equation are $e^{\kappa x}$, $e^{-\kappa x}$, $\cosh(\kappa x)$, $\sinh(\kappa x)$, provided $\kappa = \sqrt{2m(V_0 - E)/\hbar^2}$. In each case, why is the eigenfunction in that region a linear combination of at most two of the four solutions shown? What determines which linear combination and how many independent solutions are needed?

3. What is the difference in energy between the two lowest energy states of a 1.0 g particle moving in a 1.0 cm infinite square well potential? What does this tell you about measuring the energy spacing of the quantized states of a macroscopic particle?

4–5. At $t = 0$, the wave function for a particle of mass m in an infinite one-dimensional potential well located between $x = 0$ and $x = L$ is equal to

$$\psi(x,0) = \begin{cases} e^{-(x-L/2)^2/2x_0{}^2} & 0 < x < L \\ 0 & \text{otherwise} \end{cases},$$

which can be expanded as

$$\psi(x,0) = \begin{cases} \sqrt{2/L}\sum_{n=1}^{\infty} a_n \sin(n\pi x/L) & 0 < x < L \\ 0 & \text{otherwise} \end{cases},$$

where $x_0 << L$ [note that $\psi(x,0)$ is not normalized, but it is not important for this problem—it just gives an overall scaling factor to the a_n's]. In practice, you must cut off the sum at some value $n = n_{\max}$. Without formally solving the problem, estimate the value of $n_{\max}$ needed to provide a good approximation to $\psi(x,0)$. Now show that your estimate is reasonable by taking $L = 100$, $x_0 = 1$, and numerically integrating the appropriate equation to obtain the a_ns to find the maximum n that contributes significantly.

6–7. The k-state probability distribution for the eigenfunctions of the infinite square well having width L is given by [see Eq. (6.28)]

$$|\phi_n(k)|^2 = \frac{\pi L n^2 \left|1 - (-1)^n e^{-ikL}\right|^2}{\left[(kL)^2 - (n\pi)^2\right]^2}.$$

First, prove that the height of the lobes centered at $k = \pm n\pi/L$ in the limit that $n \gg 1$ is equal to $L/4\pi$. Next show that the half width at half maximum of each lobe is approximately equal to $2.79/L$ when $n \gg 1$. [Hint: Set $k = n\pi/L + \epsilon$ with $\epsilon \ll n\pi/L$ and solve the equation

$$|\phi_n(n\pi/L + \epsilon)|^2 = L/8\pi.$$

Note you *cannot* assume that $e^{iL\epsilon} \approx 1 + iL\epsilon$ since it will turn out that ϵ is of order $1/L$.]

8. For the potential step problem with energy $E > V_0$, consider the eigenfunction

$$\psi_E^R(x) = \begin{cases} \psi_{IE}^R(x) & x < 0 \\ \psi_{IIE}^R(x) & x > 0 \end{cases},$$

where

$$\psi_{IE}^R(x) = C_R e^{-ik_E x};$$
$$\psi_{IIE}^R(x) = A_R e^{-ik'_E x} + B_R e^{ik'_E x},$$

corresponds to a wave incident from the right. Calculate the amplitude reflection and transmission coefficients, R and T, and show that they can be obtained from those for the left incident wave by interchanging k_E and k'_E. Prove that the intensity reflection and transmission coefficients, $\mathcal{R}$ and $\mathcal{T}$, are unchanged. The quantities k_E and k'_E are defined in Eqs. (6.47) and (6.62), respectively.

9. One of the eigenfunctions for the step potential problem with energy $E > V_0$ is

$$\psi_{k_E}^L(x) = \begin{cases} \left(e^{ik_E x} + \frac{k_E - k'_E}{k_E + k'_E} e^{-ik_E x}\right) & x < 0 \\ \left(\frac{2k_E}{k_E + k'_E}\right) e^{ik'_E x} & x > 0 \end{cases} \qquad k_E > 0.$$

Calculate the probability current density of the entire wave function (do not break up the $x < 0$ part into incident and reflected waves) for $x < 0$ and $x > 0$ and show that the two results agree at $x = 0$. In fact, show that the current density is constant in all space. The quantities k_E and k'_E are defined in Eqs. (6.47) and (6.62), respectively.

10. The eigenfunctions for the step potential with energy $E < V_0$ is

$$\psi_{k_E}^L(x) = \begin{cases} \left(e^{ik_E x} + \frac{k_E - i\kappa_E}{k_E + i\kappa_E} e^{-ik_E x}\right) & x < 0 \\ \left(\frac{2k_E}{k_E + i\kappa_E}\right) e^{-\kappa_E x} & x > 0 \end{cases} \qquad k_E > 0.$$

Calculate the probability current density of the entire wave function and show that it vanishes in all space. The quantities k_E and κ_E are defined in Eqs. (6.47) and (6.50), respectively.

11. Obtain a graphical solution for the odd parity eigenenergies of the potential well problem. Find the minimum value of $\beta = \sqrt{\frac{2mV_0}{\hbar^2}a^2}$ needed to support an odd parity bound state.

12. Suppose a potential well having depth V_0 and width b is located inside of an *infinite* potential well having width $a > b$. Use a simple argument based on the uncertainty principle to derive an approximate condition for the existence of a bound state having $E < 0$. How does this problem differ from the one studied in the text, in which it was shown that a bound state always exists?

13–14. Now solve Problem 6.12 formally. Take both the finite well and infinite well centered at $x = 0$, such that the Hamiltonian commutes with the parity operator. Find the condition on V_0, b, and a, that will guarantee at least one bound state for $E < 0$. Show that in the limit that $a \to \infty$, there is always a bound state. Hint: What parity will the lowest energy state have? Choose a wave function that automatically satisfies the boundary condition at $x = a/2$.

15–18. In optics, when light is incident normally on a thin dielectric slab having index of refraction n and thickness d, the reflection and transmission coefficients are

$$\mathcal{R} = 1 - \mathcal{T};$$
$$\mathcal{T} = \frac{1}{\cos^2(kd) + \frac{\epsilon_n^2}{4}\sin^2(kd)},$$

where

$$\epsilon_n = n + \frac{1}{n},$$
$$k = nk_0,$$

and $k_0 = 2\pi/\lambda_0$ is the free-space propagation constant. Plot $\mathcal{T}$ as a function of $k_0 d$ for $n = 2$ and $n = 6$. Interpret your plots—that is, explain the positions of the maxima and minima in transmission on the basis of simple principles of optics.

In quantum mechanics, prove that the corresponding reflection and transmission coefficients for scattering of a particle having mass m by a potential barrier having length d and height V_0 are *identical*, provided one sets

$$n \to n_E = \sqrt{1 - V_0/E};$$
$$k \to k_E' = n_E k_E = n_E\sqrt{\frac{2mE}{\hbar^2}} = \sqrt{\frac{2m(E - V_0)}{\hbar^2}}.$$

It is assumed that $E > V_0$. The difference from the radiation problem is that as k_E is varied, the index n_E changes as well, whereas k_0 can be changed in the radiation problem without changing the index significantly. Thus it makes sense to define

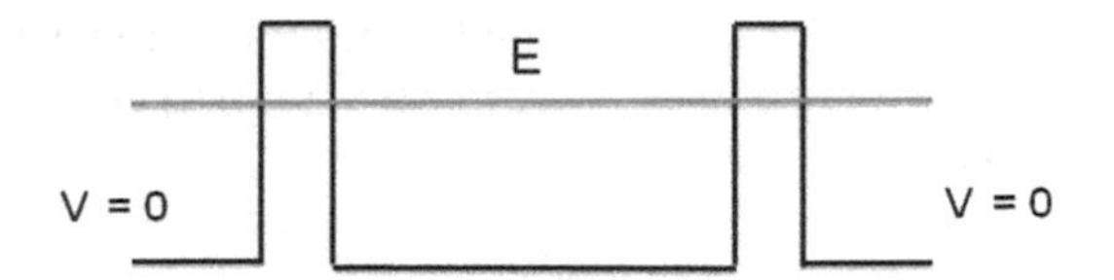

Fig. 6.20 Problem 6.21

$$k'_E d = \theta; \qquad \alpha = E/V_0;$$

$$\theta = \theta_0 \sqrt{\left(\frac{E - V_0}{V_0}\right)} = \theta_0 \sqrt{\alpha - 1};$$

$$\theta_0 = \sqrt{\frac{2mV_0}{\hbar^2}} d;$$

$$\mathcal{T} = \frac{1}{\cos^2 \theta + \frac{\epsilon_{nE}^2}{4} \sin^2 \theta},$$

so that the only parameters in the problem are θ_0 and $\alpha = E/V_0$. Plot $\mathcal{T}$ as a function of α for $\theta_0 = 10$, $\alpha = (1, 4)$ and $\theta_0 = 100$, $\alpha = (1, 1.05)$. Qualitatively how do the results differ from the radiation case? Note that with increasing θ_0 you would need better energy resolution in your incident beam to be able see the resonances. That is if you used too wide an energy bin, you could miss some of the resonances.

19. Calculate the transmission coefficient for a potential barrier having a height of 3 eV when a particle having energy 1 eV is incident, assuming the particle mass is 1 g and the barrier width is 1 cm.

20. Calculate the reflection and transmission coefficients for a particle having mass m scattered by the potential $V(x) = V_0 a\delta(x)$, where V_0 and a are positive constants.

21. For the double barrier potential shown in Fig. 6.20 it turns out that there are certain energies where the transmission is 100%, even though the transmission coefficient of *each* barrier is much less than unity. How can that be if it is very difficult for a particle incident from the left to tunnel into the space between the barriers and then tunnel out of the other side?

22. For a particle in an infinite potential well centered at the origin, prove that quantum revivals $[|\psi(x,t)|^2 = |\psi(x,0)|^2]$ occur for times t that are integral multiples of $t_r/8$ if the initial wave function is symmetric about the origin and for times t that are integral multiples of $t_r/4$ if the initial wave function is antisymmetric about the origin, where t_r is given by Eq. (6.44).

23. For a particle in an infinite potential well centered at the origin, prove that

$$|\psi(-x, t_r/2)|^2 = |\psi(x, 0)|^2$$

where t_r is given by Eq. (6.44). This result implies that there are quantum revivals at $t = t_r/2$ whenever the absolute square of the initial wave packet is either symmetric or antisymmetric about the origin.

6.7.1 Advanced Problems

1. There are two ways to calculate the expectation values of $\hat{k}^2$ and $\hat{k}^4$ $\left(\hat{k} = \hat{p}/\hbar\right)$ in an eigenstate of the particle in an infinite square well potential. One is to use $|\phi_n(k)|^2$ given in Problem 6.6–7, while the other is to use

$$\hat{k} = \frac{1}{i}\frac{d}{dx}$$

and work with the spatial eigenfunctions. Prove that both methods lead to the same value for $\left\langle \hat{k}^2 \right\rangle$. Now show that the coordinate space method leads to a finite value of $\left\langle \hat{k}^4 \right\rangle$, while the momentum space method yields an infinite result. This problem shows you that some care must be taken when the Fourier transform of the potential is not well defined. The momentum space method gives the correct (infinite) result.[6]

2. Solve Newton's equations of motion (that is, do not simply use energy conservation, solve for the dynamics) for a *classical* particle incident from the left on a potential step in one dimension. Consider both $E < V_0$ and $E > V_0$. Show that you arrive at results consistent with energy conservation. To solve this problem, replace the "step" by a ramp potential

$$V(x) = \lim_{a\to 0+} \begin{cases} V_0\,(x+a)\,/2a & -a < x < |a| \\ 0 & x \le -a \\ V_0 & x \ge a \end{cases},$$

where $\lim_{a\to 0+}$ means that a approaches zero from positive values. The slope of the ramp potential is $V_0/2a$ and approaches a potential step as $a \to 0^+$. Use this potential to solve Newton's equation for the position and velocity of the particle and then take the limit that $a \to 0^+$.

[6]See F. E. Cummings, *The particle in a box is not simple*, American Journal of Physics, Volume **45,** pp 158–160 (1977), who looks at the infinite well as the limit of a potential having steep walls. Alberto Rojo of Oakland University sent me an alternative calculation to prove that $\left\langle \hat{k}^4 \right\rangle$ diverges by considering the infinite well as the limit of a finite well whose depth is then allowed to go to infinity.

3. Normalize the wave function

$$\psi_k^L(x) = N \begin{cases} \left(e^{ikx} + \frac{k-d(k)}{k+d(k)} e^{-ikx}\right) & x < 0 \\ \frac{2k}{k+d(k)} e^{id(k)x} & x > 0 \end{cases} \qquad k > 0,$$

where

$$d(k) = \sqrt{\frac{2m(E - V_0)}{\hbar^2}} = \sqrt{k^2 - \frac{2mV_0}{\hbar^2}}.$$

To do this write

$$\psi_k^L(x) = N \lim_{\epsilon \to 0+} \begin{cases} \left(e^{ikx} + \frac{k-d(k)}{k+d(k)} e^{-ikx}\right) e^{\epsilon x/2} & x < 0 \\ \frac{2k}{k+d(k)} e^{id(k)x} e^{-\epsilon x/2} & x > 0 \end{cases} \qquad k > 0,$$

and find N such that

$$\int_{-\infty}^{\infty} dx \left[\psi_{k'}^L(x)\right]^* \psi_k^L(x) = \delta\left(k - k'\right)$$

in the limit that $\epsilon \to 0$ from positive values.

4. Consider a particle having mass m in the potential

$$V(x) = V_0 a \left[\delta(x) + \delta(x - b)\right],$$

where V_0, a, and b are positive constants. Moreover consider the limit $\beta^2/k_E a \gg 1$, where $k_E = \sqrt{2mE}/\hbar$, for which the transmission of a single barrier is small. Write the eigenfunctions as

$$\psi_E(x) = \begin{cases} e^{ik_E x} + Be^{-ik_E x} & x < 0 \\ Ce^{ik_E x} + De^{-ik_E x} & 0 \le x \le b \\ Fe^{ik_E x} & x > b \end{cases}.$$

Plot the transmission coefficient $\mathcal{T} = |F|^2$ as a function of $y = k_E b$ for $\beta' = 8$ and $0 \le y \le 20$ and show that $\mathcal{T} = 1$ when $y \approx q\pi$, for integer q. The quantity β' is defined by

$$\beta'^2 = \frac{2mV_0 ab}{\hbar^2} = \beta^2 \frac{b}{a}.$$

Also plot $|C|^2$ for the same values to show that the wave between the barriers is large compared to that outside the barriers at resonance. Interpret your result in terms of quasibound states between the barriers.

5. A particle having mass m moves in a one-dimensional potential

$$V(x) = \begin{cases} 0 & x < 0 \\ -V_0 < 0 & 0 < x < a \\ V_1 > 0 & x > a \end{cases} .$$

Write the general form of the eigenfunctions for bound states having $E < 0$ that satisfy the boundary conditions as $|x| \sim \infty$. Using the boundary conditions at $x = 0$ and $x = a$ obtain a single equation that could be solved graphically to obtain the eigenenergies. It simplifies the solution if you use sin and cos solutions for $0 < x < a$. Prove that, in the limit of a very weak well, $\beta_0^2 = 2mV_0a^2/\hbar^2 \ll 1$, a bound state exists only if $\beta_1 < \beta_0^2$, where $\beta_1^2 = \frac{2mV_1}{\hbar^2}a^2$.

Chapter 7
Simple Harmonic Oscillator: One Dimension

In the case of piecewise constant potentials, solving the Schrödinger equation was relatively easy. I obtained solutions in the various spatial regions where the potential was constant and matched the wave functions and their derivatives at places where the potential underwent a point jump discontinuity. In certain cases, additional boundary conditions had to be imposed for $x \sim \infty$ and/or $x \sim -\infty$. For *arbitrary* continuous potentials, the Schrödinger equation must be solved as an entity and, in general, such a solution must be carried out numerically. The numerical methods generally involve the use a discretized form of the Schrödinger equation, in which the kinetic energy operator and the potential are approximated on a finite grid of points. Different approaches can then be used to obtain the eigenergies and eigenfunctions.[1] For certain potentials such as the gravitational-like potential $V(x) = mgx$, the smooth potential well potential, $V(x) = -V_0 \operatorname{sech}^2(x/a)$, and the Morse potential (an anharmonic potential that is used to model intermolecular interactions),

$$V(x) = V_0 \left[1 - 2e^{-(x-x_0)/a}\right]^2, \tag{7.1}$$

it is possible to get solutions in terms of so-called *special functions* of mathematical physics. Special functions refer to quantities such as Bessel, Laguerre, hypergeometric, or Hermite functions that have been studied extensively by mathematicians. In this chapter I study the harmonic oscillator potential. As you will see, an analytic form for the eigenfunctions of the harmonic oscillator can be obtained in terms of *Hermite polynomials*.

[1]See, for example, Mohandas Pillai, Joshua Goglio, and Thad Walker, *Matrix Numerov method for solving Schrödinger's equation*, American Journal of Physics **80**, 1017–1019 (2012), and the references therein. See, also, Paolo Giannozzi, *Lecture notes Numerical Methods in Quantum Mechanics*, at http://www.fisica.uniud.it/~giannozz/Corsi/MQ/mq.html.

P.R. Berman, *Introductory Quantum Mechanics*, UNITEXT for Physics,
https://doi.org/10.1007/978-3-319-68598-4_7

The potential for a simple harmonic oscillator (SHO) associated with a particle having mass m subjected to a restoring force $-\sqrt{m\omega^2}x\mathbf{u}_x$ can be written as

$$V(x) = \frac{1}{2}m\omega^2 x^2. \tag{7.2}$$

Aside from this being the potential characterizing a particle bound by an ideal spring, it is the *approximate* potential for any interaction potential that has a point of stable equilibrium located at $x = 0$, since, in the region about $x = 0$,

$$V(x) \approx V(0) + \frac{1}{2!}\left.\frac{d^2V}{dx^2}\right|_{x=0} x^2. \tag{7.3}$$

Near a point of stable equilibrium $d^2V/dx^2\big|_{x=0}$ can be identified with $m\omega^2$ of the equivalent problem of a particle having mass m moving in a SHO potential. For example, in *optical lattices*, standing wave laser fields are used to trap atoms at the bottom of potential wells that can be approximated as harmonic oscillator potentials. The interaction potential between the nuclei of diatomic molecules can also be approximated by a harmonic oscillator potential, giving rise to *vibrational energy levels*.

7.1 Classical Problem

Most likely, you have already studied the dynamics of a classical, simple harmonic oscillator. The harmonic oscillator potential is amazing in that the motion is periodic with (angular) frequency ω, no matter how you start the oscillator. It returns to its initial position and velocity at all integral multiples of its period, $T = 2\pi/\omega$. The particle spends the least amount of time near $x = 0$ since it is moving fastest there and the most amount of time near the endpoints of its orbit,

$$\pm x_{\text{max}} = \pm\sqrt{\frac{2E}{m\omega^2}}, \tag{7.4}$$

where E is the energy of the oscillator.

The position of the oscillator as a function of time is given by

$$x = x_{\text{max}} \cos(\omega t), \tag{7.5}$$

assuming the particle starts with maximum displacement. The time-averaged probability density $P_{\text{class}}(x)$ to find the particle between x and $x + dx$ is just the fraction of a half period that the particle is located between x and $x + dx$, namely

$$P_{\text{class}}(x)dx = \left|\frac{dt}{\pi/\omega}\right| = \left|\frac{\omega\,(dt/dx)\,dx}{\pi}\right|. \tag{7.6}$$

To find dt/dx, I use the equation for conservation of energy

$$\frac{1}{2}m\left(\frac{dx}{dt}\right)^2 + \frac{1}{2}m\omega^2 x^2 = E \tag{7.7}$$

and solve for

$$\frac{dt}{dx} = \frac{1}{\frac{dx}{dt}} = \frac{1}{\sqrt{\frac{2E}{m} - \omega^2 x^2}} \tag{7.8}$$

to obtain

$$P_{\text{class}}(x) = \frac{1}{\pi\sqrt{x_{\text{max}}^2 - x^2}}, \tag{7.9}$$

provided $|x| \leq x_{\text{max}}$. For $|x| > x_{\text{max}}$, $P_{\text{class}}(x) = 0$.

As predicted, the probability density is greatest (actually infinite) at the endpoints of the motion ($x = \pm x_{\text{max}}$), and smallest at the equilibrium position ($x = 0$). The *momentum* probability density is smallest at $x = \pm x_{\text{max}}$ (where $p = 0$) and a maximum at $x = 0$ (where $p = \pm |p_{\text{max}}|$). In fact, owing to the symmetry of the Hamiltonian on exchange of momentum and position when both are expressed in dimensionless variables [see Eq. (7.13) below], the time-averaged momentum probability density $W_{\text{class}}(p)$ has the same form as $P_{\text{class}}(x)$, namely

$$W_{\text{class}}(p) = \frac{1}{\pi\sqrt{p_{\text{max}}^2 - p^2}}, \tag{7.10}$$

provided $|p| \leq p_{\text{max}} = \sqrt{2mE}$. For $|p| > p_{\text{max}}$, $W_{\text{class}}(p) = 0$. The minimum possible energy of a classical particle in the well is equal to zero.

7.2 Quantum Problem

The Hamiltonian for the quantum problem is

$$\hat{H} = \frac{\hat{p}^2}{2m} + \frac{1}{2}m\omega^2\hat{x}^2. \tag{7.11}$$

It is useful to introduce dimensionless coordinate and momentum variables defined by

$$\hat{\xi} = \sqrt{\frac{m\omega}{\hbar}}\hat{x}; \tag{7.12a}$$

$$\hat{\eta} = \sqrt{\frac{1}{\hbar m\omega}}\hat{p} = \frac{1}{i}\frac{d}{d\xi}, \tag{7.12b}$$

such that

$$\hat{H} = \frac{1}{2}\hbar\omega\left(\hat{\eta}^2 + \hat{\xi}^2\right), \tag{7.13}$$

which is obviously symmetric between the dimensionless momentum and coordinate variables. The commutator of $\hat{\xi}$ and $\hat{\eta}$ is

$$\left[\hat{\xi}, \hat{\eta}\right] = i. \tag{7.14}$$

If I measure energy in units of $\hbar\omega$, I can define a dimensionless Hamiltonian operator $\hat{H}'$ by

$$\hat{H}' = \frac{\hat{H}}{\hbar\omega} = \frac{1}{2}\left(\hat{\eta}^2 + \hat{\xi}^2\right). \tag{7.15}$$

We already know a great deal about the solutions of the oscillator problem. The energy levels are discrete and there is no energy degeneracy. Since the Hamiltonian commutes with the parity operator, the eigenfunctions must also be eigenfunctions of the parity operator, that is, they must be either even or odd functions of x. Moreover we expect the ground state wave function to be a symmetric bell-shaped curve centered at $x = 0$ (though *not* a sin function), the first excited state wave function to be an antisymmetric function with a node at the origin, the second excited state wave function to be symmetric about the origin with two nodes, etc. Without loss of generality I can label the lowest energy state by $n = 0$, the first excited state by $n = 1$, etc.

A lower bound for the ground state energy can be obtained using the uncertainty principle. Since the eigenstates have definite parity, it is clear that

$$\langle\hat{\eta}\rangle_n = \left\langle\hat{\xi}\right\rangle_n = 0 \tag{7.16}$$

for any dimensionless eigenfunction $\tilde{\psi}_n(\xi)$ of $\hat{H}'$. The notation used is

$$\left\langle\hat{O}\right\rangle_n = \int_{-\infty}^{\infty}\left[\tilde{\psi}_n(\xi)\right]^*\hat{O}\tilde{\psi}_n(\xi)d\xi, \tag{7.17}$$

where $\hat{O}$ is some arbitrary operator. From Eqs. (7.15)–(7.17), I can calculate

$$\left\langle\hat{H}'\right\rangle_0 = \frac{1}{2}\left\langle\hat{\eta}^2 + \hat{\xi}^2\right\rangle_0 = \frac{1}{2}\left(\Delta\eta^2 + \Delta\xi^2\right), \tag{7.18}$$

where $\Delta\xi^2$ is the variance of $\hat{\xi}$ and $\Delta\eta^2$ is the variance of $\hat{\eta}$ in the $n = 0$ state. It follows from Eqs. (5.94) and (7.14) that

$$\Delta\eta^2 \geq \frac{1}{4\Delta\xi^2}, \tag{7.19}$$

implying that

$$\epsilon = \frac{E}{\hbar\omega} \geq \frac{1}{2}\left(\frac{1}{4\Delta\xi^2} + \Delta\xi^2\right). \tag{7.20}$$

The minimum value of the right-hand side of this expression occurs for $\Delta\xi^2 = 1/2$; consequently

$$\epsilon \geq \frac{1}{2}\left(\frac{1}{2} + \frac{1}{2}\right) = \frac{1}{2}; \tag{7.21a}$$

$$E \geq \frac{\hbar\omega}{2}. \tag{7.21b}$$

We will see that $\hbar\omega/2$ is the *exact* ground state energy, since the ground state eigenfunction turns out to be a Gaussian (recall that the *only* minimum uncertainty wave function is a Gaussian).

I can also deduce the eigenenergies (to within a constant) by demanding that $\left|\tilde{\psi}(\xi,t)\right|^2$ is a periodic function of time having period $T = 2\pi/\omega$. Since

$$\left|\tilde{\psi}(\xi,t)\right|^2 = \sum_{n,n'} a_n a_{n'}^* \tilde{\psi}_n(\xi)\left[\tilde{\psi}_{n'}(\xi)\right]^* \exp\left[-i\left(E_n - E_{n'}\right)t/\hbar\right], \tag{7.22}$$

periodicity requires that

$$\left(E_n - E_{n'}\right)\left(2\pi/\omega\right)/\hbar = 2\pi q, \tag{7.23}$$

where q is an integer. This implies that

$$E_n = \hbar\omega\left(n + C\right), \tag{7.24}$$

where n is a positive integer or zero and C is a constant that is greater than or equal to $1/2$ [which follows from Eq. (7.21b)]. Equation (7.23) could also be satisfied if $E_n = \hbar\omega\left(n^m + C\right)$ for a positive integer m, but it is not hard to argue that m must be equal to one. We have already seen that the infinite square well energy levels vary as n^2. Since the SHO potential is less steep than the infinite well, its energy levels must vary with n less rapidly than n^2; the only integral value of m that will work is $m = 1$.

7.2.1 Eigenfunctions and Eigenenergies

For the oscillator problem the Schrödinger equation in dimensionless variables can be written as

$$\frac{1}{2}\left(-\frac{d^2}{d\xi^2}+\xi^2\right)\tilde{\psi}_n(\xi)=\epsilon_n\tilde{\psi}_n(\xi), \tag{7.25}$$

where

$$\epsilon_n=\frac{E_n}{\hbar\omega}. \tag{7.26}$$

Thus, I must solve

$$\frac{d^2\tilde{\psi}_n(\xi)}{d\xi^2}+\left(2\epsilon_n-\xi^2\right)\tilde{\psi}_n(\xi)=0. \tag{7.27}$$

As I often stress, you can solve a differential equation only if you know the solution. However you can make some progress towards a solution by building in the asymptotic form of the wave functions. We know the solution must go to zero as $|x|\rightarrow\infty$. In the case of a square well potential, the bound state eigenfunctions vary as $\exp(-\kappa|x|)$ for large $|x|$. Since the oscillator, potential increases with increasing $|x|$, we would expect a faster fall-off for the eigenfunctions.

For $|\xi|\gg 1$, I approximate Eq. (7.27) as

$$\frac{d^2\tilde{\psi}_n(\xi)}{d\xi^2}-\xi^2\tilde{\psi}_n(\xi)=0 \tag{7.28}$$

and guess a solution, $\tilde{\psi}_n(\xi)=e^{-a\xi^2}$. Then

$$\begin{aligned}\frac{d\tilde{\psi}_n(\xi)}{d\xi}&=-2a\xi e^{-a\xi^2};\\ \frac{d^2\tilde{\psi}_n(\xi)}{d\xi^2}&=4a^2\xi^2e^{-a\xi^2}-2ae^{-a\xi^2}\\ &\approx 4a^2\xi^2e^{-a\xi^2}=4a^2\xi^2\tilde{\psi}_n(\xi).\end{aligned} \tag{7.29}$$

If $a=1/2$, $\tilde{\psi}_n(\xi)=e^{-a\xi^2}$ is an approximate solution of Eq. (7.27) for $|\xi|\gg 1$. I build this dependence into the overall solution by setting

$$\tilde{\psi}_n(\xi)=e^{-\xi^2/2}H_n(\xi), \tag{7.30}$$

where $H_n(\xi)$ is a function to be determined.

I already know that $H_n(\xi)$ must be an even or odd function of ξ since the eigenfunctions are also eigenfunctions of the parity operator. I also know that the lowest energy wave function has no node in the classically allowed region, the second energy level has one node, etc. This implies that $H_n(\xi)$ could be a *polynomial* of order n.

Using Eq. (7.30) and substituting

$$\frac{d^2\tilde{\psi}_n(\xi)}{d\xi^2} = -2H_n'(\xi)\xi e^{-\xi^2/2} - H_n(\xi)e^{-\xi^2/2} + \xi^2 H_n(\xi)e^{-\xi^2/2} + H_n''(\xi)e^{-\xi^2/2} \tag{7.31}$$

into Eq. (7.27), I am led to the following differential equation for $H_n(\xi)$:

$$H_n''(\xi) - 2\xi H_n'(\xi) + (2\epsilon_n - 1) H_n(\xi) = 0, \tag{7.32}$$

where the primes indicate differentiation with respect to ξ. This is a well-known (to those who know it well) equation of mathematical physics—Hermite's differential equation. It admits polynomial solutions only if

$$2\epsilon_n = 2n + 1, \tag{7.33}$$

where n is a non-negative integer. The polynomial solutions are the only physically acceptable solutions of Hermite's equation that need concern us.[2] In *Mathematica,* the polynomial solutions of Hermite's equation,

$$H_n''(\xi) - 2\xi H_n'(\xi) + 2nH_n(\xi) = 0, \tag{7.34}$$

are designated as HermiteH[n, ξ]. Note that by limiting the solution to polynomials, I already have determined the eigenenergies, since it follows from Eq. (7.33) that

$$\epsilon_n = n + \frac{1}{2}; \qquad n = 0, 1, 2, \ldots. \tag{7.35}$$

$$E_n = \hbar\omega\left(n + \frac{1}{2}\right); \qquad n = 0, 1, 2, \ldots . \tag{7.36}$$

The energy levels are equally spaced with spacing $\hbar\omega$, a result I had already predicted based on the periodicity of the solution.

[2]The series solutions given in Eqs. (7.41) and (7.42) lead to divergent wave functions as $\xi \sim \pm\infty$ for non-integer n; however, solutions of Eq. (7.27) *do* exist that are regular as $\xi \sim \infty$ (see problems).

You can guess the first few polynomial solutions of Eq. (7.34) with little effort:

$$H_0(\xi) = 1; \tag{7.37}$$

$$H_1(\xi) = 2\xi; \tag{7.38}$$

$$H_2(\xi) = 4\xi^2 - 2; \tag{7.39}$$

$$H_3(\xi) = 8\xi^3 - 12\xi. \tag{7.40}$$

[The Hermite polynomials $H_n(\xi)$ are defined with the convention that the coefficient of the highest power of ξ is 2^n]. Others can be obtained from recursion relations, a series solution, or the so-called generating function. Hermite polynomials are one of a class of *orthogonal polynomials* that can all be treated by similar methods.

I now list several useful properties of the Hermite polynomials. The series solution for the Hermite polynomials for n even is

$$H_n(\xi) = \frac{(-1)^{n/2}\, n!}{(n/2)!}\left[1 - 2\frac{n}{2!}\xi^2 + 2^2\frac{n(n-2)}{4!}\xi^4 - 2^3\frac{n(n-2)(n-4)}{6!}\xi^6 + \cdots\right] \tag{7.41}$$

and for n odd is

$$H_n(\xi) = \frac{(-1)^{(n-1)/2}\, 2n!}{[(n-1)/2]!}\left[\xi - 2\frac{(n-1)}{3!}\xi^3 + 2^2\frac{(n-1)(n-3)}{5!}\xi^5 - \cdots\right]; \tag{7.42}$$

$H_n(\xi)$ is a polynomial of order n having even parity if n is even and odd parity if n is odd. Some recursion relations are:

$$H_n'(\xi) = 2nH_{n-1}(\xi); \tag{7.43a}$$

$$H_{n+1}(\xi) + 2nH_{n-1}(\xi) = 2\xi H_n(\xi); \tag{7.43b}$$

$$H_{n+1}(\xi) = 2\xi H_n(\xi) - H_n'(\xi). \tag{7.43c}$$

The last of these equations allows you to calculate $H_{n+1}(\xi)$ from $H_n(\xi)$ and $H_n'(\xi)$; in other words, it lets you construct all the Hermite polynomials starting from $H_0(\xi) = 1$. The recursion relations will prove very useful in calculating integrals that are needed in perturbation theory involving oscillators.

The Hermite polynomials are orthogonal (the *wave functions* must be orthogonal since there is no degeneracy) if a weighting factor is used in the integrand, namely

$$\int_{-\infty}^{\infty} e^{-\xi^2} H_n(\xi) H_m(\xi) d\xi = 2^n n! \sqrt{\pi}\,\delta_{n,m}. \tag{7.44}$$

Moreover there is a *generating function* for the Hermite polynomials: A generating function is an analytic function that can be expressed as a power series of some variable multiplied by the corresponding orthogonal polynomial. For the Hermite polynomials, the generating function is

$$F(\xi, q) = e^{-q^2+2q\xi} = \sum_{n=0}^{\infty} \frac{q^n H_n(\xi)}{n!}. \tag{7.45}$$

To evaluate the $H_n(\xi)$, the exponential is expanded and compared term by term with the series. The generating function can be used to evaluate integrals such as the one appearing in Eq. (7.44).

The normalized eigenfunctions for the SHO are

$$\tilde{\psi}_n(\xi) = \frac{1}{\sqrt{2^n n! \sqrt{\pi}}} e^{-\xi^2/2} H_n(\xi); \tag{7.46a}$$

$$\psi_n(x) = \frac{1}{\left(\frac{\hbar}{m\omega}\right)^{1/4} \sqrt{2^n n! \sqrt{\pi}}} e^{-\frac{m\omega}{2\hbar}x^2} H_n\left(\sqrt{\frac{m\omega}{\hbar}} x\right). \tag{7.46b}$$

It is often useful to rewrite the recursion relations given in Eqs. (7.43) in terms of the wave function, namely

$$\sqrt{2}\frac{d\tilde{\psi}_n(\xi)}{d\xi} = \sqrt{n}\tilde{\psi}_{n-1}(\xi) - \sqrt{n+1}\tilde{\psi}_{n+1}(\xi); \tag{7.47a}$$

$$\sqrt{2}\xi\tilde{\psi}_n(\xi) = \sqrt{n+1}\tilde{\psi}_{n+1}(\xi) + \sqrt{n}\tilde{\psi}_{n-1}(\xi); \tag{7.47b}$$

$$\sqrt{2(n+1)}\tilde{\psi}_{n+1}(\xi) = \xi\tilde{\psi}_n(\xi) - \frac{d\tilde{\psi}_n(\xi)}{d\xi}. \tag{7.47c}$$

The first few dimensionless eigenfunctions are

$$\tilde{\psi}_0(\xi) = \frac{1}{\pi^{1/4}} e^{-\xi^2/2}; \tag{7.48}$$

$$\tilde{\psi}_1(\xi) = \frac{\sqrt{2}\xi}{\pi^{1/4}} e^{-\xi^2/2}; \tag{7.49}$$

$$\tilde{\psi}_2(\xi) = \frac{\left(2\xi^2 - 1\right)}{\pi^{1/4}\sqrt{2}} e^{-\xi^2/2}; \tag{7.50}$$

$$\tilde{\psi}_3(\xi) = \frac{\left(2\xi^3 - 3\xi\right)}{\pi^{1/4}\sqrt{3}} e^{-\xi^2/2}. \tag{7.51}$$

These are graphed in Fig. 7.1. The ground state eigenfunction is a Gaussian, implying that it represents a minimum uncertainty state. It is obvious from the form of the Hamiltonian that the energy is shared equally between the kinetic ($\eta^2/2$) and potential ($\xi^2/2$) energy. Thus, in an eigenstate,

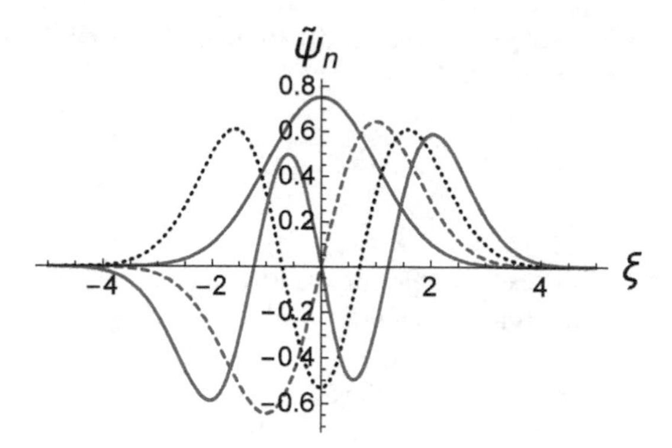

Fig. 7.1 Dimensionless oscillator wave functions, $\tilde{\psi}_n(\xi)$, for $n = 0$ (red, solid), $n = 1$ (blue, dashed), $n = 2$ (black, dotted), and $n = 3$ (brown, solid)

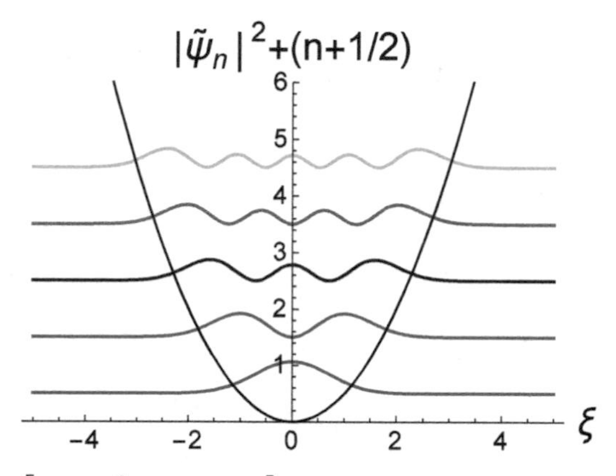

Fig. 7.2 Graphs of $\left[\left|\tilde{\psi}_n(\xi)\right|^2 + (n+1/2)\right]$ as a function of ξ for $n = 0 - 4$. The harmonic oscillator potential is superposed on the plot

$$\left\langle \hat{\xi}^2 \right\rangle_n = \left\langle \hat{\eta}^2 \right\rangle_n = \left(n + \frac{1}{2} \right). \tag{7.52}$$

In Fig. 7.2, I plot $\left[\left|\tilde{\psi}_n(\xi)\right|^2 + (n+1/2)\right]$ as a function of ξ for $n = 0 - 4$. Each curve is displaced by the corresponding eigenenergy so that you can see the probability distributions relative to the potential at the appropriate energy. The probability distributions oscillate in the classically allowed region and fall off exponentially in the classically forbidden region.

Finally I look at the eigenfunctions in the large n limit, when we would expect the spatially averaged probability density to approach the classical probability distribution in the classically allowed regime. To make a connection between the classical and quantum problems I set $x = \sqrt{\frac{\hbar}{m\omega}}\xi$, and $E_{\text{class}} = (n + 1/2)\hbar\omega$ in Eq. (7.9) to arrive at the dimensionless probability distribution,

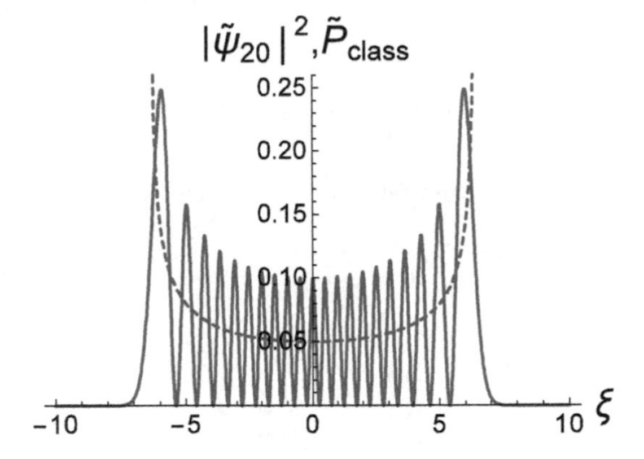

Fig. 7.3 Graphs of the dimensionless classical (blue, dashed) and quantum (red, solid) probability distributions for $n = 20$. The quantum distribution corresponds to the $n = 20$ eigenfunction, while the classical distribution corresponds to an energy $E = 20.5\hbar\omega$

$$\tilde{P}_{\text{class}}(\xi) = \sqrt{\frac{\hbar}{m\omega}} P_{\text{class}}(x) = \frac{1}{\pi\sqrt{2n + 1 - \xi^2}}, \tag{7.53}$$

which is plotted in Fig. 7.3 along with $\left|\tilde{\psi}_n(\xi)\right|^2$ for $n = 20$. You can see that the value of $\left|\tilde{\psi}_{20}(\xi)\right|^2$, averaged over oscillations, is approximately equal to $\tilde{P}_{\text{class}}(\xi)$ in the classically allowed regime.

7.2.2 *Time-Dependent Problems*

Having solved for the eigenfunctions and eigenvalues, I can construct the time-dependent solution, given some initial condition. The general time-dependent solution is

$$\psi(x, t) = \sum_n b_n e^{-i\left(n+\frac{1}{2}\right)\omega t} \psi_n(x). \tag{7.54}$$

The solution for $|\psi(x, t)|^2$ is periodic with period $T = 2\pi/\omega$, as in the classical problem (the solution for $\psi(x, t)$ is periodic with period $T = 4\pi/\omega$). In contrast to the infinite square well problem, these revivals can be viewed as *classical* in nature since the revival times are integral multiples of the oscillator period, independent of $\hbar$. The dimensionless wave function is

$$\tilde{\psi}(\xi, t) = e^{-i\omega t/2} \sum_{n=0}^{\infty} b_n e^{-in\omega t} \tilde{\psi}_n(\xi), \tag{7.55}$$

with

$$b_n = \int_{-\infty}^{\infty} d\xi \tilde{\psi}_n(\xi) \tilde{\psi}(\xi, 0). \tag{7.56}$$

One especially interesting case occurs for an initial wave function

$$\tilde{\psi}(\xi, 0) = \frac{1}{\pi^{1/4}} e^{-(\xi-\xi_0)^2/2}, \tag{7.57}$$

corresponding to the ground state wave function displaced by ξ_0. For this initial wave function,

$$b_n = \frac{1}{\sqrt{2^n n! \pi}} \int_{-\infty}^{\infty} d\xi e^{-\xi^2/2} H_n(\xi) e^{-(\xi-\xi_0)^2/2} = \frac{e^{-\xi_0^2/4} \xi_0^n}{\sqrt{2^n n!}}, \tag{7.58}$$

where the integral could be carried out using the generating function or a table of integrals. Thus, $\tilde{\psi}(\xi, 0)$ can be expanded as

$$\begin{aligned} \tilde{\psi}(\xi, 0) &= e^{-\xi_0^2/4} \sum_{n=0}^{\infty} \frac{\xi_0^n}{\sqrt{2^n n!}} \tilde{\psi}_n(\xi) \\ &= e^{-(\xi_0/\sqrt{2})^2/2} \sum_n \frac{\left(\xi_0/\sqrt{2}\right)^n}{\sqrt{n!}} \tilde{\psi}_n(\xi). \end{aligned} \tag{7.59}$$

This wave function is referred to as a *coherent state wave function* and appears in quantum optics, as well. The coherent state wave function has special properties that I will return to when I consider the oscillator using ladder operators and Dirac notation in Chap. 11. From Eqs. (7.55) and (7.58), it follows that

$$\begin{aligned} \tilde{\psi}(\xi, t) &= e^{-i\omega t/2} e^{-\xi_0^2/4} \sum_n \frac{\xi_0^n}{\sqrt{2^n n!}} e^{-in\omega t} \tilde{\psi}_n(\xi) \\ &= e^{-i\omega t/2} \frac{e^{-\xi_0^2/4}}{\pi^{1/4}} e^{-\xi^2/2} \sum_n \frac{1}{n!} \left(\frac{\xi_0}{2} e^{-i\omega t} \right)^n H_n(\xi). \end{aligned} \tag{7.60}$$

The sum is exactly that encountered with the generating function (7.45), so

$$\begin{aligned} \tilde{\psi}(\xi, t) &= e^{-i\omega t/2} \frac{e^{-\xi_0^2/4}}{\pi^{1/4}} e^{-\xi^2/2} \exp\left[2\left(\frac{\xi_0}{2} e^{-i\omega t} \right) \xi - \left(\frac{\xi_0}{2} e^{-i\omega t} \right)^2 \right] \\ &= e^{-i\omega t/2} \frac{e^{-\xi_0^2/4}}{\pi^{1/4}} e^{\xi^2/2} \exp\left\{ -\left[\xi - \left(\frac{\xi_0}{2} e^{-i\omega t} \right) \right]^2 \right\}; \end{aligned} \tag{7.61a}$$

$$\begin{aligned}\left|\tilde{\psi}\left(\xi,t\right)\right|^{2} &= \frac{e^{-\xi_{0}^{2}/2}}{\pi^{1/2}}e^{\xi^{2}}\exp\left[-2\xi^{2}+2\xi\xi_{0}\cos\left(\omega t\right)-\xi_{0}^{2}\cos\left(2\omega t\right)/2\right] \\ &= \frac{1}{\pi^{1/2}}\exp\left[-\xi^{2}+2\xi\xi_{0}\cos\left(\omega t\right)-\frac{\xi_{0}^{2}}{2}\left[1+\cos\left(2\omega t\right)\right]\right] \\ &= \frac{1}{\pi^{1/2}}\exp\left[-\xi^{2}+2\xi\xi_{0}\cos\left(\omega t\right)-\xi_{0}^{2}\cos^{2}\left(\omega t\right)\right] \\ &= \frac{1}{\pi^{1/2}}e^{-\left[\xi-\xi_{0}\cos(\omega t)\right]^{2}}. \end{aligned} \tag{7.61b}$$

The wave packet oscillates in the potential *without changing its envelope*. This is referred to as a *coherent state* since it mimics the behavior of a classical particle. Any spreading of the wave packet resulting from the various momentum components in the packet is exactly compensated by the forces acting on the particle.

7.3 Summary

In obtaining the eigenfunctions and eigenvalues of the SHO, I have solved one of the most important elementary problems in quantum mechanics. Many potentials are modeled as harmonic oscillator potentials, so these solutions are used in a wide range of applications. I did not go through a detailed derivation of the series solution of Hermite's equation, since it can be found in any standard mathematical physics text. Basically one assumes a series solution, obtains the recursion relation for the coefficients, and argues that the series must terminate, or the solution would diverge for large $|x|$. In this way, you obtain the quantization condition.

7.4 Problems

1. How do you know the eigenfunctions of the 1-D oscillator must be even or odd? How many nodes are there for $\tilde{\psi}_n(\xi)$? Based on the general solution of the time-dependent Schrödinger equation and properties of the simple harmonic oscillator, give an argument to show that the frequency difference between any two energy levels must be an integer times ω, where ω is the oscillator frequency. In general, to form a wave packet that corresponds to a particle having energy $E = 1$ J moving in a well having frequency 1 Hz, how many states would be needed?
2. Write the normalized eigenfunctions $\tilde{\psi}_n(\xi)$ of the harmonic oscillator and plot the first four normalized eigenfunctions.

3. For a dimensionless potential $V(\xi)$ varying as

$$V(\xi) = A\,|\xi|^{\mu}\,,$$

where ξ is a dimensionless variable, the (dimensionless) Hamiltonian is

$$\hat{H} = -\frac{1}{2}\frac{d^2}{d\xi^2} + A\,|\xi|^{\mu}\,.$$

For large positive ξ show that the asymptotic form of the wave function is

$$\tilde{\psi}\,(\xi) \sim \exp\left[-a\xi^{\left(\frac{\mu}{2}+1\right)}\right],$$

where a is a constant that depends on μ and A. For $\mu = 0, 2$, show the result agrees with what was obtained for piecewise potentials and the harmonic oscillator, respectively.

4. The *Virial Theorem* in mechanics states that, for closed orbits,

$$\langle T \rangle = \frac{1}{2}\,\langle \mathbf{r} \cdot \nabla V \rangle\,,$$

where T is the kinetic energy. For the 1-D oscillator prove that this implies

$$\langle T \rangle = \langle V \rangle$$

and for the electron in hydrogen $[V\,(\mathbf{r}) = -K_e/r]$ that

$$\langle T \rangle = -\frac{1}{2}\,\langle V \rangle\,.$$

5. For a 5 mW standing wave laser field having a waist area of 4 mm^2, the potential energy of a ^{85}Rb atom in a "well" of the standing wave field can be approximated as

$$V(x) = 7.9 \times 10^{-28}\,\sin^2(kx)\text{ J},$$

where $k = 2\pi/\lambda$ and $\lambda = 780$ nm. Estimate the frequency spacing of the energy levels near the bottom of the well. How cold do the atoms have to be to have most of the atoms in the ground state of the well in thermal equilibrium? Such temperatures can be achieved using techniques of laser cooling.

6. Expand the generating function

$$F(\xi, q) = e^{-q^2+2q\xi} = \sum_{n=0}^{\infty} \frac{q^n H_n(\xi)}{n!}$$

to fourth order in q and show that it gives the correct Hermite polynomials.

7. Given an initial state for the 1-D oscillator

$$\tilde{\psi}(\xi,0) = Ne^{-\xi^2/2}\left(\xi^2 + 2\right).$$

Find N such that $\tilde{\psi}(\xi,0)$ is normalized. Find $\tilde{\psi}(\xi,t)$. Calculate the average energy in this state.

8. Given a particle having mass m moving in the dimensionless potential (in units of $\hbar\omega$),

$$V(\xi) = \begin{cases} \xi^2/2 & \xi > 0 \\ 0 & \xi < 0 \end{cases},$$

show explicitly that the intensity reflection coefficient is equal to unity. [Hint, for $\xi > 0$, the time-independent Schrödinger equation is

$$\frac{d^2\tilde{\psi}_n(\xi)}{d\xi^2} + \left(2\epsilon_k - \xi^2\right)\tilde{\psi}_n(\xi) = 0,$$

where

$$\epsilon_k = \frac{E_k}{\hbar\omega} = \frac{\hbar k^2}{2m\omega} = \nu_k + \frac{1}{2} > 0.$$

A solution of this equation that goes to zero as $\xi \sim \infty$ is

$$\tilde{\psi}_{\nu_k}(\xi) = D_{\nu_k}\left(\sqrt{2}\xi\right) = 2^{-\nu_k/2}e^{-\xi^2/2}H_{\nu_k}(\xi),$$

where D_ν is a *parabolic cylinder function*. This equation defines Hermite functions H_ν when ν is non-integral. Solve Schrödinger's equation for $\xi < 0$ and $\xi > 0$. Then use the continuity of the wave function and its derivative at $\xi = 0$ to obtain an expression for the reflection coefficient. You may use the fact that $H_\nu(\xi)$ is real and that $dH_\nu/d\xi$ can be calculated using Eq. (7.43a).]

Chapter 8
Problems in Two and Three-Dimensions: General Considerations

8.1 Separable Hamiltonians in x, y, z

Going from one to two or three dimensions significantly increases the difficulty of solving Schrödinger's equation. In general the problem must be solved numerically. Moreover, the numerical solutions may be difficult to obtain. However there are classes of *separable* potentials for which the solution can be obtained easily. If the classical potential can be written as

$$V(x, y, z) = V_x(x) + V_y(y) + V_z(z), \tag{8.1}$$

then the corresponding quantum Hamiltonian for a particle having mass m moving in this potential is

$$\hat{H} = \hat{H}_x + \hat{H}_y + \hat{H}_z, \tag{8.2}$$

where

$$\hat{H}_j = \frac{\hat{p}_j^2}{2m} + \hat{V}_j; \quad j = x, y, z. \tag{8.3}$$

The eigenfunctions of $\hat{H}$ are simply the products of the eigenfunctions of $\hat{H}_x$ and $\hat{H}_y$ and $\hat{H}_z$,

$$\psi_E(x, y, z) = \psi_{E_x}(x)\,\psi_{E_y}(y)\,\psi_{E_z}(z), \tag{8.4}$$

provided

$$E = E_x + E_y + E_z, \tag{8.5}$$

P.R. Berman, *Introductory Quantum Mechanics*, UNITEXT for Physics,
https://doi.org/10.1007/978-3-319-68598-4_8

where $\psi_{E_j}(j)$ is an eigenfunction of $\hat{H}_j$ and E_j an eigenenergy of $\hat{H}_j$ $(j = x, y, z)$. You can convince yourselves that the trial solution (8.4) works if Eq. (8.5) holds. Of course you can construct an infinite number of potentials of the form given by Eq. (8.1), but I discuss only the free particle, infinite square well, and simple harmonic oscillator potential. These are the separable potentials of physical interest.

8.1.1 Free Particle

In the case of a free particle having mass m, the eigenfunctions are

$$\begin{aligned}\psi_{\mathbf{p}}(\mathbf{r}) &= \psi_{p_x}(x)\,\psi_{p_y}(y)\,\psi_{p_z}(z) = \frac{1}{(2\pi\hbar)^{3/2}}e^{ip_xx/\hbar}e^{ip_yy/\hbar}e^{ip_zz/\hbar} \\ &= \frac{1}{(2\pi\hbar)^{3/2}}e^{i\mathbf{p}\cdot\mathbf{r}/\hbar},\end{aligned} \tag{8.6}$$

with

$$E_p = \frac{p_x^2 + p_y^2 + p_z^2}{2m} = \frac{p^2}{2m}. \tag{8.7}$$

I have already discussed this solution in Chaps. 3 and 5.

8.1.2 Two- and Three-Dimensional Infinite Wells

For a particle having mass m moving in a two-dimensional, infinite height rectangular well potential located between $0 \le x \le a_x$; $0 \le y \le a_y$, the eigenfunctions are

$$\begin{aligned}\psi_{n_x,n_y}(x,y) &= \sqrt{\frac{2}{a_x}\frac{2}{a_y}}\sin\left(\frac{n_x\pi x}{a_x}\right)\sin\left(\frac{n_y\pi y}{a_y}\right); \\ n_x, n_y &= 1, 2, 3, \ldots,\end{aligned} \tag{8.8}$$

with

$$E_{n_x,n_y} = \frac{\hbar^2\pi^2}{2m}\left[\frac{n_x^2}{a_x^2} + \frac{n_y^2}{a_y^2}\right]. \tag{8.9}$$

If a_x and a_y are incommensurate, there is no energy degeneracy. If $a_x = a_y$, there is clearly at *least* a two-fold degeneracy when $n_x \neq n_y$. However, there is more degeneracy than this, in general. The problem reduces to a well-known problem in number theory to find pairs of integers n_x and n_y for which $n_x^2 + n_y^2 = n$, where n is an integer.[1] The degeneracy grows with increasing n, but very slowly, approximately as $\log \sqrt{n}$. There may be some underlying symmetry associated with this extra deneneracy, but I have yet to find it.

For a particle having mass m moving in a three-dimensional infinite well box potential, the eigenfunctions are

$$\psi_{n_x,n_y}(x, y) = \sqrt{\frac{2}{a_x}\frac{2}{a_y}\frac{2}{a_z}} \sin\left(\frac{n_x \pi x}{a_x}\right) \sin\left(\frac{n_y \pi y}{a_y}\right) \sin\left(\frac{n_z \pi z}{a_z}\right);$$
$$n_x, n_y, n_z = 1, 2, 3, \ldots, \tag{8.10}$$

with

$$E_{n_x,n_y,n_z} = \frac{\hbar^2 \pi^2}{2m}\left[\frac{n_x^2}{a_x^2} + \frac{n_y^2}{a_y^2} + \frac{n_z^2}{a_z^2}\right]. \tag{8.11}$$

If $a_x = a_y = a_z$ the degeneracy of the states increases roughly linearly with $n = \sqrt{n_x^2 + n_y^2 + n_z^2}$.

8.1.3 SHO in Two and Three Dimensions

It is a trivial matter to solve the SHO problem in two and three dimensions using rectangular coordinates. In two dimensions the Hamiltonian is

$$\hat{H} = \hat{H}_x + \hat{H}_y, \tag{8.12a}$$

$$\hat{H}_x = \frac{\hat{p}_x^2}{2m} + \frac{1}{2} m\omega_x^2 \hat{x}^2, \tag{8.12b}$$

$$\hat{H}_y = \frac{\hat{p}_y^2}{2m} + \frac{1}{2} m\omega_y^2 \hat{y}^2. \tag{8.12c}$$

The eigenenergies are

$$E_{n_x,n_y} = \hbar\omega_x\left(n_x + \frac{1}{2}\right) + \hbar\omega_y\left(n_y + \frac{1}{2}\right); \quad n_x, n_y = 0, 1, 2, 3, \ldots \tag{8.13}$$

[1] This problem in number theory is related to the so-called Ramanujan or "taxi cab" numbers. The famous number theorist Srinivasa Ramanujan is said to have commented on a taxi-cab number, 1729, as a very interesting number, since it is the smallest number expressible as the sum of two cubes in two different ways, $1^3 + 12^3$ or $9^3 + 10^3$. Generalized Ramanujan numbers are different integral solutions $\{n_1, n_2\}$ to the equation $n_1^m + n_2^m = n$, for integer n and m.

and eigenfunctions are

$$\psi_{n_x,n_y}(x,y) = \psi_{n_x}(x)\,\psi_{n_y}(y)\,, \tag{8.14}$$

where $\psi_{n_x}(x)$ and $\psi_{n_y}(y)$ are eigenfunctions of $\hat{H}_x$ and $\hat{H}_y$, respectively, given by Eq. (7.46b). If ω_x and ω_y are incommensurable, there is no degeneracy. In the limit that $\omega_x = \omega_y = \omega$, I can write the energy as

$$E_n = \hbar\omega\,(n+1)\,, \tag{8.15}$$

with

$$n = n_x + n_y. \tag{8.16}$$

In this case there is an $(n+1)$-fold degeneracy.

Similarly, in three dimensions, the eigenenergies are

$$E_{n_x,n_y,n_z} = \hbar\omega_x\left(n_x + \frac{1}{2}\right) + \hbar\omega_y\left(n_y + \frac{1}{2}\right) + \hbar\omega_z\left(n_z + \frac{1}{2}\right) \tag{8.17}$$

$\left(n_x, n_y, n_z = 0, 1, 2, 3, \ldots\right)$ and eigenfunctions are

$$\psi_{n_x,n_y,n_z}(x,y,z) = \psi_{n_x}(x)\,\psi_{n_y}(y)\,\psi_{n_z}(z)\,. \tag{8.18}$$

In the limit that $\omega_x = \omega_y = \omega_z = \omega$,

$$E_n = \hbar\omega\left(n + \frac{3}{2}\right), \tag{8.19}$$

with

$$n = n_x + n_y + n_y. \tag{8.20}$$

In this case there is an $(n+1)(n+2)/2$-fold degeneracy.

8.2 General Hamiltonians in Two and Three Dimensions

If there is no symmetry in the problem and the Hamiltonian is not a sum of the form of Eq. (8.1), you are faced with solving the Schrödinger equation numerically, in general. However, if there is symmetry, you can identify operators that commute with the Hamiltonian. For example, with cylindrical symmetry, the z component of linear momentum and z component of angular momentum commute with the

Hamiltonian; in this limit, the eigenfunctions are products of $e^{ip_z z/\hbar}e^{im\phi}$ (m must be integral for the wave function to return to itself when $\phi \rightarrow \phi + 2\pi$) times some function of the radial (cylindrical) coordinate. There is always energy degeneracy in this problem for $m \neq 0$ since states having $\pm m$ must have the same energy since the potential is invariant under a rotation about the z axis. In the case of problems with spherical symmetry, the angular momentum is conserved and can be used to classify the solutions. I turn my attention to problems with spherical symmetry in the next two chapters.

8.3 Summary

I have taken a brief excursion to look at some simple problems in two and three dimensions. In problems lacking some global symmetries, it is possible to arrive at some systematic solution of the Schrödinger equation only for separable Hamiltonians.

8.4 Problems

1. Prove that the trial solution (8.4) is an eigenfunction of the Hamiltonian (8.3) if Eq. (8.5) holds.

2. Find the eigenfunctions and eigenenergies of an infinite, two-dimensional square well having equal sides L. That is

$$V(x, y) = 0 \qquad 0 \leq x \leq L,\ 0 \leq y \leq L$$

and is infinite otherwise. In general, there is at least a two-fold degeneracy of the energy levels, but, in certain cases, show that there can be no degeneracy or greater than two-fold degeneracy.

3. Prove that the energy degeneracy for the isotropic 2-D oscillator is $(n+1)$ and that for the isotropic 3-D oscillator is $(n+1)(n+2)/2$.

4. Prove that the parity of the eigenfunctions of both the isotropic 2-D oscillator and the isotropic 3-D oscillator is $(-1)^n$.

Chapter 9
Central Forces and Angular Momentum

I have been dealing mainly with problems involving one-dimensional motion. Since nature is three-dimensional, I want to look at solutions of Schrödinger's equation in three dimensions. I consider only problems having *spherical symmetry*, that is, potentials that are a function of r only. These correspond to *central forces*. Moreover, for the most part, I restrict the discussion to bound state problems, such as the important problem of determining the bound states of the hydrogen atom. Problems related to continuum states will be discussed in the context of scattering theory in Chap. 18. Since angular momentum is conserved for central forces, we are led naturally to the quantum theory of angular momentum. *It is important to remember, however, that the results to be derived for the angular momentum operator are valid independent of the specific nature of the interaction potential, spherically symmetric or not.*

9.1 Classical Problem

In classical physics, the concept of angular momentum plays a critical role in central force motion. A particle having mass m, velocity $\mathbf{v}$, and momentum $\mathbf{p} = m\mathbf{v}$ moving in a central potential $V(r)$ experiences a force given by

$$\mathbf{F} = -\nabla V(r) = -\frac{dV(r)}{dr}\mathbf{u}_r, \tag{9.1}$$

where $\mathbf{u}_r$ is a unit vector in the $\mathbf{r}$ direction. The torque $\boldsymbol{\tau}$ on the particle vanishes,

$$\boldsymbol{\tau} = \mathbf{r} \times \mathbf{F} = \mathbf{0}, \tag{9.2}$$

P.R. Berman, *Introductory Quantum Mechanics*, UNITEXT for Physics,
https://doi.org/10.1007/978-3-319-68598-4_9

which implies that the angular momentum ,

$$\mathbf{L} = \mathbf{r} \times \mathbf{p}, \tag{9.3}$$

is a constant of the motion. Moreover,

$$\mathbf{r} \cdot \mathbf{L} = 0; \tag{9.4}$$

the motion of the particle is in a plane perpendicular to $\mathbf{L}$.

For the moment, I assume that $\mathbf{L} = L\mathbf{u}_z$, so that the motion is in the $x - y$ plane. Using polar coordinates r and ϕ in this plane, I construct the displacement and velocity vectors,

$$\mathbf{r} = r\left(\cos\phi\mathbf{u}_x + \sin\phi\mathbf{u}_y\right) = r\mathbf{u}_r, \tag{9.5a}$$

$$\begin{aligned}\mathbf{v} = \dot{\mathbf{r}} &= \dot{r}\left(\cos\phi\mathbf{u}_x + \sin\phi\mathbf{u}_y\right) + r\left(-\sin\phi\mathbf{u}_x + \cos\phi\mathbf{u}_y\right)\dot{\phi} \\ &= \dot{r}\mathbf{u}_r + r\dot{\phi}\mathbf{u}_\phi,\end{aligned} \tag{9.5b}$$

where

$$\mathbf{u}_r = \cos\phi\mathbf{u}_x + \sin\phi\mathbf{u}_y, \tag{9.6a}$$

$$\mathbf{u}_\phi = -\sin\phi\mathbf{u}_x + \cos\phi\mathbf{u}_y, \tag{9.6b}$$

and $\mathbf{u}_\phi$ is a unit vector in the direction of increasing ϕ. The kinetic energy is

$$T = \frac{1}{2}m\mathbf{v}^2 = \frac{1}{2}m\left[\dot{r}^2 + r^2\dot{\phi}^2\right]. \tag{9.7}$$

Using the relationship

$$\mathbf{L} = \mathbf{r} \times \mathbf{p} = m\mathbf{r} \times \mathbf{v} = mr^2\dot{\phi}\mathbf{u}_z, \tag{9.8}$$

I can rewrite the kinetic energy as

$$T = \frac{1}{2}m\dot{r}^2 + \frac{L^2}{2mr^2} = \frac{p_r^2}{2m} + \frac{L^2}{2mr^2}, \tag{9.9}$$

where

$$p_r = \mathbf{p} \cdot \mathbf{u}_r = m\dot{r} \tag{9.10}$$

is the *radial momentum*. Although Eq. (9.9) was derived assuming that $\mathbf{L} = L\mathbf{u}_z$, it remains valid for arbitrary directions of $\mathbf{L}$ if Eq. (9.5a) is replaced by

$$\mathbf{r} = r\left(\sin\theta\cos\phi\mathbf{u}_x + \sin\theta\sin\phi\mathbf{u}_y + \cos\theta\mathbf{u}_z\right) = r\mathbf{u}_r, \tag{9.11}$$

where (r, θ, ϕ) are now spherical coordinates.

The first term in Eq. (9.9) is the radial and the second term the angular or rotational motion contribution to the kinetic energy. The total energy is

$$E = \frac{p_r^2}{2m} + \frac{L^2}{2mr^2} + V(r), \tag{9.12}$$

such that

$$\frac{1}{2}m\dot{r}^2 = E - \left[V(r) + \frac{L^2}{2mr^2}\right]. \tag{9.13}$$

In other words, the *radial* motion is determined by the so-called *effective potential* defined by

$$V_{\text{eff}}(r) = V(r) + \frac{L^2}{2mr^2}. \tag{9.14}$$

The effective potential is extremely useful in analyzing problems involving central forces, as you shall see in Chap. 10.

9.2 Quantum Problem

9.2.1 *Angular Momentum*

The classical definition of angular momentum can be taken over to the quantum domain; that is, a Hermitian angular momentum *operator* in quantum mechanics can be defined as

$$\hat{\mathbf{L}} = \hat{\mathbf{r}} \times \hat{\mathbf{p}} = \hat{L}_x\mathbf{u}_x + \hat{L}_y\mathbf{u}_y + \hat{L}_z\mathbf{u}_z, \tag{9.15}$$

where

$$\hat{L}_x = \hat{y}\hat{p}_z - \hat{z}\hat{p}_y; \tag{9.16a}$$

$$\hat{L}_y = \hat{z}\hat{p}_x - \hat{x}\hat{p}_z; \tag{9.16b}$$

$$\hat{L}_z = \hat{x}\hat{p}_y - \hat{y}\hat{p}_x. \tag{9.16c}$$

You can show easily that $\hat{\mathbf{L}}$ is Hermitian (recall that the product of any two Hermitian operators is Hermitian if the operators commute). Since the angular momentum is a constant of the motion, we expect it to commute with the Hamiltonian. Before proving this, let me establish some basic commutation properties of the angular momentum operators. The commutator of $\hat{L}_x$ and $\hat{L}_y$ is

$$\left[\hat{L}_x, \hat{L}_y\right] = \left[\left(\hat{y}\hat{p}_z - \hat{z}\hat{p}_y\right), \left(\hat{z}\hat{p}_x - \hat{x}\hat{p}_z\right)\right]$$
$$= \hat{y}\hat{p}_x\left[\hat{p}_z, \hat{z}\right] + \hat{p}_y\hat{x}\left[\hat{z}, \hat{p}_z\right] = i\hbar\left(\hat{x}\hat{p}_y - \hat{y}\hat{p}_x\right) = i\hbar\hat{L}_z. \quad (9.17)$$

In obtaining this result you do not have to worry about the order of *commuting* operators. I can cycle this relation by letting $x \to y, y \to z, z \to x$, to obtain

$$\left[\hat{L}_y, \hat{L}_z\right] = i\hbar\hat{L}_x; \quad (9.18a)$$

$$\left[\hat{L}_z, \hat{L}_x\right] = i\hbar\hat{L}_y. \quad (9.18b)$$

The different components of the angular momentum operator do *not* commute, so it is *not* possible to measure two components simultaneously. If we knew all three components of the angular momentum simultaneously, it would imply that we could simultaneously measure both the position and momentum of the particle precisely, which would constitute a violation of the uncertainty principle.

Other useful commutation relations are:

$$\left[\hat{L}_x, \hat{x}\right] = 0; \quad \left[\hat{L}_x, \hat{y}\right] = i\hbar\hat{z}; \quad \left[\hat{L}_x, \hat{z}\right] = -i\hbar\hat{y}; \quad (9.19a)$$

$$\left[\hat{L}_x, \hat{p}_x\right] = 0; \quad \left[\hat{L}_x, \hat{p}_y\right] = i\hbar\hat{p}_z; \quad \left[\hat{L}_x, \hat{p}_z\right] = -i\hbar\hat{p}_y; \quad (9.19b)$$

$$\left[\hat{L}_x, \hat{\mathbf{p}}^2\right] = \left[\hat{L}_x, \hat{p}_x^2\right] + \left[\hat{L}_x, \hat{p}_y^2\right] + \left[\hat{L}_x, \hat{p}_z^2\right] = 0; \quad (9.19c)$$

$$\left[\hat{L}_x, \hat{V}\right] = \left[\left(\hat{y}\hat{p}_z - \hat{z}\hat{p}_y\right), \hat{V}\right] = y\left[\hat{p}_z, V(r)\right] - z\left[\hat{p}_y, V(r)\right]$$
$$= \frac{\hbar}{i}\left[y\frac{\partial V(r)}{\partial z} - z\frac{\partial V(r)}{\partial y}\right]$$
$$= \frac{\hbar}{i}\left[y\frac{dV(r)}{dr}\frac{\partial r}{\partial z} - z\frac{dV(r)}{dr}\frac{\partial r}{\partial y}\right]$$
$$= \frac{\hbar}{i}\frac{dV(r)}{dr}\left[y\frac{z}{r} - z\frac{y}{r}\right] = 0, \quad (9.19d)$$

plus terms with $x \to y, y \to z, z \to x$. The fact that $r = \sqrt{x^2 + y^2 + z^2}$ was used in deriving the last commutation relation, in which I also replaced $\hat{V}$ with $V(r)$ by assuming implicitly that each commutator acted on a function of $\mathbf{r}$. As a consequence of the commutator relations,

$$\left[\hat{\mathbf{L}}, \hat{H}\right] = \left[\hat{\mathbf{L}}, \frac{\hat{p}^2}{2m} + \hat{V}\right] = 0. \quad (9.20)$$

The angular momentum commutes with the Hamiltonian and is a constant of the motion.

Since the individual components of $\hat{\mathbf{L}}$ do not commute it is not possible to find simultaneous eigenfunctions of $\hat{L}_x, \hat{L}_y, \hat{L}_z$. As you will see, the eigenvalues of *one* component of $\hat{\mathbf{L}}$ are not sufficient to uniquely label the degenerate energy eigenfunctions of a Hamiltonian in problems having spherical symmetry. A new operator is needed that commutes with both $\hat{\mathbf{L}}$ and $\hat{H}$. An operator that satisfies these requirements is the square of the angular momentum operator, defined by

$$\hat{L}^2 = \hat{L}_x^2 + \hat{L}_y^2 + \hat{L}_z^2, \tag{9.21}$$

since

$$\left[\hat{\mathbf{L}}, \hat{L}^2\right] = 0; \tag{9.22a}$$

$$\left[\hat{L}^2, \hat{p}^2\right] = 0, \tag{9.22b}$$

in general, and

$$\left[\hat{L}^2, \hat{V}\right] = 0; \tag{9.23a}$$

$$\left[\hat{L}^2, \hat{H}\right] = 0, \tag{9.23b}$$

for spherically symmetric potentials. In effect, eigenvalues of the operator $\hat{L}^2$ determine what values of the magnitude of the angular momentum can be measured in a quantum system. Since

$$\left[\hat{L}^2, \hat{H}\right] = 0; \quad \left[\hat{\mathbf{L}}, \hat{H}\right] = 0; \quad \left[\hat{\mathbf{L}}, \hat{L}^2\right] = 0, \tag{9.24}$$

it *is* possible to find simultaneous eigenfunctions of $\hat{L}^2$, $\hat{H}$, and (any) one component of $\hat{\mathbf{L}}$ for a spherically symmetric potential. It turns out that the eigenvalues of $\hat{H}, \hat{L}^2$, and (any) one component of $\hat{\mathbf{L}}$ can be used to uniquely label the eigenfunctions of the Hamiltonian associated with spherically symmetric potentials.

9.2.1.1 Eigenfunctions of $\hat{L}^2$ and $\hat{\mathbf{L}}$

As you might imagine it is not especially convenient to get eigenfunctions of $\hat{L}^2$ and $\hat{\mathbf{L}}$ in rectangular coordinates. Spherical coordinates are the natural venue. There is still a lot of algebra involved. In coordinate space,

$$\hat{\mathbf{L}} = \hat{\mathbf{r}} \times \hat{\mathbf{p}} = \frac{\hbar}{i} \mathbf{r} \times \nabla. \tag{9.25}$$

I can express $\mathbf{r}$ and ∇ in spherical coordinates as

$$\mathbf{r} = r\mathbf{u}_r \tag{9.26}$$

and

$$\nabla = \mathbf{u}_r \frac{\partial}{\partial r} + \mathbf{u}_\theta \frac{1}{r}\frac{\partial}{\partial \theta} + \mathbf{u}_\phi \frac{1}{r\sin\theta}\frac{\partial}{\partial \phi}, \tag{9.27}$$

where the $\mathbf{u}$'s are orthogonal unit vectors,

$$\mathbf{u}_r = \sin\theta\cos\phi\mathbf{u}_x + \sin\theta\sin\phi\mathbf{u}_y + \cos\theta\mathbf{u}_z; \tag{9.28a}$$

$$\mathbf{u}_\theta = \cos\theta\cos\phi\mathbf{u}_x + \cos\theta\sin\phi\mathbf{u}_y - \sin\theta\mathbf{u}_z; \tag{9.28b}$$

$$\mathbf{u}_\phi = -\sin\phi\mathbf{u}_x + \cos\phi\mathbf{u}_y. \tag{9.28c}$$

As a consequence,

$$\begin{aligned}\hat{\mathbf{L}} &= \frac{\hbar}{i} r\mathbf{u}_r \times \left(\mathbf{u}_r \frac{\partial}{\partial r} + \mathbf{u}_\theta \frac{1}{r}\frac{\partial}{\partial \theta} + \mathbf{u}_\phi \frac{1}{r\sin\theta}\frac{\partial}{\partial \phi}\right) \\ &= \frac{\hbar}{i}\left(\mathbf{u}_\phi \frac{\partial}{\partial \theta} - \mathbf{u}_\theta \frac{1}{\sin\theta}\frac{\partial}{\partial \phi}\right),\end{aligned} \tag{9.29}$$

such that

$$\hat{\mathbf{L}} = -i\hbar\left[\begin{array}{c}\left(-\sin\phi\frac{\partial}{\partial\theta} - \cot\theta\cos\phi\frac{\partial}{\partial\phi}\right)\mathbf{u}_x \\ +\left(\cos\phi\frac{\partial}{\partial\theta} - \cot\theta\sin\phi\frac{\partial}{\partial\phi}\right)\mathbf{u}_y + \frac{\partial}{\partial\phi}\mathbf{u}_z\end{array}\right]. \tag{9.30}$$

Therefore,

$$\hat{L}_x = -\frac{\hbar}{i}\left(\sin\phi\frac{\partial}{\partial\theta} + \cot\theta\cos\phi\frac{\partial}{\partial\phi}\right); \tag{9.31a}$$

$$\hat{L}_y = \frac{\hbar}{i}\left(\cos\phi\frac{\partial}{\partial\theta} - \cot\theta\sin\phi\frac{\partial}{\partial\phi}\right); \tag{9.31b}$$

$$\hat{L}_z = \frac{\hbar}{i}\frac{\partial}{\partial\phi}. \tag{9.31c}$$

The operator $\hat{L}_z$ has a simple form in spherical coordinates owing to the fact that θ is measured from the z-axis.

I still need an expression for $\hat{L}^2$. I use Eq. (9.29) to write

$$\begin{aligned}\hat{L}^2\psi &= \hat{\mathbf{L}}\cdot\hat{\mathbf{L}}\psi = \frac{\hbar}{i}\hat{\mathbf{L}}\cdot\left(\mathbf{u}_\phi\frac{\partial\psi}{\partial\theta} - \mathbf{u}_\theta\frac{1}{\sin\theta}\frac{\partial\psi}{\partial\phi}\right) \\ &= -\hbar^2\left(\mathbf{u}_\phi\frac{\partial}{\partial\theta} - \mathbf{u}_\theta\frac{1}{\sin\theta}\frac{\partial}{\partial\phi}\right)\cdot\left(\mathbf{u}_\phi\frac{\partial\psi}{\partial\theta} - \mathbf{u}_\theta\frac{1}{\sin\theta}\frac{\partial\psi}{\partial\phi}\right).\end{aligned} \tag{9.32}$$

It is important to realize that the $\mathbf{u}_\phi$ and $\mathbf{u}_\theta$ are functions of θ and ϕ, with

$$\frac{\partial \mathbf{u}_\theta}{\partial \theta} = -\sin\theta\cos\phi\mathbf{u}_x - \sin\theta\sin\phi\mathbf{u}_y - \cos\theta\mathbf{u}_z = -\mathbf{u}_r; \tag{9.33a}$$

$$\frac{\partial \mathbf{u}_\theta}{\partial \phi} = -\cos\theta\sin\phi\mathbf{u}_x + \cos\theta\cos\phi\mathbf{u}_y = \cos\theta\mathbf{u}_\phi; \tag{9.33b}$$

$$\frac{\partial \mathbf{u}_\phi}{\partial \theta} = 0; \tag{9.33c}$$

$$\frac{\partial \mathbf{u}_\phi}{\partial \phi} = -\cos\phi\mathbf{u}_x - \sin\phi\mathbf{u}_y, \tag{9.33d}$$

such that

$$\begin{aligned}
\hat{L}^2\psi &= -\hbar^2\left(\mathbf{u}_\phi\frac{\partial}{\partial\theta} - \mathbf{u}_\theta\frac{1}{\sin\theta}\frac{\partial}{\partial\phi}\right)\cdot\left(\mathbf{u}_\phi\frac{\partial\psi}{\partial\theta} - \mathbf{u}_\theta\frac{1}{\sin\theta}\frac{\partial\psi}{\partial\phi}\right) \\
&= -\hbar^2\mathbf{u}_\phi\cdot\left[\begin{array}{c} \mathbf{u}_\phi\frac{\partial^2\psi}{\partial\theta^2} \\ +\mathbf{u}_r\frac{1}{\sin\theta}\frac{\partial\psi}{\partial\phi} - \mathbf{u}_\theta\frac{\partial}{\partial\theta}\left(\frac{1}{\sin\theta}\frac{\partial\psi}{\partial\phi}\right) \end{array}\right] \\
&\quad +\hbar^2\frac{\mathbf{u}_\theta}{\sin\theta}\cdot\left[\begin{array}{c} \left(-\cos\phi\mathbf{u}_x - \sin\phi\mathbf{u}_y\right)\frac{\partial\psi}{\partial\theta} + \mathbf{u}_\phi\frac{\partial^2\psi}{\partial\theta\partial\phi} \\ -\cos\theta\mathbf{u}_\phi\frac{1}{\sin\theta}\frac{\partial\psi}{\partial\phi} - \mathbf{u}_\theta\frac{1}{\sin\theta}\frac{\partial^2\psi}{\partial\phi^2} \end{array}\right] \\
&= -\hbar^2\left[\frac{\partial^2\psi}{\partial\theta^2} + \cot\theta\frac{\partial\psi}{\partial\theta} + \frac{1}{\sin^2\theta}\frac{\partial^2\psi}{\partial\phi^2}\right],
\end{aligned} \tag{9.34}$$

which implies that

$$\begin{aligned}
\hat{L}^2 &= -\hbar^2\left[\frac{\partial^2}{\partial\theta^2} + \cot\theta\frac{\partial}{\partial\theta} + \frac{1}{\sin^2\theta}\frac{\partial^2}{\partial\phi^2}\right] \\
&= -\hbar^2\left[\frac{1}{\sin\theta}\frac{\partial}{\partial\theta}\sin\theta\frac{\partial}{\partial\theta} + \frac{1}{\sin^2\theta}\frac{\partial^2}{\partial\phi^2}\right].
\end{aligned} \tag{9.35}$$

In deriving Eq. (9.34), I used the fact that the unit vectors is spherical coordinates are orthogonal, along with the identity $\mathbf{u}_\theta\cdot\left(\cos\phi\mathbf{u}_x + \sin\phi\mathbf{u}_y\right) = \cos\theta$.

I want to get the simultaneous eigenfunctions of $\hat{L}^2$ and one component of $\hat{\mathbf{L}}$. Since $\hat{L}_z$ has the simplest form, I choose it. It is easy to solve for eigenfunctions Φ_m of $\hat{L}_z$ using

$$\hat{L}_z\Phi_m = \frac{\hbar}{i}\frac{\partial\Phi_m}{\partial\phi} = m\hbar\Phi_m, \tag{9.36}$$

where m labels the eigenvalue of $\hat{L}_z$ having value $m\hbar$. The solution is

$$\Phi_m(\phi) = e^{im\phi}. \tag{9.37}$$

The quantity m is not arbitrary; whenever the azimuthal angle ϕ increases by 2π, $\Phi_m(\phi)$ must return to its same value. This can happen only if m is an integer (positive, negative, or zero). Thus the normalized eigenfunctions are

$$\Phi_m(\phi) = \sqrt{\frac{1}{2\pi}} e^{im\phi}; \qquad m = 0, \pm 1, \pm 2, \ldots \tag{9.38}$$

and the eigenvalues of $\hat{L}_z$ are integral multiples of $\hbar$. The quantum number m is referred to as the *magnetic quantum number*, for reasons that will become apparent when I consider the Zeeman effect in Chap. 21.

There is already an important difference from the classical case where, for a given angular momentum $\mathbf{L}$, the z component of angular momentum can take on *continuous* values from $-L$ to L. In quantum mechanics, the z-component (any component, for that matter) of angular momentum can take on only integral multiples of $\hbar$.

The eigenvalue equation for $\hat{L}^2$ is

$$\hat{L}^2 \Theta_{\ell m}(\theta, \phi) = \hbar^2 \ell(\ell+1) \Theta_{\ell m}(\theta, \phi), \tag{9.39}$$

where ℓ is totally arbitrary at this point (i.e., it need not be an integer). I assume a solution of the form

$$\Theta_{\ell m}(\theta, \phi) = G_{\ell m}(\theta) e^{im\phi}, \tag{9.40}$$

which is guaranteed to be a simultaneous eigenfunction of $\hat{L}_z$. Substituting this trial solution into Eq. (9.39) and using Eq. (9.35), I find that $G_{\ell m}(\theta)$ satisfies the ordinary differential equation

$$-\hbar^2 \left[\frac{1}{\sin\theta} \frac{d}{d\theta} \sin\theta \frac{dG_{\ell m}(\theta)}{d\theta} - \frac{m^2}{\sin^2\theta} G_{\ell m}(\theta) \right] = \hbar^2 \ell(\ell+1) G_{\ell m}(\theta). \tag{9.41}$$

By setting $x = \cos\theta$ and $G_{\ell m}(\theta) \equiv P_\ell^m(x)$, I can transform this equation into

$$\frac{d}{dx} \left[\left(1 - x^2\right) \frac{dP_\ell^m(x)}{dx} \right] + \left[\ell(\ell+1) - \frac{m^2}{1 - x^2} \right] P_\ell^m(x) = 0, \tag{9.42}$$

which is known as *Legendre's equation*. The only solutions of Legendre's equation that are regular (do not diverge) at $x = \pm 1$ $[\theta = 0, \pi]$ are the so-called *associated Legendre polynomials* $P_\ell^m(x)$ for which ℓ is a positive integer or zero that is greater than or equal to $|m|$ {Mathematica symbol LegendreP[ℓ, m, x]}. In other words, the physically acceptable solutions have

$$\ell = 0, 1, 2, \ldots. \tag{9.43}$$

and, for each value of ℓ,

$$m = -\ell, -\ell + 1, \ldots \ell - 1, \ell. \tag{9.44}$$

The eigenvalues of $\hat{L}^2$ are

$$L^2 = \hbar^2 \ell (\ell + 1), \qquad \ell = 0, 1, 2, . \tag{9.45}$$

The quantum number ℓ is referred to as the *azimuthal* or *angular momentum quantum number*.

Before looking at the eigenfunctions in more detail, I can summarize the results so far. Classically the magnitude squared of the angular momentum L^2 can take on any value from zero to infinity and L_z can take on *continuous* values from $-L$ to L. In quantum mechanics, the eigenvalues of $\hat{L}^2$ are limited to the set of discrete (quantized) values $0\hbar^2, 2\hbar^2, 6\hbar^2, \ldots \ell(\ell + 1)\hbar^2$. For each value of ℓ, the eigenvalues of $\hat{L}_z$ (or of any component of angular momentum, for that matter) vary from $-\ell\hbar$ to $\ell\hbar$ in integral steps of $\hbar$. In other words, for each value of ℓ, there $(2\ell + 1)$ values of m that are allowed.

Moreover, in quantum mechanics, we cannot know the *vector* angular momentum exactly since the components of $\hat{\mathbf{L}}$ do not commute. We can specify the magnitude of the angular momentum and one of its components, say L_z, but then there is uncertainty in both L_x and L_y. All we know about these other components is that they are constrained by

$$L_x^2 + L_y^2 = L^2 - L_z^2. \tag{9.46}$$

In other words, there is an *uncertainty cone* of L_x and L_y having radius $\left[\ell(\ell + 1) - m^2\right]^{1/2} \hbar$ for a given value of ℓ and m. I will return to a discussion of the physical interpretation of angular momentum in quantum mechanics after I look at the eigenfunctions.

You might be wondering about the result that the magnitude of the angular momentum is quantized in units of $\sqrt{\ell(\ell + 1)}\hbar$ and not $\ell\hbar$. This follows from solving the eigenvalue equation, but there does not seem to be a simple *geometric* interpretation of this result as there was for L_z. Based on the uncertainty principle, you can rule out the possibility that $L^2 = (\ell\hbar)^2$. If this were the case for integer ℓ and if the maximum value of m is equal to ℓ, then the state of maximum ℓ would have no uncertainty in $L_x^2 + L_y^2$, which is impossible since $\hat{L}_z$ does not commute with $\hat{L}_x$ and $\hat{L}_y$. You can also derive the result if you assume that ℓ is integral and that m takes on integral values from $-\ell$ to ℓ. With this assumption, for a spherically symmetric state

$$\left\langle L^2 \right\rangle = 3 \left\langle L_z^2 \right\rangle = \frac{3\hbar^2}{2\ell + 1} \sum_{-\ell}^{\ell} m^2 = \ell (\ell + 1) \hbar^2. \tag{9.47}$$

However this is a bit of a swindle since I am trying to understand how the magnitude of the angular momentum is quantized and I have already implicitly assumed it to be the case by assuming that m takes on integral values from $-\ell$ to ℓ.

The *normalized* simultaneous eigenfunctions of $\hat{L}^2$ and $\hat{L}_z$ are the so-called spherical harmonics $Y_\ell^m (\theta, \phi)$ defined by

$$Y_\ell^m (\theta, \phi) = \sqrt{\frac{2\ell + 1}{4\pi} \frac{(\ell - m)!}{(\ell + m)!}} P_\ell^m (\cos\theta) e^{im\phi} \tag{9.48}$$

(*Mathematica* symbol SphericalHarmonicY[ℓ, m, θ, ϕ]). Thus

$$\hat{L}^2 Y_\ell^m (\theta, \phi) = \hbar^2 \ell (\ell + 1) Y_\ell^m (\theta, \phi) ; \qquad \ell = 0, 1, 2, \ldots \tag{9.49a}$$

$$\hat{L}_z Y_\ell^m (\theta, \phi) = m\hbar Y_\ell^m (\theta, \phi) ; \qquad m = 0, \pm 1, \pm 2 \ldots \pm \ell. \tag{9.49b}$$

The first few $Y_\ell^m (\theta, \phi)$ are

$$Y_0^0 (\theta, \phi) = \sqrt{\frac{1}{4\pi}}; \tag{9.50a}$$

$$Y_1^0 (\theta, \phi) = \sqrt{\frac{3}{4\pi}} \cos\theta; \tag{9.50b}$$

$$Y_1^{\pm 1} (\theta, \phi) = \mp \sqrt{\frac{3}{8\pi}} \sin\theta e^{\pm i\phi}; \tag{9.50c}$$

$$Y_2^0 (\theta, \phi) = \sqrt{\frac{5}{4\pi}} \left(\frac{3\cos^2\theta - 1}{2} \right); \tag{9.50d}$$

$$Y_2^{\pm 1} (\theta, \phi) = \mp \sqrt{\frac{15}{8\pi}} \sin\theta \cos\theta e^{\pm i\phi}; \tag{9.50e}$$

$$Y_2^{\pm 2} (\theta, \phi) = \frac{1}{4} \sqrt{\frac{15}{2\pi}} \sin^2\theta e^{\pm 2i\phi}; \tag{9.50f}$$

$$Y_\ell^{-m} = (-1)^m \left(Y_\ell^{-m} \right)^* . \tag{9.50g}$$

The $Y_\ell^m (\theta, \phi)$ are orthonormal,

$$\int_0^{2\pi} d\phi \int_0^{\pi} \sin\theta \, d\theta \left[Y_\ell^m (\theta, \phi) \right]^* Y_{\ell'}^{m'} (\theta, \phi) = \delta_{\ell,\ell'} \delta_{m,m'} . \tag{9.51}$$

Under an inversion of coordinates, $\mathbf{r} \to -\mathbf{r}$, the angles change by $\theta \to \pi - \theta, \phi \to \phi + \pi$, and

$$Y_\ell^m(\pi - \theta, \phi + \pi) = (-1)^\ell Y_\ell^m(\theta, \phi). \tag{9.52}$$

Thus, the *parity* of $Y_\ell^m(\theta, \phi)$ is $(-1)^\ell$; that is, the $Y_\ell^m(\theta, \phi)$ are also simultaneous eigenfunctions of the parity operator. This must be the case since the $\hat{L}^2$, $\hat{L}_z$, and the parity operator commute and the eigenfunctions of $\hat{L}^2$ and $\hat{L}_z$ are nondegenerate.

I also list a few properties of the associated Legendre polynomials $P_\ell^m(x)$. For $m = 0$, the associated Legendre polynomials reduce to the Legendre polynomials $P_\ell(x)$ (Mathematica symbol LegendreP[ℓ, x]); that is, $P_\ell^0(x) = P_\ell(x)$, which satisfies the differential equation

$$\frac{d}{dx}\left[\left(1 - x^2\right)\frac{dP_\ell(x)}{dx}\right] + [\ell(\ell + 1)]P_\ell(x) = 0, \tag{9.53}$$

and

$$P_\ell(x) = \frac{1}{2^\ell \ell!}\frac{d^\ell}{dx^\ell}\left(x^2 - 1\right)^\ell \quad \text{(Rodrigues formula);} \tag{9.54a}$$

$$(\ell + 1)P_{\ell+1} - (2\ell + 1)xP_\ell + \ell P_{\ell-1} = 0; \tag{9.54b}$$

$$\left(1 - x^2\right)\frac{dP_\ell}{dx} = -\ell x P_\ell + \ell P_{\ell-1}; \tag{9.54c}$$

$$\int_{-1}^{1} dx P_\ell(x) P_{\ell'}(x) = \frac{2}{2\ell + 1}\delta_{\ell,\ell'}; \tag{9.54d}$$

$$\int_0^\pi \sin\theta \, d\theta P_\ell(\cos\theta) P_{\ell'}(\cos\theta) = \frac{2}{2\ell + 1}\delta_{\ell,\ell'}; \tag{9.54e}$$

$$\frac{1}{\sqrt{1 - 2xq + q^2}} = \sum_{\ell=0}^{\infty} q^\ell P_\ell(x); \quad \text{(generating function);} \tag{9.54f}$$

$$P_0(x) = 1; \quad P_1(x) = x; \quad P_2(x) = \frac{3x^2 - 1}{2}; \quad P_3(x) = \frac{5x^3 - 3x}{2}. \tag{9.54g}$$

The Legendre polynomials are defined such that

$$P_\ell(\pm 1) = (\pm 1)^\ell, \tag{9.54h}$$

implying that $P_\ell(\cos\theta) = 1$ for $\theta = 0$ and $(-1)^\ell$ for $\theta = \pi$.

The associated Legendre polynomials can be obtained from the Legendre polynomials via

$$P_\ell^m(x) = (-1)^m \left(1 - x^2\right)^{m/2} \frac{d^m}{dx^m} P_\ell(x); \quad m \geq 0; \tag{9.55a}$$

$$P_\ell^{-m}(x) = (-1)^m \frac{(\ell - m)!}{(\ell + m)!} P_\ell^m(x) \quad m \geq 0; \tag{9.55b}$$

$$\int_{-1}^{1} dx P_\ell^m(x) P_{\ell'}^m(x) = \frac{2}{2\ell + 1} \frac{(\ell + m)!}{(\ell - m)!} \delta_{\ell,\ell'}. \tag{9.55c}$$

Note that for odd m the associated Legendre "polynomials" are not polynomials at all since they contain the factor $\left(1 - x^2\right)^{m/2}$.

9.2.2 *Physical Interpretation of the Spherical Harmonics*

I now return to a discussion of the physics. It should be clear by now that the angular momentum operator plays an important role in central force problems since it is a constant of the motion. In the classical problem, suppose the angular momentum is in the positive z direction, $\mathbf{L} = L\mathbf{u}_z$. This means the orbit is in the xy plane and there is ϕ dependence in a *specific* orbit, e.g. an elliptical orbit in which the semi-major axis is along x. For a given value of energy and angular momentum, the ϕ dependence is determined by the initial conditions. If, on the other hand, we average over *all* possible initial conditions having the same energy E and angular momentum $\mathbf{L} = L\mathbf{u}_z$, there *cannot* be any ϕ dependence owing to the overall symmetry about the z-axis. In some sense, quantum mechanics does this averaging for you with respect to the eigenfunctions. This is the reason why $\left|Y_\ell^m(\theta, \phi)\right|^2$ is independent of ϕ.

You will see in the next chapter that the eigenfunctions of the Hamiltonian for spherically symmetric potentials that are simultaneous eigenfunction of $\hat{L}^2$ and $\hat{L}_z$ can be written quite generally as

$$\psi_{E\ell m}(\mathbf{r}) = R_{E\ell}(r) Y_\ell^m(\theta, \phi), \tag{9.56}$$

where $R_{E\ell}(r)$ is a radial wave function. Since $\left|Y_\ell^m(\theta, \phi)\right|^2$ is independent of ϕ , I can define an angular probability distribution for the polar angle θ by

$$W_{\ell m}(\theta) = 2\pi \sin\theta \left|Y_\ell^m(\theta, \phi)\right|^2. \tag{9.57}$$

This distribution is normalized,

$$\int_0^\pi W_{\ell m}(\theta)\, d\theta = 1. \tag{9.58}$$

It is fairly amazing that, for fixed ℓ and m, the angular probability distribution $W_{\ell m}(\theta)$ associated with an eigenfunction of a spherically symmetric potential is the *same* for *all* central forces, independent of the energy.

It is a simple matter to plot $W_{\ell m}(\theta)$ as a function of θ. Graphs of $W_{\ell m}(\theta)$ are shown as the solid red curves in Figs. 9.1, 9.2, and 9.3 for $\ell = 50$ and $m = 50, 25, 0$, respectively. There are $(\ell - |m|)$ zeroes in $W_{\ell m}(\theta)$ for $0 < \theta < \pi$, reflecting the fact that there are $(\ell - |m|)$ nodes in $P_\ell^m(\cos\theta)$. This is similar to what we found for the energy eigenfunctions for one-dimensional potentials. In this case, however, there are zero nodes for $|m| = \ell$ and a new node appears with each *decrease* in the value of $|m|$.

The question that remains, however, is "What is the physical significance of these curves?" To answer this question, I can make a comparison with the corresponding classical problem. In the limit of large quantum numbers, the quantum probability distribution, averaged over oscillations in the classically allowed region, should be approximately equal to the classical probability distribution, *averaged over all possible initial conditions consistent with the constant values of energy, magnitude of angular momentum, and z-component of angular momentum.*

To compare the classical and quantum probability distributions, I must specify the values of the angular momentum that are consistent with the conserved quantities of the quantum problem. In other words, I set

$$L = \sqrt{\ell(\ell+1)}\hbar \tag{9.59}$$

and

$$L_z = m\hbar. \tag{9.60}$$

With these values, the classical angular momentum can be located anywhere on an uncertainty cone (see Fig. 9.4) for which

$$L_x^2 + L_y^2 = L^2 - L_z^2. \tag{9.61}$$

Figure 9.4 can help you to understand the nature of the classical motion. Since the motion is in a plane perpendicular to $\mathbf{L}$, for any position of $\mathbf{L}$ on the uncertainty cone, the motion must be confined to polar angles

$$\pi/2 - \cos^{-1}(L_z/L) \le \theta \le \pi/2 + \cos^{-1}(L_z/L), \tag{9.62}$$

so that the classically allowed region is

$$\pi/2 - \cos^{-1}\left(m/\sqrt{\ell(\ell+1)}\right) \le \theta \le \pi/2 + \cos^{-1}\left(m/\sqrt{\ell(\ell+1)}\right) \tag{9.63}$$

when Eqs. (9.59) and (9.60) are used.

It can be shown (see Appendix) that the classical polar angle probability distribution is given by

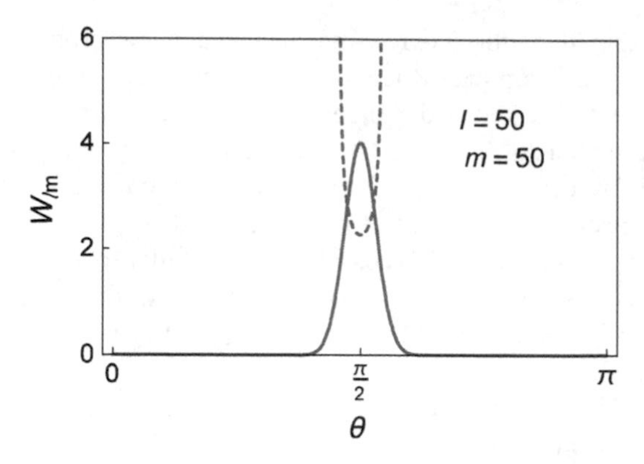

Fig. 9.1 Polar angle probability distribution $W_{\ell m}$ as a function of θ for $\ell = 50$ and $m = 50$. The solid curve is the exact, quantum result and the dashed curve is the classical probability distribution

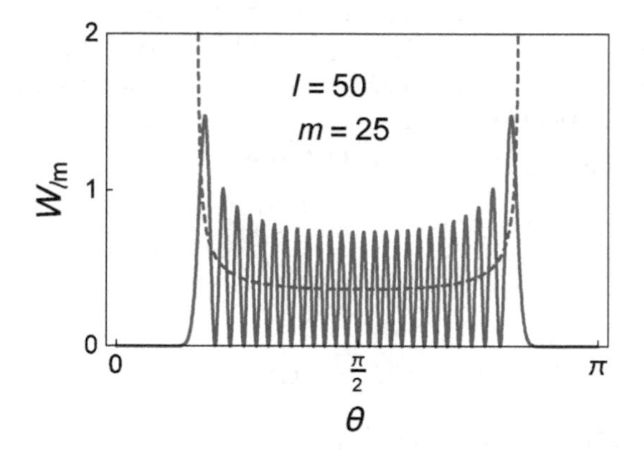

Fig. 9.2 Same as Fig. 9.1, but with $\ell = 50$ and $m = 25$

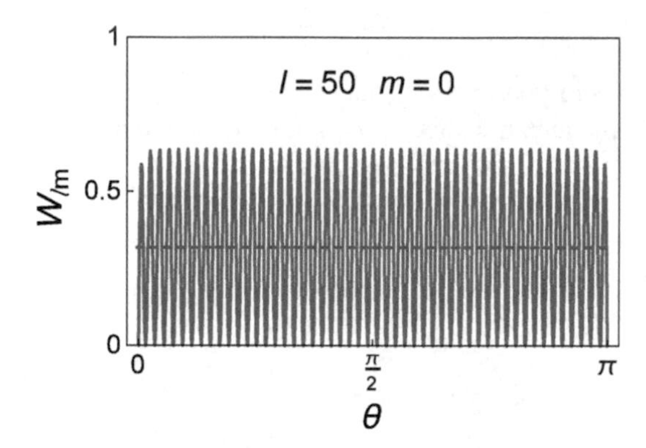

Fig. 9.3 Same as Fig. 9.1, but with $\ell = 50$ and $m = 0$

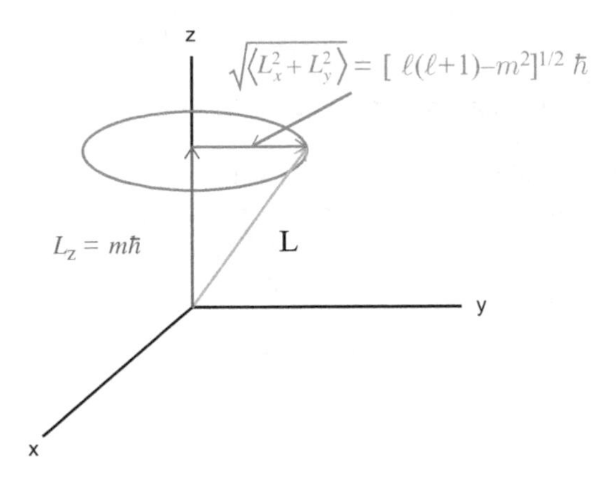

Fig. 9.4 Uncertainty cone for the x and y components of angular momentum

$$W_{\ell m}(\theta)^{\text{class}} = \frac{\sin\theta}{\pi\left\{\frac{\ell(\ell+1)-m^2}{\ell(\ell+1)} - \cos^2\theta\right\}^{1/2}}, \tag{9.64}$$

restricted to positive values of the term in curly brackets. Owing to the symmetry, the classical motion, averaged over all values of $L_x^2 + L_y^2$, consistent with Eq. (9.61), cannot depend on the azimuthal angle ϕ, even if there is a ϕ dependence for *specific* orbits.

In Figs. 9.1, 9.2 and 9.3, the classical distribution function is shown as the dashed blue curves. For $m = 50$, the motion is constrained to be very close to the xy plane ($\theta = \pi/2$) since this corresponds to $\mathbf{L} \approx L_z\mathbf{u}_z$. For $m = 25$, the values of θ are restricted to the classically allowed regime

$$0.52 \leq \theta \leq 2.62. \tag{9.65}$$

When $m = 0$, the angular momentum lies in xy plane so, for a given value of $\mathbf{L} \approx L_x\mathbf{u}_z + L_y\mathbf{u}_z$, the classical motion must be in a plane perpendicular to the xy plane. In this case, $W_{\ell 0}(\theta)^{\text{class}} = 1/\pi$; the classical angle distribution, averaged over all possible initial conditions, is constant. Since the motion is in a plane perpendicular to the xy plane and since an average over all initial conditions is taken, the resultant time-averaged angular probability distribution must be independent of θ. For values of $\ell \gg 1$, the quantum probability distribution, averaged over oscillations in the classically allowed region, is approximately equal to the classical probability distribution.

Returning to the quantum problem, I note that the eigenfunction with $m = \ell$ comes closest to the classical state in which the angular momentum is in the z direction. In other words, the eigenfunction with $m = \ell$ minimizes the angle of uncertainty cone. For this state,

$$\left|Y_{\ell}^{\ell}(\theta,\phi)\right|^{2}\sim\sin^{2\ell}\theta; \tag{9.66a}$$

$$W_{\ell m}(\theta)=2\pi\sin\theta\left|Y_{\ell}^{m}(\theta,\phi)\right|^{2}\sim\sin^{2\ell+1}\theta. \tag{9.66b}$$

Thus, $\left|Y_{\ell}^{\ell}(\theta,\phi)\right|^{2}$ is peaked about $\theta=\pi/2$, and the sharpness of the peak increases with increasing ℓ. This corresponds to the fact that the classical orbit is constrained to be very close to the *xy* plane. In fact, I can estimate how far the orbit strays from the *xy* plane by calculating

$$\langle\theta\rangle=\frac{\int_{0}^{\pi}d\theta\,\theta\sin^{2\ell+1}\theta}{\int_{0}^{\pi}d\theta\sin^{2\ell+1}\theta}=\frac{\pi}{2} \tag{9.67a}$$

$$\Delta\theta=\left(\frac{\int_{0}^{\pi}d\theta\,\theta^{2}\sin^{2\ell+1}\theta}{\int_{0}^{\pi}d\theta\sin^{2\ell+1}\theta}-\frac{\pi^{2}}{4}\right)^{1/2}\approx\frac{1}{\sqrt{2\ell+1}}. \tag{9.67b}$$

The integral in the equation for $\Delta\theta$ can be evaluated exactly in terms of hypergeometric functions, but has been approximated for large ℓ. You can also determine the dependence on ℓ by looking at the value $\Delta\theta\sim(\theta-\pi/2)$ for which the function $\sin^{2\ell+1}\theta$ is equal to $1/2$.

The uncertainty in the magnitude of $\mathbf{L}$, ΔL, is of order of the radius of the uncertainty cone,

$$\Delta L\approx\sqrt{\left\langle\hat{L}_{x}^{2}+\hat{L}_{y}^{2}\right\rangle}=\sqrt{[\ell(\ell+1)-\ell^{2}]}\hbar=\hbar\sqrt{\ell} \tag{9.68}$$

Therefore

$$\Delta L\Delta\theta\approx\hbar\sqrt{\frac{\ell}{2\ell+1}} \tag{9.69}$$

which is an *angular momentum—angle uncertainty relation*, although it is not a strict uncertainty relation since there is no Hermitian operator that corresponds to angle. For values of $m\neq\ell$, the values of both ΔL and $\Delta\theta$ increase.

The key point to remember is that the motion in the θ direction is, for the most part, restricted to a range of angles given by Eq. (9.63) when the system is in an eigenstate of $\hat{L}^2$ and $\hat{L}_z$. There is no ϕ dependence in the angular probability distribution because I have chosen eigenfunctions of $\hat{L}^2$ and $\hat{L}_z$. Had I chosen $\hat{L}^2$ and $\hat{L}_x$ the eigenfunctions would be *linear combinations* of the $Y_{\ell}^{m}(\theta,\phi)$, and the absolute square of an eigenfunction could depend on ϕ, but it would be a function only of the angle that $\mathbf{L}$ makes with the *x*-axis (see problems). The angular probability distribution associated with the simultaneous eigenfunctions of a spherically symmetric Hamiltonian and the operators $\hat{L}^2$ and $\hat{L}_z$ does not depend in any way on the specific form of the potential.

9.3 Summary

We have seen that angular momentum plays a critical role in problems having spherical symmetry. I obtained the eigenvalues and simultaneous eigenfunctions of the operators $\hat{L}^2$ and $\hat{L}_z$. Moreover I was able to make a correspondence with the analogous classical problem to help give you a physical interpretation to the spherical harmonics for problems involving spherical symmetry.

9.4 Appendix: Classical Angular Distribution

The classical polar angle probability distribution for central force motion is given by

$$W_\ell^m(\theta)^{\text{class}} = \frac{1}{T_{21}}\left|\frac{dt}{d\theta}\right| \tag{9.70}$$

where

$$T_{21} = \int_{r_{\min}}^{r_{\max}} \frac{dt}{dr}dr \tag{9.71}$$

and $r_{\min}$ is the minimum value of r and $r_{\max}$ the maximum value of r for the orbit. For *bound* orbits of a particle having mass m, it is easy to calculate $r_{\min}$ and $r_{\max}$ as roots of

$$E - V(r) - \frac{\hbar^2\ell(\ell+1)}{2mr^2} = 0. \tag{9.72}$$

On the other hand, for unbound orbits corresponding to a scattering problem, $r_{\max}$ approaches infinity. For *closed* orbits, T_{21} can be related to the period of the orbit.

The probability distribution (9.70) must be averaged over all possible classical trajectories consistent with Eqs. (9.59)–(9.61). A simple way to envision this is to first imagine the angular momentum in the z-direction. The radial coordinate r is a function of $(\phi - \phi_0)$, $r = r(\phi - \phi_0)$, where ϕ is the azimuthal angle and ϕ_0 is a constant that determines the orientation of the orbit in the $x-y$ plane. I must average the results over all values of ϕ_0 from 0 to 2π. For bound state orbits $\phi - \phi_0$ varies from 0 to 2π, but for unbound orbits $\phi - \phi_0$ may vary from some fixed angle β to $2\pi - \beta$. For example, in the Coulomb problem with positive energy, you can have

$$\beta = \cos^{-1}(1/\epsilon), \tag{9.73}$$

where ϵ is the eccentricity of the orbit; this value of β corresponds to an asymptote of a hyperbolic orbit. Once the average over ϕ_0 is performed, the entire result can be rotated such that the plane of the orbit is perpendicular to **L**. Following this procedure, I must calculate

$$W_{\ell m}(\theta)^{\text{class}} = \frac{1}{T_{21}}\left\langle\left|\frac{dt}{d\theta}\right|\right\rangle_{\phi_0}, \tag{9.74}$$

which is independent of the azimuthal angle of **L**.

To calculate $dt/d\theta$, I start from

$$\mathbf{L}\cdot\mathbf{r} = 0, \tag{9.75}$$

with

$$\mathbf{L} = L\left(\mathbf{u}_z\cos\alpha + \mathbf{u}_x\sin\alpha\right) \tag{9.76a}$$

$$\mathbf{r} = r\left(\mathbf{u}_z\cos\theta + \mathbf{u}_x\sin\theta\cos\phi + \mathbf{u}_y\sin\theta\sin\phi\right) \tag{9.76b}$$

(without loss of generality, I can take **L** in the $x-z$ plane). Clearly,

$$\cos\alpha = L_z/L. \tag{9.77}$$

From Eqs. (9.75) and (9.76), I find

$$\cos\phi = -\cot\alpha\cot\theta \tag{9.78}$$

and

$$\sin\phi = \frac{\sqrt{\sin^2\alpha - \cos^2\theta}}{\sin\alpha\sin\theta}. \tag{9.79}$$

The square of the angular momentum is given by

$$L^2 = m^2r^4\left(\dot{\theta}^2 + \sin^2\theta\dot{\phi}^2\right). \tag{9.80}$$

Using Eq. (9.78), I obtain

$$-\sin\phi\dot{\phi} = \frac{\cot\alpha}{\sin^2\theta}\dot{\theta}, \tag{9.81}$$

allowing me to calculate

$$\dot{\phi} = -\frac{\cos\alpha}{\sin\theta\sqrt{\sin^2\alpha - \cos^2\theta}}\dot{\theta}. \tag{9.82}$$

Combining Eqs. (9.80) and (9.82), I can express the square of the angular momentum as

$$L^2 = m^2 r^4 \dot{\theta}^2 \frac{\sin^2\theta}{\sin^2\alpha - \cos^2\theta}, \tag{9.83}$$

from which it follows that

$$\frac{dt}{d\theta} = \frac{mr^2}{L}\frac{\sin\theta}{\sqrt{\sin^2\alpha - \cos^2\theta}}. \tag{9.84}$$

Finally, using Eqs. (9.74) and (9.84), I arrive at

$$W_{\ell m}(\theta)^{\text{class}} = \frac{1}{T_{21}}\left\langle\left|\frac{dt}{d\theta}\right|\right\rangle_{\phi_0} = \frac{1}{T_{21}}\frac{m\left\langle r^2\right\rangle_{\phi_0}}{L}\frac{\sin\theta}{\left(\sin^2\alpha - \cos^2\theta\right)^{1/2}}. \tag{9.85}$$

Equation (9.85) is somewhat surprising. It seems to imply that

$$\frac{1}{T_{21}}\frac{m\left\langle r^2\right\rangle_{\phi_0}}{L}$$

must be independent of energy for *any* central potential. I now show that this is actually the case, starting from

$$\left\langle r^2\right\rangle_{\phi_0} = \frac{1}{2\pi}\int_0^{2\pi} r^2(\phi - \phi_0)\,d\varphi_0 = \frac{1}{2\pi}\int_0^{2\pi} r^2(\bar{\phi})d\bar{\phi}, \tag{9.86}$$

where $\bar{\phi} = \phi - \phi_0$. By writing

$$d\bar{\phi} = \frac{d\bar{\phi}}{dt}\frac{dt}{dr}dr = \dot{\phi}\frac{dt}{dr}dr, \tag{9.87}$$

I can obtain

$$\left\langle r^2\right\rangle_{\phi_0} = \frac{2}{2\pi}\int_{r_{\min}}^{r_{\max}} \dot{\phi} r^2 \frac{dt}{dr}dr = \frac{L}{\pi m}\int_{r_{\min}}^{r_{\max}}\frac{dt}{dr}dr = \frac{LT_{21}}{2\pi m}, \tag{9.88}$$

where Eq. (9.71) was used. The extra factor of 2 in Eq. (9.88) arises from the fact that as $\bar{\phi}$ varies from 0 to 2π, r varies *twice* from $r_{\min}$ to $r_{\max}$. The classical probability distribution, calculated using Eqs. (9.85) and (9.88) is

$$W_{\ell m}(\theta)^{\text{class}} = \frac{1}{T_{21}}\left\langle\left|\frac{dt}{d\theta}\right|\right\rangle_{\phi_0} = \frac{1}{\pi}\frac{\sin\theta}{\left(\sin^2\alpha - \cos^2\theta\right)^{1/2}}. \tag{9.89}$$

It is easy to verify that $W_{\ell m}(\theta)^{\text{class}}$ is normalized properly,

$$\int_{\pi/2-\alpha}^{\pi/2+\alpha} W_{\ell m}(\theta)^{\text{class}}\, d\theta = 1. \tag{9.90}$$

Using the value of α defined in Eq. (9.77),

$$\alpha = \cos^{-1}(L_z/L) = \cos^{-1}[m/\sqrt{\ell(\ell+1)}], \tag{9.91}$$

I arrive at Eq. (9.64).

9.5 Problems

1. How does quantum angular momentum differ from classical angular momentum? Why is a quantum state with $\ell = 0$ spherically symmetric, whereas a classical state with $L = 0$ has a straight line trajectory through the origin? Why is it customary to choose $\hat{L}^2$ and $\hat{L}_z$ as the commuting operators for which to find simultaneous eigenfunctions?

2. Prove that $\hat{\mathbf{L}}$ is Hermitian, that $\left[\hat{L}_y, \hat{L}_z\right] = i\hbar\hat{L}_x$, that $\left[\hat{L}_x, \hat{p}_y\right] = i\hbar\hat{p}_z$, and that $\left[\hat{L}_x, \hat{p}^2\right] = 0$.

3. Prove that $\left[\hat{L}^2, \hat{\mathbf{L}}\right] = 0$, that $\left[\hat{L}^2, V(r)\right] = 0$, and that $\left[\hat{L}^2, \hat{p}^2\right] = 0$.

4. Prove that $\hat{L}^2$ and $\hat{\mathbf{L}}$ commute with the parity operator and that the parity of $Y_\ell^m(\theta, \phi)$ is $(-1)^\ell$.

5. By making the substitutions $x = \cos\theta$, $G_{\ell m}(\theta) \to P_\ell^m(x)$, prove that the equation

$$\frac{1}{\sin\theta}\frac{d}{d\theta}\sin\theta\frac{dG_{\ell m}(\theta)}{d\theta} - \frac{m^2}{\sin^2\theta}G_{\ell m}(\theta) = -\ell(\ell+1)\,G_{\ell m}(\theta)$$

can be transformed into

$$\frac{d}{dx}\left[\left(1-x^2\right)\frac{dP_\ell^m(x)}{dx}\right] + \left[\ell(\ell+1) - \frac{m^2}{1-x^2}\right]P_\ell^m(x) = 0.$$

6–7. Classically, if the magnitude of the angular momentum is $100\hbar$ and the z-component of angular momentum is $50\hbar$, by what angle can the motion deviate from the xy plane. Plot $W_{\ell m}(\theta) = 2\pi\sin\theta\left|Y_\ell^m(\theta,\phi)\right|^2$ as a function of θ for $\ell = 100$ and $m = 50$ to see if the quantum result corresponds to the classical one. Repeat the plot for $\ell = 100$ and $m = 100$ and for $\ell = 100$ and $m = 0$ and interpret your results.

8. A rigid rotator of mass m has a Hamiltonian given by

$$\hat{H} = \frac{\hat{L}^2}{2ma^2},$$

where a is a constant. Find the eigenfunctions and eigenenergies of the rigid rotator. Are these *rotational* energy levels equally spaced? For an H_2O molecule at room temperature, estimate the number of energy levels that are occupied and the frequency spacing of the lowest rotational transition. Show that your result implies that heating in a microwave oven, which uses a frequency of about 2.4 GHz, does *not* occur by resonant absorption by the water molecules. Of course, the bond lengths in molecules are *not* rigid, giving rise to vibrations that modify the energy levels of the "rigid" rotator.

9–10. In general, what can you say about the simultaneous eigenfunctions of $\hat{L}^2$ and $\hat{L}_x$? Specifically show that the eigenfunctions of $\hat{L}_x$ for $\ell = 1$ are

$$\Phi_{\ell=1,\ell_x}(\theta,\phi) = \begin{cases} \frac{1}{2}\left[Y_1^1(\theta,\phi) + \sqrt{2}Y_1^0(\theta,\phi) + Y_1^{-1}(\theta,\phi)\right] \\ \frac{1}{\sqrt{2}}\left[Y_1^1(\theta,\phi) - Y_1^{-1}(\theta,\phi)\right] \\ \frac{1}{2}\left[Y_1^1(\theta,\phi) - \sqrt{2}Y_1^0(\theta,\phi) + Y_1^{-1}(\theta,\phi)\right] \end{cases}$$

and find the eigenvalues associated with these states (what must they be?). Note that $|\Phi_{\ell=1,\ell_x}(\theta,\phi)|^2$ now depends on ϕ in a non-trivial way. Prove, however, that $|\Phi_{\ell=1,\ell_x}(\theta,\phi)|^2$ depends only on the angle between the position vector and the x-axis.

Chapter 10
Spherically Symmetric Potentials: Radial Equation

Now that we have studied angular momentum, it is an easy matter to obtain a solution of the Schrödinger equation for spherically symmetric potentials $V(r)$. I will look at bound state solutions of Schrödinger's equation for the infinite spherical well potential, the finite spherical well potential, the Coulomb potential (hydrogen atom), and the isotropic, 3-D harmonic oscillator potential. Among these, the Coulomb potential is undoubtedly the most important, since the solution of the Coulomb problem was one of the major triumphs of quantum mechanics.

To help understand the quantum bound state radial probability distributions, it will be helpful to compare the quantum results with the classical radial probability distributions. For a particle having mass μ moving in a potential $V(r)$, the effective potential is

$$V_{\text{eff}}(r) = V(r) + L^2/\left(2\mu r^2\right), \tag{10.1}$$

where L is the magnitude of the angular momentum of the particle. For the effective potentials that I discuss (see, for example, Fig. 10.1), the bound state classical motion is always restricted to a range $r_{\min} \leq r \leq r_{\max}$, where the values of $r_{\min}$ and $r_{\max}$ are classical *radial turning points* of the orbits for a given energy and angular momentum (for $L = 0$, the classical orbit is a bounded straight line though the origin, but $r_{\min} = 0$ is still a turning point for the *radial* motion). In this chapter I use the symbol μ for the mass to distinguish it from the magnetic quantum number m—it has the added advantage that in two-body problems such as hydrogen, μ actually refers to the reduced mass of the electron.

I define a time T_{21} as the time it takes for a classical particle having mass μ to move from $r_{\min}$ to $r_{\max}$. In the case of the Coulomb and oscillator potentials, the bound orbits are *closed*. The time T_{21} is half the orbital period in the Coulomb problem and one-quarter the orbital period in the oscillator problem. For other

P.R. Berman, *Introductory Quantum Mechanics*, UNITEXT for Physics,
https://doi.org/10.1007/978-3-319-68598-4_10

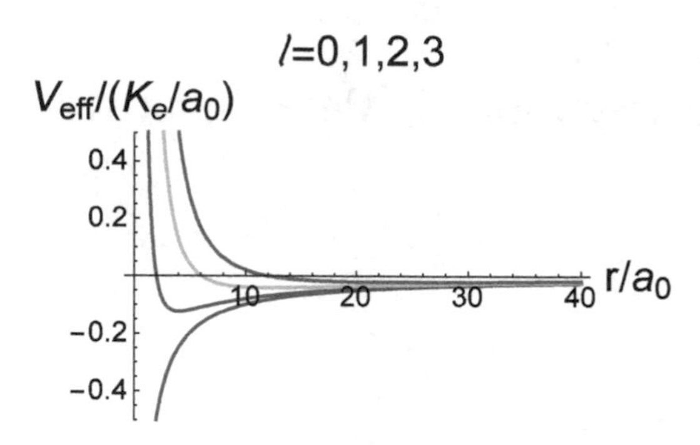

Fig. 10.1 Effective potential for the Coulomb potential, $V(r) = -K_e/r$ in units of K_e/a_0 as a function of r/a_0, where a_0 is the Bohr radius. To relate the classical and quantum problems, the angular momentum is set equal to $\hbar\sqrt{\ell(\ell+1)}$. The lowest curve has $\ell = 0$ and the other curves are in order of increasing ℓ

potentials the motion is *not* periodic and the orbits, while bound, are not closed. I assume that there is a single $r_{\min}$ and a single $r_{\max}$ in the effective potential for each energy.

The *classical radial probability distribution* is equal to the time the particle spends in an interval dt during its motion between $r_{\min}$ and $r_{\max}$, divided by T_{21}, namely

$$P^{\text{class}}(r)\,dr = \frac{|dt|}{T_{21}} = \frac{1}{T_{21}}\left|\frac{dt}{dr}\right| dr. \tag{10.2}$$

I use conservation of energy to calculate $|dt/dr|$. The energy in the classical case can be written as

$$E = \frac{\mu}{2}\left(\frac{dr}{dt}\right)^2 + \frac{L^2}{2\mu r^2} + V(r). \tag{10.3}$$

From this equation, it follows that

$$\frac{dt}{dr} = \frac{1}{\sqrt{\frac{2}{\mu}\left(E - V(r) - \frac{L^2}{2\mu r^2}\right)}}, \tag{10.4}$$

which implies that

$$T_{21} = \int_{r_{\min}}^{r_{\max}} \frac{dr}{\sqrt{\frac{2}{\mu}\left(E - V(r) - \frac{L^2}{2\mu r^2}\right)}}. \tag{10.5}$$

Combining Eqs. (10.2), (10.4), and (10.5), I find

$$P^{\text{class}}(r) = \frac{1}{T_{21}\sqrt{\frac{2}{\mu}\left(E - V(r) - \frac{L^2}{2\mu r^2}\right)}}$$
$$= \frac{1}{\sqrt{\frac{2}{\mu}\left(E - V(r) - \frac{L^2}{2\mu r^2}\right)}\int_{r_{\min}}^{r_{\max}} \frac{dr}{\sqrt{\frac{2}{\mu}\left(E - V(r) - \frac{L^2}{2\mu r^2}\right)}}}, \tag{10.6}$$

where $r_{\min}$ and $r_{\max}$ are the solutions of the equation

$$V_{\text{eff}}(r_{\min,\max}) = V(r_{\min,\max}) + \frac{L^2}{2\mu r_{\min,\max}^2} = E. \tag{10.7}$$

The effective potential is extremely useful in analyzing problems involving central forces. For example, consider the attractive Coulomb problem for which the potential energy is

$$V(r) = -\frac{K_e}{r}, \tag{10.8}$$

where K_e is a positive constant. The effective potential is drawn in Fig. 10.1 for several angular momenta. Classically, if the particle has negative energy, $E < 0$, it is always bound. For $L = 0$ the particle orbit passes through the center of force and the energy can go to $-\infty$. For any $L > 0$, the particle can never go through the origin and the particle must have a minimum energy $E_{\min}$ that can be obtained by setting $dV_{\text{eff}}(r)/dr = 0$. For $L > 0$ and $E_{\min} \leq E < 0$, the classical orbit is *bound* between the two radii $r_{\min,\max}$. At such turning points in the orbit, the *radial* kinetic energy vanishes.

If the angular momentum is zero, the orbit passes through the origin and all the kinetic energy arises from its radial component. At points where the effective potential has a minimum, the particle undergoes circular motion and all the kinetic energy arises from the angular motion. If we fix the energy in bound state problems, there is a maximum value of the angular momentum determined by

$$\frac{1}{2}\frac{L_{\max}^2}{\mu r_0^2} = E - V(r_0), \tag{10.9}$$

where r_0 is the value of the radius for which the effective potential is a minimum. On the other hand, if instead of fixing the energy we fix the angular momentum and if $L > 0$, then the energy of the particle must be greater than or equal to some minimum energy,

$$E_{\min} = V(r_0) + \frac{1}{2}\frac{L^2}{\mu r_0^2}, \tag{10.10}$$

since there is a non-vanishing component of the kinetic energy resulting from the angular motion.

In quantum mechanics there is something analogous to the radial momentum defined in Eq. (9.10), allowing us to reduce the quantum problem to an effective one-dimensional problem for the radial motion. The effective potential serves as the potential for this one-dimensional radial motion. For example, we can expect that there may be bound state motion for $E < 0$ for the Coulomb effective potential shown in Fig. 10.1. However, for $L = 0$, we do not expect the minimum energy to equal $-\infty$ and, for any $L \neq 0$, we expect that, *if* there are bound states for a given L, the minimum energy will be larger than the minimum energy of the classical problem. The particle cannot rest at the minimum of the effective potential since this would violate the uncertainty principle. As in the classical problem, for a given bound state energy, there will be a maximum value of the magnitude of the angular momentum that is allowed.

10.1 Radial Momentum

The classical Hamiltonian is given by

$$H_{\text{class}} = \frac{p_r^2}{2\mu} + \frac{L^2}{2\mu r^2} + V(r), \tag{10.11}$$

where

$$p_r = \frac{\mathbf{p} \cdot \mathbf{r}}{r}, \tag{10.12}$$

is the radial component of the momentum. On the other hand, the quantum Hamiltonian is

$$\hat{H} = \frac{\hat{p}^2}{2\mu} + \hat{V} = -\frac{\hbar^2}{2\mu}\nabla^2 + V(r). \tag{10.13}$$

Writing ∇^2 in spherical coordinates,

$$\nabla^2 = \frac{1}{r^2}\frac{\partial}{\partial r}r^2\frac{\partial}{\partial r} + \frac{1}{r^2 \sin\theta}\frac{\partial}{\partial\theta}\sin\theta\frac{\partial}{\partial\theta} + \frac{1}{r^2\sin^2\theta}\frac{\partial^2}{\partial\phi^2}, \tag{10.14}$$

and using the fact that

$$\hat{L}^2 = -\hbar^2\left[\frac{1}{\sin\theta}\frac{\partial}{\partial\theta}\sin\theta\frac{\partial}{\partial\theta} + \frac{1}{\sin^2\theta}\frac{\partial^2}{\partial\phi^2}\right], \tag{10.15}$$

I can rewrite the quantum Hamiltonian as

$$\hat{H} = -\frac{\hbar^2}{2\mu}\left(\frac{1}{r^2}\frac{\partial}{\partial r}r^2\frac{\partial}{\partial r}\right) + \frac{\hat{L}^2}{2\mu r^2} + V(r). \tag{10.16}$$

If you compare Eqs. (10.11) and (10.16), it would seem that

$$\hat{p}_r^2 = -\hbar^2 \left(\frac{1}{r^2} \frac{\partial}{\partial r} r^2 \frac{\partial}{\partial r} \right), \tag{10.17}$$

but this is not guaranteed. I must obtain an expression for the quantum-mechanical radial momentum operator corresponding to the classical variable given in Eq. (10.12) to show whether or not this is the case. To get an Hermitian operator that corresponds to this classical variable, I take the symmetrized form

$$\hat{p}_r = \frac{1}{2} \left(\frac{\mathbf{r}}{r} \cdot \hat{\mathbf{p}} + \hat{\mathbf{p}} \cdot \frac{\mathbf{r}}{r} \right) = \frac{\hbar}{2i} \left(\mathbf{u}_r \cdot \nabla + \nabla \cdot \mathbf{u}_r \right). \tag{10.18}$$

I next evaluate

$$\begin{aligned}
\hat{p}_r \psi &= \frac{\hbar}{2i} \left[\mathbf{u}_r \cdot \nabla \psi + \nabla \cdot (\mathbf{u}_r \psi) \right] \\
&= \frac{\hbar}{2i} \left[\frac{\partial \psi}{\partial r} + \psi \nabla \cdot \mathbf{u}_r + \mathbf{u}_r \cdot \nabla \psi \right] \\
&= \frac{\hbar}{2i} \left[2 \frac{\partial \psi}{\partial r} + \psi \nabla \cdot \mathbf{u}_r \right].
\end{aligned} \tag{10.19}$$

To calculate $\nabla \cdot \mathbf{u}_r$, I use spherical coordinates,

$$\nabla \cdot \mathbf{u}_r = \frac{1}{r^2} \frac{\partial}{\partial r} r^2 = \frac{2}{r}. \tag{10.20}$$

It then follows that

$$\hat{p}_r = \frac{\hbar}{i} \left(\frac{\partial}{\partial r} + \frac{1}{r} \right) \tag{10.21}$$

and

$$\begin{aligned}
\hat{p}_r^2 \psi &= -\hbar^2 \left(\frac{\partial}{\partial r} + \frac{1}{r} \right) \left(\frac{\partial \psi}{\partial r} + \frac{\psi}{r} \right) \\
&= -\hbar^2 \left[\frac{\partial^2 \psi}{\partial r^2} + \frac{2}{r} \frac{\partial \psi}{\partial r} - \frac{\psi}{r^2} + \frac{\psi}{r^2} \right] \\
&= -\hbar^2 \left[\frac{\partial^2 \psi}{\partial r^2} + \frac{2}{r} \frac{\partial \psi}{\partial r} \right] = -\hbar^2 \left(\frac{1}{r^2} \frac{\partial}{\partial r} r^2 \frac{\partial \psi}{\partial r} \right),
\end{aligned} \tag{10.22}$$

or

$$\hat{p}_r^2 = -\hbar^2 \left(\frac{1}{r^2} \frac{\partial}{\partial r} r^2 \frac{\partial}{\partial r} \right). \tag{10.23}$$

Combining Eqs. (10.16) and (10.23), I find

$$\hat{H} = \frac{\hat{p}_r^2}{2\mu} + \frac{1}{2}\frac{\hat{L}^2}{\mu r^2} + \hat{V}, \tag{10.24}$$

which mirrors the classical Hamiltonian given in Eq. (10.11).

[As an aside, I might point out the situation is *different* for problems with cylindrical symmetry about the z-axis. The classical Hamiltonian for two-dimensional motion in the xy plane is

$$H_{\text{class}} = \frac{p_\rho^2}{2\mu} + \frac{L_z^2}{2\mu\rho^2} + V(\rho), \tag{10.25}$$

where ρ is a cylindrical coordinate and $p_\rho = \mathbf{p} \cdot \mathbf{u}_\rho$. The corresponding quantum Hamiltonian is

$$\hat{H} = -\frac{\hbar^2}{2\mu}\nabla^2 + V(\rho) = -\frac{\hbar^2}{2\mu}\left(\frac{1}{\rho}\frac{\partial}{\partial\rho}\rho\frac{\partial}{\partial\rho}\right) + \frac{\hat{L}_z^2}{2\mu\rho^2} + V(\rho), \tag{10.26}$$

which would suggest that

$$\hat{p}_\rho^2 = -\hbar^2\left(\frac{1}{\rho}\frac{\partial}{\partial\rho}\rho\frac{\partial}{\partial\rho}\right) \tag{10.27}$$

in the quantum case, but this is **not** true. Instead, (see problems)

$$\hat{p}_\rho^2 = -\hbar^2\left(\frac{1}{\rho}\frac{\partial}{\partial\rho}\rho\frac{\partial}{\partial\rho}\right) + \frac{\hbar^2}{4\rho^2}, \tag{10.28}$$

and the effective potential in the quantum problem is

$$\hat{V}_{\text{eff}} = \hat{V} + \frac{\hat{L}_z^2}{2\mu\rho^2} - \frac{\hbar^2}{8\mu\rho^2}. \tag{10.29}$$

There is an attractive quantum correction, $-\hbar^2/8\mu\rho^2$, to the classical effective potential.]

10.2 General Solution of the Schrödinger Equation for Spherically Symmetric Potentials

The time-independent Schrödinger equation that must be solved is

$$\hat{H}\psi_E(\mathbf{r}) = \left[-\frac{\hbar^2}{2\mu}\nabla^2 + V(r)\right]\psi_E(\mathbf{r}) = E\psi_E(\mathbf{r}) \tag{10.30}$$

or

$$\frac{\hbar^2}{2\mu}\left(\frac{1}{r^2}\frac{\partial}{\partial r}r^2\frac{\partial\psi_E(\mathbf{r})}{\partial r}\right)+\left[E-V(r)-\frac{\hat{L}^2}{2\mu r^2}\right]\psi_E(\mathbf{r})=0. \tag{10.31}$$

Based on the fact that $\hat{H}$, $\hat{L}^2$, and $\hat{L}_z$ commute and that the $Y_\ell^m(\theta,\phi)$ are simultaneous eigenfunctions $\hat{L}^2$ and $\hat{L}_z$, I try a solution of the form

$$\psi_{E\ell m}(\mathbf{r})=R_{E\ell}(r)Y_\ell^m(\theta,\phi), \tag{10.32}$$

substitute it into Eq. (10.31), and use the fact that

$$\hat{L}^2Y_\ell^m(\theta,\phi)=\hbar^2\ell(\ell+1)Y_\ell^m(\theta,\phi) \tag{10.33}$$

to obtain

$$\frac{\hbar^2}{2\mu}\left(\frac{1}{r^2}\frac{d}{dr}r^2\frac{dR_{E\ell}(r)}{dr}\right)+\left[E-V(r)-\frac{\hbar^2\ell(\ell+1)}{2\mu r^2}\right]R_{E\ell}(r)=0;$$

$$\frac{d^2R_{E\ell}(r)}{dr^2}+\frac{2}{r}\frac{dR_{E\ell}(r)}{dr}+\frac{2\mu}{\hbar^2}\left[E-V(r)-\frac{\hbar^2\ell(\ell+1)}{2\mu r^2}\right]R_{E\ell}(r)=0. \tag{10.34}$$

Using the results of Chap. 9, I have shown that it is a simple matter to reduce all central field problems to the solution of a *one-dimensional radial equation* for the *radial wave function* $R_{E\ell}(r)$. Note that the radial wave function has units of volume$^{-3/2}$.

Let's pause for a second and appreciate the importance of Eq. (10.34). We see that, *for each value* of ℓ, there is a radial equation that must be solved for an effective potential

$$V_{\text{eff}}(r)=V(r)+\frac{\hbar^2\ell(\ell+1)}{2\mu r^2}. \tag{10.35}$$

That is, for each value of ℓ, we can determine what bound states, if any, are present. *It is helpful to remember that each value of ℓ, in effect, corresponds to a separate problem for a given central force field.* You see that the magnetic quantum number m of $Y_\ell^m(\theta,\phi)$ does not appear in Eq. (10.34). Owing to the spherical symmetry of the potential, the energy depends only on the magnitude of the angular momentum and *not* on its direction. In the classical problem, this leads to an infinite degeneracy since all directions of $\mathbf{L}$ are allowed. In quantum mechanics, however, the degeneracy is discrete since, for each value of ℓ, m can take on $(2\ell+1)$ values $[-\ell,-\ell+1,\ldots\ell-1,\ell]$. In other words, owing to spherical symmetry, the eigenfunctions $\psi_{E\ell m}(\mathbf{r})$ are at *least* $(2\ell+1)$-fold degenerate for a given value of E and ℓ.

It is sometimes convenient to introduce a function

$$u_{E\ell}(r) = rR_{E\ell}(r) \tag{10.36}$$

which transforms Eqs. (10.34) into

$$\frac{d^2u_{E\ell}(r)}{dr^2} + \frac{2\mu}{\hbar^2}\left[E - V(r) - \frac{\hbar^2\ell(\ell+1)}{2\mu r^2}\right]u_{E\ell}(r) = 0. \tag{10.37}$$

From here onwards, I usually drop the E subscript—it is implicit.

10.2.1 Boundary Conditions

As in any problem in quantum mechanics, we must examine the boundary conditions. It is assumed that $V(r) > E$ as $r \to \infty$, since the discussion is restricted to bound states. As such, I expect the radial wave function to fall off exponentially as some power of the radius as $r \to \infty$ since $r \to \infty$ corresponds to the classically forbidden region (a region where the radial contribution to the kinetic energy is negative). The exact form of the dependence depends on the nature of the potential, but the centrifugal (angular momentum term) potential does not contribute as $r \to \infty$ since it falls off as $1/r^2$.

As $r \to 0$, I require that the radial probability density, $r^2|R_\ell(r)|^2 = |u_\ell(r)|^2$, be finite at the origin. Let us first consider $\ell \neq 0$ and assume the centrifugal potential term is larger than $V(r)$ as $r \to 0$. As $r \to 0$, the radial equation can then be approximated as

$$\frac{d^2u_\ell(r)}{dr^2} - \frac{\ell(\ell+1)}{r^2}u_\ell(r) = 0, \tag{10.38}$$

which has solutions $u_\ell(r) = r^{-\ell}, r^{\ell+1}$. The $r^{-\ell}$ solution must be rejected since it leads to a radial probability density that is not finite at the origin. Thus

$$u_\ell(r) \sim r^{\ell+1} \tag{10.39}$$

as $r \to 0$. This is a general result for any potential that rises or falls less quickly than $1/r^2$ as $r \to 0$. The power law dependence in Eq. (10.39) is not surprising; the larger the angular momentum, the further away from the origin we can expect to find the particle. The origin is a classically forbidden region if $\ell \neq 0$, since a classical particle having non-zero angular momentum cannot pass through the origin.

For $\ell = 0$, the situation must be examined on a case to case basis, using

$$\frac{d^2u_0(r)}{dr^2} + \frac{2\mu}{\hbar^2}[E - V(r)]u_0(r) = 0. \tag{10.40}$$

For attractive potentials that fall off faster than $-1/r^2$ there is no solution of the radial equation that satisfies the necessary boundary condition as $r \rightarrow 0$ when $\ell = 0$. In fact, for attractive potentials that fall off faster than $-1/r^2$ as $r \rightarrow 0$, there is no normalizable solution for any value of ℓ. For attractive potentials that fall off more slowly than $1/r^2$ as $r \rightarrow 0$, it would seem that the requirement that $|u_0(r)|^2$ be finite at the origin would not rule out the possibility that $R_0(r)$ varies as $1/r$, since $r^2 |R_0(r)|^2$ would then be finite; however, $\psi(\mathbf{r}) \sim 1/r$ is *not* a solution of Schrödinger's equation since, in that limit, the $\nabla^2\psi(\mathbf{r})$ term in the Hamiltonian would give rise to a delta function that is not present in the potential. Therefore, as $r \rightarrow 0$, we must require that

$$u_0(r) = rR_0(r) \sim 0 \text{ as } r \rightarrow 0. \tag{10.41}$$

Generally speaking, for the potentials that I consider, the radial wave function $u_\ell(r)$ satisfies the boundary condition

$$u_\ell(r) = rR_\ell(r) \sim r^{\ell+1} \text{ as } r \rightarrow 0 \tag{10.42}$$

for *all* values of ℓ. In other words, $R_\ell(r)$ is finite at the origin.

To summarize, in bound state problem and for $\ell \neq 0$, there is a classically forbidden region that extends from $r = 0$ to r_{min}, a classically allowed region between r_{min} and r_{max}, and another classically forbidden region for $r > r_{\text{max}}$ (recall that r_{min} and r_{max} are the turning points of the classical orbits for a given energy). For each value of ℓ, the lowest energy state radial wave function will have zero nodes in the classically allowed region, the next higher states, one node, etc. We can expect the radial wave function to have a polynomial or sinusoidal-like dependence in the classically allowed region. In the classically forbidden regions, the radial wave function has no nodes and $R_{E\ell}(r)$ approaches zero as $r \rightarrow 0$ (for $\ell \neq 0$) as r^ℓ and as some exponential power of r as $r \rightarrow \infty$. Each state having angular momentum quantum number ℓ is $(2\ell + 1)$ fold degenerate, owing to the spherical symmetry.

I now analyze the infinite spherical potential well, finite spherical potential well, Coulomb, and isotropic oscillator potentials.

10.3 Infinite Spherical Well Potential

The infinite spherical well potential is

$$V(r) = \begin{cases} 0 & r < a \\ \infty & r > a \end{cases}. \tag{10.43}$$

The effective potential in units of $\hbar^2/2\mu a^2$, with $L^2 = \hbar^2\ell(\ell + 1)$, is shown in Fig. 10.2 as a function of r/a. Classically, there are bound states for any value of L

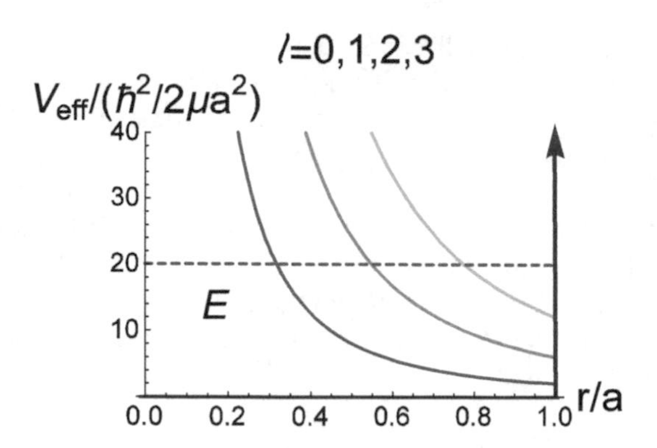

Fig. 10.2 Effective potential (in units of $\hbar^2/2\mu a^2$) for an infinite spherical well as a function of r/a. Curves corresponding to several values of angular momentum $L = \hbar\sqrt{\ell(\ell+1)}$ are shown (the $\ell = 0$ curve is along the horizontal axis). For each value of angular momentum, there is an infinity of discrete energies E possible classically. In the quantum problem, the energies are discrete. In both the classical and quantum cases, there is no upper bound to the allowed energies

and, for a given value of L, bound states occur for *all* energies

$$E \geq \frac{L^2}{2\mu a^2}, \tag{10.44}$$

where μ is the mass of the particle moving in the potential. In the quantum problem, you will see that the minimum energy for a given value of L is larger than that predicted by Eq. (10.44). Moreover the allowed energies are quantized rather than continuous. Classically, the particle is reflected each time it bounces off the spherical potential wall at $r = a$, but moves with constant velocity between bounces. There are no simple closed orbits, except for $L = 0$, when the particle moves along a diameter.

To introduce the quantum problem, let me first consider $\ell = 0$. The effective potential for $L = 0$ looks similar to that of the one-dimensional infinite square well potential (except, in the one-dimensional problem, the well width would be $2a$, going from $-a$ to a), but there is an important difference. In the three-dimensional problem, even though the potential vanishes at the center of the well, $r = 0$, there is a boundary condition that must be satisfied there. The boundary condition at $r = 0$ is $u_{k,\ell=0}(r) \equiv u_{k,0}(r) \sim r$, implying that $u_{k,0}(0) = 0$. In other words, solutions corresponding to even parity solutions of the analogous one-dimensional problem (which do not vanish at the center of the well) cannot occur in the three-dimensional case. *Remember*, the radial coordinate is *always* positive.

The solution for $\ell = 0$ is pretty simple. Equation (10.37) for the radial wave function when $r < a$ is

$$\frac{d^2 u_{k,0}(r)}{dr^2} + k^2 u_{k,0}(r) = 0, \tag{10.45}$$

where

$$k = \frac{\sqrt{2\mu E}}{\hbar}. \tag{10.46}$$

Solutions of this equation are sines and cosines of kr, but only the sine functions vanish at the origin, as required. Moreover, for the wave function to vanish at $r = a$,

$$k \to k_{n,\ell=0} \equiv k_{n0} = \frac{n\pi}{a}; \quad n = 1, 2, 3, \ldots, \tag{10.47a}$$

implying that the energy levels are quantized,

$$E_{n,\ell=0} = \frac{\hbar^2 k_{n0}^2}{2\mu}. \tag{10.47b}$$

The radial eigenfunctions are

$$R_{n0}(r) = \frac{u_{n0}(r)}{r} = \begin{cases} A\frac{\sin\left(\frac{n\pi r}{a}\right)}{r} & r < a \\ 0 & r \geq a \end{cases}, \tag{10.48}$$

where A is a normalization constant that is calculated below.

The *radial probability distribution* $P_{n\ell}(r)$ is obtained by looking at the probability to find the particle in a spherical shell between r and $r + dr$, namely

$$P_{n\ell}(r) = r^2 \int d\Omega \, |\psi_{n\ell}(\mathbf{r})|^2 = r^2 |R_{E\ell}(r)|^2 \int d\Omega \, \left|Y_\ell^m(\theta, \phi)\right|^2 = |u_{n\ell}(r)|^2. \tag{10.49}$$

[For $\ell \neq 0$, I can still label the eigenfunctions by n, although the energy is no longer given by Eqs. (10.46) and (10.47a).] For $\ell = 0$,

$$\begin{aligned} P_{n0}(r) = |u_{n0}(r)|^2 &= \begin{cases} A^2 \sin^2\left(\frac{n\pi r}{a}\right) & r < a \\ 0 & r \geq a \end{cases} \\ &= \begin{cases} \frac{2}{a} \sin^2\left(\frac{n\pi r}{a}\right) & r < a \\ 0 & r \geq a \end{cases}, \end{aligned} \tag{10.50a}$$

where the value of A was obtained using the normalization condition

$$\int_0^\infty dr P_{n0}(r) = A^2 \int_0^a dr \sin^2\left(\frac{n\pi r}{a}\right) = 1. \tag{10.50b}$$

You can view the radial probability distribution as corresponding to the *odd parity eigenstates* of the one-dimensional problem for a well of size $2a$ located between $x = -a$ and $x = a$, provided you restrict the solution to $x > 0$.

Having considered the solution for $\ell = 0$, I now discuss the solution when $\ell \neq 0$. For each value of $\ell \neq 0$, there is an infinity of quantized energy levels with some minimum energy. It might be surprising, but, for $\ell \neq 0$, it is simpler to write the equation for $r < a$ in terms of $R_{k\ell}(r)$ instead of $u_{k\ell}(r)$, since the resulting equation

$$\frac{d^2R_{k\ell}(r)}{dr^2} + \frac{2}{r}\frac{dR_{k\ell}(r)}{dr} + \left[k^2 - \frac{\ell(\ell+1)}{r^2}\right]R_{k\ell}(r) = 0 \tag{10.51}$$

is recognized as a form of Bessel's equation. The independent solutions are the so-called *spherical Bessel and Neumann functions*,

$$j_\ell(x) = \sqrt{\frac{\pi}{2x}}J_{\ell+1/2}(x); \tag{10.52a}$$

$$n_\ell(x) = \sqrt{\frac{\pi}{2x}}N_{\ell+1/2}(x), \tag{10.52b}$$

where $J_\ell(x)$ and $N_\ell(x)$ are ordinary Bessel and Neumann functions. The Bessel functions $J_n(x)$ = BesselJ[n,x] and $N_n(x)$ = BesselY[n,x] are built in functions of *Mathematica,* as are the spherical Bessel functions $j_n(x)$ = SphericalBesselJ[n,x] and $n_n(x)$ = SphericalBesselY[n,x]. The general solution of the radial equation for $r < a$ is then

$$R_{k\ell}(r) = \frac{u_{k\ell}(r)}{r} = A_\ell j_\ell(kr) + B_\ell n_\ell(kr), \tag{10.53}$$

where A_ℓ and B_ℓ are constants (that also depend implicitly of k). The solution must be consistent with the boundary condition that $R_{k\ell}(r)$ be finite at the origin. As $x \to 0$

$$j_\ell(x) \sim \frac{x^{\ell+1}}{(2\ell+1)!!}; \tag{10.54a}$$

$$n_\ell(x) \sim \begin{cases} -\frac{1}{x} & \ell = 0 \\ -\frac{(2\ell-1)!!}{x^{\ell+1}} & \ell \neq 0 \end{cases}, \tag{10.54b}$$

where

$$(2\ell+1)!! = (1)(3)\ldots(2\ell-1)(2\ell+1);$$

therefore, I must set $B_\ell = 0$ in Eq. (10.53) for all ℓ to satisfy the boundary at the origin. As a consequence, the radial wave functions are

$$R_{k\ell}(r) = \frac{u_{k\ell}(r)}{r} = \begin{cases} A_\ell j_\ell(kr) & r < a \\ 0 & r \geq a \end{cases}. \tag{10.55}$$

The first few spherical Bessel and Neumann functions are:

$$j_0(x) = \frac{\sin x}{x}; \tag{10.56a}$$

$$j_1(x) = \frac{\sin x}{x^2} - \frac{\cos x}{x}; \tag{10.56b}$$

$$j_2(x) = \left(\frac{3}{x^3} - \frac{1}{x}\right)\sin x - 3\frac{\cos x}{x^2}; \tag{10.56c}$$

$$n_0(x) = -\frac{\cos x}{x}; \tag{10.56d}$$

$$n_1(x) = -\frac{\cos x}{x^2} - \frac{\sin x}{x}; \tag{10.56e}$$

$$n_2(x) = -\left(\frac{3}{x^3} - \frac{1}{x}\right)\cos x - 3\frac{\sin x}{x^2}, \tag{10.56f}$$

and a useful asymptotic limit is

$$j_\ell(x) \sim \frac{\sin\left(x - \frac{\ell\pi}{2}\right)}{x}; \tag{10.57}$$

$$n_\ell(x) \sim \frac{-\cos\left(x - \frac{\ell\pi}{2}\right)}{x}, \tag{10.58}$$

valid for $x \gg 1$ and $x \gg \ell$.

Returning to the solution (10.55) and imposing the boundary condition that $R_{k\ell}(a) = 0$, I find

$$j_l(ka) = 0. \tag{10.59}$$

This equation can be solved numerically. For each ℓ, there is an associated effective potential that has an infinite number of energy levels. That is, for a given ℓ, the discrete energy levels can be labeled by

$$z_{n\ell} = k_{n\ell}a, \tag{10.60}$$

where $z_{n\ell}$ is the nth zero of the ℓth spherical Bessel function ($n = 1, 2, 3, \ldots$). For example, with $\ell = 1$, the lowest energy state has $z_{11} = (ka)_{11} \approx 4.5$, implying that

$$E_{11} = \frac{\hbar^2 k_{11}^2}{2\mu} = \frac{\hbar^2 k_{11}^2 a^2}{2\mu a^2} \approx 20.25\frac{\hbar^2}{2\mu a^2}. \tag{10.61}$$

The difference between the ground state energy for a given ℓ and the corresponding *classical* minimum energy $\left[\hbar^2\ell\left(\ell+1\right)/2\mu a^2\right]$, measured in units of $\hbar^2/2\mu a^2$ is equal to

$$\left(E_{1\ell} - E_\ell^{\text{class}}\right) / \left(\hbar^2/2\mu a^2\right) = \left(z_{1\ell}\right)^2 - \ell\left(\ell+1\right), \tag{10.62}$$

which is equal to π^2, 18.2, 27.2, 36.8 for $\ell = 0, 1, 2, 3$. The difference grows with increasing ℓ since Δr decreases with increasing ℓ (see Fig. 10.2), leading to larger values of Δp_r. As in the one-dimensional well, the energy levels for a given ℓ increase roughly as n^2 for large n, where n labels the corresponding zero of the spherical Bessel function.

The radial eigenfunctions are

$$R_{n\ell}(r) = \begin{cases} A_{n\ell} j_\ell(k_{n\ell} r) & r < a \\ 0 & r \geq a \end{cases} \tag{10.63}$$

and the radial probability distributions are

$$P_{n\ell}(r) = r^2 A_{n\ell}^2 j_\ell^2(k_{n\ell} r), \tag{10.64}$$

where the normalization coefficient is determined from

$$A_{n\ell} = \left[\int_0^a dr\, r^2 j_\ell^2(k_{n\ell} r) \right]^{-1/2}. \tag{10.65}$$

For each value of ℓ, the radial probability distribution has no node (other than that at $r = 0$ and $r = a$) for the lowest lying energy state, one node for the next higher energy state, etc. The dimensionless radial probability distribution $aP_{n\ell}(r)$ is plotted in Figs. 10.3, 10.4, and 10.5 as a function of r/a for $\ell = 0, 5, 10$ and $n = 1$, 2, 3. As you can see, the probability distribution is pushed further away from the origin with increasing ℓ, as would be expected from the effective potential shown in Fig. 10.2.

The classical radial probability distribution, obtained from Eqs. (10.6) and (10.7), is

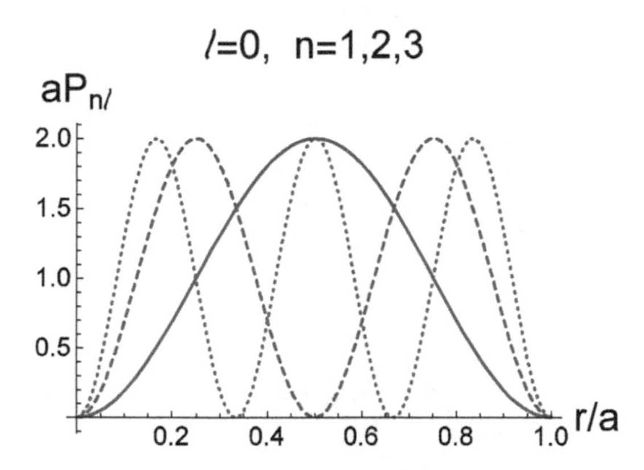

Fig. 10.3 Dimensionless radial probability distribution $aP_{n\ell}$ for the infinite well potential for $\ell = 0$ and $n = 1$ (red, solid), $n = 2$ (blue, dashed) and $n = 3$ (brown, dotted)

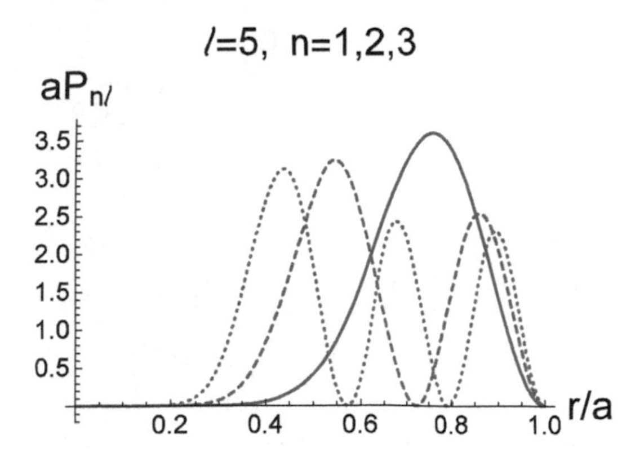

Fig. 10.4 Dimensionless radial probability distribution $aP_{n\ell}$ for the infinite well potential for $\ell = 5$ and $n = 1$ (red, solid), $n = 2$ (blue, dashed) and $n = 3$ (brown, dotted)

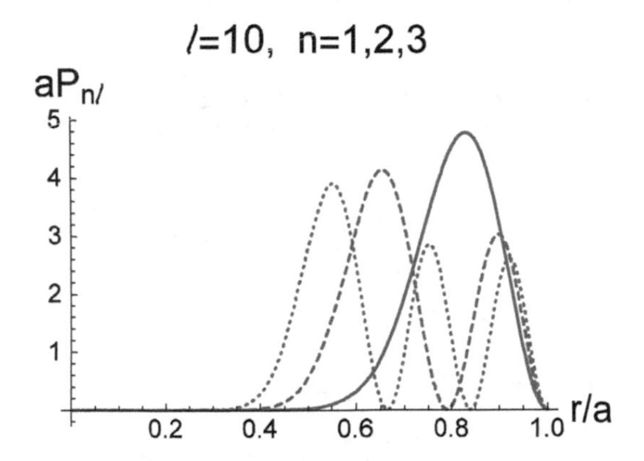

Fig. 10.5 Dimensionless radial probability distribution $aP_{n\ell}$ for the infinite well potential for $\ell = 10$ and $n = 1$ (red, solid), $n = 2$ (blue, dashed) and $n = 3$ (brown, dotted)

$$aP_{n\ell}^{\text{class}}(r) = \frac{z_{n\ell}}{\sqrt{\left(z_{n\ell}^2 - \frac{\ell(\ell+1)a^2}{r^2}\right)}\sqrt{\left(1 - \frac{\ell(\ell+1)}{z_{n\ell}^2}\right)}}. \tag{10.66}$$

To arrive at this result, I set

$$E = \frac{\hbar^2 k_{n\ell}^2}{2\mu}, \tag{10.67a}$$

$$L^2 = \hbar^2 \ell(\ell+1), \tag{10.67b}$$

in Eqs. (10.6) and (10.7) to make a correspondence with the quantum problem and used $r_{\min} = a\sqrt{\ell(\ell+1)}/z_{n\ell}$ and $r_{\max} = a$ in carrying out the integral appearing in Eq. (10.6). In Fig. 10.6, I plot $aP_{n\ell}(r)$ as a function of r/a for $\ell = 5$ and $n = 10$, along with the classical probability distribution. As you can see, the quantum radial

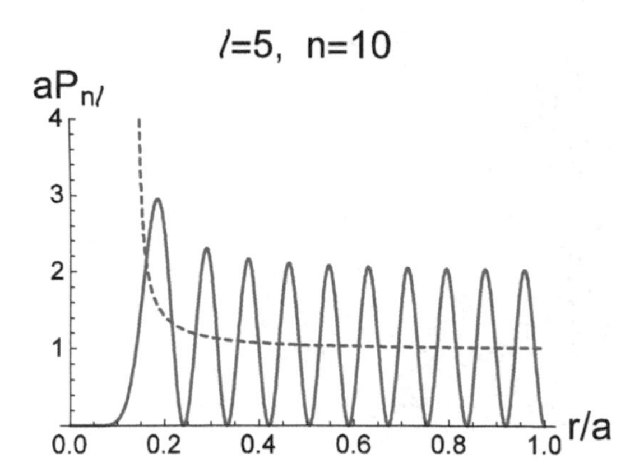

Fig. 10.6 Dimensionless radial probability distribution $aP_{n\ell}$ for the infinite well potential for $\ell = 5$ and $n = 10$; quantum distribution (red, solid), classical distribution (blue, dashed)

probability distribution in the classically allowed region, averaged over oscillations, is in good agreement with the classical distribution. The classically allowed region, obtained by solving Eq. (10.7), is defined by

$$\sqrt{\ell(\ell+1)}/z_{n\ell} < r/a < 1. \tag{10.68}$$

For $\ell = 5$ and $n = 10$, I find that $z_{10,5} = 38.9$ and that the classically allowed region is $0.141 < r/a < 1$.

Before leaving this section, I would like to return to the case of $\ell = 0$, for which

$$P_{n0}(r) = \begin{cases} \frac{2}{a}\sin^2\left(\frac{n\pi r}{a}\right) & r < a \\ 0 & r \geq a \end{cases}.$$

For large n, the radial probability density oscillates rapidly. When averaged over these oscillations, the radial probability distribution reduces to $\overline{P_{n0}(r)} = 1/a$, the *classical* probability distribution for a free particle moving along a diameter of the well. In the classical problem, the particle moves along a specific diameter (depending on the initial conditions), but in the quantum problem $|\psi_{E\ell m}(\mathbf{r})|^2 = \left|R_{E\ell}(r)Y_0^0(\theta,\phi)\right|^2 = R_{E\ell}^2(r)/4\pi$ is *spherically symmetric*. Remember that in the classical limit, quantum probability distributions correspond to classical distributions, *averaged over all possible initial conditions*. If we average the classical result over all possible initial conditions when $L = 0$, there cannot be any θ dependence since motion along every diameter of the sphere is equally likely.

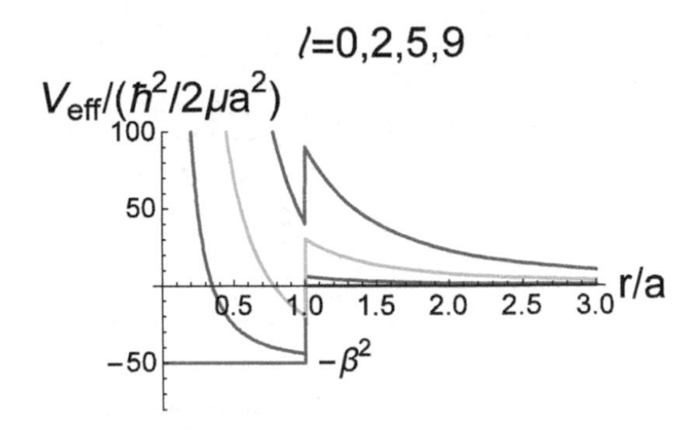

Fig. 10.7 Effective potential (in units of $\hbar^2/2\mu a^2$) for a finite spherical well as a function of r/a for $\beta^2 = 50$. Curves corresponding to several values of angular momentum $L = \hbar\sqrt{\ell(\ell+1)}$ are shown

10.4 Finite Spherical Well Potential: Bound States

Next I consider the spherical well potential,

$$V(r) = \begin{cases} -V_0 < 0 & r < a \\ 0 & r > a \end{cases}. \tag{10.69}$$

I consider only bound states, that is states for which $E < 0$. The effective potential in units of $\hbar^2/2\mu a^2$, with $L^2 = \hbar^2\ell(\ell+1)$, is shown in Fig. 10.7 as a function of r/a for $\ell = 0, 2, 5, 9$. Note that the value of V_0 in these units is β^2 (that is, $V_0/\left(\hbar^2/2\mu a^2\right) = \beta^2$). It is clear from the figure that, for fixed V_0, the number of bound states decreases with increasing angular momentum. A necessary (but not sufficient) condition for bound states to exist is

$$\ell(\ell+1) < \frac{2\mu V_0}{\hbar^2}a^2 = \beta^2; \tag{10.70}$$

otherwise, the effective potential is everywhere positive and E must be positive as well. When condition (10.70) is satisfied, the number of bound states, *if any*, depends on the values of ℓ and β.

The radial equation for $r < a$ is

$$\frac{d^2R_{E\ell}(r)}{dr^2} + \frac{2}{r}\frac{dR_{E\ell}(r)}{dr} + \left[k'^2 - \frac{\ell(\ell+1)}{r^2}\right]R_{E\ell}(r) = 0, \tag{10.71}$$

where

$$k' = \frac{\sqrt{2\mu(E+V_0)}}{\hbar} > 0, \tag{10.72}$$

while, for $r > a$, the equation is

$$\frac{d^2R_{E\ell}(r)}{dr^2}+\frac{2}{r}\frac{dR_{E\ell}(r)}{dr}+\left[-\kappa^2-\frac{\ell(\ell+1)}{r^2}\right]R_{E\ell}(r)=0, \tag{10.73}$$

where

$$\kappa=\frac{\sqrt{-2\mu E}}{\hbar}>0. \tag{10.74}$$

In both cases, the equation is a form of Bessel's equation, but I must chose the appropriate solutions consistent with the boundary conditions. The radial wave function must be finite at the origin and must not blow up as $r\rightarrow\infty$. To satisfy these boundary conditions, I take

$$R_{E\ell}(r)=\begin{cases}A_\ell j_l(k'r) & r<a\\ B_\ell h_l^{(1)}(i\kappa r) & r>a\end{cases}, \tag{10.75}$$

where

$$h_l^{(1)}(z)=j_l(z)+in_\ell(z) \tag{10.76}$$

is a *spherical Hankel function of the first kind* for which $h_l^{(1)}(i\kappa r)\sim e^{-\kappa r}$ as $r\rightarrow\infty$. The eigenenergies can then be obtained by equating the radial wave function and its derivative at $r=a$, and then solving the resulting equations graphically. In other words, I set

$$A_\ell j_l(k'a)=B_\ell h_l^{(1)}(i\kappa a); \tag{10.77a}$$

$$k'A_\ell j_l'(k'a)=i\kappa B_\ell h_l'^{(1)}(i\kappa a), \tag{10.77b}$$

where the primes on the Bessel or Hankel functions indicate derivatives that are a shorthand notation for

$$j_l'(k'a)=\left.\frac{dj_l(z)}{dz}\right|_{z=k'a} \tag{10.78a}$$

$$h_l^{(1)\prime}(i\kappa a)=\left.\frac{dh_l^{(1)}(z)}{dz}\right|_{z=i\kappa a}. \tag{10.78b}$$

Dividing Eqs. (10.77), I find

$$\frac{j_l(k'a)}{k'j_l'(k'a)}=\frac{h_l^{(1)}(i\kappa a)}{i\kappa h_l^{(1)\prime}(i\kappa a)}. \tag{10.79}$$

If you use the fact that $(k'a)^2 = \beta^2 - (\kappa a)^2$, where β is defined by Eq. (10.70), you can solve Eq. (10.79) graphically for (κa) for each value of ℓ and β. The solution determines the energy (see problems).[1]

For $\ell = 0$, I can use the fact that

$$j_0(z) = \frac{\sin z}{z}; \quad h_0^{(1)}(z) = -i\frac{e^{iz}}{z} \tag{10.80}$$

to evaluate Eq. (10.79), but it is easiest to return directly to Eq. (10.37),

$$\frac{d^2 u_{E,0}(r)}{dr^2} + k'^2 u_{E,0}(r) = 0 \tag{10.81}$$

for $r < a$ and

$$\frac{d^2 u_{E,0}(r)}{dr^2} - \kappa^2 u_{E,0}(r) = 0 \tag{10.82}$$

for $r > a$. The appropriate solutions of these equations are

$$u_{E,\ell=0}(r) = \begin{cases} A_0 \sin(k'r) & r < a \\ B_0 \exp(-\kappa r) & r > a \end{cases} \tag{10.83}$$

(only the sin solution can be taken for $r < a$ since $R_0(r) = u_0(r)/r$ must be regular at the origin). You can now solve as we did for a potential well in one dimension having width $2a$, although the solution corresponds only to the *odd* parity solutions of that problem since $u_{E,0}$ must vanish at $r = 0$, and only to the region $x > 0$ since r must be positive. There is a bound state for $\ell = 0$ only if $\beta > \pi/2$.

For values of $\ell \geq 1$, there are correspondingly higher values of β needed to support a bound state. The actual values for the eigenenergies and the number of allowed solutions are obtained by solving Eq. (10.79). For sufficiently large ℓ that violate condition (10.70), no bound states can exist in the quantum problem, even though positive energy, *classical* bound states can always be found for the effective potentials shown in Fig. 9.8 for any value of ℓ.

10.5 Bound State Coulomb Problem (Hydrogen Atom)

The electrostatic Coulomb potential is

$$V(r) = -\frac{K_e}{r}, \tag{10.84}$$

---

[1]The function $h_l^{(1)}(i\kappa a)$ is real for even ℓ and purely imaginary for odd ℓ, while the function $h_l'^{(1)}(i\kappa a)$ is real for odd ℓ and purely imaginary for even ℓ; as a consequence, the right-hand side of Eq. (10.79) is always real.

where $K_e = e^2/4\pi\epsilon_0$ is a constant. For hydrogen, the mass that appears in the Hamiltonian is not the electron mass m, but the reduced mass $\mu = mm_p/\left(m + m_p\right)$ where m_p is the proton mass. The radius r appearing in Eq. (10.84) is then the relative electron–proton separation. The effective potential is shown in Fig. 10.1; classical bound states are possible for a range of negative energies, independent of the value of ℓ. Looking at the effective potentials in Fig. 10.1, you might think that, for large ℓ, the wells are too shallow to support a bound state in the quantum problem. It turns out, however, that this is *not* the case. The slow fall off of a $1/r$ potential leads to a situation where, for any value of ℓ, there is an infinite number of bound states. For the classical problem, there is a continuum of bound state energies for each L, while in the quantum problem there is a discrete infinity of bound state energies for each ℓ.

For the potential of Eq. (10.84), the radial equation, Eq. (10.37), reduces to

$$\frac{d^2u_\ell(r)}{dr^2} + \frac{2\mu}{\hbar^2}\left[E + \frac{K_e}{r} - \frac{\hbar^2\ell\left(\ell+1\right)}{2\mu r^2}\right]u_\ell(r) = 0. \tag{10.85}$$

It is convenient to introduce dimensionless variables

$$\rho = r/a_0; \tag{10.86a}$$

$$\lambda = -E/E_R; \tag{10.86b}$$

$$u_\ell(r) \rightarrow v_\ell(\rho), \tag{10.86c}$$

where

$$a_0 = \frac{\hbar^2}{\mu K_e} = \frac{\hbar}{\mu c}\frac{1}{\alpha_{FS}} = 5.29 \times 10^{-11}\ \text{m}; \tag{10.87a}$$

$$E_R = \frac{1}{2}\frac{K_e}{a_0} = \frac{1}{2}\mu c^2\alpha_{FS}^2 = 13.6\ \text{eV}; \tag{10.87b}$$

$$\alpha_{FS} = \frac{K_e}{\hbar c} = \frac{1}{4\pi\epsilon_0}\frac{e^2}{\hbar c} \approx \frac{1}{137}. \tag{10.87c}$$

In terms of these variables, Eq. (10.85) is transformed into

$$\frac{d^2v_\ell(\rho)}{d\rho^2} + \left[-\lambda + \frac{2}{\rho} - \frac{\ell\left(\ell+1\right)}{\rho^2}\right]v_\ell(\rho) = 0. \tag{10.88}$$

This is a somewhat general dimensionless form of the radial equation. For different potentials the only term that changes is the $2/\rho$ term. I can build in the asymptotic dependence of the radial wave functions as $\rho \rightarrow 0$ and as $\rho \rightarrow \infty$. The boundary condition as $\rho \rightarrow 0$ is

$$v_\ell(\rho) \sim \rho^{\ell+1}. \tag{10.89}$$

As $\rho \to \infty$, the radial equation can be approximated as

$$\frac{d^2 v_\ell(\rho)}{d\rho^2} - \lambda v_\ell(\rho) = 0 \tag{10.90}$$

which has solutions $v_\ell(\rho) = e^{\pm\sqrt{\lambda}\rho}$. The $e^{\sqrt{\lambda}\rho}$ solution must be rejected since it leads to a radial wave function that blows up as $\rho \to \infty$. Thus

$$v_\ell(\rho) \sim e^{-\sqrt{\lambda}\rho} \tag{10.91}$$

as $\rho \to \infty$. This is a general result for the radial equation for any potential that goes to zero as $\rho \to \infty$. The exponential dependence is not surprising since the particle must penetrate into the classically forbidden.

Building in both asymptotic limits, I try a solution of the form

$$v_\ell(\rho) = \rho^{\ell+1} e^{-\sqrt{\lambda}\rho} f_\ell(\rho), \tag{10.92}$$

calculate

$$\begin{aligned} v_\ell''(\rho) &= \ell(\ell+1)\rho^{\ell-1} e^{-\sqrt{\lambda}\rho} f_\ell(\rho) \\ &\quad -2\sqrt{\lambda}(\ell+1)\rho^\ell e^{-\sqrt{\lambda}\rho} f_\ell(\rho) - 2\sqrt{\lambda}\rho^{\ell+1} e^{-\sqrt{\lambda}\rho} f_\ell'(\rho) \\ &\quad +2(\ell+1)\rho^\ell e^{-\sqrt{\lambda}\rho} f_\ell'(\rho) + \rho^{\ell+1} e^{-\sqrt{\lambda}\rho} f_\ell''(\rho) \\ &\quad +\lambda\rho^{\ell+1} e^{-\sqrt{\lambda}\rho} f_\ell(\rho), \end{aligned} \tag{10.93}$$

and substitute the result into Eq. (10.88) to arrive at

$$\rho f_\ell''(\rho) + 2\left[(\ell+1) - \rho\sqrt{\lambda}\right] f_\ell'(\rho) + 2\left[1 - (\ell+1)\sqrt{\lambda}\right] f_\ell(\rho) = 0. \tag{10.94}$$

I now make two additional changes of variable,

$$y = 2\rho\sqrt{\lambda}; \tag{10.95a}$$

$$f_\ell(\rho) \to g_\ell(y), \tag{10.95b}$$

which transforms Eq. (10.94) into

$$y\frac{d^2 g_\ell}{dy^2} + (\alpha + 1 - y)\frac{dg_\ell}{dy} + \left[\frac{1}{\sqrt{\lambda}} - (\ell+1)\right] g_\ell = 0, \tag{10.96}$$

where

$$\alpha = 2\ell + 1. \tag{10.97}$$

Equation (10.96) is known as *Laguerre's differential equation.* Only if $\alpha > -1$ do solutions of this equation exist that are regular as $\rho \to \infty$. This condition on α is a necessary but not sufficient condition for regular solutions to exist as $\rho \to \infty$. In addition it is necessary that

$$\frac{1}{\sqrt{\lambda}} - (\ell + 1) = q, \tag{10.98}$$

where q is a positive integer or zero. When both these conditions are satisfied, the physically acceptable solutions of Eq. (10.96) are

$$g_{\alpha\lambda}(y) = L_q^{\alpha}(y). \tag{10.99}$$

The $L_q^{\alpha}(y)$ are the *generalized Laguerre polynomials* that satisfy the differential equation

$$y\frac{d^2L_q^{\alpha}(y)}{dy^2} + (\alpha + 1 - y)\frac{dL_q^{\alpha}(y)}{dy} + qL_q^{\alpha}(y) = 0. \tag{10.100}$$

Some properties of the generalized Laguerre polynomials [*Mathematica* symbol LaguerreL[q, α, y]= $L_q^{\alpha}(y)$] are listed in the Appendix.

For the hydrogen atom problem,

$$\alpha = 2\ell + 1 > -1; \tag{10.101a}$$

$$q = \frac{1}{\sqrt{\lambda}} - (\ell + 1). \tag{10.101b}$$

The requirement that q be a non-negative integer leads us to the condition

$$\frac{1}{\sqrt{\lambda}} = q + \ell + 1 = n, \tag{10.102}$$

where n is defined as $(q + \ell + 1)$. From Eq. (10.102) and the fact that both q and ℓ are non-negative integers, it follows that

$$n \geq 1 \quad \text{and} \quad \ell \leq n - 1. \tag{10.103}$$

As was to be expected from classical considerations, there is a maximum angular momentum for a fixed energy.

The eigenenergies are given by

$$E_n = -\lambda E_R = -\frac{1}{2}\frac{\mu c^2 \alpha_{FS}^2}{n^2}; \qquad n = 1, 2, 3, \ldots. \tag{10.104}$$

For each value of n, ℓ can equal $0, 1, \ldots, (n-1)$, and, for each value of ℓ, m can equal $0, \pm 1, \pm 2, \ldots \pm \ell$. Thus, the energy degeneracy for a given n is

$$\sum_{\ell=0}^{n-1} (2\ell + 1) = n^2. \tag{10.105}$$

There is an "accidental degeneracy" for states having the same n, but different ℓ. This can be related to the fact that there is a conserved *dynamic* constant called the Lenz vector that points in the direction of the semi-major axis of the classical problem (the orbits are closed). In group theory the symmetry is related to the group O(4), the orthogonal group in four dimensions.

The solution of Eq. (10.96) is

$$g_{\alpha q}(y) = L_q^{\alpha}(y) = L_{n-\ell-1}^{2\ell+1}\left(2\rho\sqrt{\lambda}\right) = L_{n-\ell-1}^{2\ell+1}\left(\frac{2\rho}{n}\right). \tag{10.106}$$

To get the total radial wave function $R_{n\ell}(\rho) \sim v_\ell(\rho)/\rho$, I must multiply $L_{n-\ell-1}^{2\ell+1}\left(\frac{2\rho}{n}\right)$ by $\rho^\ell e^{-\rho/n}$ [see Eq. (10.92)]. The normalized dimensionless wave function can then be written as

$$\tilde{\psi}_{n\ell m}(\boldsymbol{\rho}) = \tilde{R}_{n\ell}(\rho) Y_\ell^m(\theta, \phi), \tag{10.107}$$

where the dimensionless radial wave function $\tilde{R}_{n\ell}(\rho)$ is

$$\begin{aligned} \tilde{R}_{n\ell}(\rho) &= \left(\frac{2}{\rho n^3}\right)^{1/2} \Lambda_{n-\ell-1}^{2\ell+1}\left(\frac{2\rho}{n}\right) \\ &= \frac{2}{n^2}\sqrt{\frac{(n-\ell-1)!}{(n+\ell)!}}\left(\frac{2\rho}{n}\right)^\ell e^{-\rho/n} L_{n-\ell-1}^{2\ell+1}\left(\frac{2\rho}{n}\right), \end{aligned} \tag{10.108}$$

and

$$\Lambda_q^{2\ell+1}(\rho) = \sqrt{\frac{q!}{(q+2\ell+1)!}} e^{-\rho/2} \rho^{\ell+1/2} L_q^{2\ell+1}(\rho) \tag{10.109}$$

is an *associated Laguerre function*. The $\tilde{\psi}_{n\ell m}(\boldsymbol{\rho})$ constitute an orthonormal set.

From the Appendix, some properties of the $L_{n-\ell-1}^{2\ell+1}$ and $\Lambda_q^{2\ell+1}$ functions are

$$L_q^{2\ell+1}(\rho) \text{ is a polynomial of order } q; \tag{10.110a}$$

$$\int_0^\infty d\rho L_q^{2\ell+1}(\rho) L_{q'}^{2\ell+1}(\rho) e^{-\rho} \rho^{2\ell+1} = \frac{(q+2\ell+1)!}{q!}\delta_{q,q'}; \tag{10.110b}$$

$$\rho\Lambda_{n-\ell-1}^{2\ell+1}(\rho) = 2n\Lambda_{n-\ell-1}^{2\ell+1}(\rho) - \sqrt{(n+\ell+1)(n-\ell)}\Lambda_{n-\ell}^{2\ell+1}(\rho) - \sqrt{(n-\ell-1)(n+\ell)}\Lambda_{n-\ell-2}^{2\ell+1}(\rho); \tag{10.110c}$$

$$\int_0^\infty d\rho\Lambda_q^\alpha(\rho)\Lambda_{q'}^\alpha(\rho) = \delta_{q,q'}; \tag{10.110d}$$

$$2\left(\frac{1}{n^3n'^3}\right)^{1/2}\int_0^\infty d\rho\,\rho\Lambda_{n-\ell-1}^{2\ell+1}\left(\frac{2\rho}{n}\right)\Lambda_{n'-\ell-1}^{2\ell+1}\left(\frac{2\rho}{n'}\right) = \delta_{n,n'}. \tag{10.110e}$$

The number of nodes in the radial wave function is $q = n - \ell - 1$.

The first few dimensionless radial wave functions are

$$\tilde{R}_{10}(\rho) = 2e^{-\rho}; \tag{10.111a}$$

$$\tilde{R}_{20}(\rho) = \frac{(2-\rho)\,e^{-\rho/2}}{2\sqrt{2}}; \tag{10.111b}$$

$$\tilde{R}_{21}(\rho) = \frac{\rho e^{-\rho/2}}{2\sqrt{6}}; \tag{10.111c}$$

$$\tilde{R}_{30}(\rho) = \frac{2\left(27-18\rho+2\rho^2\right)e^{-\rho/3}}{81\sqrt{3}}; \tag{10.111d}$$

$$\tilde{R}_{31}(\rho) = \frac{4\,(6-\rho)\,\rho e^{-\rho/3}}{81\sqrt{6}}; \tag{10.111e}$$

$$\tilde{R}_{32}(\rho) = \frac{2}{81}\sqrt{\frac{2}{15}}\rho^2e^{-\rho/3}. \tag{10.111f}$$

Dimensionless radial wave functions are plotted in Figs. 10.8, 10.9 and 10.10 for $n = 1, 2, 3$.

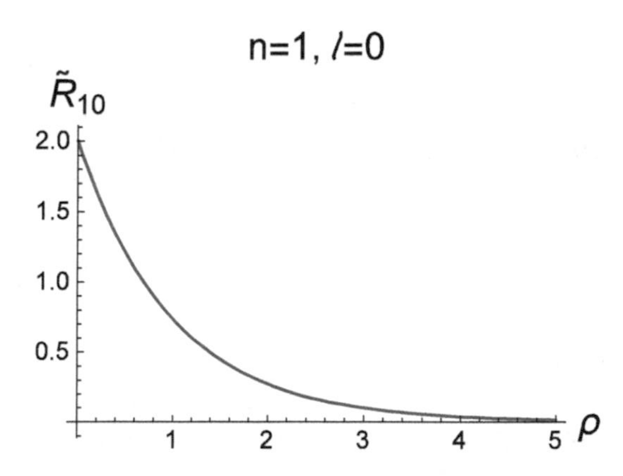

Fig. 10.8 Dimensionless hydrogenic radial wavefunction for $n = 1$

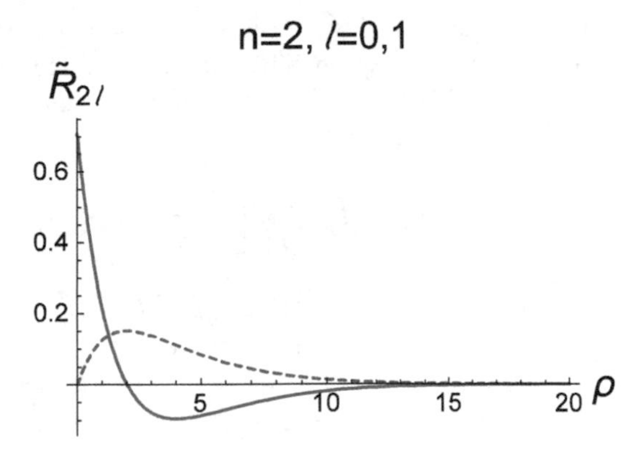

Fig. 10.9 Dimensionless hydrogenic radial wave function for $n = 2$. $\ell = 0$ (red, solid); $\ell = 1$ (blue, dashed)

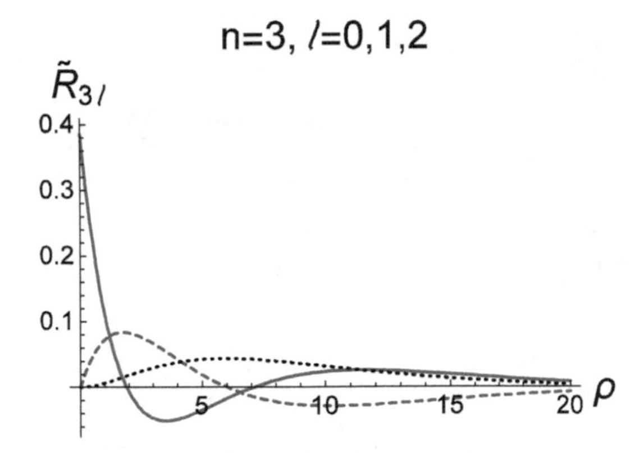

Fig. 10.10 Dimensionless hydrogenic radial wavefunction for $n = 3$. $\ell = 0$ (red, solid); $\ell = 1$ (blue, dashed); $\ell = 2$ (black, dotted)

Going from an eigenfunction that is dimensionless to one that has dimensions of $1/\sqrt{\text{volume}}$ is accomplished by taking

$$\psi_{n\ell m}(\mathbf{r}) = R_{n\ell}(r)Y_\ell^m(\theta, \phi), \tag{10.112}$$

where the radial wave function is

$$R_{n\ell}(r) = \frac{1}{a_0^{3/2}}\tilde{R}_{n\ell}\left(\frac{r}{a_0}\right) \tag{10.113}$$

or

$$R_{n\ell}(r) = \frac{1}{a_0^{3/2}}\left(\frac{2}{n^2}\right)\sqrt{\frac{(n-\ell-1)!}{(n+\ell)!}}\left(\frac{2r}{na_0}\right)^\ell e^{-r/na_0}L_{n-\ell-1}^{2\ell+1}\left(\frac{2r}{na_0}\right). \tag{10.114}$$

In order to gain some physical insight into the radial dependence of the eigenfunctions, I calculate the radial probability distribution and compare it with the corresponding classical radial probability distribution. In dimensionless units, the quantum radial probability distribution is

$$\begin{aligned}\tilde{P}_{n\ell}(\rho) = \rho^2 \tilde{R}^2_{n\ell}(\rho) &= \left(\frac{2}{n^3}\right) \rho \left[\Lambda^{2\ell+1}_{n-\ell-1}(2\rho/n)\right]^2 \\ &= \left(\frac{1}{n^2}\right) \frac{(n-\ell-1)!}{(n+\ell)!} e^{-2\rho/n} \left(\frac{2\rho}{n}\right)^{2\ell+2} \\ &\times \left[L^{2\ell+1}_{n-\ell-1}(2\rho/n)\right]^2 .\end{aligned} \tag{10.115}$$

On the other hand, the classical radial probability distribution in dimensionless units, obtained using Eq. (10.6), is given by

$$\begin{aligned}\tilde{P}^{\text{class}}_{n\ell}(\rho) &= a_0 P^{\text{class}}(a_0\rho) \\ &= \frac{a_0}{T_{21}\sqrt{\frac{2K_e}{\mu a_0}}\sqrt{\left(-\frac{1}{2n^2} + \frac{1}{\rho} - \frac{\ell(\ell+1)}{2\rho^2}\right)}},\end{aligned} \tag{10.116}$$

where a_0 is the Bohr radius and I have set $E = -E_R/n^2 = -K_e/\left(2n^2 a_0\right)$ and $L^2 = \hbar^2 \ell(\ell+1)$. Equations (10.5) and (10.7) can be used to obtain

$$T_{21} = \pi \sqrt{-\frac{\mu K_e^2}{8E^3}} = \pi n^3 \sqrt{\frac{\mu a_0^3}{K_e}}. \tag{10.117}$$

Note that T_{21} is one-half the period of the classical orbit. Combining Eqs. (10.116) and (10.117), I obtain the classical probability distribution

$$\tilde{P}^{\text{class}}_{n\ell}(\rho) = \frac{1}{\pi n^3 \sqrt{-\frac{1}{n^2} + \frac{2}{\rho} - \frac{\ell(\ell+1)}{\rho^2}}}. \tag{10.118}$$

There are two turning points, given by

$$\rho_{\text{min,max}}(n,\ell) = n^2 \left[1 \mp \sqrt{1 - \frac{\ell(\ell+1)}{n^2}}\right]. \tag{10.119}$$

These classical turning points correspond to positions in the orbit when the electron is located along the semi-major axis of the ellipse. For fixed energy (fixed n) $\rho_{\text{min}}(n,\ell)$ decreases and $\rho_{\text{max}}(n,\ell)$ increases with decreasing ℓ, a result that is deduced easily from graphs of the effective potential. The value $\ell = n-1$ corresponds most closely to circular orbits; in this limit

$$\rho_{\min,\max}(n, n-1) = n^2 \left[1 \mp \frac{1}{n}\right] \tag{10.120}$$

and the relative width of the distribution is

$$\frac{\rho_{\max}(n, n-1) - \rho_{\min}(n, n-1)}{[\rho_{\max}(n, n-1) + \rho_{\min}(n, n-1)]/2} = \frac{2n}{n^2} = \frac{2}{n}. \tag{10.121}$$

With increasing n, the classical distribution for $\ell = n - 1$ corresponds more closely to circular orbits. Of course, circular orbits *are* possible in the classical case for an energy equal to the effective potential at its minimum; however, the values I chose to simulate the quantum variables, $E_n = -K_e/\left(2n^2 a_0\right)$ and $L^2 = \hbar^2 \ell (\ell + 1)$, do not correspond to circular orbits.

I now return to the quantum probability distribution, Eq. (10.115). It possesses many of the features of the classical distribution function in the classically allowed region. For example, for $\ell = n - 1$

$$\tilde{P}_{n,n-1}(\rho) = \frac{1}{n^2 (2n-1)!} \left(\frac{2}{n}\right)^{2n} e^{-2\rho/n} \rho^{2n}. \tag{10.122}$$

This function possesses a single maximum at $\rho_c = n^2$ or $r_c = n^2 a_0$ and a relative width of order of that given by Eq. (10.121). For smaller values of ℓ, the orbits correspond to classical elliptical orbits and the electron spends less time when it is nearest to the proton, corresponding to the fact that the speed in the classical orbits is larger, the closer the electron is to the nucleus. There are $n - \ell - 1$ nodes in the classically allowed region. Using the recursion relation (10.110c), you can show that

$$\langle \rho \rangle = \int_0^\infty \rho \tilde{P}_{n\ell}(\rho) = \frac{3n^2 - \ell(\ell + 1)}{2}, \tag{10.123}$$

which is also equal to $\langle \rho \rangle$ for the classical probability distribution given by Eq. (10.118).

The (dimensionless) quantum radial probability distribution is plotted in Figs. 10.11, 10.12 and 10.13 as a function of ρ/n^2 for $n = 40$ and $\ell = 39, 25, 0$ as the solid red curves. For $\ell = n-1 = 39$ (which corresponds most closely to circular orbits), the maximum of occurs at $\rho/n^2 = 1$. For $\ell = 10$, you can see the relative maxima in the envelope of the distribution at the inner and outer turning points, $\rho_{\min,\max}(n, \ell)/n^2 = 0.23, 1.77$, with the probability largest at the outer turning point. For $\ell = 0$, the probability distribution extends to the center of force and the envelope has a single maximum near the classical turning point at $\rho/n^2 = 2$. The classical radial probability distribution $\tilde{P}_{n\ell}^{\text{class}}(\rho)$ is superimposed on the quantum distribution as the dashed blue curves in Figs. 10.11, 10.12 and 10.13. It is seen that it agrees very well with the quantum distribution, averaged over oscillations, in the classically allowed region.

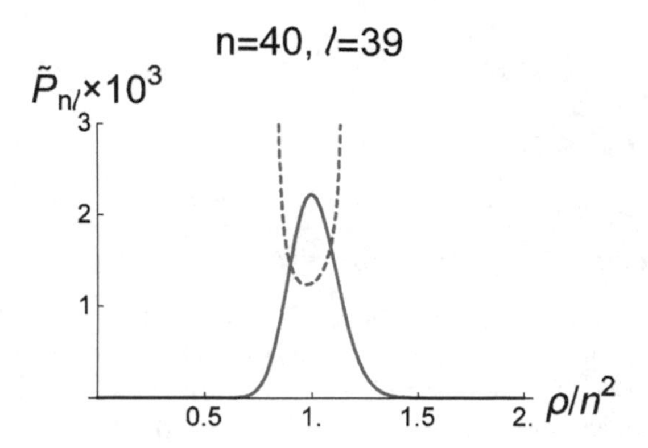

Fig. 10.11 Quantum (solid, red) and classical (blue, dashed) radial probability distributions for hydrogen for $n = 40$, $\ell = 39$

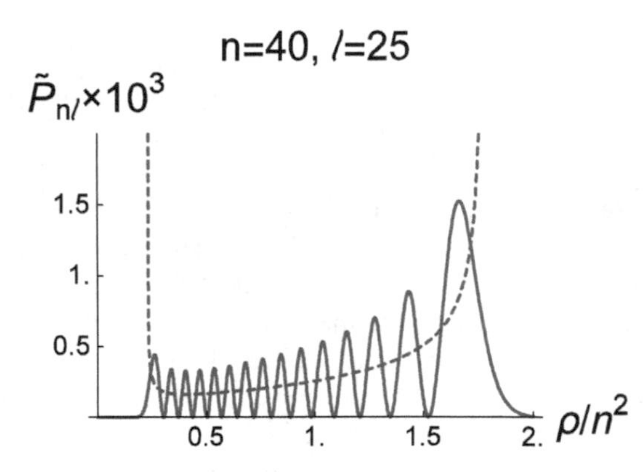

Fig. 10.12 Quantum (solid, red) and classical (blue, dashed) radial probability distributions for hydrogen for $n = 40$, $\ell = 25$

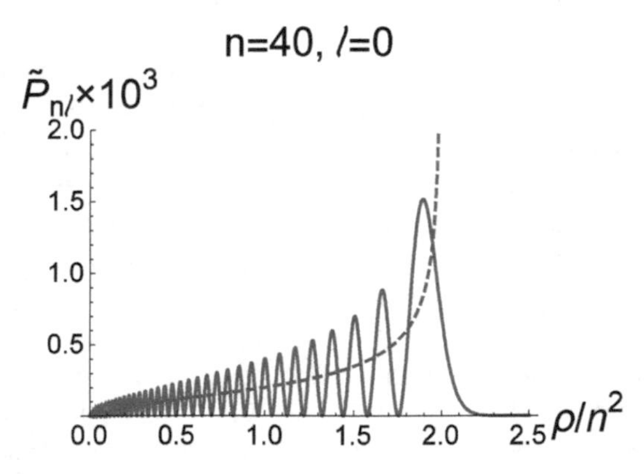

Fig. 10.13 Quantum (solid, red) and classical (blue, dashed) radial probability distributions for hydrogen for $n = 40$, $\ell = 0$

10.6 3-D Isotropic Harmonic Oscillator

The potential energy for an isotropic oscillator is

$$V(r) = \frac{1}{2}\mu\omega^2 r^2. \tag{10.124}$$

I have already solved this problem in rectangular coordinates [see Sect. 8.1.3]. The eigenenergies are

$$E_n = \left(n + \frac{3}{2}\right)\hbar\omega, \tag{10.125}$$

where n is a positive integer or zero. There is an $(n+1)(n+2)/2$ fold degeneracy for each value of n.

I can formulate a "Bohr theory" for circular orbits for the oscillator, in which

$$\mu v r = \mu\omega r^2 = (n+1)\hbar; \quad n = 0, 1, 2, \ldots. \tag{10.126}$$

$$F = \frac{\mu v^2}{r} = \mu\omega^2 r, \tag{10.127}$$

leading to

$$r_n = \sqrt{(n+1)}\sqrt{\frac{\hbar}{\mu\omega}}, \tag{10.128}$$

$$v_n = \sqrt{(n+1)}\sqrt{\frac{\hbar\omega}{\mu}}, \tag{10.129}$$

and

$$E_n = \hbar\omega(n+1). \tag{10.130}$$

The energy spacing is correct, but the levels are displaced by $-\hbar\omega/2$ from the true values.

The effective potential in units of $\hbar\omega$ is shown in Fig. 10.14 as a function of $\xi = \sqrt{\mu\omega/\hbar}r$ for $\ell = 0, 2, 5, 9$. The effective potential for $\ell = 0$ is the same as that for the one-dimensional harmonic oscillator, restricted to $x > 0$. However, since the wave function must be finite at the origin, only the odd parity solutions of the one-dimensional oscillator are allowed. In other words, for $\ell = 0$, the energies are

$$E_{n,\ell=0} = (q + 1/2)\hbar\omega; \quad q = 1, 3, 5, \ldots \tag{10.131a}$$

$$= (n + 3/2)\hbar\omega; \quad n = 0, 2, 4, \ldots \tag{10.131b}$$

Thus, $\ell = 0$ states appear only in states with n *even*. This is not a surprise. The parity of the eigenstates, obtained from Eq. (8.18) is $(-1)^{n_x+n_y+n_z} = (-1)^n$. Since

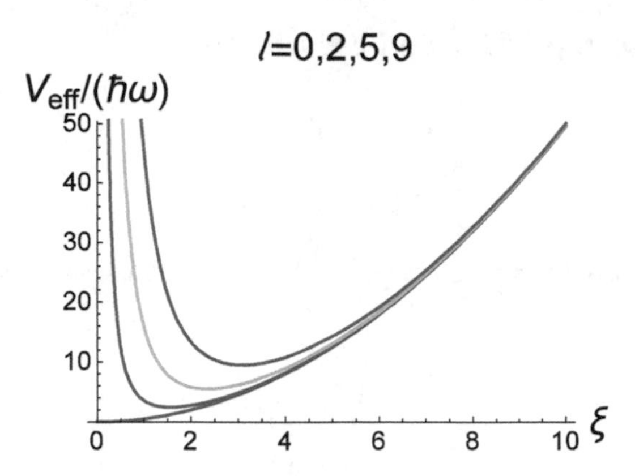

Fig. 10.14 Effective potential for the 3-D oscillator in units of $\hbar\omega$ as a function of $\xi = \sqrt{\mu\omega/\hbar}r$ for $\ell = 0, 2, 5, 9$

the parity of an $\ell = 0$ state is equal to $+1$, the value of n for all states having $\ell = 0$ must be even. As you shall see, for a given n, only those ℓ values are allowed for which $(-1)^\ell = (-1)^n$.

In spherical coordinates, the radial equation in terms of a dimensionless coordinate

$$\xi = \sqrt{\frac{\mu\omega}{\hbar}}r \tag{10.132}$$

and a dimensionless energy

$$\lambda = 2E/\hbar\omega \tag{10.133}$$

is

$$\frac{d^2u_{\lambda\ell}(\xi)}{d\xi^2} + \left[\lambda - \xi^2 - \frac{\ell(\ell+1)}{\xi^2}\right]u_{\lambda\ell}(\xi) = 0. \tag{10.134}$$

Assuming a solution of the form

$$u_{\lambda\ell}(\xi) = \xi^{\ell+1}e^{-\xi^2/2}f_{\lambda\ell}(\xi), \tag{10.135}$$

I calculate

$$\begin{aligned} u''_{\lambda\ell}(\xi) &= \ell(\ell+1)\,\xi^{\ell-1}e^{-\xi^2/2}f_{\lambda\ell}(\xi) \\ &\quad - (2\ell+3)\,\xi^{\ell+1}e^{-\xi^2/2}f_{\lambda\ell}(\xi) - 2\xi^{\ell+2}e^{-\xi^2/2}f'_{\lambda\ell}(\xi) \end{aligned}$$

$$+2\,(\ell+1)\,\xi^{\ell}e^{-\xi^2/2}f'_{\lambda\ell}(\xi)+\xi^{\ell+1}e^{-\xi^2/2}f''_{\lambda\ell}(\xi)$$
$$+\xi^{\ell+3}e^{-\xi^2/2}f_{\lambda\ell}(\xi), \tag{10.136}$$

and substitute the result into Eq. (10.134) arrive at

$$\xi f''_{\lambda\ell}(\xi)+2\left[(\ell+1)-\xi^2\right]f'_{\lambda\ell}(\xi)+[\lambda-3-2\ell]\,\xi f_{\lambda\ell}(\xi)=0. \tag{10.137}$$

In contrast to Eq. (10.94) for the hydrogen atom problem, the quantity ξ^2 appears rather than ξ in the factor multiplying $f'_{\lambda\ell}(\xi)$. With the replacements

$$y=\xi^2, \tag{10.138a}$$
$$\frac{d}{d\xi}=2\xi\frac{d}{dy}=2\sqrt{y}\frac{d}{dy}; \tag{10.138b}$$
$$\frac{d^2}{d\xi^2}=2\frac{d}{dy}+4\xi\sqrt{y}\frac{d^2}{dy^2}=2\frac{d}{dy}+4y\frac{d^2}{dy^2}; \tag{10.138c}$$
$$f_{\lambda\ell}(\xi)\rightarrow g_{\lambda\ell}(y), \tag{10.138d}$$

Eq. (10.137) is transformed into

$$yg''_{\lambda\ell}(y)+\left[\left(\ell+\frac{3}{2}\right)-y\right]g'_{\lambda\ell}(y)+\left[\frac{\lambda-3-2\ell}{4}\right]g_{\lambda\ell}(y)=0, \tag{10.139}$$

which is Laguerre's equation. For physically acceptable solutions, it is necessary that

$$\frac{\lambda-3-2\ell}{4}=q \tag{10.140}$$

where q is a positive integer or zero and

$$\alpha=\ell+\frac{3}{2}>-1. \tag{10.141}$$

The condition on α is satisfied automatically, while the condition on λ is

$$\lambda=(4q+2\ell)+3=2n+3, \tag{10.142}$$

or, using Eq. (10.133),

$$E_n=\left(n+\frac{3}{2}\right)\hbar\omega, \tag{10.143}$$

where

$$n = 2q + \ell, \tag{10.144}$$

which implies that $n = \ell, \ell + 2, \ell + 4, \ldots$ Alternatively, for a given $n \geq 0$, $\ell = n, n-2, \ldots$, with the minimum value of ℓ equal to zero if n is even and one if n is odd. Thus

$$\ell = 0, 2, 4, \ldots, n \qquad n \text{ even} \tag{10.145a}$$

$$\ell = 1, 3, 5, \ldots, n \qquad n \text{ odd.} \tag{10.145b}$$

Since there are $(2\ell + 1)$ degenerate substates for each ℓ, you can verify easily that the degeneracy of a state of given n is $(n+1)(n+2)/2$, as was found previously using rectangular coordinates. There is an "accidental" degeneracy, as in hydrogen, owing to the fact that the classical orbits are closed ellipses, but, in contrast to hydrogen, all degenerate states have the same parity. This is connected with the fact that the Lenz vector for the Coulomb problem does not commute with the parity operator, but the Lenz "vector" (actually a second rank tensor consisting of five operators) for the harmonic oscillator does commute with the parity operator. The symmetry group for the oscillator is $SU(3)$, the special unitary group in three dimensions.

The physically acceptable solution of Eq. (10.139) is

$$g_{n\ell}(y) = L_{\left(\frac{n-\ell}{2}\right)}^{\left(\ell+\frac{1}{2}\right)}(y) \tag{10.146}$$

or

$$f_{n\ell}(\xi) = L_{\left(\frac{n-\ell}{2}\right)}^{\left(\ell+\frac{1}{2}\right)}\left(\xi^2\right). \tag{10.147}$$

As a consequence, I can write the normalized (dimensionless) radial wave functions as

$$\tilde{R}_{n\ell}(\xi) = \sqrt{\frac{2}{\xi}}\Lambda_{\left(\frac{n-\ell}{2}\right)}^{\left(\ell+\frac{1}{2}\right)}\left(\xi^2\right) \tag{10.148}$$

$$= \sqrt{2}\frac{\left[\left(\frac{n-\ell}{2}\right)!\right]^{1/2}}{\left[\Gamma\left(\frac{n+\ell+3}{2}\right)\right]^{1/2}}\xi^{\ell}e^{-\xi^2/2}L_{\left(\frac{n-\ell}{2}\right)}^{\left(\ell+\frac{1}{2}\right)}\left(\xi^2\right), \tag{10.149}$$

where Λ_q^{α} is defined by Eq. (10.169) and Γ is the gamma function defined by

$$\Gamma(x) = \int_0^{\infty} t^{x-1}e^{-t}dt, \tag{10.150}$$

such that $\Gamma(n+1) = n!$ for integers $n \geq 0$. The first few dimensionless radial wave functions are

$$\tilde{R}_{00}(\xi) = \frac{2}{\pi^{1/4}} e^{-\xi^2/2}; \tag{10.151a}$$

$$\tilde{R}_{11}(\xi) = \frac{2}{\pi^{1/4}} \sqrt{\frac{2}{3}} \xi e^{-\xi^2/2}; \tag{10.151b}$$

$$\tilde{R}_{20}(\xi) = \frac{2}{\pi^{1/4}} \sqrt{\frac{2}{3}} \left(\frac{3}{2} - \xi^2\right) e^{-\xi^2/2}; \tag{10.151c}$$

$$\tilde{R}_{22}(\xi) = \frac{2}{\pi^{1/4}} \sqrt{\frac{4}{15}} \xi^2 e^{-\xi^2/2}. \tag{10.151d}$$

The radial wave functions in coordinate space are given by

$$R_{n\ell}(r) = \left(\frac{\mu\omega}{\hbar}\right)^{3/4} \tilde{R}_{n\ell}\left(\sqrt{\frac{\mu\omega}{\hbar}} r\right). \tag{10.152}$$

Qualitatively, the results are similar to those for the hydrogen atom, since the orbits are elliptical in both cases; however, the radial coordinate is a minimum on the semi-*minor* axis rather than the semi-major axis of the ellipse, since the center of force is at the origin in the case of the oscillator whereas it is at one of the foci in the Coulomb problem. The (dimensionless) radial probability distribution is

$$\tilde{P}_{n\ell}(\xi) = \xi^2 \tilde{R}_{n\ell}^2(\xi) = \frac{2\xi^{2(\ell+1)} e^{-\xi^2} \left[L_{\left(\frac{n-\ell}{2}\right)}^{\left(\ell+\frac{1}{2}\right)}\left(\xi^2\right)\right]^2 \left(\frac{n-\ell}{2}\right)!}{\Gamma\left[\frac{n}{2} + \frac{\ell}{2} + \frac{3}{2}\right]}. \tag{10.153}$$

The maximum of $\tilde{P}_{n,n}(\xi)$ (which corresponds most closely to circular orbits) occurs at $\xi_{\max} = \sqrt{(n+1)}$, the prediction of Eq. (10.128) of the Bohr theory of the oscillator. In contrast to hydrogen, $\xi_{\max}$ grows much more slowly with increasing n (as $\sqrt{n+1}$ rather than as n^2) owing to the fact that the binding force is much stronger for the oscillator than for the electron in hydrogen.

For the oscillator, it follows from Eqs. (10.5) and (10.7) that

$$T_{21} = \frac{T}{4} = \frac{\pi}{2\omega}, \tag{10.154}$$

where T is the period of the classical orbit. As a consequence, the classical radial probability distribution in dimensionless units, obtained from Eqs. (10.6) and (10.7), is

$$\tilde{P}_{n\ell}^{\text{class}}(\xi) = \sqrt{\frac{\hbar}{\mu\omega}} P^{\text{class}}\left(\sqrt{\frac{\hbar}{\mu\omega}}\xi\right)$$
$$= \frac{2}{\pi\sqrt{2n+3-\xi^2-\frac{\ell(\ell+1)}{\xi^2}}}, \tag{10.155}$$

where the argument of the square root is restricted to positive values and I have set $E = \left(n+\frac{3}{2}\right)\hbar\omega$, $L^2 = \hbar^2\ell(\ell+1)$. For $\ell = n \gg 1$ (which corresponds most closely to circular orbits), the distribution $\tilde{P}_{n\ell}^{\text{class}}(\xi)$ is confined to a small range about $\xi = \sqrt{n+1}$, as is the quantum probability distribution. For $\ell = 0$, the classical radial probability distribution extends to the center of force and there is a turning point at $\xi = \sqrt{2n+3}$. As in the Coulomb problem, the classical radial probability agrees very well with the quantum distribution, averaged over oscillations, in the classically allowed region. Comparisons are left to the problems.

10.7 Summary

The Schrödinger equation for problems with spherical symmetry can be reduced to a one-dimensional equation for the radial wave function. I have solved the radial equation for a number of problems involving spherically symmetric potentials. It is fortuitous that the solution of the spherical well potential, the Coulomb potential, and the isotropic oscillator potential can be written in terms of known functions. In most cases, it is necessary to solve the radial equation numerically. To help interpret the results, I made comparisons of the quantum probability distributions with the corresponding classical distributions.

10.8 Appendix: Laguerre Polynomials

Laguerre's equation

$$yf''(y) + [\alpha + 1 - y]f'(y) + qf(y) = 0 \tag{10.156}$$

admits polynomial solutions of order q when q is a positive integer or zero and $\alpha > -1$. The solutions of this equation with these restrictions are the generalized Laguerre polynomials

$$f(y) = L_q^\alpha(y). \tag{10.157}$$

If these conditions are not met, the solutions of Eq. (10.156) are not convergent in the interval $(0, \infty)$ and do not enter as physically acceptable solutions in the problems I am considering.

Some equations related to Eq. (10.156) and their appropriate solutions are

$$yf''(y) + [y+1]f'(y) + \left(q + \frac{\alpha}{2} + 1 - \frac{\alpha^2}{4y}\right)f(y) = 0; \tag{10.158a}$$

$$f = e^{-y}y^{\alpha/2}L_q^{\alpha}(y); \tag{10.158b}$$

$$f''(y) + \left(\frac{q + (\alpha+1)/2}{y} + \frac{1-\alpha^2}{4y^2} - \frac{1}{4}\right)f(y) = 0; \tag{10.159a}$$

$$f = e^{-y/2}y^{(\alpha+1)/2}L_q^{\alpha}(y); \tag{10.159b}$$

$$f''(y) + \left(4q + 2\alpha + 2 - y^2 + \frac{\frac{1}{4} - \alpha^2}{y^2}\right)f(y) = 0; \tag{10.160a}$$

$$f = e^{-y^2/2}y^{\alpha+1/2}L_q^{\alpha}(y^2). \tag{10.160b}$$

Some properties of the generalized Laguerre polynomials are listed below:
The orthogonality of the generalized Laguerre polynomials is expressed as

$$\int_0^{\infty} dyL_q^{\alpha}(y)L_{q'}^{\alpha}(y)e^{-y}y^{\alpha} = \frac{\Gamma(q+\alpha+1)}{q!}\delta_{q,q'}, \tag{10.161}$$

where

$$\Gamma(x) = \int_0^{\infty} t^{x-1}e^{-t}dt \tag{10.162}$$

is the gamma function; a series expansion is

$$L_q^{\alpha}(y) = \sum_{m=0}^{q}\binom{q+\alpha}{q-m}\frac{(-y)^m}{m!}, \tag{10.163}$$

where the binomial coefficient is defined as

$$\binom{a}{b} = \frac{\Gamma(a+1)}{\Gamma(b+1)\Gamma(a-b+1)}; \tag{10.164}$$

the generating function is

$$F(y,h) = \frac{1}{(1-h)^{\alpha+1}}e^{-\frac{hy}{1-h}} = \sum_{q=0}^{\infty} L_q^{\alpha}(y)h^q; \tag{10.165}$$

the identity

$$\frac{d}{dy}L_{q+1}^{\alpha}(y) = -L_q^{\alpha+1}(y) = \frac{1}{y}\left[(q+1)\,L_{q+1}^{\alpha}(y) - (q+1+\alpha)\,L_q^{\alpha}(y)\right] \tag{10.166}$$

and the fact that $L_0^{\alpha}(y) = 1$ can be used to generate all the generalized Laguerre polynomials; the first few generalized Laguerre polynomials are

$$L_0^{\alpha}(y) = 1; \tag{10.167a}$$

$$L_1^{\alpha}(y) = -y + \alpha + 1; \tag{10.167b}$$

$$L_2^{\alpha}(y) = \frac{1}{2}\left[y^2 - 2(\alpha+2)\,y + (\alpha+1)(\alpha+2)\right]; \tag{10.167c}$$

a useful recursion relation is

$$(q+1)\,L_{q+1}^{\alpha}(y) - (2q+\alpha+1-y)\,L_q^{\alpha}(y) + (q+\alpha)\,L_{q-1}^{\alpha}(y) = 0. \tag{10.168}$$

The associated Laguerre functions defined by

$$\Lambda_q^{\alpha}(y) = \sqrt{\frac{q!}{\Gamma(q+\alpha+1)}}\,e^{-y/2}y^{\alpha/2}L_q^{\alpha}(y) \tag{10.169}$$

satisfy the orthogonality relations

$$\int_0^{\infty} dy\Lambda_q^{\alpha}(y)\Lambda_{q'}^{\alpha}(y) = \delta_{q,q'}; \tag{10.170}$$

$$2\left(\frac{1}{n^3 n'^3}\right)^{1/2}\int_0^{\infty} d\rho\,\rho\Lambda_{n-\ell-1}^{2\ell+1}\left(\frac{2\rho}{n}\right)\Lambda_{n'-\ell-1}^{2\ell+1}\left(\frac{2\rho}{n'}\right) = \delta_{n,n'}, \tag{10.171}$$

and the recursion relation

$$\begin{aligned}\sqrt{(q+1)(q+\alpha+1)}\Lambda_{q+1}^{\alpha}(y) - (2q+\alpha+1-y)\,\Lambda_q^{\alpha}(y)\\ +\sqrt{q(q+\alpha)}\Lambda_{q-1}^{\alpha}(y) = 0.\end{aligned} \tag{10.172}$$

10.9 Problems

1. Find the three lowest energy states in electron volts for the $\ell = 0$ and $\ell = 1$ states of a particle having mass $\mu = 10^{-30}$ kg moving in an *infinite*, spherical potential well having radius $a = 10^{-9}$ m.

2. A particle having mass μ moves in a spherical well potential

$$V(r) = \begin{cases} -V_0 < 0 & r < a \\ 0 & r > a \end{cases}.$$

The lowest energy state has $\ell = 0$. Find the value of V_0 needed to guarantee a bound state. How does this problem differ from a three-dimensional square well potential, for which a bound state always exists?

3. Draw the effective potential for the potential of Problem 10.2. For a given energy, find values of the angular momentum for which bound states can exist classically. Classically, can bound states exist for $E > 0$? Explain. In the quantum problem, can bound states exist for $E > 0$? Explain. Find a condition on ℓ in the quantum problem that will guarantee that a bound state cannot exist.

4–5. For the potential of Problem 10.2, calculate the eigenenergies for $\ell = 1$ and $\beta = 2, 4, 6$, where $\beta = \sqrt{2\mu V_0 a^2/\hbar^2}$. In each case express the eigenenergies in terms of the dimensionless quantity $x = \kappa a = \sqrt{-2\mu E a^2/\hbar^2}$. You will have to solve Eq. (10.79) graphically to obtain the solutions. Classically, what is the minimum value of β needed to have a bound state with $E < 0$ for an angular momentum $L = \hbar\sqrt{\ell(\ell+1)}$ and $\ell = 1$? How does this compare with the minimum value of β needed to support a bound state in the quantum problem?

6. Give a very rough uncertainty principle argument to estimate the ground state energy of hydrogen. To do this replace $\langle 1/r \rangle$ by $1/\langle r \rangle$ and set $\langle p_r^2 \rangle = \hbar^2/\left(4\langle r \rangle^2\right)$ in $\left\langle \hat{H} \right\rangle$.

7. Using the effective potential for hydrogen,

$$V_{\text{eff}} = -\frac{K_e}{r} + \frac{\hbar^2 \ell(\ell+1)}{2\mu r^2},$$

where $K_e = e^2/4\pi\epsilon_0$, show that, for fixed

$$E = -\frac{1}{2}\mu c^2 \left(\frac{K_e^2}{\hbar^2 c^2}\right) \frac{1}{n^2} < 0,$$

there is a maximum value of ℓ allowed and that it corresponds *roughly* to what is found in the quantum theory, that is to $\ell_{\text{max}} = n - 1$.

8. Plot the dimensionless radial probability distribution, $\rho^2 \left|\tilde{R}_{n\ell}(\rho)\right|^2$, for the $n = 2, \ell = 1$; $n = 2, \ell = 0$; $n = 10, \ell = 9$; $n = 10, \ell = 5$; and $n = 10, \ell = 0$ states of the hydrogen atom. Interpret your results.

9. Use the recursion relation

$$\rho\Lambda_{n-\ell-1}^{2\ell+1}(\rho) = 2n\Lambda_{n-\ell-1}^{2\ell+1}(\rho) - \sqrt{(n+\ell+1)(n-\ell)}\Lambda_{n-\ell}^{2\ell+1}(\rho)$$
$$-\sqrt{(n-\ell-1)(n+\ell)}\Lambda_{n-\ell-2}^{2\ell+1}(\rho)$$

to derive

$$\langle\rho\rangle = \int_0^\infty \rho\tilde{P}_{n\ell}(\rho) = \frac{3n^2 - \ell(\ell+1)}{2},$$

where $\tilde{P}_{n\ell}(\rho)$ is the dimensionless radial probability distribution for the electron in hydrogen.

10. Construct a "Bohr theory" for the 3-D isotropic harmonic oscillator and show that the maximum of the quantum radial probability distribution for the oscillator agrees with your theory for states having $n = \ell$.

11. Plot the dimensionless radial probability distribution $\tilde{P}_{n\ell}(\xi)$ for the $n = 60$, $\ell = 60, 30, 0$ states of the 3-D isotropic harmonic oscillator given in Eq. (10.153), along with the classical distribution given by Eq. (10.155). Interpret your results.

12. Prove that the wave function for the 3-D isotropic harmonic oscillator is periodic and that $\langle\mathbf{r}\rangle$ obeys the classical equation of motion for the oscillator. Does the *radial* motion oscillate at frequency ω? Explain.

13–14. Consider the $n = 2$ state of the 3-D isotropic oscillator. Write the dimensionless wave functions in both rectangular and spherical coordinates. Write each of the rectangular wave functions in terms of the eigenfunctions in spherical coordinates.

15–16. In cylindrical coordinates (ρ, ϕ), the Hamiltonian for a particle having mass μ moving in a potential $V(\rho)$ is

$$\hat{H} = -\frac{\hbar^2}{2\mu}\nabla^2 + V(\rho)$$
$$= -\frac{\hbar^2}{2\mu}\left(\frac{1}{\rho}\frac{\partial}{\partial\rho}\rho\frac{\partial}{\partial\rho} + \frac{1}{\rho^2}\frac{\partial^2}{\partial\phi^2}\right) + V(\rho).$$

The angular momentum operator is $\hat{L}_z = \frac{\hbar}{i}\frac{\partial}{\partial\phi}$, so the Hamiltonian can be written as

$$\hat{H} = -\frac{\hbar^2}{2\mu}\left(\frac{1}{\rho}\frac{\partial}{\partial\rho}\rho\frac{\partial}{\partial\rho} - \frac{\hat{L}_z^2}{\hbar^2\rho^2}\right) + V(\rho)$$
$$= -\frac{\hbar^2}{2\mu}\left(\frac{1}{\rho}\frac{\partial}{\partial\rho}\rho\frac{\partial}{\partial\rho}\right) + \frac{\hat{L}_z^2}{2\mu\rho^2} + V(\rho).$$

On the other hand, the classical Hamiltonian is

$$H_{\text{class}} = \frac{p_\rho^2}{2\mu} + \frac{L_z^2}{2\mu\rho^2} + V(\rho) = \frac{p_\rho^2}{2\mu} + V_{\text{eff}},$$

where $p_\rho = \mathbf{p} \cdot \mathbf{u}_\rho$. This would suggest that

$$\hat{p}_\rho^2 = -\hbar^2 \left(\frac{1}{\rho} \frac{\partial}{\partial \rho} \rho \frac{\partial}{\partial \rho} \right)$$

and

$$V_{\text{eff}} = \frac{\hat{L}_z^2}{2\mu\rho^2} + V(\rho),$$

but this is **not** the case. To prove this show that $\hat{p}_\rho$ defined as

$$\begin{aligned} \hat{p}_\rho &= \frac{1}{2} \left(\hat{\mathbf{p}} \cdot \mathbf{u}_\rho + \mathbf{u}_\rho \cdot \hat{\mathbf{p}} \right) \\ &= \frac{\hbar}{2i} \left(\nabla \cdot \mathbf{u}_\rho + \mathbf{u}_\rho \cdot \nabla \right) \end{aligned}$$

is equal to

$$\hat{p}_\rho = \frac{\hbar}{i} \left(\frac{\partial}{\partial \rho} + \frac{1}{2\rho} \right)$$

and that

$$\hat{p}_\rho^2 = -\hbar^2 \left(\frac{1}{\rho} \frac{\partial}{\partial \rho} \rho \frac{\partial}{\partial \rho} \right) + \frac{\hbar^2}{4\rho^2}.$$

As a consequence show that, in the quantum problem,

$$V_{\text{eff}} = \frac{\hat{L}_z^2}{2\mu\rho^2} - \frac{\hbar^2}{8\mu\rho^2} + V(\rho).$$

In other words, there is an attractive "barrier" in the 2-D problem, even for eigenfunctions corresponding to $\left\langle \hat{L}_z^2 \right\rangle = 0$. This is the reason why there is always a bound state for a "circular" potential well in two dimensions.

17. The Hamiltonian for the previous problem is

$$\hat{H} = -\frac{\hbar^2}{2\mu} \left(\frac{1}{\rho} \frac{\partial}{\partial \rho} \rho \frac{\partial}{\partial \rho} + \frac{1}{\rho^2} \frac{\partial^2}{\partial \phi^2} \right) + V(\rho).$$

Assume a solution for the eigenfunctions of the form

$$\psi_{Em}(\boldsymbol{\rho}) = u_{Em}(\rho)e^{im\phi}/\sqrt{\rho},$$

where m is an integer, and show that the radial equation for $u_{Em}(\rho)$ is

$$u''_{Em}(\rho) + \left[\frac{2\mu}{\hbar^2}\left[E - V(\rho)\right] + \frac{1-4m^2}{4\rho^2}\right]u_{Em}(\rho) = 0.$$

For the potential $V(\rho) = \mu\rho^2\omega^2/2$ (isotropic two-dimensional harmonic oscillator), introduce dimensionless variables

$$\xi = \sqrt{\frac{\mu\omega}{\hbar}}\rho; \quad \lambda = 2E/\hbar\omega,$$

and show that the dimensionless radial function $\tilde{u}_{\lambda m}(\xi)$ obeys the differential equation

$$\tilde{u}''_{\lambda m}(\xi) + \left(\lambda + \frac{(1-4m^2)}{4\xi^2} - \xi^2\right)\tilde{u}_{\lambda m}(\xi) = 0.$$

Compare this with Eq. (10.160) to show that the eigenenergies are

$$E_n = (n+1)\hbar\omega; \quad n = 0, 1, 2, \ldots .$$

and that the dimensionless radial eigenfunctions are

$$\tilde{R}_{nm}(\xi) = \sqrt{2\frac{\left(\frac{n-|m|}{2}\right)!}{\left(\frac{n+|m|}{2}\right)!}}\xi^{|m|}e^{-\xi^2/2}L^{|m|}_{\frac{n-|m|}{2}}(\xi^2),$$

where m varies from $-n$ to n in integer steps of 2.

18–19. The classical Hamiltonian for a particle having mass μ and charge q moving in a constant magnetic field $\mathbf{B}$ is given by

$$H = \frac{(\mathbf{p} - q\mathbf{A})^2}{2\mu}$$

where

$$\mathbf{A} = -\mathbf{r} \times \mathbf{B}/2$$

is the vector potential. Prove that $\mathbf{B} = \nabla \times \mathbf{A}$ and show that the Hamiltonian can be written as

$$H = \frac{p^2}{2\mu} - \frac{q\mathbf{L} \cdot \mathbf{B}}{2\mu} + \frac{A^2}{2\mu},$$

where $\mathbf{L} = \mathbf{r} \times \mathbf{p}$ and $\mathbf{p}$ is the canonical momentum, related to the velocity by

$$\mathbf{p} = \mu \dot{\mathbf{r}} + q\mathbf{A}.$$

If the initial velocity is perpendicular to $\mathbf{B}$, prove that the motion is in a plane perpendicular to $\mathbf{B}$.

Now assume that the magnetic field is along the z-direction, $\mathbf{B} = B\mathbf{u}_z$ and consider the motion to be confined in the xy plane. Use cylindrical coordinates (ρ, ϕ). At $t = 0$, take

$$\boldsymbol{\rho}(0) = \boldsymbol{\rho}_0 = x_0\mathbf{u}_x + y_0\mathbf{u}_y;$$
$$\mathbf{v}(0) = \mathbf{v}_0 = v_{x0}\mathbf{u}_x + v_{y0}\mathbf{u}_y.$$

Use the Lorentz force equation to find $\mathbf{v}(t)$ and integrate the result to obtain $\boldsymbol{\rho}(t)$. Prove that the orbit of the particle is a circle of radius $R = v_0/\omega_c$ centered at

$$x_c = x_0 + v_{y0}/\omega_c;$$
$$y_c = y_0 - v_{x0}/\omega_c,$$

where

$$\omega_c = \frac{qB}{\mu}$$

is the *cyclotron frequency*. Prove the following relationships:

$$H = \frac{1}{2}\mu v(t)^2 = \text{constant} = \frac{1}{2}\mu v_0^2;$$
$$L_z = \text{constant} = |\boldsymbol{\rho}_0 \times \mathbf{p}_0| = \mu\left(x_0 v_{0y} - y_0 v_{0x}\right) + \omega_c \rho_0^2/2$$
$$= \omega_c \rho_c^2/2;$$
$$H = \frac{1}{2}\mu \dot{\rho}(t)^2 + V_{\text{eff}}(\rho),$$

where

$$\boldsymbol{\rho}(t) = x(t)\mathbf{u}_x + y(t)\mathbf{u}_y = \rho(t)\mathbf{u}_\rho;$$
$$\rho_c^2 = x_c^2 + y_c^2;$$

$$\mathbf{v}(t) = \dot{\boldsymbol{\rho}}(t) = v_x(t)\mathbf{u}_x + v_y(t)\mathbf{u}_y = \dot{\rho}(t)\,\mathbf{u}_\rho + \rho(t)\dot{\phi}(t)\mathbf{u}_\phi;$$

$$V_{\text{eff}}(\rho) = \frac{L_z^2}{2\mu\rho^2} - \frac{qL_zB}{2\mu} + \frac{q^2B^2}{8\mu}\rho^2.$$

Note that the distance from the origin to the center of the orbit scales as $\sqrt{L_z}$.

20–21. Now consider the corresponding quantum problem with the magnetic field in the z-direction, for which

$$\begin{aligned}\hat{H} &= \frac{\left(\hat{\mathbf{p}} - q\hat{\mathbf{A}}\right)^2}{2\mu} \\ &= \frac{\hat{p}^2}{2\mu} + \frac{q^2B^2}{8\mu}\rho^2 - \frac{q\hat{L}_zB}{2\mu}\end{aligned}$$

(the fact that $\hat{p}_j$ commutes with $\hat{A}_j$ is needed to arrive at this result). Show that the Hamiltonian is the same as that for an isotropic two-dimensional harmonic oscillator having frequency

$$\omega = \frac{\omega_c}{2} = \frac{qB}{2\mu},$$

except that there is an *additional term*,

$$-\frac{q\hbar mB}{2\mu} = -m\hbar\omega,$$

in the radial equation. As a consequence of this term, show that the equation for the radial function u_{Em} of problem 10.17 acquires an additional term, $2mu_{Em}$, and then use the results of of problem 10.17 to show that the energy levels are given by

$$E_n = \left(n + \frac{1}{2}\right)\hbar\omega_c,$$

where n is an integer ≥ 0 and m is an integer that ranges from $-n$ to infinity. Thus, there is an *infinite* degeneracy for each n, a result that can be traced to the translational symmetry of the problem. Use the effective potential to show why this degeneracy is possible for this problem, but not for the isotropic two-dimensional oscillator. The equally spaced n levels for this potential are referred to as *Landau levels* (after Lev Landau). Obtain the radial eigenfunctions and plot the dimensionless radial probability distribution for $n = 10$ and $m = 100, 400$ on the same graph. Check to see if changing the value of m results in a translation of the probability distribution and if the radial position of the center of the distribution scales approximately as $\sqrt{m}$, as predicted from classical considerations.

Chapter 11
Dirac Notation

We've reached a plateau. You now know how to solve problems in one, two, and three dimensions. Hopefully you have the basics under your belt. In this chapter, I present a somewhat more general way of specifying the state of a quantum system, based on a formalism developed by Dirac.[1]

11.1 Vector Spaces and Dirac Notation

Up to this point, I have focused on methods for obtaining the *eigenfunctions* associated with various Hamiltonians; moreover, the discussion has been limited mainly to the coordinate representation. The eigenfunctions form a complete set of functions, allowing you to expand any function as a linear superposition of the eigenfunctions. This is similar to, but not exactly identical to the situation with vectors. As you know you can expand a vector as

$$\mathbf{A} = A_x\mathbf{u}_x + A_y\mathbf{u}_y + A_z\mathbf{u}_z. \tag{11.1}$$

An *orthonormal basis set* $(\mathbf{u}_x, \mathbf{u}_y, \mathbf{u}_z)$ has been chosen for convenience. You can obtain any component of the vector by projection,

$$A_j = \mathbf{u}_j \cdot \mathbf{A}; \quad j = x, y, z. \tag{11.2}$$

In quantum mechanics, on the other hand, an arbitrary function $\psi(\mathbf{r})$ can be expanded in the set of *basis functions* corresponding to the eigenfunctions $\psi_f(\mathbf{r})$ of some Hermitian operator $\hat{F}$, namely

[1] P. A. M. Dirac, *Principles of Quantum Mechanics, Fourth Edition* (Oxford University Press, Oxford, U.K., 1958).

P.R. Berman, *Introductory Quantum Mechanics*, UNITEXT for Physics,
https://doi.org/10.1007/978-3-319-68598-4_11

$$\psi(\mathbf{r}) = \sum_f b_f \psi_f(\mathbf{r}), \tag{11.3}$$

where

$$b_f = (\psi_f, \psi) = \int d\mathbf{r} \psi_f^* (\mathbf{r})\, \psi\, (\mathbf{r})\, ; \tag{11.4}$$

the scalar product projection operation used for vectors is replaced by integration.

The analogy with vector spaces can be made exact if we deal with *eigenstates* and *state vectors* rather than eigenfunctions and wave functions. Dirac developed a powerful formalism for representing state vectors in quantum mechanics. Students leaving an introductory course in quantum mechanics often can *use* Dirac notation, but may not appreciate its significance.

11.1.1 Vector Spaces

It is probably easiest to think of Dirac notation in relation to a three-dimensional vector space. Any three dimensional vector can be written as

$$\mathbf{A} = A_x \mathbf{u}_x + A_y \mathbf{u}_y + A_z \mathbf{u}_z, \tag{11.5}$$

where A_x, A_y, A_z are the components of the vector in this x, y, z basis. I can represent the unit vectors as column vectors,

$$\mathbf{u}_x = \begin{pmatrix} 1 \\ 0 \\ 0 \end{pmatrix}; \quad \mathbf{u}_y = \begin{pmatrix} 0 \\ 1 \\ 0 \end{pmatrix}; \quad \mathbf{u}_z = \begin{pmatrix} 0 \\ 0 \\ 1 \end{pmatrix}, \tag{11.6}$$

such that the vector $\mathbf{A}$ can be written as

$$\mathbf{A} = \begin{pmatrix} A_x \\ A_y \\ A_z \end{pmatrix}. \tag{11.7}$$

Of course, the basis vectors $\mathbf{u}_x, \mathbf{u}_y, \mathbf{u}_z$ are not unique; any set of three non-collinear unit vectors would do as well. Let's call one such set $\mathbf{u}_1, \mathbf{u}_2, \mathbf{u}_3$, such that

$$\mathbf{A} = A_1 \mathbf{u}_1 + A_2 \mathbf{u}_2 + A_3 \mathbf{u}_3. \tag{11.8}$$

The vector $\mathbf{A}$ is *absolute* in the sense that it is basis-independent. For a given basis, the components of $\mathbf{A}$ change in precisely the correct manner to insure that $\mathbf{A}$ remains unchanged. The example in the problems should make this clear.

The scalar product of two vectors is defined in the usual fashion as the cosine of the angle between the vectors. Although not necessary, it is convenient to choose an *orthonormal basis*, one in which the scalar product of different basis vectors vanishes and the scalar product of a basis vector with itself is equal to unity. The basis vectors $\mathbf{u}_x, \mathbf{u}_y, \mathbf{u}_z$ constitute an orthogonal basis since

$$\begin{aligned} \mathbf{u}_x \cdot \mathbf{u}_x = \mathbf{u}_y \cdot \mathbf{u}_y = \mathbf{u}_z \cdot \mathbf{u}_z = 1; \\ \mathbf{u}_x \cdot \mathbf{u}_y = \mathbf{u}_y \cdot \mathbf{u}_z = \mathbf{u}_x \cdot \mathbf{u}_z = 0. \end{aligned} \tag{11.9}$$

In addition to a geometric interpretation to the scalar product, I can give a definition based on matrix multiplication. Even though the vectors are real quantities, in preparation for quantum mechanics, I define the *adjoint* of a vector $\mathbf{A}$ as

$$\mathbf{A}^\dagger = \begin{pmatrix} A_x^* & A_y^* & A_z^* \end{pmatrix}, \tag{11.10}$$

that is as a row matrix whose components are the complex conjugates of those of the column matrix $\mathbf{A}$. The scalar product of vectors $\mathbf{A}$ and $\mathbf{B}$ is then defined as

$$\mathbf{A} \cdot \mathbf{B} = \underline{\mathrm{A}}^\dagger \underline{\mathrm{B}} = \begin{pmatrix} A_x^* & A_y^* & A_z^* \end{pmatrix} \begin{pmatrix} B_x \\ B_y \\ B_z \end{pmatrix} = A_x^* B_x + A_y^* B_y + A_z^* B_z, \tag{11.11}$$

such that

$$\mathbf{A} \cdot \mathbf{A} = \underline{\mathrm{A}}^\dagger \underline{\mathrm{A}} = \begin{pmatrix} A_x^* & A_y^* & A_z^* \end{pmatrix} \begin{pmatrix} A_x \\ A_y \\ A_z \end{pmatrix} = |A_x|^2 + |A_y|^2 + |A_z|^2 = |\mathbf{A}|^2, \tag{11.12}$$

as desired. A line below a symbol indicates a matrix.

Expressed as matrices, the adjoints of the unit vectors given in Eq. (11.6) are

$$\underline{\mathrm{u}}_x^\dagger = \begin{pmatrix} 1 & 0 & 0 \end{pmatrix}; \quad \underline{\mathrm{u}}_y^\dagger = \begin{pmatrix} 0 & 1 & 0 \end{pmatrix}; \quad \underline{\mathrm{u}}_z^\dagger = \begin{pmatrix} 0 & 0 & 1 \end{pmatrix}. \tag{11.13}$$

You can verify that, consistent with Eq. (11.9),

$$\mathbf{u}_i \cdot \mathbf{u}_j = \underline{\mathrm{u}}_i^\dagger \underline{\mathrm{u}}_j = \delta_{i,j}; \qquad i, j = \{x, y, z\}; \tag{11.14}$$

these unit vectors form an orthonormal basis. The jth component of a vector is then obtained by projection as

$$A_j = \mathbf{u}_j \cdot \mathbf{A} = \underline{\mathrm{u}}_j^\dagger \underline{\mathrm{A}}; \qquad j = \{x, y, z\}. \tag{11.15}$$

Next, I define the action of an operator $\hat{O}$ on a vector $\mathbf{A}$ by the equation

$$\mathbf{A}' = \hat{O}\mathbf{A}; \tag{11.16}$$

that is, the operator $\hat{O}$ acting on a vector $\mathbf{A}$ produces a new vector $\mathbf{A}'$. For example, the operator $\hat{O}$ may result in a translation or a rotation of the vector $\mathbf{A}$. It won't take you too much effort to realize that a translation doesn't change a vector, but a rotation mixes up its components in some specified manner. A *linear operator* is one that produces a new vector having components that are a linear combination of the initial components of the vector. In other words, a linear operator $\hat{O}$ acting on a vector $\mathbf{A}$ produces a new vector $\mathbf{A}'$ having components

$$A'_x = O_{xx}A_x + O_{xy}A_y + O_{xz}A_z \tag{11.17a}$$

$$A'_y = O_{yx}A_x + O_{yy}A_y + O_{yz}A_z \tag{11.17b}$$

$$A'_z = O_{zx}A_x + O_{zy}A_y + O_{zz}A_z. \tag{11.17c}$$

Equation (11.16) corresponds to what is called an *active transformation*. The vector itself is operated on (e.g., the vector is rotated), but the basis vectors are left unchanged. Thus, the *new* vector is expressed in terms of the *original* basis as

$$\mathbf{A}' = A'_x\mathbf{u}_x + A'_y\mathbf{u}_y + A'_z\mathbf{u}_z. \tag{11.18}$$

It is possible to write Eq. (11.17) as a matrix equation if I define

$$\underline{O} = \begin{pmatrix} O_{xx} & O_{xy} & O_{xz} \\ O_{yx} & O_{yy} & O_{yz} \\ O_{zx} & O_{zy} & O_{zz} \end{pmatrix}; \tag{11.19}$$

that is, I represent a linear operator as a matrix, such that the transformation (11.16) can be written in matrix form as

$$\underline{A}' = \underline{O}\underline{A}. \tag{11.20}$$

This is an important result; *linear operators can be represented as matrices*. I can introduce a set of *basis matrices* of the form

$$\underline{m}_{xy} = \underline{u}_x\underline{u}_y^\dagger = \begin{pmatrix} 1 \\ 0 \\ 0 \end{pmatrix} \begin{pmatrix} 0 & 1 & 0 \end{pmatrix} = \begin{pmatrix} 0 & 1 & 0 \\ 0 & 0 & 0 \\ 0 & 0 & 0 \end{pmatrix}; \tag{11.21}$$

that is, a one for the xy element and zeroes everywhere else. If I re-label x, y, z as $1, 2, 3$, then $\underline{m}_{ij}$ has a one for the ij element and zeroes everywhere else, such that

$$\underline{O} = \sum_{i,j=1}^{3} O_{ij}\underline{m}_{ij} \tag{11.22}$$

I also replace $\mathbf{u}_x, \mathbf{u}_y, \mathbf{u}_z$ by $\mathbf{u}_1, \mathbf{u}_2, \mathbf{u}_3$. In that way, Eq. (11.22) remains valid in *any* orthonormal basis with

$$\underline{m}_{ij} = \underline{u}_i\underline{u}_j^{\dagger}; \qquad i,j = \{1,2,3\}. \tag{11.23}$$

Of course, the matrix elements of the matrix $\underline{O}$ depend on the basis. You are at liberty to represent the basis vectors as

$$\begin{pmatrix}1\\0\\0\end{pmatrix}, \begin{pmatrix}0\\1\\0\end{pmatrix}, \begin{pmatrix}0\\0\\1\end{pmatrix} \tag{11.24}$$

in any *one* orthonormal basis, but once you choose this basis, you must express all other unit vectors in terms of this specific basis. As with marriage, you make your choice and you live with it.

Note that, for orthonormal basis vectors,

$$\begin{aligned}\underline{u}_i^{\dagger}\underline{O}\underline{u}_j &= \sum_{i',j'=1}^{3} \underline{u}_i^{\dagger} O_{i'j'}\underline{m}_{i'j'}\underline{u}_j = \sum_{i',j'=1}^{3} O_{i'j'}\underline{u}_i^{\dagger}\underline{m}_{i'j'}\underline{u}_j \\ &= \sum_{i',j'=1}^{3} O_{i'j'}\underline{u}_i^{\dagger}\underline{u}_{i'}\underline{u}_{j'}^{\dagger}\underline{u}_j = \sum_{i',j'=1}^{3} O_{i'j'}\delta_{i,i'}\delta_{j,j'} = O_{ij},\end{aligned} \tag{11.25}$$

which shows you how to get matrix elements by projection.

11.1.2 Hilbert Space

I can take these ideas over to quantum mechanics. Things will be a little vague and confusing at first, but I hope that they clear up as I proceed. I consider only time-independent operators and, for the moment, only time-independent state vectors. The analogue of the vector $\mathbf{A}$ is the *state vector* $|A\rangle$, which is an abstract vector in a *Hilbert space* that can be finite or infinite-dimensional. Such a state vector is referred to as a *ket*. As in a normal vector space, I can introduce unit vectors or *basis kets* $|n\rangle$. Thus, the ket $|n\rangle$ *can be thought of as a column matrix having a 1 in the nth place and a zero everywhere else*. The state vector can be expanded as

$$|A\rangle = \sum_n A_n |n\rangle \,. \tag{11.26}$$

In other words, the state vector $|A\rangle$ is a column matrix whose nth element is A_n. As in the vector case, the state vector $|A\rangle$ has an absolute meaning. If I change the basis kets, I change the coordinates A_n but do not change $|A\rangle$. The adjoint of a ket is called a *bra* and can be represented by a row matrix. The adjoint of the basis ket $|n\rangle$ is written as

$$\langle n| = (|n\rangle)^\dagger \tag{11.27}$$

If the basis kets are orthonormal, as is often the case, then

$$\langle n |m\rangle = \delta_{m,n}. \tag{11.28}$$

You can see the origin of the bra-ket (bracket) notation. Unless stated otherwise, I will assume that the basis kets are orthonormal. The adjoint of Eq. (11.26) is defined by

$$(|A\rangle)^\dagger = \langle A| = \sum_n A_n^* \langle n| , \tag{11.29}$$

such that, for any two state vectors $|A\rangle$ and $|B\rangle$ in the same Hilbert space,

$$\langle B |A\rangle = \sum_{n,n'} A_{n'} B_n^* \langle n |n'\rangle = \sum_n A_n B_n^* = \left(\sum_n A_n^* B_n \right)^* = \langle A |B\rangle^* . \tag{11.30}$$

I can use Eq. (11.28) to obtain the A_m appearing in Eq. (11.26) by multiplying that equation by $\langle m|$ and using Eq. (11.28) to obtain

$$A_m = \langle m |A\rangle . \tag{11.31}$$

Furthermore,

$$|A\rangle = \sum_n A_n |n\rangle = \sum_n |n\rangle \langle n |A\rangle , \tag{11.32}$$

which implies that

$$\sum_n |n\rangle \langle n| = 1, \tag{11.33}$$

a statement of the completeness relation. For *continuous* kets such as $|\mathbf{r}\rangle$, the corresponding relationships are

$$\langle \mathbf{r} |\mathbf{r}'\rangle = \delta(\mathbf{r} - \mathbf{r}'); \tag{11.34a}$$

$$\int d\mathbf{r}\,|\mathbf{r}\rangle\langle\mathbf{r}| = 1. \tag{11.34b}$$

Whereas the discrete kets are dimensionless, the continuous kets $|\mathbf{r}\rangle$ have units of $1/\sqrt{\text{volume}}$.

I can also consider the effect of a linear operator $\hat{O}$ acting on the state vector $|A\rangle$. It will produce a new state vector $|A\rangle'$ that

$$|A\rangle' = \sum_n A'_n\,|n\rangle = \hat{O}\,|A\rangle = \hat{O}\sum_n A_n\,|n\rangle = \sum_n A_n\hat{O}\,|n\rangle\,. \tag{11.35}$$

If you multiply Eq. (11.35) on the left by $\langle m|$ and use the orthonormal property of the bras and kets, you can obtain

$$A'_m = \sum_n A_n\,\langle m|\,\hat{O}\,|n\rangle = \sum_n O_{mn}A_n, \tag{11.36}$$

where

$$O_{mn} = \langle m|\,\hat{O}\,|n\rangle \tag{11.37}$$

is a matrix element of the operator $\hat{O}$ in the $|n\rangle$ basis. Linear operators can be represented as matrices, just as in three-dimensional vector space. It is convenient to define *unit matrices* by

$$\underline{\mathrm{m}}_{nq} = |n\rangle\langle q|\,, \tag{11.38}$$

such that

$$\underline{\mathrm{O}} = \sum_{i,j} O_{ij}\underline{\mathrm{m}}_{ij}. \tag{11.39}$$

Remember that $|n\rangle\langle q|$ is a matrix with a one for the nq element and zeroes everywhere else. The matrix $\underline{\mathrm{O}}$ is Hermitian if $O_{nm} = (O_{mn})^*$.

I now come to the fundamental difference between a vector space and Hilbert space, as applied to quantum mechanics. In three-dimensional vector space, in general, one chooses the basis vectors without making any reference to the operators acting in the space. In quantum mechanics, a central feature is to choose a set of basis vectors that is *intimately connected* with operators acting in the Hilbert space. Moreover, it is assumed that the Hermitian operator associated with each physical observable has a matrix representation that is *diagonal* in its own basis.

That is, in a given basis $|h\rangle$ associated with a Hermitian operator $\hat{H}$, it is assumed that $\underline{\mathrm{H}}$ is diagonal,

$$\langle h|\,\hat{H}\,|h'\rangle = h\delta_{h,h'}. \tag{11.40}$$

If I multiply this equation on the left by $|h\rangle$ and sum over h using the completeness relation given in Eq. (11.33), I find

$$\hat{H}\left|h'\right\rangle = \sum_{h} h\left|h\right\rangle \delta_{h,h'} = h'\left|h'\right\rangle, \tag{11.41}$$

which is just an eigenvalue equation for $\hat{H}$! Consequently, we can use all the results that were derived in connection with the eigenvalue problem (eigenkets exist, are complete, can be chosen to form an orthonormal basis, simultaneous eigenkets can always be chosen for commuting operators, etc.). The eigenvalues h are the only possible outcomes of a measurement of the physical observable associated with $\hat{H}$ when made on a single quantum system.

I define the time-dependent state vector $|\psi(t)\rangle$ to be the solution of the time-dependent Schrödinger equation,

$$i\hbar\frac{d\left|\psi(t)\right\rangle}{dt} = \hat{H}\left|\psi(t)\right\rangle. \tag{11.42}$$

It is a simple matter to show that the solution of this equation can be written in the form

$$\left|\psi(t)\right\rangle = \sum_{E} b_E e^{-iEt/\hbar}\left|E\right\rangle, \tag{11.43}$$

provided

$$\hat{H}\left|E\right\rangle = E\left|E\right\rangle. \tag{11.44}$$

Thus, if we solve Eq. (11.44), we have a complete solution to any problem for a Hamiltonian $\hat{H}$.

Since I have assumed that an operator is diagonal in its own basis, it follows that

$$\hat{H}\left|E\right\rangle = E\left|E\right\rangle; \tag{11.45a}$$

$$\hat{\mathbf{r}}\left|\mathbf{r}\right\rangle = \mathbf{r}\left|\mathbf{r}\right\rangle; \tag{11.45b}$$

$$\hat{\mathbf{p}}\left|\mathbf{p}\right\rangle = \mathbf{p}\left|\mathbf{p}\right\rangle. \tag{11.45c}$$

Furthermore, any operator corresponding to a classical variable that is a function only of $\mathbf{r}$ has $|\mathbf{r}\rangle$ as its eigenkets and any operator corresponding to a classical variable that is a function only of $\mathbf{p}$ has $|\mathbf{p}\rangle$ as its eigenkets; that is, for operators $\hat{V}$ and $\hat{p}^2$, the matrices $\underline{V}$ and $\underline{p^2}$ associated with these operators are diagonal in the $|\mathbf{r}\rangle$ and $|\mathbf{p}\rangle$ bases, respectively,

$$\hat{V}\left|\mathbf{r}\right\rangle = V(\mathbf{r})\left|\mathbf{r}\right\rangle; \tag{11.46a}$$

$$\hat{p}^2\left|\mathbf{p}\right\rangle = p^2\left|\mathbf{p}\right\rangle. \tag{11.46b}$$

All this is well and good, but what have I accomplished? Normally in quantum mechanics we want to find the eigenenergies of the Hamiltonian, now expressed as the matrix $\underline{\mathrm{H}}$. In the $|E\rangle$ basis, $\underline{\mathrm{H}}$ is simply a diagonal matrix whose elements *are* the eigenenergies, but we have no way of calculating them yet. In other words, if we don't know the eigenenergies, the equation $\hat{H}|E\rangle = E|E\rangle$ doesn't help us to find them. It may be that we know the matrix elements of $\underline{\mathrm{H}}$ in some *other* basis, however. If that were the case, it turns out that if we diagonalize the matrix $\underline{\mathrm{H}}$, we can find the eigenenergies of $\hat{H}$, as well as the eigenkets of $\hat{H}$ in terms of the basis kets for which we know the matrix elements of $\underline{\mathrm{H}}$.

Let me give you an example. Say that the Hamiltonian is that of a free particle,

$$\underline{\mathrm{H}} = \frac{\underline{\mathrm{p}}^2}{2m}. \tag{11.47}$$

Since $\underline{\mathrm{H}}$ is diagonal in the $|\mathbf{p}\rangle$ basis, the $|\mathbf{p}\rangle$ states *are* the eigenkets of $\underline{\mathrm{H}}$ and the corresponding eigenenergies are $p^2/2m$. But what if the Hamiltonian is of the form

$$\underline{\mathrm{H}} = \frac{\underline{\mathrm{p}}^2}{2m} + \underline{\mathrm{V}}, \tag{11.48}$$

where $\underline{\mathrm{V}}$ is the matrix corresponding to the potential energy *operator* $\hat{V}$. I know the matrix elements of $\hat{p}^2$ in the $|\mathbf{p}\rangle$ basis and those of $\hat{V}$ in the $|\mathbf{r}\rangle$ basis, but not those of the *entire* Hamiltonian in any one basis.

To proceed further, I need to inject some quantum physics. Previously, one of the postulates I used was to assume that the wave functions in coordinate and momentum space were Fourier transforms of one another. This allowed me to calculate the momentum operator in coordinate space. I now replace this postulate by one in which the *Poisson bracket* of classical mechanics for two variables is replaced by the commutator of the matrices in quantum mechanics corresponding to those variables, multiplied by $(i\hbar)^{-1}$. The Poisson bracket of two arbitrary functions F and G with respect to canonical variables q and p is defined as

$$[F, G]_{q,p} = \frac{\partial H}{\partial q}\frac{\partial G}{\partial p} - \frac{\partial H}{\partial p}\frac{\partial G}{\partial q}. \tag{11.49}$$

As a consequence, the Poisson bracket of x and p_x is

$$[x, p_x]_{x,p_x} = \frac{\partial x}{\partial x}\frac{\partial p_x}{\partial p_x} - \frac{\partial x}{\partial p_x}\frac{\partial p_x}{\partial x} = 1. \tag{11.50}$$

To arrive at the analogous equation for quantum mechanics, I replace $[x, p_x]_{x,p_x}$ by $(i\hbar)^{-1}[\hat{x}, \hat{p}_x]$ to arrive at the commutator relation $[\hat{x}, \hat{p}_x] = i\hbar$. This will allow me to evaluate matrix elements of $\hat{\mathbf{p}}$ in the $|\mathbf{r}\rangle$ basis.

Armed with this commutator, I can try to evaluate matrix elements of

$$\hat{H} = \frac{\hat{p}^2}{2m} + \hat{V} \tag{11.51}$$

in the $|\mathbf{r}\rangle$ basis. Matrix elements of the second term are easy to obtain since

$$\langle \mathbf{r}| \hat{V} |\mathbf{r}'\rangle = V(\mathbf{r})\,\delta(\mathbf{r} - \mathbf{r}'); \tag{11.52}$$

the potential energy matrix is diagonal in the $|\mathbf{r}\rangle$ basis. But what about matrix elements of $\hat{p}^2$ in the $|\mathbf{r}\rangle$ basis? I start from the assumed form for the commutator,

$$[\hat{x}, \hat{p}_x] = i\hbar, \tag{11.53}$$

and evaluate

$$\begin{aligned} \langle x| [\hat{x}, \hat{p}_x] |x'\rangle &= i\hbar \langle x |x'\rangle = i\hbar\delta(x-x'); \\ \langle x| \hat{x}\hat{p}_x - \hat{p}_x\hat{x} |x'\rangle &= i\hbar\delta(x-x'). \end{aligned} \tag{11.54}$$

Since $\hat{x}$ is a Hermitian operator and since $\hat{x}|x'\rangle = x'|x'\rangle$, it follows that

$$\langle x| \hat{x} = (\hat{x}|x\rangle)^\dagger = x(|x\rangle)^\dagger = x\langle x|, \tag{11.55}$$

allowing me to rewrite Eq. (11.54) as

$$(x - x') \langle x| \hat{p}_x |x'\rangle = i\hbar\delta(x-x'). \tag{11.56}$$

Equation (11.56) is of the form $xf(x) = a\delta(x)$, which has as solution

$$f(x) = -a\frac{d}{dx}\delta(x). \tag{11.57}$$

To prove that this is a solution, I work backwards starting from

$$xf(x) = -ax\frac{d}{dx}\delta(x) = -a\frac{d}{dx}[x\delta(x)] + a\delta(x) = a\delta(x), \tag{11.58}$$

having used the relation $x\delta(x) = 0$. Thus,

$$\langle x| \hat{p}_x |x'\rangle = \frac{\hbar}{i}\frac{d}{dx}\delta(x-x'); \tag{11.59a}$$

$$\langle \mathbf{r}| \hat{p}_x |\mathbf{r}'\rangle = \frac{\hbar}{i}\frac{\partial}{\partial x}\delta(\mathbf{r} - \mathbf{r}'); \tag{11.59b}$$

$$\langle \mathbf{r}| \hat{\mathbf{p}} |\mathbf{r}'\rangle = \frac{\hbar}{i}\nabla_r\delta(\mathbf{r} - \mathbf{r}'). \tag{11.59c}$$

Similarly,

$$\langle \mathbf{p}| \hat{\mathbf{r}} |\mathbf{p}'\rangle = i\hbar \nabla_p \delta(\mathbf{p} - \mathbf{p}'). \tag{11.60}$$

It is clear that $\hat{p}_x$ is not diagonal in the $|x\rangle$ basis since the derivative of a delta function is *not* proportional to $\delta(x-x')$; that is, it is non-zero for values of $x \neq x'$. You can understand this easily. A delta function $\delta(x)$ is like a narrow Gaussian centered at $x = 0$ whose derivative vanishes at $x = 0$, has a sharp maximum for $x < 0$, and a sharp minimum for $x > 0$—it is non-zero for values of $x \neq 0$. Of course we knew beforehand that $\hat{\mathbf{p}}$ could not be diagonal in the $|\mathbf{r}\rangle$ basis since two operators can have simultaneous eigenkets if and only if the operators commute.

I have made some progress. Using the assumed commutation relation between $\hat{x}$ and $\hat{p}_x$, I found a matrix representation of $\hat{\mathbf{p}}$ in the $|\mathbf{r}\rangle$ basis. The matrix is not diagonal, nor will the matrix $\underline{\mathrm{p}}^2$ be diagonal. As a consequence, to get the eigenvalues of $\underline{\mathrm{H}}$, I must diagonalize $\underline{\mathrm{H}}$. In other words, I seek a linear combination of energy kets in the $|\mathbf{r}\rangle$ basis,

$$|E\rangle = \int d\mathbf{r} \, \langle \mathbf{r} \, |E\rangle \, |\mathbf{r}\rangle \tag{11.61}$$

that diagonalizes $\underline{\mathrm{H}}$, that is, a basis for which

$$\hat{H} \, |E\rangle = E \, |E\rangle \, . \tag{11.62}$$

If I can find the expansion coefficients $\langle \mathbf{r} \, |E\rangle$, I will have accomplished this task. In other words, I will have found a basis in which $\hat{H}$ is diagonal—the eigenenergies are simply the diagonal elements of the matrix $\underline{\mathrm{H}}$ in the new basis.

To get the expansion components, I start from Eq. (11.62) and multiply on the left by $\langle \mathbf{r}|$ to obtain

$$\langle \mathbf{r}| \left(\frac{\hat{p}^2}{2m} + \hat{V} \right) |E\rangle = E \, \langle \mathbf{r} \, |E\rangle \, . \tag{11.63}$$

I now use the completeness relation

$$\int d\mathbf{r} \, |\mathbf{r}\rangle \, \langle \mathbf{r}| = 1 \tag{11.64}$$

and insert complete sets at will. In fact, when dealing with Dirac notation and you are lost about what to do, you can always insert some complete sets and see what happens! Inserting Eq. (11.64) into Eq. (11.63), I find

$$\int d\mathbf{r}' \, \langle \mathbf{r}| \left(\frac{\hat{p}^2}{2m} + \hat{V} \right) |\mathbf{r}'\rangle \langle \mathbf{r}' \, |E\rangle = E \, \langle \mathbf{r} \, |E\rangle \, , \tag{11.65a}$$

$$\int d\mathbf{r}'\left(\frac{\langle\mathbf{r}|\,\hat{p}^2\,|\mathbf{r}'\rangle}{2m}+\langle\mathbf{r}|\,\hat{V}\,|\mathbf{r}'\rangle\right)\langle\mathbf{r}'\,|E\rangle=E\,\langle\mathbf{r}\,|E\rangle\,, \tag{11.65b}$$

$$\int d\mathbf{r}'\left(\frac{\langle\mathbf{r}|\,\hat{p}^2\,|\mathbf{r}'\rangle}{2m}+V(\mathbf{r})\delta(\mathbf{r}-\mathbf{r}')\right)\langle\mathbf{r}'\,|E\rangle=E\,\langle\mathbf{r}\,|E\rangle\,, \tag{11.65c}$$

$$\int d\mathbf{r}'\frac{\langle\mathbf{r}|\,\hat{p}^2\,|\mathbf{r}'\rangle}{2m}\langle\mathbf{r}'\,|E\rangle+V(\mathbf{r})\,\langle\mathbf{r}\,|E\rangle=E\,\langle\mathbf{r}\,|E\rangle\,, \tag{11.65d}$$

and the problem reduces to evaluating

$$\begin{aligned}\int d\mathbf{r}'\,\langle\mathbf{r}|\,\hat{p}^2\,|\mathbf{r}'\rangle\langle\mathbf{r}'\,|E\rangle&=\int d\mathbf{r}''\int d\mathbf{r}'\,\langle\mathbf{r}|\,\hat{\mathbf{p}}\,|\mathbf{r}''\rangle\cdot\langle\mathbf{r}''|\,\hat{\mathbf{p}}\,|\mathbf{r}'\rangle\langle\mathbf{r}'\,|E\rangle\\&=\int d\mathbf{r}''\int d\mathbf{r}'\frac{\hbar}{i}\nabla_r\delta(\mathbf{r}-\mathbf{r}'')\cdot\langle\mathbf{r}''|\,\hat{\mathbf{p}}\,|\mathbf{r}'\rangle\langle\mathbf{r}'\,|E\rangle\,.\end{aligned} \tag{11.66}$$

Since ∇_r acts only on $\mathbf{r}$, I can take it out of the integral and obtain

$$\begin{aligned}&\int d\mathbf{r}'\,\langle\mathbf{r}|\,\hat{p}^2\,|\mathbf{r}'\rangle\langle\mathbf{r}'\,|E\rangle\\&=\frac{\hbar}{i}\nabla_r\int d\mathbf{r}''\int d\mathbf{r}'\delta(\mathbf{r}-\mathbf{r}'')\cdot\langle\mathbf{r}''|\,\hat{\mathbf{p}}\,|\mathbf{r}'\rangle\langle\mathbf{r}'\,|E\rangle\\&=\frac{\hbar}{i}\nabla_r\cdot\int d\mathbf{r}'\,\langle\mathbf{r}|\,\hat{\mathbf{p}}\,|\mathbf{r}'\rangle\langle\mathbf{r}'\,|E\rangle=\frac{\hbar}{i}\nabla_r\cdot\int d\mathbf{r}'\frac{\hbar}{i}\nabla_r\delta(\mathbf{r}-\mathbf{r}')\,\langle\mathbf{r}'\,|E\rangle\\&=-\hbar^2\nabla_r\cdot\nabla_r\int d\mathbf{r}'\delta(\mathbf{r}-\mathbf{r}')\,\langle\mathbf{r}'\,|E\rangle=-\hbar^2\nabla_r^2\,\langle\mathbf{r}\,|E\rangle\end{aligned} \tag{11.67}$$

or

$$-\frac{\hbar^2}{2m}\nabla_r^2\,\langle\mathbf{r}\,|E\rangle+V(\mathbf{r})\,\langle\mathbf{r}\,|E\rangle=E\,\langle\mathbf{r}\,|E\rangle\,. \tag{11.68}$$

But this is nothing more (or less) than Schrödinger's equation if I identify

$$\langle\mathbf{r}\,|E\rangle=\psi_E(\mathbf{r}). \tag{11.69}$$

Diagonalizing $\underline{\mathrm{H}}$ *in the* $|\mathbf{r}\rangle$ *basis is equivalent to solving the Schrödinger equation!* I now have the connection between Dirac notation and the wave function. In general, the bra-ket notation doesn't simplify the problem, since we still have to solve the Schrödinger equation. Nevertheless Dirac notation provides a powerful formalism that lets us express results in a basis-independent fashion.

Often, I will need to calculate the matrix elements of an operator $\hat{A}$ in the energy basis, namely $\langle E|\,\hat{A}\,|E'\rangle$. Suppose that the operator $\hat{A}$ corresponds to a physical variable $A(\mathbf{r})$ that is a function of coordinates only—$\hat{A}$ is diagonal in the $|\mathbf{r}\rangle$ basis.

Then

$$\begin{aligned}\langle E|\hat{A}\left|E'\right\rangle &= \int d\mathbf{r}d\mathbf{r}'\left\langle E\left|\mathbf{r}\right\rangle\left\langle\mathbf{r}\right|\hat{A}\left|\mathbf{r}'\right\rangle\right\langle\mathbf{r}'\left|E'\right\rangle \\ &= \int d\mathbf{r}d\mathbf{r}'\psi_E^*(\mathbf{r})A(\mathbf{r})\delta\left(\mathbf{r}-\mathbf{r}'\right)\psi_{E'}(\mathbf{r}') \\ &= \int d\mathbf{r}\psi_E^*(\mathbf{r})\hat{A}(\mathbf{r})\psi_{E'}(\mathbf{r}) \qquad (11.70)\end{aligned}$$

is the way in which matrix elements are most often evaluated.

11.1.3 Schrödinger's Equation in Momentum Space

As an example of the use of Dirac notation, I derive the time-independent Schrödinger's equation in momentum space. As in coordinate space, I expand

$$|E\rangle = \int d\mathbf{p}\left\langle\mathbf{p}\left|E\right\rangle\right|\mathbf{p}\rangle \qquad (11.71)$$

and try to find the expansion coefficients $\langle\mathbf{p}\,|E\rangle$. I start from $\hat{H}\,|E\rangle = E\,|E\rangle$ and multiply on the left by $\langle\mathbf{p}|$ to get

$$\langle\mathbf{p}|\left(\frac{\hat{p}^2}{2m}+\hat{V}\right)|E\rangle = E\left\langle\mathbf{p}\left|E\right\rangle\right. . \qquad (11.72)$$

I proceed as before and obtain

$$\begin{aligned}\int d\mathbf{p}'\,\langle\mathbf{p}|\left(\frac{\hat{p}^2}{2m}+\hat{V}\right)\left|\mathbf{p}'\right\rangle\left\langle\mathbf{p}'\left|E\right\rangle\right. &= E\left\langle\mathbf{p}\left|E\right\rangle\right.; \\ \int d\mathbf{p}'\left(\frac{p^2}{2m}\delta(\mathbf{p}-\mathbf{p}')+\langle\mathbf{p}|\hat{V}\left|\mathbf{p}'\right\rangle\right)\left\langle\mathbf{p}'\left|E\right\rangle\right. &= E\left\langle\mathbf{p}\left|E\right\rangle\right.; \\ \frac{p^2}{2m}\left\langle\mathbf{p}\left|E\right\rangle\right. + \int d\mathbf{p}'\,\langle\mathbf{p}|\hat{V}\left|\mathbf{p}'\right\rangle\left\langle\mathbf{p}'\left|E\right\rangle\right. &= E\left\langle\mathbf{p}\left|E\right\rangle\right. . \qquad (11.73)\end{aligned}$$

However, I do not know matrix elements of $\hat{V}$ in the $|\mathbf{p}\rangle$ basis, but do know them in the $|\mathbf{r}\rangle$ basis. So I add some more completeness relations:

$$\begin{aligned}&\int d\mathbf{p}'\,\langle\mathbf{p}|\hat{V}\left|\mathbf{p}'\right\rangle\left\langle\mathbf{p}'\left|E\right\rangle\right. \\ &= \int d\mathbf{r}'\int d\mathbf{r}\int d\mathbf{p}'\left\langle\mathbf{p}\left|\mathbf{r}\right\rangle\left\langle\mathbf{r}\right|\hat{V}\left|\mathbf{r}'\right\rangle\right\langle\mathbf{r}'\left|\,\mathbf{p}'\right\rangle\left\langle\mathbf{p}'\left|E\right\rangle\right.\end{aligned}$$

$$= \int d\mathbf{r}' \int d\mathbf{r} \int d\mathbf{p}' \langle \mathbf{p} | \mathbf{r} \rangle V(\mathbf{r}') \delta(\mathbf{r} - \mathbf{r}') \langle \mathbf{r}' | \mathbf{p}' \rangle \langle \mathbf{p}' | E \rangle$$

$$= \int d\mathbf{r} V(\mathbf{r}) \int d\mathbf{p}' \langle \mathbf{p} | \mathbf{r} \rangle \langle \mathbf{r} | \mathbf{p}' \rangle \langle \mathbf{p}' | E \rangle . \tag{11.74}$$

To evaluate this I need to know the value of $\langle \mathbf{p} | \mathbf{r} \rangle = \langle \mathbf{r} | \mathbf{p} \rangle^*$.

To evaluate $\langle \mathbf{r} | \mathbf{p} \rangle$, I start from the eigenvalue equation

$$\hat{\mathbf{p}} |\mathbf{p}\rangle = \mathbf{p} |\mathbf{p}\rangle ; \tag{11.75}$$

$$\langle \mathbf{r} | \hat{\mathbf{p}} |\mathbf{p}\rangle = \mathbf{p} \langle \mathbf{r} |\mathbf{p}\rangle ; \tag{11.76}$$

$$\int d\mathbf{r}' \langle \mathbf{r} | \hat{\mathbf{p}} |\mathbf{r}'\rangle \langle \mathbf{r}' | \mathbf{p}\rangle = \mathbf{p} \langle \mathbf{r} |\mathbf{p}\rangle ; \tag{11.77}$$

$$\frac{\hbar}{i} \int d\mathbf{r}' \nabla_r \delta(\mathbf{r} - \mathbf{r}') \langle \mathbf{r}' | \mathbf{p}\rangle = \mathbf{p} \langle \mathbf{r} |\mathbf{p}\rangle ; \tag{11.78}$$

$$\frac{\hbar}{i} \nabla_r \int d\mathbf{r}' \delta(\mathbf{r} - \mathbf{r}') \langle \mathbf{r}' | \mathbf{p}\rangle = \mathbf{p} \langle \mathbf{r} |\mathbf{p}\rangle ; \tag{11.79}$$

$$\frac{\hbar}{i} \nabla_r \langle \mathbf{r} | \mathbf{p}\rangle = \mathbf{p} \langle \mathbf{r} |\mathbf{p}\rangle ; \tag{11.80}$$

$$\langle \mathbf{r} | \mathbf{p}\rangle = \frac{1}{(2\pi\hbar)^{3/2}} e^{i\mathbf{p}\cdot\mathbf{r}/\hbar} = \langle \mathbf{p} | \mathbf{r}\rangle^* , \tag{11.81}$$

where a normalization factor has been included to ensure that

$$\langle \mathbf{r} | \mathbf{r}'\rangle = \int d\mathbf{p} \langle \mathbf{r} |\mathbf{p}\rangle \langle \mathbf{p} |\mathbf{r}'\rangle = \frac{1}{(2\pi\hbar)^3} \int d\mathbf{p} e^{i\mathbf{p}\cdot(\mathbf{r}-\mathbf{r}')/\hbar} = \delta\left(\mathbf{r} - \mathbf{r}'\right) . \tag{11.82}$$

Substituting Eq. (11.81) into Eq. (11.74) I obtain

$$\int d\mathbf{p}' \langle \mathbf{p} | \hat{V} |\mathbf{p}'\rangle \langle \mathbf{p}' |E\rangle = (2\pi\hbar)^{-3} \int d\mathbf{p}' \int d\mathbf{r} V(\mathbf{r}) e^{-i(\mathbf{p}-\mathbf{p}')\cdot\mathbf{r}/\hbar} \langle \mathbf{p}' |E\rangle$$

$$= (2\pi\hbar)^{-3/2} \int d\mathbf{p}' \tilde{V}\left(\mathbf{p} - \mathbf{p}'\right) \langle \mathbf{p}' |E\rangle , \tag{11.83}$$

where

$$\tilde{V}(\mathbf{q}) = (2\pi\hbar)^{-3/2} \int d\mathbf{r} V(\mathbf{r}) e^{-i\mathbf{q}\cdot\mathbf{r}/\hbar} \tag{11.84}$$

is the Fourier transform of $V(\mathbf{r})$. Combining Eqs. (11.73), (11.74), and (11.83), I arrive at

$$\frac{p^2}{2m}\Phi_E(\mathbf{p}) + \frac{1}{(2\pi\hbar)^{3/2}}\int d\mathbf{p}'\tilde{V}\left(\mathbf{p}-\mathbf{p}'\right)\Phi_E(\mathbf{p}') = E\Phi_E(\mathbf{p}), \tag{11.85}$$

which is Schrödinger's equation in momentum space for the wave function

$$\Phi_E(\mathbf{p}) = \langle \mathbf{p} \,|E\rangle\,. \tag{11.86}$$

In momentum space, Schrödinger's equation is an integral equation.

Also, given the fact that

$$|\mathbf{r}\rangle = \int d\mathbf{p}\,\langle \mathbf{p}\,|\mathbf{r}\rangle\,|\mathbf{p}\rangle\,, \tag{11.87}$$

I can multiply on the left by $\langle E|$ and use Eq. (11.81) to obtain

$$\langle E\,|\mathbf{r}\rangle = \frac{1}{(2\pi\hbar)^{3/2}}\int d\mathbf{p}\,\langle E\,|\mathbf{p}\rangle\,e^{-i\mathbf{p}\cdot\mathbf{r}/\hbar}. \tag{11.88}$$

Taking the complex conjugate of this equation and using Eqs. (11.69) and (11.86), I find

$$\psi_E(\mathbf{r}) = \frac{1}{(2\pi\hbar)^{3/2}}\int d\mathbf{p}\Phi_E(\mathbf{p})e^{i\mathbf{p}\cdot\mathbf{r}/\hbar}. \tag{11.89}$$

The energy eigenfunctions in coordinate and momentum space are Fourier transforms of one another. Earlier I postulated this result and was able to derive the commutation relations for the position and momentum operators. Here I postulated the commutation relations and was led to the result that $\psi_E(\mathbf{r})$ and $\Phi_E(\mathbf{p})$ are Fourier transforms of one another.

Although I have shown that diagonalizing the Hamiltonian in the coordinate representation is equivalent to solving the Schrödinger equation, there are some cases where it is possible to get the eigenvalues and eigenfunctions working directly from the Dirac notation formalism. I now turn my attention to two such cases, the simple harmonic oscillator and the angular momentum operator.

11.2 Simple Harmonic Oscillator

Let us reconsider the SHO in 1-D which has a Hamiltonian in units of $\hbar\omega$ given by

$$\hat{H}' = \frac{\hat{H}}{\hbar\omega} = \frac{1}{2}\left(\hat{\eta}^2 + \hat{\xi}^2\right), \tag{11.90}$$

where

$$\hat{\xi} = \sqrt{\frac{m\omega}{\hbar}}\hat{x}; \tag{11.91a}$$

$$\hat{\eta} = \sqrt{\frac{1}{\hbar m\omega}}\hat{p} = \frac{1}{i}\frac{\partial}{\partial\xi}. \tag{11.91b}$$

It will prove convenient to introduce operators a and $a^\dagger$ (I don't put hats on them even though they are operators) that are defined by

$$a = \sqrt{\frac{m\omega}{2\hbar}}\left(\hat{x} + i\frac{\hat{p}}{m\omega}\right) = \frac{\hat{\xi} + i\hat{\eta}}{\sqrt{2}}; \tag{11.92a}$$

$$a^\dagger = \sqrt{\frac{m\omega}{2\hbar}}\left(\hat{x} - i\frac{\hat{p}}{m\omega}\right) = \frac{\hat{\xi} - i\hat{\eta}}{\sqrt{2}}. \tag{11.92b}$$

In terms of these operators,

$$\hat{H}' = a^\dagger a + \frac{1}{2} = \hat{n} + \frac{1}{2}, \tag{11.93}$$

where

$$\hat{n} = a^\dagger a. \tag{11.94}$$

The operators a and $a^\dagger$ satisfy the commutation relations

$$\left[a, a^\dagger\right] = -\left[a^\dagger, a\right] = 1; \ \ [a, a] = \left[a^\dagger, a^\dagger\right] = 0. \tag{11.95}$$

It turns out, by being a bit clever, you can find the eigenenergies and eigenfunctions of the SHO without solving the Schrödinger equation. For reasons that will become obvious, the operators a and $a^\dagger$ are often referred to as *ladder operators* and $\hat{n}$ as the *number operator*.

To start, I label the eigenkets of $\hat{n}$ by $|n\rangle$ without any restriction on n (it need not be an integer) such that

$$\hat{H}'|n\rangle = \left(\hat{n} + \frac{1}{2}\right)|n\rangle = \epsilon_n|n\rangle \tag{11.96}$$

where

$$\epsilon_n = E_n/\hbar\omega = (n + 1/2). \tag{11.97}$$

and the energy E_n is given by

$$E_n = (n + 1/2)\hbar\omega. \tag{11.98}$$

This is perfectly arbitrary since there is no restriction on n.

The operator a is a *lowering* or *destruction* operator, which can be proved as follows: First I note that

$$\hat{H}'(a|n\rangle) = \hat{H}'a|n\rangle = \left(\left[\hat{H}',a\right] + a\hat{H}'\right)|n\rangle. \quad (11.99)$$

But

$$\left[\hat{H}',a\right] = \left[a^\dagger a, a\right] = a^\dagger [a,a] + \left[a^\dagger, a\right]a = -a, \quad (11.100)$$

implying that

$$\begin{aligned}\hat{H}'(a|n\rangle) &= \left(-a + a\hat{H}'\right)|n\rangle = [-a + a(n+1/2)]|n\rangle \\ &= [(n-1)+1/2](a|n\rangle). \quad (11.101)\end{aligned}$$

In other words, $(a|n\rangle)$ is an eigenket of $\hat{H}'$ having eigenvalue $n-1$.

By successively applying the operator a, I keep lowering the eigenvalue by one. However, the expectation value of the number operator is always positive or zero, since

$$\langle n|\hat{n}|n\rangle = \langle n|a^\dagger a|n\rangle = (a|n\rangle)^\dagger (a|n\rangle) \geq 0. \quad (11.102)$$

Thus as I keep applying the operator a, I must come to a lowest state $|n_{\min}\rangle$ for which $a|n_{\min}\rangle = 0$. As a consequence,

$$a^\dagger (a|n_{\min}\rangle) = 0 = \hat{n}|n_{\min}\rangle = n_{\min}|n_{\min}\rangle, \quad (11.103)$$

which requires that $n_{\min} = 0$. The value of $n_{\min} = 0$ was reached by continually lowering n by one; therefore, n must be a positive integer or zero and the (dimensionless) eigenenergies are

$$\epsilon_n = (n+1/2); \quad n = 0,1,2,3,\ldots. \quad (11.104)$$

I have obtained the eigenenergies without having solved Schrödinger's equation.

Since $(a|n\rangle)$ is an eigenket of $\hat{H}$ having eigenvalue $n-1$, I can write

$$a|n\rangle = c_n|n-1\rangle, \quad (11.105)$$

where c_n is a constant. Thus

$$\begin{aligned}n &= \langle n|a^\dagger a|n\rangle = (a|n\rangle)^\dagger (a|n\rangle) = (c_n|n-1\rangle)^\dagger (c_n|n-1\rangle) \\ &= |c_n|^2 \langle n-1|n-1\rangle = |c_n|^2. \quad (11.106)\end{aligned}$$

If I choose the phase of c_n equal to zero, then $c_n = \sqrt{n}$ and

$$a\,|n\rangle = \sqrt{n}\,|n-1\rangle\,. \tag{11.107}$$

In a similar manner I can show that $a^\dagger$ is a *raising* or *creation* operator and that

$$a^\dagger\,|n\rangle = \sqrt{n+1}|n+1\rangle\,, \tag{11.108}$$

such that the number operator leaves the ket unchanged,

$$\hat{n}\,|n\rangle = a^\dagger a\,|n\rangle = n|n\rangle, \tag{11.109}$$

as required since the Hamiltonian is diagonal in the $|n\rangle$ basis.

From Eq. (11.108), it follows that

$$|1\rangle = \frac{a^\dagger}{\sqrt{1!}}|0\rangle; \qquad |2\rangle = \frac{a^\dagger}{\sqrt{1!}}|1\rangle = \frac{\left(a^\dagger\right)^2}{\sqrt{2!}}|0\rangle, \tag{11.110}$$

and so forth, leading to

$$|n\rangle = \frac{\left(a^\dagger\right)^n}{\sqrt{n!}}|0\rangle. \tag{11.111}$$

The (dimensionless) wave function in terms of dimensionless variables is given by

$$\psi_n(\xi) = \langle \xi\;|n\rangle\,. \tag{11.112}$$

Using ladder operators I have obtained the eigenenergies and expressions for all the eigenkets in terms of the eigenket $|0\rangle$, but have not yet found explicit expressions for the wave functions.

To do so, I look at the equation

$$a\,|0\rangle = 0, \tag{11.113}$$

put in a complete set via

$$\int_{-\infty}^{\infty} d\xi'\,a\,|\xi'\rangle\langle\xi'\;|0\rangle = 0, \tag{11.114}$$

and multiply on the left by $\langle\xi|$,

$$\int_{-\infty}^{\infty} d\xi'\,\langle\xi|\,a\,|\xi'\rangle\,\psi_0(\xi') = 0; \tag{11.115a}$$

$$\int_{-\infty}^{\infty} d\xi'\,\langle\xi|\left(\hat{\xi} + i\hat{\eta}\right)|\xi'\rangle\,\psi_0(\xi') = 0; \tag{11.115b}$$

$$\int_{-\infty}^{\infty} d\xi' \left[\xi\delta\left(\xi-\xi'\right) + \frac{d\delta\left(\xi-\xi'\right)}{d\xi} \right] \psi_0(\xi') = 0; \tag{11.115c}$$

$$\xi\psi_0(\xi) + \frac{d}{d\xi}\int_{-\infty}^{\infty} d\xi'\delta\left(\xi-\xi'\right)\psi_0(\xi') = 0, \tag{11.115d}$$

which leads to

$$\xi\psi_0(\xi) + \frac{d\psi_0(\xi)}{d\xi} = 0. \tag{11.116}$$

The solution of this equation is

$$\psi_0(\xi) = N\exp\left(-\xi^2/2\right) \tag{11.117}$$

where $N = \pi^{-1/4}$ is a normalization factor. Thus I have found the ground state wave function. Higher order wave functions can be calculated using the recursion relations that are given in Chap. 7. In fact I could derive the needed recursion relation by starting from Eq. (11.108) and multiplying on the left by $\langle\xi|$; that is,

$$\langle\xi|\, a^{\dagger}|n\rangle = \sqrt{n+1}\,\langle\xi\;|n+1\rangle = \sqrt{n+1}\psi_{n+1}(\xi) = \langle\xi|\left(\hat{\xi} - i\hat{\eta}\right)|n\rangle/\sqrt{2}. \tag{11.118}$$

Following the same procedure that led to Eq. (11.116), I find

$$\sqrt{2\left(n+1\right)}\psi_{n+1}(\xi) = \xi\psi_n(\xi) - \frac{d\psi_n(\xi)}{d\xi}, \tag{11.119}$$

which is Eq. (7.47c) and allows you to calculate all the wave functions if you know $\psi_0(\xi)$. The ladder operators a and $a^{\dagger}$ are particularly useful for obtaining matrix elements of operators of the form ξ^m for the harmonic oscillator. I will use them in applications of perturbation theory.

11.2.1 Coherent State

I want to return to the coherent state that I have already introduced in my discussion of the harmonic oscillator in Chap. 7, where I found that, for an initial wave function

$$\psi_{\text{coh}}\left(\xi,0\right) = \frac{1}{\pi^{1/4}}e^{-\left(\xi-\xi_0\right)^2/2} = \sum_{n=0}^{\infty} \frac{\left(\xi_0/\sqrt{2}\right)^n e^{-\left(\xi_0/\sqrt{2}\right)^2/2}}{\sqrt{n!}}\psi_n\left(\xi\right), \tag{11.120}$$

the wave packet envelope does not change its shape as a function of time. In terms of Dirac notation this state is

$$|\psi(0)\rangle_{\text{coh}} = \sum_{n=0}^{\infty} \frac{\left(\xi_0/\sqrt{2}\right)^n e^{-\left(\xi_0/\sqrt{2}\right)^2/2}}{\sqrt{n!}} |n\rangle , \tag{11.121}$$

such that

$$\begin{aligned} |\psi(t)\rangle_{\text{coh}} &= e^{-i\omega t/2} \sum_{n=0}^{\infty} \frac{\left(\xi_0/\sqrt{2}\right)^n e^{-\left(\xi_0/\sqrt{2}\right)^2/2} e^{-in\omega t}}{\sqrt{n!}} |n\rangle \\ &= e^{-i\omega t/2} |\psi_{\text{coh}}(t)\rangle , \end{aligned} \tag{11.122}$$

where

$$|\psi_{\text{coh}}(t)\rangle = \sum_{n=0}^{\infty} \frac{\left(\xi_0/\sqrt{2}\right)^n e^{-\left(\xi_0/\sqrt{2}\right)^2/2} e^{-in\omega t}}{\sqrt{n!}} |n\rangle . \tag{11.123}$$

is the conventional form for a coherent state vector. In taking expectation values of operators, it makes no difference whether I use $|\psi(t)\rangle_{\text{coh}}$ or $|\psi_{\text{coh}}(t)\rangle$.

It is now a relatively simple matter to calculate

$$\left\langle \hat{\xi} \right\rangle_{\text{coh}} = \frac{1}{\sqrt{2}} \langle \psi_{\text{coh}}(t)| \left(a + a^\dagger\right) |\psi_{\text{coh}}(t)\rangle \tag{11.124}$$

and

$$\begin{aligned} \left\langle \hat{\xi}^2 \right\rangle_{\text{coh}} &= \frac{1}{2} \langle \psi_{\text{coh}}(t)| \left(a + a^\dagger\right)^2 |\psi_{\text{coh}}(t)\rangle \\ &= \frac{1}{2} \langle \psi_{\text{coh}}(t)| \left(a^2 + a^{\dagger 2} + 2a^\dagger a + 1\right) |\psi_{\text{coh}}(t)\rangle \end{aligned} \tag{11.125}$$

and show they are consistent with a wave packet that oscillates in the potential without changing its shape.

Instead, I adopt somewhat different method that is useful in oscillator problems and in quantum optics. I define

$$\xi(t) = \xi_0 e^{-i\omega t}/\sqrt{2} \tag{11.126}$$

and operate with the destruction operator on $|\psi_{\text{coh}}(t)\rangle$,

$$a|\psi_{\text{coh}}(t)\rangle = \sum_{n=0}^{\infty} \frac{[\xi(t)]^n e^{-|\xi(t)|^2/2}}{\sqrt{n!}} \sqrt{n}\, |n-1\rangle$$

$$= \xi(t) \sum_{n=1}^{\infty} \frac{[\xi(t)]^{n-1} e^{-|\xi(t)|^2/2}}{\sqrt{(n-1)!}} |n-1\rangle$$
$$= \xi(t) \sum_{n=0}^{\infty} \frac{[\xi(t)]^{n} e^{-|\xi(t)|^2/2}}{\sqrt{n!}} |n\rangle$$
$$= \xi(t) |\psi_{\text{coh}}(t)\rangle . \tag{11.127}$$

Thus, the coherent state at time t is an eigenstate of the destruction operator with eigenvalue $\xi(t)$.[2] I change the notation by writing

$$|\xi(t)\rangle_{\text{coh}} \equiv |\psi_{\text{coh}}(t)\rangle = \sum_{n=0}^{\infty} \frac{[\xi(t)]^{n} e^{-|\xi(t)|^2/2}}{\sqrt{n!}} |n\rangle , \tag{11.128a}$$

such that

$$a |\xi(t)\rangle_{\text{coh}} = \xi(t) |\xi(t)\rangle_{\text{coh}} . \tag{11.128b}$$

Somewhat more generally, the coherent state of an oscillator, $|\alpha\rangle$, can be defined by

$$a |\alpha\rangle = \alpha |\alpha\rangle \tag{11.129}$$

with

$$|\alpha\rangle = \sum_n \frac{\alpha^n}{\sqrt{n!}} e^{-|\alpha|^2/2} |n\rangle$$
$$= \sum_n \frac{(\alpha a^\dagger)^n}{n!} e^{-|\alpha|^2/2} |0\rangle = e^{\alpha a^\dagger} e^{-|\alpha|^2/2} |0\rangle . \tag{11.130}$$

Equation (11.129) is an alternative definition of a coherent state. Thus for the coherent state $|\psi_{\text{coh}}(t)\rangle$ having

$$\alpha(t) = \left(\xi_0 e^{-i\omega t}/\sqrt{2}\right), \tag{11.131}$$

with ξ_0 real,

$$\langle\xi\rangle_{\text{coh}} = \frac{1}{\sqrt{2}} \langle\alpha(t)| \left(a + a^\dagger\right) |\alpha(t)\rangle = \frac{\alpha(t) + \alpha^*(t)}{\sqrt{2}} = \xi_0 \cos(\omega t) \tag{11.132}$$

---

[2]Since $\hat{a}$ is not a Hermitian operator, there is no guarantee that it possesses an orthonormal set of eigenkets (in fact, the eigenkets are *not* orthogonal). If you try to follow a procedure similar to the one that led to Eq. (11.127) to arrive at an eigenvalue equation for $\hat{a}^\dagger$, you will find that it is not possible—a set of normalizable eigenkets does not exist for the operator $\hat{a}^\dagger$.

and

$$\begin{aligned}\left\langle \xi^2 \right\rangle_{\text{coh}} &= \frac{1}{2}\left\langle \alpha(t)\right| \frac{\left(a^2 + a^{\dagger 2} + 2a^{\dagger}a + 1\right)}{2} \left|\alpha(t)\right\rangle \\ &= \frac{1}{2}\left[\alpha^2(t) + \left[\alpha^*(t)\right]^2 + 2\left|\alpha(t)\right|^2 + 1\right] \\ &= \left(\frac{\xi_0}{\sqrt{2}}\right)^2 \frac{e^{2i\omega t} + e^{-2i\omega t} + 2}{2} + \frac{1}{2} \\ &= \xi_0^2 \cos^2(\omega t) + 1/2. \end{aligned} \tag{11.133}$$

From Eqs. (11.132) and (11.133), I find that the variance

$$\Delta\xi^2 = \left\langle \xi^2 \right\rangle - \left\langle \xi \right\rangle^2 = \frac{1}{2} \tag{11.134}$$

is constant in time—the wave packet does not spread.[3] The coherent state with $\xi_0 = 0$ is the *vacuum state* of the oscillator and is simply the ground state of the oscillator. For $\xi_0 \neq 0$, the wave packet corresponds to the ground state eigenfunction displaced by an amount ξ_0 and oscillates in the potential without changing its shape.

11.3 Angular Momentum Operator

Angular momentum can also be analyzed using ladder operators. I designate the simultaneous eigenkets of $\hat{L}^2$ and $\hat{L}_z$ by $|\gamma\beta\rangle$; that is,

$$\hat{L}^2 \left|\gamma\beta\right\rangle = \hbar^2\gamma \left|\gamma\beta\right\rangle \tag{11.135a}$$

$$\hat{L}_z \left|\gamma\beta\right\rangle = \hbar\beta \left|\gamma\beta\right\rangle. \tag{11.135b}$$

At this point, γ and β are totally arbitrary. Given that $\hat{\mathbf{L}}$ is Hermitian, $\hat{L}^2 = \hat{\mathbf{L}}^{\dagger} \cdot \hat{\mathbf{L}}$ must have non-negative eigenvalues; $\gamma \geq 0$. Moreover, since the expectation value of $\hat{L}_x^2 + \hat{L}_y^2$ in any state must also be non-negative, it follows that

$$\left\langle\gamma\beta\right| \left(\hat{L}^2 - \hat{L}_z^2\right) \left|\gamma\beta\right\rangle = \left\langle\gamma\beta\right| \left(\hat{L}_x^2 + \hat{L}_y^2\right) \left|\gamma\beta\right\rangle = \hbar^2\left(\gamma - \beta^2\right) \geq 0, \tag{11.136}$$

[3] I have proved only that the variance is constant, not that the absolute square of the wave function does not change its shape. That result was proved in Chap. 7. However, you can prove that all moments of the coordinate for the oscillator are constant, which is equivalent to proving the wave packet does not change its shape.

or

$$\gamma \geq \beta^2. \tag{11.137}$$

This is not a surprising result—we expect the z-component of angular momentum to be less than or equal to the magnitude of the angular momentum.

I now form ladder operators

$$\hat{L}_\pm = \hat{L}_x \pm i\hat{L}_y. \tag{11.138}$$

The fact that these are ladder operators follows from the commutation relations

$$\left[\hat{L}_z, \hat{L}_\pm\right] = \left[\hat{L}_z, \hat{L}_x \pm i\hat{L}_y\right] = i\hbar\left(\hat{L}_y \mp i\hat{L}_x\right) = \pm\hbar\hat{L}_\pm. \tag{11.139}$$

Using these commutation relations, I find

$$\hat{L}_\pm\hat{L}_z\left|\gamma\beta\right\rangle = \hbar\beta\hat{L}_\pm\left|\gamma\beta\right\rangle = \left(\mp\hbar\hat{L}_\pm + \hat{L}_z\hat{L}_\pm\right)\left|\gamma\beta\right\rangle, \tag{11.140}$$

or

$$\hat{L}_z\left[\hat{L}_\pm\left|\gamma\beta\right\rangle\right] = \hbar\left(\beta \pm 1\right)\left[\hat{L}_\pm\left|\gamma\beta\right\rangle\right]. \tag{11.141}$$

In other words, $\hat{L}_\pm\left|\gamma\beta\right\rangle$ is an eigenket of $\hat{L}_z$ having eigenvalue $\hbar\left(\beta \pm 1\right)$, namely

$$\hat{L}_\pm\left|\gamma\beta\right\rangle = \hbar C_\pm\left(\gamma, \beta\right)\left|\gamma, \beta \pm 1\right\rangle, \tag{11.142}$$

where $C_\pm\left(\gamma, \beta\right)$ is a constant.

Thus, by applying $\hat{L}_\pm$ to $\left|\gamma\beta\right\rangle$ successively, I keep raising or lowering the eigenvalues by one unit of $\hbar$. This cannot go on forever, however, since I know that $\gamma \geq \beta^2$. Thus there must be a maximum value β_{max} and a minimum value β_{min} for which

$$\hat{L}_+\left|\gamma, \beta_{\text{max}}\right\rangle = 0; \tag{11.143a}$$

$$\hat{L}_-\left|\gamma, \beta_{\text{min}}\right\rangle = 0, \tag{11.143b}$$

implying that

$$\left(\beta_{\text{max}} - \beta_{\text{min}}\right) = \text{ positive integer or zero.} \tag{11.144}$$

I now calculate

$$\left\langle\gamma, \beta_{\text{max}}\right|\hat{L}_-\hat{L}_+\left|\gamma\beta_{\text{max}}\right\rangle = 0; \tag{11.145}$$

$$\left\langle\gamma, \beta_{\text{max}}\right|\left(\hat{L}_x - i\hat{L}_y\right)\left(\hat{L}_x + i\hat{L}_y\right)\left|\gamma\beta_{\text{max}}\right\rangle = 0; \tag{11.146}$$

$$\langle\gamma,\beta_{\max}|\,\hat{L}_x^2+\hat{L}_y^2-i\left[\hat{L}_y,\hat{L}_x\right]|\gamma\beta_{\max}\rangle=0; \tag{11.147}$$

$$\langle\gamma,\beta_{\max}|\,\hat{L}^2-\hat{L}_z^2-\hbar\hat{L}_z\,|\gamma\beta_{\max}\rangle=0; \tag{11.148}$$

$$\gamma-\beta_{\max}^2-\beta_{\max}=0; \tag{11.149}$$

$$\gamma=\beta_{\max}\left(\beta_{\max}+1\right). \tag{11.150}$$

Similarly by considering $\langle\gamma,\beta_{\min}|\,\hat{L}_+\hat{L}_-\,|\gamma\beta_{\min}\rangle=0$, I find

$$\gamma=\beta_{\min}\left(\beta_{\min}-1\right), \tag{11.151}$$

from which it follows that $\beta_{\min}=-\beta_{\max}$ and, using Eq. (11.144), that

$$2\beta_{\max}=\text{ positive integer or zero.} \tag{11.152}$$

From Eqs. (11.150) and (11.144) you see that $\gamma=\ell\left(\ell+1\right)$, where $\ell=\beta_{\max}$ is a positive half-integer, integer, or zero, and β, which I now denote by m, can take on values from $-\ell$ to ℓ in integer steps. Thus, as for the SHO, I obtained the eigenvalues of the operators $\hat{L}^2$ and $\hat{L}_z$ without solving a differential equation.

To obtain the values of $C_+\left(\ell,m\right)$ appearing in Eq. (11.142), I evaluate

$$\begin{aligned}\langle\ell,m|\,\hat{L}_-\hat{L}_+\,|\ell,m\rangle&=\left(\hat{L}_+\,|\ell,m\rangle\right)^\dagger\hat{L}_+\,|\ell,m\rangle\\&=\hbar^2\left|C_+\left(\ell,m\right)\right|^2\langle\ell,m+1\,|\ell,m+1\rangle=\hbar^2\left|C_+\left(\ell,m\right)\right|^2\\&=\langle\ell,m|\,\hat{L}^2-\hat{L}_z^2-\hbar\hat{L}_z\,|\ell,m\rangle=\hbar^2\left[\ell\left(\ell+1\right)-m^2-m\right],\end{aligned} \tag{11.153}$$

having used the identity $\hat{L}_-\hat{L}_+=\hat{L}^2-\hat{L}_z^2-\hbar\hat{L}_z$. Consequently,

$$C_+\left(\ell,m\right)=\sqrt{\ell\left(\ell+1\right)-m^2-m} \tag{11.154}$$

and

$$\begin{aligned}\hat{L}_+\,|\ell,m\rangle&=\hbar C_+\left(\ell,m\right)|\ell,m+1\rangle=\hbar\sqrt{\ell\left(\ell+1\right)-m^2-m}\,|\ell,m+1\rangle\\&=\hbar\sqrt{\left(\ell-m\right)\left(\ell+m+1\right)}\,|\ell,m+1\rangle.\end{aligned} \tag{11.155}$$

Similarly, you can show that

$$C_-\left(\ell,m\right)=\sqrt{\ell\left(\ell+1\right)-m^2+m} \tag{11.156}$$

and

$$\begin{aligned}\hat{L}_-\,|\ell,m\rangle&=\hbar C_-\left(\ell,m\right)|\ell,m-1\rangle=\hbar\sqrt{\ell\left(\ell+1\right)-m^2+m}\,|\ell,m-1\rangle\\&=\hbar\sqrt{\left(\ell+m\right)\left(\ell-m+1\right)}\,|\ell,m-1\rangle.\end{aligned} \tag{11.157}$$

These are equations that will prove useful in later chapters.

Finally, I need to say something about the half-integral values of ℓ and m, which are not ruled out by this discussion. Half-integral values of angular momentum *are* possible when electron spin is included. However, without spin I can expand the $|\ell m\rangle$ kets in terms of the kets in coordinate space as

$$|\ell m\rangle = \int d\mathbf{r} \left\langle \mathbf{r} \left| \ell m \right\rangle \right| \mathbf{r}\rangle . \tag{11.158}$$

In Appendix B, I obtain matrix elements of $\hat{\mathbf{L}}$ in the $|\mathbf{r}\rangle$ basis and, starting from

$$\hat{L}_z |\ell m\rangle = m\hbar |\ell m\rangle ; \tag{11.159a}$$

$$\hat{L}^2 |\ell m\rangle = \hbar^2 \ell (\ell + 1) |\ell m\rangle , \tag{11.159b}$$

I prove that the expansion coefficients $\langle \mathbf{r} | \ell m\rangle$ are solutions of the differential equation for the spherical harmonics. I have already shown that the only physically acceptable solutions for the spherical harmonics that are regular at both $\theta = 0, \pi$ and are unchanged when $\phi \to \phi + 2\pi$ correspond to integral values of ℓ and m with $\ell \geq |m|$. As a consequence, the half-integral values of ℓ and m must be rejected.

The expansion coefficient $\langle \mathbf{r} | \ell m\rangle$ is proportional to, but not equal to $Y_\ell^m (\theta, \phi)$ (it has the wrong units). To identify the spherical harmonic with an inner product in Dirac notation, I write the ket $|\mathbf{r}\rangle$ as

$$|\mathbf{r}\rangle = \frac{|r\rangle \, |\mathbf{u}_r\rangle}{r}, \tag{11.160}$$

where $|\mathbf{u}_r\rangle$ satisfies the orthogonality condition

$$\langle \mathbf{u}_r | \mathbf{u}_r' \rangle = \delta \left(\cos\theta - \cos\theta' \right) \delta \left(\phi - \phi' \right) \tag{11.161}$$

and can be interpreted as a solid-angle ket since the unit vector $\mathbf{u}_r$ depends only on the spherical angles θ and ϕ. Equations (11.160) and (11.161) are consistent with the fact that

$$\langle \mathbf{r} | \mathbf{r}' \rangle = \delta \left(\mathbf{r} - \mathbf{r}' \right) = \frac{\delta \left(r - r' \right) \delta \left(\cos\theta - \cos\theta' \right) \delta \left(\phi - \phi' \right)}{r^2}. \tag{11.162}$$

In terms of these solid angle kets (see Appendix B)

$$\langle \mathbf{u}_r | \ell m\rangle = Y_\ell^m (\theta, \phi) , \tag{11.163}$$

which relates the spherical harmonics to inner products in Dirac notation.

11.4 Solving Problems Using Dirac Notation

Things have been a bit formal in discussing Dirac notation, so let me summarize where and when you use it in practical situations relative to the Schrödinger equation. In general, you can often solve problems involving a discrete subspace using Dirac notation, but need the Schrödinger equation to solve problems involving an infinite number of levels. I give a few examples below to help illustrate these points.

11.4.1 Hydrogen Atom

I have already solved Schrödinger's equation for the hydrogen atom potential and found that, if $E < 0$, there is an infinite number of bound states. I also obtained the wave functions associated with these bound states. For $E > 0$, there is an infinite number of continuum states along with their associated eigenfunctions.

If you try to solve this problem using Dirac notation, you start from

$$\hat{H}\left|E\right\rangle = E\left|E\right\rangle, \tag{11.164}$$

but this is useless for obtaining the eigenvalues since you don't already know them! To proceed, you expand

$$\left|E\right\rangle = \int d\mathbf{r}\left\langle \mathbf{r}\left|E\right\rangle\right|\mathbf{r}\right\rangle \tag{11.165}$$

and try to find the expansion coefficients $\left\langle \mathbf{r}\left|E\right\rangle\right.$. I have already shown that these expansion coefficients satisfy the Schrödinger equation, so there is no advantage in using Dirac notation to obtain the eigenvalues and eigenfunctions of the hydrogen atom. On the other hand, you can label the eigenkets of the hydrogen atom by the eigenvalues of the commuting operators $\hat{H}, \hat{L}^2, \hat{L}_z$, namely $|n, \ell, m\rangle$, where n, ℓ, m are the quantum numbers corresponding to these physical quantities.

11.4.2 Harmonic Oscillator

This is a mixed case. You can solve for the eigenfunctions and eigenvalues using the Schrödinger equation without much trouble. However, as we have seen, it is also possible to introduce ladder operators and solve for the eigenvalues and eigenfunctions without having ever solved the Schrödinger equation directly. Calculation of matrix elements is easiest using the ladder operators. Although not so obvious, it is also possible to solve for the eigenvalues of the hydrogen atom without solving the Schrödinger equation based on the group $O(4)$ and the fact that both angular momentum and the Lenz vector are conserved.[4]

11.4.3 Angular Momentum

I will discuss formal aspects of angular momentum in Chaps. 19–20. Dirac notation is the preferred notation for dealing with angular momentum since you can label the simultaneous eigenkets of $\hat{L}^2$ and $\hat{L}_z$ by $|\ell m_\ell\rangle$ and write matrices for both L^2 and $\hat{L}_z$. For a given integral value of ℓ, each of these diagonal matrices has dimension $(2\ell + 1) \times (2\ell + 1)$.

To illustrate the power of Dirac notation, I would like to discuss the eigenkets of $\hat{L}^2$ and either $\hat{L}_x$ and $\hat{L}_z$. I represent the eigenkets of $\hat{L}^2$ and $\hat{L}_x$ by $|\ell, \ell_x\rangle$ and those of $\hat{L}^2$ and $\hat{L}_z$ by $|\ell, \ell_z\rangle$. Clearly the *eigenvalues* must be identical, since there is nothing *physically* that can distinguish $\hat{L}_x$ from $\hat{L}_z$. I can take the basis kets (a column vector with a 1 in one place and zeroes everywhere else) to be those of $\hat{L}_x$ *or* $\hat{L}_z$, *but not both.* For example, let me take the basis kets to be $|\ell, \ell_z\rangle$ and take $\ell = 1$. In this $\ell = 1$ subspace

$$\underline{\mathrm{L}}_z = \begin{pmatrix} 1 & 0 & 0 \\ 0 & 0 & 0 \\ 0 & 0 & -1 \end{pmatrix} \hbar, \tag{11.166}$$

where the order is $\ell_z = 1, 0, -1$. Moreover, using ladder operators, it is easy to calculate

$$\underline{\mathrm{L}}_x = \begin{pmatrix} 0 & 1 & 0 \\ 1 & 0 & 1 \\ 0 & 1 & 0 \end{pmatrix} \frac{\hbar}{\sqrt{2}}. \tag{11.167}$$

By diagonalizing $\underline{\mathrm{L}}_x$ I find the eigenvalues $\ell_x = 1, 0, -1$ as expected. The eigenkets (with the $\ell = 1$ label suppressed) are given by

$$\begin{aligned}
|\ell_x = 1\rangle &= \frac{1}{2}\left(|\ell_z = 1\rangle + \sqrt{2}\,|\ell_z = 0\rangle + |\ell_z = -1\rangle\right); \\
|\ell_x = 0\rangle &= \frac{1}{\sqrt{2}}\left(|\ell_z = 1\rangle - |\ell_z = -1\rangle\right); \\
|\ell_x = -1\rangle &= \frac{1}{2}\left(|\ell_z = 1\rangle - \sqrt{2}\,|\ell_z = 0\rangle + |\ell_z = -1\rangle\right).
\end{aligned} \tag{11.168}$$

[4]For a concise, excellent discussion, see Chap. 7 in *Quantum Mechanics,* Third Edition (McGraw Hill, New York, 1968) by L. Schiff.

The difference between $\hat{L}_x$ and $\hat{L}_z$ becomes apparent when I consider the eigenfunctions rather than the eigenkets. The reason for this is that the spherical coordinate system is one in which the polar angle is measured from the z axis, making the eigenfunctions of $\hat{L}_z$ simpler than those of $\hat{L}_x$. If I take the inner product of Eqs. (11.168) with $\langle \mathbf{u}_r|$ and use the fact that $\langle \mathbf{u}_r |\ell, \ell_z\rangle \equiv \langle \mathbf{u}_r |\ell m\rangle = Y_\ell^m(\theta, \phi)$, I find

$$\begin{aligned}\Phi_1^1(\theta,\phi) &= \tfrac{1}{2}\left[Y_1^1(\theta,\phi) + \sqrt{2}Y_1^0(\theta,\phi) + Y_1^{-1}(\theta,\phi)\right] \\ \Phi_1^0(\theta,\phi) &= \tfrac{1}{\sqrt{2}}\left[Y_1^1(\theta,\phi) - Y_1^{-1}(\theta,\phi)\right] \\ \Phi_1^{-1}(\theta,\phi) &= \tfrac{1}{2}\left[Y_1^1(\theta,\phi) - \sqrt{2}Y_1^0(\theta,\phi) + Y_1^{-1}(\theta,\phi)\right]\end{aligned} \tag{11.169}$$

are the eigenfunctions of $\hat{L}_x$. You see that, although the eigenkets are essentially identical, the eigenfunctions are most simply expressed if you use $\hat{L}_z$ rather than $\hat{L}_x$. To solve for the eigenfunctions of $\hat{L}_x$ directly using $\hat{L}_x\Phi_1^m(\theta,\phi) = m\hbar\Phi_1^m(\theta,\phi)$ with $\hat{L}_x$ given by Eq. (9.31a) would be more difficult.

11.4.4 Limited Subspaces

Sometimes an atom–field interaction is limited to a discrete subspace. Examples are an atom interacting with a field that is resonant with its ground to first excited state transition frequency or a magnetic field acting on an atom in its ground state, including spin. Whenever there are just a few levels in the problem, Dirac notation is usually the method of choice for attacking the problem.

11.5 Connection with Linear Algebra

Suppose you have a Hamiltonian matrix of the form

$$\underline{\mathrm{H}} = \begin{pmatrix} 1 & 0.5 & 2 \\ 0.5 & 5 & 1 \\ 2 & 1 & 8 \end{pmatrix}. \tag{11.170}$$

You know this is not the Hamiltonian in the energy basis since it would be *diagonal* in that basis. Let us imagine that Eq. (11.170) represents the Hamiltonian in the u basis with

$$\underline{\mathrm{u}}_1 = \begin{pmatrix} 1 \\ 0 \\ 0 \end{pmatrix}; \quad \underline{\mathrm{u}}_2 = \begin{pmatrix} 0 \\ 1 \\ 0 \end{pmatrix}; \quad \underline{\mathrm{u}}_3 = \begin{pmatrix} 0 \\ 0 \\ 1 \end{pmatrix}, \tag{11.171}$$

How can I get the eigenvalues of $\underline{\mathrm{H}}$ and its eigenvectors expressed in the u basis? To do so I expand

$$|E\rangle = \sum_{n=1}^{3} \langle u_n |E\rangle |u_n\rangle \tag{11.172}$$

and show that obtaining the $\langle u_n |E\rangle$ is equivalent to diagonalizing the $\underline{\mathrm{H}}$ matrix. I start from

$$\underline{\mathrm{H}} |E\rangle = \sum_{n=1}^{3} \underline{\mathrm{H}} |u_n\rangle \langle u_n |E\rangle = E |E\rangle \tag{11.173}$$

and multiply by $\langle u_m|$ to obtain

$$\sum_{n=1}^{3} \langle u_m| \underline{\mathrm{H}} |u_n\rangle \langle u_n |E\rangle = E \langle u_m |E\rangle . \tag{11.174}$$

This is just the equation that you encountered in linear algebra to diagonalize a matrix and find its eigenvectors, that is

$$(\underline{\mathrm{H}}_{11} - E) \langle u_1 |E\rangle + \underline{\mathrm{H}}_{21} \langle u_2 |E\rangle + \underline{\mathrm{H}}_{31} \langle u_3 |E\rangle = 0; \tag{11.175a}$$

$$\underline{\mathrm{H}}_{21} \langle u_1 |E\rangle + (\underline{\mathrm{H}}_{22} - E) \langle u_2 |E\rangle + \underline{\mathrm{H}}_{23} \langle u_3 |E\rangle = 0; \tag{11.175b}$$

$$\underline{\mathrm{H}}_{31} \langle u_1 |E\rangle + \underline{\mathrm{H}}_{32} \langle u_2 |E\rangle + (\underline{\mathrm{H}}_{22} - E) \langle u_3 |E\rangle = 0. \tag{11.175c}$$

The equation has a non-trivial solution only if the determinant of the coefficients vanishes. By setting the determinant of the coefficients equal to zero, I can use this result to obtain the three energy eigenvalues $E_n \{n = 1, 2, 3\}$. For each eigenenergy E_m, Eqs. (11.175) are solved for the $\langle u_n |E_m\rangle \{n = 1, 2, 3\}$, allowing me to obtain the eigenkets as

$$|E_m\rangle = \sum_{n=1}^{3} \langle u_n |E_m\rangle |u_n\rangle = \sum_{n=1}^{3} a_n^{(m)} |u_n\rangle , \tag{11.176}$$

where

$$a_n^{(m)} = \langle u_n |E_m\rangle . \tag{11.177}$$

In matrix form,

$$|E_1\rangle = \begin{pmatrix} \langle u_1 |E_1\rangle \\ \langle u_2 |E_1\rangle \\ \langle u_3 |E_1\rangle \end{pmatrix}; \quad |E_2\rangle = \begin{pmatrix} \langle u_1 |E_2\rangle \\ \langle u_2 |E_2\rangle \\ \langle u_3 |E_2\rangle \end{pmatrix}; \quad |E_3\rangle = \begin{pmatrix} \langle u_1 |E_3\rangle \\ \langle u_2 |E_3\rangle \\ \langle u_3 |E_3\rangle \end{pmatrix}. \tag{11.178}$$

Thus, finding the expansion coefficients of $|E\rangle$ in the u basis is equivalent to diagonalizing the Hamiltonian.

I form a matrix $\underset{\sim}{\mathrm{S}}^{\dagger}$ by placing the eigenvectors in *columns*, such that

$$\underset{\sim}{\mathrm{S}}^{\dagger} = \begin{pmatrix} \langle u_1 | E_1 \rangle & \langle u_1 | E_2 \rangle & \langle u_1 | E_3 \rangle \\ \langle u_2 | E_1 \rangle & \langle u_2 | E_2 \rangle & \langle u_2 | E_3 \rangle \\ \langle u_3 | E_1 \rangle & \langle u_3 | E_2 \rangle & \langle u_3 | E_3 \rangle \end{pmatrix}. \tag{11.179}$$

The eigenvectors are then given by

$$\begin{pmatrix} |E_1\rangle \\ |E_2\rangle \\ |E_3\rangle \end{pmatrix} = \left(\underset{\sim}{\mathrm{S}}^{\dagger}\right)^T \begin{pmatrix} |u_1\rangle \\ |u_2\rangle \\ |u_3\rangle \end{pmatrix} = \underset{\sim}{\mathrm{S}}^{*} \begin{pmatrix} |u_1\rangle \\ |u_2\rangle \\ |u_3\rangle \end{pmatrix} \tag{11.180}$$

For *normalized* eigenkets, $\underset{\sim}{\mathrm{S}}^{\dagger}$ is a unitary matrix, having inverse of $\left(\underset{\sim}{\mathrm{S}}^{\dagger}\right)^{-1} = \underset{\sim}{\mathrm{S}}$ and

$$\underset{\sim}{\mathrm{S}}\underset{\sim}{\mathrm{H}}\underset{\sim}{\mathrm{S}}^{\dagger} = \underset{\sim}{\mathrm{E}} \tag{11.181}$$

where $\underset{\sim}{\mathrm{E}}$ is a diagonal matrix whose elements are the eigenvalues of $\underset{\sim}{\mathrm{H}}$. To get the matrix $\underset{\sim}{\mathrm{S}}^{\dagger}$ in Mathematica, use Transpose[Orthogonalize[Eigenvectors[H]]], where H={{h_{11},h_{12},h_{13}},{h_{21},h_{22},h_{23}},{h_{31},h_{32},h_{33}}}. To get the eigenvalues, use Eigenvalues[H].

It is easy to show that $\underset{\sim}{\mathrm{S}}\underset{\sim}{\mathrm{H}}\underset{\sim}{\mathrm{S}}^{\dagger} = \underset{\sim}{\mathrm{E}}$, in general; that is, for matrices having arbitrary dimension. Since $\underset{\sim}{\mathrm{H}}|E_m\rangle = E_m |E_m\rangle$, it follows that

$$\langle E_n| H |E_m\rangle = E_n\delta_{n,m}; \tag{11.182}$$

$$\sum_{p,q} \langle E_n |u_p\rangle\langle u_p| H |u_q\rangle\langle u_q |E_m\rangle = E_n\delta_{n,m}; \tag{11.183}$$

$$\sum_{p,q} \left(a_p^{(n)}\right)^* H_{pq}a_q^{(m)} = E_n\delta_{n,m}; \tag{11.184}$$

$$\sum_{p,q} \left(S^{\dagger}\right)^*_{pn} H_{pq} \left(S^{\dagger}\right)_{qm} = \sum_{p,q} S_{np}H_{pq} \left(S^{\dagger}\right)_{qm} = E_n\delta_{n,m}; \tag{11.185}$$

$$\left(\underset{\sim}{\mathrm{S}}\underset{\sim}{\mathrm{H}}\underset{\sim}{\mathrm{S}}^{\dagger}\right)_{nm} = E_n\delta_{n,m}. \tag{11.186}$$

As an example, consider

$$\underset{\sim}{\mathrm{H}} = \begin{pmatrix} 2 & -2i \\ 2i & 5 \end{pmatrix} \tag{11.187}$$

in the $|1\rangle, |2\rangle$ basis in some arbitrary units. To diagonalize $\underset{\sim}{\mathrm{H}}$, I set

$$|\underset{\sim}{\mathrm{H}}| = \begin{vmatrix} 2-E & -2i \\ 2i & 5-E \end{vmatrix} = 0 \tag{11.188}$$

and evaluate the determinant to obtain

$$E^2 - 7E + 6 = 0; \qquad E = 6, 1. \tag{11.189}$$

For $E = 6$

$$\begin{aligned} -4a_1^{(1)} - 2ia_2^{(1)} &= 0; \\ a_2^{(1)} &= 2ia_1^{(1)} \end{aligned} \tag{11.190}$$

Therefore

$$|E_1\rangle = a_1^{(1)} \left(|1\rangle + 2i\,|2\rangle\right). \tag{11.191}$$

I normalize by taking

$$\begin{aligned} \left|a_1^{(1)}\right|^2 (1+4) &= 1; \\ a_1^{(1)} &= \frac{1}{\sqrt{5}}, \end{aligned} \tag{11.192}$$

such that

$$|E_1\rangle = \frac{1}{\sqrt{5}} \left(|1\rangle + 2i\,|2\rangle\right). \tag{11.193}$$

Similarly, for $E = 1$

$$\begin{aligned} a_1^{(2)} - 2ia_2^{(2)} &= 0; \\ a_1^{(2)} &= 2ia_2^{(2)} \end{aligned} \tag{11.194}$$

and

$$|E_2\rangle = a_2^{(2)} \left(2i\,|1\rangle + |2\rangle\right) = \frac{1}{\sqrt{5}} \left(2i\,|1\rangle + |2\rangle\right). \tag{11.195}$$

Therefore

$$\underset{\sim}{S}^\dagger = \frac{1}{\sqrt{5}} \begin{pmatrix} 1 & 2i \\ 2i & 1 \end{pmatrix}; \quad \underset{\sim}{S} = \frac{1}{\sqrt{5}} \begin{pmatrix} 1 & -2i \\ -2i & 1 \end{pmatrix}, \tag{11.196}$$

$$\begin{pmatrix} |E_1\rangle \\ |E_2\rangle \end{pmatrix} = \underset{\sim}{S}^* \begin{pmatrix} |1\rangle \\ |2\rangle \end{pmatrix} = \frac{1}{\sqrt{5}} \begin{pmatrix} 1 & 2i \\ 2i & 1 \end{pmatrix} \begin{pmatrix} |1\rangle \\ |2\rangle \end{pmatrix}, \tag{11.197}$$

and

$$\underline{S}\underline{H}\underline{S}^{\dagger} = \begin{pmatrix} 6 & 0 \\ 0 & 1 \end{pmatrix}. \tag{11.198}$$

Note that for any matrix that is 3×3 or larger, you need to diagonalize numerically rather than analytically since the characteristic equation to find the eigenvalues will be cubic or higher. In Mathematica, simply put a decimal point in one of the entries. Thus for

$$\underline{H} = \begin{pmatrix} 1 & 1 & 2 \\ 1 & 5 & 3 \\ 2 & 3 & 8 \end{pmatrix}, \tag{11.199}$$

you can write

$$h = \{\{1., 1, 2\}, \{1, 5, 3\}, \{2, 3, 8\}\} \tag{11.200}$$

and Mathematica gives eigenvalues

$$\{10.383, 3.1598, 0.457203\}$$

and eigenvectors {{0.231057, 0.50604, 0.830985}, {0.0690456, -0.860472, 0.504798}, {0.970487, -0.0592615, -0.233758}}, corresponding to

$$|E_1\rangle = 0.231057\,|1\rangle + 0.50604\,|2\rangle + 0.830985\,|3\rangle\,; \tag{11.201a}$$

$$|E_2\rangle = 0.0690456\,|1\rangle - 0.860472\,|2\rangle + 0.504798\,|3\rangle\,; \tag{11.201b}$$

$$|E_3\rangle = 0.970487\,|1\rangle - 0.0592615\,|2\rangle - 0.233758\,|3\rangle\,, \tag{11.201c}$$

with energies from highest to lowest.

11.5.1 Time Dependence

The time-dependent Schrödinger equation is solved in the same manner used in the wave function approach, once the eigenvectors and eigenkets are determined. The only tricky problem is that the initial conditions are often given in terms of the *original* basis rather than the eigenket basis. A simple example illustrates this point. Consider the matrix

$$\underline{H} = \hbar\omega \begin{pmatrix} 2 & -2i \\ 2i & 5 \end{pmatrix} \tag{11.202}$$

in the $|1\rangle, |2\rangle$ basis; the variable ω has units of frequency. Suppose at $t = 0$ the system is in state $|1\rangle$. What is $|\psi(t)\rangle$? I have already calculated the eigenenergies and eigenkets of this Hamiltonian in the previous section. Using Eq. (11.43), I expand $|\psi(t)\rangle$ as

$$|\psi(t)\rangle = b_1 e^{-iE_1 t/\hbar} |E_1\rangle + b_2 e^{-iE_2 t/\hbar} |E_2\rangle . \tag{11.203}$$

The initial condition is

$$|\psi(0)\rangle = |1\rangle = b_1 |E_1\rangle + b_2 |E_2\rangle . \tag{11.204}$$

Taking inner products with the eigenkets and using Eqs. (11.193) and (11.195), I find the expansion coefficients

$$b_1 = \langle E_1 | \psi(0)\rangle = \langle E_1 | 1\rangle = \frac{1}{\sqrt{5}} \left(\langle 1| - 2i \langle 2|\right) 1\rangle = \frac{1}{\sqrt{5}}; \tag{11.205a}$$

$$b_2 = \langle E_2 | \psi(0)\rangle = \langle E_2 | 1\rangle = \frac{1}{\sqrt{5}} \left(-2i \langle 1| + \langle 2|\right) 1\rangle = \frac{-2i}{\sqrt{5}}, \tag{11.205b}$$

and then use Eq. (11.203) to obtain the state vector

$$|\psi(t)\rangle = \frac{1}{\sqrt{5}} \left(e^{-6i\omega t} |E_1\rangle - 2ie^{-i\omega t} |E_2\rangle\right) . \tag{11.206}$$

The state vector can be re-expressed in terms of the *original* basis as

$$\begin{aligned} |\psi(t)\rangle &= \frac{1}{\sqrt{5}} \left(\frac{e^{-6i\omega t}}{\sqrt{5}} \left(|1\rangle + 2i|2\rangle\right) - \frac{2i}{\sqrt{5}} e^{-i\omega t} \left(2i|1\rangle + |2\rangle\right) \right) \\ &= \frac{1}{5} \left[\left(e^{-6i\omega t} + 4e^{-i\omega t}\right) |1\rangle + 2i \left(e^{-6i\omega t} - e^{-i\omega t}\right) |2\rangle \right] . \end{aligned} \tag{11.207}$$

As a consequence the state probabilities in terms of the original basis states are

$$P_1(t) = \frac{1}{25} \left[17 + 8 \cos(5\omega t)\right] ; \tag{11.208a}$$

$$P_2(t) = \frac{8}{25} \left[1 - \cos(5\omega t)\right] , \tag{11.208b}$$

and both these probabilities oscillate in time.

In general, for an arbitrary system, the state vector is

$$|\psi(t)\rangle = \sum_n b_n e^{-iE_n t/\hbar} |E_n\rangle = \sum_n \langle n | \psi(0)\rangle e^{-iE_n t/\hbar} |E_n\rangle , \tag{11.209}$$

where the eigenkets $|E_n\rangle$ are the eigenkets of a Hamiltonian $\underline{H}$ having eigenenergies E_n. Suppose, however, that we know matrix elements of the Hamiltonian only in some other basis denoted by $|n\rangle$ and that, although $\underline{H}$ is not diagonal in this basis, the initial state vector $|\psi(0)\rangle$ can be expressed in this basis as

$$|\psi(0)\rangle = \sum_n a_n |n\rangle . \tag{11.210}$$

You can then calculate the state vector at any time using Eq. (11.209) to be

$$|\psi(t)\rangle = \sum_{m,n} a_n \langle E_m| n\rangle e^{-iE_m t/\hbar} |E_m\rangle . \tag{11.211}$$

However, from Eq. (11.180), you know that

$$|E_m\rangle = \sum_{n'} (S^*)_{mn'} |n'\rangle . \tag{11.212}$$

The adjoint of this equation is

$$\langle E_m| = \sum_{m'} (S)_{mm'} \langle m'| \tag{11.213}$$

(note the use of different dummy variables). Substituting these expressions into Eq. (11.211), I find

$$\begin{aligned} |\psi(t)\rangle &= \sum_{n,n',m,m'} a_n (S)_{mm'} \langle m'| n\rangle e^{-iE_m t/\hbar} (S^*)_{mn'} |n'\rangle \\ &= \sum_{n,n',m} a_n (S)_{mn} e^{-iE_m t/\hbar} (S^*)_{mn'} |n'\rangle \\ &= \sum_{n,n',m} a_n (S^\dagger)_{n'm} (S)_{mn} e^{-iE_m t/\hbar} |n'\rangle , \end{aligned} \tag{11.214}$$

which is the desired result.

In our previous example, $a_1 = 1, a_2 = 0$, and

$$\underline{S}^\dagger = \frac{1}{\sqrt{5}} \begin{pmatrix} 1 & 2i \\ 2i & 1 \end{pmatrix}; \quad \underline{S} = \frac{1}{\sqrt{5}} \begin{pmatrix} 1 & -2i \\ -2i & 1 \end{pmatrix}, \tag{11.215}$$

such that

$$\begin{aligned}
|\psi(t)\rangle &= \sum_{n,m} \left(S^{\dagger}\right)_{nm} (S)_{m1}\, e^{-iE_m t/\hbar}\, |n\rangle \\
&= \left[\left(S^{\dagger}\right)_{11} (S)_{11}\, e^{-iE_1 t/\hbar} + \left(S^{\dagger}\right)_{12} (S)_{21}\, e^{-iE_2 t/\hbar}\right] |1\rangle \\
&\quad + \left[\left(S^{\dagger}\right)_{21} (S)_{11}\, e^{-iE_1 t/\hbar} + \left(S^{\dagger}\right)_{22} (S)_{21}\, e^{-iE_2 t/\hbar}\right] |2\rangle \\
&= \frac{1}{5}\left[\left(e^{-6i\omega t} + 4e^{-i\omega t}\right)|1\rangle + 2i\left(e^{-6i\omega t} - e^{-i\omega t}\right)|2\rangle\right],
\end{aligned} \tag{11.216}$$

in agreement with Eq. (11.207).

11.6 Summary

I have shown that Dirac notation is a powerful method for dealing with problems in quantum mechanics. Rather than specify a given representation, you can write state vectors in a basis-independent manner. Connection with specific representations such as the coordinate or momentum representation can then be obtained by taking inner products. In Dirac notation, operators are represented by matrices. As such Dirac notation is closely related to Heisenberg's matrix formulation of quantum mechanics.

11.7 Appendix A: Matrix Properties

In this Appendix, I list some matrix properties. I simply use standard type for all matrices, rather than underlined quantities. The *identity matrix* I has $I_{ij} = \delta_{i,j}$, that is, ones along the diagonal and zeroes everywhere else. The *inverse* A^{-1} of a matrix A satisfies

$$A^{-1}A = AA^{-1} = 1. \tag{11.217}$$

The *transpose* A^T of a matrix A is defined by

$$\left(A^T\right)_{ij} = A_{ji}. \tag{11.218}$$

The Hermitian adjoint $A^{\dagger}$ of a matrix A is defined by

$$\left(A^{\dagger}\right)_{ij} = A_{ji}^{*} \text{ or } A^{\dagger} = \left(A^T\right)^{*} \tag{11.219}$$

and

$$(AB)^{-1} = B^{-1}A^{-1}; \tag{11.220a}$$
$$(AB)^T = B^T A^T; \tag{11.220b}$$
$$(AB)^\dagger = B^\dagger A^\dagger. \tag{11.220c}$$

A matrix U is *unitary* if

$$U^\dagger U = UU^\dagger = 1 \tag{11.221}$$

and a matrix O is *orthogonal* if

$$O^T O = OO^T = 1. \tag{11.222}$$

For a unitary matrix $U^{-1} = U^\dagger$ and the determinant of U is a complex number having unit magnitude. For an orthogonal matrix $O^{-1} = O^T$ and the determinant of O is unity.

A matrix H is Hermitian if

$$H^\dagger = H. \tag{11.223}$$

For any Hermitian matrix it is always possible to find a unitary matrix U such that

$$UHU^\dagger = E \tag{11.224}$$

where E is a diagonal matrix. The columns of $U^\dagger$ are the eigenvectors and the diagonal elements of E are the eigenvalues. If H is real, then U is also real so it is an orthogonal matrix.

11.8 Appendix B: Spherical Harmonics in Dirac Notation

To express the spherical harmonics in Dirac notation, I first obtain matrix elements of $\hat{\mathbf{L}}$ in the $|\mathbf{r}\rangle$ basis. To do so, I write $\hat{\mathbf{L}} = \hat{\mathbf{r}} \times \hat{\mathbf{p}}$ and use Eq. (11.59c) to arrive at

$$\langle \mathbf{r}| \hat{L}_x |\mathbf{r}'\rangle = \left(y\frac{\partial}{\partial z} - z\frac{\partial}{\partial y} \right) \delta\left(\mathbf{r} - \mathbf{r}'\right), \tag{11.225}$$

along with its cyclical permutations. I next use the the relationships $r = \sqrt{x^2 + y^2 + z^2}$, $\theta = \cos^{-1}(z/r)$, $\phi = \tan^{-1}(y/x)$, $x = r\sin\theta\cos\phi$, and $y = r\sin\theta\sin\phi$ to express the partial derivatives in spherical coordinates as

$$\frac{\partial}{\partial x} = \frac{\partial r}{\partial x}\frac{\partial}{\partial r} + \frac{\partial \theta}{\partial x}\frac{\partial}{\partial \theta} + \frac{\partial \phi}{\partial x}\frac{\partial}{\partial \phi}$$

$$= \frac{x}{r}\frac{\partial}{\partial r} + \frac{\cos\theta\cos\phi}{r}\frac{\partial}{\partial \theta} - \frac{\sin\phi}{r\sin\theta}\frac{\partial}{\partial \phi}; \tag{11.226a}$$

$$\frac{\partial}{\partial y} = \frac{y}{r}\frac{\partial}{\partial r} + \frac{\cos\theta\sin\phi}{r}\frac{\partial}{\partial \theta} + \frac{\cos\phi}{r\sin\theta}\frac{\partial}{\partial \phi}; \tag{11.226b}$$

$$\frac{\partial}{\partial z} = \frac{z}{r}\frac{\partial}{\partial r} - \frac{\sin\theta}{r}\frac{\partial}{\partial \theta}. \tag{11.226c}$$

It then follows that

$$\langle \mathbf{r}| \hat{L}_x |\mathbf{r}'\rangle = -\frac{\hbar}{i}\left(\sin\phi\frac{\partial}{\partial\theta} + \cot\theta\cos\phi\frac{\partial}{\partial\phi}\right)\delta\left(\mathbf{r}-\mathbf{r}'\right); \tag{11.227a}$$

$$\langle \mathbf{r}| \hat{L}_y |\mathbf{r}'\rangle = \frac{\hbar}{i}\left(\cos\phi\frac{\partial}{\partial\theta} - \cot\theta\sin\phi\frac{\partial}{\partial\phi}\right)\delta\left(\mathbf{r}-\mathbf{r}'\right); \tag{11.227b}$$

$$\langle \mathbf{r}| \hat{L}_z |\mathbf{r}'\rangle = \frac{\hbar}{i}\frac{\partial}{\partial\phi}\delta\left(\mathbf{r}-\mathbf{r}'\right) \tag{11.227c}$$

and

$$\langle \mathbf{r}| \hat{L}^2 |\mathbf{r}'\rangle = \int d\mathbf{r}'' \langle \mathbf{r}| \hat{L}_x |\mathbf{r}''\rangle\langle \mathbf{r}''| \hat{L}_x |\mathbf{r}'\rangle$$

$$+ \int d\mathbf{r}'' \langle \mathbf{r}| \hat{L}_y |\mathbf{r}''\rangle\langle \mathbf{r}''| \hat{L}_y |\mathbf{r}'\rangle$$

$$+ \int d\mathbf{r}'' \langle \mathbf{r}| \hat{L}_z |\mathbf{r}''\rangle\langle \mathbf{r}''| \hat{L}_z |\mathbf{r}'\rangle$$

$$= -\hbar^2\left[\frac{\partial^2}{\partial\theta^2} + \cot\theta\frac{\partial}{\partial\theta} + \frac{1}{\sin^2\theta}\frac{\partial^2}{\partial\phi^2}\right]\delta\left(\mathbf{r}-\mathbf{r}'\right). \tag{11.228}$$

Using the eigenvalue equation

$$\hat{L}_z |\ell m\rangle = m\hbar |\ell m\rangle, \tag{11.229}$$

I insert a complete set and multiply on the left by $\langle \mathbf{r}|$ to transform this equation into

$$\int d\mathbf{r}' \langle \mathbf{r}| \hat{L}_z |\mathbf{r}'\rangle\langle \mathbf{r}' |\ell m\rangle = m\hbar \langle \mathbf{r} |\ell m\rangle \tag{11.230}$$

and then use Eq. (11.227c) to obtain

$$\int d\mathbf{r}' \frac{\partial}{\partial \phi} \delta\left(\mathbf{r}-\mathbf{r}'\right) \left\langle \mathbf{r}' \left| \ell m \right\rangle = im \left\langle \mathbf{r} \left| \ell m \right\rangle ;\right.\right.$$

$$\frac{\partial}{\partial \phi} \left\langle \mathbf{r} \left| \ell m \right\rangle = im \left\langle \mathbf{r} \left| \ell m \right\rangle .\right.\right. \tag{11.231}$$

Similarly, starting from

$$\hat{L}^2 \left| \ell m \right\rangle = \hbar^2 \ell \left(\ell + 1\right) \left| \ell m \right\rangle , \tag{11.232}$$

and using Eq. (11.228), I find

$$-\left[\frac{\partial^2}{\partial \theta^2} + \cot\theta \frac{\partial}{\partial \theta} + \frac{1}{\sin^2\theta} \frac{\partial^2}{\partial \phi^2}\right] \left\langle \mathbf{r} \left| \ell m \right\rangle = \ell\left(\ell+1\right) \left\langle \mathbf{r} \left| \ell m \right\rangle .\right.\right. \tag{11.233}$$

To relate the $\langle \mathbf{r} | \ell m\rangle$ to the spherical harmonics I write the ket $|\mathbf{r}\rangle$ as

$$\left|\mathbf{r}\right\rangle = \frac{\left|r\right\rangle \left|\mathbf{u}_r\right\rangle}{r}, \tag{11.234}$$

multiply Eqs. (11.231) and (11.233) by $r^2 |r\rangle$, integrate over r, and use the completeness relation

$$\int_0^\infty dr\, r^2 \left|r\right\rangle \left\langle r\right| = 1 \tag{11.235}$$

to arrive at

$$\frac{\partial}{\partial \phi} \left\langle \mathbf{u}_r \left| \ell m \right\rangle = im \left\langle \mathbf{u}_r \left| \ell m \right\rangle ;\right.\right. \tag{11.236a}$$

$$-\left[\frac{\partial^2}{\partial \theta^2} + \cot\theta \frac{\partial}{\partial \theta} + \frac{1}{\sin^2\theta} \frac{\partial^2}{\partial \phi^2}\right] \left\langle \mathbf{u}_r \left| \ell m \right\rangle = \ell\left(\ell+1\right) \left\langle \mathbf{u}_r \left| \ell m \right\rangle ,\right.\right. \tag{11.236b}$$

implying that

$$\left\langle \mathbf{u}_r \left| \ell m \right\rangle = Y_\ell^m \left(\theta, \phi\right) .\right. \tag{11.237}$$

Note that Eq. (11.237) has the correct normalization, since

$$\begin{aligned} \left\langle \ell' m' \left| \ell m \right\rangle \right. &= \int d\Omega \left\langle \ell' m' \left| \mathbf{u}_r \right\rangle \left\langle \mathbf{u}_r \left| \ell m \right\rangle \right.\right. \\ &= \int d\Omega \left[Y_{\ell'}^{m'}\left(\theta,\phi\right)\right]^* Y_\ell^m\left(\theta,\phi\right) = \delta_{\ell,\ell'}\delta_{m,m'} . \end{aligned} \tag{11.238}$$

11.9 Problems

1. In Dirac notation, how are Hermitian operators represented? What does it mean to say that the matrix elements of an operator depend on the basis? Give an example by considering the matrix elements of the operator $\hat{x}$ in the coordinate and momentum bases. Is it diagonal in the coordinate basis? in the momentum basis? Why is the diagonalization of the Hamiltonian in the coordinate basis equivalent to solving the Schrödinger equation?

2. Consider the two-dimensional vector $\mathbf{A} = \mathbf{u}_x + 2\mathbf{u}_y$. Suppose you want to use different orthonormal basis vectors defined by

$$\mathbf{u}_{1,2} = \frac{\mathbf{u}_x \pm \mathbf{u}_y}{\sqrt{2}}.$$

Express the unit vectors $\mathbf{u}_x$ and $\mathbf{u}_y$ as column vectors in this basis and find the coordinates of $\mathbf{A}$ in this basis. Show explicitly that $A_1\mathbf{u}_1 + A_2\mathbf{u}_2 = A_x\mathbf{u}_x + A_y\mathbf{u}_y$.

3. The adjoint or Hermitian conjugate $A^\dagger$ of a matrix A is defined by $\left(A^\dagger\right)_{mn} = (A_{nm})^*$. Show that for two matrices A and B for which matrix multiplication can be defined, $(AB)^\dagger = B^\dagger A^\dagger$. As a consequence prove that $\hat{A}\hat{B}$ is Hermitian only if $\left[\hat{A}, \hat{B}\right] = 0$.

4. Using Dirac notation, prove that the eigenvalues of a Hermitian operator are real and the eigenkets having nondegenerate eigenvalues are orthogonal.

5. Using Dirac notation prove that two Hermitian operators can possess simultaneous eigenkets if, and only if, the operators commute.

6. Suppose that in the $|g\rangle$ basis, a Hamiltonian has the form

$$H = \begin{pmatrix} 3 & 1 \\ 1 & 2 \end{pmatrix}.$$

Find the eigenvalues and (normalized) eigenvectors (express the eigenvectors in the $|g\rangle$ basis). Do the calculation yourself and then check the result on a computer using, for example, the Eigenvalues and Eigenvectors operations in Mathematica (e.g., Orthogonalize[Eigenvectors[{{3.,1},{1,2}}]]—putting the decimal point in will give you numerical values).

7. In Problem 11.6, find a matrix $\underline{S}$ such that $\underline{S}\underline{H}\underline{S}^\dagger$ is a diagonal matrix having diagonal elements equal to the eigenvalues of $\underline{H}$. Check your answer using a computer program.

8. In Problem 11.6, suppose that at $t = 0$ a particle is in the state $|1\rangle$ in the $|g\rangle$ basis. Find the wave function as a function of time in terms of the $|g\rangle$ basis. Show that the probability to be in state $|1\rangle$ in the $|g\rangle$ basis oscillates as a function of time.

9. The time-dependent Schrödinger equation in Dirac notation can be written as

$$i\hbar \frac{d\,|\psi(t)\rangle}{dt} = \hat{H}\,|\psi(t)\rangle\,.$$

For an operator $\hat{A}$ that has no explicit time dependence, prove

$$d\left\langle \psi(t)\left|\hat{A}\right|\psi(t)\right\rangle/dt = \frac{1}{i\hbar}\left\langle \psi(t)\left|\left[\hat{A},\hat{H}\right]\right|\psi(t)\right\rangle.$$

What can you conclude about operators that commute with the Hamiltonian?

10. Using ladder operators, evaluate matrix elements $\langle n|\,\hat{x}\,|q\rangle$ and $\langle n|\,\hat{x}^2\,|q\rangle$ for the 1-D harmonic oscillator. What is the general structure of the matrices $\underline{\mathrm{H}}$, $\underline{\mathrm{x}}$, and $\underline{\mathrm{x}}^2$ in the $|n\rangle$ basis? That is, are these matrices diagonal or, if not, which elements are non-vanishing?

11. Using ladder operators, evaluate matrix elements $\langle n|\,\hat{p}_x\,|m\rangle$ and $\langle n|\,\hat{p}_x^2\,|m\rangle$ for the 1-D harmonic oscillator. Moreover, prove that

$$\left(\frac{\langle n|\,\hat{p}_x^2\,|q\rangle}{2m} + \frac{m\omega^2\,\langle n|\,\hat{x}^2\,|q\rangle}{2}\right) = \hbar\omega\left(n+\frac{1}{2}\right)\delta_{n,q} = E_n\delta_{n,q}.$$

12. Using ladder operators, evaluate matrix elements,

$$\langle n|\,\hat{p}_x\hat{x}\,|q\rangle \text{ and } \langle n|\,\hat{x}\hat{p}_x\,|q\rangle\,,$$

for the 1-D harmonic oscillator and show explicitly that

$$\langle n|\,\hat{x}\hat{p}_x\,|q\rangle - \langle n|\,\hat{p}_x\hat{x}\,|q\rangle = i\hbar\delta_{n,q}.$$

13. At $t = 0$, a particle having mass m moving in a one-dimensional oscillator potential having associated frequency ω is in the state

$$|\psi(0)\rangle = N\left(|0\rangle + 2\,|1\rangle\right),$$

where N is a normalization constant. Find the expectation value of the position operator as a function of time.

14–15. Consider the one-dimensional harmonic oscillator in dimensionless coordinates for which

$$\hat{H} = \frac{\hbar\omega}{2}\left(\hat{\eta}^2 + \xi^2\right) = \hbar\omega\left(a^\dagger a + \frac{1}{2}\right).$$

Suppose that the normalized state vector at time t is given by

$$|\psi(t)\rangle = e^{-i\omega t/2} \sum_{n=0}^{\infty} a_n e^{-in\omega t} |n\rangle ,$$

where $|n\rangle$ are the eigenkets of $\hat{H}$.

(a) Using this state vector, evaluate $\left\langle \hat{\xi}^2 \right\rangle = \langle \psi(t)| \hat{\xi}^2 |\psi(t)\rangle$ and show that it has a time-independent component and a component that oscillates with frequency 2ω. How does this dependence compare with that of $x(t)^2$ for a classical oscillator?
(b) Evaluate $\left\langle \hat{H} \right\rangle$.
(c) Obtain a differential equation for $d^2 \left\langle \hat{\xi}^2 \right\rangle / dt^2$ that could be solved in terms of the initial conditions and the value of $\left\langle \hat{H} \right\rangle$. [Hint: First obtain an equation for $d\left\langle \hat{\xi}^2 \right\rangle / dt$ that will involve some product of operators. Then obtain an equation of motion for this product of operators and use the fact that $\left\langle \hat{H} \right\rangle = \frac{\hbar\omega}{2} \left\langle \hat{\eta}^2 + \hat{\xi}^2 \right\rangle$ to eliminate $\left\langle \hat{\eta}^2 \right\rangle$.]
(d) Show that the solution of the differential equation of part (c) has the correct form found in part (a).

Note: The answers to parts (a) and (b) will be in the form of sums.

16–17. Starting from the coherent state

$$|\psi(t)\rangle = e^{-i\omega t/2} \sum_{n=0}^{\infty} \frac{\left(\xi_0/\sqrt{2}\right)^n e^{-\left(\xi_0/\sqrt{2}\right)^2/2} e^{-in\omega t}}{\sqrt{n!}} |n\rangle ,$$

use Eqs. (11.107) and (11.108) to calculate

$$\langle \xi \rangle = \langle \psi(t)| \left(a + a^{\dagger}\right) |\psi(t)\rangle / \sqrt{2}$$

and

$$\left\langle \xi^2 \right\rangle = \langle \psi(t)| \left(a + a^{\dagger}\right)^2 |\psi(t)\rangle / 2.$$

Show that the results agree with Eqs. (11.132) and (11.133).

18. (a) Using your knowledge of Dirac notation, evaluate

$$\int d\Omega \left[Y_{\ell'}^{m'}(\theta, \phi) \right]^* \hat{L}_x Y_{\ell}^{m}(\theta, \phi),$$

where the integral is over solid angle.

(b) Given an *arbitrary* operator $\hat{A}$ that commutes with $\hat{L}^2$. Prove that

$$\langle \ell' m' | \hat{A} | \ell m \rangle = 0$$

unless $\ell = \ell'$.

19. Evaluate $\langle \ell m | \hat{L}_y | \ell' m' \rangle$ using ladder operators and find explicit expressions for the eigenkets of $\hat{L}_y$ for $\ell = 1$, in terms of the $|\ell m\rangle$ basis.

20–21. Show that for a spherically symmetric potential, the momentum space eigenfunctions in spherical (momentum) coordinates can be written as

$$\Phi_E(\mathbf{p}) = P_{E\ell}(p) Y_\ell^m(\theta_p, \phi_p),$$

where the "radial" wave function $P_{E\ell}(p)$ is a solution of

$$\left(p^2 - 2mE\right) P_{E\ell}(p) = -\frac{2m}{(2\pi\hbar)^{3/2}} \int_0^\infty p'^2 dp' \tilde{V}_\ell\left(p, p'\right) P_{E\ell}(p'),$$

and

$$\tilde{V}_\ell\left(p, p'\right) = \frac{(4\pi)^2}{(2\pi\hbar)^{3/2}} \int_0^\infty r^2 dr V(r) j_\ell(pr/\hbar) j_\ell(p'r/\hbar).$$

Find $P_{E\ell}(p)$ for a free particle.

To solve this problem you will need to use the expansion

$$e^{i\mathbf{k}\cdot\mathbf{r}} = 4\pi \sum_{\ell=0}^{\infty} \sum_{m=-\ell}^{\ell} i^\ell j_\ell(kr) \left[Y_\ell^m(\theta, \phi)\right]^* Y_\ell^m(\theta_k, \phi_k),$$

in which $j_\ell(kr)$ is a spherical Bessel function.

Chapter 12
Spin

12.1 Classical Magnetic Moment

I have completed most of the material typically covered in a one-semester course on quantum mechanics. There is one additional topic, however, that needs to be discussed, that of intrinsic angular momentum or *spin*. Spin arises naturally when one considers the relativistic version of the Schrödinger equation, namely the Dirac equation. In that equation the wave function for a free electron has four components, two of which correspond to positive energy solutions. The two positive energy solutions can be associated with two spin components of the electron (the negative energy solutions correspond to anti-particles). Historically, spin was deduced from atomic spectra. In turned out that the spectra could be explained if one assigned an intrinsic angular momentum to the electron. Today we know that all elementary particles possess such an intrinsic angular momentum (which can be zero for some particles). *Fermions* (named after Enrico Fermi) are particles whose intrinsic angular momentum quantum number corresponds to half integral values of intrinsic spin (measured in units of $\hbar$) while *bosons* (named after Satyendra Nath Bose) are particles whose intrinsic angular momentum quantum number corresponds to integral values of intrinsic spin.

Before discussing spin, it is useful to recall the relationship between the magnetic moment of a charged particle and its angular momentum relative to some origin. The magnetic moment of an electron circulating about the origin is given by

$$\begin{aligned}\mu &= \frac{1}{2}\int \mathbf{r}'\times\mathbf{J}(\mathbf{r}')d\mathbf{r}' = -\frac{e}{2}\int \mathbf{r}'\times\mathbf{v}(\mathbf{r}')\delta(\mathbf{r}-\mathbf{r}')d\mathbf{r}' = -\frac{e\mathbf{r}\times\mathbf{v}}{2} \\ &= -\frac{e\mathbf{r}\times m_e\mathbf{v}}{2m_e} = -\frac{e\mathbf{r}\times\mathbf{p}}{2m_e} = -\frac{e\mathbf{L}}{2m_e},\end{aligned} \tag{12.1}$$

where $\mathbf{J}(\mathbf{r}') = -e\mathbf{v}(\mathbf{r})\delta(\mathbf{r}-\mathbf{r}')$ is the (electric) current density associated with an electron located at position $\mathbf{r}$ moving with velocity $\mathbf{v}$. Remember that e is the

P.R. Berman, *Introductory Quantum Mechanics*, UNITEXT for Physics,
https://doi.org/10.1007/978-3-319-68598-4_12

magnitude of the charge of the electron—the magnetic moment of an electron and its angular momentum are in opposite directions. When a particle having a magnetic moment $\boldsymbol{\mu}$ is placed in a magnetic field characterized by a field induction vector **B**, there is an energy

$$V = -\boldsymbol{\mu} \cdot \mathbf{B} = \frac{e\mathbf{L} \cdot \mathbf{B}}{2m_e} \tag{12.2}$$

associated with the magnetic moment–magnetic field interaction. If the field is taken to be along the z axis, then

$$V = -\boldsymbol{\mu} \cdot \mathbf{B} = \frac{eBL_z}{2m_e} = \beta_0 B \frac{L_z}{\hbar}, \tag{12.3}$$

where

$$\beta_0 = \frac{e\hbar}{2m_e} = 9.27 \times 10^{-24}\ \mathrm{J\,T^{-1}} = 5.79 \times 10^{-5}\ \mathrm{eV\,T^{-1}} \tag{12.4}$$

is the *Bohr magneton.* In frequency units,

$$\frac{\beta_0}{h} = 14.0\ \mathrm{GHz/T} = 1.40\ \mathrm{MHz/Gauss.} \tag{12.5}$$

In the quantum problem, the magnetic moment becomes an operator and the potential is replaced by the operator

$$\hat{V} = -\hat{\boldsymbol{\mu}} \cdot \mathbf{B} = \frac{eB\hat{L}_z}{2m_e} = \beta_0 B \frac{\hat{L}_z}{\hbar}. \tag{12.6}$$

For the electron in hydrogen, the energy levels characterized by a given value of ℓ are $(2\ell + 1)$-fold degenerate in the absence of the magnetic field. In the presence of an external magnetic field these degenerate levels are split into $(2\ell + 1)$ distinct energy levels, with the spacing between adjacent levels of order $\Delta E_m = \beta_0 B$. I will derive explicit expressions for this so-called *Zeeman splitting* in Chap. 21. In the Earth's magnetic field, which is typically of order 0.5 Gauss, the splittings are of order 1 MHZ.

12.2 Spin Magnetic Moment

Since ℓ is integral, each state of a given ℓ splits into an *odd* number of states in a magnetic field. This is referred to as the *normal* Zeeman effect. However, in some cases it is observed that the splitting is into an *even* number of levels. This is referred to as the *anomalous* Zeeman effect. In order to explain this and other observed

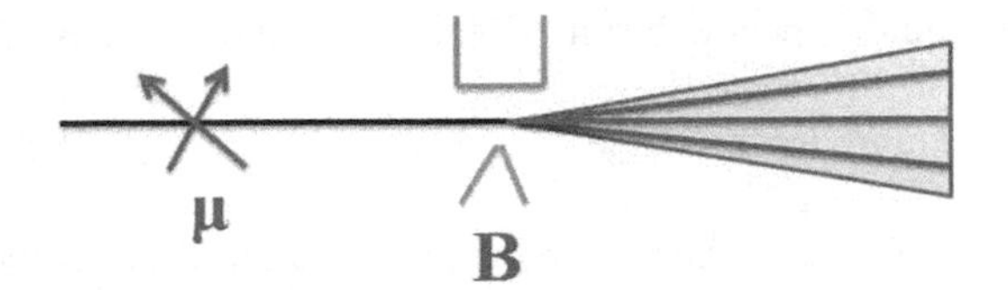

Fig. 12.1 Schematic representation of the Stern-Gerlach experiment. A beam of atoms having randomly oriented magnetic moments are passed through a region where there is a magnetic field gradient. Classically, there is a continuous range of deflection angles. For atoms having quantized angular momentum characterized by the quantum number ℓ, one would expect $(2\ell + 1)$ distinct traces

features of atomic structure, Goudsmit and Uhlenbeck proposed the concept of an intrinsic angular momentum or *spin* for the electron.[1] This also helped to explain an apparent anomaly in another experiment.

The other experiment is the Stern-Gerlach experiment carried out in 1921 (see Fig. 12.1).[2] If a beam of neutral particles, each having a magnetic moment $\boldsymbol{\mu}$, is sent into an *inhomogeneous* magnetic field $\mathbf{B}(\mathbf{r})$, there is a force on each particle given by

$$\mathbf{F}(\mathbf{r}, t) = \nabla [\boldsymbol{\mu}(t) \cdot \mathbf{B}(\mathbf{r})] = \nabla \left[\mu_x(t) B_x(\mathbf{r}) + \mu_y(t) B_y(\mathbf{r}) + \mu_z(t) B_z(\mathbf{r}) \right]. \tag{12.7}$$

Owing to its interaction with the field, the magnetic moment becomes a function of time. In the experiment of Stern and Gerlach, neutral atoms moving in the x direction pass through a region in which there is a magnetic field having a large homogeneous component in the z direction and smaller spatially varying components in all directions (the field components must satisfy $\nabla \cdot \mathbf{B} = 0$). An atom can possess a magnetic moment owing to the motion of its electrons about the nucleus. The magnetic moment precesses about the large homogeneous field component in the z direction, resulting in values of $\mu_x(t)$ and $\mu_y(t)$ that effectively average to zero and to a value of μ_z that is approximately constant, such that the resulting force on an atom produced by the field gradient is given approximately by

$$\mathbf{F} \approx \mu_z \frac{\partial B_z}{\partial z} \mathbf{u}_z \approx \mu_z \frac{\partial B}{\partial z} \mathbf{u}_z. \tag{12.8}$$

[1]See, for example, George E. Uhlenbeck and Samuel A. Goudsmit, *Spinning Electrons and the Structure of Spectra*, Nature **117,** 264–265 (1926). For an interesting paper on this discovery, see the article by Abraham Pais, *George Uhlenbeck and the discovery of electron spin,* Physics Today **42**, 34–40 (1989).

[2]Walter Gerlach and Otto Stern, *Der experimentelle Nachweis der Richtungsquantelung im Magnetfeld (Experimental Proof of Space Quantization in a Magnetic Field)*, Zeitschrift für Physik. **9**, 349–352 (1922); an English translation can be found at http://www.applet-magic.com/sterngerlach.htm.

Classically, we would expect a beam of particles having a magnetic moment $\boldsymbol{\mu}$ oriented in random directions to be deflected over a continuous range in the z-direction corresponding to the range of values $-\mu \leq \mu_z \leq \mu$ in the incoming beam.

In the quantum case, μ_z is replaced by the expectation value of the z-component of the magnetic moment operator which, in turn, is proportional to the expectation value of $\hat{L}_z$. Thus, owing to the gradient force, atoms having an angular momentum quantum number ℓ should be split into $(2\ell + 1)$ *discrete* paths after passing through the magnetic field region, each path corresponding to a different value of the z-component of angular momentum or magnetic moment. It was found, however, that when silver atoms were sent through the apparatus, *two* paths were detected even though ground state silver atoms were known to have $\ell = 0$.

Compton and Pauli had previously suggested that the electron could possess an intrinsic magnetic moment, but the credit for the explanation usually goes to Goudsmit and Uhlenbeck (although they never won the Nobel prize for the discovery of spin). To fit the experimental data, the operator $\hat{\boldsymbol{\mu}}_s$ associated with intrinsic magnetic moment of the electron and the operator $\hat{\mathbf{S}}$ associated with the intrinsic spin angular momentum of the electron must be related by

$$\boldsymbol{\mu}_s = -g_e \frac{e\mathbf{S}}{2m_e}; \tag{12.9}$$

moreover, it is necessary that the spin operator $\hat{\mathbf{S}}$ appearing in Eq. (12.9) corresponds to an angular momentum quantum number $s = 1/2$. The quantity g_e appearing in Eq. (12.9) is referred to as the *electron g-factor*. To fit the spectroscopic data, Goudsmit and Uhlenbeck took $g_e = 2$. It turns out that g_e is *approximately* equal to 2, but there are corrections related to the interaction of the electron with the vacuum field. The electron g-factor is one of the most precisely determined constants in physics with

$$g_e = 2.0023193043622 \pm 0.0000000000015, \tag{12.10}$$

and is in agreement with theoretical calculations. I will take $g_e = 2$.

It is natural to talk about spin using Dirac notation. The spin state of an electron can be described by the eigenket $|sm_s\rangle$ where $s = 1/2$, $m_s = \pm 1/2$. The state with $m_s = +1/2$ is referred to as *spin up* and that with $m_s = -1/2$ as *spin down*. As for any angular momentum,

$$\hat{S}^2 \left|sm_s\right\rangle = \hbar^2 s\left(s+1\right)\left|sm_s\right\rangle = \frac{3}{4}\hbar^2 \left|sm_s\right\rangle; \tag{12.11a}$$

$$\hat{S}_z \left|\pm 1/2\right\rangle = m_s\hbar \left|\pm 1/2\right\rangle = \pm\frac{1}{2}\hbar \left|\pm 1/2\right\rangle; \tag{12.11b}$$

$$\left[\hat{S}_x, \hat{S}_y\right] = i\hbar \hat{S}_z; \tag{12.11c}$$

$$\hat{S}_+ \left|-1/2\right\rangle = \left(\hat{S}_x + i\hat{S}_y\right)\left|-1/2\right\rangle = \hbar \left|1/2\right\rangle; \tag{12.11d}$$

$$\hat{S}_- \left|1/2\right\rangle = \left(\hat{S}_x - i\hat{S}_y\right)\left|1/2\right\rangle = \hbar \left|-1/2\right\rangle; \tag{12.11e}$$

$$\hat{S}_+ \left|1/2\right\rangle = 0; \quad \hat{S}_- \left|-1/2\right\rangle = 0. \tag{12.11f}$$

(the s quantum number is suppressed in most of these equations since it always equals 1/2). It is important to remember that electron spin is an *intrinsic* spin and has nothing to do with motion of charge—any classical models based on a rotating charge give meaningless results. Thus, spin acts in its own *abstract* space, independent of the orbital angular momentum of the particle. The intrinsic magnetic moment of the spin is a direction opposite to its intrinsic spin angular momentum.

In the $|m_s\rangle$ basis, the spin operators are 2×2 matrices

$$\underset{\sim}{S}^2 = \frac{3}{4}\hbar^2 \begin{pmatrix} 1 & 0 \\ 0 & 1 \end{pmatrix}; \tag{12.12a}$$

$$\underset{\sim}{S}_z = \frac{1}{2}\hbar \begin{pmatrix} 1 & 0 \\ 0 & -1 \end{pmatrix}; \tag{12.12b}$$

$$\underset{\sim}{S}_x = \frac{1}{2}\hbar \begin{pmatrix} 0 & 1 \\ 1 & 0 \end{pmatrix}; \tag{12.12c}$$

$$\underset{\sim}{S}_y = \frac{1}{2}\hbar \begin{pmatrix} 0 & -i \\ i & 0 \end{pmatrix}; \tag{12.12d}$$

$$\underset{\sim}{S}_+ = \hbar \begin{pmatrix} 0 & 1 \\ 0 & 0 \end{pmatrix}; \quad \underset{\sim}{S}_- = \hbar \begin{pmatrix} 0 & 0 \\ 1 & 0 \end{pmatrix}, \tag{12.12e}$$

where the order of the matrix indices is

$$\left|1/2\right\rangle = \left|\uparrow\right\rangle; \qquad \left|-1/2\right\rangle = \left|\downarrow\right\rangle \tag{12.13}$$

(that is, $\left(\underset{\sim}{S}_z\right)_{11} = \left(\underset{\sim}{S}_z\right)_{\uparrow\uparrow}$) and $|\uparrow\rangle$ and $|\downarrow\rangle$ are referred to as spin up and spin down eigenkets, respectively. It is customary to write

$$\underset{\sim}{S} = \hbar\boldsymbol{\sigma}/2, \tag{12.14}$$

where

$$\boldsymbol{\sigma} = \sigma_x \mathbf{u}_x + \sigma_y \mathbf{u}_y + \sigma_z \mathbf{u}_z, \tag{12.15}$$

$$\sigma_x = \begin{pmatrix} 0 & 1 \\ 1 & 0 \end{pmatrix}; \tag{12.16a}$$

$$\sigma_y = \begin{pmatrix} 0 & -i \\ i & 0 \end{pmatrix}; \tag{12.16b}$$

$$\sigma_z = \begin{pmatrix} 1 & 0 \\ 0 & -1 \end{pmatrix}, \tag{12.16c}$$

and the components of $\boldsymbol{\sigma}$ are referred to as *Pauli matrices*. Together with the identity,

$$1 = \begin{pmatrix} 1 & 0 \\ 0 & 1 \end{pmatrix}, \tag{12.17}$$

the Pauli matrices form a complete basis set for all 2×2 Hermitian matrices. Note that

$$\mathrm{Tr}\,(\sigma_x) = \sum_{i=1}^{3} (\sigma_x)_{ii} = 0; \tag{12.18a}$$

$$\sigma_x^2 = 1; \tag{12.18b}$$

$$\sigma_x \sigma_y + \sigma_y \sigma_x = 0; \tag{12.18c}$$

$$\left[\sigma_x, \sigma_y\right] = 2i\sigma_z; \tag{12.18d}$$

$$\sigma_x \sigma_y = i\sigma_z, \tag{12.18e}$$

along with all cyclic variations of these equations with $x \to y$, $y \to z$, and $z \to x$. Since they satisfy Eq. (12.18c), the Pauli matrices are said to *anti-commute*.

A general spin state can be written as

$$|\chi\rangle = \alpha\,|\uparrow\rangle + \beta\,|\downarrow\rangle = \begin{pmatrix} \alpha \\ \beta \end{pmatrix}. \tag{12.19}$$

The corresponding wave functions for spin up and spin down can be written as χ_+ and χ_-, respectively, although these wave functions *have no meaning in coordinate space*. They are defined only in the abstract spin vector space. Thus, the eigenfunctions of the electron in the hydrogen atom can be written as $\psi_{n\ell m}(\mathbf{r})\,\chi_\pm$ and the eigenkets as

$$|n\ell m; s = 1/2, m_s = \pm 1/2\rangle = |n\ell m\rangle\,|sm_s\rangle\,. \tag{12.20}$$

Spin is a strange quantity. To help you understand why, I will look at the way in which the spin components transform under a rotation of the coordinate system. Although the spin wave functions are not functions in coordinate space, they *will* change under a rotation of the coordinate system since the x, y, z, components of the spin are inter-mixed. Imagine that we have a Stern-Gerlach experiment that picks out spin up relative to some quantization axis. We now pass these atoms through a second Stern-Gerlach magnet that is oriented at an angle $\theta \leq \pi$ relative to the first (that is, the spin is now defined relative to a new axis that is at an angle θ relative to the first). The question is, "What is the spin state after passing through the second magnet?"

Before answering this question let us think about what happens in a similar situation with the electric field vector of light, which is a vector quantity. After passing through a polarizer whose axis is in the z direction (the light is propagating in the y direction) and an *analyzer* in the $x - z$ plane whose axis is at an angle $\theta \leq \pi$ relative to the z-axis, the field intensity after passing through the analyzer is proportional to $\cos^2\theta$. That is, if the analyzer is rotated by π relative to the polarizer, the intensity is unchanged. If the analyzer is rotated by $\pi/2$ (crossed polarizers), no light gets through.

But what happens with spin? In the spin experiment, the first magnet has $\mathbf{B} = B_0\mathbf{u}_z$ and the second

$$\mathbf{B} = B_0\left(\mathbf{u}_z\cos\theta + \mathbf{u}_x\sin\theta\right). \tag{12.21}$$

To see what occurs, I must express the original state vector in terms of the new state vectors defined relative to the new direction of the field. In other words, I need to look at the interaction of the spin magnetic moment with the new field,

$$\hat{V} = -\hat{\boldsymbol{\mu}}_s\cdot\mathbf{B} = \frac{e\hat{\mathbf{S}}\cdot\mathbf{B}}{m_e} = \frac{eB_0}{m_e}\left(\hat{S}_x\sin\theta + \hat{S}_z\cos\theta\right); \tag{12.22}$$

$$\underset{\sim}{\mathrm{V}} = \frac{eB_0\hbar}{2m_e}\begin{pmatrix}\cos\theta & \sin\theta\\ \sin\theta & -\cos\theta\end{pmatrix} = \beta_0 B_0\begin{pmatrix}\cos\theta & \sin\theta\\ \sin\theta & -\cos\theta\end{pmatrix}. \tag{12.23}$$

It is a simple matter to diagonalize this matrix. The eigenvalues are $\pm\beta_0 B_0$ and the eigenkets are (work this out)

$$|\uparrow\rangle' = \cos\left(\frac{\theta}{2}\right)|\uparrow\rangle + \sin\left(\frac{\theta}{2}\right)|\downarrow\rangle; \tag{12.24a}$$

$$|\downarrow\rangle' = -\sin\left(\frac{\theta}{2}\right)|\uparrow\rangle + \cos\left(\frac{\theta}{2}\right)|\downarrow\rangle. \tag{12.24b}$$

Expressing the original ket in terms of the new ket, I find

$$|\uparrow\rangle = \cos\left(\frac{\theta}{2}\right)|\uparrow\rangle' - \sin\left(\frac{\theta}{2}\right)|\downarrow\rangle'. \tag{12.25}$$

Thus the probability to find the atom in spin up after the second magnet is $\cos^2(\theta/2)$ instead of being proportional to $\cos^2\theta$ as I found for the electric field case. It takes a rotation of π to block the passage of the spin up component emerging from the second magnet [only a spin down component (with respect to the new quantization axis) is transmitted], whereas it takes a rotation of $\pi/2$ to (totally) block the emerging optical field. It is not difficult to understand why a rotation of π leads to no transmission of a spin up component—when it enters the second magnet, there is no longer any spin up component with respect to the new quantization axis.

In order to see the effect of rotation on the spin state vector from a somewhat different perspective, imagine an electron having its magnetic moment aligned along a strong homogeneous magnetic field, taken to be in the z-direction. Now slowly rotate the magnetic field about the y-axis. For sufficiently slow rotations, the magnetic moment will stay aligned with the field. As a consequence, after a rotation of π the spin has been flipped from down to up (recall that spin angular momentum and magnetic moment differ in sign for an electron) relative to the original quantization axis. After a rotation of 2π, the magnetic moment is again aligned along the positive z-direction, *but the sign of the spin state vector has changed* [see Eq. (12.25) with $\theta = 2\pi$]. It takes a rotation of 4π to return the state vector to its original value. The state vector is said to transform as a *spinor* under rotation. Rotations are discussed in more detail in Chaps. 19 and 20.

12.3 Spin-Orbit Coupling in Hydrogen

If we put ourselves in the rest frame of the electron in a hydrogen atom, we would see the proton undergoing orbital motion. This orbital motion produces a magnetic field at the position of the electron. Since the electron has a magnetic moment, there is *spin–orbit interaction* in hydrogen that adds a term to the Hamiltonian that is given classically by

$$H_{\mathrm{so}} = -\boldsymbol{\mu} \cdot \mathbf{B}_{\mathrm{proton}}. \tag{12.26}$$

There are two ways to evaluate H_{so} that both give the same (wrong) answer. First we can look at the field produced by the proton at the position $\mathbf{r}_e$ of the electron. From the Biot-Savart law, the magnetic induction can be calculated as

$$\mathbf{B}(\mathbf{r}_e) = \frac{\mu_0}{4\pi} \int d\mathbf{r}' \frac{\mathbf{J}_p(\mathbf{r}') \times (\mathbf{r}_e - \mathbf{r}')}{|\mathbf{r}_e - \mathbf{r}'|^3}, \tag{12.27}$$

where the current density $\mathbf{J}_p(\mathbf{r}')$ of the proton is given by

$$\mathbf{J}_p(\mathbf{r}') = e\left(\mathbf{v}_p - \mathbf{v}_e\right) \delta\left(\mathbf{r}' - \mathbf{r}_p\right) = -e\mathbf{v}\delta\left(\mathbf{r}' - \mathbf{r}_e + \mathbf{r}\right), \tag{12.28}$$

$\mathbf{v}_p$ is the velocity of the proton, $\mathbf{v}_e$ is the electron velocity, $\mathbf{r}_p$ is proton's coordinate, and

$$\mathbf{r} = \mathbf{r_e} - \mathbf{r_p}; \qquad \mathbf{v} = \mathbf{v_e} - \mathbf{v}_p \tag{12.29}$$

are the position and velocity of the electron relative to the proton. As a consequence,

$$\mathbf{B}(\mathbf{r}_e) = -e\frac{\mu_0 \mathbf{v}\times\mathbf{r}}{4\pi r^3} = \frac{e}{4\pi\epsilon_0}\frac{\mathbf{r}\times\mathbf{v}}{r^3c^2}. \tag{12.30}$$

This leads to

$$\begin{aligned} H_{\text{so}} &= -\boldsymbol{\mu}\cdot\mathbf{B}(\mathbf{r_e}) = \frac{1}{4\pi\epsilon_0}\frac{e^2\mathbf{S}}{m_e}\cdot\frac{\mathbf{r}\times\mathbf{v}}{r^3c^2} \\ &= \frac{1}{4\pi\epsilon_0}\frac{e^2\mathbf{S}\cdot\mathbf{L}}{m_e^2r^3c^2}. \end{aligned} \tag{12.31}$$

Alternatively, we can use the relativistic formula for the magnetic induction in the electron's instantaneous rest frame produced by the static electric field $\mathbf{E}_p(\mathbf{r})$ in the proton's rest frame, namely

$$\mathbf{B}(\mathbf{r}) \approx -\frac{\mathbf{v}\times\mathbf{E}_p(\mathbf{r})}{c^2} = -\frac{e}{4\pi\epsilon_0}\frac{\mathbf{v}\times\mathbf{r}}{r^3c^2}, \tag{12.32}$$

a result that agrees with Eq. (12.30).

Both methods give a result which is *twice* the correct result. There is a correction, known as the *Thomas Precession*,[3] connected with the fact that the electron is in an accelerating reference frame. Why the correction factor is *exactly* one/half remains somewhat of a mystery to me.[4] In any event the correct spin–orbit interaction is

$$\hat{H}_{\text{so}} = \frac{1}{8\pi\epsilon_0}\frac{e^2\hat{\mathbf{S}}\cdot\hat{\mathbf{L}}}{m_e^2r^3c^2}, \tag{12.33}$$

where classical variables have been replaced by quantum-mechanical operators. As I have mentioned, the correct spin-orbit coupling emerges naturally in Dirac's relativistic treatment of the bound states of hydrogen. In Chap. 21, I use perturbation theory within the framework of the Schrödinger equation to calculate the changes in the energy levels resulting from the spin-orbit coupling.

12.4 Coupling of Orbital and Spin Angular Momentum

Since the spin and orbital angular momentum act in different Hilbert spaces, they commute,

$$\left[\hat{S}_i, \hat{L}_j\right] = 0. \tag{12.34}$$

[3]It is discussed in Sect. 11.8 in Jackson *Classical Electrodynamics*, but the derivation is too involved to reproduce here.

[4]But see this link, http://aether.lbl.gov/www/classes/p139/homework/seven.pdf.

Thus, I can label the eigenkets as

$$|\ell, s; m_\ell, m_s\rangle = |\ell, m_\ell\rangle\, |s, m_s\rangle\,, \tag{12.35}$$

which are eigenkets of $\hat{L}^2, \hat{L}_z, \hat{S}^2, \hat{S}_z$. For an isolated atom, however, it turns out that $\hat{L}_z$ and $\hat{S}_z$ need not be constants of the motion when the spin-orbit interaction given by Eq. (12.33) is included. On the other hand, the *total angular momentum*

$$\hat{\mathbf{J}} = \hat{\mathbf{L}} + \hat{\mathbf{S}} \tag{12.36}$$

is a constant of the motion. As a result, it is often convenient to use eigenkets of the commuting operators $\hat{L}^2, \hat{S}^2, \hat{J}^2, \hat{J}_z$, denoted by $|\ell, s, ; j, m\rangle$. They are related to the product state eigenkets by

$$|\ell, s, ; j, m\rangle = \sum_{m_\ell, m_s} \langle \ell, m_\ell; s, m_s\, |\ell, s; j, m\rangle\, |\ell, s; m_\ell, m_s\rangle\,, \tag{12.37}$$

where the expansion coefficients,

$$\langle \ell, s; m_\ell, m_s\, |\ell, s; j, m\rangle \equiv \begin{bmatrix} \ell & s & j \\ m_\ell & m_s & m \end{bmatrix}, \tag{12.38}$$

are referred to as *Clebsch-Gordan coefficients*. These coefficients, as well as the associated 3-J symbols, are discussed in detail in Chap. 20. The Clebsch-Gordan coefficients are built-in functions in many symbolic mathematical programs, which use a closed form expression to evaluate them.

For $s = 1/2$, owing to the properties of the Clebsch-Gordan coefficients, only two terms enter the sum,

$$\begin{aligned}
\left|\ell, \frac{1}{2}, ; j = \ell \pm \frac{1}{2}, m\right\rangle &= \begin{bmatrix} \ell & 1/2 & \ell \pm \frac{1}{2} \\ m - 1/2 & 1/2 & m \end{bmatrix} \left|\ell, \frac{1}{2}; m - \frac{1}{2}, \frac{1}{2}\right\rangle \\
&+ \begin{bmatrix} \ell & 1/2 & \ell \pm \frac{1}{2} \\ m + 1/2 & -1/2 & m \end{bmatrix} \left|\ell, \frac{1}{2}; m + \frac{1}{2}, -\frac{1}{2}\right\rangle \\
&= \pm\sqrt{\frac{\ell \pm m + 1/2}{2\ell + 1}} \left|\ell, \frac{1}{2}; m - \frac{1}{2}, \frac{1}{2}\right\rangle \\
&+ \sqrt{\frac{\ell \mp m + 1/2}{2\ell + 1}} \left|\ell, \frac{1}{2}; m + \frac{1}{2}, -\frac{1}{2}\right\rangle.
\end{aligned} \tag{12.39}$$

It is not really possible to write an equivalent expression for the wave functions since the ket $\left|\ell, \frac{1}{2}, ; j = \ell \pm \frac{1}{2}, m\right\rangle$ represents a mixture of orbital and spin states. Of course you could define the wave function by taking the inner product of this equation with $\langle \mathbf{r}|$. In this way the eigenfunctions of hydrogen, including spin can be written as

$$\psi_{n,\ell,j=\ell\pm\frac{1}{2},m}(\mathbf{r}) = \left\langle \mathbf{r} \left| n,\ell,\frac{1}{2},;j=\ell\pm\frac{1}{2},m \right.\right\rangle$$

$$= \pm\sqrt{\frac{\ell \pm m + 1/2}{2\ell+1}}\psi_{n\ell,m-1/2}(\mathbf{r})\,\chi_+$$

$$+\sqrt{\frac{\ell \mp m + 1/2}{2\ell+1}}\psi_{n\ell,m+1/2}(\mathbf{r})\,\chi_-, \tag{12.40}$$

where the $\psi_{n\ell m_\ell}(\mathbf{r})$ are eigenfunctions of the hydrogen atom without spin.

States of hydrogen are labeled as nL_J where n is the energy or electronic quantum number, L is the ℓ quantum number in an old-fashioned scheme in which $\ell = 0$ is S (sharp), $\ell = 1$ is P (principal), $\ell = 2$ is D (diffuse), $\ell = 3$ is F (fundamental), $\ell = 4$ is G, etc., and J is the j value. Thus $2P_{3/2}$ is the state having $n = 2$, $\ell = 1$, and $j = 3/2$.

12.5 Spin and Statistics

Most of the discussion to date has focused on the quantum mechanics of a single particle. I did look at the eigenfunctions of a Hamiltonian that was the sum of two commuting Hamiltonians in Chap. 5. That result is strictly valid only if the particles governed by each of the Hamiltonians are distinguishable. But what happens if they are indistinguishable, such as the two electrons in the helium atom?

The eigenfunctions of a two-particle system can be written as $\psi_E(\mathbf{r}_1,\mathbf{r}_2)$ where $\mathbf{r}_j$ is the coordinate of particle j. If the particles are indistinguishable, there is an *exchange degeneracy* in this system. I can define an operator that interchanges the coordinates of the two particles. This exchange or *permutation operator* is Hermitian and commutes with the Hamiltonian so it is possible to find simultaneous eigenfunctions of the Hamiltonian and the permutation operator. Let me denote the exchange or permutation operator that interchanges particle 1 and 2 by $\hat{P}_{12}$ and its eigenfunctions by $\psi_{E,P}(\mathbf{r}_1,\mathbf{r}_2)$. Clearly

$$\hat{P}_{12}^2\psi_{E,P}(\mathbf{r}_1,\mathbf{r}_2) = \psi_{E,P}(\mathbf{r}_1,\mathbf{r}_2) \tag{12.41}$$

and, since $\hat{P}_{12}$ is Hermitian, this implies that the eigenvalues of $\hat{P}_{12}$ must equal ± 1. This, in turn, implies that either

$$\psi_{E,P}(\mathbf{r}_2,\mathbf{r}_1) = \psi_{E,P}(\mathbf{r}_1,\mathbf{r}_2) \tag{12.42}$$

or

$$\psi_{E,P}(\mathbf{r}_2, \mathbf{r}_1) = -\psi_{E,P}(\mathbf{r}_1, \mathbf{r}_2). \tag{12.43}$$

It turns out that the wave functions of all elementary particles satisfy either Eq. (12.42) or (12.43). Particles having integral intrinsic spin that satisfy Eq. (12.42) and are called *bosons* (photon, graviton, mesons) while particles having half-integral intrinsic spin satisfy Eq. (12.43) and are called *fermions* (electron, neutrino, quarks, baryons).

For multi-particle systems, the eigenfunctions for bosons must be symmetric on the interchange of any two particles and that for fermions must be antisymmetric on the interchange of any two particles. In practice you can show that you need not worry about this symmetrization of the wave function if the spatial wave functions for the particles never overlap.

For two spin 1/2 particles moving in a potential, it is always possible to write the eigenfunctions as a product of a symmetric (antisymmetric) spatial wave function multiplied by an antisymmetric (symmetric) spin state. The spin eigenkets of the two particles are eigenkets of the total spin operator, $\hat{\mathbf{S}} = \hat{\mathbf{S}}_1 + \hat{\mathbf{S}}_2$, and can be expressed in terms of Clebsch-Gordan coefficients as

$$|s_1, s_2, ; s, m\rangle = \sum_{m_1, m_2} \begin{bmatrix} s_1 & s_2 & s \\ m_1 & m_2 & m \end{bmatrix} |s_1 m_1\rangle \, |s_2 m_2\rangle , \tag{12.44}$$

where the ket $|s_1, s_2, ; s, m\rangle$ is a simultaneous eigenket of the operators $\hat{S}_1^2, \hat{S}_2^2, \hat{S}^2, \hat{S}_z$. Using this equation with $s_1 = s_2 = 1/2$, I find that the coupled spin eigenkets of two electrons consist of an antisymmetric or spin *singlet* state

$$\left| \frac{1}{2}, \frac{1}{2}; s = 0, m_s = 0 \right\rangle = \frac{1}{\sqrt{2}} \left(|\uparrow\downarrow\rangle - |\downarrow\uparrow\rangle \right), \tag{12.45}$$

and a spin *triplet* state

$$\left| \frac{1}{2}, \frac{1}{2}; s = 1, m_s = 1 \right\rangle = |\uparrow\uparrow\rangle ; \tag{12.46a}$$

$$\left| \frac{1}{2}, \frac{1}{2}; s = 1, m_s = 0 \right\rangle = \frac{1}{\sqrt{2}} \left(|\uparrow\downarrow\rangle + |\downarrow\uparrow\rangle \right); \tag{12.46b}$$

$$\left| \frac{1}{2}, \frac{1}{2}; s = 1, m_s = 1 \right\rangle = |\downarrow\downarrow\rangle , \tag{12.46c}$$

where the first arrow in the kets on the right-hand side of the equation refers to particle 1 and the second to particle 2. For N spin 1/2 particles, with $N > 2$, it is no longer possible to construct total spin eigenkets all of which are symmetric or antisymmetric on the exchange of any two particles. There is no totally antisymmetric spin state for $N > 2$. Of the 2^N total spin eigenkets, only $(N + 1)$ are totally symmetric on the interchange of any two particles and these correspond the the spin state having total spin $s = N/2$.

There is a profound difference in the allowed energy levels of a system of indistinguishable particles, depending on whether the particles are fermions or bosons. A simple example will suffice to illustrate this idea. Imagine we have three non-interacting particles in a one-dimensional infinite square well potential. You are asked to obtain the ground state energy and eigenfunction of the system assuming (1) the particles are bosons and (2) the particles are spin 1/2 fermions. If the particles are bosons they can all be in the same energy state so the ground state energy is simply $3E_0$, where E_0 is the ground state energy of a single particle in the well. The ground state wave function is simply a product of the ground state wave functions for each particle, multiplied by a symmetric state of the spins of the particles.

On the other hand, if the particles are spin 1/2 fermions, they cannot all have the same ground state spatial wave function. Such a wave function is symmetric on exchange of any two particles, which would require that the spin wave function be antisymmetric on the exchange of any two particles. It is impossible to form such a state for three particles that is non-vanishing. Thus the third particle must be in the first excited state of the well having energy E_1. The total energy of this 3-particle system is $E = 2E_0 + E_1$, while the wave function is

$$\psi_E(x_1, x_2, x_3) = \frac{1}{\sqrt{3!}} \begin{vmatrix} \psi_0(x_1)\,\chi_\uparrow(1) & \psi_0(x_1)\,\chi_\downarrow(1) & \psi_1(x_1)\,\chi_\uparrow(1) \\ \psi_0(x_2)\,\chi_\uparrow(2) & \psi_0(x_2)\,\chi_\downarrow(2) & \psi_1(x_2)\,\chi_\uparrow(2) \\ \psi_0(x_3)\,\chi_\uparrow(3) & \psi_0(x_3)\,\chi_\downarrow(3) & \psi_1(x_3)\,\chi_\uparrow(3) \end{vmatrix}, \tag{12.47}$$

where the determinant is referred to as a *Slater determinant* and is manifestly antisymmetric on the exchange of any two particles. (The third state could have equally well been spin down rather than spin up.)

As a second example of the role of spin and statistics, consider a conductor that is modeled as a sea of free electrons confined to a volume. If I take the volume to be a cube having side L and assume that the wave function vanishes on the surfaces of the cube, the energy levels of a *single* electron in this volume are given by

$$E_n = \frac{\hbar^2 k_n^2}{2m_e} = \frac{\hbar^2 n^2 \pi^2}{2m_e L^2} = \frac{\hbar^2 \left(n_x^2 + n_y^2 + n_z^2\right) \pi^2}{2m_e L^2}, \tag{12.48}$$

where m_e is the electron mass and n_x, n_y, n_z take on all positive integral values, with $n = \sqrt{n_x^2 + n_y^2 + n_z^2}$. If there are N electrons in the volume, then two electrons can occupy the lowest energy level ($n_x = n_y = n_z = 1$), *six* electrons can occupy the next highest energy level [($n_x = n_y = 1$; $n_z = 2$) or ($n_x = n_z = 1$; $n_y = 2$) or ($n_y = n_z = 1$; $n_x = 2$)], etc. If N is large and the electrons occupy the lowest possible energy state, then the total number of particles in the *Fermi sphere* is given by

$$N = 2 \times \frac{1}{8} \frac{4\pi n_F^3}{3} \tag{12.49}$$

where the factor of 2 is for two spin states for each value of (n_x, n_y, n_z), while the factor of $1/8$ is required since only one quadrant of the sphere is permitted for positive (n_x, n_y, n_z). The quantity n_F is the smallest value of n that is needed to accommodate the N electrons. For an electron density $\mathfrak{N} = N/L^3$, the corresponding k value and energy corresponding to n_F are given by

$$k_f = \frac{\pi n_F}{L} = \left(3\pi^2\mathfrak{N}\right)^{1/3}; \tag{12.50a}$$

$$E_f = \frac{\hbar^2 k_f^2}{2m_e} = \frac{\hbar^2\left(3\pi^2\mathfrak{N}\right)^{2/3}}{2m_e}. \tag{12.50b}$$

The quantities k_f and E_f are referred to as the magnitude of the *Fermi k-vector* and *Fermi energy*, respectively. The total energy of all the electrons is equal to

$$E_{\text{total}} = \int_0^N E_f(N')dN' = E_f \int_0^N (N'/N)^{2/3} dN' = \frac{3}{5} N E_f \tag{12.51}$$

and the average energy of an electron is

$$\langle E \rangle = \frac{E_{\text{total}}}{N} = \frac{3}{5} E_f. \tag{12.52}$$

12.6 Summary

The concept of electron spin was introduced. The electron possesses an intrinsic angular momentum corresponding to a spin angular momentum quantum number $s = 1/2$. Some of the properties of spin were discussed, as were the modifications to the energy levels of hydrogen resulting from spin. Finally, systems of identical particles were considered and were found to obey either Bose or Fermi statistics, depending on their intrinsic spin angular momentum. No two identical fermions can possess all the same quantum labels since their wave function must be antisymmetric on exchange of particles. The fact that no two *electrons* can possess all the same quantum numbers is often referred to as the *Pauli Exclusion Principle*, based on a proposal of Pauli to explain the shell structure of atoms.

12.7 Problems

1. Consider an electron in a magnetic induction $\mathbf{B} = B_0\mathbf{u}_x$. Find the eigenkets and eigenvalues for the Hamiltonian $\hat{H} = 2\beta_0 \mathbf{B} \cdot \hat{\mathbf{S}}/\hbar$. Calculate the state vector as a function of time if the electron is in its spin down state at $t = 0$. Neglect any center-of-mass motion of the electron—consider only its spin components.

2. Suppose that the lowest energy eigenvalue for a single electron moving in a one-dimensional infinite potential well is 1.0 eV. The well is located between $x = 0$ and $x = a$ so the eigenfunctions for a single electron are given by

$$\psi_n(x) = \sqrt{\frac{2}{a}} \sin\left(\frac{n\pi x}{a}\right); \qquad n = 1, 2, 3, \ldots.$$

(a) For two *non-interacting* electrons moving in this potential find the ground state eigenfunction and eigenenergy. You must include the fact that electrons are fermions.
(b) Calculate the next highest energy and explain why there are four eigenfunctions having this energy.
(c) Calculate the third highest energy and explain why this energy state is nondegenerate.

3. Two *non-interacting* particles, each having mass m move in a one-dimensional oscillator potential characterized by frequency ω.

(a) If the particles have spin zero write the eigenfunctions and eigenenergies for the three lowest energy state manifolds of the two-particle system.
(b) If the particles have spin $1/2$ write the eigenfunctions and eigenenergies for the three lowest energy state manifolds of the two-particle system.

In each case, you can express your answers in terms on the single particle eigenfunctions $\psi_{n_1}(x_1)$ and $\psi_{n_2}(x_2)$. By "three lowest energy state manifolds" I mean all and any degenerate states having one of the three lowest energies. Explain your reasoning.

4. Consider the protons in water as independent spin 1/2 particles. The spin of the proton is equal to 1/2, its g factor of 5.59, and its magnetic moment operator is

$$\hat{\mu}_p = 5.59\frac{e}{2m_p}\hat{\mathbf{S}}_p,$$

where $m_p = 1.67 \times 10^{-27}$ kg is the proton mass and $\hat{\mathbf{S}}_p$ is its spin operator. Note, the nuclear Bohr magneton $\beta_n = \beta_0\frac{m_e}{m_p} = 5.05 \times 10^{-27}$ J·T^{-1}

(a) Calculate the energy level splitting in frequency units for the spin states of protons in a static magnetic field of 1.0 T.
(b) Assume the protons are in thermal equilibrium at room temperature (about 300 °K). Estimate the population ratio of the two spin states according to Boltzmann's law, $W(E) \sim \exp(-E/k_B T)$. Even though the population difference is small, it is still sufficient to use for *magnetic resonance imaging* (MRI) since there are so many protons present. In MRI, a radio frequency magnetic field is applied to drive transitions between the spin states.

5. The Hamiltonian for a spin in a magnetic induction $\mathbf{B}$ is $\hat{H} = 2\beta_0 \mathbf{B} \cdot \hat{\mathbf{S}}/\hbar$, where β_0 is the Bohr magneton and $\hat{\mathbf{S}}$ is the spin operator. Prove that

$$\frac{d\left\langle \hat{\mathbf{S}} \right\rangle}{dt} = \boldsymbol{\omega} \times \left\langle \hat{\mathbf{S}} \right\rangle$$

and obtain an expression $\boldsymbol{\omega}$.

6. Write all the eigenfunctions for the electron in the $n = 3$ state of hydrogen using the $\left|n\ell j m_j\right\rangle$ basis.

7–8. Prove that the eigenkets of $\hat{S}_x$ and $\hat{S}_y$ can be written in terms of the eigenkets of $\hat{S}_z$ as

$$|\pm\rangle_x = \frac{1}{\sqrt{2}} \left(|\uparrow\rangle \pm |\downarrow\rangle\right);$$

$$|\pm\rangle_y = \frac{1}{\sqrt{2}} \left(|\uparrow\rangle \pm i\,|\downarrow\rangle\right),$$

where the + refers to spin up and − to spin down. For a quantization axis defined by

$$\mathbf{u}\left(\theta, \phi, \psi\right) = \cos\psi \mathbf{u}_\theta + \sin\psi \mathbf{u}_\varphi,$$

prove that the eigenkets of the operator $\hat{S}_{\theta,\phi,\psi} = \hat{\mathbf{S}} \cdot \mathbf{u}\left(\theta, \phi, \psi\right)$ can be written in terms of the eigenkets of $\hat{S}_z$ as

$$|+\rangle_{\theta,\phi,\psi} = \sqrt{\frac{1 - \cos\psi \sin\theta}{2}}\, |\uparrow\rangle + e^{i\phi} \frac{\cos\psi \cos\theta + i\sin\theta}{\sqrt{2\left(1 - \cos\psi \sin\theta\right)}}\, |\downarrow\rangle;$$

$$|-\rangle_{\theta,\phi,\psi} = e^{-i\phi} \frac{\cos\psi \cos\theta - i\sin\theta}{\sqrt{2\left(1 - \cos\psi \sin\theta\right)}}\, |\uparrow\rangle - \sqrt{\frac{1 - \cos\psi \sin\theta}{2}}\, |\downarrow\rangle.$$

The angles θ and ϕ are the polar and azimuthal angles of a spherical coordinate system, while ψ is the angle of the quantization axis relative to the $\mathbf{u}_\theta$ direction in a plane perpendicular to $\mathbf{u}_r$.

9–10. (a) Use Eq. (12.44) with $s = s_{12}$ to prove Eqs. (12.45) and (12.46) for a two-electron system.

(b) Now add a third electron to the system and form total spin eigenkets of a three-electron system defined by

$$|s_1, s_2, s_3, s_{12}; s, m\rangle = \sum_{m_{12}, m_3} \begin{bmatrix} s_{12} & s_3 & s \\ m_{12} & m_3 & m \end{bmatrix} |s_{12}, s_3, ; m_{12}, m_3\rangle,$$

where the ket $|s_1, s_2, s_3, s_{12}; s, m\rangle$ is a simultaneous eigenket of the operators $\hat{S}_1^2, \hat{S}_2^2, \hat{S}_3^2, \hat{S}_{12}^2, \hat{S}^2, \hat{S}_z$ and

$$\hat{\mathbf{S}} = \hat{\mathbf{S}}_1 + \hat{\mathbf{S}}_2 + \hat{\mathbf{S}}_3 = \hat{\mathbf{S}}_{12} + \hat{\mathbf{S}}_3$$

is the total spin operator. Express your answer in terms of the eigenkets $|s_1, s_2, s_3; m_1, m_2, m_3\rangle$ and show that eigenkets having $s = 3/2$ are totally symmetric on the interchange of any two particles, but the other eigenkets are neither symmetric nor antisymmetric on the interchange of any two particles. You will have to look up the needed values of the Clebsch-Gordan coefficients.

Chapter 13
(A) Review of Basic Concepts (B) Feynman Path Integral Approach (C) Bell's Inequalities Revisited

Chapters 1–12 form what could be the basis for a one semester course in quantum mechanics. Before moving on to advanced topics such as time-independent perturbation theory, the variational method, the WKB approximation, and irreducible tensor operators, it can't hurt to review the basic concepts of quantum mechanics that I have covered up to this point. I also use this chapter to discuss an alternative formulation of quantum mechanics given by Feynman based on path-integrals. Finally I present a proof of Bell's theorem and see how it can be tested.

13.1 Review of Basic Concepts

13.1.1 *Postulates*

The postulates of quantum mechanics are:

1. The absolute square of the wave function $|\psi(\mathbf{r},t)|^2$ corresponds to the probability density of finding the particle at position $\mathbf{r}$ at time t.
2. To each physical observable in classical mechanics, there corresponds a Hermitian operator in quantum mechanics.
3. The time dependence of $\psi(\mathbf{r},t)$ is governed by the Schrödinger equation,

$$i\hbar\frac{\partial\psi(\mathbf{r},t)}{\partial t} = \hat{H}\psi(\mathbf{r},t) \tag{13.1}$$

   where $\hat{H}$ is the energy operator of the system.
4. The only possible outcome of a measurement on a *single* quantum system of a physical observable associated with a given Hermitian operator is one of the eigenvalues of the operator.

P.R. Berman, *Introductory Quantum Mechanics*, UNITEXT for Physics,
https://doi.org/10.1007/978-3-319-68598-4_13

5. The fifth postulate can take different forms. One way to state the postulate is that the Poisson brackets of two dynamic variables of classical mechanics are replaced by the $(i\hbar)^{-1}$ times the commutator of the corresponding Hermitian operators in quantum mechanics, the postulate I used when discussing Dirac notation. An alternative and equivalent postulate is that the Fourier transform of $\psi(\mathbf{r}, t)$, denoted by $\Phi(\mathbf{p}, t)$, corresponds to the wave function in momentum space, the postulate I used in the wave function approach.

The fundamental equation of quantum mechanics is the time-dependent Schrödinger equation,

$$i\hbar\frac{\partial\psi(\mathbf{r},t)}{\partial t} = \hat{H}\psi(\mathbf{r},t) = \left[\frac{\hat{p}^2}{2m} + \hat{V}\right]\psi(\mathbf{r},t) = \left[\frac{-\hbar^2\nabla^2}{2m} + V(\mathbf{r})\right]\psi(\mathbf{r},t); \tag{13.2a}$$

$$i\hbar\frac{\partial\psi(x,t)}{\partial t} = \hat{H}\psi(x,t) = \left[\frac{\hat{p}_x^2}{2m} + \hat{V}\right]\psi(x,t) = \left[\frac{-\hbar^2}{2m}\frac{\partial^2}{\partial x^2} + V(x)\right]\psi(x,t). \tag{13.2b}$$

That is, matter is described by a wave equation, in which $|\psi(\mathbf{r},t)|^2$ is given the interpretation of the probability density to find a particle at position $\mathbf{r}$ at time t. I list equations in both one and three dimensions.

13.1.2 Wave Function Approach

A solution of the time-dependent Schrödinger equation is

$$\psi(\mathbf{r},t) = e^{-iEt/\hbar}\psi(\mathbf{r}); \tag{13.3a}$$

$$\psi(x,t) = e^{-iEt/\hbar}\psi(x), \tag{13.3b}$$

provided

$$\hat{H}\psi(\mathbf{r}) = E\psi(\mathbf{r}); \tag{13.4a}$$

$$\hat{H}\psi(x) = E\psi(x). \tag{13.4b}$$

Thus, if you solve the *time-independent* Schrödinger equation and find the eigenfunctions $\psi_E(\mathbf{r})$ $[\psi_E(x)]$ and the eigenvalues E, you have a complete solution to the problem. In other words,

$$\psi(\mathbf{r},t)=\sum_E b_E e^{-iEt/\hbar}\psi_E(\mathbf{r}); \tag{13.5a}$$

$$\psi(x,t)=\sum_E b_E e^{-iEt/\hbar}\psi_E(x) \tag{13.5b}$$

is a general solution of the time-dependent Schrödinger equation, provided Eqs. (13.4) hold. I can equally write this as

$$\psi(\mathbf{r},t)=\sum_E b_E(t)\psi_E(\mathbf{r}); \tag{13.6a}$$

$$\psi(x,t)=\sum_E b_E(t)\psi_E(x), \tag{13.6b}$$

with

$$b_E(t)=e^{-iEt/\hbar}b_E. \tag{13.7}$$

In this form it is clear that the probability to be in an eigenstate,

$$P_E=|b_E(t)|^2=|b_E|^2, \tag{13.8}$$

is independent of time. The quantity $b_E(t)$ is referred to as the *state amplitude.*

Although the probability to be in a given eigenstate is constant in time, the wave function squared or probability density evolves in time, since

$$|\psi(\mathbf{r},t)|^2=\sum_{E,E'} b_E^* b_{E'}\psi_E^*(\mathbf{r})\,\psi_{E'}(\mathbf{r})\,e^{i(E-E')t/\hbar}; \tag{13.9a}$$

$$|\psi(x,t)|^2=\sum_{E,E'} b_E^* b_{E'}\psi_E^*(x)\,\psi_{E'}(x)\,e^{i(E-E')t/\hbar}, \tag{13.9b}$$

is a function of time, in general. The relative phases of the state amplitudes give rise to the time dependence.

To solve a dynamic problem in which you are given $\psi(\mathbf{r},0)$ [$\psi(x,0)$], you first find the eigenfunctions and eigenenergies of the Hamiltonian. Then you set

$$\psi(\mathbf{r},0)=\sum_E b_E\psi_E(\mathbf{r}); \tag{13.10a}$$

$$\psi(x,0)=\sum_E b_E\psi_E(x), \tag{13.10b}$$

which lets you calculate

$$b_E=(\psi_E,\psi(\mathbf{r},0))=\int d\mathbf{r}\psi_E^*(\mathbf{r})\,\psi(\mathbf{r},0); \tag{13.11a}$$

$$b_E=(\psi_E,\psi(x,0))=\int dx\psi_E^*(x)\,\psi(x,0) \tag{13.11b}$$

and

$$\psi(\mathbf{r},t) = \sum_E b_E e^{-iEt/\hbar}\psi_E(\mathbf{r}); \tag{13.12a}$$

$$\psi(x,t) = \sum_E b_E e^{-iEt/\hbar}\psi_E(x). \tag{13.12b}$$

Example 1 A particle having mass m moves in an infinite square well potential located between $x = 0$ and $x = L$. At $t = 0$,

$$\psi(x,0) = \begin{cases} \sqrt{\frac{1}{a}}; & \frac{L}{2} - \frac{a}{2} \le x \le \frac{L}{2} + \frac{a}{2} \\ 0; & \text{otherwise} \end{cases}, \tag{13.13}$$

with $a < L$. Find $\psi(x,t)$.

The eigenfunctions and eigenvalues are

$$\psi_n(x) = \sqrt{\frac{2}{L}}\sin\left(\frac{n\pi x}{L}\right); \qquad n = 1,2,3,\ldots \tag{13.14a}$$

$$E_n = \frac{\hbar^2 n^2 \pi^2}{2mL^2}. \tag{13.14b}$$

Therefore

$$\begin{aligned} b_n &= \int_0^L dx\psi_n^*(x)\,\psi(x,0) = \sqrt{\frac{1}{a}}\sqrt{\frac{2}{L}}\int_{\frac{L}{2}-\frac{a}{2}}^{\frac{L}{2}+\frac{a}{2}} dx\sin\left(\frac{n\pi x}{L}\right) \\ &= -\sqrt{\frac{2L}{a}}\frac{1}{n\pi}\left\{\cos\left[\frac{n\pi}{2}\left(1+\frac{a}{L}\right)\right] - \cos\left[\frac{n\pi}{2}\left(1-\frac{a}{L}\right)\right]\right\} \end{aligned} \tag{13.15}$$

and

$$\psi(x,t) = \sqrt{\frac{2}{L}}\sum_{n=1}^{\infty} b_n \exp\left(-i\frac{\hbar^2 n^2\pi^2}{2mL^2}t\right)\sin\left(\frac{n\pi x}{L}\right). \tag{13.16}$$

Example 2 Sometimes you can read off the expansion coefficients by inspection. Imagine in the previous problem that

$$\psi(x,0) = \frac{1}{\sqrt{2}}\sqrt{\frac{2}{L}}\left[\sin\left(\frac{\pi x}{L}\right) + \sin\left(\frac{2\pi x}{L}\right)\right]. \tag{13.17}$$

Clearly $b_1 = b_2 = \frac{1}{\sqrt{2}}$ and all other b_n are zero, such that

$$\psi(x,t) = \frac{1}{\sqrt{2}}\sqrt{\frac{2}{L}}\begin{bmatrix} \exp\left(-i\frac{\hbar^2\pi^2}{2mL^2}t\right)\sin\left(\frac{\pi x}{L}\right) \\ +\exp\left(-i\frac{4\hbar^2\pi^2}{2mL^2}t\right)\sin\left(\frac{2\pi x}{L}\right) \end{bmatrix}. \tag{13.18}$$

Note that $|\psi(x,t)|^2$ is a function of time,

$$|\psi(x,t)|^2 = \frac{1}{L}\left[\sin^2\left(\frac{\pi x}{L}\right) + \sin^2\left(\frac{2\pi x}{L}\right) + 2\sin\left(\frac{\pi x}{L}\right)\sin\left(\frac{2\pi x}{L}\right)\cos\left(\frac{3\hbar^2\pi^2}{2mL^2}t\right)\right]. \quad (13.19)$$

13.1.3 Quantum-Mechanical Current Density

The total probability is conserved for a single-particle quantum system. The probability current density,

$$\mathbf{J}(\mathbf{r},t) = \frac{i\hbar}{2m}\left[\psi(\mathbf{r},t)\nabla\psi^*(\mathbf{r},t) - \psi^*(\mathbf{r},t)\nabla\psi(\mathbf{r},t)\right]; \quad (13.20a)$$

$$J_x(x,t) = \frac{i\hbar}{2m}\left[\psi(x,t)\frac{\partial\psi^*(x,t)}{\partial x} - \psi^*(x,t)\frac{\partial\psi(x,t)}{\partial x}\right], \quad (13.20b)$$

satisfies an equation of continuity

$$\frac{\partial}{\partial t}\rho(\mathbf{r},t) + \nabla\cdot\boldsymbol{J}(\mathbf{r},t) = 0; \quad (13.21a)$$

$$\frac{\partial}{\partial t}\rho(x,t) + \frac{\partial J_x(x,t)}{\partial x} = 0, \quad (13.21b)$$

where

$$\rho(\mathbf{r},t) = |\psi(\mathbf{r},t)|^2; \quad (13.22a)$$

$$\rho(x,t) = |\psi(x,t)|^2 \quad (13.22b)$$

is the probability density.

13.1.4 Eigenfunctions and Eigenenergies

It is important to recognize that *each potential energy function gives rise to its own set of eigenfunctions*. Thus, the eigenfunctions of a free particle are infinite, plane mono-energetic waves, while those of any bound state problem are localized in space. But what about the unbound state eigenfunctions of the hydrogen atom? They are not localized states, nor are they plane wave states. Instead they must be found by solving the Schrödinger equation for positive energies. Similarly, for the problem of a one-dimensional well and particle energies $E > 0$, the solutions of Shcrödinger's equation are sines and cosines in each region, but the eigenfunction is not a single plane wave state. In three-dimensional problems with spherical symmetry, there is a separate effective potential problem to solve for each value of ℓ.

13.1.5 Operators

The expectation value of an operator for a quantum system is defined by

$$\left\langle \hat{A} \right\rangle = \int d\mathbf{r}\, \psi^* (\mathbf{r}, t)\, \hat{A} \psi (\mathbf{r}, t)\,. \tag{13.23}$$

For operators with no explicit time dependence,

$$i\hbar \frac{\partial \left\langle \hat{A} \right\rangle}{\partial t} = \left\langle \left[\hat{A}, \hat{H} \right] \right\rangle \tag{13.24}$$

where $\hat{H}$ is the Hamiltonian operator. The time dependence in $\left\langle \hat{A} \right\rangle$ arises from the time dependence in $\psi (\mathbf{r}, t)$. If $\left[\hat{A}, \hat{H} \right] = 0$, this implies that the dynamic variable associated with $\hat{A}$ is a constant of the motion. For an operator that doesn't commute with the Hamiltonian, its expectation value is not constant, in general.

The expectation values obey the laws of classical physics. This is known as *Ehrenfest's theorem*. That is

$$\frac{d \langle \hat{\mathbf{r}} \rangle}{dt} = \frac{1}{i\hbar} \left\langle \left[\hat{\mathbf{r}}, \hat{H} \right] \right\rangle = \frac{1}{i\hbar} \left\langle \left[\hat{\mathbf{r}}, \frac{\hat{p}^2}{2m} \right] \right\rangle = \frac{\langle \hat{\mathbf{p}} \rangle}{m}; \tag{13.25a}$$

$$\begin{aligned} \frac{d \langle \hat{\mathbf{p}} \rangle}{dt} &= \frac{1}{i\hbar} \left\langle \left[\hat{\mathbf{p}}, \hat{H} \right] \right\rangle = \frac{1}{i\hbar} \langle [\hat{\mathbf{p}}, V(\mathbf{r})] \rangle \\ &= - \langle \nabla V(\mathbf{r}) \rangle = \left\langle \hat{\mathbf{F}} \right\rangle, \end{aligned} \tag{13.25b}$$

where $\hat{\mathbf{F}} = - \nabla V(\mathbf{r})$ can be thought of the operator associated with the force acting on the particle.

13.1.5.1 Commuting Operators: Simultaneous Eigenfunctions

Two Hermitian operators possess simultaneous eigenfunctions if and only if they commute. Moreover, any operator that commutes with the Hamiltonian is a constant of the motion. Constants of the motion can be used to label eigenfunctions of the Hamiltonian. In general, whenever there is a dynamic constant of the motion such as momentum or angular momentum, there is an energy degeneracy associated with that conserved quantity. In that case the eigenvalues of the conserved operators can be used to label different eigenfunctions that have the same energy. For example, for the free particle, there is infinite degeneracy for any non zero energy since the direction of the momentum can be in any direction. However, by using simultaneous eigenfunctions of both the energy and momentum (momentum is conserved), you can uniquely label each energy eigenfunction. Similarly, in problems with spherical symmetry, angular momentum is conserved and you can label states by the eigenvalues associated with the commuting operators $\hat{L}^2$, $\hat{L}_z$, and $\hat{H}$. There is always a $(2\ell + 1)$ energy degeneracy associated with each state of a given ℓ. In general, the energy in problems with spherical symmetry depends on both the ℓ quantum number and some additional quantum number—only in problems with some extra symmetry such as hydrogen and the 3-d harmonic oscillator does the energy depend only on a single quantum number n. For example, in an infinite spherical potential well, for each ℓ, there is a set of possible energies for each ℓ that can be labeled by the zeroes of the spherical Bessel functions. In that case, the energy is determined by both ℓ and a quantum number n for which $j_\ell(k_n a) = 0$, where $n = 1, 2, 3, \ldots$ labels the zeroes of the Bessel function.

13.1.6 Measurement in Quantum Mechanics

What truly sets quantum mechanics apart from classical physics is the existence of a single quantum system. There is no classical analogue of a single quantum system in a superposition state. In contrast to a closed classical system of particles and fields for which the energy is constant, it is impossible to assign a unique energy to a single quantum system that is in a superposition state of two or more energies. A measurement of the energy will yield one of the energies in the superposition state, but we don't know which one. If you measure the dynamic variable associated with a Hermitian operator for any single quantum system in a superposition state of eigenfunctions of that operator, you get one and only one eigenvalue of that operator. Large numbers of measurements on identically prepared quantum systems are needed to map out the probability function for each of the eigenvalues.

13.1.7 Dirac Notation

Although the wave function usually gives us some information about the spatial probability distribution, it is possible to formulate quantum mechanics in a somewhat more abstract formalism using Dirac notation. In Dirac notation each Hermitian operator has its own set of eigenkets. These kets can be represented as column vectors with a 1 in one position and zeroes everywhere else. Each operator is diagonal in its own basis with the diagonal elements simply the eigenvalues of the operator. In some cases, as for the harmonic oscillator or for angular momentum, it is possible to define ladder operators and obtain the eigenenergies without solving Schrödinger's equation. However in most cases, this is impossible and one reverts to solving eigenvalue equations using the wave function formalism. The real import of Dirac notation is that different representations, such as momentum and coordinate, appear on an equal footing.

By calculating matrix elements of the momentum operator in the coordinate representation I could establish a connection between the wave function formalism and Dirac notation. To calculate matrix elements of the momentum operator in the coordinate representation, I used the fifth postulate and the Poisson bracket of x and p_x to obtain the commutation relation between $\hat{x}$ and $\hat{p}_x$ and used the commutator to show that

$$\langle \mathbf{r} | \hat{\mathbf{p}} | \mathbf{r}' \rangle = \frac{\hbar}{i} \nabla_r \delta(\mathbf{r} - \mathbf{r}'). \tag{13.26}$$

As a consequence, matrix elements of the entire Hamiltonian could be evaluated in the coordinate basis. The Hamiltonian is not diagonal in the coordinate representation; however when the energy eigenkets are expanded in terms of the coordinate representation, the expansion coefficients turn out to be the wave function in coordinate space,

$$\psi_E(\mathbf{r}) = \langle \mathbf{r} | E \rangle . \tag{13.27}$$

I will often use Dirac notation. Remember, however, that one must often revert to the wave function formalism to carry out any calculations. The time-dependent Schrödinger equation can be written in Dirac notation as

$$i\hbar \frac{d |\psi(t)\rangle}{dt} = \hat{H} |\psi(t)\rangle = \left[\frac{\hat{p}^2}{2m} + \hat{V} \right] |\psi(t)\rangle . \tag{13.28}$$

The most general solution is

$$|\psi(t)\rangle = \sum_n b_n e^{-iE_n t/\hbar} |E_n\rangle , \tag{13.29}$$

provided

$$\hat{H} |E_n\rangle = E_n |E_n\rangle . \tag{13.30}$$

Remember that the eigenkets are just column vectors with a 1 in one position and zeroes everywhere else and that $|\psi(t)\rangle$ is a column vector with $b_n e^{-iE_n t/\hbar}$ in the nth position.

In contrast to the wave function approach, if you are given $|\psi(0)\rangle$ in terms of the eigenkets,

$$|\psi(0)\rangle = \sum_n b_n |E_n\rangle , \tag{13.31}$$

you already know all the b_n's. If you were given the initial wave function, you must calculate the b_n's as before. In other words,

$$b_n = \langle E_n | \psi(0) \rangle = \int d\mathbf{r} \langle E_n | \mathbf{r} \rangle \langle \mathbf{r} | \psi(0) \rangle = \int d\mathbf{r} \psi^*_{E_n}(\mathbf{r}) \psi(\mathbf{r}, 0). \tag{13.32}$$

Example: As an example of the use of Dirac notation, consider a hydrogen atom (neglecting spin) in an external magnetic field. The Hamiltonian for the electron is

$$\hat{H} = \hat{H}_0 + \beta_0 \mathbf{B} \cdot \hat{\mathbf{L}}/\hbar, \tag{13.33}$$

where $\hat{H}_0$ is the Hamiltonian in the absence of the magnetic field and β_0 is the Bohr magneton. As long as the field is along the z axis, it is simple to find the eigenfunctions and eigenenergies since the Y_ℓ^m's are still eigenfunctions. The energies are simply shifted by $\beta_0 m B$, where m is the magnetic quantum number. However if the field is not along the z axis, it is not at all obvious how to solve for the eigenfunctions.

In Dirac notation, it is trivial to obtain the eigenkets. The eigenkets are simply $|n\ell m_B\rangle$, where $m_B\hbar$ is the eigenvalue associated with the component of $\mathbf{L}$ along the direction of the magnetic field. The eigenenergies are shifted by $\beta_0 m_B B$. Of course, the eigenfunctions are still not simple to obtain since $\langle \hat{\mathbf{r}} | \ell m_B \rangle$ is not a spherical harmonic. However, if we express the Hamiltonian in terms of the (complete) set of eigenkets $|n\ell m_z\rangle$ which is easy to do, then the eigenkets in the $|n\ell m_B\rangle$ basis can be obtained by diagonalizing the Hamiltonian, just as was done in Sect. 11.4.3 for the operator $\hat{L}_x$. The eigenfunctions are obtained as linear combinations of the spherical harmonics having the same ℓ, but different m.

13.1.8 Heisenberg Representation

As a second example of the use of Dirac notation, I would like to discuss the *Heisenberg representation* or *Heisenberg picture.* The Schrödinger operators I have been using are time-independent. The expectation values of these operators are time-dependent, in general, owing to the time-dependence of the wave functions or state vectors associated with a quantum system. In the Heisenberg representation, these roles are reversed. The state vector of the system becomes constant in time, whereas the Heisenberg operators are functions of time, in general.

There is a unitary operator that connects the two representations. For a quantum system characterized by a Hamiltonian $\hat{H}$, I define the state vector in the Heisenberg representation by

$$|\psi\rangle^H = e^{\frac{i}{\hbar}\hat{H}t} |\psi(t)\rangle = e^{\frac{i}{\hbar}\hat{H}t} e^{-\frac{i}{\hbar}\hat{H}t} |\psi(0)\rangle = |\psi(0)\rangle, \tag{13.34}$$

which is constant in time. The expectation value of an operator $\hat{O}$ is

$$\langle \hat{O} \rangle = \langle \psi(t) | \hat{O} | \psi(t) \rangle = \langle \psi(0) | e^{\frac{i}{\hbar}\hat{H}t} \hat{O} e^{-\frac{i}{\hbar}\hat{H}t} | \psi(0) \rangle$$
$$= \langle \psi(0) | e^{\frac{i}{\hbar}\hat{H}t} \hat{O} e^{-\frac{i}{\hbar}\hat{H}t} | \psi(0) \rangle =^{H} \langle \psi | \hat{O}^{H}(t) | \psi \rangle^{H}, \quad (13.35)$$

where

$$\hat{O}^{H}(t) = e^{\frac{i}{\hbar}\hat{H}t} \hat{O} e^{-\frac{i}{\hbar}\hat{H}t}. \quad (13.36)$$

The Heisenberg operator $\hat{O}^{H}(t)$ is time-dependent, in general and obeys the equation of motion

$$i\hbar \frac{d\hat{O}^{H}(t)}{dt} = -e^{\frac{i}{\hbar}\hat{H}t} \hat{H} \hat{O} e^{-\frac{i}{\hbar}\hat{H}t} + e^{\frac{i}{\hbar}\hat{H}t} \hat{O} \hat{H} e^{-\frac{i}{\hbar}\hat{H}t} = [\hat{O}^{H}(t), \hat{H}]. \quad (13.37)$$

The advantage of the Heisenberg picture is that the operators often obey equations of motion that are identical to their classical counterparts. Thus, Ehrenfest's theorem holds directly for the Heisenberg operators. It is important to note that commutation laws such as $[\hat{x}^{H}(t), \hat{p}^{H}(t)] = i\hbar$ remain valid in the Heisenberg picture, but that $[\hat{x}^{H}(t), \hat{p}^{H}(t')] \neq i\hbar$, in general, if $t \neq t'$. The time-independent operator $\hat{H}$ is the same in both the Schrödinger and Heisenberg representations.

13.2 Feynman Path-Integral Approach

In 1948 Richard Feynman published a paper in *Reviews of Modern Physics* entitled *Space-Time Approach to Non-Relativistic Quantum Mechanics.*[1] In this paper, Feynman formulated an alternative theory of quantum mechanics based on *propagators*, rather than the Schrödinger equation. As Feynman noted in his introduction, "The formulation is equivalent to the more usual formulations. There are, therefore, no fundamentally new results. However, there is a pleasure in recognizing old things from a new point of view." It turns out that the propagator or *path-integral* approach is now used routinely in both non-relativistic and relativistic quantum mechanics.[2] The approach is related to the WKB method that is discussed in Chap. 16 in that both effectively involve what is called a *stationary phase approximation*.

Feynman's approach is based on the classical *action*. The classical action S for a particle having mass m moving in a potential $V(\mathbf{r})$ from position $\mathbf{r}_a$ at time t_a to position $\mathbf{r}_b$ at time t_b is defined by

[1] R. P. Feynman, Reviews of Modern Physics, Vol. 20, pp. 367–387 (1948).

[2] A general discussion of the path-integral approach can be found in L. S. Schulman, *Techniques and Applications of Path Integration* (Dover Publications, Mineola, N.Y., 2005).

$$S(\mathbf{r}_a, t_a; \mathbf{r}_b, t_b) = \int_{t_a}^{t_b} L(\mathbf{r}, \dot{\mathbf{r}})\, dt, \tag{13.38}$$

where

$$L(\mathbf{r}, \dot{\mathbf{r}}) = \frac{m\dot{\mathbf{r}} \cdot \dot{\mathbf{r}}}{2} - V(\mathbf{r}), \tag{13.39}$$

is the associated Lagrangian. The position $\mathbf{r}$ and velocity $\dot{\mathbf{r}}$ appearing in these equations are implicit functions of time. The *Principle of Least Action* states that of all the possible classical paths that take the particle from position $\mathbf{r}_a$ at time t_a to position $\mathbf{r}_b$ at time t_b, the path that makes S an extremum corresponds to the actual dynamics of the particle. It is not overly difficult to show that the requirement $\delta S = 0$, where δ corresponds to a variational derivative over particle trajectories having fixed endpoints $(\mathbf{r}_a, t_a)$ and $(\mathbf{r}_b, t_b)$, leads to Lagrange's equation for the particle.

In analogy with this concept, Feynman postulated that the wave function in quantum mechanics evolves as

$$\psi(\mathbf{r}, t) = \int_{-\infty}^{\infty} K(\mathbf{r}, t; \mathbf{r}', t')\psi(\mathbf{r}', t')d\mathbf{r}', \tag{13.40}$$

where the *propagator* is $K(\mathbf{r}_b, t_b; \mathbf{r}_a, t_a)$ is *assumed* to be given by

$$\begin{aligned} K(b,a) \equiv K(\mathbf{r}_b, t_b; \mathbf{r}_a, t_a) &= \frac{1}{N} \int_{\mathbf{r}_a}^{\mathbf{r}_b} \exp\left(\frac{i}{\hbar} S(\mathbf{r}_a, t_a; \mathbf{r}_b, t_b)\right) \mathcal{D}\mathbf{r}(t) \\ &\equiv \frac{1}{N} \int_{\mathbf{r}_a}^{\mathbf{r}_b} \exp\left(\frac{i}{\hbar} S(b,a)\right) \mathcal{D}\mathbf{r}(t), \end{aligned} \tag{13.41}$$

where N is a normalization constant. The *functional differential* $\mathcal{D}\mathbf{r}(t)$ is *not* an integral over position; it defines an operation in which the integrand is summed over all *classical* paths from $(\mathbf{r}_a, t_a)$ to $(\mathbf{r}_b, t_b)$.

Of course, you can postulate anything you want. Only if the postulates are consistent with experiment can they form the basis for a useful theory. In Feynman's case, he showed (by an argument reproduced in the Appendix) that his formulation was totally equivalent to that described by Schrödinger's equation. Moreover his formalism can offer some computational advantages for calculations in both non-relativistic and relativistic quantum mechanics. I will show how the propagator can be calculated and then evaluate it for a free particle.

The Feynman approach is similar to the Principle of Least Action, but there is a fundamental difference. In the Principle of Least Action, one path is picked out by demanding that the action is an extremum. On the other hand, the propagator in Eq. (13.41) involves a sum over *all* classical paths from $(\mathbf{r}_a, t_a)$ to $(\mathbf{r}_b, t_b)$. In quantum mechanics it is impossible to deterministically define the path of a particle. However, owing to the fact that $\hbar$ appears in the denominator of the phase in the exponent in Eq. (13.41), the phase is expected to be large. In that case, the major contribution to the integral will be along the path for which the phase is an extremum (path of stationary phase), a condition on the action similar to that encountered in the classical case. For all other paths, the exponential factor oscillates rapidly and leads to destructive interference when the integral in Eq. (13.41) is evaluated.

In order to evaluate the propagator, the time integral can be broken down into multiple steps, with the number of steps eventually approaching infinity. From the definition given in Eq. (13.38), it is clear that

$$S(b,a) = S(c,a) + S(b,c). \tag{13.42}$$

Since $S(b,a)$ is the *classical* action, it contains no quantum-mechanical operators. As a consequence

$$\begin{aligned} \exp\left(\frac{i}{\hbar}S(b,a)\right) &= \exp\left(\frac{i}{\hbar}\left[S(c,a)+S(b,c)\right]\right) \\ &= \exp\left(\frac{i}{\hbar}S(b,c)\right)\exp\left(\frac{i}{\hbar}S(c,a)\right). \end{aligned} \tag{13.43}$$

The integral in Eq. (13.41) can be written as an integral from the $\mathbf{r}_a$ to $\mathbf{r}_c$ multiplied by an integral from $\mathbf{r}_c$ to $\mathbf{r}_b$. To allow for all classical paths, I must integrate over all possible intermediate locations $\mathbf{r}_c$; that is,

$$\begin{aligned} K(b,a) &= \frac{1}{N}\int_{\mathbf{r}_a}^{\mathbf{r}_b} \exp\left(\frac{i}{\hbar}S(b,a)\right)\mathcal{D}\mathbf{r}(t) \\ &= \frac{1}{N}\int d\mathbf{r}_c \int_{\mathbf{r}_c}^{\mathbf{r}_b}\mathcal{D}\mathbf{r}_2(t)\int_{\mathbf{r}_a}^{\mathbf{r}_c}\mathcal{D}\mathbf{r}_1(t)\exp\left(\frac{i}{\hbar}S(b,c)\right)\exp\left(\frac{i}{\hbar}S(c,a)\right) \\ &= \frac{1}{N}\int d\mathbf{r}_c K(b,c)K(c,a). \end{aligned} \tag{13.44}$$

The integral is over all space.

I can now continue this process, dividing the interval from $\mathbf{r}_a$ to $\mathbf{r}_b$ into n sections, each having a temporal duration ϵ. At the end of the calculation a limit is taken in which $n \to \infty$ and $\epsilon \to 0$, with the product $n\epsilon \to (t_b - t_a)$. Since each time interval interval becomes infinitely small, I can use Eqs. (13.38) and (13.41) to estimate the propagator in the interval from $(\mathbf{r}_i, t_i)$ to $(\mathbf{r}_{i+1}, t_{i+1})$ as

$$\begin{aligned} K(i+1,i) &= \frac{1}{N}\int_{\mathbf{r}_i}^{\mathbf{r}_{i+1}} \exp\left(\frac{i}{\hbar}\int_{t_i}^{t_{i+1}} L\left(\mathbf{r},\dot{\mathbf{r}}\right)dt\right)\mathcal{D}\mathbf{r}(t) \\ &\approx \frac{1}{N}\exp\left[\frac{i\epsilon}{\hbar}L\left(\frac{\mathbf{r}_i+\mathbf{r}_{i+1}}{2},\frac{\mathbf{r}_{i+1}-\mathbf{r}_i}{\epsilon}\right)\right], \end{aligned} \tag{13.45}$$

since the classical path from $(\mathbf{r}_i, t_i)$ to $(\mathbf{r}_{i+1}, t_{i+1})$ reduces to an infinitesimal interval for which $\mathbf{r}\left(t_j\right) \approx (\mathbf{r}_i + \mathbf{r}_{i+1})/2$ and $\dot{\mathbf{r}}\left(t_j\right) = \mathbf{v}\left(t_j\right) \approx (\mathbf{r}_{i+1} - \mathbf{r}_i)/\epsilon$. It then follows that the total propagator is equal to

$$K(b,a) = \lim_{\substack{\epsilon\to 0\\ n\to\infty\\ n\epsilon\to(t_b-t_a)}} \left(\frac{2\pi i\hbar\epsilon}{m}\right)^{-\frac{3n}{2}} \int d\mathbf{r}_1 \ldots d\mathbf{r}_{n-1} \times \prod_{j=0}^{n-1} \exp\left[\frac{i\epsilon}{\hbar} L\left(\frac{\mathbf{r}_j + \mathbf{r}_{j+1}}{2}, \frac{\mathbf{r}_{j+1} - \mathbf{r}_j}{\epsilon}\right)\right], \tag{13.46}$$

where $\mathbf{r}_0 = \mathbf{r}_a$, $\mathbf{r}_n = \mathbf{r}_b$. The normalization factor that was used,

$$N = \left(\frac{2\pi i\hbar\epsilon}{m}\right)^{3/2}, \tag{13.47}$$

is derived in the Appendix.

As an example, I recalculate the one-dimensional free particle propagator already obtained in Eq. (3.48). The one-dimensional propagator is given by

$$K(b,a) = \lim_{\substack{\epsilon\to 0\\ n\to\infty\\ n\epsilon\to(t_b-t_a)}} \left(\frac{2\pi i\hbar\epsilon}{m}\right)^{-\frac{n}{2}} \int_{-\infty}^{\infty} dx_1 \ldots dx_{n-1} \times \prod_{j=0}^{n-1} \exp\left[\frac{i\epsilon}{\hbar} L\left(\frac{x_j + x_{j+1}}{2}, \frac{x_{j+1} - x_j}{\epsilon}\right)\right], \tag{13.48}$$

where the free-particle Lagrangian for a particle having mass m is

$$L\left(\frac{x_j + x_{j+1}}{2}, \frac{x_{j+1} - x_j}{\epsilon}\right) = m\frac{v_j^2}{2} = m\frac{\left(x_{j+1} - x_j\right)^2}{2\epsilon^2}. \tag{13.49}$$

Using the fact that

$$\int_{-\infty}^{\infty} dx_j \exp\left[\frac{im}{2\hbar\epsilon}\left(x_j - x_{j-1}\right)^2\right] \exp\left[\frac{im}{2(n-j)\hbar\epsilon}\left(x_n - x_j\right)^2\right] = \left(\frac{2\pi i\hbar\epsilon}{m}\frac{n-j}{n-j+1}\right)^{1/2} \exp\left[\frac{im}{2\hbar\epsilon\left(n-j+1\right)}\left(x_n - x_{j-1}\right)^2\right], \tag{13.50}$$

you can show by successive integrations starting with the integral over x_{n-1} that

$$K(b,a) = \lim_{\substack{\epsilon \to 0 \\ n \to \infty \\ n\epsilon \to (t_b - t_a)}} \left(\frac{2\pi i\hbar\epsilon}{m}\right)^{-\frac{n}{2}} \prod_{j=1}^{n-1} \left(\frac{2\pi i\hbar\epsilon}{m} \frac{n-j}{n-j+1}\right)^{1/2} \times \exp\left[\frac{im}{2n\epsilon\hbar}(x_n - x_0)^2\right]. \tag{13.51}$$

Since $x_0 = x_a$, $x_n = x_b$, and

$$\prod_{j=1}^{n-1} \left(\frac{n-j}{n-j+1}\right)^{1/2} = \sqrt{\frac{1}{n}}, \tag{13.52}$$

it then follows that

$$K(b,a) = \left(\frac{m}{2\pi i\hbar\,(t_b - t_a)}\right)^{1/2} \exp\left[\frac{im}{2\hbar}\frac{(x_b - x_a)^2}{(t_b - t_a)}\right], \tag{13.53}$$

which agrees with Eq. (3.48).

13.3 Bell's Theorem

13.3.1 Proof of Bell's Theorem

In Chap. 1, I promised a more detailed discussion of Bell's inequalities, once electron spin was introduced. I now make good on that promise. The proof of Bell's Theorem has nothing to do directly with quantum mechanics. It is based solely on measurements on correlated systems. I shall refer to each system as a "particle," but this need not be the case. I assume that we have two particles (A and B) and two detectors (1 and 2). The same property of each particle is measured at each detector, particle A is measured at detector 1 and particle B at detector 2. The measurements are assumed to be correlated and the correlation is assumed to occur as the result of some *hidden variable* encoded in the particles. The detectors are separated by a large distance, insofar as the measurement of one of the particles at one of the detectors cannot influence the measurement of the other particle at the other detector (this assumption implies that only *local hidden variables,* created in the particles at their creation, are used to explain the correlations between the particles).

I now assume that each detector can result only in a measurement of ± 1 when measuring the property of the particle. The value that is measured will depend on the angular orientation of the detector. Moreover, I assume that the measurements are perfectly anti-correlated for the same orientation of the detectors. That is, if detector 1 measures a value $+1$ for an orientation α of detector 1, then detector 2 must measure a value of -1 for the same orientation α of detector 2. This correlation is assumed to occur as a result of a hidden variable λ.

The proof of Bell's Theorem then follows directly.[3] Let me denote the orientation angle α of detector 1 by $\theta_{1\alpha}$ and the orientation angle β of detector 2 by $\theta_{2\beta}$. Then, according to my assumptions, the only possible measurements are

$$A(\theta_{1\alpha},\lambda) = \pm 1; \qquad B(\theta_{2\beta},\lambda) = \pm 1, \tag{13.54}$$

with

$$A(\theta_{1\alpha},\lambda) = -B(\theta_{2\alpha},\lambda). \tag{13.55}$$

It is assumed that the measurements are controlled by a hidden variable λ, with $0 \le \lambda \le 1$, governed by some distribution $\rho(\lambda)$ with

$$\int_0^1 \rho(\lambda)\, d\lambda = 1. \tag{13.56}$$

I now define $E\left(\theta_{1\alpha},\theta_{2\beta}\right)$ as the expectation value of the product of the two measurements. (It is important to recognize that $E\left(\theta_{1\alpha},\theta_{2\beta}\right)$ is *not* a probability—it can be negative.) Then

$$\begin{aligned} E\left(\theta_{1\alpha},\theta_{2\beta}\right) &= \int_0^1 d\lambda\, \rho(\lambda) A(\theta_{1\alpha},\lambda) B(\theta_{2\beta},\lambda) \\ &= -\int_0^1 d\lambda\, \rho(\lambda) A(\theta_{1\alpha},\lambda) A(\theta_{1\beta},\lambda), \end{aligned} \tag{13.57}$$

where the second line is obtained using Eq. (13.55). As a consequence, it follows that[4]

$$\begin{aligned} &E\left(\theta_{1\alpha},\theta_{2\beta}\right) - E\left(\theta_{1\alpha},\theta_{2\gamma}\right) \\ &\quad = -\int_0^1 d\lambda\, \rho(\lambda) \left[A(\theta_{1\alpha},\lambda)A(\theta_{1\beta},\lambda) - A(\theta_{1\alpha},\lambda)A(\theta_{1\gamma},\lambda)\right]. \end{aligned} \tag{13.58}$$

Using the fact that $\left[A(\theta_{1\beta},\lambda)\right]^2 = 1$, I can rewrite this equation as

$$\begin{aligned} &E\left(\theta_{1\alpha},\theta_{2\beta}\right) - E\left(\theta_{1\alpha},\theta_{2\gamma}\right) \\ &\quad = \int_0^1 d\lambda\, \rho(\lambda) A(\theta_{1\alpha},\lambda)A(\theta_{1\beta},\lambda) \left[A(\theta_{1\beta},\lambda)A(\theta_{1\gamma},\lambda) - 1\right]. \end{aligned} \tag{13.59}$$

[3]John Bell, *On the Einstein Podolsky Rosen Paradox*, Physics **1**, 195–200 (1964).

[4]Some authors object to Bell using the same value of λ for different measurements. See, for example, Karl Hess, *Einstein Was Right!* (CRC Press, Boca Raton, FL, 2015).

Since the product $\left|A(\theta_{1\alpha},\lambda)A(\theta_{1\beta},\lambda)\right| \leq 1$ and $1 - A(\theta_{1\beta},\lambda)A(\theta_{1\gamma},\lambda) \geq 0$, I know that

$$\left|\int_0^1 d\lambda\, \rho(\lambda)\, A(\theta_{1\alpha},\lambda)A(\theta_{1\beta},\lambda)\left[A(\theta_{1\beta},\lambda)A(\theta_{1\gamma},\lambda) - 1\right]\right|$$
$$\leq \int_0^1 d\lambda\, \rho(\lambda)\left[1 - A(\theta_{1\beta},\lambda)A(\theta_{1\gamma},\lambda)\right] = 1 + E\left(\theta_{1\beta},\theta_{2\gamma}\right), \quad (13.60)$$

which, when combined with Eq. (13.59), yields

$$\left|E\left(\theta_{1\alpha},\theta_{2\beta}\right) - E\left(\theta_{1\alpha},\theta_{2\gamma}\right)\right| - E\left(\theta_{1\beta},\theta_{2\gamma}\right) \leq 1. \quad (13.61)$$

Equation (13.61) is a statement of Bell's Theorem. If measurements on any perfectly anti-correlated system of two particles violate this inequality, then the correlations cannot be attributed to a local hidden variable.

Often a modified form of Bell's inequalities is used,[5]

$$\left|E(\theta_{1\alpha},\theta_{2\beta}) - E(\theta_{1\alpha},\theta_{2\beta'}) + E(\theta_{1\alpha'},\theta_{2\beta}) + E(\theta_{1\alpha'},\theta_{2\beta'})\right| \leq 2. \quad (13.62)$$

which does not depend on the condition that the events at similarly oriented detectors be perfectly correlated or anti-correlated.

13.3.2 Electron Spin Measurements

Electron spin measurements satisfy all the requirements given above for perfectly anti-correlated measurements if the two electrons are prepared in the spin singlet state,

$$|\psi\rangle = \frac{1}{\sqrt{2}}\left(|\uparrow\downarrow\rangle - |\downarrow\uparrow\rangle\right), \quad (13.63)$$

where the first arrow in each ket refers to electron 1 and the second to electron 2.[6] The electrons are assumed to move in the $\pm z$ directions. The expectation value of the product of spin measurements when detector 1 is oriented at an angle $\theta_{1\alpha}$ relative to the x-axis and detector 2 is oriented at an angle $\theta_{2\beta}$ relative to the x-axis is given by

[5]J.F. Clauser, M.A. Horne, A. Shimony, R.A. Holt (1969), *Proposed experiment to test local hidden-variable theories*, Physical Review Letters **23**, 880–4 (1969).

[6]Although the singlet state is written for a quantization axis in the z direction, the same state is realized for *any* quantization axis.

$$E\left(\theta_{1\alpha},\theta_{2\beta}\right)=\left\langle\begin{array}{c}\boldsymbol{\sigma}_A\cdot\left(\cos\theta_{1\alpha}\mathbf{u}_x+\sin\theta_{1\alpha}\mathbf{u}_y\right)\\ \times\boldsymbol{\sigma}_B\cdot\left(\cos\theta_{2\beta}\mathbf{u}_x+\sin\theta_{2\beta}\mathbf{u}_y\right)\end{array}\right\rangle$$
$$=\left\langle\left(\sigma_{Ax}\cos\theta_{1\alpha}+\sigma_{Ay}\sin\theta_{1\alpha}\right)\left(\sigma_{Bx}\cos\theta_{2\beta}+\sigma_{By}\sin\theta_{2\beta}\right)\right\rangle, \quad (13.64)$$

where

$$\boldsymbol{\sigma}_j=\sigma_{jx}\mathbf{u}_x+\sigma_{jy}\mathbf{u}_y+\sigma_{jz}\mathbf{u}_z; \quad j=A,B, \quad (13.65)$$

and $\sigma_{j\nu}$ $\{\nu=x,y,z\}$ is a Pauli spin matrix acting in the space of spin j $\{j=A,B\}$. In other words, the expectation value of spin (in units of $\hbar/2$) for a single spin system measured relative to a quantization direction $\mathbf{u}$ is $\langle\boldsymbol{\sigma}\cdot\mathbf{u}\rangle$. Calculation of the expectation value given in Eq. (13.64) for the state vector given in Eq. (13.63) is straightforward. For example,

$$\begin{aligned}\langle\sigma_{Ax}\sigma_{Bx}\rangle&=\frac{1}{2}\left(\langle\uparrow\downarrow|-\langle\downarrow\uparrow|\right)\sigma_{Ax}\sigma_{Bx}\left(|\uparrow\downarrow\rangle-|\downarrow\uparrow\rangle\right)\\&=\frac{1}{2}\left(\langle\uparrow\downarrow|-\langle\downarrow\uparrow|\right)\sigma_{Ax}\left(|\uparrow\uparrow\rangle-|\downarrow\downarrow\rangle\right)\\&=\frac{1}{2}\left(\langle\uparrow\downarrow|-\langle\downarrow\uparrow|\right)\left(|\downarrow\uparrow\rangle-|\uparrow\downarrow\rangle\right)=-1. \end{aligned}\quad (13.66)$$

Similarly,

$$\left\langle\sigma_{Ax}\sigma_{By}\right\rangle=0; \quad (13.67a)$$
$$\left\langle\sigma_{Ay}\sigma_{Bx}\right\rangle=0; \quad (13.67b)$$
$$\left\langle\sigma_{Ay}\sigma_{By}\right\rangle=-1. \quad (13.67c)$$

Substituting these results into Eq. (13.64), I find

$$E\left(\theta_{1\alpha},\theta_{2\beta}\right)=-\left(\cos\theta_{1\alpha}\cos\theta_{2\beta}+\sin\theta_{1\alpha}\sin\theta_{2\beta}\right)=-\cos\left(\alpha-\beta\right). \quad (13.68)$$

By combining Eqs. (13.61) and (13.68), I arrive at Bell's inequality for the spin system,

$$\left|\cos\left(\alpha-\beta\right)-\cos\left(\alpha-\gamma\right)\right|+\cos\left(\beta-\gamma\right)\leq 1. \quad (13.69)$$

If I take $\alpha=0$, $\beta=3\pi/4$, $\gamma=\pi/4$, then the left-hand side of this equation is equal to $\sqrt{2}$ which violates the inequality. The correlations that occur in quantum mechanics cannot be explained by a local hidden variable theory.

In experiments, the correlation functions $E\left(\theta_{1\alpha},\theta_{2\beta}\right)$ are not measured directly. What is measured are individual and coincidence counts at the two detectors. However it is possible to relate the coincidence counts to the correlation functions to obtain a form of Bell's inequalities for the measurable quantities. A violation of

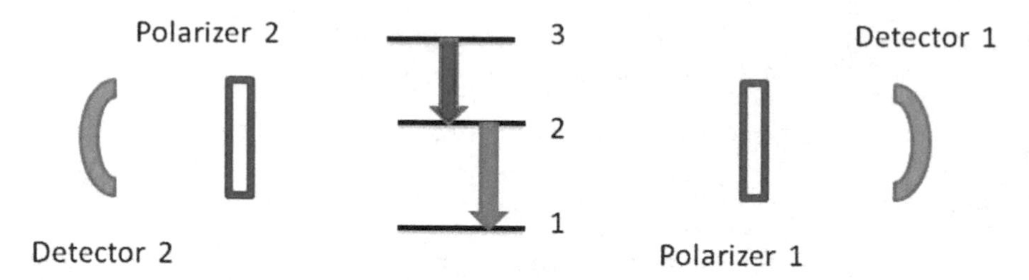

Fig. 13.1 Cascade emission to test Bell's inequalities. The emission on the 3–2 transition propagating in one direction is correlated with the radiation emitted on the 2–1 transition propagating in the opposite direction. Measurements are made for different settings of the polarizers

Bell's inequality has been observed using electron spin states in nitrogen vacancy centers in diamond with detectors placed 1.3 km apart.[7] The experimental spin measurements agree with the quantum predictions.

13.3.3 Photon Polarization Measurements

Experimentally it proved easier to use photons rather than spin to look for violations of Bell's inequalities. To understand how an experiment using photons can be used in a Bell's experiment, look at the atomic level scheme shown in Fig. 13.1. Atoms are prepared in level three and then undergo *spontaneous emission* that takes them to levels 2 and 1 via a cascade process in which a single photon is emitted on each transition (it is actually better to say that a *single photon state* is emitted on each transition since the radiation emerges as a pulse, not as a monochromatic plane wave state). There is no classical analogue of a single photon state, which makes it ideal to use in experiments to test Bell's inequalities.

The polarization of the radiation depends on the angular momentum associated with each level and the direction of emission. If an $\ell = 0 \rightarrow 1 \rightarrow 0$ cascade is chosen, the radiation emitted on *each* transition is *unpolarized* in *all* directions, but the polarization of the successively emitted single photon states is *correlated.* In other words, if the polarization of the radiation emitted on the 3–2 transition is not measured, then measurements of the radiation on the 2–1 transition will show it to be unpolarized in any direction. The situation changes if polarizers are placed between the atoms and the detectors, and the outcome of the measurement on the 2–1 transition is correlated with that on the 3–2 transition. For example, when the detectors are placed along opposite directions from the source and the polarizers

[7]B. Hensen, H. Bernien, A. E. Dréau, A. Reiserer, N. Kalb, M. S. Blok, J. Ruitenberg, R. F. L. Vermeulen, R. N. Schouten, C. Abellán, W. Amaya, V. Pruneri, M. W. Mitchell, M. Markham, D. J. Twitchen, D. Elkouss, S. Wehner, T. H. Taminiau and R. Hanson, *Loophole-free Bell inequality violation using electron spins separated by 1.3 kilometres,* Nature **526**, 682–686 (2015).

have the same orientation relative to the axis between the detectors, if a photon is detected at detector 1 then there is a 100% probability that a photon detected at detector 2 will have the same polarization, assuming perfect detectors (e.g. in a given direction, a linearly polarized photon emitted on the 3–2 transition is correlated with a photon emitted in the opposite direction on the 3–2 transition having the *same* linear polarization). In an ideal Bell's experiment, the direction of the polarizers is chosen *after* the radiation is emitted so the orientation of the detectors cannot affect the emission process.[8]

With ideal measurements,

$$E(\theta_{1\alpha}, \theta_{2\beta}) = \cos\left[2\left(\theta_{1\alpha} - \theta_{2\beta}\right)\right]. \tag{13.70}$$

For a relative angle of zero or π between the polarizers there is perfect correlation between the detectors, $E = 1$, while for relative polarizer angles of $\pi/2$ and $3\pi/2$ the events at the detectors are perfectly anti-correlated, $E = -1$. Substituting Eq. (13.70) into Eq. (13.62), I obtain

$$\left|\begin{array}{l} \cos\left[2\left(\theta_{1\alpha} - \theta_{2\beta}\right)\right] - \cos\left[2(\theta_{1\alpha} - \theta_{2\beta'})\right] \\ + \cos\left[2(\theta_{1\alpha'} - \theta_{2\beta})\right] + \cos\left[2(\theta_{1\alpha'} - \theta_{2\beta'})\right] \end{array}\right| \leq 2. \tag{13.71}$$

An optimal geometry for violating Bell's inequality is one in which

$$\phi = \left|\theta_{2\beta} - \theta_{1\alpha}\right| = \left|\theta_{2\beta'} - \theta_{1\alpha'}\right| = \left|\theta_{2\beta} - \theta_{1\alpha'}\right| = \left|\theta_{2\beta'} - \theta_{1\alpha}\right|/3. \tag{13.72}$$

For $\phi = \pi/8$ or $3\pi/8$, the left-hand side of Eq. (13.71) takes on its maximum value of $2\sqrt{2}$ and violates the inequality. The experimental measurements agree with the quantum predictions and violate Bell's inequalities.

There is another way to see the inconsistency of local hidden variable theories that does not make direct use of the inequality given in Eq. (13.62).[9] Suppose that the radiation is emitted in opposite directions along the x-axis and that each detector is oriented in the $y - z$ plane so that its axis makes an angle of 0, $2\pi/3$ or $4\pi/3$ relative to the z-axis; that is, each detector has three possible positions. If a photon gets through the detector, I denote this by a y (yes) and if it does not get through

[8]S. J. Freedman and J. F. Clauser, *Experimental Test of Local Hidden-Variable Theories*, Physical Review Letters **28**, 938–941 (1972); E. S. Fry and R. C. Thompson, R. C. (1976), *Experimental Test of Local Hidden Variables Theories*, Physical Review Letters **37**, 465-468 (1976); A. Aspect, P. Grangier, and G. Roger, *Experimental Realization of Einstein-Podolsky-Rosen-Bohm Gedankenexperiment: A New Violation of Bell's Inequalities*, Physical Review Letters **49**, 91–94 (1982); W. Tittel, J. Brendel, H. Zbinden, N. Gisin, *Violation of Bell inequalities by photons more than 10 km apart*, Physical Review Letters **81**, 3563–3566 (1998); J-A. Larsson, M. Giustina, J. Kofler, B. Wittmann, R. Ursin, and S. Ramelow, *Bell violation with entangled photons, free of the coincidence-time loophole*. Physical. Review A **90,** 032107 (2014).

[9]This argument follows that given by Robert Adair in *The Great Design, Particles, Fields, and Creation* (Oxford University Press, New York, 1987), pp. 185–187.

by an n (no). For a given orientation of a single detector, 50% of the photons get through, on average. Thus if the detectors are at the same angles, there will be 50% yy and 50% nn and 0% yn or ny. When the detectors are at different angles, if the first photon gets through the first detector (which has a 50% probability), the second photon must have the same polarization as the first and there is a 25% $[\cos^2(2\pi/3)$ or $\cos^2(4\pi/3)]$ chance that it gets through the second detector and a 75% that it does not. Thus the probabilities are 0.125 (0.5×0.25) for yy and 0.375 (0.5×0.75) for yn. On the other hand, if the first photon does not get through the first detector (which has a 50% probability), the second photon must have its polarization perpendicular to that of the first photon and there is a 75% $[\cos^2(\pi/6)$ or $\cos^2(5\pi/6)]$ chance that it gets through the second detector and a 25% that it does not. Thus the probabilities are 0.375 (0.5×0.75) for ny and 0.125 (0.5×0.25) for nn. I now show how these results lead to a contradiction if a local hidden variables theory is used.

Assume now that each photon that is emitted already has its polarization encoded. When the detectors are aligned, there is a 100% correlation between the two photons, which implies that the photons carry the same code. Since there are three detector positions and two possible outcomes for each position (y or n), there are eight possible codes for *each* photon ($yyy, yyn, yny, ynn, nyy, nny, nyn, nnn$). For example, yny implies that a photon will get through the detector for $\theta = 0, 4\pi/3$ but not for $\theta = 2\pi/3$. Since n and y are equally likely for a given position, this implies that the probability for each of $yyn, yny, ynn, nyy, nny, nyn$ must be equal (call it β) and that for each of yyy, nnn must be equal (call it α). Set the detectors at a relative angle of $2\pi/3$ and try to calculate α and β. Let $W(a, b, c)$ represent the probability that we get the result a when the first detector is at 0, b when it is at $2\pi/3$, and c when it is at $4\pi/3$. In this example, only the a and b indices are relative—the third index can be either y or n and must be summed over for a specific choice of the first two indices. Thus,

$$W(yy) = W(yyy) + W(yyn) = 0.125 = \alpha + \beta, \tag{13.73a}$$

$$W(nn) = W(nny) + W(nnn) = 0.125 = \alpha + \beta, \tag{13.73b}$$

$$W(yn) = W(yny) + W(ynn) = 0.375 = 2\beta, \tag{13.73c}$$

$$W(ny) = W(nyy) + W(nyn) = 0.375 = 2\beta. \tag{13.73d}$$

For example, the first line asks what is the probability that both photons get through the detectors (which was found to be 0.125 when the detectors are at a relative angle of $2\pi/3$) and the third line that the first photon gets through and the second one doesn't (which was found to be 0.375 when the detectors are at a relative angle of $2\pi/3$). Solving for α and β gives $\beta = 0.375/2$ and $\alpha = -0.125/2$. Since negative probabilities are not allowed, we are led to a contradiction. Of course, inequality (13.62) can be viewed as a statement that negative probabilities are not allowed.

13.3.4 Quantum Teleportation

Entangled Bell states can be used as the basis for *quantum teleportation*.[10] In classical digital communication, messages to be transmitted are encoded into a stream of zeroes or ones, so-called classical bits of information. Such communication is not secure insofar as it can be intercepted by an "eavesdropper." To avoid such problems, it is now possible to send encoded messages using quantum bits or *qubits* of information over distances as large as 1200 km.[11]

A qubit is a quantum superposition state. For example, suppose Alice (in such examples, it seems that it is *Alice* who always sends information to *Bob*, trying to avoid eavesdropping by *Eve*) wants to send the superposition state

$$|\psi\rangle_1 = (\alpha\,|H\rangle_1 + \beta\,|V\rangle_1) \tag{13.74}$$

to Bob. You can think of the states $|H\rangle_1$ and $|V\rangle_1$ as two orthogonal polarization states (horizontal and vertical) of a single photon state; the subscript 1 indicates that this is the first photon state in this teleportation scheme. To initiate the teleportation protocol, she creates an entangled Bell state of two other single photon states, using a method such as cascade emission. For example, the entangled Bell state might be

$$|\psi\rangle_{23} = \frac{1}{\sqrt{2}}\left(|H\rangle_2\,|H\rangle_3 + |V\rangle_2\,|V\rangle_3\right). \tag{13.75}$$

The subscripts 2 and 3 indicate that these states belong to the second and third single photon states. The third single photon state is sent to Bob and the second is kept by Alice. Alice now has two, single photon states (1 and 2), while Bob has one (3).

The entire state of the system of quantum bit and Bell state is

$$|\psi\rangle_T = \frac{1}{\sqrt{2}}\left(|H\rangle_2\,|H\rangle_3 + |V\rangle_2\,|V\rangle_3\right)\left(\alpha\,|H\rangle_1 + \beta\,|V\rangle_1\right). \tag{13.76}$$

[10] See, for example, *The Physics of Quantum Information*, edited by Dirk Bouwmeester, Artur Ekert, and Anton Zeilinger (Springer-Verlag, Berlin, 2000), and references therein.

[11] See, for example, Xiao-Song Ma, Thomas Herbst, Thomas Scheidl, Daqing Wang, Sebastian Kropatschek, William Naylor, Bernhard Wittmann, Alexandra Mech, Johannes Kofler, Elena Anisimova, Vadim Makarov, Thomas Jennewein, Rupert Ursin and Anton Zeilinger, *Quantum teleportation over 143 kilometres using active feed-forward*, Nature **489**, 269–273 (2012); Raju Valivarthi, Marcel.li Grimau Puigibert, Qiang Zhou, Gabriel Aguilar, Varun Verma, Francesco Marsili, Matthew D. Shaw, Sae Woo Nam, Daniel Oblak and Wolfgang Tittel, *Quantum teleportation across a metropolitan fibre network,* Nature Photonics **10**, 676–680 (2016); J. Yin et al., *Satellite-based entanglement distribution over 1200 kilometers*, Science **356**, 1140-1144 (2017).

Alice now makes a measurement of the state of her two single photon states in a Bell-state basis.[12] That is, she measures *one* of the four states

$$|\psi\rangle_{A1} = \frac{1}{\sqrt{2}}\left(|H\rangle_1 |H\rangle_2 + |V\rangle_1 |V\rangle_2\right); \tag{13.77a}$$

$$|\psi\rangle_{A2} = \frac{1}{\sqrt{2}}\left(|H\rangle_1 |H\rangle_2 - |V\rangle_1 |V\rangle_2\right); \tag{13.77b}$$

$$|\psi\rangle_{A3} = \frac{1}{\sqrt{2}}\left(|H\rangle_1 |V\rangle_2 + |V\rangle_1 |H\rangle_2\right); \tag{13.77c}$$

$$|\psi\rangle_{A4} = \frac{1}{\sqrt{2}}\left(|H\rangle_1 |V\rangle_2 - |V\rangle_1 |H\rangle_2\right). \tag{13.77d}$$

Alice's measurement projects Bob's single photon state into one of the states (see problems)

$$|\psi\rangle_{B1} = \alpha |H\rangle_3 + \beta |V\rangle_3 ; \tag{13.78a}$$

$$|\psi\rangle_{B2} = \alpha |H\rangle_3 - \beta |V\rangle_3 ; \tag{13.78b}$$

$$|\psi\rangle_{B3} = \beta |H\rangle_3 + \alpha |V\rangle_3 ; \tag{13.78c}$$

$$|\psi\rangle_{B4} = \beta |H\rangle_3 - \alpha |V\rangle_3 . \tag{13.78d}$$

Bob and Alice share a publicly accessible *key distribution* that correlates each Bell state in Eqs. (13.77) with the corresponding single photon state in Eqs. (13.78). Alice sends Bob two classical bits of information over a public line to tell him which Bell state measurement she made (for example, $\{0,0\}$ could correspond to an $A1$ measurement, $\{0,1\}$ to an $A2$ measurement, etc.) Bob then knows what to do to recover the initial quantum bit. For example, if Alice measured the state given in Eq. (13.77a), he already has the desired qubit. If she measured the state given in Eq. (13.77c), he carries out the unitary *quantum gate operation*

$$|\psi\rangle'_{B1} = \sigma_x |\psi\rangle_{B3} = \begin{pmatrix} 0 & 1 \\ 1 & 0 \end{pmatrix} \begin{pmatrix} \beta \\ \alpha \end{pmatrix} = \alpha |H\rangle_3 + \beta |V\rangle_3 , \tag{13.79}$$

in which he swaps the horizontal and vertical polarization components of his state. Other states are treated in a similar fashion.

[12] For example, she can make such a Bell state measurement by entangling her single photon states using beam splitters [Dik Bouwmeester, Jian-Wei Pan, Klaus Mattle, Manfred Eibl, Harald Weinfurter and Anton Zeilinger, *Experimental quantum teleportation,* Nature **390**, 575–579 (1997)] or nonlinear crystals [Yoon-Ho Kim, Sergei Kulik, and Yanhua Shih, *Quantum Teleportation of a Polarization State with a Complete Bell State Measurement*, Physical Review Letters **86**, 1370–1374 (2001)].

If the eavesdropper Eve intercepts the classical communication between Alice and Bob, it is useless to her since she doesn't have the single photon state that was sent to Bob. Moreover, if she intercepts the single photon state and makes a measurement on it, she will necessarily corrupt that superposition state. It will be evident to Alice and Bob that someone has been listening in.

13.3.5 Why the Big Fuss?

There are thousands of papers written on Bell's inequalities and more appear regularly. Many people believe in the validity of Bell's theorem and some don't. The carrying out of "loophole-free" experiments demonstrating violations of Bell's inequalities has enhanced or even made some scientific careers. The New York Times seems to be enamored with experiments that prove "Einstein was wrong" and confirm "spooky action at a distance." It seems that people are either fascinated, disturbed, or dissatisfied (or some combination of these) with experiments that have shown violations of Bell's inequalities.

I am not of the opinion that all this attention is merited. In the spin experiment, for example, each observer detects spin up or spin down 50% of the time, on average, independent of the orientation of her detector. It is only when the observers use classical communication channels to compare the results of their measurements do they find that there are correlations. How do they explain such correlations? To say that one measurement *influenced* the other is not particularly meaningful. In effect, the best answer to this question is "the reason there are correlations is because there are correlations." It is the same as asking people why two neutral objects attract one another. Attributing this attraction to "gravity" does not in any way explain the attraction. The laws of nature are what they are. People have become familiar with gravitational attraction, but do not experience the effects of quantum correlations in their everyday lives. If quantum mechanics is a correct theory, there will be violations of Bell's inequalities even if the proof of Bell's theorem has flaws. This does not prevent people from looking for alternative theories, as Einstein did in formulating an alternative theory of gravity in his Theory of General Relativity, but so far there has not yet been a theory to replace quantum mechanics. I believe that quantum mechanics remains in incomplete theory insofar as it does not address the dynamics of wave function collapse.

13.4 Summary

This chapter serves as a bridge between the introductory course material and the more advanced applications of the basic theory. A brief review was presented. I have also taken the opportunity to present two additional topics. The first was an alternative approach to the quantum theory based on the Feynman's path integrals and the second a more detailed treatment of Bell's theorem. These are both interesting topics, but the results that were derived are not used in the remainder of this text.

13.5 Appendix: Equivalence of Feynman Path Integral and Schrödinger Equation

In this Appendix, I show that Schrödinger's equation can be derived from the Feynman propagator. I work in one-dimension to simplify matters, but the results can be generalized easily to three dimensions. In the Feynman approach, the wave function at time t is related to that at time t' by the integral equation

$$\psi(x,t) = \int_{-\infty}^{\infty} K(x,t;x',t')\psi(x',t')dx', \tag{13.80}$$

where the propagator is defined in Eq. (13.41). To derive a differential equation for $\psi(x,t)$ I consider an infinitesimal time interval ϵ with $t = t' + \epsilon$ for which the propagator can be approximated as

$$\begin{aligned} K(x,t;x',t') &= \frac{1}{N}\exp\left[\frac{i}{\hbar}\int_{t'}^{t'+\epsilon} L(x,\dot{x})\,dt''\right] \\ &\approx \frac{1}{N}\exp\left[\frac{i\epsilon}{\hbar}L\left(\frac{x+x'}{2},\frac{x-x'}{\epsilon}\right)\right], \end{aligned} \tag{13.81}$$

since $\dot{x}(t'') \approx [x(t'+\epsilon) - x(t')]/\epsilon$ and $x(t'') \approx [x(t'+\epsilon) + x(t')]/2$. The quantity N is a normalization constant, whose value is derived below. In this limit

$$\psi(x,t+\epsilon) = \frac{1}{N}\int_{-\infty}^{\infty} \exp\left[\frac{i\epsilon}{\hbar}L\left(\frac{x+x'}{2},\frac{x-x'}{\epsilon}\right)\right]\psi(x',t)dx'. \tag{13.82}$$

I choose a specific form for the Lagrangian consistent with the Hamiltonian I have used in discussing the Schrödinger equation, namely

$$L(x,\dot{x}) = \frac{m\dot{x}^2}{2} - V(x), \tag{13.83}$$

which corresponds to a particle having mass m moving in the potential $V(x)$. With this Lagrangian,

$$\psi(x,t+\epsilon) = \frac{1}{N}\int_{-\infty}^{\infty} \exp\left[\frac{i}{\hbar}\left(\frac{m(x-x')^2}{2\epsilon} - \epsilon V\left(\frac{x+x'}{2}\right)\right)\right]\psi(x',t)dx', \tag{13.84}$$

or, setting $x - x' = y$,

$$\psi(x, t + \epsilon) = \frac{1}{N} \int_{-\infty}^{\infty} \exp\left(i\frac{my^2}{2\hbar\epsilon}\right) \exp\left[-i\epsilon V\left(x - \frac{y}{2}\right)/\hbar\right] \psi(x - y, t) dy. \tag{13.85}$$

The first exponent blows up as $\epsilon \to 0$. As a consequence the exponential will oscillate rapidly and lead to destructive interference for the integral except in a small interval about $y = 0$. I keep the first exponential intact, but expand the remaining factors in a power series in both ϵ and y to arrive at

$$\begin{aligned}\psi(x, t + \epsilon) &= \frac{1}{N} \int_{-\infty}^{\infty} \exp\left(i\frac{my^2}{2\hbar\epsilon}\right) [1 - i\epsilon V(x)/\hbar] \\ &\quad \times \left[\psi(x, t) - \frac{\partial \psi(x, t)}{\partial x} y + \frac{1}{2} \frac{\partial^2 \psi(x, t)}{\partial x^2} y^2\right] dy. \end{aligned} \tag{13.86}$$

All higher order terms in the expansion vanish in the limit that $\epsilon \to 0$. If I keep only the lead term in the expansion, I find

$$\psi(x, t + \epsilon) \approx \frac{1}{N} \int_{-\infty}^{\infty} \exp\left(i\frac{my^2}{2\hbar\epsilon}\right) \psi(x, t) dy = \frac{1}{N} \sqrt{\frac{2\pi i\hbar\epsilon}{m}} \psi(x, t), \tag{13.87}$$

where the integral can be evaluated using contour integrals in the complex plane or taken from integral tables. For this equation to be valid as $\epsilon \to 0$, I must require that

$$N = \sqrt{\frac{2\pi i\hbar\epsilon}{m}}. \tag{13.88}$$

Using the relationships

$$\int_{-\infty}^{\infty} \exp\left(i\frac{my^2}{2\hbar\epsilon}\right) dy = \sqrt{\frac{2\pi i\hbar\epsilon}{m}}; \tag{13.89a}$$

$$\int_{-\infty}^{\infty} \exp\left(i\frac{my^2}{2\hbar\epsilon}\right) y dy = 0; \tag{13.89b}$$

$$\begin{aligned}\int_{-\infty}^{\infty} \exp\left(i\frac{my^2}{2\hbar\epsilon}\right) y^2 dy &= \frac{2\hbar\epsilon}{i} \frac{d}{dm} \int_{-\infty}^{\infty} \exp\left(i\frac{my^2}{2\hbar\epsilon}\right) dy \\ &= -\frac{\hbar\epsilon}{i} \sqrt{\frac{2\pi i\hbar\epsilon}{m^{3/2}}}, \end{aligned} \tag{13.89c}$$

and Eq. (13.88) in Eq. (13.86), and taking the limit $\epsilon \to 0$, I find

$$\frac{\partial\psi(x,t)}{\partial t} = \lim_{\epsilon\to 0}\frac{\psi(x,t+\epsilon)-\psi(x,t)}{\epsilon} = -\frac{1}{2}\frac{\hbar}{im}\frac{\partial^2\psi(x,t)}{\partial x^2} - \frac{i}{\hbar}V(x)\,\psi(x,t), \tag{13.90}$$

or

$$i\hbar\frac{\partial\psi(x,t)}{\partial t} = -\frac{\hbar^2}{2m}\frac{\partial^2\psi(x,t)}{\partial x^2} + V(x)\,\psi(x,t), \tag{13.91}$$

which is Schrödinger's equation.

13.6 Problems

1–2. Write an expression in the form of Eq. (13.48) for the propagator of a particle having mass m moving in a uniform gravitational field for which the classical Hamiltonian is $H = p^2/2m + mgx$. In this case the calculation is more complicated than it is for a free particle, but it can be shown that[13]

$$K(b,a) = \left(\frac{m}{2\pi i\hbar\tau}\right)^{1/2} \times \exp\left\{i\frac{m\tau}{2\hbar}\left[\left(\frac{x_b-x_a}{\tau}\right)^2 - g\,(x_b+x_a) - \frac{g^2\tau^2}{12}\right]\right\},$$

where $\tau = (t_b - t_a)$. Show that the kernel is equal to

$$K(b,a) = \left(\frac{m}{2\pi i\hbar\tau}\right)^{1/2}\exp\left[iS_{cl}(b,a)/\hbar\right],$$

where $S_{cl}(b,a)$ is the classical action.

Using this kernel with

$$\psi(x,0) = \frac{1}{\pi^{1/4}\sigma^{1/2}}e^{-(x-x_0)^2/(2\sigma^2)}e^{ik_0x},$$

prove that

$$|\psi(x,t)|^2 = \left(\frac{1}{\pi\sigma(t)^2}\right)^{1/2} e^{-\left(x-x_0-v_0t-gt^2/2\right)^2/\sigma(t)^2},$$

[13]See, for example, S. Huerfano, S. Sahu, and M. Socolovsky, *Quantum Mechanics and the Weak Equivalence Principle,* International Journal of Pure and Applied Mathematics **49**, 153–166 (2008).

where

$$\sigma(t)^2 = \sigma^2 + \left(\frac{\hbar t}{m\sigma}\right)^2$$

and $v_0 = \hbar k_0/m$. In other words, in a uniformly accelerating reference frame, the particle spreads as if it were a free particle.

3–4. Write an expression in the form of Eq. (13.48) for the propagator of a particle having mass m moving in a 1-D simple harmonic potential for which the classical Hamiltonian is $H = p^2/2m + m\omega^2 x^2/2$. Again, the calculation is more complicated than it is for a free particle, but it can be shown that[14]

$$K(b,a) = \left(\frac{m\omega}{2\pi i\hbar \sin(\omega\tau)}\right)^{1/2} \times \exp\left[\frac{im\omega}{2\hbar \sin(\omega\tau)}\left[\left(x_b^2 + x_a^2\right)\cos(\omega\tau) - 2x_b x_a\right]\right],$$

where $\tau = (t_b - t_a)$. Show that the kernel is equal to

$$K(b,a) = \left(\frac{m\omega}{2\pi i\hbar \sin(\omega\tau)}\right)^{1/2} \exp\left[iS_{cl}(b,a)/\hbar\right],$$

where $S_{cl}(b,a)$ is the classical action.

Using this kernel, find $|\psi(\xi,t)|^2$, given

$$\psi(\xi,0) = \frac{1}{\pi^{1/4}} e^{-(\xi-\xi_0)^2/2},$$

where $\xi = \sqrt{m\omega/\hbar}\,x$ is a dimensionless variable, and show that your answer agrees with Eq. (7.61b).

5. Suppose that two electrons are emitted in the correlated spin state

$$|\psi\rangle = \frac{1}{\sqrt{2}}\left(|\uparrow\downarrow\rangle - |\downarrow\uparrow\rangle\right)$$

and propagate in opposite directions, $\pm\mathbf{u}_r$, where

$$\mathbf{u}_r = \sin\theta\cos\phi\,\mathbf{u}_x + \sin\theta\sin\phi\,\mathbf{u}_y + \cos\theta\,\mathbf{u}_z$$

[14] See, for example, K. Hira, *Derivation of the harmonic oscillator propagator using the Feynman path integral and recursive relations*, European Journal of Physics **34**, 777–785 (2013).

is a unit vector, and r, θ, and ϕ are spherical coordinates. If the detectors are located in planes perpendicular to $\mathbf{u}_r$ such that the orientation of the detectors is given by

$$\mathbf{u}_A(\theta,\phi,\psi_A) = \cos\psi_A \mathbf{u}_\theta + \sin\psi_A \mathbf{u}_\varphi;$$
$$\mathbf{u}_B(\theta,\phi,\psi_B) = \cos\psi_B \mathbf{u}_\theta + \sin\psi_B \mathbf{u}_\varphi.$$

Prove that

$$\langle(\boldsymbol{\sigma}_A\cdot\mathbf{u}_A)(\boldsymbol{\sigma}_B\cdot\mathbf{u}_B)\rangle = -\cos(\psi_A-\psi_B),$$

where

$$\boldsymbol{\sigma}_j = \sigma_{jx}\mathbf{u}_x + \sigma_{jy}\mathbf{u}_y + \sigma_{jz}\mathbf{u}_z; \quad j = A, B,$$

and $\sigma_{j\nu}$ $\{\nu = x, y, z\}$ is a Pauli spin matrix acting in the space of spin j $\{j = A, B\}$.

6. Under a rotation of the quantization axis, the state vector

$$|\psi\rangle = \frac{1}{\sqrt{2}}(|\uparrow\downarrow\rangle - |\downarrow\uparrow\rangle) = \frac{1}{\sqrt{2}}(|\uparrow\rangle_1|\downarrow\rangle_2 - |\downarrow\rangle_1|\uparrow\rangle_2)$$

is transformed into

$$|\psi\rangle' = \frac{1}{\sqrt{2}}(|\uparrow\rangle_1'|\downarrow\rangle_2' - |\downarrow\rangle_1'|\uparrow\rangle_2')$$

where

$$\begin{pmatrix}|\uparrow\rangle_{1,2}' \\ |\downarrow\rangle_{1,2}'\end{pmatrix} = \underset{\sim}{U}\begin{pmatrix}|\uparrow\rangle_{1,2} \\ |\downarrow\rangle_{1,2}\end{pmatrix}$$

and $\underset{\sim}{U}$ is a unitary matrix having determinant equal to 1. Prove that, $|\psi\rangle' = |\psi\rangle$; that is, the spins are always anti-correlated for the singlet state, independent of the choice of quantization axis, as you would expect.

7. Make a contour plot of the left side of Eq. (13.69) with axes $x = \beta - \alpha$ and $y = \gamma - \alpha$ to determine the range of detector angles for which the Bell's inequality is violated.

8. Make a contour plot of the left side of Eq. (13.71) with axes $\theta_{1\alpha'}$ and $\theta_{2\beta}$ for $\theta_{1\alpha} = 0$ and $\theta_{2\beta'} = 0, \pi/8, \pi/4, 3\pi/8, \pi/2$ to determine the range of detector angles for which the Bell's inequality is violated.

9. The excited state of a *quantum dot* can decay into a superposition of two, nearly degenerate ground states $|1\rangle$ and $|2\rangle$. If the decay is to state $|1\rangle$, the radiation has horizontal (H) polarization and, if the decay is to state $|2\rangle$, the radiation has

vertical (V) polarization along a given direction of emission. Following emission, the radiation and the quantum dot are in the entangled state[15]

$$|\psi\rangle = \frac{1}{\sqrt{2}} \left(|1, H\rangle - i\,|2, V\rangle \right).$$

If a polarizer is placed in the path of the emitted radiation at an angle θ relative to the x axis and a signal is detected, what is the resultant state vector for the quantum dot? Experimentally, an effective polarization angle can be associated with the superposition state of the quantum dot and the state of the quantum dot can be read out using auxiliary laser pulses. By using different rotation angles for the polarizer and using different rotations of the quantum dot's polarization, it is possible to use the entangled state to demonstrate a violation of a Bell's inequality.

10. Show that when Alice makes measurements of the quantum state given in Eq. (13.76) using the Bell states given in Eq. (13.77), Bob's single photon state is projected into the states given in Eq. (13.78).

[15] J. R. Schaibley, A. P. Burgers, G. A. McCracken, L.-M. Duan, P. R. Berman, A. S. Bracker, D. Gammon, L. Sham, and D. G. Steel, *Entanglement between a Single Electron Spin Confined to an InAs Quantum Dot and a Photon*, Physical Review Letters **110**, 167401 pp. 1–5 (2013).

Chapter 14
Perturbation Theory

Unfortunately, it is impossible to obtain analytic solutions of the Schrödinger equation for most potentials. However there is a large class of problems in quantum mechanics where the Hamiltonian consists of two parts. The first part corresponds to an isolated quantum system, such as a hydrogen atom, for which we know the exact eigenfunctions and eigenenergies. The second part of the Hamiltonian corresponds to the interaction of the isolated quantum system with some external potential. If the second potential is weak (and I have to define what I mean by weak), it is possible to develop a systematic procedure to approximate the eigenfunctions and eigenvalues of the total Hamiltonian. This procedure is referred to as *perturbation theory*.

Perturbation theory is covered in all standard texts. I will stress some points that are not often discussed in the standard treatments of the problem. The basic idea is simple. Suppose we can write the Hamiltonian of a system as

$$\hat{H} = \hat{H}_0 + \lambda \hat{H}' \tag{14.1}$$

or, in matrix form, as

$$\underline{H} = \underline{H}_0 + \lambda \underline{H}'. \tag{14.2}$$

The quantity λ is just for bookkeeping—I will eventually set it equal to unity. It lets us keep track of the fact that $\hat{H}'$ is supposed to be a small perturbation to $\hat{H}_0$, whatever that means. Furthermore I assume that we know the eigenfunctions $\psi_n^{(0)}(\mathbf{r})$ of $\hat{H}_0$ and the corresponding eigenvalues $E_n^{(0)}$,

$$\hat{H}_0 \psi_n^{(0)}(\mathbf{r}) = E_n^{(0)} \psi_n^{(0)}(\mathbf{r}). \tag{14.3}$$

In Dirac notation,

$$\underline{H}_0 \left| E_n \right\rangle^{(0)} = E_n^{(0)} \left| E_n \right\rangle^{(0)} \tag{14.4}$$

P.R. Berman, *Introductory Quantum Mechanics*, UNITEXT for Physics,
https://doi.org/10.1007/978-3-319-68598-4_14

(remember that $|E_n\rangle^{(0)}$ is a column matrix with 1 in the nth location and zeroes everywhere else). The problem is then straightforward—how can we approximate the energies and eigenfunctions or eigenkets for the total Hamiltonian. In Dirac notation, we are simply trying to develop methods for approximately diagonalizing a Hamiltonian.

14.1 Non-degenerate Perturbation Theory

I will derive the perturbation theory results using Dirac notation and then simply list the corresponding results for the wave functions. In *non-degenerate perturbation theory*, it is assumed that any states coupled by the perturbation (that is, states $|n\rangle$ and $|n'\rangle$ for which $\underline{H}'_{nn'} \neq 0$) have different energies.

I start from

$$\underline{H} = \underline{H}_0 + \lambda \underline{H}', \tag{14.5a}$$

$$\underline{H}_0 |E_n\rangle^{(0)} = E_n^{(0)} |E_n\rangle^{(0)}, \tag{14.5b}$$

$$\underline{H} |E_n\rangle = E_n |E_n\rangle, \tag{14.5c}$$

and try a solution

$$E_n = E_n^{(0)} + \lambda E_n^{(1)} + \lambda^2 E_n^{(2)} + \cdots \tag{14.6a}$$

$$|E_n\rangle = |E_n\rangle^{(0)} + \lambda |E_n\rangle^{(1)} + \lambda^2 |E_n\rangle^{(2)} + \cdots, \tag{14.6b}$$

substitute it into Eq. (14.5c) and equate equal powers of λ. After that, I can set $\lambda = 1$, since it was used only for bookkeeping. I calculate corrections to the energy to order $E_n^{(2)}$ and to the wave function to order $|E_n\rangle^{(1)}$.

Equating coefficients of λ^0 simply reproduces Eq. (14.5b). Equating coefficients of λ yields

$$\underline{H}_0 |E_n\rangle^{(1)} + \underline{H}' |E_n\rangle^{(0)} = E_n^{(0)} |E_n\rangle^{(1)} + E_n^{(1)} |E_n\rangle^{(0)}. \tag{14.7}$$

Multiplying Eq. (14.7) by ${}^{(0)}\langle E_n|$, I obtain

$${}^{(0)}\langle E_n| \underline{H}_0 |E_n\rangle^{(1)} + {}^{(0)}\langle E_n| \underline{H}' |E_n\rangle^{(0)} = E_n^{(0)}\,{}^{(0)}\langle E_n |E_n\rangle^{(1)} + E_n^{(1)}. \tag{14.8}$$

In the ${}^{(0)}\langle E_n|\underline{H}_0 |E_n\rangle^{(1)}$ term, I can let $\underline{H}_0$ act to the left,

$${}^{(0)}\langle E_n| \underline{H}_0 = \left[\underline{H}_0 |E_n\rangle^{(0)}\right]^{\dagger} = E_n^{(0)}\,{}^{(0)}\langle E_n| \tag{14.9}$$

to arrive at

$$E_n^{(0)\,(0)}\langle E_n|E_n\rangle^{(1)} + H'_{nn} = E_n^{(0)\,(0)}\langle E_n|E_n\rangle^{(1)} + E_n^{(1)}, \tag{14.10}$$

or

$$E_n^{(1)} = {}^{(0)}\langle E_n|\underline{\mathrm{H}}'|E_n\rangle^{(0)} = H'_{nn}, \tag{14.11}$$

where all matrix elements are taken with respect to the unperturbed basis. The first order corrections to the energy are simply the diagonal elements of $\underline{\mathrm{H}}'$.

Multiplying Eq. (14.7) by ${}^{(0)}\langle E_m|$ with $m \neq n$ and using Eq. (14.9), I find

$$E_m^{(0)\,(0)}\langle E_m|E_n\rangle^{(1)} + H'_{mn} = E_n^{(0)\,(0)}\langle E_m|E_n\rangle^{(1)}, \tag{14.12}$$

or

$${}^{(0)}\langle E_m|E_n\rangle^{(1)} = \frac{H'_{mn}}{E_n^{(0)} - E_m^{(0)}}. \tag{14.13}$$

This can now be used to calculate $|E_n\rangle$ to first order, since

$$\begin{aligned}
|E_n\rangle &= \sum_m {}^{(0)}\langle E_m|E_n\rangle\,|E_m\rangle^{(0)} \\
&\approx |E_n\rangle^{(0)} + \sum_m {}^{(0)}\langle E_m|E_n\rangle^{(1)}\,|E_m\rangle^{(0)} \\
&\approx |E_n\rangle^{(0)} + \sum_{m\neq n} \frac{H'_{mn}}{E_n^{(0)} - E_m^{(0)}}|E_m\rangle^{(0)} + {}^{(0)}\langle E_n|E_n\rangle^{(1)}\,|E_n\rangle^{(0)} \\
&= \left[1 + {}^{(0)}\langle E_n|E_n\rangle^{(1)}\right]|E_n\rangle^{(0)} + \sum_{m\neq n} \frac{H'_{mn}}{E_n^{(0)} - E_m^{(0)}}|E_m\rangle^{(0)}.
\end{aligned} \tag{14.14}$$

The ${}^{(0)}\langle E_n|E_n\rangle^{(1)}$ term in this expression is often confusing to students. In mth order perturbation theory, the value of ${}^{(0)}\langle E_n|E_n\rangle^{(m)}$ is completely arbitrary since it cannot be determined from Eqs. (14.5) and (14.6). This is not tragic, however, since the mth order value of the energy and mth order expectation values of all operators in a state $|\psi\rangle$ are *independent* of the values of ${}^{(0)}\langle E_n|E_n\rangle^{(q)}$ for $q < m$. In effect, different choices for ${}^{(0)}\langle E_n|E_n\rangle^{(m)}$ simply result in different normalization factors for the eigenkets.

Conventionally the value of ${}^{(0)}\langle E_n|E_n\rangle^{(m)}$ is set equal to zero for all m. With this choice, the eigenkets calculated in each order of perturbation theory are determined uniquely and the mth order eigenket does not depend on the values of ${}^{(0)}\langle E_n|E_n\rangle^{(q)}$ for $q < m$ (clearly, since they have been set equal to zero). The price you pay is that the eigenkets are not normalized, in general, in each order of perturbation theory. The first order eigenkets *are* normalized, however, since

$$\langle E_n | E_n \rangle \approx \left| 1 +^{(0)} \langle E_n | E_n \rangle^{(1)} \right|^2 = 1, \tag{14.15}$$

where terms of order $H_{mn}'^2$ have been neglected. With $^{(0)} \langle E_n | E_n \rangle^{(1)} = 0$,

$$|E_n\rangle^{(1)} = \sum_{m \neq n} \frac{H_{mn}'}{E_n^{(0)} - E_m^{(0)}} |E_m\rangle^{(0)} . \tag{14.16}$$

Equating coefficients of λ^2 in Eq. (14.5c) yields

$$\underline{\mathrm{H}}_0 |E_n\rangle^{(2)} + \underline{\mathrm{H}}' |E_n\rangle^{(1)} = E_n^{(0)} |E_n\rangle^{(2)} + E_n^{(1)} |E_n\rangle^{(1)} + E_n^{(2)} |E_n\rangle^{(0)} . \tag{14.17}$$

Multiplying Eq. (14.17) by $^{(0)} \langle E_n |$, Eq. (14.9), and canceling terms, I obtain

$$\begin{aligned} E_n^{(2)} &= {}^{(0)} \langle E_n | \underline{\mathrm{H}}' |E_n\rangle^{(1)} = \langle E_n | \underline{\mathrm{H}}' \sum_{m \neq n} \frac{H_{mn}'}{E_n^{(0)} - E_m^{(0)}} |E_m\rangle^{(0)} \\ &= \sum_{m \neq n} \frac{H_{nm}' H_{mn}'}{E_n^{(0)} - E_m^{(0)}} = \sum_{m \neq n} \frac{\left| H_{nm}' \right|^2}{E_n^{(0)} - E_m^{(0)}} . \end{aligned} \tag{14.18}$$

This completes the derivation. If wave functions rather than kets are used, all the results are still valid. In terms of the wave function, Eq. (14.16) is replaced by

$$\psi_n^{(1)}(\mathbf{r}) = \sum_{m \neq n} \frac{H_{mn}'}{E_n^{(0)} - E_m^{(0)}} \psi_m^{(0)}(\mathbf{r}) \tag{14.19}$$

and the matrix elements of $\underline{\mathrm{H}}'$ are evaluated as

$$H_{mn}' =^{(0)} \langle E_m | \underline{\mathrm{H}}' |E_n\rangle^{(0)} = \int d\mathbf{r} \left[\psi_m^{(0)}(\mathbf{r}) \right]^* \hat{H}' \psi_n^{(0)}(\mathbf{r}). \tag{14.20}$$

The combined results are

$$E_n = E_n^{(0)} + H_{nn}' + \sum_{m \neq n} \frac{\left| H_{nm}' \right|^2}{E_n^{(0)} - E_m^{(0)}} + \cdots \tag{14.21a}$$

$$|E_n\rangle = |E_n\rangle^{(0)} + \sum_{m \neq n} \frac{H_{mn}'}{E_n^{(0)} - E_m^{(0)}} |E_m\rangle^{(0)} + \cdots \tag{14.21b}$$

$$\psi_n(\mathbf{r}) = \psi_m^{(0)}(\mathbf{r}) + \sum_{m \neq n} \frac{H_{mn}'}{E_n^{(0)} - E_m^{(0)}} \psi_m^{(0)}(\mathbf{r}) + \cdots . \tag{14.21c}$$

For perturbation theory to be valid, the off diagonal matrix elements of $\left|\underline{H}'_{mn}\right|$ must be small compared with the *difference* in eigenvalues $\left|E_n^{(0)} - E_m^{(0)}\right|$ *and* the ratio $\left|E_n^{(1)} - E_m^{(1)}\right| / \left|E_n^{(0)} - E_m^{(0)}\right|$ must be much less than unity. Clearly this fails for any degenerate states that are coupled by the perturbation.

14.1.1 Examples

(1) Find the eigenvalues to order 0.01 and the eigenvectors to order 0.1 of the following matrix:

$$\underline{H} = \begin{pmatrix} 1.1 & 0.1 & -0.2 & 0.2 \\ 0.1 & 4.2 & 0.1 & -0.1 \\ -0.2 & 0.1 & 8 & -0.2 \\ 0.2 & -0.1 & -0.2 & 13.9 \end{pmatrix}, \tag{14.22}$$

written in the $\left|E_m\right\rangle^{(0)}$ basis ($m = 1-4$). I set

$$\underline{H}_0 = \begin{pmatrix} 1 & 0 & 0 & 0 \\ 0 & 4 & 0 & 0 \\ 0 & 0 & 8 & 0 \\ 0 & 0 & 0 & 14 \end{pmatrix}; \tag{14.23a}$$

$$\underline{H}' = \begin{pmatrix} 0.1 & 0.1 & -0.2 & 0.2 \\ 0.1 & 0.2 & 0.1 & -0.1 \\ -0.2 & 0.1 & 0 & -0.2 \\ 0.2 & -0.1 & -0.2 & -0.1 \end{pmatrix}, \tag{14.23b}$$

with $E_1^{(0)} = 1$, $E_2^{(0)} = 4$, $E_3^{(0)} = 8$, $E_2^{(0)} = 12$. Therefore, from Eq. (14.21a), I find

$$\begin{aligned} E_1 \approx 1 + 0.1 + \frac{0.01}{1-4} + \frac{0.04}{1-8} + \frac{0.04}{1-14} &= 1.08788; \\ E_1(\text{exact}) = 1.08777\,; \end{aligned} \tag{14.24a}$$

$$\begin{aligned} E_2 \approx 4 + 0.2 + \frac{0.01}{4-1} + \frac{0.01}{4-8} + \frac{0.01}{4-14} &= 4.19983; \\ E_2(\text{exact}) = 4.20013\,; \end{aligned} \tag{14.24b}$$

$$\begin{aligned} E_3 \approx 8 + 0 + \frac{0.04}{8-1} + \frac{0.01}{8-4} + \frac{0.04}{8-14} &= 8.00155; \\ E_3(\text{exact}) = 8.00093\,; \end{aligned} \tag{14.24c}$$

$$E_4 \approx 14 - 0.1 + \frac{0.04}{14-1} + \frac{0.01}{14-4} + \frac{0.04}{14-8} = 13.9107;$$
$$E_4(\text{exact}) = 13.9112\,, \tag{14.24d}$$

where the exact values were obtained by numerically diagonalizing the Hamiltonian. The eigenvectors are obtained from Eq. (14.21a) as

$$|E_1\rangle \approx \begin{pmatrix}1\\0\\0\\0\end{pmatrix} + \begin{pmatrix}0\\ \frac{0.1}{1-4}\\ \frac{-0.2}{1-8}\\ \frac{0.2}{1-14}\end{pmatrix} = \begin{pmatrix}1\\-0.0333\\0.0286\\-0.0154\end{pmatrix};$$
$$|E_1\rangle\,(\text{exact}) = \begin{pmatrix}0.999\\-0.0335\\0.0289\\-0.0154\end{pmatrix}; \tag{14.25a}$$

$$|E_2\rangle \approx \begin{pmatrix}0\\1\\0\\0\end{pmatrix} + \begin{pmatrix}\frac{0.1}{4-1}\\ 0\\ \frac{0.1}{4-8}\\ \frac{-0.1}{4-14}\end{pmatrix} = \begin{pmatrix}0.0333\\1\\-0.025\\0.010\end{pmatrix};$$
$$|E_2\rangle\,(\text{exact}) = \begin{pmatrix}0.0344\\0.999\\-0.0240\\0.0091\end{pmatrix}; \tag{14.25b}$$

$$|E_3\rangle \approx \begin{pmatrix}0\\0\\1\\0\end{pmatrix} + \begin{pmatrix}\frac{-0.2}{8-1}\\ \frac{0.1}{8-4}\\ 0\\ \frac{-0.2}{8-14}\end{pmatrix} = \begin{pmatrix}-0.0286\\0.025\\1\\0.0333\end{pmatrix};$$
$$|E_3\rangle\,(\text{exact}) = \begin{pmatrix}-0.0276\\0.0246\\0.999\\0.0352\end{pmatrix}; \tag{14.25c}$$

$$|E_4\rangle \approx \begin{pmatrix}0\\0\\0\\1\end{pmatrix} + \begin{pmatrix}\frac{0.2}{14-1}\\ \frac{-0.1}{14-4}\\ \frac{-0.2}{14-8}\\ 0\end{pmatrix} = \begin{pmatrix}0.0154\\-0.01\\-0.0333\\1\end{pmatrix};$$

$$|E_4\rangle\,(\text{exact}) = \begin{pmatrix} 0.0161 \\ -0.0105 \\ -0.0345 \\ 0.999 \end{pmatrix}. \tag{14.25d}$$

(2) An infinite square well of width a has a potential bump at its center of the form

$$H'(x) = \begin{cases} V_0 & |x| \le b/2 \\ 0 & a/2 > |x| \ge b/2 \end{cases}, \tag{14.26}$$

with $V_0 > 0$ and $a > b > 0$. Find corrections to the energy and the eigenfunctions to order V_0. I have positioned the well between $-a/2$ and $a/2$ to exploit the symmetry of the problem. The unperturbed eigenfunctions and eigenenergies are

$$E_n^{(0)} = \frac{\hbar^2 n^2 \pi^2}{2ma^2}; \tag{14.27a}$$

$$\psi_n^{(0)}(x) = \sqrt{\frac{2}{a}} \begin{cases} \cos\left(\frac{n\pi x}{a}\right) & n \text{ odd} \\ \sin\left(\frac{n\pi x}{a}\right) & n \text{ even} \end{cases}. \tag{14.27b}$$

Therefore,

$$\begin{aligned} E_n &\approx E_n^{(0)} + E_n^{(1)} \\ &= \frac{\hbar^2 n^2 \pi^2}{2ma^2} + \frac{2V_0}{a} \int_{-b/2}^{b/2} dx \begin{cases} \cos^2\left(\frac{n\pi x}{a}\right) & n \text{ odd} \\ \sin^2\left(\frac{n\pi x}{a}\right) & n \text{ even} \end{cases} \\ &= \frac{\hbar^2 n^2 \pi^2}{2ma^2} + \frac{V_0}{a}\left[b \pm \frac{a \sin\left(\frac{bn\pi}{a}\right)}{n\pi}\right], \end{aligned} \tag{14.28}$$

where Eqs. (14.11) and (14.20) were used, and the plus (minus) sign is for n odd (even). The energy of each state is raised by the potential, as is expected from the positive energy bump, but odd n (even parity) states are raised more since the wave function squared is a maximum at the center of the perturbation. The new eigenfunctions are

$$\begin{aligned} \psi_n(x) &\approx \psi_n^{(0)}(x) + \psi_n^{(1)}(x) \\ &= \psi_n^{(0)}(x) + \sum_{q \ne n} \frac{H'_{qn}}{E_n^{(0)} - E_q^{(0)}} \psi_q^{(0)}(x) \\ &= \psi_n^{(0)}(x) + \sum_{q \ne n} \frac{V_0 \int_{-b/2}^{b/2} dx \psi_q^{(0)}(x) \psi_n^{(0)}(x)}{\frac{\hbar^2 \pi^2}{2ma^2}\left(n^2 - q^2\right)} \psi_q^{(0)}(x). \end{aligned} \tag{14.29}$$

For n odd, I find

$$\psi_n(x) \approx \sqrt{\frac{2}{a}} \cos\left(\frac{n\pi x}{a}\right) - \frac{4ma^2V_0}{\hbar^2\pi^3} \times \sum_{q\neq n;\, q \text{ odd}} \left[\frac{(q+n)\sin\left(\frac{b\pi(q-n)}{2a}\right) + (q-n)\sin\left(\frac{b\pi(q+n)}{2a}\right)}{\left(n^2-q^2\right)^2} \right] \times \sqrt{\frac{2}{a}} \cos\left(\frac{q\pi x}{a}\right) \tag{14.30}$$

and, for n even,

$$\psi_n(x) \approx \sqrt{\frac{2}{a}} \sin\left(\frac{n\pi x}{a}\right) - \frac{4ma^2V_0}{\hbar^2\pi^3} \times \sum_{q\neq n;\, q \text{ even}} \left[\frac{(q+n)\sin\left(\frac{b\pi(q-n)}{2a}\right) - (q-n)\sin\left(\frac{b\pi(q+n)}{2a}\right)}{\left(n^2-q^2\right)^2} \right] \times \sqrt{\frac{2}{a}} \sin\left(\frac{q\pi x}{a}\right). \tag{14.31}$$

The lowest energy eigenfunctions are pushed away from the origin by the potential, but less so for the even n (odd parity) eigenfunctions, since the odd-parity eigenfunctions vanished at the origin (where the perturbation is a maximum). The series converges rapidly for $q \gg n$.

(3) Consider a 1-D harmonic oscillator in dimensionless units having

$$\hat{H}_0 = a^\dagger a + 1/2, \tag{14.32}$$

subjected to a perturbation

$$\hat{H}' = \epsilon\hat{\xi}^4 = \epsilon\frac{\left(a + a^\dagger\right)^4}{4}, \tag{14.33}$$

where ϵ is a positive constant having magnitude much less than unity. Find the corrections to the energy to order ϵ.

The first order correction to the energy is

$$E_n^{(1)} = H'_{nn} = \frac{\epsilon}{4}\langle n| \left(a + a^\dagger\right)^4 |n\rangle. \tag{14.34}$$

When you expand $\left(a + a^\dagger\right)^4$ the only nonvanishing diagonal matrix elements result from terms having equal powers of a and $a^\dagger$ (why?). That is,

$$
\begin{aligned}
E_n^{(1)} &= H'_{nn} = \frac{\epsilon}{4} \langle n| \left(a + a^{\dagger}\right)^4 |n\rangle \\
&= \frac{\epsilon}{4} \langle n| a^2 a^{\dagger 2} + a^{\dagger 2} a^2 + a a^{\dagger 2} a + a^{\dagger} a^2 a^{\dagger} + a^{\dagger} a a^{\dagger} a + a a^{\dagger} a a^{\dagger} |n\rangle \\
&= \frac{\epsilon}{4} \left[(n+1)(n+2) + n(n-1) + 2n(n+1) + n^2 + (n+1)^2 \right] \\
&= \frac{\epsilon}{4} \left(6n^2 + 6n + 3\right). \qquad (14.35)
\end{aligned}
$$

For sufficiently large n, perturbation theory breaks down since $E_n^{(0)}$ varies linearly with n and $E_n^{(1)}$ varies quadratically with n.

(4) *van der Waals interaction*: Finally consider two hydrogen atoms in their ground states separated by a distance $R > a_0$, where a_0 is the Bohr radius. Imagine they both lie on the z axis. The electrostatic interaction energy in lowest non-vanishing order is the dipole–dipole interaction for which

$$
\begin{aligned}
\hat{H}' &= -\frac{3\left(\hat{\mathbf{p}}_{e1} \cdot \mathbf{u}_z\right)\left(\hat{\mathbf{p}}_{e2} \cdot \mathbf{u}_z\right) - \hat{\mathbf{p}}_{e1} \cdot \hat{\mathbf{p}}_{e2}}{4\pi\epsilon_0 R^3} \\
&= -e^2 \frac{2\hat{z}_1\hat{z}_2 - \hat{x}_1\hat{x}_2 - \hat{y}_1\hat{y}_2}{4\pi\epsilon_0 R^3}, \qquad (14.36)
\end{aligned}
$$

where $\hat{\mathbf{p}}_{ej} = -e\hat{\mathbf{r}}_j$ $(j = 1, 2)$ is the dipole moment operator of atom j. I want to calculate the change in the ground state energy

$$
\Delta E_g = E_g - E_g^{(0)}
$$

resulting from the electrostatic interaction, an energy associated with the van der Waals interaction. The unperturbed Hamiltonian is the sum of the Coulomb potential Hamiltonians for each atom, implying that the unperturbed eigenkets are

$$
|n_1\ell_1 m_1; n_2\ell_2 m_2\rangle = |n_1\ell_1 m_1\rangle\, |n_2\ell_2 m_2\rangle \qquad (14.37)
$$

and the unperturbed eigenenergies are

$$
E_{n_1,n_2}^{(0)} = -\frac{e^2}{8\pi\epsilon_0 a_0}\left(\frac{1}{n_1^2} + \frac{1}{n_2^2}\right), \qquad (14.38)
$$

where a_0 is the Bohr radius. I denote the ground state eigenket of the noninteracting two-atom system by

$$
|g\rangle = |100; 100\rangle. \qquad (14.39)
$$

It is easy to prove that the first order energy change is zero. To second order, with carets on the operators suppressed,

$$\Delta E_g = E_g - E_{1,1}^{(0)} = E_g^{(2)} = \sum_{\substack{n_1,\ell_1,m_1 \\ n_2,\ell_2,m_2 \\ n_1,n_2 \neq 1,1}}' \frac{\left|\langle g|\, H' \,|n_1\ell_1 m_1; n_2\ell_2 m_2\rangle\right|^2}{E_{1,1}^{(0)} - E_{n_1,n_2}^{(0)}}$$

$$= \sum_{\substack{n_1,\ell_1,m_1 \\ n_2,\ell_2,m_2 \\ n_1,n_2 \neq 1,1}}' \frac{\langle g|\, H' \,|n_1\ell_1 m_1; n_2\ell_2 m_2\rangle \langle n_1\ell_1 m_1; n_2\ell_2 m_2|\, H' \,|g\rangle}{E_{1,1}^{(0)} - E_{n_1,n_2}^{(0)}}. \quad (14.40)$$

The prime on the summation symbol in Eq. (14.40) is a shorthand notation for a sum over bound states and an integral over *continuum* states of the hydrogen atoms. Since *all* intermediate states are included in the sum in Eq. (14.21a), the continuum states have to be included as well. The sum (and integral) can be evaluated using hydrogenic wave functions, but I can get a rough estimate of the sum by evaluating the denominator in Eq. (14.40) at specific values of n_1 and n_2. In Chap. 15, I will get a lower bound on ΔE_g by using only the $n_1 = 2$, $n_2 = 2$ intermediate state, but here I evaluate the denominator at the ionization energy $n_1 = \infty$, $n_2 = \infty \left[E_{n_1,n_2}^{(0)} = 0\right]$, for which

$$\Delta E_g \approx \frac{1}{E_{1,1}^{(0)}} \sum_{\substack{n_1,\ell_1,m_1 \\ n_2,\ell_2,m_2 \\ n_1,n_2 \neq 1,1}}' \langle g|\, H' \,|n_1\ell_1 m_1; n_2\ell_2 m_2\rangle$$

$$\times \langle n_1\ell_1 m_1; n_2\ell_2 m_2|\, H' \,|g\rangle$$

$$= \sum_{\substack{n_1,\ell_1,m_1 \\ n_2,\ell_2,m_2}}' \frac{\langle g|\, H' \,|n_1\ell_1 m_1; n_2\ell_2 m_2\rangle \langle n_1\ell_1 m_1; n_2\ell_2 m_2|\, H' \,|g\rangle}{E_{1,1}^{(0)}}$$

$$-\frac{1}{E_{1,1}^{(0)}} \left|\langle g|\, H' \,|g\rangle\right|^2 = \frac{1}{E_{1,1}^{(0)}} \langle g|\, H'^2 \,|g\rangle, \quad (14.41)$$

where the completeness relation was used, as well as the fact that

$$\langle g|\, H' \,|g\rangle = 0. \quad (14.42)$$

Evaluating the denominator at the ionization energy is not such a bad idea since it is at the border between the discrete and continuum states, that is, somewhere in the "middle" of the summation.

In evaluating the matrix elements needed in Eq. (14.41), the only non-vanishing terms are

$$\Delta E_g \approx \frac{e^4}{(4\pi\epsilon_0)^2 R^6 E_{1,1}^{(0)}} \langle g|\, 4z_1^2 z_2^2 + x_1^2 x_2^2 + y_1^2 y_2^2 \,|g\rangle$$

$$= \frac{e^4}{(4\pi\epsilon_0)^2 R^6 E_{1,1}^{(0)}}$$

$$\times \langle g | r_1^2 r_2^2 \begin{pmatrix} 4\cos^2\theta_1 \cos^2\theta_2 \\ +\sin^2\theta_1 \sin^2\theta_2 \cos^2\phi_1 \cos^2\phi_2 \\ +\sin^2\theta_1 \sin^2\theta_2 \sin^2\phi_1 \sin^2\phi_2 \end{pmatrix} | g \rangle$$

$$= \frac{e^4}{(4\pi\epsilon_0)^2 R^6 E_{1,1}^{(0)}} \left| \int_0^\infty r^4 |R_{10}(r)|^2 dr \right|^2 \left(4\frac{1}{3}\frac{1}{3} + \frac{1}{3}\frac{1}{3} + \frac{1}{3}\frac{1}{3} \right)$$

$$= \frac{6e^4}{9(4\pi\epsilon_0)^2 R^6 E_{1,1}^{(0)}} \left| \int_0^\infty r^4 |R_{10}(r)|^2 dr \right|^2$$

$$= \frac{6e^4 \pi}{9(4\pi\epsilon_0)^2 R^6 E_{1,1}^{(0)}} \left| \int_0^\infty \frac{4}{a_0^3} r^4 e^{-2r/a_0} dr \right|^2 = \frac{6a_0^4 e^4}{(4\pi\epsilon_0)^2 R^6 E_{1,1}^{(0)}}$$

$$= \frac{6a_0^4 e^4}{(4\pi\epsilon_0)^2 R^6 \left(-2e^2/8\pi\epsilon_0 a_0\right)} = -\frac{6a_0^5 e^2}{4\pi\epsilon_0 R^6}. \tag{14.43}$$

Since a_0 is proportional to $\hbar^2$, the van der Waals interaction is of quantum-mechanical origin, although the quantum-mechanical dependence is linked to the quantum states of the individual atoms through a_0, rather than the quantum nature of the interaction between the atoms. The value from an accurate variational calculation[1] (see Chap. 15 for a discussion of variational calculations) is

$$\Delta E_g \approx -\frac{6.50a_0^5 e^2}{4\pi\epsilon_0 R^6}, \tag{14.44}$$

so I didn't do too bad.

In the intermediate steps used to arrive at Eqs. (14.43), I needed to use the ground state wave function

$$\psi_g(r) = \psi_{100}(r) = R_{10}(r) Y_0^0(\theta, \phi) = \frac{2}{(a_0)^{3/2}} e^{-r/a_0} \sqrt{\frac{1}{4\pi}}, \tag{14.45}$$

the integral identities

$$\frac{1}{4\pi} \int_0^{2\pi} d\phi \int_0^\pi \cos^2\theta \sin\theta d\theta = \frac{1}{3}, \tag{14.46a}$$

$$\frac{1}{4\pi} \int_0^\pi \sin^2\theta d\theta \int_0^{2\pi} \sin^2\phi d\phi$$
$$= \frac{1}{4\pi} \int_0^\pi \sin^2\theta d\theta \int_0^{2\pi} \cos^2\phi d\phi = \frac{1}{3}, \tag{14.46b}$$

and the value of $E_{1,1}^{(0)}$ given by Eq. (14.38).

[1] Linus Pauling and J. Y. Beach, *The van der Waals Interaction of Hydrogen Atoms*, Physical Review **47**, 686–692 (1935).

14.2 Degenerate Perturbation Theory

Clearly nondegenerate perturbation theory breaks down whenever there is energy degeneracy, *if* the degenerate levels are coupled by the perturbation. As an example, consider the 2×2 matrix

$$\underset{\sim}{\mathrm{H}} = \begin{pmatrix} a & \epsilon \\ \epsilon & a \end{pmatrix}, \tag{14.47}$$

with $\epsilon \ll a$ and both a and ϵ positive. The eigenvalues are

$$E_1 = a - \epsilon; \qquad E_2 = a + \epsilon, \tag{14.48}$$

and the eigenkets are

$$|E_1\rangle \approx \frac{1}{\sqrt{2}} \begin{pmatrix} 1 \\ -1 \end{pmatrix}; \qquad |E_2\rangle \approx \frac{1}{\sqrt{2}} \begin{pmatrix} 1 \\ 1 \end{pmatrix}. \tag{14.49}$$

In other words, although the eigenenergies are changed only slightly if $\epsilon \ll a$, the eigenkets are changed to *zeroth order* in ϵ. Sometimes it is possible to guess the correct eigenkets by building in some physics and using "good" quantum numbers. I will look at this more carefully after I have reviewed angular momentum and spin. However without some physics to fall back on, you must diagonalize any degenerate submatrix *exactly* and then proceed using normal degenerate perturbation theory. Even then, there can be some traps.

Before giving some examples, let me put the problem in some perspective. Consider the energy levels in hydrogen, neglecting spin. Each state characterized by the quantum number n is n^2 fold degenerate. If a perturbation couples states *within* a given manifold designated by n, then nondegenerate perturbation theory fails. In essence, each submatrix corresponding to the degenerate manifold must be diagonalized *exactly*. Then, using the new eigenkets, you can carry out *non-degenerate* perturbation theory between *different* n manifolds. Often, the small additional corrections produced by such coupling are of little interest. *In that case, degenerate perturbation theory consists simply of diagonalizing each degenerate manifold of levels.* The examples should make this clear.

Example 1 Approximate matrix diagonalization.

Suppose that a Hamiltonian has the form

$$\underset{\sim}{\mathrm{H}} = \begin{pmatrix} 1 & 0.1 & 0.2 \\ 0.1 & 1 & -0.1 \\ 0.2 & -0.1 & 4 \end{pmatrix} \tag{14.50}$$

in the $|E_1\rangle^{(0)}$, $|E_2\rangle^{(0)}$, $|E_3\rangle^{(0)}$ basis. The basis kets,

$$|E_1\rangle^{(0)} = \begin{pmatrix} 1 \\ 0 \\ 0 \end{pmatrix}; \quad |E_2\rangle^{(0)} = \begin{pmatrix} 0 \\ 1 \\ 0 \end{pmatrix}; \quad |E_3\rangle^{(0)} = \begin{pmatrix} 0 \\ 0 \\ 1 \end{pmatrix} \tag{14.51}$$

are eigenkets of

$$\underline{H}_0 = \begin{pmatrix} 1 & 0 & 0 \\ 0 & 1 & 0 \\ 0 & 0 & 4 \end{pmatrix}. \tag{14.52}$$

I want to find the approximate eigenenergies and eigenkets of $\underline{H}$. Since there is energy degeneracy and coupling between the degenerate levels, I must diagonalize the degenerate subblock. To this end, I define

$$\widetilde{\underline{H}}_0 = \begin{pmatrix} 1 & 0.1 & 0 \\ 0.1 & 1 & 0 \\ 0 & 0 & 4 \end{pmatrix}; \tag{14.53a}$$

$$\underline{H}' = \begin{pmatrix} 0 & 0 & 0.2 \\ 0 & 0 & -0.1 \\ 0.2 & -0.1 & 0 \end{pmatrix}. \tag{14.53b}$$

I first diagonalize $\widetilde{\underline{H}}_0$ and find the eigenvalues and eigenkets,

$$E_1 = 0.9; \qquad |\widetilde{E_1}\rangle = \frac{1}{\sqrt{2}} \begin{pmatrix} 1 \\ -1 \\ 0 \end{pmatrix}; \tag{14.54a}$$

$$E_2 = 1.1; \qquad |\widetilde{E_2}\rangle = \frac{1}{\sqrt{2}} \begin{pmatrix} 1 \\ 1 \\ 0 \end{pmatrix}; \tag{14.54b}$$

$$E_3 = 4; \qquad |\widetilde{E_3}\rangle = |E_3\rangle = \begin{pmatrix} 0 \\ 0 \\ 1 \end{pmatrix}. \tag{14.54c}$$

As expected, the eigenkets are changed to zeroth order. The energy of the symmetric ket is increased and that of the antisymmetric ket is decreased (if the coupling were negative, as in an attractive interaction, the symmetric state would be lowered in energy).

The procedure for diagonalizing a matrix is reviewed in Sect. 11.3. I form the matrix

$$\underline{S}^{\dagger} = \begin{pmatrix} \frac{1}{\sqrt{2}} & \frac{1}{\sqrt{2}} & 0 \\ -\frac{1}{\sqrt{2}} & \frac{1}{\sqrt{2}} & 0 \\ 0 & 0 & 1 \end{pmatrix}, \tag{14.55}$$

with the eigenvectors put in as *columns*. In other words,

$$\widetilde{|E_n\rangle} = \sum_m S^{\dagger}_{mn} |E_m\rangle = \sum_m S^{*}_{nm} |E_m\rangle, \tag{14.56}$$

$$S^{\dagger}_{mn} = \langle E_m \widetilde{|E_n\rangle}, \tag{14.57}$$

[note the order of the indices in Eq. (14.56)] such that

$$\underline{S}\widetilde{\underline{H}}_0\underline{S}^{\dagger} = \begin{pmatrix} 0.9 & 0 & 0 \\ 0 & 1.1 & 0 \\ 0 & 0 & 4 \end{pmatrix}. \tag{14.58}$$

In terms of the new eigenkets

$$\widetilde{\underline{H}} = \underline{S}\widetilde{\underline{H}}_0\underline{S}^{\dagger} + \underline{S}\underline{H}'\underline{S}^{\dagger} \tag{14.59a}$$

$$= \begin{pmatrix} 0.9 & 0 & 0 \\ 0 & 1.1 & 0 \\ 0 & 0 & 4 \end{pmatrix} + \frac{1}{\sqrt{2}} \begin{pmatrix} 0 & 0 & 0.3 \\ 0 & 0 & 0.1 \\ 0.3 & 0.1 & 0 \end{pmatrix} \tag{14.59b}$$

$$= \widetilde{\underline{H}}_0 + \widetilde{\underline{H}}', \tag{14.59c}$$

where I have *redefined* the eigenkets of $\widetilde{\underline{H}}_0$ as having a 1 in one row and zeroes everywhere else, that is

$$\widetilde{|E_1\rangle}^{(0)} = \begin{pmatrix} 1 \\ 0 \\ 0 \end{pmatrix}; \quad \widetilde{|E_2\rangle}^{(0)} = \begin{pmatrix} 0 \\ 1 \\ 0 \end{pmatrix}; \quad \widetilde{|E_3\rangle}^{(0)} = \begin{pmatrix} 0 \\ 0 \\ 1 \end{pmatrix}. \tag{14.60}$$

Now I can use *nondegenerate* perturbation theory to find

$$E_1 \approx 0.9 + \frac{0.3^2/2}{0.9-4} = 0.8855; \quad E_1(\text{exact}) = 0.88544; \tag{14.61a}$$

$$E_2 \approx 1.1 + \frac{0.1^2/2}{1.1-4} = 1.0983; \quad E_2(\text{exact}) = 1.0984; \tag{14.61b}$$

$$E_3 \approx 4 + \frac{0.3^2/2}{4-0.9} + \frac{0.1^2/2}{4-1.1} = 4.01624; \quad E_3(\text{exact}) = 4.01616. \tag{14.61c}$$

The eigenkets are

$$|E_1\rangle \approx \begin{pmatrix}1\\0\\0\end{pmatrix} + \begin{pmatrix}0\\0\\\frac{0.3/\sqrt{2}}{0.9-4}\end{pmatrix} = \begin{pmatrix}1\\0\\-0.0684\end{pmatrix}; \tag{14.62a}$$

$$|E_2\rangle \approx \begin{pmatrix}0\\1\\0\end{pmatrix} + \begin{pmatrix}0\\0\\\frac{0.1/\sqrt{2}}{1.1-4}\end{pmatrix} = \begin{pmatrix}1\\0\\-0.0244\end{pmatrix}; \tag{14.62b}$$

$$|E_3\rangle \approx \begin{pmatrix}0\\0\\1\end{pmatrix} + \begin{pmatrix}\frac{0.3/\sqrt{2}}{4-0.9}\\\frac{0.1/\sqrt{2}}{4-1.1}\\0\end{pmatrix} = \begin{pmatrix}0.0684\\0.0244\\1\end{pmatrix}. \tag{14.62c}$$

To get the eigenkets in terms of the original basis, I use Eq. (14.54) and find

$$\begin{aligned}|E_1\rangle &\approx \widetilde{|E_1\rangle}^{(0)} - 0.0711\widetilde{|E_3\rangle}^{(0)}\\ &= \frac{1}{\sqrt{2}}\left(|E_1\rangle^{(0)} - |E_2\rangle^{(0)}\right) - 0.0684\,|E_3\rangle^{(0)}\\ &= \begin{pmatrix}0.707\\-0.707\\-0.0684\end{pmatrix}; \quad |E_1\rangle\,(\text{exact}) = \begin{pmatrix}0.721\\-0.689\\-0.0684\end{pmatrix},\end{aligned} \tag{14.63a}$$

$$\begin{aligned}|E_2\rangle &\approx \widetilde{|E_2\rangle}^{(0)} - 0.0244\widetilde{|E_3\rangle}^{(0)}\\ &= \frac{1}{\sqrt{2}}\left(|E_1\rangle^{(0)} + |E_2\rangle^{(0)}\right) - 0.0244\,|E_3\rangle^{(0)}\\ &= \begin{pmatrix}0.707\\0.707\\-0.0244\end{pmatrix}; \quad |E_2\rangle\,(\text{exact}) = \begin{pmatrix}0.690\\0.724\\-0.0226\end{pmatrix},\end{aligned} \tag{14.63b}$$

$$\begin{aligned}|E_3\rangle &\approx 0.0684\widetilde{|E_1\rangle}^{(0)} + 0.0244\widetilde{|E_2\rangle}^{(0)} + \widetilde{|E_3\rangle}^{(0)}\\ &= \frac{0.0684}{\sqrt{2}}\left(|E_1\rangle^{(0)} - |E_2\rangle^{(0)}\right)\\ &\quad + \frac{0.0244\left(|E_1\rangle^{(0)} + |E_2\rangle^{(0)}\right)}{\sqrt{2}} + |E_3\rangle^{(0)}\\ &= \begin{pmatrix}0.0656\\-0.0311\\1\end{pmatrix}; \quad |E_3\rangle\,(\text{exact}) = \begin{pmatrix}0.0651\\-0.0309\\0.997\end{pmatrix}.\end{aligned} \tag{14.63c}$$

The eigenenergies are good approximations to the exact eigenenergies as is eigenket $|E_3\rangle$, but eigenkets $|E_1\rangle$ and $|E_2\rangle$ are not as good as expected. Where did I go wrong? It turns out that if you calculate *second order* corrections to the wave function, corrections to $|E_1\rangle$ and $|E_2\rangle$ are of order of the coupling matrix elements (of order 0.1) divided by $\tilde{E}_2^{(0)} - \tilde{E}_1^{(0)} = 0.2$, multiplied by first order corrections to $|E_3\rangle$. That is, you can show that the second order corrections to the eigenket $|E_n\rangle$ are given by

$$|E_n\rangle^{(2)} = -\sum_{m\neq n} \frac{\widetilde{H}'_{mn}\widetilde{H}'_{nn}\widetilde{|E_m\rangle}^{(0)}}{\left[\tilde{E}_n^{(0)} - \tilde{E}_m^{(0)}\right]^2} + \sum_{k,m\neq n} \frac{\widetilde{H}'_{km}\widetilde{H}'_{mn}\widetilde{|E_k\rangle}^{(0)}}{\left[\tilde{E}_n^{(0)} - \tilde{E}_m^{(0)}\right]\left[\tilde{E}_n^{(0)} - \tilde{E}_k^{(0)}\right]}. \tag{14.64}$$

A glance at Eq. (14.59b) will reveal that the first term does not contribute since $\widetilde{H}'_{nn} = 0$. However, if I evaluate $|E_1\rangle^{(2)}$, the second term yields [recall that $\widetilde{H}'_{12} = 0$]

$$\begin{aligned} |E_1\rangle^{(2)} &= \frac{\widetilde{H}'_{23}\widetilde{H}'_{31}\widetilde{|E_2\rangle}^{(0)}}{\left[\tilde{E}_1^{(0)} - \tilde{E}_3^{(0)}\right]\left[\tilde{E}_1^{(0)} - \tilde{E}_2^{(0)}\right]} = \frac{1}{2}\frac{0.1\times 0.3}{(-3.1)\times(-0.2)}\widetilde{|E_2\rangle}^{(0)} \\ &= 0.024\widetilde{|E_2\rangle}^{(0)}, \end{aligned} \tag{14.65}$$

whereas

$$|E_3\rangle^{(1)} = 0.0684\widetilde{|E_1\rangle}^{(0)} + 0.0244\widetilde{|E_2\rangle}^{(0)}. \tag{14.66}$$

Thus the second order correction to $|E_1\rangle$ is of the same order as the *first* order corrections of $|E_3\rangle$. Thus, if you want eigenkets $|E_1\rangle$ and $|E_2\rangle$ correct to order 0.001 you must diagonalize the *entire* matrix.

To be a bit more specific, if the matrix elements that couple the degenerate states in a given manifold of levels is *large* compared to those that couple these levels to levels *outside* the degenerate subblock, then the procedure of diagonalizing the degenerate subblock and then applying degenerate perturbation theory is valid. However when these couplings are comparable, the method fails to give the new eigenkets of the *degenerate* states to first order in the coupling, it gives it only to zeroth order in the coupling. It *does* give the correct energies. Usually in degenerate perturbation theory we are content to have the degenerate eigenkets correct to *zeroth* order in the perturbation. In other words, in degenerate perturbation theory, the states *outside* the degenerate subspace are often not considered *at all*, since they produce small modifications to the eigenkets, provided the coupling strength is much less than the energy spacing between nondegenerate energy manifolds.

Example 2 Linear Stark shift in hydrogen for $n = 2$.

A classic example of degenerate perturbation theory is the linear Stark effect in hydrogen. In the $n = 2$ state of hydrogen, neglecting spin and the Lamb shift [the modifications of the energy levels resulting from spin and the Lamb shift (produced by fluctuations of the vacuum field) are discussed in Chap. 21], there are four degenerate levels corresponding to $n = 2, \ell = 0$ and $n = 2, \ell = 1, m_\ell = 0, \pm 1$. In the presence of a static electric field, $\mathbf{E} = \mathcal{E}\mathbf{u}_z$ directed along the z axis,

$$\hat{H}' = -\hat{\mathbf{p}}_e \cdot \mathbf{E} = e\mathcal{E}\widehat{r\cos\theta}, \tag{14.67}$$

where $\hat{\mathbf{p}}_e = -e\hat{\mathbf{r}}$ is the electric dipole moment operator, e the magnitude of the charge of the electron, and z the coordinate of the electron relative to the proton. This interaction couples only states having different parity. It couples only the $n = 2, \ell = 0$ and $\ell = 1$ states. Moreover, the matrix element of $\hat{z}$ is nonvanishing only between the $n = 2, \ell = 0$ and $n = 2, \ell = 1, m_\ell = 0$ states. The matrix element that is needed, $\langle 210 | e\mathcal{E}\widehat{r\cos\theta} | 200 \rangle$, can be calculated using the hydrogenic wave functions and one finds

$$\langle 210 | e\mathcal{E}\widehat{r\cos\theta} | 200 \rangle = -3e\mathcal{E}a_0, \tag{14.68}$$

where a_0 is the Bohr radius. Thus the symmetric state,

$$\psi_s = \frac{1}{\sqrt{2}} \left(\psi_{200} + \psi_{210} \right), \tag{14.69}$$

is lowered in energy by $3e\mathcal{E}a_0$ and the antisymmetric state,

$$\psi_a = \frac{1}{\sqrt{2}} \left(\psi_{200} - \psi_{210} \right), \tag{14.70}$$

is raised by this amount. There are higher order corrections to the energy and wave functions of the $n = 2$ state that arise from couplings to states *outside* the $n = 2$ manifold. These can be calculated using nondegenerate perturbation theory with the eigenfunctions ψ_s and ψ_a.

The mixing of different parity states of hydrogen produced by an external electric field is important in problems involving spontaneous emission. The ψ_{200} state is metastable, living for about 1/7 of a second, since it cannot decay to the ground state via a dipole transition. However, if an electric field mixes in some of the ψ_{210} state, the transition becomes electric-dipole allowed and the atom decays rapidly, on the order on nanoseconds. For atoms other than hydrogen, there is no "accidental" degeneracy of states having different parity and there is no linear Stark shift, only a quadratic Stark shift that can be calculated using second order perturbation theory.

14.3 Summary

We have seen that it is possible to develop a systematic approach for approximating the eigenenergies and eigenfunctions of a quantum system for which the Hamiltonian can be written as the sum of an exactly solvable part and a "small" perturbation. Different techniques are required when there is degeneracy in the unperturbed energy levels; effectively, part of the Hamiltonian corresponding to coupled degenerate states must be diagonalized exactly. The perturbation theory outlined in this chapter can be applied to a wide range of problems in quantum mechanics. However there are other classes of problems where approximate solutions can still be obtained even though the Hamiltonian cannot be written as the sum of an exactly solvable part and a "small" perturbation. It is to such cases that I now turn my attention.

14.4 Problems

1–2. To see the effect of the normalization term $\alpha_n =^{(0)} \langle E_n | E_n \rangle^{(1)}$ on the expectation value of operators, suppose there is a two-state quantum system whose eigenkets in the absence of any perturbations are $|1\rangle^{(0)}$ and $|2\rangle^{(0)}$. Moreover, assume that there is a perturbation having matrix elements $H'_{21} = H'_{12}$ such that, to first order in H'_{21},

$$
\begin{aligned}
|1\rangle &\approx (1+\alpha_1)\,|1\rangle^{(0)} - \beta\,|2\rangle^{(0)}\,;\\
|2\rangle &\approx (1+\alpha_2)\,|2\rangle^{(0)} + \beta\,|1\rangle^{(0)}\,,
\end{aligned}
$$

where $\beta = H'_{12}/\left(E_2^{(0)} - E_1^{(0)}\right)$. To simplify matters, neglect all terms of order β^2 in this problem (first order perturbation theory) and take α_1, α_2, as real (β is assumed real since $H'_{21} = H'_{12}$ and $\hat{H}'$ is Hermitian). Take the state vector for the quantum system as $|\psi\rangle = |1\rangle^{(0)}$ and assume that there is a Hermitian operator $\hat{A}$ associated with this system having matrix elements in the *unperturbed* basis given by A_{ij} ($i = 1,2; j = 1,2$). Clearly the expectation value of $\hat{A}$ in the state $|\psi\rangle$ is equal to A_{11}.

(a) Expand the state vector $|\psi\rangle$ as

$$
|\psi\rangle = b_1\,|1\rangle + b_2\,|2\rangle
$$

and evaluate b_1 and b_2 in terms of α_1, α_2, and β to first order in β. This shows you that the new state vectors depend on the choice of α_1 and α_2.

(b) Evaluate $\langle 1 | 1 \rangle$ and $\langle 2 | 2 \rangle$ in terms of α_1, α_2, and β to first order in β.

(c) Finally calculate

$$
\langle \psi | \hat{A} | \psi \rangle = \frac{\sum_{i,j=1,2} b_i b_j^* \langle j | \hat{A} | i \rangle}{\sum_{i=1,2} |b_i|^2 \langle i | i \rangle}
$$

and show that it is equal to A_{11}. In other words, the expectation values of operators in a given quantum state cannot depend on the choice of α_1 and α_2.

3. Show that in second order perturbation theory, the value of $E_n^{(2)}$ is independent of $\alpha_n =^{(0)} \langle E_n | E_n \rangle^{(1)}$.

4–5. Use perturbation theory to find the approximate eigenvalues (correct to order 0.01) and eigenvectors (correct to order 0.1) for the Hamiltonian

$$\underset{\sim}{\mathrm{H}} = \begin{pmatrix} 2.1 & 0.1 & -0.3 & 0.2 \\ 0.1 & 5 & 0.15 & -0.1 \\ -0.3 & 0.15 & 9 & 0 \\ 0.2 & -0.1 & 0 & 11.9 \end{pmatrix}.$$

Compare your answer with the exact results obtained from a computer solution. For example, you can use the Eigenvalues and Eigenvectors operations in Mathematica. Mathematica also has an operation *Orthogonalize* that will orthogonalize the vectors for you [e.g., Orthogonalize[Eigenvectors[{{2.1,0.1,–0.3,0.2},{0.1,5,0.15,–0.1}, {–0.3,0.15,9,0},{0.2,–0.1,0,11.9}}]]—putting the decimal point in will give you numerical values].

6–7. Consider a particle having mass m moving in a one-dimensional infinite potential well located between $-a/2$ and $a/2$, subject to the perturbation

$$H'(x) = \begin{cases} V_0 & -b/2 \le x \le b/2 \\ 0 & \text{otherwise} \end{cases},$$

given in Eq. (14.26). Under what conditions are the perturbation solutions given in Eqs. (14.28), (14.30), and (14.31) valid? Take $V_0/E_1^{(0)} = 0.1$ and $b/a = 0.8$ and plot the *change* in the (dimensionless) eigenfunctions defined by

$$\sqrt{a/2}\delta\psi_n(x) = \sqrt{a/2}\left[\psi_n(x) - \psi_n^{(0)}(x)\right],$$

for $n = 1, 2$ as a function of x/a, where $\psi_n(x)$ is given by Eqs. (14.30), and (14.31). Show that both states are pushed away from the origin, but the symmetric state more than the antisymmetric one. Why is this so?

Now obtain *exact* equations from which the eigenenergy of the ground state can be calculated. Graphically, find the change in $k_1 a$,

$$\delta(k_1 a) = k_1(exact)a - \pi,$$

for $V_0/E_1^{(0)} = 0.1$ and $b/a = 0.8$. Compare your result with the perturbative solution that can be obtained from Eq. (14.28). [Hint: Use the fact that the ground state has even parity and choose a wave function that builds in the boundary condition at $x = a$.]

8–9. Consider a one-dimensional harmonic oscillator having $\hat{H}_0 = \frac{\hat{\eta}^2}{2} + \frac{\hat{\xi}^2}{2}$ in dimensionless variables subjected to a perturbation $\hat{H}' = b\hat{\xi}$ with $b \ll 1$. The total

Hamiltonian is $\hat{H} = \frac{\hat{\eta}^2}{2} + \frac{\hat{\xi}^2}{2} + b\hat{\xi}$. To what simple physical system does this system correspond? Classically what is the motion of a particle governed by this Hamiltonian?

Using perturbation theory, calculate the new eigenenergies to second order in b and the new wave functions to first order in b. Now solve the problem exactly and compare the results with perturbation theory (you may have to use some of the recursion relations given in Chap. 7). Do you think higher order perturbation theory will converge to the exact values?

Using the fact that the exact solution for $\left\langle\hat{\xi}\right\rangle$ as a function of time is the same as that of classical mechanics, obtain an expression for $\left\langle\hat{\xi}\right\rangle$ as a function of time, given that $\left\langle\hat{\xi}\right\rangle = \langle\hat{\eta}\rangle = 0$ at $t = 0$ and show that the solution depends linearly on b. What does this imply about all higher contributions to $\left\langle\hat{\xi}\right\rangle$ using perturbation theory?

10–11. A one-dimensional harmonic oscillator having $\hat{H}_0 = \frac{\hat{\eta}^2}{2} + \frac{\hat{\xi}^2}{2}$ in dimensionless variables is subjected to a perturbation $\hat{H}' = \frac{b\hat{\xi}^2}{2}$ with $b \ll 1$. Using perturbation theory, calculate the new eigenenergies to second order in b and the new wave functions to first order in b. Now solve the problem exactly and compare the results with perturbation theory. Do you think higher order perturbation theory will converge to the exact values?

12–13. Given the matrix

$$\begin{pmatrix} 2 & 0.1 & -0.3 & 0.2 \\ 0.1 & 2 & 0.15 & -0.1 \\ -0.3 & 0.15 & 6 & 0 \\ 0.2 & -0.1 & 0 & 10 \end{pmatrix}.$$

Use degenerate perturbation theory to find the eigenvalues correct to order 0.01 and the eigenvectors correct to order 0.1. Check your answer using a computer solution.

14. Consider the simple harmonic oscillator in two dimensions. The eigenfunctions can be written as $\psi_{n_x n_y}(x, y) = \psi_{n_x}(x)\psi_{n_y}(y)$, where $\psi_{n_x}(x)$ and $\psi_{n_y}(y)$ are eigenfunctions for the one-dimensional simple harmonic oscillator. The corresponding eigenenergies are $E_{n_x n_y} = \hbar\omega(n_x + n_y + 1)$. The ground state is nondegenerate, but the first excited state is two-fold degenerate. Suppose the oscillator is subject to a perturbation $\hat{H}' = g\hat{x}\hat{y}$, where g is a constant. Find corrections to the energies of the first excited states using degenerate perturbation theory and find the wave functions for the first excited states correct to *zero* order in g. Even though you diagonalized the matrix exactly for the first excited states, why is your result for the energies not exact? Without doing any detailed calculations, estimate the error in the energies.

15. Prove that

$$\langle 210|\,\widehat{r\cos\theta}\,|200\rangle = -3a_0.$$

16. For the linear Stark effect in hydrogen, what external field strength is needed to produce a frequency splitting of 1.0 GHz (equal to the Lamb shift) between the $2S$ and $2P$ levels (in this problem, neglect the Lamb shift and assume the levels are degenerate in the absence of the field). Is this a large or small field compared with the breakdown voltage of air ($\approx 10^6$ V/m)? Recall that the Stark coupling matrix element is $-3ea_0\mathcal{E}$.

17–18. In a "real" hydrogen atom, the $2P$ level decays via *spontaneous emission* at a rate $\Gamma \approx 6\times 10^8\ \text{s}^{-1}$ (corresponding to a lifetime of about 1.6 ns) while the $2S$ level is metastable and its decay can be neglected in this problem (the lifetime of the $2S$ level is about $1/7$ s). In some experiments, hydrogen atoms are prepared in their $2S$ levels and a "quenching" Stark field is applied to measure the $2S$ level population by observing the radiation emitted from the $2P$ levels back to the ground state. Assume that a hydrogen atom is prepared in the $2S$ level and a Stark field is applied along the z axis at $t = 0$ such that the Stark coupling matrix element, $-3ea_0\mathcal{E}$, is equal to $10\hbar\Gamma$. Neglect spin and the Lamb shift (which modifies this result). Decay adds a term $-(\Gamma/2)\,b_{2P}$ to the equation for the time derivative $\dot{b}_{2P}$ of the $2P$ state amplitude. With this knowledge, proceed as you should in any quantum problem. Use the approximate correct eigenkets for the $n = 2$ manifold,

$$|a,s\rangle = \frac{1}{\sqrt{2}}\left(|200\rangle \pm |210\rangle\right),$$

and obtain equations for $\dot{b}_{a,s}$. Show that these equations are

$$\dot{b}_s(t) = i\epsilon b_s(t) - \frac{\Gamma}{4}\left[b_s(t) - b_a(t)\right]$$

$$\dot{b}_a(t) = -i\epsilon b_a(t) + \frac{\Gamma}{4}\left[b_s(t) - b_a(t)\right],$$

where $\epsilon = 10\Gamma$. In this problem the unperturbed energy of the $n = 2$ manifold has been set arbitrarily to zero since it does not enter into the solution.

Given the initial condition, solve these equations (you can use a program such as DSolve in Mathematica or solve them on your own) and plot the total excited state population, $|b_a|^2 + |b_s|^2$, as a function of Γt to show that it exhibits oscillations. These are not the *quantum beats* of quantum optics. In a standard quantum beat experiment the *total* upper state population does not oscillate, whereas the total radiated emission does exhibit such beats owing to interference in the radiation emitted from each state. In this problem the oscillations are caused by the coupling of the symmetric and antisymmetric states via spontaneous emission.

Chapter 15
Variational Approach

The variational approach is a powerful technique for obtaining ground state (and, to a lesser extent, excited state) energies. The idea behind the method is simple—guess a trial wave function. That is, choose a trial function

$$\psi_t(\mathbf{r}) \equiv \psi_t(\mathbf{r}, \alpha_1, \alpha_2, \ldots \alpha_N), \tag{15.1}$$

with as many free parameters $(\alpha_1, \alpha_2, \ldots \alpha_N)$ as you wish and use it to calculate the ground state energy. The trial function should be consistent with the boundary conditions of the problem. In that case, $\psi_t(\mathbf{r})$ can be expanded in terms of the *exact* eigenfunctions of a Hamiltonian $\hat{H}$ as

$$\psi_t(\mathbf{r}) = \sum_n c_n \psi_n(\mathbf{r}). \tag{15.2}$$

Now form

$$\begin{aligned} E &= \frac{\int \psi_t^*(\mathbf{r}) \hat{H} \psi_t(\mathbf{r}) d\mathbf{r}}{\int \psi_t^*(\mathbf{r}) \psi_t(\mathbf{r}) d\mathbf{r}} \\ &= \frac{\sum_{n,n'} \int \psi_{n'}^*(\mathbf{r}) \hat{H} \psi_n(\mathbf{r}) c_{n'}^* c_n d\mathbf{r}}{\sum_{n,n'} \int \psi_{n'}^*(\mathbf{r}) \psi_n(\mathbf{r}) c_{n'}^* c_n d\mathbf{r}} \\ &= \frac{\sum_n E_n |c_n|^2}{\sum_n |c_n|^2} \geq E_0, \end{aligned} \tag{15.3}$$

where E_0 is the ground state energy [i.e., the sum starts from $n = 0$)]. The denominator can be set equal to unity if the trial wave function is normalized. By varying the parameters $\alpha_1, \alpha_2, \ldots \alpha_N$ such that E is a minimum, you can get an *upper* bound on the ground state energy of the system since $E_0 \leq E$. If you know something about the wave functions and can choose a wave function that is

P.R. Berman, *Introductory Quantum Mechanics*, UNITEXT for Physics,
https://doi.org/10.1007/978-3-319-68598-4_15

orthogonal to the ground state wave function, then you can also use this method to estimate the energy of the first excited state. For example, if the ground state is an $\ell = 0$ state, you can choose a trial wave function for the first excited state that corresponds to a $\ell = 1$ state, implying that $c_0 = 0$. Most of the time the variational method is used to obtain ground state energies.

The advantage of the variational approach over perturbation theory is that the exact Hamiltonian is used. That is, sometimes the Hamiltonian cannot be written as $\hat{H} = \hat{H}_0 + \hat{H}'$ in any obvious fashion. The disadvantage of the method is that it depends on how good your guess is. There is no simple way to estimate its accuracy. Generally speaking you keep adding parameters to your trial wave function and see if the energy converges.

Another advantage of the variational approach is that if you are a fairly good guesser, you are rewarded with an excellent approximation to the energy. Imagine you guess

$$\psi_t(\mathbf{r}) = \psi_n(\mathbf{r}) + \delta\psi_n(\mathbf{r}), \tag{15.4}$$

where $\psi_n(\mathbf{r})$ is the exact wave function. Then

$$\begin{aligned} E &= \frac{\int \psi_t^*(\mathbf{r})\hat{H}\psi_t(\mathbf{r})d\mathbf{r}}{\int \psi_t^*(\mathbf{r})\psi_t(\mathbf{r})d\mathbf{r}} \\ &= \frac{\int [\psi_n(\mathbf{r}) + \delta\psi_n(\mathbf{r})]^* \hat{H} [\psi_n(\mathbf{r}) + \delta\psi_n(\mathbf{r})] \, d\mathbf{r}}{\int [\psi_n(\mathbf{r}) + \delta\psi(\mathbf{r})]^* [\psi_n(\mathbf{r}) + \delta\psi_n(\mathbf{r})] \, d\mathbf{r}} \\ &= \frac{E_n + E_n \int \{\psi_n^*(\mathbf{r})\delta\psi_n(\mathbf{r}) + [\delta\psi_n(\mathbf{r})]^* \psi_n(\mathbf{r})\} \, d\mathbf{r}}{1 + \int \{\psi_n^*(\mathbf{r})\delta\psi_n(\mathbf{r}) + [\delta\psi_n(\mathbf{r})]^* \psi_n(\mathbf{r})\} \, d\mathbf{r}} + O\left([\delta\psi_n(\mathbf{r})]^2\right) \\ &= E_n + O\left([\delta\psi_n(\mathbf{r})]^2\right); \end{aligned} \tag{15.5}$$

that is, the error is of order $[\delta\psi_n(\mathbf{r})]^2$.

Of course there may be no way you can guess the wave function. However, if the Hamiltonian can be written as

$$\hat{H} = \hat{H}_0 + \hat{H}', \tag{15.6}$$

where $\hat{H}'$ is a small perturbation to $\hat{H}_0$, you can use the wave function obtained from perturbation theory that is correct to second order in $\hat{H}'$ as a trial function to calculate the energy correct to order $\hat{H}'^4$.

I will give two examples of calculations using the variational approach. First, suppose that

$$\hat{H} = \frac{\hbar\omega}{2}\left(-\frac{d^2}{d\xi^2} + \xi^2\right), \tag{15.7}$$

where

$$\xi = \sqrt{\frac{m\omega}{\hbar}}x \tag{15.8}$$

is a dimensionless variable. I guess

$$\psi_t = e^{-\alpha\xi^2} \tag{15.9}$$

since I know the ground state must have even parity. Then I can use

$$\frac{\int_{-\infty}^{\infty} e^{-\alpha\xi^2}\xi^2 e^{-\alpha\xi^2} d\xi}{\int_{-\infty}^{\infty} e^{-2\alpha\xi^2} d\xi} = \frac{1}{4\alpha} \tag{15.10}$$

and calculate

$$\begin{aligned}
E &= \frac{\int_{-\infty}^{\infty} \psi_t^*(x)\hat{H}\psi_t(x)dx}{\int_{-\infty}^{\infty} \psi_t^*(x)\psi_t(x)dx} \\
&= \frac{\frac{\hbar\omega}{2}\int_{-\infty}^{\infty} e^{-\alpha\xi^2}\left(-\frac{d^2}{d\xi^2} + \xi^2\right) e^{-\alpha\xi^2} d\xi}{\int_{-\infty}^{\infty} e^{-2\alpha\xi^2} d\xi} \\
&= \frac{\frac{\hbar\omega}{2}\int_{-\infty}^{\infty} e^{-\alpha\xi^2}\left(2\alpha - 4\alpha^2\xi^2 + \xi^2\right) e^{-\alpha\xi^2} d\xi}{\int_{-\infty}^{\infty} e^{-2\alpha\xi^2} d\xi} \\
&= \hbar\omega\left(\frac{\alpha}{2} + \frac{1}{8\alpha}\right) \geq E_0.
\end{aligned} \tag{15.11}$$

I minimize this expression with respect to α,

$$\frac{dE}{d\alpha} = \hbar\omega\left(\frac{1}{2} - \frac{1}{8\alpha^2}\right) = 0; \tag{15.12a}$$

$$\alpha = 1/2. \tag{15.12b}$$

For this value of α,

$$\psi_t = e^{-\xi^2/2} \tag{15.13a}$$

$$E = \hbar\omega/2 \geq E_0. \tag{15.13b}$$

I ended up with the exact wave function and the exact ground state energy since I made a perfect guess! On the other hand, the results associated with some other guesses are

$$\psi_t = e^{-\beta\xi^4}; \quad \beta = 1/6; \quad E = 0.585\hbar\omega; \tag{15.14a}$$

$$\psi_t = \operatorname{sech}(\beta\xi); \quad \beta = \sqrt{\frac{\pi}{2}} = 1.2533; \quad E = 0.524\hbar\omega; \tag{15.14b}$$

$$\psi_t = \frac{1}{\beta^2 + \xi^2}; \quad \beta = \frac{1}{2^{1/4}} = 0.841; \quad E = 0.707\hbar\omega, \tag{15.14c}$$

where the value of β in each case is chosen to minimize the energy. Thus, how well I do depends on my guess. There is no easy way to estimate the error.

As a second example, I consider

$$\hat{H} = -\frac{\hbar^2}{2m}\frac{d^2}{dx^2} + mg\,|x| \tag{15.15}$$

and use a trial wave function

$$\psi_t = \left(\frac{2\alpha}{\pi}\right)^{1/4} e^{-\alpha x^2}, \tag{15.16}$$

which is normalized. Then

$$\begin{aligned} E &= \int_{-\infty}^{\infty} \psi_t^*(x)\hat{H}\psi_t(x)dx \\ &= \left(\frac{2\alpha}{\pi}\right)^{1/2} \int_{-\infty}^{\infty} e^{-\alpha x^2}\left(-\frac{\hbar^2}{2m}\frac{d^2}{dx^2} + mg\,|x|\right) e^{-\alpha x^2} dx \\ &= \frac{\hbar^2\alpha}{2m} + \frac{mg}{\sqrt{2\pi\alpha}} \geq E_0. \end{aligned} \tag{15.17}$$

Setting

$$\frac{dE}{d\alpha} = \left(\frac{\hbar^2}{2m} - \frac{mg}{2\alpha\sqrt{2\pi\alpha}}\right) = 0; \tag{15.18a}$$

$$\alpha = \left(\frac{m^2 g}{\sqrt{2\pi}\hbar^2}\right)^{2/3}, \tag{15.18b}$$

which leads to

$$E = \frac{3}{2\,(2\pi)^{1/3}}\left(mg^2\hbar^2\right)^{1/3} = 0.813\left(mg^2\hbar^2\right)^{1/3} \geq E_0, \tag{15.19}$$

whereas the exact eigenvalue (obtained from solving the Schrödinger equation in terms of Airy functions) is $E_0 = 0.809\left(mg^2\hbar^2\right)^{1/3}$. The calculated upper bound for the ground state energy is almost equal to the exact eigenenergy, since the Gaussian trial function is a good approximation to the exact ground state wave function.

15.1 Combining Perturbation Theory and the Variational Approach

The variational approach gives an upper bound for the ground state energy of a quantum system. It is sometimes possible to use conventional perturbation theory to get a *lower* bound for the ground state energy. In this way you can get both upper and lower bounds for the energy. Suppose you have a quantum system for which the Hamiltonian is

$$\hat{H} = \hat{H}_0 + \hat{H}'. \tag{15.20}$$

It is assumed that you know the eigenenergies and eigenfunctions of $\hat{H}_0$ and that $\hat{H}'$ can be considered as a perturbation. I denote the ground state energy by $E_g^{(0)}$ and the change in the ground state energy resulting from the perturbation by ΔE_g. Moreover I assume that the first order correction resulting from the perturbation vanishes, $H'_{gg} = 0$. Then the change in the ground state energy, calculated using perturbation theory, is given approximately by

$$\begin{aligned}\Delta E_g &= \sum\nolimits'_{n \neq g} \frac{\left|H'_{gn}\right|^2}{E_g^{(0)} - E_n^{(0)}} \geq \frac{1}{E_g^{(0)} - E_1^{(0)}} \sum\nolimits'_{n \neq g} \left|H'_{gn}\right|^2 \\ &= \frac{1}{E_g^{(0)} - E_1^{(0)}} \left[\left(\hat{H}'\right)^2\right]_{gg}.\end{aligned} \tag{15.21}$$

where $E_1^{(0)}$ is the energy of the lowest energy state for which $H'_{g1} \neq 0$ (this choice leads to the most restrictive lower bound). The completeness relation, $\sum_n \left|H'_{gn}\right|^2 = 1$, along with the fact that $H'_{gg} = 0$, was used in the last step of the derivation. Recall that the prime on the summation symbol is a shorthand notation for a sum over bound states and an integral over continuum states. To illustrate the technique, I calculate both upper and lower bounds for the van der Waals interaction energy between two ground state hydrogen atoms.

The Hamiltonian in this case is given by Eq. (15.20) with

$$\hat{H}_0 = -\frac{\hbar^2 \nabla_1^2}{2m_e} - \frac{e^2}{4\pi\epsilon_0 r_1} - \frac{\hbar^2 \nabla_2^2}{2m_e} - \frac{e^2}{4\pi\epsilon_0 r_2} \tag{15.22}$$

and

$$H' = -e^2 \frac{2z_1 z_2 - x_1 x_2 - y_1 y_2}{4\pi\epsilon_0 R^3} \tag{15.23}$$

(even though H' is an operator, I leave the "hat" off of it since all calculations are carried out in the coordinate representation where the operators in $\hat{H}'$ are replaced by their functional values). The ground state has $n_1 = n_2 = 1$ and the lowest energy state for which $H'_{gn} \neq 0$ has $n_1 = n_2 = 2$.

The perturbation theory calculation is virtually identical to that in Chap. 14 with

$$E_g^{(0)} = E_{1,1}^{(0)} = -\frac{e^2}{4\pi\epsilon_0 a_0} \tag{15.24a}$$

$$E_1^{(0)} = E_{2,2}^{(0)} = -\frac{e^2}{4\pi\epsilon_0 a_0}\frac{1}{4}, \tag{15.24b}$$

where $E_{n_1,n_2}^{(0)}$ is the unperturbed energy for one atom having quantum number n_1 and the other n_2. The only difference from Chap. 14 is that I need to replace $1/\left(E_{1,1}^{(0)}\right)$ appearing in Eq. (14.41) by $1/\left(E_{1,1}^{(0)} - E_{2,2}^{(0)}\right)$, which multiplies the final result, Eq. (14.44), by a factor 4/3, namely

$$\Delta E_g = \Delta E_{1,1}^{(2)} \geq -\frac{8a_0^5 e^2}{4\pi\epsilon_0 R^6}. \tag{15.25}$$

Using a variational approach, I assume an (unnormalized) wave function of the form

$$\psi_t(\mathbf{r}_1, \mathbf{r}_2) = \psi_g(r_1)\psi_g(r_2)\left(1 + \beta H'\right) \tag{15.26}$$

where $\psi_g(r) = \psi_{100}(r)$ is the ground state wave function of an isolated hydrogen atom and H' is the perturbation energy, which is a function of $\mathbf{r}_1$ and $\mathbf{r}_2$. From the variational principle I know that

$$E_g + \Delta E_g \leq \frac{\int \psi_t^*(\mathbf{r}_1, \mathbf{r}_2)\left(\hat{H}_0 + H'\right)\psi_t(\mathbf{r}_1, \mathbf{r}_2)d\mathbf{r}_1 d\mathbf{r}_2}{\int \psi_t^*(\mathbf{r}_1, \mathbf{r}_2)\psi_t(\mathbf{r}_1, \mathbf{r}_2)d\mathbf{r}_1 d\mathbf{r}_2}. \tag{15.27}$$

I substitute Eqs. (15.22) and (15.23) into Eq. (15.27) and use the fact that

$$\int \left|\psi_g(r_1)\psi_g(r_2)\right|^2 H' d\mathbf{r}_1 d\mathbf{r}_2 = 0 \tag{15.28}$$

and that the unperturbed ground state wave functions are real to obtain

$$\begin{aligned}
E_g &= E_g^{(0)} + \Delta E_g \\
&\leq \frac{\int \left(1 + \beta H'\right)\psi_g(r_1)\psi_g(r_2)\left(\hat{H}_0 + H'\right)\psi_g(r_1)\psi_g(r_2)\left(1 + \beta H'\right)d\mathbf{r}_1 d\mathbf{r}_2}{\int \left|\psi_g(r_1)\psi_g(r_2)\right|^2 (1 + \beta H')^2 d\mathbf{r}_1 d\mathbf{r}_2} \\
&= \frac{E_g^{(0)} + \int \psi_g(r_1)\psi_g(r_2)\left[2\beta H'^2 + \beta^2 H'\hat{H}_0 H'\right]\psi_g(r_1)\psi_g(r_2)d\mathbf{r}_1 d\mathbf{r}_2}{\int \left|\psi_g(r_1)\psi_g(r_2)\right|^2 \left(1 + \beta^2 H'^2\right)d\mathbf{r}_1 d\mathbf{r}_2}.
\end{aligned} \tag{15.29}$$

keeping terms up to order H'^2 only. The contributions to

$$\int \psi_g(r_1)\psi_g(r_2)\left[\beta^2 H'\hat{H}_0 H'\right]\psi_g(r_1)\psi_g(r_2)d\mathbf{r}_1 d\mathbf{r}_2, \tag{15.30}$$

from all the cross terms in the integrand (that is, terms varying as $z_1z_2x_1x_2$, $z_1z_2y_1y_2$, $x_1x_2y_1y_2$) vanish. Thus, Eq. (15.30) consists of a sum of three terms and each term contains a factor of the form

$$\int \psi_g(r)r_\alpha\left(-\frac{\hbar^2\nabla^2}{2m_e}-\frac{e^2}{4\pi\epsilon_0 r}\right)\left[r_\alpha\psi_g(r)\right]d\mathbf{r}, \tag{15.31}$$

where $r_\alpha = x, y$, or z. It is possible to show that all such integrals vanish as well. For example,

$$\begin{aligned}
&\int \psi_g(r)z\left(-\frac{\hbar^2\nabla^2}{2m_e}-\frac{e^2}{4\pi\epsilon_0 r}\right)\left[z\psi_g(r)\right]d\mathbf{r}\\
&=\frac{1}{\pi a_0^3}\int r\cos\theta e^{-r/a_0}\left(-\frac{\hbar^2\nabla^2}{2m_e}-\frac{e^2}{4\pi\epsilon_0 r}\right)\left[r\cos\theta e^{-r/a_0}\right]d\mathbf{r}\\
&=\frac{2}{a_0^3}\int_0^\infty r^2dr\int_0^\pi \sin\theta d\theta e^{-r/a_0}r\cos\theta\\
&\quad\times\left[-\frac{\hbar^2\cos\theta}{2m_e}\left(\frac{r-4a_0}{a_0^2}\right)e^{-r/a_0}-\frac{e^2\cos\theta}{4\pi\epsilon_0}e^{-r/a_0}\right]\\
&=\left(\frac{\hbar^2}{2m_e}-\frac{e^2a_0}{8\pi\epsilon_0}\right)=0.
\end{aligned} \tag{15.32}$$

As a consequence, Eq. (15.29) reduces to

$$\begin{aligned}
E_g^{(0)}+\Delta E_g &\le \frac{E_g^{(0)}+2\beta\int\psi_g(r_1)\psi_g(r_2)H'^2\psi_g(r_1)\psi_g(r_2)d\mathbf{r}_1d\mathbf{r}_2}{1+\beta^2\int\left|\psi_g(r_1)\psi_g(r_2)\right|^2H'^2d\mathbf{r}_1d\mathbf{r}_2}\\
&\approx E_g^{(0)}(1-\beta^2K)+2\beta K,
\end{aligned} \tag{15.33}$$

where

$$K=\int\psi_g(r_1)\psi_g(r_2)H'^2\psi_g(r_1)\psi_g(r_2)d\mathbf{r}_1d\mathbf{r}_2=\frac{6a_0^6\left(E_g^{(0)}\right)^2}{R^6}. \tag{15.34}$$

Minimizing with respect to β, I find

$$\beta = 1/E_g^{(0)}, \tag{15.35}$$

such that

$$\Delta E_g \leq K/E_g^{(0)} = \frac{6a_0^6 E_g^{(0)}}{R^6} = -\frac{6e^2 a_0^5}{4\pi\epsilon_0 R^6}. \tag{15.36}$$

The perturbation and variational results can be combined to provide both upper and lower limits for the shift,

$$-\frac{8e^2 a_0^5}{4\pi\epsilon_0 R^6} \leq \Delta E_g \leq -\frac{6e^2 a_0^5}{4\pi\epsilon_0 R^6}. \tag{15.37}$$

The result from a multi-parameter variational approach is $\Delta E \approx -\frac{6.50e^2 a_0^5}{4\pi\epsilon_0 R^6}$.[1]

15.2 Helium Atom

The Hamiltonian for the helium atom is

$$\hat{H} = \frac{\hat{p}_1^2}{2m_e} + \frac{\hat{p}_2^2}{2m_e} - \frac{e^2}{4\pi\epsilon_0}\left(\frac{2}{r_1} + \frac{2}{r_2} - \frac{1}{|\mathbf{r}_2 - \mathbf{r}_1|}\right) \tag{15.38}$$

where the two electrons are denoted by 1 and 2. Although the last term, which represents the electron–electron interaction is not much smaller than the attractive terms, I can still try to treat it by perturbation theory setting

$$\hat{H}_0 = \frac{\hat{p}_1^2}{2m_e} + \frac{\hat{p}_2^2}{2m_e} - \frac{Ze^2}{4\pi\epsilon_0}\left(\frac{1}{r_1} + \frac{1}{r_2}\right); \tag{15.39a}$$

$$H' = \frac{e^2}{4\pi\epsilon_0 |\mathbf{r}_2 - \mathbf{r}_1|}, \tag{15.39b}$$

where $Z = 2$ is the nuclear charge. The Hamiltonian $\hat{H}_0$ is the sum of two Hamiltonians, one for each particle. Each of these Hamiltonians is identical to that of hydrogen if the replacement $e^2 \rightarrow Ze^2 = 2e^2$ is made. This implies that the Bohr radius a_0 (which varies as $1/e^2$) must be replaced by

$$a = a_0/2 \tag{15.40}$$

[1]Linus Pauling and J. Y. Beach, *The van der Waals Interaction of Hydrogen Atoms*, Physical Review **47**, 686–692 (1935).

in the hydrogenic wave functions. As a consequence, the unperturbed ground state wave function is

$$\psi^{(0)}(r_1, r_2) = \psi_g(r_1)\,\psi_g(r_2) = \frac{1}{\pi a^3} e^{-(r_1+r_2)/a} \tag{15.41}$$

and the unperturbed ground state energy is

$$E_g^{(0)} = 2\left(\frac{-Ze^2}{8\pi\epsilon_0 a}\right) = 2\left(\frac{-Ze^2}{4\pi\epsilon_0 a_0}\right) = -108.8\text{ eV} = 8E_0, \tag{15.42}$$

where E_0 is the ground state energy of hydrogen. The ground state energy of helium, calculated using perturbation theory, is

$$E_g \approx 8E_0 + \frac{e^2}{4\pi\epsilon_0}\int d\mathbf{r}_1 \int d\mathbf{r}_2 \frac{1}{\pi^2 a^6} e^{-2(r_1+r_2)/a} \frac{1}{|\mathbf{r}_2 - \mathbf{r}_1|}. \tag{15.43}$$

The integral in Eq. (15.43) can be evaluated by expanding $\frac{1}{|\mathbf{r}_2-\mathbf{r}_1|}$ in terms of spherical harmonics as[2]

$$\frac{1}{|\mathbf{r}_2 - \mathbf{r}_1|} = 4\pi \sum_{\ell=0}^{\infty}\sum_{m=-\ell}^{\ell} \frac{1}{2\ell+1}\frac{r_<^{\ell}}{r_>^{\ell+1}}\left[Y_\ell^m(\theta_1, \phi_1)\right]^* Y_\ell^m(\theta_2, \phi_2), \tag{15.44}$$

where $r_<$ is the lesser of r_1 and r_2 while $r_>$ is the greater of r_1 and r_2. With this expansion, only the $\ell = 0$ term in the sum in Eq. (15.44) contributes to the integrals in Eq. (15.43) owing to the fact that

$$\int d\Omega\, Y_\ell^m(\theta, \phi) = \sqrt{4\pi}\,\delta_{\ell,0}\delta_{m,0}. \tag{15.45}$$

As a consequence, using Eqs. (15.43), (15.44), and (15.40), I find

$$\begin{aligned}
E_g &\approx 8E_0 + \frac{16e^2}{4\pi\epsilon_0 a^6}\int_0^\infty r_1^2 dr_1 \int_0^\infty r_2^2 dr_2 \frac{e^{-2(r_1+r_2)/a}}{r_>} \\
&= 8E_0 + \frac{16e^2}{4\pi\epsilon_0 a^6}\int_0^\infty r_1^2 dr_1 \int_0^{r_1} r_2^2 dr_2 \frac{e^{-2(r_1+r_2)/a}}{r_1} \\
&\quad + \frac{16e^2}{4\pi\epsilon_0 a^6}\int_0^\infty r_1^2 dr_1 \int_{r_1}^\infty r_2^2 dr_2 \frac{e^{-2(r_1+r_2)/a}}{r_2} \\
&= 8E_0 + \frac{16e^2}{4\pi a\epsilon_0}\left(\frac{5}{256} + \frac{5}{256}\right) = E_g^{(0)} + \frac{5}{8}\frac{e^2}{4\pi a\epsilon_0} \\
&= 8E_0 - 2.5E_0 = -74.8\text{ eV}.
\end{aligned} \tag{15.46}$$

[2]See, for example, John David Jackson, *Classical Electrodynamics, Third Edition* (John Wiley and Sons, Inc., New York, 1999) Sect. 3.6.

The experimental result is −78.9 eV so I have done better than might be expected.

I can use the variational method to get an even better estimate. Since each electron is partially shielded from the nucleus by the other electron, I try a (normalized) trial wave function of the form

$$\psi_t(\mathbf{r}_1, \mathbf{r}_2) = \frac{1}{\pi a^3} e^{-(r_1+r_2)/a} \tag{15.47}$$

where

$$a = a_0/Z_{\text{eff}} \tag{15.48}$$

and Z_{eff} is the parameter to be varied. I expect to find $1 < Z_{\text{eff}} < 2$, since Z_{eff} is the shielded nuclear charge seen by each electron. The Hamiltonian (15.38) can be written as $\hat{H} = \hat{H}_1 + \hat{H}_2$, with

$$\hat{H}_1 = \frac{\hat{p}_1^2}{2m_e} + \frac{\hat{p}_2^2}{2m_e} - \frac{e^2}{4\pi\epsilon_0}\left(\frac{Z_{\text{eff}}}{r_1} + \frac{Z_{\text{eff}}}{r_2}\right); \tag{15.49a}$$

$$H_2 = \frac{e^2}{4\pi\epsilon_0 \left|\mathbf{r}_2 - \mathbf{r}_1\right|} - \frac{e^2}{4\pi\epsilon_0}\left(\frac{2 - Z_{\text{eff}}}{r_1} + \frac{2 - Z_{\text{eff}}}{r_2}\right), \tag{15.49b}$$

In this form, the the trial wave function is the ground state eigenfunction of the Hamiltonian $\hat{H}_1$, while H_2 can be considered as a small perturbation. Since $\hat{H}_1$ corresponds to a Hamiltonian in which each electron moves independently in the field of a nucleus having change Z_{eff}, it follows immediately that

$$\iint d^3r_1 d^3r_2 \psi_t^*(\mathbf{r}_1, \mathbf{r}_2) \hat{H}_1 \psi_t(\mathbf{r}_1, \mathbf{r}_2) = -\frac{2Z_{\text{eff}}e^2}{8\pi\epsilon_0 a} = -\frac{2Z_{\text{eff}}^2 e^2}{8\pi\epsilon_0 a_0}. \tag{15.50}$$

Moreover, I have already shown in Eq. (15.46) that

$$\iint d^3r_1 d^3r_2 \psi_t^*(\mathbf{r}_1, \mathbf{r}_2) \frac{e^2}{4\pi\epsilon_0 \left|\mathbf{r}_2 - \mathbf{r}_1\right|} \psi_t(\mathbf{r}_1, \mathbf{r}_2) = \frac{5}{8}\frac{e^2}{4\pi a\epsilon_0} = \frac{5}{8}\frac{Z_{\text{eff}}e^2}{4\pi\epsilon_0 a_0}. \tag{15.51}$$

Thus I need only calculate

$$\begin{aligned} &-\frac{e^2}{4\pi\epsilon_0}\iint d^3r_1 d^3r_2 \psi_t^*(\mathbf{r}_1, \mathbf{r}_2)\left(\frac{2 - Z_{\text{eff}}}{r_1} + \frac{2 - Z_{\text{eff}}}{r_2}\right)\psi_t(\mathbf{r}_1, \mathbf{r}_2) \\ &= -\frac{2e^2\left(2 - Z_{\text{eff}}\right)}{4\pi\epsilon_0 a} = -\frac{e^2 Z_{\text{eff}}\left(2 - Z_{\text{eff}}\right)}{2\pi\epsilon_0 a_0}. \end{aligned} \tag{15.52}$$

Combining Eqs. (15.50)–(15.52), I find

$$\begin{aligned} E &= \iint d^3r_1 d^3r_2 \psi_t^* (\mathbf{r}_1, \mathbf{r}_2) \hat{H} \psi_t (\mathbf{r}_1, \mathbf{r}_2) \\ &= \frac{e^2}{8\pi a_0 \epsilon_0} \left(-2Z_{\text{eff}}^2 + \frac{5}{4} Z_{\text{eff}} - 4Z_{\text{eff}} (2 - Z_{\text{eff}}) \right) \\ &= |E_0| \left(2Z_{\text{eff}}^2 - \frac{27}{4} Z_{\text{eff}} \right). \end{aligned} \tag{15.53}$$

Minimizing this expression with respect to Z_{eff}, I find $Z_{\text{eff}} = 27/16$ and

$$E_g \leq 5.695 E_0 = -77.46 \text{ eV}, \tag{15.54}$$

which is about 2% from the experimental value of −78.9 eV. By writing the Hamiltonian as the sum of the two terms given in Eqs. (15.49), I was able to obtain a good guess for the ground state wave function.

15.3 Summary

The variational method is an extremely powerful method for obtaining the ground state energy of quantum systems. It is the preferred method for approximating the ground state energy of atoms such as helium. In certain cases, the variational method can be combined with perturbation theory to obtain both lower and upper bounds for the ground state energy.

15.4 Problems

1. Use the variational method to estimate the ground state energy E_0 of the potential $V = mgz$, $z > 0$; $V = \infty$, $z < 0$. Use the (normalized) trial wave function

$$\psi_t = \left(\frac{128\alpha^3}{\pi} \right)^{1/4} z e^{-\alpha z^2},$$

which satisfies the boundary condition at $z = 0$. Compare your answer with the value, $E_0 = 2.3381 (\hbar^2 m g^2 / 2)^{1/3}$, obtained from an exact solution of the Schrödinger equation.

2–3. Use the variational method to estimate an upper bound $E_{0\mu}$ to the ground state energy of a particle having mass m moving in a potential $V(x) = V_0 |x/a|^\mu$, using a Gaussian trial wave function, $\psi_t = e^{-x^2/b^2}$. You can assume that μ, V_0, and a are greater than zero and use the fact that

$$\int_0^\infty z^\mu e^{-z^2} dz = \Gamma\left(\frac{1+\mu}{2}\right),$$

where Γ is the gamma function. Show that your result yields $E_{0\mu} = \hbar\omega_0/2$ when $\mu = 2$ and $V_0 = m\omega_0^2/2$. Why does the upper bound give the exact result in this case? Plot E_u/V_0 as a function of μ for $\beta^2 = 2mV_0a^2/\hbar^2 = 1$ and $0 \leq \mu \leq 5$. As $\mu \to \infty$, show that the potential approaches that of an infinite square well having width $2a$, but that the variational upper bound for the energy diverges. This shows that the choice of a Gaussian trial function for large values of μ is not a good choice. Why not? [Note: The exact ground state energies are given by $(E_{\text{exact}}/V_0)\,\beta^{\mu/(\mu+2)} = 1.0188, 1, 1.02295, 1.06036, 1.1023$ for $\mu = 1, 2, 3, 4, 5$.]

4. Repeat the previous calculation to estimate an upper bound $E_{1\mu}$ to the first excited state energy of a particle having mass m moving in the potential $V(x) = V_0\,|x/a|^\mu$, where μ is a positive integer. Now you must take a trial wave function that is consistent with the parity of the first excited state.

5. The Hamiltonian of a quantum system is given by $\hat{H} = \hat{H}_0 + \hat{H}'$, where $\hat{H}'$ can be considered as a perturbation. If the first order perturbation theory contribution to the ground state energy $E_0^{(1)}$ vanishes, prove that the second order correction satisfies

$$E_0^{(2)} \geq \frac{1}{E_0^{(0)} - E_1^{(0)}} \langle 0|\,\hat{H}'^2\,|0\rangle\,.$$

where $E_1^{(0)}$ is the first excited state energy. Use this result to get a lower bound for the Stark shift of the ground state of hydrogen when the atom is put into an external, constant electric field $\mathcal{E}$ along the z axis for which $\hat{H}' = e\mathcal{E}\hat{z}$. It turns out that the second order correction can be evaluated exactly and is given by

$$\Delta E = -\frac{9}{4}\,(4\pi\epsilon_0)\,\mathcal{E}^2 a_0^3.$$

How does your bound compare?

6–7. Now do a variational calculation to get an upper bound for the ground state energy of hydrogen in a static electric field. Since the field couples states with opposite parity, you can try a trial wave function of the form

$$\psi_t = \frac{e^{-r/a_0}}{\sqrt{\pi}a_0^{3/2}}(1 + \beta H'),$$

with $H' = e\mathcal{E}z$, and vary β to get the upper bound. Combine this result with that of Problem 15.5 and show that the exact solution fits between your two bounds.

[Hint: Follow the same procedure used in the text for the variational calculation of the van der Waals energy. You can use the fact that

$$\int \psi_t^* H' \hat{H}_0 \left[H' \psi_t\right] d\mathbf{r} = 0$$

(you can try to prove this if you wish, but it is not required)].

8. Starting with a (normalized) trial wave function $\psi_t = \left(\frac{\alpha}{\pi}\right)^{1/4} e^{-\alpha x^2/2}$ with $\alpha > 0$, use the variational method to prove that, for a one-dimensional potential that is attractive in the sense that

$$\int_{-\infty}^{\infty} dx V(x) < 0$$

and for which $V(x) \rightarrow 0$ as $|x| \rightarrow \infty$, there is always a bound state (that is, an eigenstate with $E < 0$). [Hint: Consider the limit that $\alpha \sim 0$ for the energy associated with the trial wave function.]

9–10. Obtain an upper bound for the ground state energy of a particle having mass m confined to a square well potential having depth V_0 located between $-a/2$ and $a/2$. Use normalized trial wave functions

$$\psi_t^{(1)} = \left(\frac{\alpha}{\pi}\right)^{1/4} e^{-\alpha x^2/2};$$

$$\psi_t^{(2)} = \left(\frac{\alpha}{2}\right)^{1/2} \operatorname{sech}(\alpha x),$$

with $\alpha > 0$. In the limit that $\beta^2 = 2mV_0 a^2/\hbar^2 \ll 1$, show that the upper bound to the energy is always greater than that given in Eq. (6.96). Why does the second trial function provide a better limit for the exact energy?

Chapter 16
WKB Approximation

16.1 WKB Wave Functions

The WKB (Wentzel, Kramers, Brillouin) approximation is covered in most standard quantum mechanics texts. Basically, the WKB approximation is a type of *eikonal* approximation in which it is assumed the potential varies very slowly compared with the average de Broglie wavelength of a particle. In some sense, the quantum particle is acting somewhat like a classical particle. This is not quite the case, since the particle is still described by a wave function. It is like trying to have the best of both the classical and quantum worlds. Generally it is valid for high energies such that the de Broglie wavelength is large compared with distances over which the potential varies significantly.

Since this is a type of *semi-classical* approximation, it makes sense to look for a solution of Schrödinger's equation as a power series in $\hbar$. Terms to zeroth order in $\hbar$ correspond to a classical limit and the higher order terms provide quantum corrections. The WKB method generally works only in 1-D problems, but can be used in problems in 3-D with spherical symmetry, since the radial equation is an equation in one variable. I consider only one-dimensional motion in the x direction in this chapter, but return to the radial equation in discussing scattering theory.

The time independent Schrödinger equation is

$$\psi'' + k^2(x)\psi = 0 \tag{16.1}$$

where

$$k(x) = \sqrt{\frac{2m}{\hbar^2}[E - V(x)]} = p(x)/\hbar \tag{16.2}$$

and primes indicate derivatives with respect to x. For the moment I assume that $E > V(x)$ so that $k(x)$ is real and positive. If k were a constant, then the solution of Eq. (16.1) would be $\psi(x) = e^{\pm ikx}$. This suggests that I try a solution of the form

P.R. Berman, *Introductory Quantum Mechanics*, UNITEXT for Physics,
https://doi.org/10.1007/978-3-319-68598-4_16

$$\psi(x) = e^{iS(x)/\hbar} \tag{16.3}$$

and obtain equations for $S(x)$ that can be solved to give $S(x)$ as a power series in $\hbar$. If I do so, I need retain only those terms varying as $\hbar^0$ or $\hbar$, since higher order terms do not contribute to Eq. (16.3) in the limit that $\hbar \to 0$.

To obtain such a series solution for $S(x)$, I first calculate the derivatives of $\psi(x)$,

$$\psi' = \left[iS'(x)/\hbar\right] e^{iS(x)/\hbar}; \tag{16.4a}$$

$$\psi'' = -\left[S'(x)/\hbar\right]^2 e^{iS(x)/\hbar} + \left[iS''(x)/\hbar\right] e^{iS(x)/\hbar}, \tag{16.4b}$$

and substitute the expression for ψ'' into Eq. (16.1) to arrive at

$$\left[S'(x)\right]^2 = \hbar^2 k^2(x) + i\hbar S''(x). \tag{16.5}$$

Note that

$$p^2(x) = \hbar^2 k^2(x) \tag{16.6}$$

is independent of $\hbar$.

It looks like I haven't accomplished much since I started from a linear differential equation for $\psi(x)$ and ended up with a highly nonlinear differential equation for $S(x)$. The idea is to solve Eq. (16.5) iteratively, assuming that $S(x)$ is slowly varying so that the second derivative term in Eq. (16.5) can be treated as a small correction. This is equivalent to solving for $S(x)$ as a power series in $\hbar$. To lowest order

$$S'(x) = \pm\hbar k(x) = \pm p(x). \tag{16.7}$$

I use this to approximate

$$S''(x) = \pm p'(x), \tag{16.8}$$

and substitute this result back into Eq. (16.5) to obtain

$$\begin{aligned}
\left[S'(x)\right]^2 &= p^2(x) \pm i\hbar p'(x); \\
S'(x) &= \pm p(x)\sqrt{1 \pm i\hbar p'(x)/p^2(x)} \\
&\approx \pm p(x) + i\hbar\frac{p'(x)}{2p(x)} \\
&= \pm\hbar k(x) + i\hbar\frac{k'(x)}{2k(x)},
\end{aligned} \tag{16.9}$$

having used the fact that the sign of the $i\hbar p'(x)$ term is correlated with the sign of the $\pm p(x)$ factor.

The solution of this equation, neglecting integration constants, is

$$\begin{aligned} S(x) &= \pm\hbar \int k(x)dx + i\hbar \int \frac{k'(x)}{2k(x)}dx \\ &= \pm\hbar \int k(x)dx + \frac{i\hbar}{2} \int \frac{d}{dx} \ln[k(x)]\, dx \\ &= \pm\hbar \int k(x)dx + \frac{i\hbar}{2} \ln[k(x)], \end{aligned} \tag{16.10}$$

leading to

$$\begin{aligned} \psi_{WKB}(x) &= e^{iS(x)/\hbar} = \exp\left\{\pm i \int k(x)dx - \frac{1}{2}\ln[k(x)]\right\} \\ &= \frac{C}{\sqrt{k(x)}} \exp\left\{\pm i \int k(x)dx\right\}, \end{aligned} \tag{16.11}$$

where the integration constants have been absorbed into the normalization constant C.

If the energy is less than $V(x)$, this equation is replaced by

$$\psi_{WKB}(x) = \frac{C}{\sqrt{\kappa(x)}} \exp\left\{\pm \int \kappa(x)dx\right\}, \tag{16.12}$$

where

$$\kappa(x) = \sqrt{\frac{2m}{\hbar^2}[V(x) - E]} > 0. \tag{16.13}$$

It is not too difficult to estimate the validity conditions for the WKB approximation. From Eq. (16.10) a necessary condition for the validity of the approach requires the second term be smaller than the first, or

$$\left|\frac{k'(x)}{2k^2(x)}\right| \ll 1; \tag{16.14a}$$

$$\hbar\left|\frac{p'(x)}{2p^2(x)}\right| \ll 1. \tag{16.14b}$$

Since

$$p'(x) = -\frac{\sqrt{2m}}{2}\frac{dV/dx}{\sqrt{E-V}}, \tag{16.15}$$

the validity condition can be written as

$$\hbar\frac{\sqrt{2m}}{4}\left|\frac{dV/dx}{\sqrt{E-V}}\right| \ll \left|p^2(x)\right| = \frac{h^2}{\lambda_{dB}^2(x)};$$

$$\lambda_{dB}(x)\left|\frac{dV}{dx}\right| \ll \sqrt{\frac{|E-V|}{2m}}\frac{8\pi h}{\lambda_{dB}(x)} = 8\pi\frac{\left|p^2(x)\right|}{2m}, \tag{16.16}$$

where $\lambda_{dB}(x) = h/\left|p(x)\right|$. The potential must change slowly compared to the kinetic energy over distances of order of a wavelength for the WKB approximation to be valid.

Equation (16.16) is a necessary condition for the validity of the WKB approximation, but it is not sufficient. If higher order terms in the expansion used to solve Eq. (16.5) are included, they lead to additional phases in the exponent appearing in Eq. (16.3). As I have stressed, when terms appear in an exponent, they must have an absolute value much less than unity for them to be neglected. If you carry out the expansion to next order, you will see that the condition

$$\left[\lambda_{dB}(x)\right]^2\left|\frac{d^2V}{dx^2}\right| \ll \frac{\left|p^2(x)\right|}{2m} \tag{16.17}$$

must also be satisfied.

16.2 Connection Formulas

As long as conditions (16.16) and (16.17) are satisfied, the WKB method can be used to approximate the eigenfunctions for a particle moving in a potential $V(x)$. Let's check to see when we can expect this to be the case for the potentials shown in Fig. 16.1. (a) Assuming that conditions (16.16) and (16.17) hold, then the WKB approximation is valid for all x, and the WKB eigenfunctions for a given energy correspond to particles moving to the right or left. If you construct a wave packet having average energy E greater than the barrier height incident from the left, you will find that there is no reflected wave, the wave packet and simply moves to the right, adjusting its kinetic energy to changes in the barrier height. (b) In this case, the WKB approximation necessarily breaks down at the point discontinuities in the potential since $dV/dx = \infty$ at these points. In the exact quantum problem, there is always a reflected wave. We have seen already that the reflection coefficient is independent of $\hbar$ for step potentials in the high energy limit. In Fig. 16.1c–f, the WKB approximation necessarily fails at the *classical turning points* a and b since the momentum $p(x) = 0$ at such points.

Even when there are classical turning points for a given energy, it is still possible to use the WKB approximation. One proceeds by calculating the WKB wave functions in the regions away from the turning points and piecing together the solutions using the *exact* solutions in the vicinity of the turning points. At each

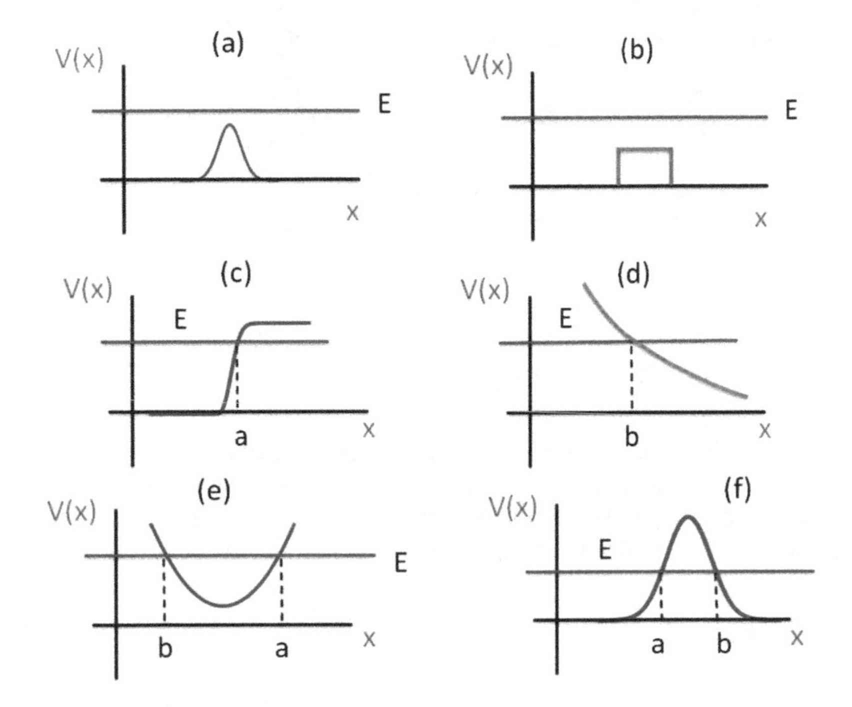

Fig. 16.1 The WKB approximation fails at point discontinuities in the potential and at the classical turning points. The classical turning point a has the classical region to the left and the classical turning point b has the classical region to the right. (**a**) smooth barrier (**b**) square barrier (**c**) classical region to left of turning point (**d**) classical region to right of turning point (**e**) bound states (**f**) barrier with tunneling

classical turning point, the spatially dependent de Broglie wavelength $\lambda_{dB}(x) = h/p(x) = 2\pi/k(x)$ is infinite. It is generally assumed that the potential varies linearly with x in the region of a turning point and that the slope is sufficiently small to insure that the exact solution for a linear potential (which are so-called Airy functions) extends into the region where the WKB wave functions are valid. In this manner, one arrives at a number of *connection formulas* which tell you how to piece together the solutions. I give a derivation of one of the connection formulas in the Appendix; derivations can also be found in most textbooks on quantum mechanics.

The connection formulas must be written for the two possible cases of a turning point with the classically allowed region to the right or to the left of the turning point:

Turning point at $x = b$ (classically allowed region to the right):

$$\begin{aligned}&\frac{A}{\sqrt{\kappa(x)}}\exp\left\{-\int_x^b \kappa(x')dx'\right\} + \frac{B}{\sqrt{\kappa(x)}}\exp\left\{\int_x^b \kappa(x')dx'\right\} \\ &\leftrightarrow \frac{2A}{\sqrt{k(x)}}\cos\left\{\int_b^x k(x')dx' - \frac{\pi}{4}\right\} - \frac{B}{\sqrt{k(x)}}\sin\left\{\int_b^x k(x')dx' - \frac{\pi}{4}\right\}. \quad (16.18)\end{aligned}$$

Turning point at $x = a$ (classically allowed region to the left):

$$\frac{2A}{\sqrt{k(x)}}\cos\left\{\int_x^a k(x')dx' - \frac{\pi}{4}\right\} - \frac{B}{\sqrt{k(x)}}\sin\left\{\int_x^a k(x')dx' - \frac{\pi}{4}\right\}$$
$$\leftrightarrow \frac{A}{\sqrt{\kappa(x)}}\exp\left\{-\int_a^x \kappa(x')dx'\right\} + \frac{B}{\sqrt{\kappa(x)}}\exp\left\{\int_a^x \kappa(x')dx'\right\}. \quad (16.19)$$

16.2.1 Bound State Problems

If you look at Fig. 16.1e, you can deduce that there is a discrete infinity of bound states for this potential. You can estimate the bound state energies using the WKB approximation. The answer should be good for the high energy states. To do so, begin on the left ($x < b$) with an exponentially decreasing function and use

$$\frac{1}{\sqrt{\kappa(x)}}\exp\left\{-\int_x^b \kappa(x)dx\right\} \to \frac{2}{\sqrt{k(x)}}\cos\left\{\int_b^x k(x)dx - \frac{\pi}{4}\right\} \quad (16.20)$$

to connect the solution to the region $b < x < a$. To connect the wave function in the region $b < x < a$ to one in the region $x > a$, I rewrite the right-hand side of Eq. (16.20) as

$$\begin{aligned}
&\frac{2}{\sqrt{k(x)}}\cos\left\{\int_b^x k(x)dx - \frac{\pi}{4}\right\} \\
&= \frac{2}{\sqrt{k(x)}}\cos\left\{-\int_x^a k(x)dx + \int_b^a k(x)dx - \frac{\pi}{4}\right\} \\
&= \frac{2}{\sqrt{k(x)}}\cos\left\{\int_x^a k(x)dx - \int_b^a k(x)dx - \frac{\pi}{4} + \frac{\pi}{2}\right\} \\
&= \frac{2}{\sqrt{k(x)}}\cos\left\{\int_x^a k(x)dx - \frac{\pi}{4}\right\}\cos\left\{\int_b^a k(x)dx - \frac{\pi}{2}\right\} \\
&\quad + \frac{2}{\sqrt{k(x)}}\sin\left\{\int_x^a k(x)dx - \frac{\pi}{4}\right\}\sin\left\{\int_b^a k(x)dx - \frac{\pi}{2}\right\}. \quad (16.21)
\end{aligned}$$

I can now use the connection formula (16.19) to extend the solution to the $x > a$ region; however, in doing so, the sin term leads to an exponentially increasing wave function, which is not physical. The only way to avoid this is to have

$$\sin\left\{\int_b^a k(x)dx - \frac{\pi}{2}\right\} = 0, \quad (16.22)$$

which implies that

$$\int_b^a k(x)dx = \left(n + \frac{1}{2}\right)\pi \tag{16.23}$$

(n is a positive integer or zero) or

$$\int_b^a p(x)dx = \left(n + \frac{1}{2}\right)\pi\hbar; \quad n = 0, 1, 2 \ldots. \tag{16.24}$$

This is the *semiclassical quantization condition*. Note that the WKB approximation breaks down near the turning points, which is the reason that it may not be possible to normalize the entire wave function.

If the potential is infinite for $x < 0$, then you start from the right of the turning point at $x = a$ and work backwards towards the turning point at $x = b = 0$ to obtain

$$\frac{2}{\sqrt{k(x)}}\cos\left\{\int_x^a k(x)dx - \frac{\pi}{4}\right\} \leftarrow \frac{1}{\sqrt{\kappa(x)}}\exp\left\{-\int_a^x \kappa(x)dx\right\}. \tag{16.25}$$

Since the wave function must vanish at $x = 0$, you are led to the requirement that

$$\int_0^a k(x)dx - \frac{\pi}{4} = \left(n + \frac{1}{2}\right)\pi \tag{16.26}$$

or

$$\int_0^a k(x)dx = \left(n + \frac{3}{4}\right)\pi; \quad n = 0, 1, 2 \ldots. \tag{16.27}$$

If the potential is infinite at *both* $x = a$ and $x = b$, this equation is replaced by

$$\int_b^a k(x)dx = (n + 1)\pi; \quad n = 0, 1, 2 \ldots. \tag{16.28}$$

16.3 Examples

Equation (16.24) may not give the correct energies, but it always gives a good idea of how the energy levels scale with n at high energy. For example, consider the potential

$$V(x) = V_0 \left|x/x_0\right|^q = \alpha \left|x\right|^q, \tag{16.29}$$

with $q > 0$, $V_0 > 0$, $x_0 > 0$, and

$$\alpha = V_0/x_0^q. \tag{16.30}$$

The classical energy for a particle having mass m moving in this potential is

$$E = \frac{p^2}{2m} + \alpha\,|x|^q \tag{16.31}$$

and the classical turning points $[p(x) = 0]$ occur at

$$a = -b = (E/\alpha)^{1/q}, \tag{16.32}$$

leading to the quantization condition

$$\begin{aligned}&\int_{-(E/\alpha)^{1/q}}^{(E/\alpha)^{1/q}} \sqrt{2m\left(E - \alpha\,|x|^q\right)}dx \\ &= 2\int_0^{(E/\alpha)^{1/q}} \sqrt{2m\left(E - \alpha\,|x|^q\right)}dx = \left(n + \frac{1}{2}\right)\pi\hbar. \end{aligned} \tag{16.33}$$

The integral is tabulated and one finds

$$\sqrt{2m\pi}\frac{E^{\left(\frac{1}{q}+\frac{1}{2}\right)}}{\alpha^{1/q}}\frac{\Gamma(1+1/q)}{\Gamma(3/2+1/q)} = \left(n + \frac{1}{2}\right)\pi\hbar, \tag{16.34}$$

where Γ is the gamma function. Solving for the energy, I find

$$\frac{E_{q,n}}{V_0} = C_{q,n}\left(\frac{1}{\beta^2}\right)^{\frac{q}{q+2}}, \tag{16.35}$$

where

$$C_{q,n} = \left[\frac{\Gamma(3/2+1/q)\sqrt{\pi}}{\Gamma(1+1/q)}\left(n + \frac{1}{2}\right)\right]^{2q/(q+2)} \tag{16.36a}$$

and

$$\beta^2 = \frac{2mV_0x_0^2}{\hbar^2}. \tag{16.36b}$$

For the harmonic oscillator potential with $q = 2$ and $\alpha = m\omega^2/2$, it follows that $V_0 = m\omega^2x_0^2/2$,

$$\beta^2 = \frac{4V_0^2}{\hbar^2\omega^2}, \tag{16.37}$$

$$C_{2,n} = \frac{\Gamma(2)\sqrt{\pi}}{\Gamma(3/2)}\left(n+\frac{1}{2}\right) = 2\left(n+\frac{1}{2}\right), \tag{16.38}$$

and

$$E_{2,n} = \left(n+\frac{1}{2}\right)\hbar\omega. \tag{16.39}$$

The fact that I got the exact answer is an "accident." In the classically allowed region, the WKB wave function looks nothing like the true wave function for the $n = 0$ state. With increasing n, by choosing the WKB wave functions to agree with the exact wave functions at the origin (with this choice the WKB wave functions are not normalized), you will find that the WKB and exact wave functions are in good agreement as long as you stay away from the turning points.

For arbitrary q, the energies vary as

$$E_{q,n} \sim (n+1/2)^{2q/(q+2)}. \tag{16.40}$$

For $q = 4$, $C_{4,n} = 2.185\,(n+1/2)^{4/3} = 0.867, 3.75, 7.41, 11.6$ for $n = 0, 1, 2, 3$, while the numerical solution of the Schrödinger equation[1] yields $C_{4,n} = 1.06$, 3.80, 7.46, 11.6. Not so great for $n = 0$, but not too bad for $n \geq 1$. As $q \to \infty$, the potential approximates an infinite square well potential having width $2x_0$ and the WKB energies vary as $(n+1/2)^2$. The WKB energies given by Eq. (16.35) of the high-lying n states nearly coincide with the exact results, but the $n = 0$ state energy is off by a factor of 4, since a better quantization condition to use is the one given in Eq. (16.28), rather than that given in Eq. (16.23), because the potential walls approximate those of an infinite potential well. In certain cases, the WKB approximation may provide a lower bound for the ground state energy.[2]

The condition given in Eq. (16.16) for the WKB approximation to be valid reduces to

$$\frac{q}{4\beta}\left|\frac{x}{x_0}\right|^{q-1}\left|C_{q,n}\left(\frac{1}{\beta^2}\right)^{\frac{q}{q+2}} - \left|\frac{x}{x_0}\right|^q\right|^{-3/2} \ll 1. \tag{16.41}$$

[1] I took these values from the lecture notes of Professor Klaus Schulten on the web site http://www.ks.uiuc.edu/Services/Class/PHYS480/.

[2] L.F. Barrágan-Gil and A. Camacho, Modern Physics Letters **22**, 2675–2687 (2007) prove that a lower bound to the ground state energy can be obtained using WKB considerations for potentials that vary as x^q for $x > 0$ and are infinite for $x < 0$, provided $1 < q < 5/2$. However the lower bound they obtain is lower than that which would be calculated using the WKB quantization condition.

It turns out that inequality (16.41) holds over a large range of parameter space. It is violated near $x/x_0 = 0$ if $q \ll 1$ (the derivative of the potential diverges at the origin if $q < 1$) and near the turning points. The worst violations (that is, over the largest range of x/x_0) occur for $n = 0$ and for q of order unity. It is not readily apparent to me how the validity condition for the WKB *wave function* given in Eq. (16.41) translates into one for the accuracy of the *energy levels* calculated using the WKB approximation. That is, although the violation of inequality (16.41) for $n = 0$ decreases with increasing q for fixed β, the relative error in the $n = 0$ energy calculated in the WKB approximation increases with increasing q. As was mentioned previously, this can be associated with using the wrong quantization condition as the potential begins to approximate the infinite square well potential.[3] For fixed q and β, the WKB approximation becomes better with increasing n.

The connection formulas can also be used to calculate the reflection and transmission coefficients for the barrier shown in Fig. 16.1f. You start with the WKB wave function for $x > b$ in the form

$$\frac{2F}{\sqrt{k(x)}} \cos\left\{\int_b^x k(x)dx - \frac{\pi}{4}\right\} - \frac{G}{\sqrt{k(x)}} \sin\left\{\int_b^x k(x)dx - \frac{\pi}{4}\right\} \tag{16.42}$$

and choose G such that this is in the form of a wave moving to the right only, that is, something varying as $T \exp\left\{i\left[\int_b^x k(x)dx - \frac{\pi}{4}\right]\right\}$. You then propagate this solution all the way back to $x < a$ using the connection formulas and write the wave function for $x < a$ in the form of a wave moving to the right plus one moving to the left as

$$A \exp\left\{i\left[\int_x^a k(x)dx - \frac{\pi}{4}\right]\right\} + R \exp\left\{-i\left[\int_x^a k(x)dx - \frac{\pi}{4}\right]\right\}. \tag{16.43}$$

The (intensity) transmission coefficient is then $|T/A|^2$ and the reflection coefficient is $|R/A|^2$. The result for the transmission coefficient is the same as that given by Eq. (6.120) for a square barrier, if the quantity κd is replaced by $\int_a^b \kappa(x)dx$, where d is the width for the square barrier and a and b are the classical turning points. You are asked to prove this in the problems.

16.4 Summary

The WKB approximation is generally referred to as a semi-classical approximation since it is valid in the limit of large energies, where the de Broglie wave length is large over distances in which the potential varies significantly. It always fails at classical turning points, but connection formulas can be used to connect the WKB

[3]See H. Friedtich and J. Trost, *Nonintegral Maslov indices*, Physical Review A **54**, 1136–1145 (1996). I thank R. Shakeshaft for pointing out this reference to me.

wave functions in regions on both sides of the turning point. The WKB method provides a relatively simple way for estimating the energies of all but the lowest energy bound states in one-dimensional problems. The equation that allows you to do this, Eq. (16.24), is essentially the Bohr quantization condition.

16.5 Appendix: Connection Formulas

To illustrate how the connection formulas can be derived, I consider the case where the turning point is at $x = a$ such that the classical region is to the left of the turning point. The basic idea is to expand the potential in the region of the turning point and keep only the term that is linear in $x - a$. That is, in the region of the turning point I approximate

$$V(x) \approx V(a) + \left.\frac{dV}{dx}\right|_{x=a}(x-a), \tag{16.44}$$

such that

$$E - V(x) \approx -\left.\frac{dV}{dx}\right|_{x=a}(x-a), \tag{16.45}$$

since $E = V(a)$. From Fig. 16.1 you can see that the slope is positive at the turning point at $x = a$. The next step is to obtain an exact solution of Schrödinger's equation for this potential and hope that the solution remains valid at distances sufficiently far from $x = a$ to join with the WKB solutions. This will normally be the case if the general validity conditions for the WKB approximation given in Eqs. (16.16) and (16.17) are satisfied, provided the first derivative of the potential at the turning point does not vanish.

In the region of the turning point, Schrödinger's equation is

$$\frac{d^2\psi}{dx^2} - K^2(x)\psi = 0, \tag{16.46}$$

where

$$K^2(x) = \frac{2m}{\hbar^2}\left.\frac{dV}{dx}\right|_{x=a}(x-a) = \begin{cases} -k^2(x) & x < a \\ \kappa^2(x) & x > a \end{cases} \tag{16.47}$$

is positive for $x > a$ and negative for $x < a$. In terms of a dimensionless variable z defined by

$$z = \left[\frac{2m}{\hbar^2}\left.\frac{dV}{dx}\right|_{x=a}\right]^{1/3}(x-a), \tag{16.48}$$

Eq. (16.46) is transformed into

$$\frac{d^2\psi}{dz^2} - z\psi = 0. \tag{16.49}$$

This is a well-known (to those who know it well) differential equation of mathematical physics known as Airy's equation, having independent solutions denoted by Ai(z) and Bi(z) (Mathematica symbols AiryAi[z] and AiryBi[z]). The general solution of Eq. (16.49) is

$$\psi(z) = C_1\mathrm{Ai}(z) + C_2\mathrm{Bi}(z). \tag{16.50}$$

The idea is to use the asymptotic forms of the Airy functions to join the solution for $\psi(z)$ to the WKB solutions in the regions $x < a$ and $x > a$. The WKB validity condition given in Eq. (16.16) corresponds to the requirement that $|z| \gg 1$. To see this, I use Eq. (16.45) and assume that $dV/dx > 0$ is constant in the region near the turning point to write

$$\frac{dV}{dx} = \left|\frac{E - V(x)}{|x-a|}\right| = \frac{|p(x)|^2}{2m\,|x-a|}. \tag{16.51}$$

Condition (16.16) then reduces to

$$\frac{dV}{dx} = \frac{|p(x)|^2}{2m\,|x-a|} \ll \frac{8\pi\,|p(x)|^2}{2m\lambda_{dB}(x)} \tag{16.52}$$

or

$$|x-a| \gg \hbar/4\,|p(x)|, \tag{16.53}$$

which corresponds to

$$|z|^3 = \frac{2m}{\hbar^2}\frac{dV}{dx}|x-a|^3 = \frac{2m}{\hbar^2}\frac{|p(x)|^2\,|x-a|^3}{2m\,|x-a|} = \frac{|p(x)|^2\,|x-a|^2}{\hbar^2} \gg \frac{1}{16}. \tag{16.54}$$

I need the asymptotic forms of Ai(z) and Bi(z) for $|z| \gg 1$.

The needed asymptotic forms are

$$\mathrm{Ai}(z) \sim \frac{1}{2\sqrt{\pi}}z^{-1/4}e^{-y} \quad z \gg 1; \tag{16.55a}$$

$$\mathrm{Ai}(z) \sim \frac{1}{\sqrt{\pi}}(-z)^{-1/4}\cos\left(y - \frac{\pi}{4}\right) \quad z \ll -1; \tag{16.55b}$$

$$\mathrm{Bi}(z) \sim \frac{1}{\sqrt{\pi}} z^{-1/4} e^{y} \qquad z \gg 1; \tag{16.55c}$$

$$\mathrm{Bi}(z) \sim -\frac{1}{\sqrt{\pi}} (-z)^{-1/4} \sin\left(y - \frac{\pi}{4}\right) \qquad z \ll -1, \tag{16.55d}$$

where

$$y = \frac{2}{3} |z|^{3/2}. \tag{16.56}$$

I am now in a position to connect the WKB solution for $x < a$ to that for $x > a$. To simplify matters I write the WKB solution in the form

$$\psi_{WKB}(x) = \begin{cases} \frac{A_1}{\sqrt{k(x)}} \cos\left(\int_x^a k(x')dx' - \frac{\pi}{4}\right) & \\ +\frac{A_2}{\sqrt{k(x)}} \sin\left(\int_x^a k(x')dx' - \frac{\pi}{4}\right) & x < a \\ \frac{B_1}{\sqrt{\kappa(x)}} e^{-\int_a^x \kappa(x')dx'} + \frac{B_2}{\sqrt{\kappa(x)}} e^{\int_a^x \kappa(x')dx'} & x > a \end{cases}, \tag{16.57}$$

while the exact asymptotic solution (for $z \gg 1$) near the turning point, obtained using Eqs. (16.50) and (16.55), is

$$\psi_{\mathrm{Airy}}(z) = \begin{cases} C_1 \frac{1}{\sqrt{\pi}} (-z)^{-1/4} \cos\left(y - \frac{\pi}{4}\right) & \\ -C_2 \frac{1}{\sqrt{\pi}} (-z)^{-1/4} \sin\left(y - \frac{\pi}{4}\right) & x < a \\ C_1 \frac{1}{2\sqrt{\pi}} z^{-1/4} e^{-y} + C_2 \frac{1}{\sqrt{\pi}} z^{-1/4} e^{y} & x > a \end{cases}. \tag{16.58}$$

In the range where the potential is linear, if $x < a$,

$$\int_x^a k(x')dx' = \sqrt{\frac{2m}{\hbar^2}\frac{dV}{dx}} \int_x^a \sqrt{a - x'}dx' = \frac{2}{3}\sqrt{\frac{2m}{\hbar^2}\frac{dV}{dx}} (a - x)^{3/2} = y, \tag{16.59}$$

and, if $x > a$,

$$\int_a^x \kappa(x')dx' = \sqrt{\frac{2m}{\hbar^2}\frac{dV}{dx}} \int_a^x \sqrt{x' - a}dx' = \frac{2}{3}\sqrt{\frac{2m}{\hbar^2}\frac{dV}{dx}} (x - a)^{3/2} = y. \tag{16.60}$$

Comparing Eqs. (16.57) and (16.58), and using Eqs. (16.47) and (16.48) to show that $-z = c\sqrt{k(x)}$ for $x < a$ and $z = c\sqrt{\kappa(x)}$ for $x > a$ (c = constant), I find that

$$B_1 = A_1/2; \quad B_2 = -A_2, \tag{16.61}$$

giving rise to the connection formula given in Eq. (16.19). The calculation proceeds in the same manner for the turning point at $x = b$, but the slope of the potential is negative at this turning point.

16.6 Problems

1. In the WKB approximation, what is the reflection coefficient when a wave packet is sent into a one-dimensional barrier with the energy above the barrier height? How and why does this differ from the problem of a rectangular barrier?

2. Use the WKB method to estimate the energy levels of a particle having mass m in the potential

$$V(z) = \begin{cases} mgz & z > 0 \\ \infty & z < 0 \end{cases} .$$

In this case, because of the infinite wall at the origin, the semi-classical quantization condition is given by Eq. (16.27). Compare your answer with the solutions $E_n = 2.338, 4.088, 5.521, 6.787$ for $n = 0, 1, 2, 3$ obtained from an exact solution of the Schrödinger equation, where E_n is expressed in units of $(\hbar^2 mg^2/2)^{1/3}$. Show that the exact solution for the ground state energy lies between the WKB value and the variational upper bound of $E_0 = 2.34477$ calculated in Problem 15.2–3

3. Evaluate Eq. (16.35) for the potential $V(x) = V_0 |x/x_0|^q$ when $n = 0$ to get the WKB approximation to the ground state energy. Compare your result with the exact solutions $(E/V_0)\,\beta^{q/(q+2)} = 1.0188,\ 1,\ 1.02295,\ 1.06036,\ 1.1023$ for $q = 1, 2, 3, 4, 5$, with $\beta^2 = 2mV_0x_0^2/\hbar^2$. Also compare your result with the variational solution of Problem 15.2–3 by plotting the ratio of the variational to WKB solutions as a function of q. Show that the WKB solution is greater than the variational solution for $q < 2$ and less than the variational solution for $q > 2$. Also show that the WKB solution deviates more and more from the variational result with increasing q when $q > 2$. You may use the fact that the variational solution is

$$E/V_0 = \beta^{-\frac{q}{q+2}} \left[\frac{1}{\xi^2} + \frac{1}{2^{\frac{q}{2}}} \frac{\Gamma\left(\frac{1+q}{2}\right)}{\sqrt{\pi}} \xi^q \right]$$

with

$$\xi = \left[\frac{2^{\frac{q}{2}+1}}{q} \frac{\sqrt{\pi}}{\Gamma\left(\frac{1+q}{2}\right)} \right]^{\frac{1}{q+2}} .$$

4. Show that the WKB estimate of the energy levels of a particle having mass m in the potential

$$V(x) = \begin{cases} V_0 \left| x/x_0 \right|^{\mu} & x > 0 \\ \infty & x < 0 \end{cases}$$

can be obtained from Eq. (16.36a) by replacing $(n + 1/2)$ with $(2n + 3/2)$. With this substitution, verify that the results are consistent with Problem 16.2 for $n = 0, 1, 2, 3$. For this potential, show that the variational approximation to the ground state energy is identical to that of Problem 15.4 if you use the same trial function. Plot the ratio of the variational to WKB solutions for the ground state energy as a function of q and determine the range of q for which it is possible (but not guaranteed) that the WKB ground state energy is a lower bound to the true energy.

5. For $q \to \infty$, show that the potential $V(x) = V_0 \left| x/x_0 \right|^q$ approaches that of an infinite square well. Show that, as $q \to \infty$, Eq. (16.35) gives the correct energy for the infinite square well potential for $n \gg 1$ and 1/4 of the exact result for $n = 0$.

6. The Hamiltonian for the harmonic oscillator in dimensionless coordinates is $H' = \left(\eta^2 + \xi^2\right)/2$. The semiclassical quantization condition for the dimensionless energy levels ϵ_n of this Hamiltonian is

$$\int_{-\sqrt{2\epsilon}}^{\sqrt{2\epsilon}} \sqrt{2\left(\epsilon_n - \frac{\xi^2}{2}\right)} d\xi = \left(n + \frac{1}{2}\right)\pi; \qquad n = 0, 1, \ldots,$$

where the dimensionless value of $k_n(x)$ is $\tilde{k}(\xi) = 2\epsilon_n - \xi^2$. Show that the WKB quantization condition gives the exact energies, $\epsilon_n = n + 1/2$. Plot the WKB wave function

$$\left(\tilde{\psi}_n\right)_{WKB} = \frac{C_n}{\sqrt{\tilde{k}_n(\xi)}} \cos\left[\int_{\xi}^{\sqrt{2n+1}} d\xi' \sqrt{\left(2n + 1 - \xi'^2\right)} - \frac{\pi}{4}\right]$$

in the classically allowed regime and compare it with the exact wave function

$$\tilde{\psi}_n(\xi) = \frac{1}{\sqrt{2^n n! \sqrt{\pi}}} e^{-\xi^2/2} H_n(\xi)$$

for $n = 0$ and $n = 24$. Choose the constant C_n of the WKB wave function so it agrees with the exact wave function at $\xi = 0$ and plot the $n = 0$ case from $0 \leq \xi \leq 0.7$ and the $n = 24$ case from $0 \leq \xi \leq 6.7$ (that is to within about 0.3 from the classical turning point). The WKB energies are exact, but are the WKB wave functions also exact?

7. Use the connection formulas to extend the WKB solution of the previous problem to the region to the right of the classical turning point. Plot the approximate wave function for $n = 24$ using the WKB solutions in the regions $0 \le \xi \le 6.4$ and $\xi > 7.6$ and the Airy function solution given by Eq. (16.50) in the region $6.4 \le \xi \le 7.6$; on the same graph, plot the exact eigenfunction of the oscillator. [Hint: Show that the variable z defined by Eq. (16.48) is

$$z \to z_n = (2\xi)^{1/3} \left(\xi - \sqrt{(2n+1)} \right),$$

express $\tilde{k}_n(\xi)$ in terms of z_n, and compare Eqs. (16.57) and (16.58) to obtain the constants C_1 and C_2 appearing in Eq. (16.50).]

8–9. Use the WKB method to estimate the transmission coefficient for a smooth potential barrier when the energy is less than the barrier height. Show that the result is the same as that for a square barrier if the quantity $e^{-\kappa d}$ is replaced by $e^{-\int_a^b \kappa(x)dx}$, where d is the width for the square barrier and a and b are the classical turning points for the smooth barrier.

Chapter 17
Scattering: 1-D

Most of what we know about the structure of matter comes from scattering experiments. When I discuss scattering in 3-D, I will review classical scattering theory, but for the time being, I want to discuss the scattering problem in one-dimension. Scattering is simple in principle—send something in and see what comes out. I will give a detailed analysis of scattering in one-dimension for the step potential shown in Fig. 17.1 and then give a qualitative discussion for other potentials. The step potential can be written as

$$V(x) = V_0 \Theta(x), \tag{17.1}$$

where $\Theta(x)$ is the Heaviside step function which is zero for $x < 0$ and one for $x \geq 0$.

Classically, this is a simple problem. If the energy of a particle incident from the left is less than V_0, it is reflected at the potential step with a change in the sign of its velocity; if the energy is greater than V_0 there is no reflection and the particle is transmitted with reduced energy. However, for a true step potential, the problem is *always* quantum-mechanical in nature since the potential changes over a distance that is small compared to a de Broglie wavelength of the particle. In that case, part of the wave packet is always reflected if $E > V_0$. Of course, a *real* potential can only approximate a step function; quantum mechanics is needed if the potential changes over a distance that is small compared to a de Broglie wavelength of the particle. As you will see, there is an additional quantum effect when the energy is less than the height of the step.

There are two basic approaches to solving a scattering problem. In the *steady-state* approach, it is assumed that some type of steady-state has been reached in which a continuous, mono-energetic beam of particles is incident on the scattering region. You can then identify a current density for both the incident and scattered particles. This is the easiest way to do scattering theory and corresponds to the idealized situation in which there is a single energy in the problem, since the incoming beam is mono-energetic. In effect we deal with a *single* energy

P.R. Berman, *Introductory Quantum Mechanics*, UNITEXT for Physics,
https://doi.org/10.1007/978-3-319-68598-4_17

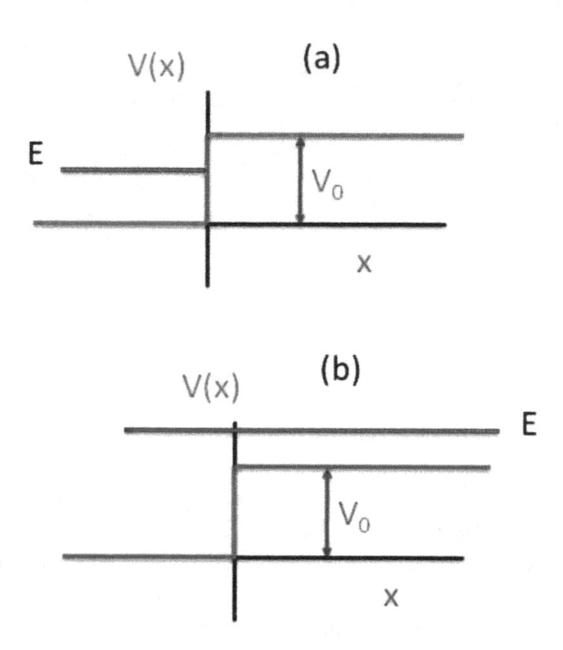

Fig. 17.1 Step potential: (**a**) Energy less than the barrier height. (**b**) Energy greater than the barrier height

eigenstate of the system. The second approach is more difficult mathematically, but more interesting from a physical viewpoint. In this *time-dependent* approach, one sends in a wave packet to the scattering region and looks at the emerging wave packets. Both methods give identical results for the reflection and transmission coefficients when the spatial width of the wave packet is taken to be arbitrarily large (approaching a mono-energetic wave). Scattering is usually a uniquely well-defined problem only in this limit.

Before we start, let me consider the classical scattering problem with one particle a second having energy $E > V_0$ moving towards the step potential from the left with a velocity of one meter per second. The situation is depicted in Fig. 17.2 for successive times differing by one second. As can be seen, the density per unit length *increases* for particles once they are past the barrier, and the speed of each particle *decreases*. However the current density $J = \mathcal{N}v$, where $\mathcal{N}$ is the density per unit length, remains *constant*. The ratio of the transmitted to incident particle current density is a measure of the transmission coefficient, which is unity in the classical case.

Returning to the quantum problem, I proceed as in any problem in quantum mechanics by finding the stationary state eigenfunctions. If $E_k = \hbar^2 k^2/2m > V_0$ and m is the mass of the particle, independent (normalized) eigenfunctions for the potential step may be taken as

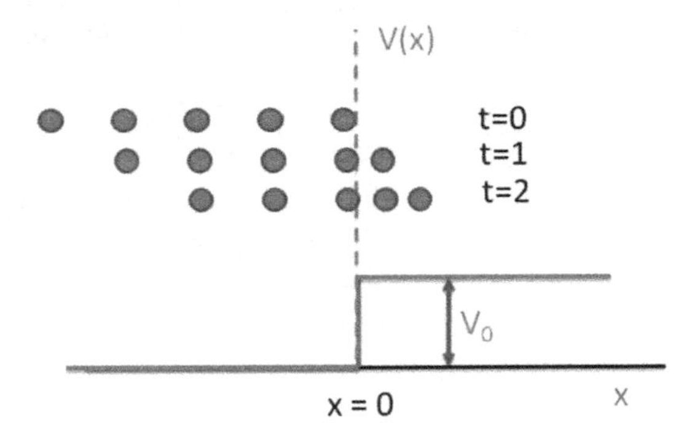

Fig. 17.2 Particles passing the barrier slow down, but their density increases in such a manner that the particle current density remains constant

$$\psi_k^+(x) = \frac{\Theta(k)}{\sqrt{2\pi}} \begin{cases} e^{ikx} + a(k)e^{-ikx} & x < 0 \\ b(k)e^{ik'x} & x \geq 0 \end{cases}; \tag{17.2a}$$

$$\psi_k^-(x) = \frac{\Theta(k - k_b)}{\sqrt{2\pi}} \sqrt{\frac{k}{k'}} \begin{cases} e^{-ik'x} + a_-(k)e^{ik'x} & x \geq 0 \\ b_-(k)e^{-ikx} & x < 0 \end{cases}, \tag{17.2b}$$

where

$$a(k) = \frac{k - k'}{k + k'}; \quad a_-(k) = -\frac{k - k'}{k + k'}, \tag{17.3}$$

$$b(k) = \frac{2k}{k + k'}; \quad b_-(k) = \frac{2k'}{k + k'}, \tag{17.4}$$

$$k' = \sqrt{\frac{2m(E_k - V_0)}{\hbar^2}} = \sqrt{k^2 - k_b^2}, \tag{17.5}$$

and

$$k_b = \sqrt{\frac{2mV_0}{\hbar^2}}. \tag{17.6}$$

There are two independent eigenfunctions for a given value of

$$k = \sqrt{\frac{2mE_k}{\hbar^2}}. \tag{17.7}$$

The $\psi_k^+(x)$ eigenfunction corresponds to a wave incident from the left and the $\psi_k^-(x)$ eigenfunction to one incident from the right.

If $E_k < V_0$, then the eigenfunctions are non-degenerate and given by

$$\psi_k^+(x) = \frac{\Theta(k)}{\sqrt{2\pi}} \begin{cases} e^{ikx} + c(k)e^{-ikx} & x < 0 \\ d(k)e^{-\kappa x} & x \geq 0 \end{cases}, \tag{17.8}$$

where

$$\kappa = \sqrt{k_b^2 - k^2} \tag{17.9}$$

and

$$c(k) = \frac{k - i\kappa}{k + i\kappa} = e^{-2i\alpha(k)}; \quad \alpha(k) = \tan^{-1}\frac{\kappa}{k}, \tag{17.10}$$

$$d(k) = \frac{2k}{k + i\kappa} = \frac{2k}{\sqrt{k^2 + \kappa^2}} e^{-i\alpha(k)}. \tag{17.11}$$

17.1 Steady-State Approach

I assume that there is a single energy eigenfunction, corresponding to a wave incident from the left. The trick is to break up the probability current density associated with $\psi_k^+(x)$ into three parts, the e^{ikx} part for $x < 0$ corresponding to the incident wave, the e^{-ikx} part for $x < 0$ corresponding to the reflected wave, and the e^{ikx} or $e^{-\kappa x}$ part for $x > 0$ corresponding to the transmitted wave. Thus, the incident probability current density is calculated using $\psi_i(x) = e^{ikx}/\sqrt{2\pi}$ as

$$\begin{aligned} J_i &= \frac{\hbar}{2mi}\left[\psi_i^*(x)\frac{d\psi_i(x)}{dx} - \psi_i(x)\frac{d\psi_i^*(x)}{dx}\right] \\ &\approx \frac{\hbar}{2mi}\left[e^{-ikx}ike^{ikx} - e^{ikx}\left(-ike^{-ikx}\right)\right]\rho = \rho v_k, \end{aligned} \tag{17.12}$$

where $\rho = 1/(2\pi)$ is the probability density associated with the incident wave and

$$v_k = \frac{\hbar k}{m}. \tag{17.13}$$

For $E > V_0$, the reflected probability current density is calculated using $\psi_r(x) = a(k)e^{-ikx}/\sqrt{2\pi}$ as

$$J_r = -\left|a(k)\right|^2 \rho v_k, \tag{17.14}$$

and the transmitted probability current density is calculated using $\psi_t(x) = b(k)e^{ik'x}/\sqrt{2\pi}$ as

$$J_t = \left|b(k)\right|^2 \rho v_k', \tag{17.15}$$

where

$$v'_k = \frac{\hbar k'}{m} = \frac{\hbar}{m}\sqrt{k^2 - k_b^2}. \quad (17.16)$$

The reflection coefficient is

$$\mathcal{R} = \frac{-J_r}{J_i} = |a(k)|^2 = \left(\frac{k-k'}{k+k'}\right)^2 \quad (17.17)$$

and the transmission coefficient is

$$\mathcal{T} = \frac{J_t}{J_i} = \frac{v'_k}{v_k}|b(k)|^2 = \frac{k'}{k}|b(k)|^2 = \frac{4kk'}{(k+k')^2}, \quad (17.18)$$

with

$$\mathcal{R} + \mathcal{T} = 1. \quad (17.19)$$

For $E_k < V_0$, the reflected probability current density is calculated using $\psi_r(x) = c(k)e^{ikx}/\sqrt{2\pi}$ as

$$J_r = -|c(k)|^2 \rho v_k = \rho v_k. \quad (17.20)$$

The reflection coefficient is

$$\mathcal{R} = \frac{-J_r}{J_i} = 1. \quad (17.21)$$

The probability current density in the barrier vanishes since the wave function is real for $x > 0$. All these results were derived previously in Chap. 6.

17.2 Time-Dependent Approach

In the time-dependent approach, the initial state wave function is expanded in terms of the stationary state eigenfunctions $\psi_k^{+,-}(x)$ to obtain the expansion coefficients $\Phi(k)$, which are then used to calculate $\psi(x,t)$ as

$$\psi(x,t) = \int_0^\infty dk \left[\Phi_+(k)\psi_k^+(x) + \Phi_-(k)\psi_k^-(x)\right] e^{-i\hbar k^2 t/2m}. \quad (17.22)$$

The integral is restricted to positive values of k since the eigenfunctions $\psi_k^{+,-}(x)$ given in Eqs. (17.2) and (17.8) are so restricted. The key point in solving the scattering problem in this fashion is to choose a wave packet incident from the left that is sufficiently broad to insure that it does not spread much on the time scale of the scattering. Moreover, it should be much broader than the distance over which the potential varies significantly (in this case the variation occurs as a step, so the packet is *always* larger in extent than the interval over which the potential changes). In other words, you want the central energy E_0 associated with the incident wave packet to be fairly well-defined. The same type of formalism can then be used for both $E_0 > V_0$ and $E_0 < V_0$.

Since the incident wave packet is localized to the left of the step and moving to the right, $\Phi_-(k) \approx 0$. Moreover, to calculate the expansion coefficients from the initial wave packet, I can neglect any of the e^{-ikx} components of the eigenfunctions $\psi_k^+(x)$ given in Eqs. (17.2a) and (17.8), respectively, since they correspond to a wave packet moving to the left. To a good approximation, the initial wave function, centered at $x = -x_0 < 0$ at time $t = 0$ can be expanded as

$$\psi(x,0) \approx \int_0^\infty dk\, \Phi_+(k)\psi_k^+(x) \approx \frac{1}{\sqrt{2\pi}}\int_{-\infty}^\infty dk\, \Phi(k)e^{ikx}, \tag{17.23}$$

where $\Phi(k)$ is a real function, sharply peaked about $k = k_0$, corresponding to energy

$$E_0 = \hbar^2 k_0^2/2m, \tag{17.24}$$

and the integral is extended to $-\infty$ based on the assumption that $\Phi(k) \approx 0$ for $k < 0$. In other words, the initial wave packet doesn't yet know about the step potential it is going to encounter, so it can be expanded in terms of *free-particle plane wave eigenfunctions*. This is a key step in solving the problem.

After a time $t \gg x_0/v_0$, with v_0 defined by

$$v_0 = \hbar k_0/m, \tag{17.25}$$

the scattering is finished. I now look at the wave function for times $t \gg x_0/v_0$.

There is one final point to note before starting the calculation. For a wave packet moving to the right with a fairly well-defined energy, I can write Eq. (17.23) as

$$\psi(x,0) \approx \frac{1}{\sqrt{2\pi}}\int_{-\infty}^\infty dk\, \Phi(k)e^{ikx} = \frac{1}{\sqrt{2\pi}}e^{ik_0x}\int_{-\infty}^\infty dk\, \Phi(k)e^{i(k-k_0)x}. \tag{17.26}$$

If $\Phi(k)$ is *symmetric* about $k = k_0$, as I shall assume, then

$$\psi(x,0) \approx \frac{1}{\sqrt{2\pi}}e^{ik_0x}\int_{-\infty}^\infty dk\, \Phi(k)\cos[(k-k_0)\,x]. \tag{17.27}$$

Without loss of generality, I can assume that the integral is positive, implying that

$$\psi(x,0) = |\psi(x,0)|\, e^{ik_0x}. \tag{17.28}$$

The phase factor gives rise to the motion of the wave packet to the right with speed $v_0 = \hbar k_0/m$.

17.2.1 $E_0 > V_0$ for Step Potential

I expand the wave function at time t in terms of the *exact* eigenfunctions, but neglect any contributions from $\psi_k^-(x)$. For $E_0 > V_0$

$$\psi(x,t) \approx \frac{1}{\sqrt{2\pi}} \int_0^{\infty} dk\, \Phi(k)\psi_k^+(x)e^{-i\hbar k^2 t/2m}$$
$$= \frac{1}{\sqrt{2\pi}} \int_{-\infty}^{\infty} dk\, \Phi(k)e^{-i\hbar k^2 t/2m} \begin{cases} e^{ikx} + a(k)e^{-ikx} & x < 0 \\ b(k)e^{ik'x} & x \geq 0 \end{cases}, \quad (17.29)$$

where $\Phi(k)$ is a sharply peaked, real function centered at $k = k_0$. The calculation proceeds exactly as in Chap. 3. That is, *neglecting spreading* [i.e., approximating $k^2 = [k_0 + (k - k_0)]^2 \approx k_0^2 + 2k_0(k - k_0) = 2kk_0 - k_0^2$], I find

$$\psi(x,t) = \int_{-\infty}^{\infty} dk\, \Phi(k)\psi_k(x)e^{-i\hbar k^2 t/2m} \approx \frac{1}{\sqrt{2\pi}} e^{i\hbar k_0^2 t/2m}$$
$$\times \int_{-\infty}^{\infty} dk\, \Phi(k)e^{-ikv_0 t} \begin{cases} e^{ikx} + a(k_0)e^{-ikx} & x < 0 \\ b(k_0)e^{ik'x} & x \geq 0 \end{cases}, \quad (17.30)$$

where $a(k)$ and $b(k)$ are evaluated at $k = k_0$. The only extra feature I have to deal with is the $e^{ik'x}$ term.

Since k' appears in an exponent and is a function of k, as is evident from Eq. (17.5), I cannot simply evaluate it at $k = k_0$, but must include a correction term as well, to insure that each term in the exponent has a magnitude much less than unity. Consequently, I expand k' around $k = k_0$ in the exponent,

$$e^{ik'x} \approx \exp\left[i\left(k_0' + \left.\frac{dk'}{dk}\right|_{k=k_0}(k - k_0)\right)x\right], \quad (17.31)$$

where

$$k_0' = \sqrt{k_0^2 - k_b^2}.$$

I then use the relationship

$$\frac{dk'}{dk} = \frac{k}{k'} \quad (17.32)$$

to evaluate

$$k_0' + \left.\frac{dk'}{dk}\right|_{k=k_0} (k - k_0) = k_0' + \frac{k_0 k}{k_0'} - \frac{k_0^2}{k_0'}, \tag{17.33}$$

and substitute this result into Eqs. (17.31) and (17.30) to arrive at

$$\psi(x,t) \approx \frac{1}{\sqrt{2\pi}} e^{i\hbar k_0^2 t/2m} \int_{-\infty}^{\infty} dk\, \Phi(k) e^{-ikv_0 t} \times \begin{cases} e^{ikx} + a(k_0)e^{-ikx} & x < 0 \\ e^{i\left(k_0' - k_0^2/k_0'\right)x} b(k_0) e^{ikk_0 x/k_0'} & x \geq 0 \end{cases}. \tag{17.34}$$

Comparing Eq. (17.34) with Eq. (17.23), I find that

$$\psi(x,t) = e^{i\hbar k_0^2 t/2m} \begin{cases} \psi(x - v_0 t, 0) + a(k_0)\psi(-x - v_0 t, 0) & x < 0 \\ b(k_0) e^{i\left(k_0' - k_0^2/k_0'\right)x} \psi\left[\left(\frac{k_0}{k_0'}\right)x - v_0 t, 0\right] & x \geq 0 \end{cases}. \tag{17.35}$$

Equation (17.35) gives the time-dependent wave function, neglecting spreading. I should note that the expansion of k' around $k = k_0$ breaks down when $E_k \approx V_0$. In that case, dispersion leads to a wave packet that no longer propagates without distortion. A necessary condition to neglect this dispersion is $k_0'\sigma \gg 1$, where σ is the width of the incident packet. This condition may not be sufficient, however, for certain types of wave packets and values of x.

I look at each term in Eq. (17.35) separately and remember that the original wave packet is centered at $x = -x_0 < 0$ at $t = 0$. I consider only those times $t \gg x_0/v_0$, for which the scattering is complete. The first term in the first line of Eq. (17.35), $\psi(x - v_0 t, 0)$, is peaked at

$$\begin{aligned} x_c - v_0 t &= -x_0; \\ x_c &= v_0 t - x_0, \end{aligned} \tag{17.36}$$

which corresponds to positive $x_c \gg 0$ for $t \gg x_0/v_0$. In other words,

$$\psi(x - v_0 t, 0) \approx 0 \quad \text{for } x < 0 \text{ and } t \gg x_0/v_0, \tag{17.37}$$

implying that the first term in the first line of Eq. (17.35) is approximately equal to zero for times $t \gg x_0/v_0$. The second term in the first line of Eq. (17.35), $a(k_0)\psi(-x - v_0 t, 0)$, is peaked at

$$\begin{aligned} -x_c - v_0 t &= -x_0; \\ x_c &= x_0 - v_0 t \end{aligned} \tag{17.38}$$

and corresponds to the reflected wave for $t \gg x_0/v_0$. The term in the second line of Eq. (17.35), $b(k_0)\psi\left[\left(\frac{k_0}{k_0'}\right)x - v_0 t, 0\right]$, is peaked at

$$\left(\frac{k_0}{k_0'}\right)x_c - v_0 t = -x_0;$$

$$x_c = v_0'\left(t - \frac{x_0}{v_0}\right), \tag{17.39}$$

where

$$v_0' = \frac{\hbar k_0'}{m} \tag{17.40}$$

is the speed of the transmitted wave. For $t \gg x_0/v_0$, this corresponds to the transmitted wave packet, moving with speed v_0'. Note that it takes a time

$$t_0 = x_0/v_0 \tag{17.41}$$

for the center of the initial wave packet to reach the origin.

I now return to the final result, Eq. (17.35), which gives an approximation to $\psi(x,t)$ for *all* times. I can use Eq. (17.28) to transform this equation into

$$\psi(x,t) \approx e^{-i\hbar k_0^2 t/2m}\begin{cases} e^{ik_0 x}\left|\psi(x - v_0 t, 0)\right| & \\ +a(k_0)e^{-ik_0 x}\left|\psi(-x - v_0 t, 0)\right| & x < 0 \\ b(k_0)e^{ik_0' x}\left|\psi\left[\left(\frac{k_0}{k_0'}\right)x - v_0 t, 0\right]\right| & x \ge 0 \end{cases}. \tag{17.42}$$

To illustrate the physical content of Eq. (17.42), I choose an initial wave packet of the form

$$\psi(x,0) = \frac{1}{(\pi\sigma^2)^{1/4}}e^{-(x+x_0)^2/2\sigma^2}e^{ik_0 x} \tag{17.43}$$

and substitute it into Eq. (17.42) to arrive at

$$\psi(x,t) \approx \frac{e^{-i\hbar k_0^2 t/2m}}{(\pi\sigma^2)^{1/4}}\begin{cases} e^{ik_0 x}e^{-(x+x_0-v_0 t)^2/2\sigma^2} & \\ +a(k_0)e^{-ik_0 x}e^{-(-x+x_0-v_0 t)^2/2\sigma^2} & x < 0 \\ b(k_0)e^{ik_0' x}e^{-\left[x - v_0'(t - x_0/v_0)\right]^2/2\sigma'^2} & x \ge 0 \end{cases}, \tag{17.44}$$

where

$$\sigma' = \sigma k_0'/k_0 < \sigma. \tag{17.45}$$

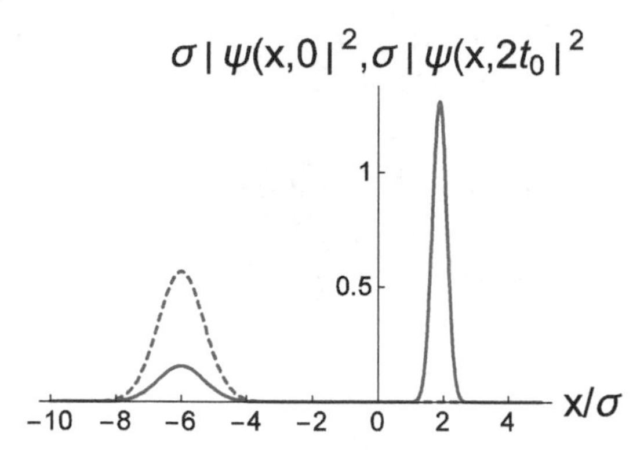

Fig. 17.3 Graphs of $\sigma\left|\psi(x,0)\right|^2$ (blue, dashed curve), along with $\left|\psi(x,t=2t_0)\right|^2$ (red, solid curve) for $k_0\sigma = 200$, $k_b\sigma = 190$, $k_0'\sigma = 62.5$, $v_0t_0/\sigma = 6$

In Fig. 17.3 I plot $\sigma\left|\psi(x,0)\right|^2$, along with $\sigma\left|\psi(x,t=2t_0)\right|^2$. The reflected and transmitted packets are seen clearly in the figure. For the dimensionless parameters $k_0\sigma = 200$, $k_b\sigma = 190$, $k_0'\sigma = 62.5$, $v_0t_0/\sigma = 6$, the peak value of $\left|\psi(x,t)\right|^2$ of the reflected wave is reduced by a factor of $\left|a(k_0)\right|^2 = 0.274$ and that of the transmitted wave is increased by a factor of $\left|b(k_0)\right|^2 = 2.32$. Moreover, since $\sigma' < \sigma$, the transmitted packet is compressed, as was the classical particle density shown in Fig. 17.3, and the speed of the transmitted wave packet is reduced by a factor of $k_0'/k_0 = 0.31$. Although the amplitude of the transmitted packet is larger than that of the incident packet, the transmitted probability current density is less than that of the initial packet, as I now show.

I can use Eqs. (17.42) and (5.139) to show that the probability current density associated with the incident wave packet is approximately

$$\begin{aligned} J_i &= \frac{\hbar}{2mi}\left[\psi_i^*(x,t)\frac{d\psi_i(x,t)}{dx} - \psi_i(x,t)\frac{d\psi_i^*(x,t)}{dx}\right] \\ &= \frac{\hbar}{2mi}\left[\left|\psi^*(x-v_0t,0)\right|e^{-ik_0x}\frac{d\left(\left|\psi(x-v_0t,0)\right|e^{ik_0x}\right)}{dx}\right] + \text{c.c.} \\ &\approx \frac{\hbar}{2mi}\left[ik_0\left|\psi(x-v_0t,0)\right|^2\right] + \text{c.c.} = v_0\left|\psi(x-v_0t,0)\right|^2, \end{aligned} \tag{17.46}$$

where $\psi_i(x,t)$ corresponds to the incident wave packet *before* it reaches the barrier. In deriving Eq. (17.46), I assumed that $k_0\sigma \gg 1$, consistent with the assumption that the incident wave packet is sharply peaked in momentum space. Similarly,the reflected probability current density for times well after the packet has scattered from the barrier is approximately

$$J_r = -\left|a(k_0)\right|^2 v_0\left|\psi\left[x+v_0t,0\right]\right|^2. \tag{17.47}$$

and the transmitted probability current density is

$$J_t = |b(k_0)|^2 v_0' \left| \psi \left[\left(\frac{k_0}{k_0'} \right) x - v_0 t, 0 \right] \right|^2 . \tag{17.48}$$

Since $|b(k_0)|^2 v_0' < v_0$, the maximum transmitted current density is less than the maximum incident current density.

The reflection coefficient is equal to the magnitude of time-integrated reflected probability current density (this is what a detector placed in the path of the particle would measure) divided by the time-integrated initial probability current density

$$\mathcal{R} = \frac{-\int J_r dt}{\int J_i dt} = \frac{|a(k_0)|^2 v_0}{v_0} = |a(k_0)|^2 = \left(\frac{k_0 - k_0'}{k_0 + k_0'} \right)^2 \tag{17.49}$$

and the transmission coefficient is

$$\mathcal{T} = \frac{\int J_t dt}{\int J_i dt} = \frac{|a(k_0)|^2 v_0'}{v_0} = \frac{k_0'}{k_0} |b(k_0)|^2 = \frac{4k_0 k_0'}{(k_0 + k_0')^2} < 1. \tag{17.50}$$

These results agree with the steady-state approach.

17.2.2 $E_0 < V_0$ for Step Potential

For $E_0 < V_0$

$$\begin{aligned} \psi(x,t) &\approx \int_0^\infty dk\, \Phi(k) \psi_k^+(x) e^{-i\hbar k^2 t/2m} \\ &\approx \int_{-\infty}^\infty dk\, \Phi(k) e^{-i\hbar k^2 t/2m} \begin{cases} e^{ikx} + c(k)e^{-ikx} & x < 0 \\ d(k)e^{-\kappa x} & x \geq 0 \end{cases} . \end{aligned} \tag{17.51}$$

The calculation proceeds as before, except that there are a few wrinkles. You might think that all that is necessary is to replace k' by $i\kappa$ and k_0' by $i\kappa_0$ but you would be wrong on two counts. First, for $x > 0$, the major contribution to the integral over k may no longer be centered at $k = k_0$, even though $\Phi(k)$ is peaked at $k = k_0$. The reason for this is that the eigenfunction for $x > 0$ is an exponentially decreasing function of $\kappa x = \sqrt{k_b^2 - k^2} x$, so it is possible to get a relatively larger contribution to the integral over k near $k = k_b$ rather than $k = k_0$. In general, the integral in Eq. (17.51) must be done numerically for $x > 0$. On the other hand, for the scattering problem I am interested in calculating the wave function only when $t \gg t_0 = x_0/v_0$, times for which the wave function in the region $x > 0$ is negligibly small. Thus I neglect the region $x > 0$ in my analysis of the problem, at least for the moment.

The second problem with replacing k' by $i\kappa$ and k_0' by $i\kappa_0$ is linked to the fact that, while $a(k)$ and $b(k)$ are real, $c(k)$ and $d(k)$ are complex; that is,

$$c(k) = e^{-i\phi(k)}; \tag{17.52a}$$

$$d(k) = \frac{2k}{\sqrt{k^2+\kappa^2}} e^{-i\phi(k)/2}; \tag{17.52b}$$

$$\phi(k) = 2\alpha(k) = 2\tan^{-1}(\kappa/k) = 2\tan^{-1}\left(\sqrt{k_b^2 - k^2}/k\right). \tag{17.52c}$$

When a function appears in an exponent and you expand the function, only terms that are much less than unity can be neglected. I cannot simply replace $\phi(k)$ by $\phi(k_0)$. Instead, I must expand it as

$$\phi(k) \approx \phi(k_0) + \frac{d\phi}{dk_0}(k - k_0) \tag{17.53}$$

where

$$\frac{d\phi}{dk_0} = \left.\frac{d\phi}{dk}\right|_{k=k_0} = -\frac{2}{\kappa_0} \tag{17.54}$$

and

$$\kappa_0 = \sqrt{\frac{2m(V_0 - E_0)}{\hbar^2}}. \tag{17.55}$$

Using this result and Eq. (17.621) in Eq. (17.51) with $k^2 \approx -k_0^2 + 2k_0 k$, I find for $x < 0$,

$$\begin{aligned}
\psi(x,t) &\approx e^{i\hbar k_0^2 t/2} m\psi(x - v_0 t, 0) + e^{i\hbar k_0^2 t/2} e^{-i\phi(k_0) + ik_0 \frac{d\phi}{dk_0}} \\
&\quad \times \int_{-\infty}^{\infty} dk\, \Phi(k) e^{-ikv_0 t} e^{-ikx} e^{-ik\frac{d\phi}{dk_0}} \\
&= e^{i\hbar k_0^2 t/2} \psi(x - v_0 t, 0) \\
&\quad + e^{i\hbar k_0^2 t/2} e^{-i\phi(k_0) + ik_0 \frac{d\phi}{dk_0}} \psi\left(-x - v_0 t - \frac{d\phi}{dk_0}, 0\right) \\
&= e^{-i\hbar k_0^2 t/2} e^{ik_0 x} \left|\psi(x - v_0 t, 0)\right| \\
&\quad + e^{-i\hbar k_0^2 t/2} e^{-i\phi(k_0)} e^{-ik_0 x} \left|\psi\left(-x - v_0 t - \frac{d\phi}{dk_0}, 0\right)\right|.
\end{aligned} \tag{17.56}$$

Since

$$\frac{d\phi}{dk_0}=\frac{d\phi}{dE_0}\frac{dE_0}{dk_0}=\frac{\hbar^2k_0}{m}\frac{d\phi}{dE_0}=\hbar v_0\frac{d\phi}{dE_0}=-\frac{2}{\kappa_0}, \tag{17.57}$$

it follows that the reflected wave packet is centered at

$$-x_c-v_0\left(t+\hbar\frac{d\phi}{dE}\right)=-x_0;$$

$$x_c=x_0-v_0\left(t+\hbar\frac{d\phi}{dE}\right), \tag{17.58}$$

or

$$x_c=x_0-v_0\left(t-\frac{2}{\kappa_0 v_0}\right). \tag{17.59}$$

There is a *time delay* in the scattering given by

$$t_d=-\frac{1}{v_0}\frac{d\phi}{dk_0}=-\hbar\frac{d\phi}{dE_0}=\frac{2}{\kappa_0 v_0}=\frac{\hbar}{\sqrt{E(V_0-E)}}. \tag{17.60}$$

The delay is proportional to $\hbar$—this is a quantum effect. How can we interpret this time delay? In the stationary state approach to the scattering problem, the probability current density vanishes for $x>0$. In the wave packet approach, however, it cannot vanish at all times. There must be a positive probability current density up until the time the center of the wave packet reaches the step and a negative probability current density as the (time delayed) wave packet is reflected. It is as if the packet is checking out the potential barrier and then decides it does not have enough energy to be transmitted so it goes back from whence it came, but it takes a time delay for the wave packet to figure this out. The probability current density at the origin, given by

$$J(x=0,t)=\frac{i\hbar}{2m}\left[\psi(x,t)\frac{\partial\psi^*(x,t)}{\partial x}-\psi^*(x,t)\frac{\partial\psi(x,t)}{\partial x}\right]_{x=0}, \tag{17.61}$$

with $\psi(x,t)$ given by Eq. (17.56) and $\psi(x,0)$ by Eq. (17.43), is plotted in Fig. 17.4 in units of $v_0/\left(\sqrt{\pi}\right)$ as a function of v_0t/σ for $x_0/\sigma=6$, $k_b\sigma=200$ and $k_0\sigma=190, 100$. You can see that the curve has a "dispersion-like" form, even though it is not exactly proportional to the derivative of the wave packet envelope. The peak amplitude of the curve and the time delay decrease with increasing k_b/k_0. In the low energy limit $k_b/k_0\sim\infty$, $J(0,t)\sim0$ and $t_d\sim0$. The time delay is largest when $E\approx V_0$; in this limit, the wave function penetrates significantly into the barrier.

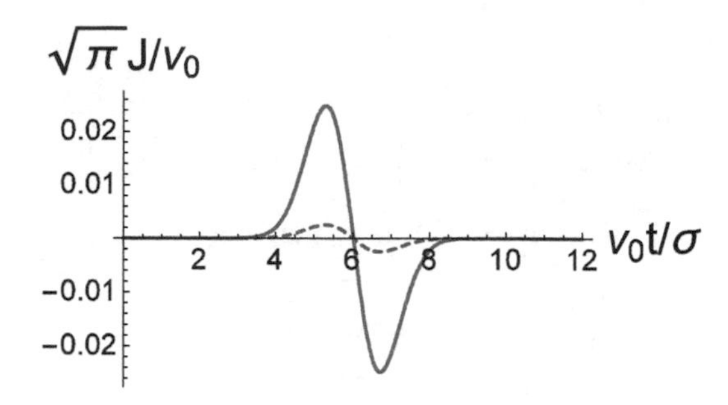

Fig. 17.4 Probability current density at $x = 0$ in dimensionless units as a function of v_0t/σ for $x_0/\sigma = 6$ and $k_b\sigma = 200$. The red solid curve is for $k_0\sigma = 190$ and the blue dashed curve for $k_0\sigma = 100$

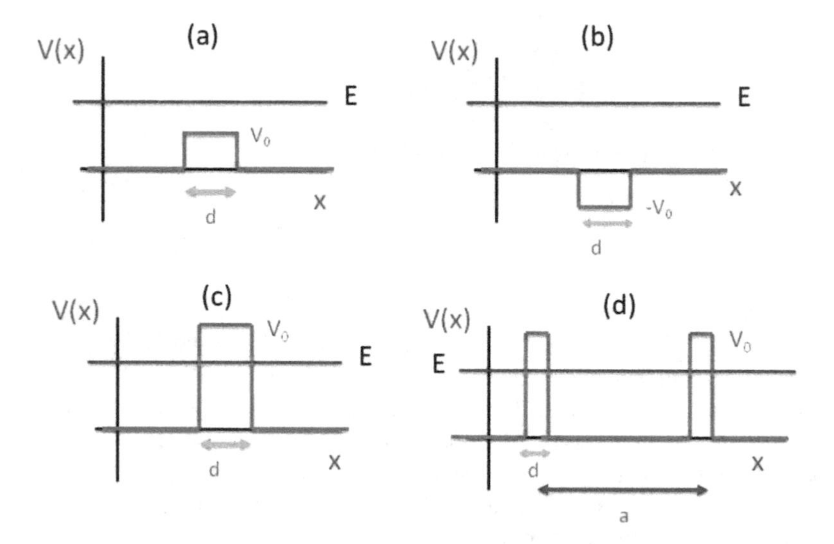

Fig. 17.5 Different piecewise constant potentials and incident scattering energies

Similar analyses can be used for different potentials. I will describe in qualitative terms what happens in each of the cases shown in Fig. 17.5.

Cases (a) and (b) are analogous to an optical field incident normally on a thin dielectric film. There is constructive interference in reflection when twice the film thickness, $2d$, is a half integral number of wavelengths and destructive interference when it is an integral number of wavelengths. The transmission goes to unity at positions of destructive interference in reflection. The resonances can be very narrow if the index of refraction is high. In the quantum-mechanical problem, you can also have very narrow transmission resonances as a function of $k_0'd$. In the scattering problem, such resonances are accompanied by large time delays. To avoid wave packet spreading and to see these resonances, you must take the width of the initial wave packet to be much larger than the barrier width, *multiplied* by the number of "bounces" the packet makes in the region of the potential. The number of "bounces" is simply the barrier width d divided by $v_0' t_d$.

In case (c), there can be quantum-mechanical tunneling through the barrier. There is also a time delay in this problem, but no resonance phenomena. The wave function simply builds up and decays inside the barrier.

Case (d) is perhaps the most interesting. It corresponds to a Fabry-Perot filter for light. If each barrier has a very small transmission coefficient separately, *the transmission coefficient for the double barrier can approach unity when the incoming wavelength corresponds to a standing wave pattern between the two barriers*! This is the result found from the steady-state approach. How can this be? How can the wave penetrate significantly through the first barrier since it doesn't even *know* about the second barrier? The answer to this question is "It can't" if the width of the packet is much less than a, the separation of the barriers. In that case, the wave packet is partially transmitted by the barrier, but mostly reflected by it. The part that is transmitted bounces back and forth between the barriers, leaking out a little bit each time. Thus, the transmitted wave is a *series* of packets, as is the reflected wave, with the time between reflected or transmitted packets equal to the round-trip time in the cavity.

How then does a Fabry-Perot filter work? In order to see the narrow resonances, you must choose an initial wave packet that is much larger than the distance a between the barriers, multiplied by number of bounces in the cavity, $a/(v_0 t_d)$, which can be enormous. The time delay near resonance is inversely proportional to the transmission coefficient for a *single* barrier. In other words, you need an incoming packet that is quasi-monochromatic to see the narrow resonances. Part of the packet penetrates through the barrier, is reflected from the second barrier, and interferes with subsequent transmission through the barrier. After a long time, a steady state standing wave pattern is formed in the cavity having an intensity that is much larger than that of the incident wave—the part that leaks out the other end has an intensity equal to the initial wave intensity. After *very* long times, the initial wave packet can be almost totally transmitted by the double barrier.

17.3 Summary

We have seen several interesting features of scattering from one-dimensional potentials. Both steady-state and time-dependent approaches were used. In some sense, the time-dependent approach provides a justification for the equations for the reflection and transmission coefficients used in the steady-state approach and provides a prescription for obtaining the time delay from the phase of the steady-state eigenfunctions. I now turn my attention to scattering from spherically symmetric potentials.

17.4 Problems

1. How does one-dimensional scattering by a step potential differ for classical and quantum-mechanical scattering of a particle? What is the optical analogue of scattering by an attractive square well? Why would you expect resonances in this case?

2. In the time-dependent scattering approach I assumed that a wave packet with a fairly well-defined energy was incident on the potential step from the left. If the energy is greater than the step height, there could be contributions to the wave function from the eigenfunctions given in Eq. (17.2b). Prove that such contributions are negligible at all times. Moreover, prove that the contribution for the e^{-ikx} part of the $\psi_k^+(x)$ eigenfunction makes a negligible contribution to $\Phi_+(k)$.

3–5. (a) Consider classical scattering of a particle having mass m by a square well potential for which $V(x) = -V_0 < 0$ for $|x| \leq a/2$ and is zero otherwise. A particle is incident from the left with energy $E = mv_0^2/2$. Show that, compared to the case when there is no potential, there is a negative time delay (that is the particle reaches a point $x > 0$ faster than it would in the absence of the potential) given by

$$t_d^{cl} = -\frac{a}{v_0} + \frac{a}{v_0'},$$

where $v_0' = \sqrt{2m(E + V_0)}$

(b) Now consider the analogous quantum problem for the scattering of a quasi-monoenergetic wave packet having energy centered at $E_0 = mv_0^2/2 = \hbar^2 k_0^2/2m$. The (amplitude) reflection and transmission coefficients for a mono-energetic wave packet having energy $E = \hbar k$ are given in Eqs. (6.107) with k_E replaced by k. If you write these coefficients as

$$R(k) = |R(k)|\, e^{i\phi_R(k)}; \qquad T(k) = |T(k)|\, e^{i\phi_T(k)},$$

show that the quantum time delays for the reflected and transmitted packets are given by

$$t_d^R = -\frac{1}{v_0}\frac{d\phi_R}{dk_0};$$

$$t_d^T = \frac{1}{v_0}\frac{d\phi_T}{dk_0}.$$

(c) Plot the intensity transmission coefficient as a function of $y = k_0 a$ for $\beta = \sqrt{2mV_0/\hbar^2}a = 500$ and $0 < y < 120$. Show that there are narrow resonances when

$$k_0' a = \sqrt{\frac{2m\left(E_0 + V_0\right)}{\hbar^2}} a = n\pi$$

for integer $n > \beta/\pi$.

(d) Plot the classical time delay and the quantum transmission time delay on the same graph in units of $a/v_0 = ma^2/\hbar y$ as a function of $y = k_0 a$ for $\beta = 500$ and $0 < y < 120$. Show that, at the position of the resonances, the quantum delay is *greater* than the classical delay. This supports the contention that the particle "bounces" back and forth in as it is scattered near resonance.

(e) In the low energy limit, $y \ll 1$, show that the time delay is given by

$$t_d^T = -\frac{ma^2}{\hbar y}\left(1 + \frac{2\cot\beta}{\beta}\right).$$

It follows from Eqs. (6.88) and (6.89) that, for $y \ll 1$, a new bound state appears near zero energy whenever $\beta = n\pi$. Thus the time delay diverges near such resonances. Since the phase varies rapidly in such regions, there is considerable dispersion and the picture of a transmitted, time-delayed, undistorted wave packet is no longer valid.[1]

6. Calculate the time delay in transmission of the scattering of a quasi-monoenergetic wave packet of a particle having mass m by a delta function potential barrier,

$$V(x) = V_0 a\delta\left(x\right)$$

where V_0, d, and a are positive constants. Show that it vanishes in the limit $\hbar \to 0$.

7–8. Consider scattering of a particle having mass m by a one-dimensional potential

$$V(x) = V_0 a\left[\delta\left(x\right) + \delta\left(x - d\right)\right],$$

where V_0, d, and a are positive constants.

(a) Calculate and plot the intensity transmission coefficient $\mathcal{T} = |T|^2$ as a function of $y = kd$ ($k = \sqrt{2mE/\hbar^2}$) for $\beta'^2 = 2mV_0 ad/\hbar^2 = 100$ in the strong barrier limit,

$$\frac{\beta^2}{2ka} = \frac{mV_0 a^2}{ka\hbar^2} = \frac{\beta'^2}{2y} \gg 1.$$

[1]The dependence of the time delay on β is similar to that encountered in the dependence of the scattering length (the scattering length is discussed in Chap. 18) on magnetic field strength when the field is used to tune the energy in an open scattering channel to that of a bound state in a closed channel of the intermolecular potentials. Such *Feshbach resonances* play an important role in controlling interactions in Bose-Einstein condensates [for a review, see the article by C. Chin, R. Grimm, P. Julienne, and E. Tiesinga, *Feshbach resonances in ultracold gases,* Reviews of Modern Physics **82**, 1225–1286 (2010)].

Show that there are resonances where the transmission goes to unity and interpret your result.

(b) From the expression for the transmission amplitude T calculate and plot the scattering time delay, in units of $d/v = md^2/\hbar y$, experienced by a quasi-monoenergetic wave packet as a function of y. Show that, at the position of the resonances, the time delay goes through a maximum. Interpret your results. The time delay in these units is the number of bounces the particle makes before being scattered.

9. Consider the scattering of a quasi-monoenergetic wave packet having central energy $E = \hbar^2 k^2/2m$ by the potential

$$V(x) = \begin{cases} V_0 a \delta(x) & x < d \\ \infty & x > d \end{cases},$$

where V_0 and a are positive constants and m is the particle's mass. Plot the time delay in units of $d/v = md^2/\hbar y$ as a function of $y = ka$ for $\beta'^2 = 2mV_0 ad/\hbar^2 = 100$ in the strong barrier limit, $\beta'^2/2y \gg 1$, and interpret your results. The time delay in these units is the number of bounces the particle makes before being scattered.

Chapter 18
Scattering: 3-D

In order to appreciate the intricacies of quantum-mechanical scattering, you have to know something about classical scattering. When two particles collide and undergo elastic scattering, both the classical and quantum scattering problems can be very difficult to solve. The first step in the analysis of this problem is to make a transformation to the center-of-mass frame of the two particles. Then the interaction can be reduced to the scattering of a particle having reduced mass μ from a center of potential having relative coordinate $\mathbf{r}$. One calculates the scattering in the center-of-mass frame and then must transform back to laboratory coordinates. I will discuss only the problem of scattering in the center-of-mass frame or, equivalently, scattering of a particle having mass μ from a potential $V(r)$ that is assumed to possess spherical symmetry. The particle is incident along the z axis and has energy E.

18.1 Classical Scattering

Let me first review classical scattering, indicated schematically in Fig. 18.1. The scattering depends on the *impact parameter* b and energy E of the incoming particle, as well as on the nature of $V(r)$. If the particle has initial speed $v_0 = \sqrt{2E/\mu}$, the magnitude of the particle's angular momentum, impact parameter, and energy are related by

$$|\mathbf{L}| = L = \mu v_0 b = \sqrt{2\mu E}b; \tag{18.1a}$$

$$E = \frac{L^2}{2\mu b^2}. \tag{18.1b}$$

P.R. Berman, *Introductory Quantum Mechanics*, UNITEXT for Physics,
https://doi.org/10.1007/978-3-319-68598-4_18

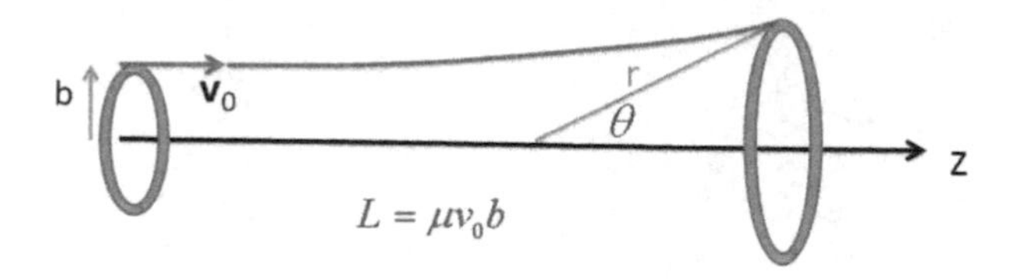

Fig. 18.1 Scattering by a spherically symmetric potential. Particles having initial velocity $\mathbf{v}_0 = v_0\mathbf{u}_z$ that are incident with an impact parameter between b and $b + db$ are scattered into a ring having area $2\pi r^2 \sin\theta d\theta$

Since the potential is spherically symmetric the angular momentum is a constant of the motion.

Moreover, as shown in Fig. 18.1, any particle incident within a ring having radius b and width db is scattered into a ring on a spherical surface having polar angle θ $(0 \le \theta \le \pi)$, circumference $2\pi r \sin\theta$ and width $rd\theta$. Thus, the area of the ring into which the particle is scattered $d\sigma/d\Omega$ is $2\pi r^2 \sin\theta d\theta$. The *differential cross section* $d\sigma/d\Omega$ is then defined by

$$\frac{d\sigma}{d\Omega}d\Omega = \frac{\text{number of particles scattered into } d\Omega \text{ per unit time}}{I}, \tag{18.2}$$

where I is the number of particles incident per unit area per unit time and

$$d\Omega = 2\pi \sin\theta d\theta \tag{18.3}$$

is an element of solid angle. Since each particle in the ring having radius b and width db is scattered into $d\Omega$,

$$\frac{d\sigma}{d\Omega}2\pi \sin\theta d\theta = \frac{I 2\pi b db}{I} \tag{18.4}$$

or

$$\frac{d\sigma}{d\Omega} = \left|\frac{b}{\sin\theta}\frac{db}{d\theta}\right| = \left|\frac{b}{\sin\theta \frac{d\theta}{db}}\right|, \tag{18.5}$$

where, by convention, $\frac{d\sigma}{d\Omega}$ is always positive. If more than one impact parameter b_j gives rise to scattering at the *same* θ, then Eq. (19.1) is replaced by

$$\frac{d\sigma}{d\Omega} = \sum_j \left|\frac{b_j}{\sin\theta \frac{d\theta}{db_j}}\right|. \tag{18.6}$$

The problem is reduced to finding θ as a function of b.

The *total cross section* σ is defined as

$$\sigma = \frac{\text{number of particles scattered per unit time}}{I} = \int \frac{d\sigma}{d\Omega} d\Omega. \tag{18.7}$$

For any infinite range potential, the classical cross section is infinite, since particles are scattered no matter how large their impact parameter. For potentials having a finite range $r_{\max}$, $\sigma = \pi r_{\max}^2$.

Calculating θ as a function of b is a standard problem in classical mechanics. Since the potential is spherically symmetric, the orbit is in a plane perpendicular to the direction of the angular momentum and the total energy,

$$E = \frac{1}{2}\mu \dot{r}^2 + \frac{L^2}{2\mu r^2} + V(r), \tag{18.8}$$

is the sum of the potential energy and both radial and angular contributions to the kinetic energy. The *radial* motion is determined by the effective potential

$$V_{\text{eff}}(r) = V(r) + \frac{L^2}{2\mu r^2}, \tag{18.9}$$

already discussed in Chaps. 9 and 10.

Equation (18.8) can be integrated directly to obtain r as a function of t or of the scattering angle θ; however, I need to distinguish the *scattering angle* θ measured in the laboratory from the *deflection angle* Θ, which is the final value of the orbit angle relative to the z-axis (see Fig. 18.2). The scattering angle is always restricted to lie between 0 and π. If the potential is repulsive, the deflection angle must also be between 0 and π since the particle either goes straight through or is deflected away from the scattering center. In this case the scattering and deflection angles are identical. For attractive potentials, the particle is attracted so the deflection angle is negative (remember Apollo 13). Moreover the attraction can be so great that

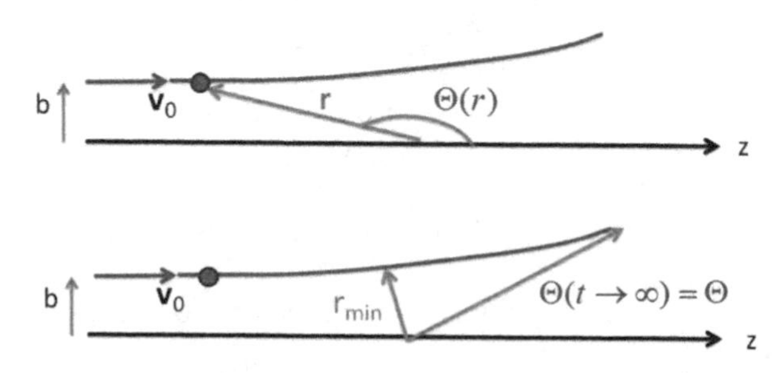

Fig. 18.2 At any point in the scattering the orbit angle can be considered as a function of the radial distance r. The final value of the orbit angle as $t \sim \infty$ is designated as the deflection angle

the deflection angle can even approach negative infinity under certain conditions (orbiting). The relation between the two angles is given by

$$\begin{aligned} \theta &= |\Theta|; & -\pi &\leq \Theta \leq \pi; \\ \theta &= \Theta + 2q\pi; & -2q\pi &\leq \Theta \leq (-2q+1)\,\pi; \\ \theta &= -\Theta - 2q\pi; & -(2q+1)\,\pi &\leq \Theta \leq -2q\pi, \end{aligned} \tag{18.10}$$

where q is a non-negative integer. Since $d\Theta/db = \pm d\theta/db$ and $\sin\Theta = \pm\sin\theta$, θ can be replaced by Θ in Eq. (18.5).

In any scattering problem involving a spherically symmetric potential, there is a radius of closest approach $r = r_{\min}$ about which the orbit is symmetric. The time origin is chosen such that $t = 0$ when $r = r_{\min}$. For spherically symmetric potentials the orbit is in a plane perpendicular to $\mathbf{L}$ and the orbit at any time can be specified by the radial coordinate and the orbit angle Θ, which can be taken to be a function of time or radial coordinate. Since the orbit is symmetric about $r = r_{\min}$, $\Theta(r)$ is a double valued function of r; for each value of r, there are two values of Θ. However, if you restrict r such that $r < r_{\min}$ or $r > r_{\min}$, it becomes a single-valued function. The particle is incident along the *negative* z-axis at $t = -\infty$, with $\Theta(r \to \infty, t \to -\infty) = \pi$. Owing to the symmetry of the orbit

$$\begin{aligned} \Theta(r = r_{\min}, t = 0) &= \pi + \frac{\Theta(r \to \infty, t \to \infty) - \pi}{2} \\ &= \frac{\pi + \Theta}{2}, \end{aligned} \tag{18.11}$$

where

$$\Theta \equiv \Theta(r \to \infty, t \to \infty) \tag{18.12}$$

is the deflection angle (see Fig. 18.2).

Equation (18.8) can be solved to yield

$$\dot{r} = \pm\sqrt{\frac{2}{\mu}}\sqrt{E - \frac{L^2}{2\mu r^2} - V(r)} \tag{18.13}$$

with

$$L = -\mu r^2 \dot{\Theta} \tag{18.14}$$

(note that $\dot{r} < 0$ for $r < r_{\min}$, $\dot{r} > 0$ for $r > r_{\min}$, and $\dot{\Theta} < 0$). Therefore, for $r > r_{\min}$

$$\frac{d\Theta}{dr} = \frac{\dot{\Theta}}{\dot{r}} = \frac{-L}{\mu r^2 \sqrt{\frac{2}{\mu}}\sqrt{E - \frac{L^2}{2\mu r^2} - V(r)}}. \tag{18.15}$$

Since Θ is a single-valued function of r for $r > r_{\min}$, I can integrate Eq. (18.15) to obtain

$$\Theta - \Theta(r = r_{\min}) = -\frac{L}{\sqrt{2\mu}} \int_{r_{\min}}^{\infty} \frac{dr}{r^2\sqrt{E - \frac{L^2}{2\mu r^2} - V(r)}};$$

$$\Theta - \left[\frac{\pi + \Theta}{2}\right] = -\frac{L}{\sqrt{2\mu}} \int_{r_{\min}}^{\infty} \frac{dr}{r^2\sqrt{E - \frac{L^2}{2\mu r^2} - V(r)}};$$

$$\Theta = \pi - 2b \int_{r_{\min}}^{\infty} \frac{dr}{r^2\sqrt{1 - \frac{b^2}{r^2} - \frac{V(r)}{E}}}, \tag{18.16}$$

where Eq. (18.1b) was used.

As an example, consider scattering from a hard sphere having radius a (see Fig. 18.3), for which $r_{\min} = a$ and $V(r) = 0$ for $r > a$. Then, for $b < a$

$$\Theta = \pi - 2b \int_{a}^{\infty} \frac{dr}{r^2\sqrt{1 - \frac{b^2}{r^2}}} = 2\cos^{-1}(b/a), \tag{18.17}$$

which can also be read from the figure. For $b > a$, $\Theta = 0$. With

$$\frac{d\Theta}{db} = -\frac{2}{a\sqrt{1 - (b/a)^2}}; \tag{18.18a}$$

$$\sin\Theta = 2\cos\frac{\Theta}{2}\sin\frac{\Theta}{2} = 2\frac{b}{a}\sqrt{1 - \frac{b^2}{a^2}}, \tag{18.18b}$$

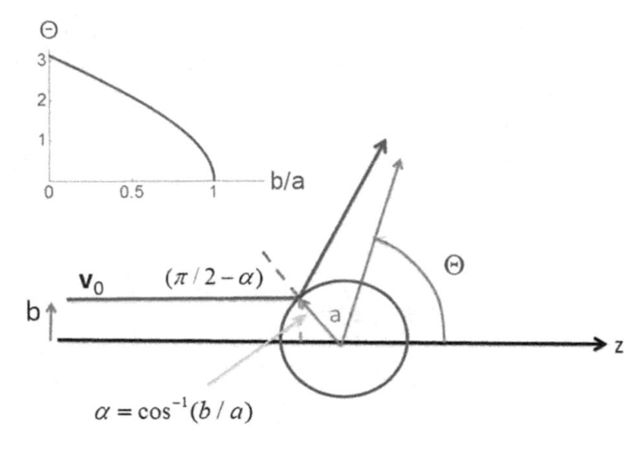

Fig. 18.3 Hard sphere scattering. The graph insert shows the deflection angle as a function of impact parameter

it follows that

$$\frac{d\sigma}{d\Omega} = \left| \frac{b}{\sin\Theta \frac{d\Theta}{db}} \right|$$

$$= \frac{b}{\frac{2b}{a}\sqrt{1-(b/a)^2}\frac{2}{a\sqrt{1-(b/a)^2}}} = \frac{a^2}{4}. \tag{18.19}$$

The scattering is isotropic! Moreover the total cross section is

$$\sigma = \int \frac{d\sigma}{d\Omega} d\Omega = \frac{a^2}{4} \int d\Omega = \frac{a^2}{4} 4\pi = \pi a^2, \tag{18.20}$$

as expected.

To discuss arbitrary potentials, it is useful to look at the effective potential. Remember that

$$L = \sqrt{2\mu E} b; \tag{18.21}$$

for a given energy, *specifying the angular momentum is equivalent to specifying the impact parameter*. Using the effective potential for scattering problems is a little different than using it for bound state problems, as I did in Chap. 10. In bound state problems, the radial motion is restricted between a minimum and maximum radius. In scattering problems, however, the particle is incident from $r = \infty$, enters the scattering region, reaches a minimum radial distance $r = r_{\min}$, and then exits the scattering region with its radial distance again approaching $r = \infty$ as $t \sim \infty$. Remember that the effective potential is for the *radial motion only* and does not give a complete description of the orbit. At $r = r_{\min}$, $\dot{r} = 0$ and all contributions to the kinetic energy of the particle come from the angular motion.

Let us first look at the case where there is *no* potential. Of course, there is no preferred origin in this problem, but we can pick one at random. For a free particle the path is a straight line having impact parameter b relative to the origin, as shown in Fig. 18.4. However, relative to this arbitrarily chosen origin, the particle has both a radial and angular contribution to its kinetic energy if $L \neq 0$ (both the radial distance and polar angle of the particle vary during its motion). The effective potential for a free particle is

$$V_{\text{eff}} = \frac{L^2}{2\mu r^2} = \frac{Eb^2}{r^2}. \tag{18.22}$$

and the minimum radial distance is

$$r_{\min} = \sqrt{\frac{L^2}{2\mu E}} = b. \tag{18.23}$$

These features are seen clearly in Fig. 18.4.

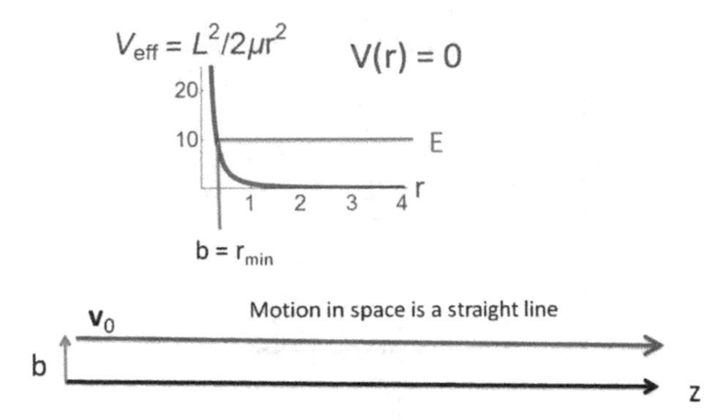

Fig. 18.4 "Scattering" in the absence of a potential. The effective potential is shown for $L \neq 0$. The minimum radius is $r_{min} = b$. In coordinate space, the particle path in space is a straight line parallel to the z-axis

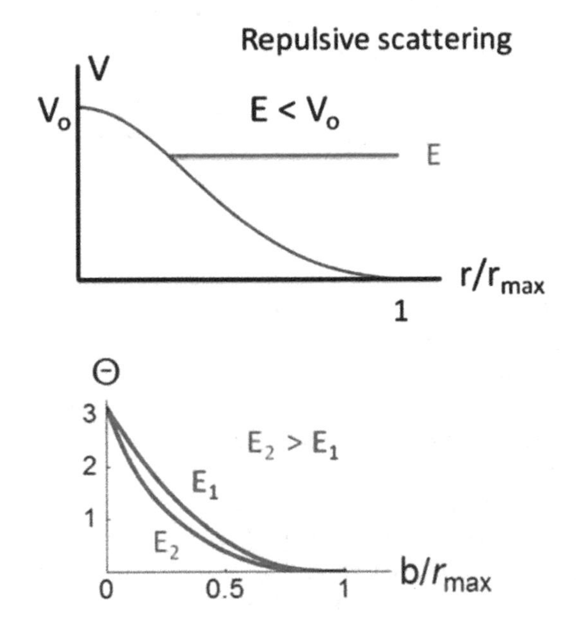

Fig. 18.5 Scattering by a finite range repulsive potential when the energy is less than the maximum of the potential

Next consider scattering by a monotonically decreasing repulsive potential having finite range r_{max}, such as that shown in Fig. 18.5, when the energy E of the particle is *less* than the maximum height V_0 of the potential. In this case, for $b = 0$ ($L = 0$) the particle is repelled by the potential and goes back along its original direction, $\Theta = \pi$. With increasing impact parameter b, the strength of the potential and the scattering angle decreases monotonically, ultimately reaching zero when the impact parameter is larger than the range r_{max}. If $b > r_{max}$, the particle moves on a

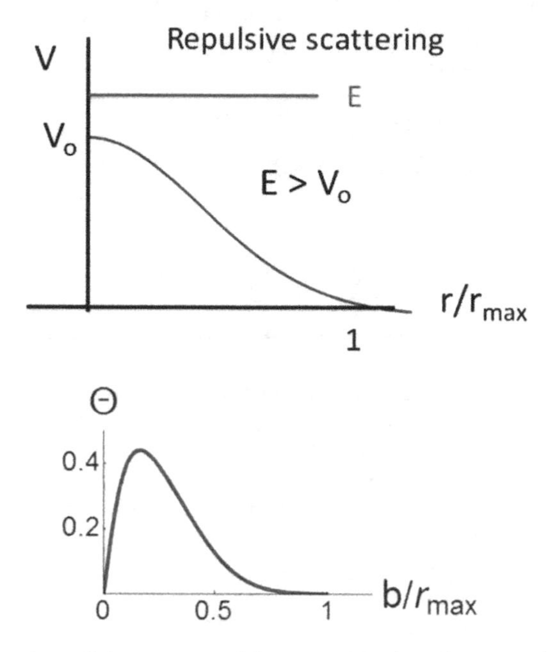

Fig. 18.6 Scattering by a finite range repulsive potential when the energy is greater than the maximum of the potential

straight line and does not encounter the potential at all. Graphs of Θ as a function of b are shown in Fig. 18.5. For a given impact parameter $b < r_{\max}$ the deflection angle decreases with increasing energy, as would be expected.

The scattering is more interesting when the energy is greater than the maximum of the potential energy (see Fig. 18.6). In this case, for $b = 0$ ($L = 0$) the particle slows down when it reaches the potential, passes through the origin, and speeds up as it leaves the potential; the deflection angle is $\Theta = 0$. But the scattering angle must also go to zero when the impact parameter is larger than the range $r_{\max}$ of the potential. As a consequence, the scattering angle must pass through a maximum. A graph of Θ as a function of b is shown in Fig. 18.6. Since $d\Theta/db = 0$ at some scattering angle, the differential scattering cross section becomes *infinite* at this impact parameter, analogous to *rainbow scattering* (actually rainbow scattering actually depends on the *wave* nature of light—see below).

For attractive potentials new features appear in the scattering since the effective potential is no longer a monotonic function of r. I will assume that the potential falls off faster than $1/r^2$ for large r and that the potential is finite at $r = 0$. With these restrictions, the effective potential for different values of angular momentum takes on the general structure indicated in Fig. 18.7. It is positive at $r = 0$ (except for $L = 0$) and also positive as $r \to \infty$ (except for $L = 0$). For sufficiently small values of $L = \sqrt{2\mu E}b \neq 0$ there is a *maximum* in the effective potential (see Fig. 18.7),

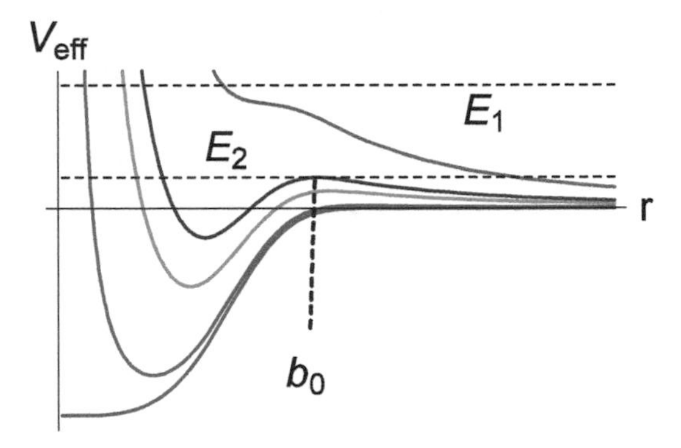

Fig. 18.7 Effective potentials for an attractive potential. The lowest curve is for $L = 0$ and each subsequent curve corresponds to a higher value of angular momentum. For sufficiently large L, there is no longer a local maximum

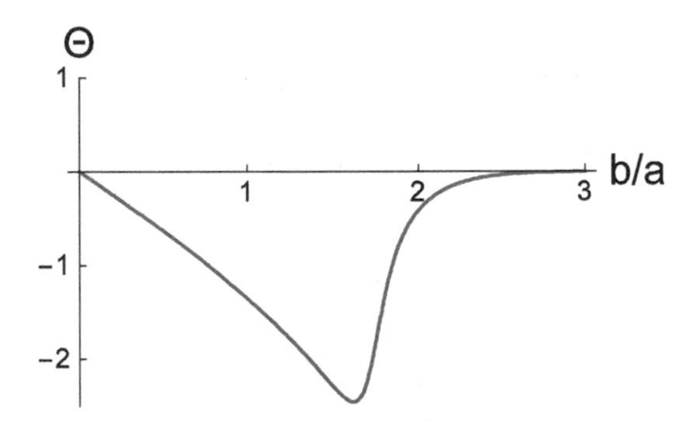

Fig. 18.8 Deflection angle for the attractive Gaussian potential of Eq. (18.24) for $V_0/E = 4$, corresponding to an energy such as E_1 shown in Fig. 18.7

but for larger values of L, no maximum occurs and the effective potential simply decays monotonically.

How is a particle scattered by such a potential? For impact parameter $b = 0$ ($L = 0$) the particle passes through the origin, $\Theta = 0$. The deflection angle also goes to zero for large values of b, but from *negative* values, since the potential is attractive. For intermediate values of b, the nature of the scattering depends on the energy of the incoming particle. For an energy such as E_1 shown in Fig. 18.7 there is *no* value of the angular momentum for which the energy intersects the effective potential at a local maximum. In this limit, a graph of the deflection angle versus impact parameter is shown in Fig. 18.8 for a potential of the form

$$V(r) = -V_0 e^{-r^2/a^2}. \tag{18.24}$$

There is rainbow-like scattering where $d\Theta/db = 0$, as there was for repulsive scattering with $E > V_0$.

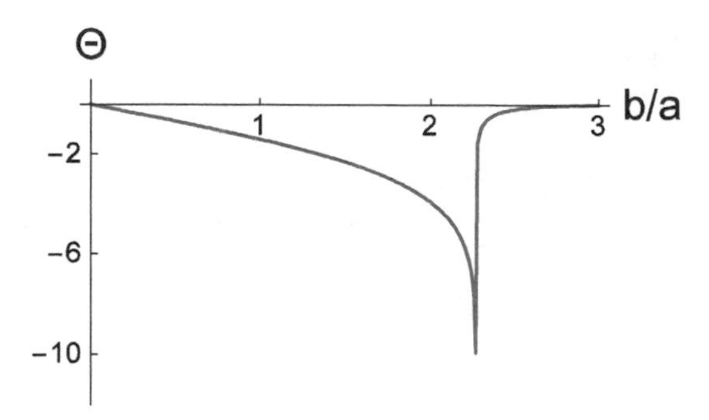

Fig. 18.9 Deflection angle for the attractive Gaussian potential of Eq. (18.24) for $V_0/E = 16$, corresponding to an energy such as E_2 shown in Fig. 18.7. Orbiting occurs for $b/a = 2.27$

On the other hand, for energies such as E_2 in Fig. 18.7 there is *always* a value of angular momentum for which the energy intersects the effective potential at its local maximum. At this angular momentum or impact parameter, the particle approaches the radius $r = b_0$ corresponding to the local maximum, but it takes an *infinite* time to get there. This corresponds to *orbiting* and the deflection angle can take on infinitely many values an infinite number of times. Although the deflection angle diverges, the differential cross section is finite. In Fig. 18.9, the deflection angle Θ is plotted as a function of b/a for the potential given by Eq. (18.24) with $V_0/E = 16$. Near $b/a = 2.27$, which corresponds to the impact parameter where $dV(r)/dr = 0$ for this potential and energy, the deflection angle is greater than 12π, indicating that orbiting has occurred.[1]

When orbiting occurs, there can be several impact parameters $b \neq 0$ for which $-\Theta_N$ is an even multiple of π (forward scattering) and for which $-\Theta_N$ is an odd multiple of π (backscattering). Since the scattering angle θ is equal to either 0 or π for these values of the deflection angle Θ_N, the differential scattering cross section, $d\sigma/d\Omega = |b/(\sin\theta d\theta/db)|$, become large or even infinite at such deflection angles (depending on the value of $d\theta/db$ at these points). This enhanced scattering is referred to as *glory scattering*. Road signs have a coating that gives rise to glory backscattering; the backscattering of light from water droplets is another example of the glory effect.

Although I said that there is no longer a local maximum with increasing L for attractive potentials, there is an exception to this general result. For potentials having a sharp cutoff, such as the *spherical well potential*,

[1] If you expand the square root in the denominator of Eq. (18.16) about $r = r_{\min}$ at the impact parameter corresponding to orbiting, you will find it varies as $(r - r_{\min})$, giving a deflection angle that diverges as $\ln[(r - r_{\min})/a]$. Numerically, it is very hard to reproduce this very slow divergence. There are very special values, $V_0/E = e^2$, $B_0/E = 4$, $r_{\min}/a = \sqrt{2}$, for which the *second* derivative also vanishes when $E = E_0$; in this case, the divergence is more rapid, varying as $[(r - r_{\min})/a]^{-1/2}$.

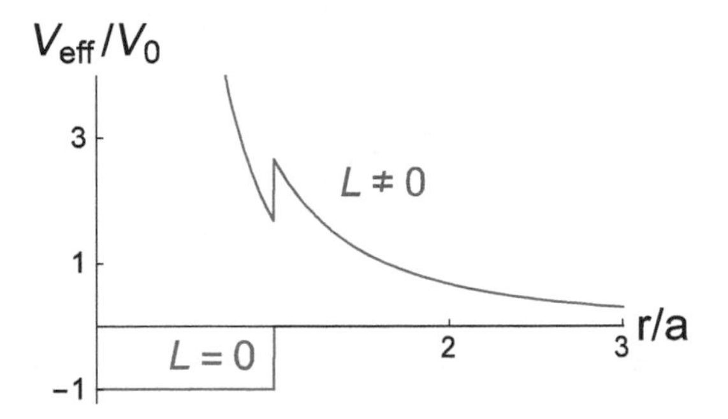

Fig. 18.10 Effective potential for a spherical well potential

$$V(r) = \begin{cases} -V_0 & r \leq a \\ 0 & r > a \end{cases}, \tag{18.25}$$

there is *always* a local maximum in the effective potential for $L \neq 0$ (see Fig. 18.10), independent of the value of L. The spherical well potential is attractive in the sense that any particle incident with impact parameter $b < a$ undergoes scattering with a negative deflection angle. It gives rise to scattering that is analogous to the scattering of electromagnetic radiation by a dielectric sphere having index of refraction,

$$n = \sqrt{1 + V_0/E}, \tag{18.26}$$

but with *no* reflections off the sphere (only transmission). For the spherical well potential, the deflection angle as a function of impact parameter can be calculated from Eq. (18.16) as

$$\Theta = 2\left[-\sin^{-1}\left(\frac{b}{a}\right) + \sin^{-1}\left(\frac{b}{na}\right)\right] \text{Heaviside}(1 - b/a), \tag{18.27}$$

where Heaviside$(x) = 1$ if $x > 0$ and is zero otherwise. The deflection angle is shown in Fig. 18.11 as a function of b/a for $n = 1.4$. The differential cross section, which can be calculated using Eqs. (18.5) and (18.27), is given by

$$\frac{d\sigma}{d\Omega} = \frac{n^2 a^2}{4\cos(\theta/2)} \frac{[n - \cos(\theta/2)][n\cos(\theta/2) - 1]}{\left[1 + n^2 - 2n\cos(\theta/2)\right]^2} \times \text{Heaviside}\left[n\cos(\theta/2) - 1\right] \tag{18.28}$$

and vanishes for deflection angles having $\Theta < \Theta_{\min} = -2\cos^{-1}(1/n)$. There is no rainbow-like or glory scattering in this classical problem.

I finish this section with a brief discussion of scattering of light rays by a dielectric sphere. In contrast to classical particle scattering by a spherical well potential, the scattering of light rays by a dielectric sphere, such as a water droplet,

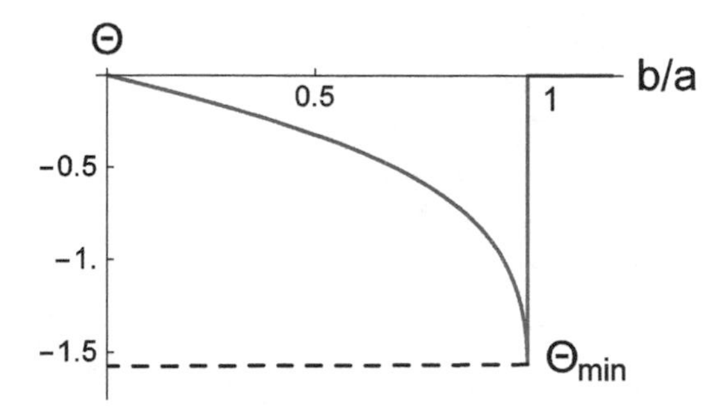

Fig. 18.11 Deflection angle for a spherical well potential with $n = \sqrt{1 + V_0/E} = 1.4$

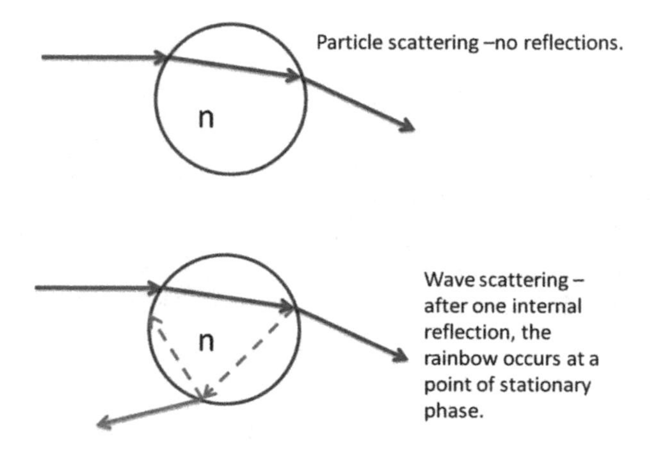

Fig. 18.12 Classical scattering by a spherical potential and rainbow scattering by a dielectric sphere (the wave reflected off the outer surface of the dielectric sphere is not shown)

can lead to rainbow and glory scattering as a result of reflections of the light on the *inner* surface of the water droplet (see Fig. 18.12). Even though the rainbow represents a geometrical optics limit for light (as does refraction at a dielectric interface), the formation of a rainbow still relies on the fact that light is a wave. Since there is a sudden change of index of refraction at the air–water droplet interface, there are wave-like effects (just as the reflection coefficient at a potential step in the quantum problem is independent of $\hbar$—it is a geometrical, wave-like effect). The scattering cross section can be thought to result from *rays* of light incident with different impact parameters. A light ray that is incident on the sphere with impact parameter b is partially reflected and partially transmitted at the outer surface of the sphere. It then undergoes an infinite number of reflections (and transmissions) on the *inner* surface of the sphere. You can show using geometric optics that the deflection angle Θ_N for the light that emerges after N reflections on the inner surface of the sphere is given by

$$\Theta_N = \left\{\left[\pi - 2\sin^{-1}(b/a)\right] - N\left[\pi - 2\sin^{-1}(b/na)\right]\right\} \text{Heaviside}(1 - b/a), \tag{18.29}$$

where n is the index of refraction of the sphere relative to the medium surrounding it. Equation (18.29) also holds for the reflection of an incoming ray off the *outer* surface of the sphere if you set $N = 0$.

For $N = 1$ the result is similar to that of classical scattering by a spherical well potential having depth V_0, with $V_0/E = n^2 - 1$—there is no contribution to rainbow or glory scattering. For $N \geq 2$, you can use Eq. (18.29) to search for contributions to rainbows (where $d\Theta_N/db = 0$) or glories (values of $-\Theta_N$ that are an integral multiple of π for $b \neq 0$). In this manner, you can show that, for $n = 4/3$ (water), the principal rainbow angle, corresponding to $N = 2$, occurs at $\Theta_N = -138°$ or $42°$ from the back scattering angle. Note that there is no *total* internal reflection on the inner surface, simply reflection. Contributions to glory scattering are possible for $N \geq 3$.

Equation (18.29) is also valid in the case where light goes from a higher to lower index as from scattering off an air bubble in water, provided $b/a < n$ ($n < 1$ is the relative index of refraction of the bubble to the medium surrounding the bubble). On the other hand, if $n < b/a < 1$, there is total reflection from the outer surface of the sphere with $\Theta = 2\cos^{-1}(b/a)$. In general, for $n < 1$, Eq. (18.29) is replaced by

$$\begin{aligned}\Theta_N = &\left\{\left[\pi - 2\sin^{-1}(b/a)\right] - N\left[\pi - 2\sin^{-1}(b/na)\right]\right\} \text{Heaviside}(n - b/a) \\ &+2\cos^{-1}(b/a)\text{Heaviside}(1 - b/a)\delta_{N,0}.\end{aligned} \tag{18.30}$$

In this case you can show that there is no rainbow scattering, but contributions to glories are still possible for $N \geq 3$. For $N = 0$ the result corresponds to the classical problem of scattering by a repulsive, *spherical barrier potential* when the energy of the particle is greater than the height V_0 of the well, with $V_0/E = 1 - n^2$.

18.2 Quantum Scattering

I now turn my attention to quantum scattering. This is a really important problem, since many experiments must be analyzed in terms of the differential cross sections associated with the scattering process. It is clear that quantum effects become important as soon as the scattering potential varies significantly over distances of a de Broglie wavelength of the incident particle (wave). If there is a sharp change in the potential, such as in hard sphere scattering, there are always diffraction effects. In this limit, no matter what the energy of the incident particle, there is a "wave-like" contribution to the scattering. We have already seen such effects in the scattering of a wave packet from a one-dimensional potential barrier or well. Aside from the sharp cutoff effects, there can be additional quantum effects when the de Broglie wavelength of the incident particle is comparable to distances over which the potential changes. Thus, for low energy scattering, a quantum treatment is often needed. The rainbow-like, orbiting, and glory effects that occur for fixed impact parameters in classical scattering are "washed out" in quantum scattering since it is not possible to specify both a precise impact parameter and scattering angle owing to the uncertainty principle, but analogous processes still occur. Moreover, we will see that tunneling also occurs in certain limits, leading to very sharp resonances in the scattering.

Quantum-mechanical scattering represents a very complex problem in most cases. There are many books devoted to this subject. In this introductory presentation, I limit the discussion to scattering of structureless particles by spherically symmetric potentials. The use of the effective potential greatly aids the analysis of the problem. As for one-dimensional problems, the calculations can be given in terms of a steady-state or time-dependent approach. I will mention the time-dependent approach briefly, but concentrate on the steady-state approach. In both cases, the theory is based on the prediction (I will prove this later on) that the *asymptotic* form of the (unnormalized) eigenfunctions as $r \to \infty$ is of the form

$$\psi_k(\mathbf{r}) \sim \left(e^{ikz} + f_k(\theta) \frac{e^{ikr}}{r} \right). \tag{18.31}$$

The first term in Eq. (18.31) corresponds to a plane wave incident from the left in the positive z direction, while the second term corresponds to a *spherically scattered* wave that is weighted with the *scattering amplitude* $f_k(\theta)$. Had I carried out a time-dependent treatment, I would find that an initial wave packet having spatial extent w in the z direction and propagating in the z direction is transformed into the original wave packet propagating as if no scattering occurred, plus an outgoing *shell* having thickness w that is centered at radius $r = -z_0 + v_0 t$, where $-z_0$ is the initial location of the center of the packet and v_0 is the average speed of the initial packet. The spherical shell is weighted by $f_k(\theta)$. To arrive at this picture of the scattering, it is necessary to assume that the initial wave packet has a k-space amplitude $\Phi(\mathbf{k})$ that is sharply peaked about $\mathbf{k} = k_0 \mathbf{u}_z$ and that the phase of $f_k(\theta)$ is approximately constant for values of $\mathbf{k}$ close to $k_0 \mathbf{u}_z$. If the phase is not constant, but is slowly varying, there is a time delay associated with the scattering process that can depend on the scattering angle. For more rapid variations of the phase, such as those that occur near resonances, the outgoing wave packet has a complicated structure owing to dispersion associated with the phase of the scattering amplitude.

As in the 1-D case, I can get the differential scattering cross section by considering the probability current density in all but the forward direction (I return to what happens in the forward direction below). The incident probability current density is that associated with the e^{ikz} term in Eq. (18.31) and is equal to

$$\mathbf{J}_i = v_k \mathbf{u}_z = (\hbar k / \mu)\, \mathbf{u}_z, \tag{18.32}$$

where $v_k = \hbar k / \mu$. The scattered current density is that associated with the $\psi_s(\mathbf{r}) = f_k(\theta) e^{ikr}/r$ term in Eq. (18.31) and is given by

$$\mathbf{J}_s(\mathbf{r}) = \frac{i\hbar}{2\mu}\left[\psi_s(x)\nabla\psi_s^*(\mathbf{r}) - \psi_s^*(\mathbf{r})\nabla\psi_s(\mathbf{r})\right]$$
$$= \frac{i\hbar}{2\mu}\left[f_k(\theta)\frac{e^{ikr}}{r}\nabla\left\{f_k^*(\theta)\frac{e^{-ikr}}{r}\right\} - f_k^*(\theta)\frac{e^{-ikr}}{r}\nabla\left\{f_k(\theta)\frac{e^{ikr}}{r}\right\}\right]. \quad (18.33)$$

I use the expression for the gradient operator in spherical coordinates,

$$\nabla = \frac{\partial}{\partial r}\mathbf{u}_r + \frac{1}{r}\frac{\partial}{\partial\theta}\mathbf{u}_\theta + \frac{1}{r\sin\theta}\frac{\partial}{\partial\phi}\mathbf{u}_\phi, \quad (18.34)$$

to calculate

$$\nabla\left\{f_k(\theta)\frac{e^{ikr}}{r}\right\} = \left(ik - \frac{1}{r}\right)f_k(\theta)\frac{e^{ikr}}{r}\mathbf{u}_r + \frac{e^{ikr}}{r^2}\frac{df_k(\theta)}{d\theta}\mathbf{u}_\theta. \quad (18.35)$$

To get the outgoing probability flux into a solid angle $d\Omega$, I multiply $\mathbf{J}_s(\mathbf{r})$ by $r^2 d\Omega$, take the scalar product with $\mathbf{u}_r$, and let $r \to \infty$,

$$\lim_{r\to\infty} \mathbf{J}_s \cdot \mathbf{u}_r r^2 d\Omega = v_k \left|f_k(\theta)\right|^2 d\Omega. \quad (18.36)$$

The differential cross section is then calculated as

$$\frac{d\sigma}{d\Omega}d\Omega = \frac{v_k \left|f_k(\theta)\right|^2 d\Omega}{\mathbf{J}_i \cdot \mathbf{u}_z} = \left|f_k(\theta)\right|^2 d\Omega. \quad (18.37)$$

That is,

$$\frac{d\sigma}{d\Omega} = \left|f_k(\theta)\right|^2 \quad (18.38)$$

is the differential scattering cross section and

$$\sigma = \int \left|f_k(\theta)\right|^2 d\Omega = 2\pi\int_0^\pi \left|f_k(\theta)\right|^2 \sin\theta \, d\theta \quad (18.39)$$

is the total scattering cross section.

There are effectively two ways to calculate $f_k(\theta)$. One involves a formal solution of the Schrödinger equation and gives rise to the *Born series*. The second involves a solution of the radial equation for a given value of angular momentum ℓ and is referred to as the method of partial waves, for reasons to be discussed in a moment. Of course, if you can find the exact eigenfunctions for a given potential, you need only expand them for large r and compare the results with Eq. (18.31) to identify $f_k(\theta)$.

18.2.1 Method of Partial Waves

In the classical problem, for a given energy, each incident impact parameter or value of the angular momentum is associated with a single scattering angle. In quantum mechanics this picture is no longer valid since it would imply that we could simultaneously specify both the position (impact parameter) and momentum (given by the scattering angle) of the particle. Nevertheless, it is possible to write the differential scattering amplitude as arising from contributions from each value of ℓ, since ℓ is a conserved quantum number. This is the origin of the terminology, *method of partial waves.* Several (in principle, *all*) values of ℓ give rise to scattering at the same angle, and there can be interference among the various contributions. Nevertheless, it is convenient to view the quantum-mechanical scattering as arising from contributions from different impact parameters.

An uncertainty principle argument can also help you to understand why quantum total cross sections can be finite for infinite range potentials, whereas they are infinite in the corresponding classical case. The question is, when does the uncertainty principle lead to a breakdown of a classical approach. The uncertainty principle requires that the transverse momentum Δp_t imparted to the particle by the potential multiplied by the impact parameter that gives rise to this scattering must be greater than or of order $\hbar$; that is

$$b\Delta p_t \gtrsim \hbar. \tag{18.40}$$

For a potential V varying as C/r^s, the change in transverse momentum can be estimated as

$$\Delta p_t = \int F dt \approx \frac{dV}{dr}\frac{b}{v} \approx \frac{C}{b^{s+1}}\frac{b}{v} = \frac{C}{b^s}\frac{b}{v}, \tag{18.41}$$

where v is the incident speed. Therefore, the maximum impact parameter $b_{\max}$ for which a classical picture can remain valid is determined from the condition

$$b_{\max}\frac{C}{vb_{\max}^s} \approx \hbar, \tag{18.42}$$

which yields

$$b_{\max} \approx \left(\frac{C}{\hbar v}\right)^{\frac{1}{s-1}}. \tag{18.43}$$

As a consequence, we might expect that the total quantum cross section is of order

$$\pi b_{max}^2 = \pi \left(\frac{C}{\hbar v} \right)^{\frac{2}{s-1}}, \tag{18.44}$$

a quantity that is finite for $s > 1$. Note that Eq. (18.44) predicts an infinite cross section for Coulomb scattering, which is actually the case, even quantum-mechanically. For potentials that fall off faster than $1/r$ as $r \to \infty$ there is always an *effective* maximum range of the potential given by $r_{max} = b_{max}$ and an associated diffractive scattering component in the differential scattering cross section that would arise from a potential having a *sharp* cut-off at $r = r_{max}$.

To apply the method of partial waves, I must solve the Schrödinger equation exactly. This is no easy task, but at least I can solve it for large r. If I assume that the scattering potential falls off faster than $1/r^2$, the effective potential at large r is dominated by the angular momentum term—in other words, the effective potential is that of a free particle. I have considered free particle solutions in spherical coordinates in Chap. 10. The general solution for the radial wave function of a free particle for $r \neq 0$ can be written as (see Appendix 1)

$$R_\ell(r) = ru_\ell(r) = A_\ell j_\ell(kr) + B_\ell n_\ell(kr), \tag{18.45}$$

where

$$k = \sqrt{\frac{2\mu E}{\hbar^2}}, \tag{18.46}$$

are $j_\ell(kr)$ and $n_\ell(kr)$ are spherical Bessel and Neumann functions.

Thus, for spherically symmetric potentials, the solution for the wave function, valid at large r, is

$$\psi_k(\mathbf{r}) \sim \sum_{\ell=0}^{\infty} [A_\ell j_\ell(kr) + B_\ell n_\ell(kr)] P_\ell(\cos\theta). \tag{18.47}$$

Owing to the azimuthal symmetry of the scattering process about the z-axis, the wave function has no ϕ dependence, which is why the Legendre polynomials appear in the summation rather than the spherical harmonics. The actual values of A_ℓ and B_ℓ can be obtained only if you solve the radial equation *exactly* for *all* r. To obtain an expression for the scattering amplitude, I must expand Eq. (18.47) for large kr. Using the asymptotic form of the spherical Bessel and Neumann functions,

$$j_\ell(x) \sim \frac{\sin\left(x - \frac{\ell\pi}{2}\right)}{x}; \tag{18.48}$$

$$n_\ell(x) \sim \frac{-\cos\left(x - \frac{\ell\pi}{2}\right)}{x}, \tag{18.49}$$

valid for $x \gg 1$ and $x \gg \ell$, I find

$$\psi_k(\mathbf{r}) \sim \sum_{\ell=0}^{\infty} \frac{A_\ell \sin\left(kr - \frac{\ell\pi}{2}\right) - B_\ell \cos\left(kr - \frac{\ell\pi}{2}\right)}{kr} P_\ell(\cos\theta). \tag{18.50}$$

Since I have already assumed that

$$\psi_k(\mathbf{r}) \sim \left(e^{ikz} + f_k(\theta)\frac{e^{ikr}}{r}\right), \tag{18.51}$$

I must now equate Eqs. (18.50) and (18.51) to find the appropriate $f_k(\theta)$.

To do so, I first use the identity

$$e^{ikz} = e^{ikr\cos\theta} = \sum_{\ell=0}^{\infty} i^\ell (2\ell+1) j_\ell(kr) P_\ell(\cos\theta), \tag{18.52}$$

which follows from the solution of the free particle Schrödinger equation in spherical coordinates (see Appendix 1). For large kr I use the asymptotic form of the Bessel function to rewrite this equation as

$$e^{ikz} \sim \sum_{\ell=0}^{\infty} i^\ell (2\ell+1) \frac{\sin(kr - \frac{\ell\pi}{2})}{kr} P_\ell(\cos\theta). \tag{18.53}$$

I next set

$$A_\ell = D_\ell \cos\delta_\ell; \tag{18.54a}$$

$$B_\ell = -D_\ell \sin\delta_\ell; \tag{18.54b}$$

$$\tan\delta_\ell = -\frac{B_\ell}{A_\ell}, \tag{18.54c}$$

such that

$$A_\ell \sin\left(kr - \frac{\ell\pi}{2}\right) - B_\ell \cos\left(kr - \frac{\ell\pi}{2}\right) = D_\ell \sin\left(kr - \frac{\ell\pi}{2} + \delta_\ell\right). \tag{18.55}$$

The quantity δ_ℓ is referred to as a *partial wave phase.* By combining Eqs. (18.50), (18.51), (18.53), and (18.55), I find

$$\begin{aligned} \psi_k(\mathbf{r}) &\sim \frac{1}{kr}\sum_{\ell=0}^{\infty} D_\ell \sin\left(kr - \frac{\ell\pi}{2} + \delta_\ell\right) P_\ell(\cos\theta) \\ &= \frac{1}{kr}\sum_{\ell=0}^{\infty} i^\ell (2\ell+1) \sin\left(kr - \frac{\ell\pi}{2}\right) P_\ell(\cos\theta) + \frac{f_k(\theta)}{r} e^{ikr} \end{aligned} \tag{18.56}$$

or

$$f_k(\theta)e^{ikr} = \frac{1}{k}\sum_{\ell=0}^{\infty}\left[D_\ell \sin\left(kr - \frac{\ell\pi}{2} + \delta_\ell\right) - i^\ell(2\ell+1)\sin\left(kr - \frac{\ell\pi}{2}\right)\right] \times P_\ell(\cos\theta). \quad (18.57)$$

To solve this equation, I equate coefficients of $e^{\pm ikr}$ appearing on both sides of the equation. Equating coefficients of e^{-ikr} I find

$$0 = \frac{1}{k}\sum_{\ell=0}^{\infty} i^\ell \left[D_\ell e^{-i\delta_\ell} - i^\ell(2\ell+1)\right] P_\ell(\cos\theta). \quad (18.58)$$

Since θ is arbitrary, the only possible solution is

$$D_\ell = i^\ell e^{i\delta_\ell}(2\ell+1). \quad (18.59)$$

Equating coefficients of e^{ikr} then yields

$$f_k(\theta) = \frac{1}{2ik}\sum_{\ell=0}^{\infty}(-i)^\ell \left[D_\ell e^{i\delta_\ell} - i^\ell(2\ell+1)\right] P_\ell(\cos\theta). \quad (18.60)$$

Combining Eqs. (18.59) and (18.60), I obtain the scattering amplitude

$$f_k(\theta) = \frac{1}{2ik}\sum_{\ell=0}^{\infty}(2\ell+1)\left(e^{2i\delta_\ell} - 1\right) P_\ell(\cos\theta), \quad (18.61)$$

which can also be written as

$$f_k(\theta) = \frac{1}{k}\sum_{\ell=0}^{\infty}(2\ell+1)e^{i\delta_\ell}\sin(\delta_\ell)\, P_\ell(\cos\theta). \quad (18.62)$$

The problem is solved once you calculate the partial wave phases δ_ℓ.

To find the δ_ℓ, you can proceed as follows:

- For each value of ℓ, solve Schrödinger's equation *exactly* for the given potential to obtain the eigenfunctions.

- As $r \to \infty$, express the radial eigenfunctions in the form

$$\begin{aligned} u_\ell(r) &= R_\ell(r)/r = A_\ell j_\ell(kr) + B_\ell n_\ell(kr) \\ &\sim A_\ell \sin\left(kr - \frac{\ell\pi}{2}\right) - B_\ell \cos\left(kr - \frac{\ell\pi}{2}\right) \\ &= D_\ell \sin\left(kr - \frac{\ell\pi}{2} + \delta_\ell\right) \end{aligned} \tag{18.63}$$

from which the δ_ℓ can be extracted.

Of course it is not easy to solve the Schrödinger equation for most potentials. If you need only a few partial waves, then the method of partial waves is most useful. If the range of the potential is of order a, then impact parameters of order a are needed. That is

$$b_{\max} = \frac{L_{\max}}{\mu v} \approx \frac{\hbar \ell_{\max}}{\mu v} \approx a; \tag{18.64a}$$

$$\ell_{\max} \approx \frac{\mu v a}{\hbar} = \frac{pa}{\hbar} = ka. \tag{18.64b}$$

Therefore, the partial wave expansion is especially useful when $ka \lesssim 1$. On the other hand, even for high energies, $ka \gg 1$, it is often possible to solve the radial equation using the WKB method and then extract the phase shifts. In this limit, the method of partial waves can also be used at high energies.

18.2.1.1 Differential and Total Cross Sections: Optical Theorem

Before discussing some specific examples and examining the meaning of the partial wave phases, I derive an expression for the total cross section and relate it to the scattering amplitude. In doing so, I can address the question of the nature of the scattering in the forward direction. From Eqs. (18.38) and (18.62), I find

$$\frac{d\sigma}{d\Omega} = \frac{1}{k^2}\left|\sum_{\ell=0}^{\infty}(2\ell+1)e^{i\delta_\ell}\sin\delta_\ell P_\ell(\cos\theta)\right|^2 \tag{18.65}$$

and

$$\begin{aligned} \sigma = \frac{2\pi}{k^2}\sum_{\ell,\ell'=0}^{\infty}\int_{-1}^{1} & d(\cos\theta)(2\ell+1)e^{i\delta_\ell}\sin\delta_\ell P_\ell(\cos\theta)(2\ell'+1)e^{-i\delta_{\ell'}} \\ & \times \sin\delta_{\ell'} P_{\ell'}(\cos\theta) \end{aligned}$$

$$= \frac{2\pi}{k^2} \sum_{\ell,\ell'=0}^{\infty} (2\ell+1) e^{i\delta_\ell} \sin\delta_\ell (2\ell'+1) e^{-i\delta_{\ell'}} \sin\delta_{\ell'} \frac{2\delta_{\ell,\ell'}}{(2\ell+1)}$$

$$= \frac{4\pi}{k^2} \sum_{\ell=0}^{\infty} (2\ell+1) \sin^2\delta_\ell. \tag{18.66}$$

Since $P_\ell(1) = P_\ell\left[\cos(\theta = 0)\right] = 1$, this can be re-expressed as

$$\sigma = \frac{4\pi}{k} \mathrm{Im}\left[\frac{1}{k} \sum_{\ell=0}^{\infty} (2\ell+1) e^{i\delta_\ell} \sin\delta_\ell P_\ell(\cos 0)\right]$$

$$= \frac{4\pi}{k} \mathrm{Im} f_k(0), \tag{18.67}$$

which is known as the *optical theorem*. It relates the total cross section to the forward scattering amplitude. What is the physical meaning of this equation?

Some insight into the answer to this question can be obtained by returning to the asymptotic form of the wave function

$$\psi_k(\mathbf{r}) \sim \left(e^{ikz} + f_k(\theta)\frac{e^{ikr}}{r}\right). \tag{18.68}$$

The initial probability current density is

$$\mathbf{J}_i = \frac{\hbar k}{\mu} \mathbf{u}_z \tag{18.69}$$

and the final probability current density is

$$\begin{aligned} \mathbf{J}_f &= \frac{i\hbar}{2\mu}\left[e^{ikz} + f_k(\theta)\frac{e^{ikr}}{r}\right] \\ &\quad \times \left[-ike^{-ikz}\mathbf{u}_z - f_k^*(\theta)\frac{e^{-ikr}}{r}\left(ik + \frac{1}{r}\right)\mathbf{u}_r + \frac{e^{-ikr}}{r^2}\frac{df_k^*(\theta)}{d\theta}\mathbf{u}_\theta\right] \\ &\quad + \mathrm{c.c.} \\ &= \frac{\hbar k}{\mu}\mathbf{u}_z + \frac{\hbar k}{\mu}\frac{|f_k(\theta)|^2}{r^2}\mathbf{u}_r \\ &\quad + \left\{\frac{i\hbar}{2\mu}\left[e^{ikz} + f_k(\theta)\frac{e^{ikr}}{r}\right]\frac{e^{-ikr}}{r^2}\frac{df_k^*(\theta)}{d\theta}\mathbf{u}_\theta + \mathrm{c.c.}\right\} \end{aligned}$$

$$+\frac{\hbar k}{2\mu}\left[e^{-ikz}f_k(\theta)\frac{e^{ikr}}{r}\mathbf{u}_z+\text{c.c.}\right]$$
$$+\frac{\hbar k}{2\mu}\left[e^{ikz}f_k^*(\theta)\left(1+\frac{1}{ikr}\right)\frac{e^{-ikr}}{r}\mathbf{u}_r+\text{c.c.}\right]. \tag{18.70}$$

For probability to be conserved, the *change* in the normal component of the probability density integrated over a spherical surface must vanish as the radius R of the sphere approaches infinity. Using the facts that $z = r\cos\theta$, $\mathbf{u}_z\cdot\mathbf{u}_r = \cos\theta$, $\oint\cos\theta d\Omega = 0$, $\mathbf{u}_r\cdot\mathbf{u}_\theta = 0$, $d\Omega = d\phi d(\cos\theta)$, and carrying out the integration over ϕ, I find

$$0=\frac{\mu}{\hbar k}\lim_{R\to\infty}\oint R^2 d\Omega\left(\mathbf{J}_f-\mathbf{J}_i\right)\cdot\mathbf{u}_r=\int d\Omega\left|f_k(\theta)\right|^2+2\pi$$
$$\times\lim_{R\to\infty}\int_{-1}^{1}d(\cos\theta)\left\{\begin{array}{c}\frac{R}{2}\left[e^{-ikR\cos\theta}\cos\theta f_k(\theta)e^{ikR}+\text{c.c.}\right]\\+\frac{R}{2}\left[e^{ikR\cos\theta}f_k^*(\theta)\left(1+\frac{1}{ikR}\right)e^{-ikR}+\text{c.c.}\right]\end{array}\right\}. \tag{18.71}$$

There are two types of terms, the first involves the radial flow of the scattered wave and the second and third represent interference of the incident and scattered waves. Integrating the second and third terms by parts, I obtain

$$\int d\Omega\left|f_k(\theta)\right|^2+\pi\left[\frac{f_k(0)+f_k(\pi)e^{-2ikR}}{-ik}+\text{c.c.}\right]$$
$$+\pi\left[\frac{f_k^*(0)-f_k^*(\pi)e^{2ikR}}{ik}+\text{c.c.}\right]=0. \tag{18.72}$$

Additional terms in the integration by parts are neglected since they vanish in the limit $R\to\infty$. The $e^{\pm 2ikR}$ terms cancel and I am left with

$$\sigma=\int d\Omega\left|f_k(\theta)\right|^2=\frac{2\pi}{ik}\left[f_k(0)-f_k^*(0)\right]=\frac{4\pi}{k}\operatorname{Im}f_k(0). \tag{18.73}$$

This result now has a very simple physical interpretation. The incident wave (or wave packet) is scattered in all directions. In the forward direction the incident wave *interferes* with the scattered wave to reduce the probability current in the forward direction in precisely the amount that corresponds to scattering in all the other directions. In other words, the optical theorem is just a statement of conservation of probability.

18.2.1.2 Interpretation of the Partial Wave Phases

There is also a simple interpretation that can be given to the partial wave phases. For a free particle

$$R_\ell(r) = j_\ell(kr) \sim \frac{\sin\left(kr - \frac{\ell\pi}{2}\right)}{kr} \tag{18.74}$$

as $r \to \infty$. For a particle scattered by a potential, $\sin\left(kr - \frac{\ell\pi}{2}\right)$ is replaced by $\sin\left(kr - \frac{\ell\pi}{2} + \delta_\ell\right)$ [see Eqs. (18.50) and (18.55)]. Thus the partial wave phases are simply phase *shifts* in the asymptotic wave function produced by the potential. For example, for an infinitely repulsive hard sphere potential having radius a, the wave function is displaced from the origin. For $\ell = 0$ it is displaced by a, giving rise to a phase shift $\delta_0 = -ka$. For higher ℓ, the shift depends on ℓ but is always negative. Any purely repulsive potential always produces negative phase shifts, analogous to the fact that classical repulsive scattering always leads to positive deflection angles.

For scattering by attractive potentials, new and interesting phenomena can occur. The phase shifts are positive, in general, because the wave function is pulled in by the potential, but for strong potentials the wave function can be pulled in so much that an extra oscillation occurs and gives rise to negative phase shifts of the type you would normally associate with a repulsive interaction. Although the deflection angle is negative for classical scattering by attractive potentials, deflection angles such as those satisfying $-\pi < \Theta < -2\pi$ correspond to scattering that can mimic that of a repulsive potential. It is often stated that the interatomic potential in Bose condensates is repulsive since repulsive interactions are needed to produce the condensates. Actually the true interaction is attractive, but the *effective* interaction is repulsive.

There is another interpretation that can be given to the phase shifts that is valid for large energy and angular momenta. The radial wave equation for $u_\ell(r) = rR_\ell(r)$ is

$$\frac{d^2u_\ell}{dr^2} + k^2u_\ell - \frac{2\mu V(r)u_\ell}{\hbar^2} - \frac{\ell(\ell+1)}{r^2}u_\ell = 0. \tag{18.75}$$

Asymptotically, as $r \to \infty$,

$$u_\ell(r) \sim \sin\left(kr - \frac{\ell\pi}{2} + \delta_\ell\right). \tag{18.76}$$

Equation (18.75) can be solved using the WKB method in the limit of high energies and high angular momentum, since the effective potential does not vary significantly over a de Broglie wavelength in this limit. By solving Eq. (18.75) as I did in Chap. 16, you can show that the WKB wave function for $r > a$, where a is the classical turning point, is given by

$$u_\ell^{WKB}(r) = \frac{C}{\sqrt{k(r)}}\sin\left[\int_a^r k(r')dr' + \pi/4\right] \tag{18.77}$$

where C is a constant and

$$\begin{aligned} k(r) &= \sqrt{k^2 - \frac{2\mu V(r)}{\hbar^2} - \frac{\ell(\ell+1)}{r^2}} \\ &= \sqrt{\frac{2\mu}{\hbar^2}}\sqrt{E - V(r) - \frac{\hbar^2 \ell(\ell+1)}{2\mu r^2}}. \end{aligned} \tag{18.78}$$

As $r \to \infty$, the radial function can be written as

$$u_\ell^{WKB}(r) = \frac{C}{\sqrt{k(r)}} \sin\left[kr + \int_a^\infty dr' \left[k(r') - k\right] - ka + \pi/4\right]. \tag{18.79}$$

By comparing Eqs. (18.76) and (18.79), you can see that the partial wave phase shift is given by

$$\begin{aligned} \delta_\ell^{WKB} &= \frac{\ell\pi}{2} + \int_a^\infty dr \left[k(r') - k\right] - ka + \pi/4 \\ &= \frac{\ell\pi}{2} + \sqrt{\frac{2\mu E}{\hbar^2}} \int_a^\infty dr \left(\sqrt{1 - \frac{V(r)}{E} - \frac{\hbar^2 \ell(\ell+1)}{2\mu E r^2}} - 1\right) \\ &\quad -ka + \pi/4. \end{aligned} \tag{18.80}$$

For a classical limit in which $\ell \gg 1$, $\ell(\ell+1) \approx \ell^2$, the derivative of the phase shift with respect to ℓ is given by

$$\frac{d}{d\ell}\delta_\ell^{WKB} \approx \frac{\pi}{2} - \hbar^2 \ell \sqrt{\frac{1}{2\mu E \hbar^2}} \int_a^\infty \frac{dr}{r^2\left(\sqrt{1 - \frac{V(r)}{E} - \frac{\hbar^2 \ell^2}{2\mu E r^2}}\right)}. \tag{18.81}$$

If I set $L = \hbar\ell = \mu v b = \sqrt{2\mu E} b$ and compare Eq. (18.81) with Eq. (18.16), I find

$$2\frac{d}{d\ell}\delta_\ell^{WKB} = \pi - 2b \int_a^\infty \frac{dr}{r^2 \sqrt{1 - \frac{b^2}{r^2} - \frac{V(r)}{E}}} = \Theta; \tag{18.82}$$

the derivative of the phase shift is related to the deflection angle for scattering with impact parameter b! Using this approximation, one can show that it is possible to recover the classical limit for the differential scattering cross section in the high energy limit.

18.2.1.3 Calculation of the Partial Phase Shifts

I consider only two examples: scattering by a hard sphere and scattering by a spherical well potential, but first I discuss some general qualitative features of the partial wave approach. The scattering associated with the first few partial waves is designated by the standard letter scheme of angular momentum. Thus, $\ell = 0$ corresponds to S wave scattering, $\ell = 1$ to P wave scattering, $\ell = 2$ to D wave scattering, etc. The largest angular momentum quantum number $\ell_{\max}$ that can be expected to contribute to the sum over ℓ for scattering by a potential having finite range $r_{\max}$ is

$$\ell_{\max} = \frac{L_{\max}}{\hbar} = \frac{\mu v r_{\max}}{\hbar} \approx \frac{r_{\max}}{\lambda_{dB}} \approx k r_{\max}. \tag{18.83}$$

At low energies, $k r_{\max} \ll 1$, only S wave scattering is important. In that case

$$\frac{d\sigma}{d\Omega} \approx \frac{\sin^2 \delta_0}{k^2}; \tag{18.84a}$$

$$\sigma = \frac{4\pi \sin^2 \delta_0}{k^2} = \frac{\sin^2 \delta_0}{\pi} \lambda_{dB}^2. \tag{18.84b}$$

The scattering is isotropic and, since the de Broglie wavelength is larger than the scattering range, the incoming wave "surrounds" the scattering center and is scattered by the total surface area of the potential rather than the cross-sectional area.

In the opposite limit of high energy scattering, where

$$\ell_{\max} = \frac{L_{\max}}{\hbar} = \frac{\mu v r_{\max}}{\hbar} \approx \frac{r_{\max}}{\lambda_{dB}} \approx k r_{\max} \gg 1, \tag{18.85}$$

many partial waves contribute to the scattering. One can convert the sum over ℓ to an integral over impact parameters. For each impact parameter, there is a scattering angle that corresponds to the maximum contribution to the integral (point of stationary phase). In some sense, we recapture the classical limit in which scattering at a given impact parameter gives rise to scattering at a particular angle. The calculation is reminiscent of the Feynman path integral approach where the classical path makes the major contribution to an integral of the action between two fixed points. There is a difference from the classical case however, since contributions from different ℓ can interfere, giving rise to rapid oscillations in the differential cross section as a function of angle—it is only when these oscillations are assumed to average to zero (as would be the case if an experiment could not resolve the oscillations as a function of angle) does one recover the classical limit of the differential cross section.

Hard Sphere Scattering

I first consider the hard sphere potential

$$V(r) = \begin{cases} \infty & r \leq a \\ 0 & r > a \end{cases} . \tag{18.86}$$

Recall that in the classical case, the differential scattering cross section is $a^2/4$ (isotropic) and the total cross section is πa^2. The maximum impact parameter giving rise to scattering is $b_{\max} = a$. For an incident energy $E = \mu v^2/2$, there is no scattering for $L > \mu v a$. In the quantum problem we would expect that partial waves having $\ell < \mu v a/\hbar = ka$ provide the major contribution to the scattering.

In the quantum problem, the radial wave function vanishes for $r \leq a$. For $r > a$ the potential is that of a free particle, giving a general solution

$$R_\ell(r) = A_\ell j_\ell(kr) + B_\ell n_\ell(kr). \tag{18.87}$$

The radial wave function must vanish at $r = a$, leading to the condition

$$0 = A_\ell j_\ell(ka) + B_\ell n_\ell(ka) \tag{18.88}$$

or

$$\tan \delta_\ell = -\frac{B_\ell}{A_\ell} = \frac{j_\ell(ka)}{n_\ell(ka)}. \tag{18.89}$$

Some care must be used in interpreting δ_ℓ if you use a computer program to evaluate it, since most programs return the principal value of $\tan^{-1}$.

On the other hand there is no ambiguity in calculating the value of $\sin \delta_\ell$ using $\delta_\ell = \tan^{-1} [j_\ell(ka)/n_\ell(ka)]$. The result is

$$\sin \delta_\ell = \frac{j_\ell(ka)}{\sqrt{[j_\ell(ka)]^2 + [n_\ell(ka)]^2}}. \tag{18.90}$$

As a consequence, the differential cross section can then be obtained using Eq. (18.65) as

$$\frac{d\sigma}{d\Omega} = \frac{1}{k^2} \left| \sum_{\ell=0}^{\infty} (2\ell + 1) e^{i \tan^{-1}\left(\frac{j_\ell(ka)}{n_\ell(ka)}\right)} \frac{j_\ell(ka)}{\sqrt{[j_\ell(ka)]^2 + [n_\ell(ka)]^2}} P_\ell(\cos\theta) \right|^2 , \tag{18.91}$$

and the total cross section using Eq. (18.66) as

$$\sigma = \frac{4\pi}{k^2}\sum_{\ell=0}^{\infty}(2\ell+1)\frac{[j_\ell(ka)]^2}{[j_\ell(ka)]^2+[n_\ell(ka)]^2}. \tag{18.92}$$

Recall that $\lambda_{dB} = 2\pi/k$ and $k^2 = 2\mu E/\hbar^2$, such that

$$ka = 2\pi\frac{a}{\lambda_{dB}} = \sqrt{\frac{2\mu E a^2}{\hbar^2}} \tag{18.93}$$

is a dimensionless measure of the energy. If $ka \ll 1$, the de Broglie wavelength is larger than the scattering radius and quantum effects are important. For $ka \gg 1$ you might expect that the quantum result reduces to the classical one, but this is only partially true. Let me analyze the scattering in the low and high energy regions, $ka \ll 1$ and $ka \gg 1$, respectively.

Low Energy - $ka \ll 1$

In this limit, only $\ell = 0$ contributes significantly. For $\ell = 0$,

$$\tan\delta_0 = \frac{\sin(ka)}{-\cos(ka)} = -\tan ka; \tag{18.94a}$$

$$\delta_0 = -ka, \tag{18.94b}$$

the wave is shifted so that there is a node in the radial wave function at $r = a$. As I already mentioned, the radial wave function is displaced by a distance a when $\ell = 0$ for hard sphere scattering. The differential and total cross sections are given approximately by

$$\frac{d\sigma}{d\Omega} \approx \frac{1}{k^2}\frac{[j_0(ka)]^2}{[j_0(ka)]^2+[n_0(ka)]^2} = \frac{\sin^2(ka)}{k^2} \approx a^2; \tag{18.95a}$$

$$\sigma \approx 4\pi a^2. \tag{18.95b}$$

The scattering is isotropic and the total cross section is four times the geometrical one—in some sense the wave "sees" the entire surface area of the sphere, instead of just the cross section.

High Energy - $ka \gg 1$

In this limit, values of $\ell \lesssim ka$ contribute to the cross section. For $ka \gg 1$, the differential and total cross sections can be approximated by methods that are discussed in Appendix 2. Near $\theta = 0$, there is a peak resulting from *diffractive scattering* at the sharp boundary of the hard sphere. For larger θ, the differential cross section settles down to a value close to the classical limit of $a^2/4$. The diffraction peak has an amplitude of order

$$\frac{d\sigma\left(\theta=0\right)}{d\Omega}\approx\frac{a^{2}}{4}\left(ka\right)^{2}, \tag{18.96}$$

width of order $\Delta\theta \approx 4/ka$, and contributes πa^2 to the total cross section. The classical part of the scattering also contributes πa^2 to the total cross section, such that the total cross section is

$$\sigma = 2\pi a^{2}, \tag{18.97}$$

twice the classical limit. This is another example where a sharp change in the potential gives rise to a wave effect, regardless of the incident energy. Of course you never see such diffractive scattering in classical experiments since you would have to prepare a coherent wave to scatter off another particle. In some laser spectroscopy experiments, evidence for diffractive scattering of Yb (atomic mass =178) was shown explicitly.[2] Diffractive scattering also contributes significantly to the diffusion coefficients of atoms in vapors.

For arbitrary *ka*, you can simply use Eq. (18.89) to evaluate the phase shifts and then carry out the sums in Eqs. (18.65) and (18.66) to obtain the differential and total cross sections. This may take some time to do on a computer, but not that long if $ka < 100$. Some graphs are shown in Figs. 18.13, 18.14 and 18.15, the last of which corresponds to high energy scattering, where the forward scattering diffractive peak is seen clearly. A graph $\sigma/\pi a^2$ as a function of *ka* is shown in Fig. 18.16. The cross section varies from $4\pi a^2$ in the low energy limit to $2\pi a^2$ in the high energy limit.

Scattering by a Spherical Potential Well

I now consider scattering by the spherical well potential

$$V(r)=\begin{cases}-V_{0} & r\leq a\\ 0 & r>a\end{cases}. \tag{18.98}$$

Although I concentrate mainly on the case of positive V_0, the analysis is equally valid for a spherical barrier potential, with $V_0 < 0$. This is an extraordinarily rich problem and I cannot discuss many aspects of the solution, but you can examine the numerical solutions at your leisure. In this case, many features of the scattering are *not* related to the sharp discontinuity in the potential; they occur for *any* finite range potential. That is what makes a study of the spherical well potential so important.

Recall that in the corresponding case of electromagnetic scattering by a dielectric sphere, there were many interesting effects possible, such as rainbow and glory

[2]See, for example, R. A. Forber, L. Spinelli, J. E. Thomas, and M. S. Feld, *Observation of Quantum Diffractive Velocity-Changing Collisions by Use of Two-Level Heavy Optical radiators*, Physical Review Letters **50**, 331–334 (1983).

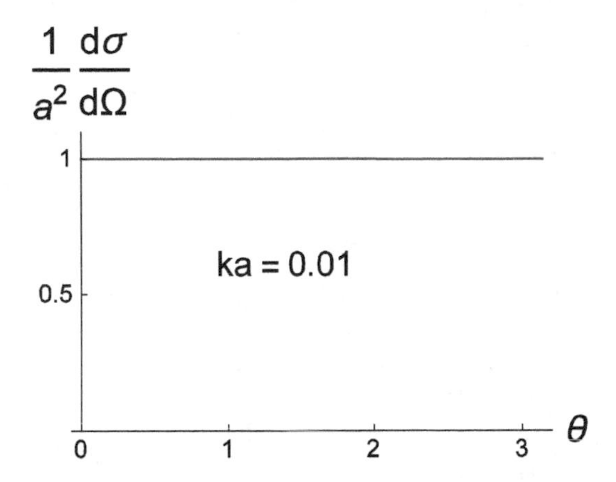

Fig. 18.13 Hard sphere scattering for $ka = 0.01$. The total scattering cross section is $\sigma = 4.0\pi a^2$

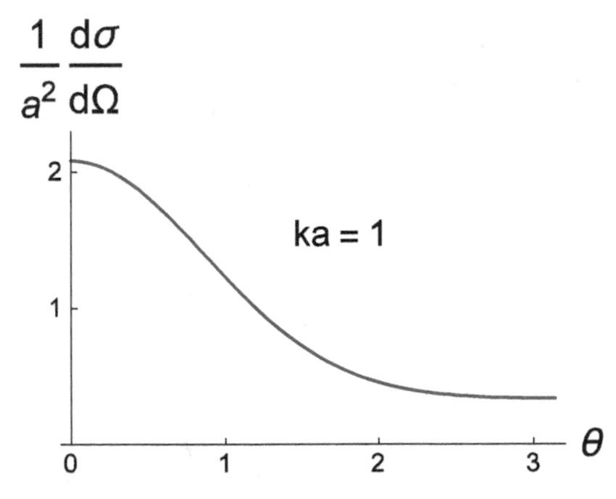

Fig. 18.14 Hard sphere scattering for $ka = 1$. The total scattering cross section is $\sigma = 3.4\pi a^2$

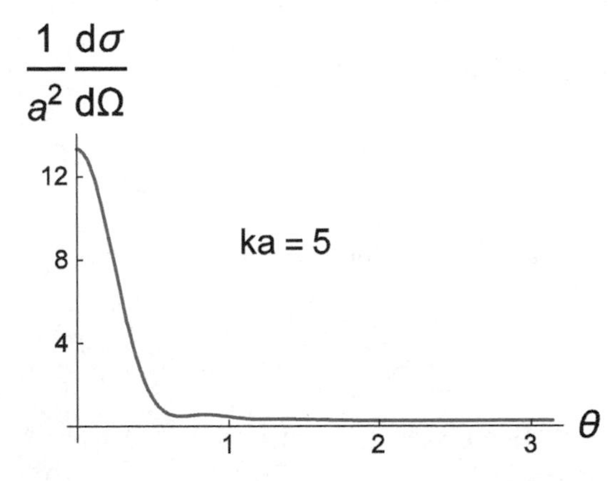

Fig. 18.15 Hard sphere scattering for $ka = 5$, showing a forward diffraction peak and isotropic scattering outside the peak. The total scattering cross section is $\sigma = 2.6\pi a^2$

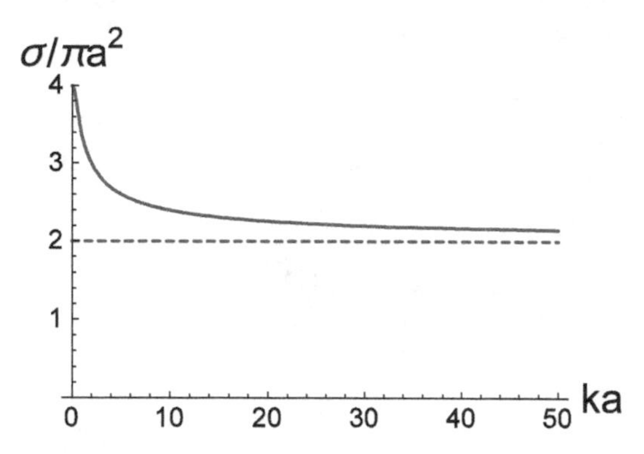

Fig. 18.16 Total cross section σ (in units of πa^2) as a function of ka for hard sphere scattering. At high energies the cross section approaches $\sigma/\pi a^2 \sim 2$, twice the classical value

scattering. We can try to look for some of these effects in the quantum case. The maximum impact parameter giving rise to scattering is $b_{\max} = a$ in the classical problem so, as in the case of hard sphere scattering, we can expect that partial waves having $\ell \lesssim ka$ to provide the major contribution to the scattering.

The potential is constant for both $r < a$ and $r > a$, but the radial wave function must be finite as $r \to 0$. As such, the general solution for the radial wave function is

$$R_\ell(r) = \begin{cases} C_\ell j_\ell(k_1 r) & r < a \\ A_\ell j_\ell(kr) + B_\ell n_\ell(kr) & r > a \end{cases}, \tag{18.99}$$

where

$$k = \sqrt{\frac{2\mu E}{\hbar^2}}; \qquad k_1 = \sqrt{\frac{2\mu\,(E+V_0)}{\hbar^2}}. \tag{18.100}$$

[For $V_0 < 0$, k_1 is replaced by $k_2 = \sqrt{2\mu\,(E-V_0)/\hbar^2}$. Although k_2 is purely imaginary for $E < V_0$, $j_\ell(k_2 r)$ is still real.] Matching the radial wave function and its derivatives at $r = a$, I find

$$A_\ell j_\ell(x) + B_\ell n_\ell(x) = C_\ell j_\ell(x_1) \tag{18.101a}$$

$$A_\ell k j_\ell'(x) + B_\ell k n_\ell'(x) = C_\ell k_1 j_\ell'(x_1) \tag{18.101b}$$

where $j_\ell'(x)$ is a shorthand notation for $dj_\ell(y)/dy|_{y=x}$, $n_\ell'(x)$ is a shorthand notation for $dn_\ell(y)/dy|_{y=x}$, and

$$x = ka; \qquad x_1 = k_1 a. \tag{18.102}$$

The partial wave phase shifts are given by Eq. (18.54c), namely

$$\tan\delta_\ell = -\frac{B_\ell}{A_\ell}. \tag{18.103}$$

Solving Eqs. (18.101), I find

$$\tan\delta_\ell = -\frac{B_\ell}{A_\ell} = \frac{xj_\ell(x_1)j'_\ell(x) - x_1 j_\ell(x)j'_\ell(x_1)}{xn'_\ell(x)j_\ell(x_1) - x_1 j'_\ell(x_1)n_\ell(x)}. \tag{18.104}$$

In effect I have solved the problem since all the phase shifts can be evaluated easily using a simple computer program. Once I have calculated the phase shifts from Eq. (18.104), I can construct

$$f_k(\theta) = \frac{1}{k}\sum_{\ell=0}^{\infty}(2\ell+1)e^{i\delta_\ell}\sin\delta_\ell P_\ell(\cos\theta); \tag{18.105a}$$

$$\sigma_k = \frac{4\pi}{k^2}\sum_{\ell=0}^{\infty}(2\ell+1)\sin^2\delta_\ell. \tag{18.105b}$$

The dimensionless variables that enter are

$$x = ka; \qquad x_1 = \sqrt{x^2+\beta^2}; \quad \beta = \sqrt{\frac{2\mu V_0 a^2}{\hbar^2}}. \tag{18.106}$$

From your work on the eigenenergies of a particle in a spherical potential well, you will recognize that β is a measure of the number of bound states in the potential.

Low Energy - $x = ka \ll 1$

In this limit, only $\ell = 0$ contributes significantly and

$$\frac{d\sigma}{d\Omega} \approx \frac{\sin^2\delta_0}{k^2}; \tag{18.107a}$$

$$\sigma \approx 4\pi\frac{\sin^2\delta_0}{k^2}. \tag{18.107b}$$

For spherical barrier potentials, the dependence of the cross section on β is somewhat boring (see problems) and need not be discussed.

Spherical well potentials present a whole new story. There are values of β for which $\delta_0 = 0$ and S wave scattering is totally suppressed—this is known as the *Ramsauer-Townsend effect*. On the other hand, there are values of $\beta = (n + 1/2)\pi$ for which bound states in the potential well occur near zero energy—the first such state occurs when $\beta = \pi/2$. For these values of β there is a *resonance* near $x = ka = 0$ and the wave function "bounces back and forth many times" as it is scattered. As a result the scattering cross section has a narrow, sharply peaked resonance as a function of β for fixed x.

These features are seen clearly in Figs. 18.17 and 18.18 where both the resonances and Ramsauer-Townsend effect can be seen. Note that this does not necessarily imply there is a resonance in the cross section as a function of *energy*

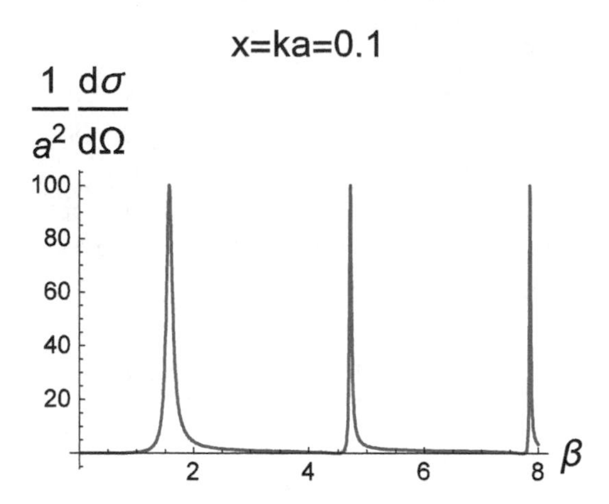

Fig. 18.17 Differential scattering cross section (in units of a^2) as a function of well strength parameter β

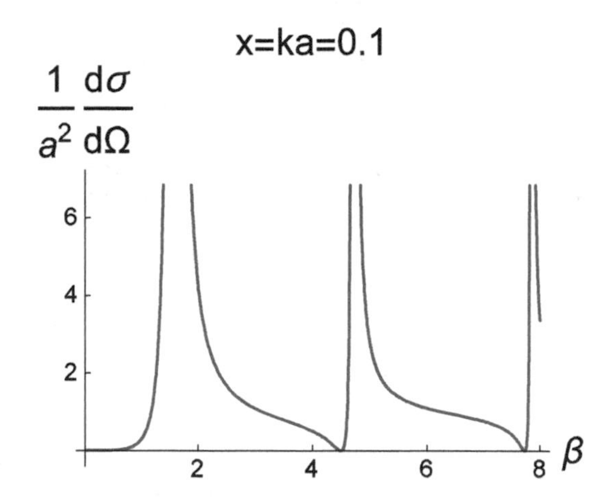

Fig. 18.18 Blow-up of Fig. 18.17 showing the Ramsauer-Townsend effect

for fixed β. The resonance effect at low energy is related to that associated with the perfect transmission for a one-dimensional square-well potential having width $2a$ [see Eq. (6.110) with a replaced by $2a$].

An expression for the differential cross section can be obtained without difficulty by substituting in the values of the spherical Bessel and Neumann functions and their derivatives for $\ell = 0$. Alternatively you can solve the radial equation directly for $\ell = 0$. Using either method, and setting $x_1 \approx \beta$ since $x \ll 1$, you will find

$$\tan \delta_0 = x \left(\frac{\tan \beta}{\beta} - 1 \right), \tag{18.108a}$$

$$\sin\delta_0 = \frac{x\left(\frac{\tan\beta}{\beta}-1\right)}{\sqrt{1+x^2\left(\frac{\tan\beta}{\beta}-1\right)^2}}, \tag{18.108b}$$

and

$$\frac{d\sigma}{d\Omega} \approx \frac{\sin^2\delta_0}{k^2} = \frac{a^2\left(\frac{\tan\beta}{\beta}-1\right)^2}{1+x^2\left(\frac{\tan\beta}{\beta}-1\right)^2}. \tag{18.109}$$

Note that, as a function of β, there are resonances whenever $\beta = (n+1/2)\pi$. At the resonance positions, $a^{-2}d\sigma/d\Omega = 1/x^2$ and the width $\Delta\beta$ of the resonances is of order x. There is a Ramsauer-Townsend effect whenever $\tan\beta/\beta = 1$.

Since $x \ll 1$, for all values of β except those corresponding to resonance,

$$\frac{d\sigma}{d\Omega} \approx a^2\left(\frac{\tan\beta}{\beta}-1\right)^2 = a_s^2 \tag{18.110}$$

where

$$a_s = -a\left(\frac{\tan\beta}{\beta}-1\right) = \lim_{k\to 0}\left(-\frac{1}{k}\tan\delta_0\right) \tag{18.111}$$

is referred to as the *scattering length.* Thus, low energy scattering can be characterized by a single parameter a_s, except in the region of a resonance (where a second parameter called the effective range r_{eff} is also needed). This is a very general result for any type of low energy scattering. In other words, the specific form of the potential is unimportant; the low energy cross section depends only on the scattering length which, in turn, depends on some *integral* property of the potential, such as the strength parameter β.

There is a simple geometric interpretation that can be given to the scattering length. Recall that the radial equation for $\ell = 0$ is simply

$$\begin{cases} \frac{d^2u_0(r)}{dr^2} + k_1^2 u_0(r) = 0 & r < a \\ \frac{d^2u_0(r)}{dr^2} + k^2 u_0(r) = 0 & r > a \end{cases}. \tag{18.112}$$

The general solution of these equations satisfying the boundary condition that $u_0(0) = 0$ is

$$\begin{cases} u_0(r) = A\sin(k_1 r) & r < a \\ u_0(r) = B\sin(kr+\delta_0) & r > a \end{cases}. \tag{18.113}$$

Matching the radial wave functions and their derivatives at $r = a$, and using the fact that $x = ka \ll 1$, I obtain the solution

$$u_0(r) \approx \begin{cases} A\sin(\beta r/a) & r < a \\ \frac{A\sin\beta}{x\cos\delta_0 + \sin\delta_0}\sin(xr/a + \delta_0) & r > a \end{cases}, \tag{18.114}$$

where δ_0 is determined from Eq. (18.108a). The value of β determines how many oscillations there are in $u_0(r)$ in the region $r < a$ [there are no oscillations for a *spherical barrier* potential since $\sin(\beta r/a) \rightarrow i\sinh(\beta r/a)$ which increases monotonically for $r < a$]. The slope of the radial wave function at $r = a$,

$$\left.\frac{du_0(r)}{dr}\right|_{r=a} = \frac{A\beta}{a}\cos(\beta), \tag{18.115}$$

is associated with the tangent to the function $u_0(r)$ at that point. I extend this tangent and look for the point r_s where it crosses the r-axis (see Fig. 18.19). I can calculate r_s using

$$u_0(a) - \left.\frac{du_0(r)}{dr}\right|_{r=a}(a - r_s) = u_0(r_s) = 0, \tag{18.116}$$

which has as solution

$$r_s = a\left(1 - \frac{\tan\beta}{\beta}\right) = -\frac{\tan(\delta_0)}{k} = a_s. \tag{18.117}$$

In other words, the scattering length is the radius at which the tangent to the radial wave function $u_0(r)$ at $r = a$ crosses the r-axis. For spherical well potentials with $\beta < \pi/2$, the slope is positive but always less than $u_0(a)/a$; consequently, $a_s < 0$. Since $a_s \neq 0$ there is no Ramsauer-Townsend when $\beta < \pi/2$. At the first resonance, $\beta = \pi/2$, $a_s = -\infty$ (the tangent is parallel to the r-axis), but for a value of β slightly larger than $\pi/2$ the wave function "turns over" and the scattering length becomes positive, mimicking the effects of a repulsive potential. This is how the scattering in a Bose condensate can appear to be repulsive, even if the actual potential is attractive. For still larger values of β, the scattering length decreases and eventually passes through $a_s = 0$, giving rise to the Ramsauer-Townsend effect, just before the next resonance. These features are illustrated in Fig. 18.19. [For spherical barrier potentials the radial wave function varies as $\sinh(\beta r/a)$ and the slope at $r = a$, $A\beta\cosh(\beta)/a$, is positive and always greater than $u_0(a)/a = A\tanh(\beta)/a$; consequently, $a_s > 0$.]

High Energy - $x = ka \gg 1$

When I consider scattering at high energies, a new range of phenomena can appear. I have already noted that the scattering is analogous to scattering by a dielectric sphere having index of refraction

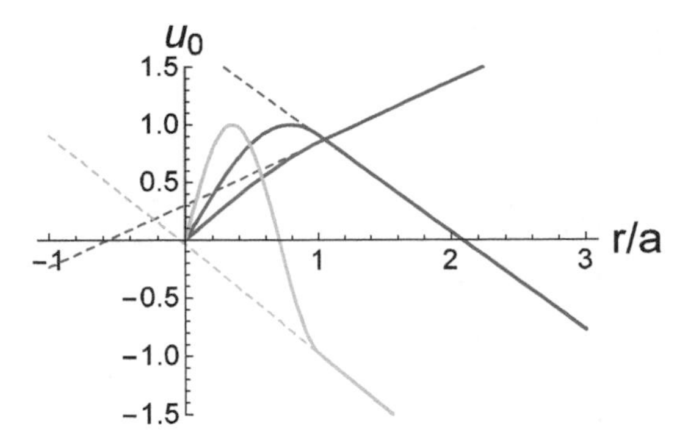

Fig. 18.19 Graphs of u_0 as a function of r/a for a spherical well potential with $A = 1, x = 0.1$ and $\beta = 1$ (red curve), 2 (blue curve), 4.5 (green curve). The dashed curves intersect the horizontal axis at a value corresponding to the scattering length. For $\beta = 1$, the scattering length is negative, for $\beta = 2$, the scattering length is positive, and for $\beta = 4.5$, the scattering length is zero (Ramsauer-Townsend effect)

$$n_{\text{eff}} = \sqrt{1 + \frac{V_0}{E}} = \sqrt{1 + \frac{\beta^2}{x^2}}. \tag{18.118}$$

Therefore, for $x = ka \gg 1$ and β of order unity, the index approaches unity and there is very little scattering. The more interesting regime is when x and β are comparable, so that n_{eff} is on the order or 1.4 or so. In that limit, one may see such effects as rainbow and glory scattering, although it may be necessary to go to values of x as large as 1000 to clearly see the rainbow effects.

I need to consider three ranges of angular momenta values to get a qualitative understanding of the scattering. First consider $\ell < ka$ which corresponds to impact parameters $b < a$ [see Fig. 18.20 with energy E_3]. This is analogous to the situation in optics where a ray enters the dielectric sphere and undergoes internal reflections in the sphere. Both rainbow and glory scattering occur. You can almost "see" the result. If a "particle" comes in, it is reflected and transmitted at $r = a$, reaches a point of closest approach r_0 and then is transmitted and reflected on its way out at $r = a$. The wave that exits the sphere is the "normal" refracted wave, but that wave is also reflected on the inner surface of the sphere and can emerge at some other direction and correspond to rainbow scattering for some incident impact parameter. Since an infinite number of reflections are possible, all scattering angles can occur and glory scattering occurs a well.

For $\ell > x_1 = \sqrt{x^2 + \beta^2}$, any penetration into the barrier is weak and there is very little scattering since $b \gg a$ [see Fig. 18.20 with energy E_1]. No tunneling into a classically allowed region is possible. However for $x_1 > \ell > x$, such tunneling is possible even though the impact parameter $b > a$ (this is a wave-like effect) [see Fig. 18.20 with energy E_2]. Most of the time the scattering is negligible, but if the

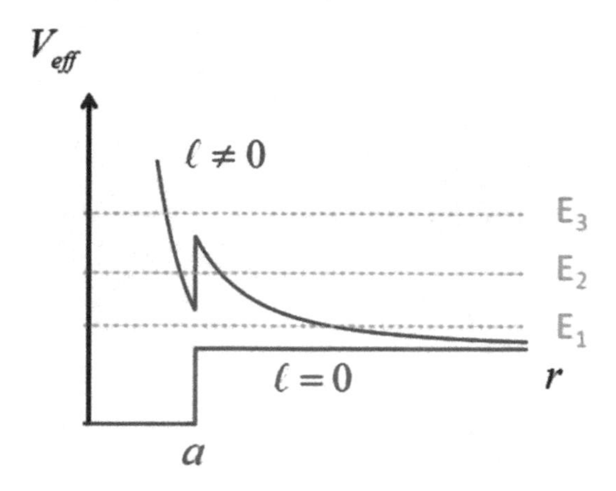

Fig. 18.20 Effective potential for a spherical well potential

incident energy corresponds to the energy of the quasi-bound states in the effective potential, there can be very narrow resonances in the scattering at these energies. Of course, the differential scattering cross section involves a sum over all values of ℓ, so all three types of effects are present at once.

The differential scattering cross section is shown in Figs. 18.21, 18.22, and 18.23, for $x = ka = \beta = 5, 20, 100$. There is a forward diffraction peak as in scattering by a hard sphere, but now there is evidence for glory scattering at $\theta = \pi$. Moreover, for this value of the effective index of refraction, $n_{\text{eff}} = \sqrt{2}$, the rainbow angle is predicted to be $\theta \approx 2.6\,\text{rad}$; the peak at this value in the $ka = 100$ graph, which sharpens with increasing ka, may be an indication that rainbow scattering is occurring. The oscillations in the graph correspond to interference effects arising from contributions from different ℓ to scattering at a given angle.

The classical differential cross section, given in Eq. (18.28) with $n = n_{\text{eff}} = \sqrt{1 + V_0/E}$, is also plotted in Figs. 18.21, 18.22 and 18.23 as the dashed curves. The classical cross section does not have a forward diffraction peak, but otherwise is in good agreement with the quantum distribution, averaged over oscillations, in the classically allowed region, which extends to $\theta_{\max} = 2\cos^{-1}(1/n_{\text{eff}})$. For $n_{\text{eff}} = \sqrt{2}$, $\theta_{\max} = \pi/2 = 1.57$.

For the case of scattering by a spherical barrier potential I can carry over the results of those for the spherical well potential by replacing V_0 with $-|V_0|$. In the limit that $E \ll |V_0|$, the results are similar to that for hard sphere scattering. For $E > |V_0|$, the net change from the attractive case is that the effective index is now $n_{\text{eff}} = \sqrt{1 - \frac{|V_0|}{E}}$ and the quantity β becomes pure imaginary. There is no longer any rainbow scattering and the interference effects that are present for scattering by a spherical well potential are diminished, since there are not as many impact

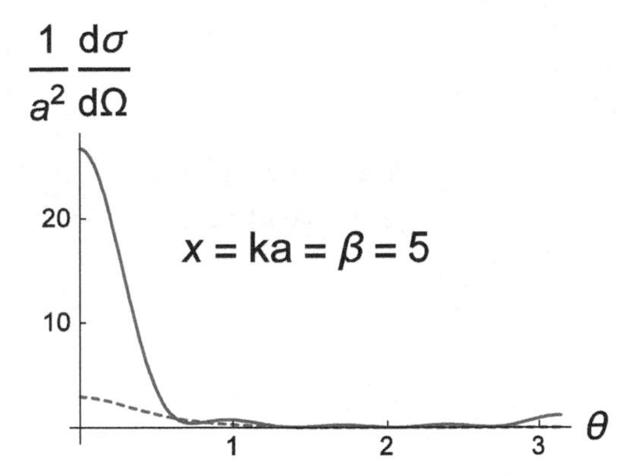

Fig. 18.21 Scattering by a spherical well potential showing the forward diffraction peak. The dashed, blue curve is the corresponding result for classical particle scattering, for which the diffraction peak is absent

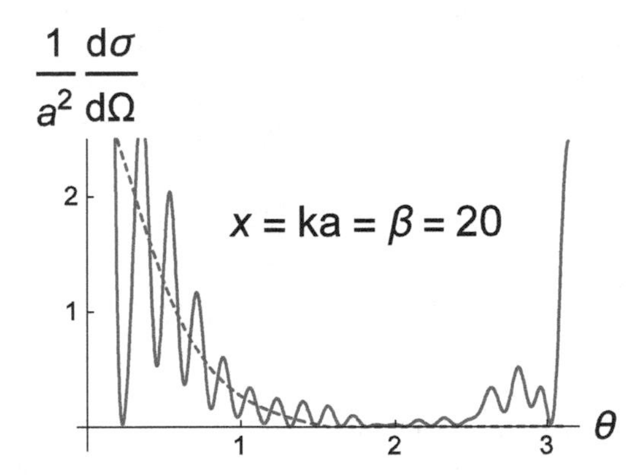

Fig. 18.22 Scattering by a spherical well potential showing the glory. The dashed, blue curve is the classical result

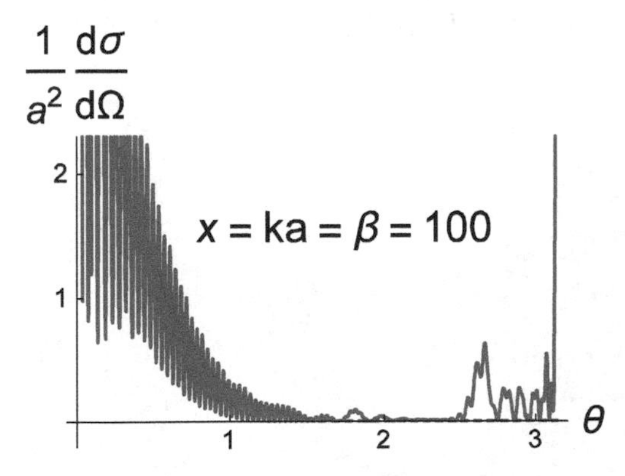

Fig. 18.23 Scattering by a spherical well potential showing the glory and a possible rainbow. The dashed, blue curve is the classical result

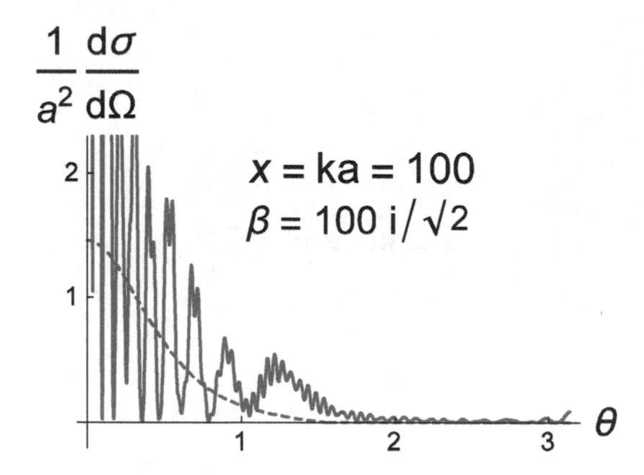

Fig. 18.24 Scattering by a spherical barrier potential. The dashed, blue curve is the classical result

parameters that give rise to the same scattering angle for repulsive scattering as there are for attractive scattering. In Fig. 18.24, where $x = 100$ and $\beta = 100i/\sqrt{2}$ ($n_{\text{eff}} = 1/\sqrt{2}$), you can see that there are fewer oscillations than in the corresponding case of scattering by a spherical well potential and that there is no rainbow, but there does appear to be a slight glory.

18.2.2 Born Approximation

For potentials requiring numerical solutions, the method of partial waves is especially useful in the low energy scattering limit, when only a few partial waves are needed. It would be good to have a result that is valid in the high energy limit, other than the WKB partial wave result. The *Born series* provides such a solution to this problem. The Born series is effectively a perturbative approach, valid when the scattering cross section can be expanded in a power series in the potential.

To derive the Born series, I start from the Schrödinger equation written in the form

$$\nabla^2\psi + k^2\psi = U(\mathbf{r})\psi, \tag{18.119}$$

where

$$U(\mathbf{r}) = \frac{2\mu}{\hbar^2}V(\mathbf{r}). \tag{18.120}$$

The idea is to treat the right-hand side as a perturbation. I look for a solution to this equation in the form

$$\psi_k(\mathbf{r}) = e^{i\mathbf{k}\cdot\mathbf{r}} - \frac{1}{4\pi}\int d\mathbf{r}' G(\mathbf{r}-\mathbf{r}')U(\mathbf{r}')\psi(\mathbf{r}'), \tag{18.121}$$

substitute it into Eq. (18.119) and find that, for this to be a solution, I must require that

$$\left(\nabla^2 + k^2\right) G(\mathbf{r}-\mathbf{r}') = -4\pi\delta(\mathbf{r}-\mathbf{r}'). \tag{18.122}$$

The quantity $G(\mathbf{r}-\mathbf{r}')$ is known as a *Green function*. It is not difficult to verify by direct substitution that

$$G(\mathbf{r}-\mathbf{r}') = \frac{e^{\pm ik|\mathbf{r}-\mathbf{r}'|}}{|\mathbf{r}-\mathbf{r}'|} \tag{18.123}$$

is a solution of Eq. (18.122). Which sign in the exponent to take depends on the physics. If the scattering is to correspond to *outgoing* spherical waves, the positive sign must be taken.

Thus, the formal solution for $\psi_k(\mathbf{r})$, obtained from Eqs. (18.121) and (18.123), is

$$\psi_k(\mathbf{r}) = e^{i\mathbf{k}\cdot\mathbf{r}} - \frac{1}{4\pi}\int d\mathbf{r}' \frac{e^{ik|\mathbf{r}-\mathbf{r}'|}}{|\mathbf{r}-\mathbf{r}'|}U(\mathbf{r}')\psi_k(\mathbf{r}'). \tag{18.124}$$

It looks like I haven't accomplished *anything* but to go from a differential to an integral equation. However, this is precisely the form that is useful for an iterative or perturbation series solution in powers of $U(\mathbf{r})$. In other words, to zeroth order in $U(\mathbf{r})$

$$\psi_k^{(0)}(\mathbf{r}) = e^{i\mathbf{k}\cdot\mathbf{r}} \tag{18.125}$$

and to first order

$$\psi_k^{(1)}(\mathbf{r}) = e^{i\mathbf{k}\cdot\mathbf{r}} - \frac{1}{4\pi}\int d\mathbf{r}' \frac{e^{ik|\mathbf{r}-\mathbf{r}'|}}{|\mathbf{r}-\mathbf{r}'|}U(\mathbf{r}')e^{i\mathbf{k}\cdot\mathbf{r}'}, \tag{18.126}$$

which is known as the *first Born approximation*. You can continue iterating the solution to obtain the Born series as a power series in the potential, but it is usually difficult to go beyond the first term. The first Born approximation is the most important result in high energy scattering.

To get the scattering amplitude, I must compare Eq. (18.126) in the limit $r \to \infty$ with Eq. (18.31). To do so, I take the z axis along $\mathbf{k}$, and expand

$$e^{ik|\mathbf{r}-\mathbf{r}'|} \approx e^{ikr(1-\hat{\mathbf{r}}\cdot\mathbf{r}')} = e^{ikr}e^{-i\mathbf{k}'\cdot\mathbf{r}'}, \tag{18.127}$$

where

$$\mathbf{k}' = k\mathbf{u}_r \tag{18.128}$$

gives the direction of scattering. In this manner, I find that as $r \to \infty$,

$$\psi_k^{(1)}(\mathbf{r}) \sim e^{ikz} - \frac{e^{ikr}}{4\pi r} \int d\mathbf{r}' e^{-i\mathbf{q}\cdot\mathbf{r}'} U(\mathbf{r}'), \tag{18.129}$$

where

$$\mathbf{q} = \mathbf{k}' - \mathbf{k}. \tag{18.130}$$

Comparing Eqs. (18.31) and (18.129), I obtain the scattering amplitude

$$f_k(\theta) = -\frac{1}{4\pi} \int d\mathbf{r} e^{-i\mathbf{q}\cdot\mathbf{r}} U(\mathbf{r}) = -\frac{\mu}{2\pi\hbar^2} \int d\mathbf{r} e^{-i\mathbf{q}\cdot\mathbf{r}} V(\mathbf{r}). \tag{18.131}$$

For spherically symmetric potentials, $V(\mathbf{r})$ is a function of r only. In this limit, to evaluate the integral, I take z along $\mathbf{q}$ such that

$$\mathbf{q} \cdot \mathbf{r} = qr \cos\theta, \tag{18.132}$$

enabling me to compute

$$\begin{aligned}
\int d\mathbf{r} e^{-i\mathbf{q}\cdot\mathbf{r}} V(r) &= \int d\mathbf{r} e^{-iqr\cos\theta} V(r) \\
&= 2\pi \int_0^\infty r^2 dr \int_{-1}^1 d(\cos\theta) e^{-iqr\cos\theta} V(r) \\
&= \frac{4\pi}{q} \int_0^\infty dr \sin(qr)\, rV(r),
\end{aligned} \tag{18.133}$$

such that

$$f_k(\theta) = -\frac{2\mu}{q\hbar^2} \int_0^\infty dr \sin(qr)\, rV(r). \tag{18.134}$$

The magnitude of $\mathbf{q}$ is given by

$$q = \left|\mathbf{k}' - \mathbf{k}\right| = k\left|\mathbf{u}_r - \mathbf{u}_z\right| = k\sqrt{2(1-\cos\theta)} = 2k \sin(\theta/2). \tag{18.135}$$

Roughly speaking, Eq. (18.134) is the Fourier sine transform of r times the potential. The hope is that the dependence of the Born cross section on q will reveal something about the nature of the potential giving rise to the scattering.

It is difficult to give a rigorous estimate of the corrections to the first Born approximation. Sometimes the Born approximation is used without justification since it is the only simple calculation that can be carried out. By looking at the second order term in the Born series and evaluating the terms in the integrand at $r = 0$, one can estimate that the first Born approximation is valid if

$$\left| \frac{\mu}{k\hbar^2} \int_0^\infty drV(r) \left[e^{2ikr} - 1 \right] \right| \ll 1. \tag{18.136}$$

If there is some effective range a_0 to the potential and if the maximum of the potential is V_0, then the validity condition given by Eq. (18.136) reduces to

$$\mu V_0 a_0^2/\hbar^2 \ll 1 \text{ if } ka_0 \ll 1; \tag{18.137a}$$

$$\mu V_0 a_0^2 / \left(ka_0 \hbar^2 \right) \ll 1 \text{ if } ka_0 \gg 1. \tag{18.137b}$$

Looking at Eq. (18.137b), you see that the first Born approximation is generally valid for high energy scattering, $ka_0 \gg 1$. It can be valid even for low energy scattering if $\beta^2 = 2\mu V_0 a_0^2/\hbar^2 \ll 1$; that is, for values of the strength β that are sufficiently small to insure that *no bound states could exist in the case of an attractive potential*. In fact, since condition (18.137a) is generally satisfied when condition (18.137b) holds, a sufficient condition for the Born series to converge is that $-|V(r)|$ is not strong enough to support a bound state (along with the additional requirements that both $\int_0^\infty dr\,|V(r)|\,r$ and $\int_0^\infty dr\,|V(r)|\,r^2$ are finite). Although conditions (18.137) are sufficient, they may not be *necessary* for validity of the first Born approximation.

Examples:

If

$$V(r) = V_0 e^{-r^2/a^2}, \tag{18.138}$$

then

$$\begin{aligned} f_k(\theta) = f(q) &= -\frac{2\mu V_0}{q\hbar^2} \int_0^\infty dr \sin(qr)\, re^{-r^2/a^2} \\ &= -\frac{2\mu V_0}{q\hbar^2} \left[\frac{\sqrt{\pi}}{4} a^3 q e^{-a^2q^2/4} \right] = -\frac{\mu V_0 \sqrt{\pi} a^3 e^{-a^2q^2/4}}{2\hbar^2} \end{aligned} \tag{18.139}$$

and

$$\frac{d\sigma}{d\Omega} = |f_k(\theta)|^2 = \pi a^2 \left[\left(\frac{\mu V_0 a^2}{2\hbar^2} \right)^2 e^{-2k^2a^2 \sin^2(\theta/2)} \right], \tag{18.140}$$

where Eq. (18.135) was used. For $ka \gg 1$, the differential cross section exhibits a diffraction-like central peak with a maximum at $\theta = 0$, but, in contrast to the diffraction pattern of an opaque disk, there are no oscillations for larger values of θ. Note that, if $ka \ll 1$, $d\sigma/d\Omega$ is constant, as for any low energy S-wave scattering.

The first Born approximation is valid if condition (18.136) holds

$$\left| \frac{\mu V_0}{k\hbar^2} \int_0^\infty dr e^{-r^2/a^2} \left[e^{2ikr} - 1 \right] \right| \ll 1, \tag{18.141}$$

which can be rewritten as

$$\left| \frac{\beta^2}{2x} \int_0^\infty d\rho e^{-\rho^2} \left[e^{2ix\rho} - 1 \right] \right| \ll 1, \tag{18.142}$$

where $x = ka$. The integral can be evaluated in terms of known functions, but can also be done numerically. The inequality is satisfied for $\beta \ll 1$ if $x \leq 1$ and for $\beta^2 \ll x$ if $x \gg 1$.

As a second example, I take

$$V(r) = \begin{cases} -V_0 & r \leq a \\ 0 & r > a \end{cases}. \tag{18.143}$$

For this potential

$$\begin{aligned} f_k(\theta) = f(q) &= \frac{2\mu V_0}{q\hbar^2} \int_0^a dr \sin(qr)\, r \\ &= \frac{2\mu V_0}{q\hbar^2} \left[\frac{\sin(aq) - aq\cos(aq)}{q^2} \right] \end{aligned} \tag{18.144}$$

and

$$\frac{d\sigma}{d\Omega} = |f(q)|^2 = a^2 \beta^4 \left[\frac{\sin(aq) - aq\cos(aq)}{q^3 a^3} \right]^2, \tag{18.145}$$

where

$$\beta = \sqrt{\frac{2\mu V_0 a^2}{\hbar^2}} \tag{18.146}$$

is a measure of the number of bound states supported by the potential. Recall that $q = 2k \sin(\theta/2)$.

If $ka \ll 1$, then

$$\frac{d\sigma}{d\Omega} \sim \frac{\beta^4 a^2}{9} \tag{18.147}$$

and the scattering is isotropic. This low energy result agrees with Eq. (18.110) obtained using the method of partial phases, provided $\beta \ll 1$ [in that limit, $(\tan\beta/\beta - 1)^2 \sim \beta^4/9$].

Next consider the limit that $qa \gg 1$, that is, high energy scattering outside the diffraction cone defined by $\theta_d \approx 1/ka$. In this case,

$$f(q) \sim -\beta^2 a \frac{\cos(aq)}{q^2a^2} \quad (18.148)$$

and

$$\frac{d\sigma}{d\Omega} \sim \beta^4 a^2 \frac{\cos^2(aq)}{q^4a^4}. \quad (18.149)$$

Inside the diffraction cone, that is, for $\theta_d \ll 1/ka$, the result is still given by Eq. (18.147). The overall pattern is similar to that for Fraunhofer diffraction of optical radiation by a circular aperture.

According to conditions (18.136), the first Born approximation should be valid if

$$\left|\frac{\mu V_0}{k\hbar^2}\int_0^a dr\left[e^{2ikr}-1\right]\right| = \frac{\beta^2}{2x}\left|\left[\frac{e^{2ix}-1}{2ix}-1\right]\right| \ll 1, \quad (18.150)$$

where $x = ka$. Thus the first Born approximation is valid for $\beta^2 \ll 1$ if $x \ll 1$ and for $\beta^2/x \ll 1$ if $x \gg 1$. However these are sufficient, but not necessary conditions. By comparing the exact solution given by Eqs. (18.104)–(18.105), you can show that if $\beta/x = \sqrt{V_0/E} \ll 1$, then the first Born approximation is valid for all θ. This limit corresponds to scattering by a sphere whose effective index of refraction is approximately equal to unity.

The total cross section is given by

$$\begin{aligned}\sigma_k &= 2\pi\int_0^\pi |f(q)|^2 \sin\theta d\theta \\ &= \frac{2\pi a^2\beta^4}{k^2a^2}\int_0^{2ka} dy\frac{[\sin y - y\cos y]^2}{y^5} \\ &= \pi\beta^4a^2\left[\frac{32x^4 - 8x^2 - 1 + 4x\sin(4x) + \cos(4x)}{64x^6}\right],\end{aligned} \quad (18.151)$$

where Eq. (18.145) was used along with the substitutions $y = qa = 2ka\sin(\theta/2) = 2x\sin(\theta/2)$ and $\sin\theta = 2\sin(\theta/2)\cos(\theta/2)$.

Note that the Born approximation can fail even in the limit $\hbar \sim 0$ ($ka \sim \infty$), owing to the sharp boundary of the spherical potential well. That is, in the limit $ka = x \gg 1$,

$$\sigma \sim \frac{\pi a^2 \beta^4}{2x^2} = \frac{\pi a^2 \beta^2}{2} \left(\frac{\beta}{x} \right)^2, \tag{18.152}$$

which is a good approximation to the total cross section when $\beta/x = \sqrt{V_0/E} \ll 1$. However, if $\beta = x \gg 1$, this result is not valid since it gives $\sigma \sim \pi a^2 \beta^2/2$ whereas the correct cross section, calculated from Eqs. (18.105) and (18.104), is of order $2\pi a^2$.

18.3 Summary

I have given a rather detailed discussion of elementary quantum scattering theory, trying to emphasize the underlying physics. To facilitate the discussion, I reviewed aspects of classical scattering of particles and electromagnetic radiation, since many features encountered in classical scattering, such as rainbows and glories, resurface in quantum scattering theory. There are two different methods that can be used to solve the quantum scattering problem. The method of partial waves is especially useful for low energy scattering while the Born approximation is basically a high energy, perturbative approach. Scattering by various potentials was considered and different types of resonance phenomena were explored, as were diffractive effects associated with sharp changes in the scattering potential.

18.4 Appendix A: Free Particle Solution in Spherical Coordinates

For a free particle, we know that the (unnormalized) eigenfunctions are simply

$$\psi_k(\mathbf{r}) = e^{i\mathbf{k}\cdot\mathbf{r}}, \tag{18.153}$$

where $k^2 = 2\mu E/\hbar^2$. Why should we go the trouble of solving this in spherical coordinates when there is no natural origin to the problem? The reason for this is that, although the particle is not free in the scattering problem, the functional form of the eigenfunctions for the region $r > r_{\max}$ is the same as that of a free particle, provided the potential vanishes for $r > r_{\max}$.

To solve the free particle problem in spherical coordinates, I recall, that for any spherically symmetric potential, the eigenfunctions have the form

$$\psi_{k\ell m}(\mathbf{r}) = R_\ell(r) Y_\ell^m(\theta, \phi), \tag{18.154}$$

where the $Y_\ell^m(\theta, \phi)$s are spherical harmonics and $R_\ell(r) = u_\ell(r)/r$ and $u_\ell(r)$ satisfy the radial equations

$$\frac{1}{r^2}\frac{d}{dr}\left(r^2\frac{dR_\ell}{dr}\right) + \left[k^2 - \frac{2\mu V(r)}{\hbar^2} - \frac{\ell(\ell+1)}{r^2}\right]R_\ell = 0; \tag{18.155a}$$

$$\frac{d^2u_\ell}{dr^2} + \left[k^2 - \frac{2\mu V(r)}{\hbar^2} - \frac{\ell(\ell+1)}{r^2}\right]u_\ell = 0. \tag{18.155b}$$

For $V(r) = 0$, the equations are

$$\frac{d}{dr}\left(r^2\frac{dR_\ell}{dr}\right) + k^2r^2R_\ell - \ell(\ell+1)R_\ell = 0; \tag{18.156a}$$

$$\frac{d^2u_\ell}{dr^2} + k^2u_\ell - \frac{\ell(\ell+1)}{r^2}u_\ell = 0, \tag{18.156b}$$

having linearly independent solutions that are spherical Bessel and Neumann functions,

$$R_\ell^{(1)}(kr) = j_\ell(x) = \sqrt{\frac{\pi}{2x}}J_{\ell+1/2}(x); \tag{18.157a}$$

$$R_\ell^{(2)}(kr) = n_\ell(x) = \sqrt{\frac{\pi}{2x}}N_{\ell+1/2}(x), \tag{18.157b}$$

where $J_\ell(x)$ and $N_\ell(x)$ are ordinary Bessel and Neumann functions and $x = kr$. For the free particle, the Neumann function solutions must be rejected because they are not regular at the origin.

As a consequence, the free particle eigenfunction $e^{i\mathbf{k}\cdot\mathbf{r}}$ can be expanded in terms of spherical harmonics and spherical Bessel functions as

$$\psi_k(\mathbf{r}) = e^{i\mathbf{k}\cdot\mathbf{r}} = \sum_{\ell=0}^{\infty}\sum_{m=-\ell}^{\ell} A_{\ell m}j_\ell(kr)Y_\ell^m(\theta,\phi). \tag{18.158}$$

It is not trivial to find the expansion coefficients $A_{\ell m}$ directly from Eq. (18.158), but it can be shown that[3]

$$A_{\ell m} = 4\pi i^\ell\left[Y_\ell^m(\theta_k,\phi_k)\right]^*, \tag{18.159}$$

[3]Actually, Eq. (18.159) can be derived from Eq. (18.160) using the addition theorem for spherical harmonics. For a derivation of the addition theorem, see George Arfken, *Mathematical Methods for Physicists*, Third Edition (Academic Press, San Diego, 1985).

where (θ_k, ϕ_k) are the polar and azimuthal angles of the vector $\mathbf{k}$. If the z axis is taken along $\mathbf{k}$, then $\theta_k = 0$ and Eq. (18.158) reduces to

$$e^{ikz} = e^{ikr\cos\theta} = \sum_{\ell=0}^{\infty} i^{\ell}(2\ell+1)j_{\ell}(kr)P_{\ell}(\cos\theta), \tag{18.160}$$

the result I needed to obtain the scattering amplitude using the method of partial waves.

I can give a simple derivation of Eq. (18.160). In Eq. (18.158) with the z axis taken along $\mathbf{k}$, the left-hand side is independent of ϕ so I can set $\phi = 0$ in the right-hand side. Then the spherical harmonics reduce to Legendre polynomials and the sum over m can be carried out, enabling me to expand

$$e^{ikz} = e^{ikr\cos\theta} = \sum_{\ell'=0}^{\infty} B_{\ell'}j_{\ell'}(kr)P_{\ell'}(\cos\theta), \tag{18.161}$$

where the $B_{\ell'}$s are some new expansion coefficients. I multiply by $P_{\ell}(\cos\theta)$ and integrate over $\cos\theta$ using the orthogonality of the Legendre polynomials to obtain

$$B_{\ell}j_{\ell}(kr) = \frac{2\ell+1}{2}\int_{-1}^{1} dx e^{ikrx}P_{\ell}(x)dx, \tag{18.162}$$

where $x = \cos\theta$. I now take the limit that $r \to \infty$ on both sides. On the right-hand side I can integrate by parts using $u = P_{\ell}(x)$ and $dv = e^{ikrx}$ and on the left-hand side I use Eq. (18.48) to arrive at

$$B_{\ell}\frac{\sin\left(kr - \frac{\ell\pi}{2}\right)}{kr} = \frac{2\ell+1}{2}\left[\frac{P_{\ell}(1)e^{ikr} - P_{\ell}(-1)e^{ikr}}{ikr}\right]. \tag{18.163}$$

I have neglected the integral term in the integration by parts since it vanishes in the limit $r \to \infty$ (you can see this by carrying out additional integration by parts on the integral term). Since $P_{\ell}(1) = 1$ and $P_{\ell}(-1) = (-1)^{\ell}$, it follows from Eq. (18.163) that

$$\begin{aligned}
B_{\ell}\frac{e^{ikr}e^{-i\ell\pi/2} - e^{-ikr}e^{i\ell\pi/2}}{2ikr} &= \frac{2\ell+1}{2}\left[\frac{e^{ikr} - (-1)^{\ell}e^{-ikr}}{ikr}\right];\\
B_{\ell}e^{-i\ell\pi/2}\left(e^{ikr} - e^{-ikr}(-1)^{\ell}\right) &= (2\ell+1)\left(e^{ikr} - (-1)^{\ell}e^{-ikr}\right);\\
B_{\ell} &= e^{i\ell\pi/2}(2\ell+1) = i^{\ell}(2\ell+1),
\end{aligned} \tag{18.164}$$

and the proof is complete.

18.5 Appendix B: Hard Sphere Scattering when $ka \gg 1$

In this Appendix, I give some of the mathematical details for hard sphere scattering in the limit that $ka \gg 1$. The maximum value of ℓ that contributes to the sum over partial waves in Eq. (18.62) is approximately equal to ka. To apply asymptotic methods, I break the scattering into two regions, $\theta \ll (ka)^{-1/3}$ and $\theta \gg (ka)^{-1/3}$ in all cases I assume that $ka \gg 1$.

If $\theta \ll (ka)^{-1/3}$ and $\ell \lesssim ka$,

$$P_\ell(\cos\theta) \sim J_0\left[(\ell + 1/2)\,\theta\right], \tag{18.165}$$

where $J_0\left[(\ell + 1/2)\,\theta\right]$ is a normal Bessel function, leading to a scattering amplitude [Eq. (18.62)]

$$f_k(\theta) \approx \frac{1}{2ik}\sum_{\ell=0}^{\infty}(2\ell+1)\left(e^{2i\delta_\ell}-1\right)J_0\left[(\ell + 1/2)\,\theta\right]. \tag{18.166}$$

The δ_ℓ are determined from Eq. (18.89). The function $e^{2i\delta_\ell}$ oscillates rapidly for $\ell < ka$ and goes to unity for $\ell > ka$, allowing me to write

$$f_k(\theta) \approx \frac{-1}{2ik}\sum_{\ell=0}^{ka}(2\ell+1)J_0\left[(\ell + 1/2)\,\theta\right]. \tag{18.167}$$

I can set $z = \ell + 1/2$ and replace the sum by the integral to arrive at

$$f_k(\theta) \approx \frac{-1}{ik}\int_0^{ka} dz J_0\left[z\theta\right] z = \frac{ia}{\theta}J_1\left[ka\theta\right], \tag{18.168}$$

such that

$$\frac{d\sigma}{d\Omega} = |f_k(\theta)|^2 \approx \frac{a^2}{4}\left[\frac{2J_1\left[ka\theta\right]}{ka\theta}\right]^2 k^2a^2. \tag{18.169}$$

Equation (18.169) is valid for $0 \le \theta \ll (ka)^{-1/3}$, when $ka \gg 1$. The first minimum in the differential cross section occurs at $\theta \approx 3.83/ka < (ka)^{-1/3}$. As such, Eq. (18.169) correctly describes the forward diffraction peak associated with hard sphere scattering. Since $J_1(z) \sim z/2$ as $z \to 0$, the diffraction peak has an amplitude

$$\frac{d\sigma}{d\Omega}(\theta = 0) \approx \frac{a^2}{4}(ka)^2. \tag{18.170}$$

The contribution to the *total* cross section from the diffractive component can be obtained by assuming that most of the contributions come from small angles $\theta \lesssim 1/ka \ll (ka)^{-1/3}$. This allows me to replace $\sin\theta$ by θ and to replace the upper integration limit $(ka)^{-1/3}$ by ∞; that is,

$$\begin{aligned}\sigma_{\text{diff}} &= 2\pi \frac{k^2a^4}{4} \int_0^{(ka)^{-1/3}} \left[\frac{2J_1\left[ka\theta\right]}{ka\theta}\right]^2 \sin\theta d\theta \\ &\approx 2\pi \frac{k^2a^4}{4} \int_0^{\infty} \left[\frac{2J_1\left[ka\theta\right]}{ka\theta}\right]^2 \theta d\theta \\ &= 2\pi a^2 \int_0^{\infty} \left[\frac{J_1(z)}{z}\right]^2 zdz = 2\pi a^2 \left(\frac{1}{2}\right) = \pi a^2. \end{aligned} \tag{18.171}$$

To obtain the differential cross section outside the diffraction cone, that is, for $\theta \gg (ka)^{-1/3}$, I can approximate

$$P_\ell(\cos\theta) \sim \sqrt{\frac{1}{\pi\ell\sin\theta}} \cos\left[(\ell+1/2)\,\theta - \frac{\pi}{4}\right], \tag{18.172}$$

such that

$$f_k(\theta) \approx \frac{1}{2ik} \sum_{\ell=0}^{\infty} (2\ell+1)\left(e^{2i\delta_\ell}-1\right) \sqrt{\frac{1}{\pi\ell\sin\theta}} \cos\left[(\ell+1/2)\,\theta - \frac{\pi}{4}\right]. \tag{18.173}$$

Equation (18.172) is valid only for $\ell \gg 1$, but large values of ℓ provide the major contribution to the sum given in Eq. (18.173). The sum can be replaced by an integral and the method of stationary phase can be used to evaluate the integral. That is, after converting to an integral, for a given θ, there is a major contribution to the integral from the angular momentum (impact parameter) giving rise to the scattering at this angle; moreover, this impact parameter corresponds to the *classical* impact parameter giving rise to the scattering at this angle. This is a general result for high-energy scattering for any potential. Explicitly for the hard sphere potential, one finds

$$f_k(\theta) \approx \frac{a}{2} e^{i\alpha}, \tag{18.174}$$

where α is some phase that can be calculated using a WKB type approximation. Thus, *outside* the diffraction cone,

$$\frac{d\sigma}{d\Omega} = |f_k(\theta)|^2 \approx \frac{a^2}{4}, \tag{18.175}$$

the classical result.

The total cross section can be obtained directly from Eq. (18.168) as

$$\sigma = \frac{4\pi}{k}\,\mathrm{Im}f_k(0) = \frac{4\pi}{k}\frac{ka^2}{2} = 2\pi a^2. \tag{18.176}$$

The diffractive and classical scattering each contribute πa^2 to the total cross section.

18.6 Problems

1. Explain why specifying the impact parameter and the energy is equivalent to specifying the angular momentum and the energy in classical scattering by a spherically symmetric potential. Under what conditions can there be rainbow-like and glory scattering of classical particles? Why is the total cross section for classical scattering infinite for a potential having infinite range?

2. This problem refers to classical scattering.

(a) For the potential

$$V(r) = \begin{cases} 0, & r \geq a \\ \infty, & r < a \end{cases},$$

find the angular momentum for which the scattering angle goes to zero.

(b) For the potential

$$V(r) = -V_0 e^{-r^2/a^2},$$

where V_0 is positive, plot the effective potential as a function of r/a when

$$\frac{L^2}{2\mu a^2} = \frac{V_0}{3} = 100$$

in some appropriate energy units. For a particle of mass μ having energy $E > 0$, show that both bound and free orbits are possible. Indicate on the graph the energy for which orbiting can occur.

3. Outline the procedures that are used for solving a scattering problem using the method of partial waves and using the Born approximation. In general, when is each method most useful? Are there cases where only the $\ell = 0$ partial wave contributes, yet the first Born approximation gives a result that agrees with $\ell = 0$ scattering? Explain. Why can't the Born approximation give a good result for hard sphere scattering?

4–6. Consider the hard sphere potential,

$$V(r) = \begin{cases} 0 & r \geq a; \\ \infty & r < a \end{cases}.$$

Numerically obtain the *total* cross section for $x = ka = 0.1, 10, 100, 500$ and compare it with πa^2. You need only go from $\ell = 0$ to $\ell = ka + 25$ in evaluating the sum over partial waves in each case (why?). Interpret your result.

For $ka = 75$, numerically calculate and plot the differential scattering cross section for $\theta < 0.25$ and in the range $\pi/4 < \theta < \pi$. Interpret your results.

Plot $\cos(2\delta_\ell)$ as a function of ℓ for $ka = 50$ to illustrate that, for $ka \gg 1$, $\cos(2\delta_\ell)$ oscillates rapidly for $\ell < ka$ and is approximately equal to unity for $\ell > ka$. Use this result to show that the total cross section for $ka \gg 1$ is equal to $2\pi a^2$.

7–8. (a) In scattering of light rays by a sphere having index of refraction $n > 1$ and radius a, you can show using simple geometric optics (you are not asked to do this), that the scattering angle Θ_N after N internal reflections is given by Eq. (18.29), where b is the impact parameter of the incident light ray. Plot Θ_1 as a function of b/a, which mirrors classical scattering by a spherical well potential with $V_0/E = n^2 - 1$. Find the maximum value of $|\Theta_1|$. Also show that a primary rainbow ($N = 2$) exists for $n = \sqrt{2}$ and find the rainbow angle. Find the rainbow angle for $n = 4/3$ (water) and translate this result into an angle from which the rainbow is seen from Earth. For $n = \sqrt{2}$ show that contributions to glory scattering are possible, but only for $N \geq 4$.

(b) In the case of scattering by a sphere whose index of refraction relative to its surroundings is less than unity, Θ_N is given by Eq. (18.30). In this case, show that for $n = 1/\sqrt{2}$ that there is no rainbow scattering but contributions to glories are still possible for $N \geq 3$.

9–11. Now look at the quantum problem of scattering of a particle having mass μ by the spherical well potential,

$$V(r) = \begin{cases} -V_0 & r \leq a \\ 0 & r > a \end{cases},$$

with $V_0 > 0$. The scattering is analogous to scattering of electromagnetic radiation by a dielectric sphere having relative index of refraction

$$n_{\text{eff}} = \sqrt{1 + \frac{V_0}{E}}.$$

Take $V_0/E = 1$ and plot the differential cross section as a function of θ for $ka = 15$. Identify the structures in the cross section.

Next repeat the calculation for scattering by the spherical barrier potential

$$V(r) = \begin{cases} V_0 & r \leq a \\ 0 & r > a \end{cases},$$

with $V_0/E = 1/2$. In this case, the scattering is analogous to the scattering of electromagnetic radiation by a bubble in a dielectric medium, when the index of refraction of the bubble relative to that of its surrounding medium is given by

$$n_{\text{eff}} = \sqrt{1 - \frac{V_0}{E}}.$$

Why does the qualitative structure of the differential cross sections differ?

In this problem you must sum the contributions from the phase shifts numerically. Do you see rainbow and glories in each case?

12. Derive Eq. (18.108a). Show that, for low energy scattering by a spherical barrier potential, the differential scattering cross section does not exhibit any of the resonant structures found for scattering by a spherical well potential.

13. Derive Eq. (18.114) and the corresponding result for scattering by a spherical barrier potential. Prove that the scattering length is always positive for scattering by a spherical barrier potential.

14. Prove that $G(\mathbf{r}) = e^{ikr}/r$ is a solution of the equation

$$\left(\nabla^2 + k^2\right) G(\mathbf{r}) = -4\pi\delta(\mathbf{r}).$$

You can use the fact that $\nabla^2(1/r) = -4\pi\delta(\mathbf{r})$. Hint: You can also use the fact that, for any two functions f and g,

$$\nabla^2 (fg) = \boldsymbol{\nabla} \cdot \boldsymbol{\nabla} (fg) = \boldsymbol{\nabla} \cdot (f\boldsymbol{\nabla} g + g\boldsymbol{\nabla} f) = f\nabla^2 g + g\nabla^2 f + 2\boldsymbol{\nabla} f \cdot \boldsymbol{\nabla} g.$$

15. Calculate the differential scattering cross section in first Born approximation for a potential $V(r) = V_0 \text{sech}(r/a)$. Plot your results as a function of θ for $\beta = \left(2\mu V_0/\hbar^2\right)^{1/2} a = 0.5$ and $x = ka = \left(2\mu E/\hbar^2\right)^{1/2} a = 2$. In general, when would you expect the Born approximation to be a good approximation for this potential?

16–17. Consider scattering by the spherical well potential,

$$V(r) = \begin{cases} -V_0 < 0 & r \leq a \\ 0 & r > a \end{cases}.$$

Using the validity condition (18.136), show that the Born approximation is valid provided $\beta^2/x \ll 1$, when $x = ka \gg 1$.

Compare the exact solution [obtained using Eqs. (18.104) and (18.105a)] and the Born approximation result [Eq. (18.145)] for the differential cross section for $x = 10$

and $\beta = 1, x = 10$ and $\beta = 2$, and $x = 10$ and $\beta = 6$. Do you think the condition $\beta^2/x \ll 1$ is necessary for the validity of the Born approximation?

18. The total cross section for scattering by a spherical well potential in Born approximation is given by Eq. (18.151). You can try to reproduce this result, but it is not part of the problem. Find σ in the limit that $ka \ll 1$ and $ka \gg 1$. For $ka = 0.1, 1, 30$ compare the total cross section with that calculated "exactly" (that is, summing the partial waves) using the method of partial waves. In each case take a value of $\beta \ll ka$, $\beta = ka$, and $\beta \gg ka$. Under what conditions is the Born approximation valid?

19–20. Return to Problem 18.9–11 and plot two graphs of the total cross section (in units of πa^2) for $\beta = 10$ as a function of $x = ka$. In one graph take $0 \leq x \leq 20$ and in the second graph, take $20 \leq x \leq 100$. Show that there are resonances in the total cross section. The positions of the resonances should coincide with the energies of the quasibound states of the effective potential. To check this in a very *rough* manner, calculate the energies of the quasibound states for $\ell \leq 15$ using the WKB approximation,

$$\int_{r_1}^{r_2} k(r)dr = (n + 1/2)\,\pi,$$

where

$$k(r)a = \sqrt{x^2 + \beta^2 - \frac{\ell(\ell+1)}{r^2/a^2}},$$

and r_1 and r_2 are the classical turning points of the classical bound states of the effective potential for energy $E > 0$ (the WKB approximation cannot be expected to provide accurate positions of the resonances since there is at most one bound state for each value of ℓ). For $x \gg 1$, compare your result with that of the Born approximation, $\sigma = \pi\beta^4/2x^2$. If you take larger values of β, there will be more quasibound states since the depth of the well in the effective potential and the number of quasibound states grows with increasing β.

Chapter 19
Symmetry and Transformations: Rotation Matrices

Now that I have looked at approximation techniques and scattering theory, I return to some more formal aspects of quantum mechanics. I begin with a discussion of symmetry and see how this leads to a somewhat more sophisticated picture of angular momentum. We have seen already that energy degeneracy arises in problems for which there is an associated symmetry. Moreover, in most of these cases, there is also a conserved dynamic variable that was connected with the symmetry (e.g., in central field potentials, angular momentum is conserved and the potential is spherically symmetric).

Symmetry is central to modern physics. When you learn about the standard model of particle physics, you will see the critical role symmetry plays in the classification of particles and forces. Even in elementary quantum mechanics, symmetry considerations are important. I will try to give you a very brief introduction to this topic. I might say from the outset that I often encounter sign problems in considering either active or passive transformations, but I have learned to live with it. I need to consider transformations of coordinates, vectors, scalar functions, vector functions, and operators. These quantities transform differently under symmetry operations. Before applying the ideas of symmetry operations to quantum mechanics, I review what happens to coordinates, vectors, and functions under translation and rotation.

19.1 Active Versus Passive Transformations

In general, it is possible to define scalar, vector, and tensor operators by their transformation properties under a given symmetry operation such as translation or rotation. To make matters confusing, one can look at changes from either a passive (coordinate axes change) or an active (axes remain fixed, but vectors and functions are transformed) point of view. I will review these concepts for translation in one-dimension and rotations in both two and three dimensions. The study of translations

P.R. Berman, *Introductory Quantum Mechanics*, UNITEXT for Physics,
https://doi.org/10.1007/978-3-319-68598-4_19

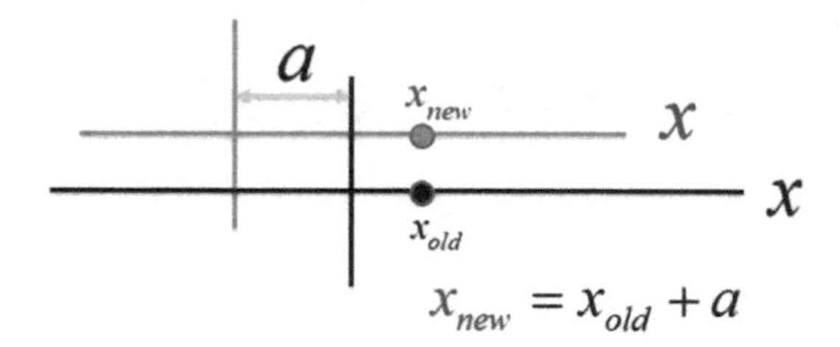

Fig. 19.1 Passive transformation of the coordinate axis to the left by a units. A point that has coordinate x_{old} in the original coordinate system has $x_{\text{new}} = x + a$ in the new coordinate system. The upper axis in the figure is the translated one

and rotations is important from a practical standpoint. For example, if you need to average collision cross sections over different orientations of the colliding particles, you can calculate the result for a given orientation of the colliding particles and then use rotation matrices to transform the result to account for arbitrary orientations.

19.1.1 Passive and Active Translations

19.1.1.1 Passive Translation

In a passive translation of the coordinate system the coordinate axes are translated. Imagine that the x-axis in Fig. 19.1 is translated a units to the *left*. If I label the coordinate of a point in the original coordinate system by x_{old} and in the new coordinate system by x_{new}, then $x_{\text{new}} = x_{\text{old}} + a$. I define a translation operator $\hat{T}^P(-a)$ for a passive transformation to the left by a units as one that changes the coordinate of a point from x_{old} to x_{new} according to

$$x_{\text{new}} = \hat{T}^P(-a)x_{\text{old}} = x_{\text{old}} + a. \tag{19.1}$$

Under this passive transformation, functions $f(x)$ are not translated; however, since the coordinate axis is translated, the *functional form* of $f(x)$ changes. The new functional form is denoted by $f_P(x; -a)$, where the minus sign is associated with a translation of the coordinate axis to the left. Since the value of the function at a fixed point in space is unchanged, the relationship between $f_P(x; -a)$ and $f(x)$ is given by

$$f_P(x; -a) = f\left[\left(\hat{T}^P(-a)\right)^{-1} x\right] = f\left[\hat{T}^P(a)x\right] = f(x - a), \tag{19.2}$$

where

$$\left(\hat{T}^P(-a)\right)^{-1} = \hat{T}^P(a) \tag{19.3}$$

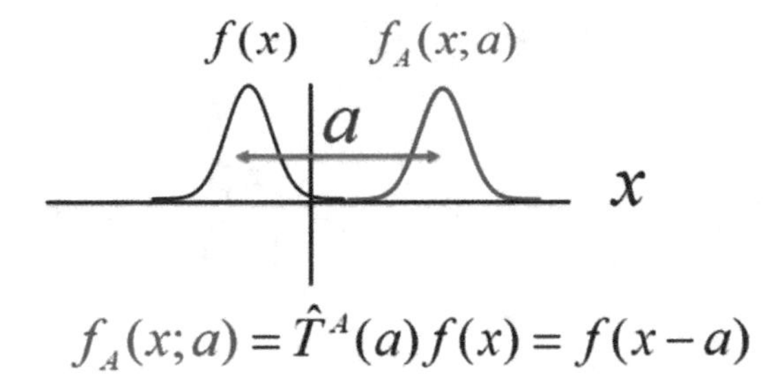

Fig. 19.2 Active transfomation of the function $f(x)$ to the right by a units

is the *inverse transformation*. For example, suppose the original function is $f(x) = e^{-x^2}$, which is centered at $x = 0$. The new function, $f_P(x; -a) = e^{-(x-a)^2}$, is now centered at $x = a$ since $x_{\text{new}} = a$ is the transformed coordinate of $x_{\text{old}} = 0$ in the original coordinate system. In effect, Eq. (19.2) defines the properties of a scalar function f_P under translation.

19.1.1.2 Active Translation

I *define* an active translation $\hat{T}^A(a)$ of a *function* to the right by a units as producing the same function I would get by translating the axis by a units to the *left* in a passive transformation (see Fig. 19.2), that is,

$$f_A(x; a) = \hat{T}^A(a) f(x) = f_P(x; -a) = f\left[\left(\hat{T}^P(a)\right) x\right] = f(x-a). \tag{19.4}$$

If the original function was centered at $x = 0$, the transformed function is centered at $x = a$. A translation of the function to the right is equivalent to a transformation of the coordinate axis to the left. Note that the passive transformation acts on coordinates while the active transformation acts on functions. In other words, $\hat{T}^A(a)$ translates functions to the right by a units while $\hat{T}^P(a)$ translates the coordinate axis to the right by a units.

19.1.2 Passive and Active Rotations

19.1.2.1 Passive Rotations

Consider a rotation of the coordinate axes in which the position vector $\mathbf{r}$ remains fixed. The axes are rotated by an angle $-\phi$ (see Fig. 19.3) about the z-axis. The unit vectors are denoted by $\mathbf{u}_1 = \mathbf{u}_x$, $\mathbf{u}_2 = \mathbf{u}_y$. Under the rotation, coordinates of the position vector change (even though the vector remains fixed in space) according to

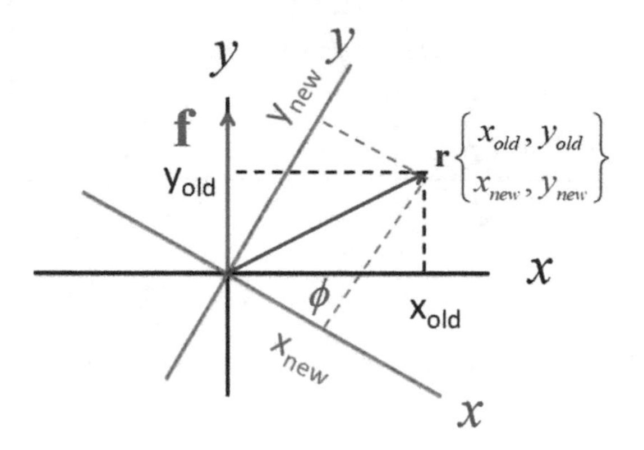

Fig. 19.3 A passive rotation of the coordinate axes by an angle $-\phi$ about the z axis. The position vector **r** and the constant vector **f** are unchanged in an absolute sense, but their coordinates change. On the other hand, the unit vectors, which are aligned along the coordinate axes change in this passive rotation. The position vector shown has coordinates $(x_{\text{old}}, y_{\text{old}})$ in the original coordinate system and $(x_{\text{new}}, y_{\text{new}})$ in the new coordinate system

$$\mathbf{r}_{\text{new}} = \begin{pmatrix} x_{\text{new}} \\ y_{\text{new}} \end{pmatrix} = \underset{\sim}{\mathrm{R}}^P(-\phi)\mathbf{r}_{\text{old}} = \begin{pmatrix} \cos\phi & -\sin\phi \\ \sin\phi & \cos\phi \end{pmatrix} \begin{pmatrix} x_{\text{old}} \\ y_{\text{old}} \end{pmatrix}, \tag{19.5}$$

where

$$\underset{\sim}{\mathrm{R}}^P(\phi) = \begin{pmatrix} \cos\phi & \sin\phi \\ -\sin\phi & \cos\phi \end{pmatrix} \tag{19.6}$$

is the *passive rotation matrix*. In other words, under this passive transformation of $-\phi$, the position vector $\mathbf{r}_{\text{old}} = (x_{\text{old}}, y_{\text{old}})$ in the old system has coordinates in the new system given by

$$x_{\text{new}} = x_{\text{old}} \cos\phi - y_{\text{old}} \sin\phi; \tag{19.7a}$$

$$y_{\text{new}} = x_{\text{old}} \sin\phi + y_{\text{old}} \cos\phi. \tag{19.7b}$$

Moreover, the basis vectors have also changed since they are aligned along the new axes, with

$$(\mathbf{u}_i)_{\text{new}} = \sum_{j=1}^{3} \underset{\sim}{\mathrm{R}}^P_{ij}(-\phi)\,(\mathbf{u}_j)_{\text{old}}, \tag{19.8}$$

where $\underset{\sim}{\mathrm{R}}^P_{ij}$ are elements of the passive rotation matrix (19.6).

The effect of a passive rotation on a scalar function is defined in an analogous manner to that for translations, namely

$$f_P(\mathbf{r}; -\phi) = f\left[\left(\underset{\sim}{\mathrm{R}}^P(-\phi)\right)^{-1}\mathbf{r}\right] = f\left[\left(\underset{\sim}{\mathrm{R}}^P(-\phi)\right)^{\dagger}\mathbf{r}\right] = f\left[\underset{\sim}{\mathrm{R}}^P(\phi)\mathbf{r}\right], \tag{19.9}$$

where I used the fact that $\underset{\sim}{\mathrm{R}}^{P}(\phi)$ is a unitary matrix whose inverse is simply its adjoint. Things can get a little more complicated when you look at the effect of rotations on vector functions (such as the electric field). For a vector function $\mathbf{E}(\mathbf{r})$, under a passive rotation of $-\phi$,

$$\mathbf{E}_P(\mathbf{r};-\phi) = \underset{\sim}{\mathrm{R}}^{P}(-\phi)\mathbf{E}\left[\underset{\sim}{\mathrm{R}}^{P}(\phi)\mathbf{r}\right], \tag{19.10}$$

the components are changed and they are evaluated at the new coordinates.

19.1.2.2 Active Rotations

As in the case of translations I can define an active rotation of a scalar function by an operator $\hat{R}^{A}(\phi)$ that, when acting on a function $f(\mathbf{r})$, produces a new function according to

$$f_A(\mathbf{r};\phi) = \hat{R}^{A}(\phi)f(\mathbf{r}) = f_P(\mathbf{r};-\phi) = f\left[\underset{\sim}{\mathrm{R}}^{P}(\phi)\mathbf{r}\right]. \tag{19.11}$$

Now here's where things get really confusing. Note that $\hat{R}^{A}(\phi)$ is an *operator* while $\underset{\sim}{\mathrm{R}}^{P}(\phi)$ is a matrix. The operator $\hat{R}^{A}(\phi)$ operates on functions and the matrix $\underset{\sim}{\mathrm{R}}^{P}(\phi)$ operates on the position vector. However I can *define* an active transformation *matrix* by

$$\underset{\sim}{\mathrm{R}}^{A}(\phi) = \underset{\sim}{\mathrm{R}}^{P}(-\phi) = \left[\underset{\sim}{\mathrm{R}}^{P}(\phi)\right]^{\dagger} = \begin{pmatrix} \cos\phi & -\sin\phi \\ \sin\phi & \cos\phi \end{pmatrix}. \tag{19.12}$$

With this definition, Eq. (19.11) becomes

$$f_A(\mathbf{r};\phi) = \hat{R}^{A}(\phi)f(\mathbf{r}) = f\left[\left(\underset{\sim}{\mathrm{R}}^{A}(\phi)\right)^{\dagger}\mathbf{r}\right] = f\left[\left(\underset{\sim}{\mathrm{R}}^{A}(\phi)\right)^{-1}\mathbf{r}\right]. \tag{19.13}$$

A vector function $\mathbf{E}(\mathbf{r})$ transforms under an active rotation by ϕ as

$$\mathbf{E}_A(\mathbf{r};\phi) = \hat{R}^{A}(\phi)\mathbf{E}(\mathbf{r}) = \underset{\sim}{\mathrm{R}}^{A}(\phi)\mathbf{E}\left[\left(\underset{\sim}{\mathrm{R}}^{A}(\phi)\right)^{-1}\mathbf{r}\right]. \tag{19.14}$$

The operation of $\hat{R}^{A}(\phi)$ on a vector function is to change the components and evaluate the coordinates at the rotated values, leaving the length of the vector unchanged. For a *constant* vector function $\mathbf{f}$,

$$\mathbf{f}_A(\phi) = \hat{R}^{A}(\phi)\mathbf{f} = \underset{\sim}{\mathrm{R}}^{A}(\phi)\mathbf{f}. \tag{19.15}$$

For constant vectors, the rotation operator acts as a matrix and we can think of the rotation operator as rotating the vector (see Fig. 19.4).

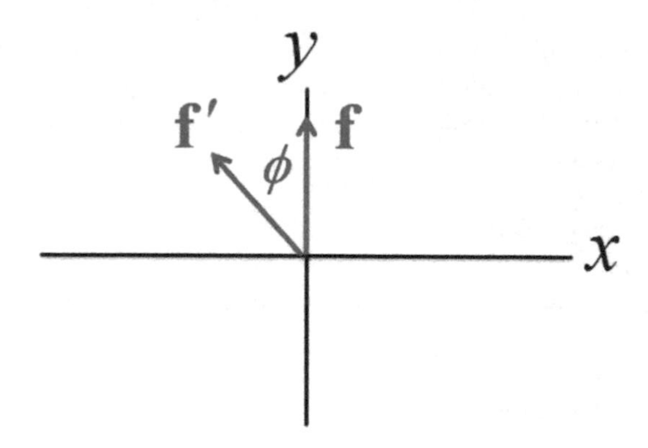

Fig. 19.4 An active rotation of a *constant* vector **f** by angle ϕ about the z axis. The axes and unit vectors remain constant, but the coordinates of the vector change

19.1.3 Extension to Three Dimensions

To specify a rotation in three dimensions, I need to specify the axis of rotation $\mathbf{u_n}(\theta,\phi)$ and the angle of rotation ω about this axis. The quantity $\mathbf{u_n}(\theta,\phi)$ is a unit vector in a spherical coordinate system having polar angle θ and azimuthal angle ϕ. Thus, *three* angles have to be specified.

Instead of specifying the axis of rotation and the rotation angle, I can equally well represent any rotation by the *Euler angles* (α,β,γ). The convention chosen for a passive rotation is one that takes the axes (x, y, z) into the axes (X, Y, Z) (see Fig. 19.5) using

- a rotation by α about the z axis
- a rotation by β about the *new* y axis
- a rotation by γ about the *new* z axis.

This leads to the passive rotation matrix

$$
\begin{aligned}
\underline{\mathrm{R}}^{P}(\alpha,\beta,\gamma) &= \begin{pmatrix} \cos\gamma & \sin\gamma & 0 \\ -\sin\gamma & \cos\gamma & 0 \\ 0 & 0 & 1 \end{pmatrix} \begin{pmatrix} \cos\beta & 0 & -\sin\beta \\ 0 & 1 & 0 \\ \sin\beta & 0 & \cos\beta \end{pmatrix} \\
&\quad \times \begin{pmatrix} \cos\alpha & \sin\alpha & 0 \\ -\sin\alpha & \cos\alpha & 0 \\ 0 & 0 & 1 \end{pmatrix} \\
&= \begin{pmatrix} \cos\alpha\cos\beta\cos\gamma - \sin\gamma\sin\alpha & \cos\gamma\cos\beta\sin\alpha + \cos\alpha\sin\gamma & -\sin\beta\cos\gamma \\ -\sin\gamma\cos\beta\cos\alpha - \sin\alpha\cos\gamma & -\sin\alpha\cos\beta\sin\gamma + \cos\gamma\cos\alpha & \sin\beta\sin\gamma \\ \cos\alpha\sin\beta & \sin\alpha\sin\beta & \cos\beta \end{pmatrix}.
\end{aligned}
\tag{19.16}
$$

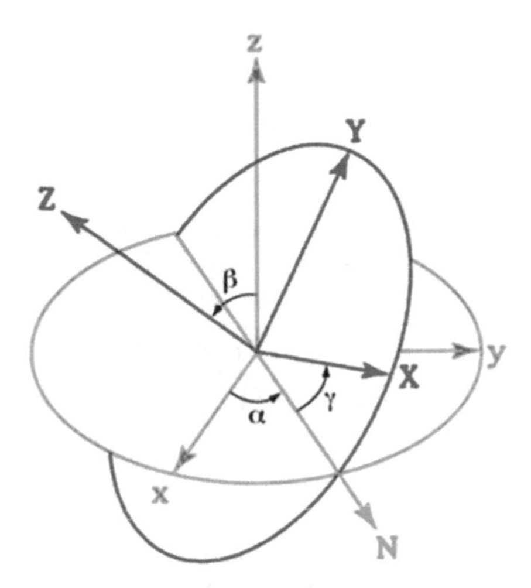

Fig. 19.5 Euler angles. A passive rotation of α about the z axis takes the axes (x, y, z) into (x', y', z). This is followed by a passive rotation of β about the y' axis that takes the axes (x', y', z) into (x'', y', Z) and a passive rotation of γ about the Z axis that takes the axes (x'', y', Z) into (X, Y, Z). The vector **N** that lies along the intersection of the xy and XY planes is referred to as the *line of nodes*

The active rotation *matrix* is defined as the inverse of this passive rotation matrix,

$$\begin{aligned}\underline{\mathrm{R}}^{A}(\alpha,\beta,\gamma) &= \left[\underline{\mathrm{R}}^{P}(\alpha,\beta,\gamma)\right]^{-1} = \underline{\mathrm{R}}_{z}^{A}(\alpha)\underline{\mathrm{R}}_{y}^{A}(\beta)\underline{\mathrm{R}}_{z}^{A}(\gamma)\\ &= \begin{pmatrix} \cos\alpha\cos\beta\cos\gamma-\sin\gamma\sin\alpha & -\cos\alpha\cos\beta\sin\gamma-\sin\alpha\cos\gamma & \cos\alpha\sin\beta\\ \sin\alpha\cos\beta\cos\gamma+\cos\alpha\sin\gamma & -\sin\alpha\cos\beta\sin\gamma+\cos\alpha\cos\gamma & \sin\alpha\sin\beta\\ -\sin\beta\cos\gamma & \sin\beta\sin\gamma & \cos\beta \end{pmatrix},\end{aligned} \tag{19.17}$$

where the rotations are now relative to the *fixed* axes. The matrix $\underline{\mathrm{R}}_{z}^{A}(\alpha)$ is the active rotation matrix for a rotation about the z axis by an angle α given by Eq. (19.12); note that in active rotations, *the γ rotation is carried out first*. Under an active rotation,

$$f_{A}(\mathbf{r};\alpha,\beta,\gamma) = \hat{R}^{A}(\alpha,\beta,\gamma)f(\mathbf{r}) = f\left[\left[\underline{\mathrm{R}}^{A}(\alpha,\beta,\gamma)\right]^{-1}\mathbf{r}\right] \tag{19.18}$$

for scalar functions and

$$\mathbf{E}_{A}(\mathbf{r};\alpha,\beta,\gamma) = \hat{R}^{A}(\alpha,\beta,\gamma)\mathbf{E}(\mathbf{r}) = \underline{\mathrm{R}}^{A}(\alpha,\beta,\gamma)\mathbf{E}\left[\left[\underline{\mathrm{R}}^{A}(\alpha,\beta,\gamma)\right]^{-1}\mathbf{r}\right] \tag{19.19}$$

for vector functions. The rotation matrix acting on a vector must leave the length of the vector unchanged. Since the rotation matrix is real, the rotation matrix must be an orthogonal matrix to preserve the length of vectors. In fact, the matrix given in Eq. (19.17) is the most general orthogonal 3×3 matrix having determinant equal to $+1$.

For a *constant* vector field,

$$\mathbf{f}_A(\alpha,\beta,\gamma) = \hat{R}^A(\alpha,\beta,\gamma)\mathbf{f} = \underset{\sim}{\mathrm{R}}^A(\alpha,\beta,\gamma)\mathbf{f}. \tag{19.20}$$

As a simple example consider the vector $\mathbf{f} = \left(\mathbf{u}_x + \mathbf{u}_y\right)/\sqrt{2}$. I want to rotate this vector so it is aligned along the z axis. I can do this by a passive transformation $(\alpha,\beta,\gamma) = (\pi/4, \pi/2, 0)$. Under such a passive transformation

$$\mathbf{f}_P(\pi/4,\pi/2,0) = \underset{\sim}{\mathrm{R}}^P(\pi/4,\pi/2,0)\mathbf{f} \begin{pmatrix} 0 & 0 & -1 \\ -\frac{1}{\sqrt{2}} & \frac{1}{\sqrt{2}} & 0 \\ \frac{1}{\sqrt{2}} & \frac{1}{\sqrt{2}} & 0 \end{pmatrix} \begin{pmatrix} \frac{1}{\sqrt{2}} \\ \frac{1}{\sqrt{2}} \\ 0 \end{pmatrix} = \begin{pmatrix} 0 \\ 0 \\ 1 \end{pmatrix}. \tag{19.21}$$

The vector is now along the *new* z axis.

Alternatively I can use an active rotation with $(\alpha,\beta,\gamma) = (0,-\pi/2,-\pi/4)$ (recall in active rotations the γ rotation is carried out first). Under this active rotation,

$$\begin{aligned}\mathbf{f}_A\left(0,-\pi/2,-\pi/4\right) &= \underset{\sim}{\mathrm{R}}^A\left(0,-\pi/2,-\pi/4\right)\mathbf{f} \\ &= \begin{pmatrix} 0 & 0 & -1 \\ -1/\sqrt{2} & 1/\sqrt{2} & 0 \\ 1/\sqrt{2} & 1/\sqrt{2} & 0 \end{pmatrix} \begin{pmatrix} 1/\sqrt{2} \\ 1/\sqrt{2} \\ 0 \end{pmatrix} = \begin{pmatrix} 0 \\ 0 \\ 1 \end{pmatrix}.\end{aligned} \tag{19.22}$$

The vector has been rotated to lie along the (original) z axis. Additional examples are given below.

Although the unit vectors used to expand vectors are unchanged by an active transformation, I can nevertheless see how the unit vectors, *considered as vectors*, transform in this case. Since the unit vectors are constant vectors, they transform as

$$\mathbf{u}_A(\alpha,\beta,\gamma)_i = \underset{\sim}{\mathrm{R}}^A(\alpha,\beta,\gamma)\mathbf{u}_i, \tag{19.23}$$

so that, for example,

$$\mathbf{u}_A(\alpha,\beta,\gamma)_1 = \underset{\sim}{\mathrm{R}}^A(\alpha,\beta,\gamma)\begin{pmatrix} 1 \\ 0 \\ 0 \end{pmatrix} = \begin{pmatrix} \underset{\sim}{\mathrm{R}}^A_{11}(\alpha,\beta,\gamma) \\ \underset{\sim}{\mathrm{R}}^A_{21}(\alpha,\beta,\gamma) \\ \underset{\sim}{\mathrm{R}}^A_{31}(\alpha,\beta,\gamma) \end{pmatrix}. \tag{19.24}$$

In other words, the transformation can be written as

$$\mathbf{u}_A(\alpha,\beta,\gamma)_i = \sum_{j=1}^{3} \underset{\sim}{\mathrm{R}}^A_{ji}(\alpha,\beta,\gamma)\mathbf{u}_j. \tag{19.25}$$

Note the order of the indices on the rotation matrix elements.

19.2 Quantum Mechanics

All these ideas can now be taken over to quantum mechanics. In fact we can use our understanding of the physics associated with Hamiltonians for the free particle and spherically symmetric potentials to obtain expressions for operators such as the translation and rotation operators. To this point, I have defined these operators only implicitly through Eqs. (19.4) and (19.11), in terms of their actions on functions.

Before starting any formalism, it might be helpful to remind you of what you already know about symmetry operations. For the moment I consider only continuous symmetry operations such as translation or rotation, but will discuss discrete symmetry operations as well. So what do you know?

1. If an operator commutes with the Hamiltonian, the expectation value of the physical observable associated with that operator is a constant of the motion.
2. If an operator $\hat{A}$ commutes with a Hamiltonian $\hat{H}$, and if $|E\rangle$ is an eigenket of $\hat{H}$ with eigenenergy E, then $\hat{A}\,|E\rangle$ is also an eigenket of $\hat{H}$ with eigenenergy E. We are guaranteed that $\hat{A}\,|E\rangle$ is equal to a constant times $|E\rangle$ *only* if there is no degeneracy in the eigenenergies of $\hat{H}$. When there is degeneracy, $\hat{A}\,|E\rangle$ is not *necessarily* equal to a constant times $|E\rangle$, this implies that there can be more than one eigenket with the same energy—when an operator commutes with the Hamiltonian, there is often energy degeneracy.
3. An operator $\hat{A}$ acting on an eigenket $|E\rangle$ of a given Hamiltonian $\hat{H}$ produces a new ket $|E\rangle'$, in general, that is not necessarily an eigenket of $\hat{H}$. Suppose that $\hat{A}$ corresponds to a symmetry operation such as a translation or rotation in Hilbert space. If the Hamiltonian $\hat{H}$ is invariant under this symmetry operation, the *eigenenergy* of the transformed state $|E\rangle'$ cannot change; it must also be an eigenket of $\hat{H}$. Since $|E\rangle' \neq |E\rangle$, in general, this again implies some energy degeneracy.
4. Combining these ideas, we see that if a Hamiltonian is invariant under some symmetry operation, there is associated with the symmetry operation a Hermitian operator that commutes with the Hamiltonian. As you will see, the operator is a generator of infinitesimal transformations on the wave function or eigenkets that leave the Hamiltonian invariant.

A simple example will serve to illustrate these ideas. Let us look at the case of a free particle having mass m in one-dimension for which the Hamiltonian is

$$\hat{H} = \frac{\hat{p}^2}{2m} \tag{19.26}$$

where $\hat{p} = -i\hbar d/dx$. The (unnormalized) eigenfunctions can be taken as

$$\psi_k(x) = e^{\pm ikx} \tag{19.27}$$

with

$$k = \sqrt{\frac{2mE}{\hbar^2}} > 0. \tag{19.28}$$

Equally well, the eigenfunctions can be taken as

$$\psi_k(x) = \cos(kx)\,; \qquad \sin(kx)\,. \tag{19.29}$$

The eigenfunctions are two-fold degenerate.

How is all this related to the momentum operator? Since $\left[\hat{p}, \hat{H}\right] = 0$, if $\psi_k(x)$ is an eigenfunction, then $\hat{p}\psi_k(x)$ is also an eigenfunction having the same energy. If I take e^{ikx} as an eigenfunction, then $\hat{p}\psi_k(x) = \hbar k e^{ikx}$, which is the same eigenfunction. That is, if I use the *simultaneous* eigenfunctions of $\hat{p}$ and $\hat{H}$, I do not generate a new eigenfunction by applying $\hat{p}$ to an eigenfunction. On the other hand, if I choose $\cos(kx)$ as an eigenfunction of $\hat{H}$ then $\hat{p}\psi_k(x) = i\hbar k \sin(kx)$. It generates a *new* eigenfunction with the same energy, demonstrating the energy degeneracy.

19.2.1 Translation Operator

Since the momentum operator commutes with the Hamiltonian and since the free particle Hamiltonian is invariant under translation, we might expect that the momentum operator is somehow connected with the translation operator in quantum mechanics. To demonstrate that this is indeed the case, I start by considering an infinitesimal displacement ϵ of an arbitrary wave function $\psi(x)$, for which

$$\psi(x-\epsilon) \approx \left(1 - \epsilon\frac{d}{dx}\right)\psi(x) = \left(1 - \frac{i}{\hbar}\epsilon\hat{p}\right)\psi(x). \tag{19.30}$$

The momentum operator is said to be *the generator of infinitesimal translations.* I can build up a finite translation a by taking N infinitesimal translations, each having $\epsilon = a/N$, and then take the limit as $N \to \infty$. In this manner I obtain

$$\begin{aligned}
\psi(x;a)_T &= \hat{T}(a)\psi(x) = \lim_{N\to\infty}\left(1 - e^{-\frac{i}{\hbar}\frac{a}{N}\hat{p}}\right)^N \psi(x) \\
&= \lim_{N\to\infty}\left(1 - e^{-\frac{i}{\hbar}\frac{a}{N}\hat{p}}\right)^{N-1} \psi(x - a/N) \\
&= \lim_{N\to\infty}\left(1 - e^{-\frac{i}{\hbar}\frac{a}{N}\hat{p}}\right)^{N-2} \psi(x - a/N - a/N) = \cdots \\
&= \lim_{N\to\infty}\psi(x - Na/N) = \psi(x-a),
\end{aligned} \tag{19.31}$$

where

$$\hat{T}(a) = \lim_{N\to\infty}\left(1 - e^{-\frac{i}{\hbar}\frac{a}{N}\hat{p}}\right)^N = e^{-i\hat{p}a/\hbar} = e^{-a\frac{d}{dx}} \tag{19.32}$$

is the translation operator and $\psi(x;a)_T$ is the translated function. The translation operator is unitary.

If the exponential in Eq. (19.32) is expanded in a Taylor series about $a = 0$ and operates on a function $\psi(x)$, the following series is generated:

$$\begin{aligned} e^{-a\frac{d}{dx}}\psi(x) &= \left(1 - a\frac{d}{dx} + \frac{(-a)^2}{2!}\frac{d^2}{dx^2} - \cdots\right)\psi(x) \\ &= \psi(x) - a\frac{d\psi(x)}{dx} + \frac{(-a)^2}{2!}\frac{d^2\psi(x)}{dx^2} - \cdots . \end{aligned} \tag{19.33}$$

The right side of this equation is a Taylor expansion of $\psi(x-a)$ about $a = 0$, but the Taylor expansion *does not necessarily converge*. Thus,

$$e^{-i\hat{p}a/\hbar}\psi(x) = \psi(x-a) \tag{19.34}$$

only if the series expansion of $\psi(x-a)$ about $a = 0$ converges for *all* x. This is always the case for any infinitesimal translation if $\psi(x)$ is an analytic function, for polynomial $\psi(x)$, and for an $\psi(x)$ that is an exponential of an analytic function, but it is not true in general (see the Appendix for more details).

[Here is a simple example where it works. Take

$$\psi(x) = 2x^2 + x \tag{19.35}$$

and let the translation of the function be two units to the right. The function is transformed as

$$\begin{aligned} \psi(x;2)_T &= e^{-\frac{i}{\hbar}2\hat{p}}\psi(x) = e^{-2\frac{d}{dx}}\left(2x^2 + x\right) \\ &= \left(1 - 2\frac{d}{dx} + \frac{(-2)^2}{2!}\frac{d^2}{dx^2} - \cdots\right)\left(2x^2 + x\right) \\ &= 2x^2 + x - 8x - 2 + 8 = 2x^2 - 7x + 6, \end{aligned} \tag{19.36}$$

which can be written as

$$\psi(x;2)_T = 2\,(x-2)^2 + (x-2)\,; \tag{19.37}$$

the original function is simply translated to the right by two units.]

If the momentum operator commutes with the Hamiltonian, then the Hamiltonian is invariant under translation, $\hat{T}(a)\hat{H}\hat{T}^{\dagger}(a) = \hat{H}$. This is a general result, continuous operators that commute with a Hamiltonian are the generators of infinitesimal transformations under which the Hamiltonian remains invariant. In this case, the momentum operator generates infinitesimal translations. *By looking at the infinitesimal transformations that are generated by operators that commute with the Hamiltonian, we can determine the symmetry properties of the Hamiltonian!*

The situation is a bit different for discrete operators such as the parity operator defined by

$$\hat{\mathcal{P}}\psi(x) = \psi(-x). \tag{19.38}$$

There can be no infinitesimal transformations associated with a discrete operator of this form. On the other hand, if an operator commutes with the parity operator, this could imply degeneracy. You can see this if you choose the eigenfunctions of the free particle Hamiltonian that are *not* eigenfunctions of the parity operator, namely $e^{\pm ikx}$. Clearly,

$$\hat{\mathcal{P}}e^{ikx} = e^{-ikx} \tag{19.39}$$

produces a new eigenfunction with the same energy. Other discrete operators of importance are the *time-reversal operator* and the *charge-conjugation operator* (see problems). All physical processes are believed to be invariant under the product of the parity, charge-conjugation, and time-reversal operators (CPT theorem).

19.2.2 Rotation Operator

Now let us consider a Hamiltonian with a spherically symmetric potential. You know that the angular momentum operator commutes with this Hamiltonian so you can see what type of infinitesimal transformation it generates. Of course, you can probably guess that it will produce a rotation since the Hamiltonian is invariant under rotations, but let's see. I need only consider $\hat{L}_z$ since the results can be generalized for the other components. I look at

$$\begin{aligned}
\left(1 - \frac{i}{\hbar}\delta\phi\hat{L}_z\right)\psi(\mathbf{r}) &= \left[1 - \frac{i}{\hbar}\delta\phi\left(x\hat{p}_y - y\hat{p}_x\right)\right]\psi(\mathbf{r}) \\
&= \left[1 - \delta\phi\left(x\frac{\partial}{\partial y} - y\frac{\partial}{\partial x}\right)\right]\psi(\mathbf{r}) \\
&= \psi(\mathbf{r}) - \delta\phi\left(x\frac{\partial\psi(\mathbf{r})}{\partial y} - y\frac{\partial\psi(\mathbf{r})}{\partial x}\right)
\end{aligned}$$

$$= \psi(x,y,z) + \frac{\partial\psi(x,y,z)}{\partial x}(y\delta\phi) - \frac{\partial\psi(x,y,z)}{\partial y}(x\delta\phi)$$
$$\approx \psi(x + y\delta\phi, y - x\delta\phi, z), \tag{19.40}$$

where $\delta\phi$ is an infinitesimal angle. But under an active rotation $\hat{R}$ of the vector $\mathbf{r}$ by an infinitesimal angle $\delta\phi$ about the z axis,

$$\hat{R}_z(\delta\phi)\,\mathbf{r} = \underline{\mathrm{R}}_z(\delta\phi)\,\mathbf{r} = (x - y\delta\phi)\,\mathbf{u}_x + (y + x\delta\phi)\,\mathbf{u}_y + \mathbf{u}_z \tag{19.41}$$

Therefore,

$$\psi(\mathbf{r};\delta\phi)_R = \left(1 - \frac{i}{\hbar}\delta\phi\hat{L}_z\right)\psi(\mathbf{r}) \approx \psi(x + y\delta\phi, y - x\delta\phi, z) \approx \psi\left[\underline{\mathrm{R}}_z(\delta\phi)^{-1}\,\mathbf{r}\right], \tag{19.42}$$

where $\underline{\mathrm{R}}_z(\delta\phi)$ is a rotation matrix corresponding to a rotation about the z-axis and $\psi(\mathbf{r};\delta\phi)_R$ is the rotated function. A comparison with Eq. (19.13) allows us to conclude that the operator $\hat{L}_z$ is the generator of infinitesimal rotations about the z axis. For an arbitrary infinitesimal rotation $\delta\omega$ about the $\mathbf{u_n}$ direction,

$$\psi(\mathbf{r};\mathbf{u_n},\delta\omega)_R = \left(1 - \frac{i}{\hbar}\mathbf{u_n}\cdot\hat{\mathbf{L}}\delta\omega\right)\psi(\mathbf{r}) \approx \psi(\underline{\mathrm{R}}(\mathbf{u_n},\delta\omega)^{-1}\,\mathbf{r}) \tag{19.43}$$

and, for a finite rotation ω about the $\mathbf{u_n}$ direction, the rotation operator is

$$\hat{R}(\mathbf{u_n},\omega) = \lim_{N\to\infty}\left(1 - \frac{i}{\hbar}\mathbf{u_n}\cdot\hat{\mathbf{L}}\frac{\omega}{N}\right)^N = e^{-\frac{i}{\hbar}\omega\mathbf{u_n}\cdot\hat{\mathbf{L}}}. \tag{19.44}$$

In this case, the transformed function is

$$\psi(\mathbf{r};\mathbf{u_n},\omega)_R = e^{-\frac{i}{\hbar}\omega\mathbf{u_n}\cdot\hat{\mathbf{L}}}\psi(\mathbf{r}) = \psi(\underline{\mathrm{R}}(\mathbf{u_n},\omega)^{-1}\,\mathbf{r}). \tag{19.45}$$

We have seen already that rotations can be described as active rotations about the z, y, and z axes by the Euler angles (γ,β,α); as a consequence, the rotation operator can also be written as

$$\hat{R}(\alpha,\beta,\gamma) = e^{-\frac{i}{\hbar}\alpha\hat{L}_z}e^{-\frac{i}{\hbar}\beta\hat{L}_y}e^{-\frac{i}{\hbar}\gamma\hat{L}_z} \tag{19.46}$$

and the transformed wave function as

$$\psi(\mathbf{r})_{\alpha,\beta,\gamma} = e^{-\frac{i}{\hbar}\alpha\hat{L}_z}e^{-\frac{i}{\hbar}\beta\hat{L}_y}e^{-\frac{i}{\hbar}\gamma\hat{L}_z}\psi(\mathbf{r}) = \psi(\underline{\mathrm{R}}(\alpha,\beta,\gamma)^{-1}\,\mathbf{r}). \tag{19.47}$$

Since the $\exp\left(i\hat{H}\right)$ is unitary if $\hat{H}$ is Hermitian, the rotation operator is unitary; that is, the value of $|\psi(\mathbf{r})|$ is unchanged when acted upon by the rotation operator.

19.2.2.1 Transformation of Eigenkets Under Rotation

So far I have been looking at the effect of transformations on wave functions. In some sense, the eigenkets replace the unit vectors, since they can be represented as column vectors. However, I cannot simply use the three-dimensional rotation matrix to transform the infinite dimensional basis vectors in Hilbert space. Instead I need a generalization of Eq. (19.25). To do this I *define* an active rotation of the eigenkets $|\ell m\rangle$ by

$$\begin{aligned}|\ell m\rangle_R &= \hat{R}(\alpha,\beta,\gamma)|\ell m\rangle = \sum_{\ell',m'}|\ell' m'\rangle\langle\ell' m'|\hat{R}(\alpha,\beta,\gamma)|\ell m\rangle \\ &= \sum_{m'}|\ell m'\rangle\langle\ell m'|\hat{R}(\alpha,\beta,\gamma)|\ell m\rangle = \sum_{m'}\mathcal{D}^{(\ell)}_{m'm}(\alpha,\beta,\gamma)|\ell m'\rangle, \end{aligned} \quad (19.48)$$

where $|\ell m\rangle_R$ is the transformed basis ket and

$$\mathcal{D}^{(\ell)}_{m'm}(\alpha,\beta,\gamma) = \langle\ell m'|\hat{R}(\alpha,\beta,\gamma)|\ell m\rangle = \langle\ell m'|e^{-\frac{i}{\hbar}\alpha\hat{L}_z}e^{-\frac{i}{\hbar}\beta\hat{L}_y}e^{-\frac{i}{\hbar}\gamma\hat{L}_z}|\ell m\rangle \quad (19.49)$$

is a matrix element of the rotation operator in the $|\ell m\rangle$ basis, which is diagonal in the ℓ quantum number. Note the order of the subscripts in Eq. (19.48), which is consistent with Eq. (19.25). For a state vector

$$|\psi\rangle = \sum_{\ell' m'} a_{\ell m}|\ell m\rangle, \quad (19.50)$$

the rotated state vector is

$$\begin{aligned}|\psi\rangle_R &= \hat{R}(\alpha,\beta,\gamma)|\psi\rangle = \sum_{\ell,m} a_{\ell m}\hat{R}(\alpha,\beta,\gamma)|\ell m\rangle \\ &= \sum_{\ell,m,m'} a_{\ell m}\mathcal{D}^{(\ell)}_{m'm}(\alpha,\beta,\gamma)|\ell m'\rangle \\ &= \sum_{\ell,m,m'}(a_{\ell m'})_R|\ell m'\rangle, \end{aligned} \quad (19.51)$$

where

$$(a_{\ell m'})_R = \sum_{m'}\mathcal{D}^{(\ell)}_{m'm}(\alpha,\beta,\gamma)(a_{\ell m}). \quad (19.52)$$

The components of the state vector transform as would be expected for any vector.

I will usually refer to the $\mathcal{D}^{(\ell)}_{m'm}(\alpha,\beta,\gamma)$ as *rotation matrices* (even though they are actually matrix elements of the rotation operator), to be distinguished from the three-dimensional rotation matrix $\underline{R}(\alpha,\beta,\gamma)$. Some properties of the rotation matrices are:

$$\mathcal{D}^{(\ell)}_{m'm}(\alpha,\beta,\gamma)=\left[\mathcal{D}^{(\ell)}_{mm'}(-\gamma,-\beta,-\alpha)\right]^*; \tag{19.53a}$$

$$\left[\mathcal{D}^{(\ell)}_{m'm}(\alpha,\beta,\gamma)\right]^*=(-1)^{m-m'}\mathcal{D}^{(\ell)}_{-m',-m}(\alpha,\beta,\gamma); \tag{19.53b}$$

$$\left[\mathcal{D}^{(\ell)}(\alpha,\beta,\gamma)\right]^{-1}=\left[\mathcal{D}^{(\ell)}(\alpha,\beta,\gamma)\right]^{\dagger}=\mathcal{D}^{(\ell)}(-\gamma,-\beta,-\alpha); \tag{19.53c}$$

$$\sum_m \mathcal{D}^{(\ell)}_{mm'}\left[\mathcal{D}^{(\ell)}_{mm''}\right]^*=\delta_{m',m''}=\sum_m \mathcal{D}^{(\ell)}_{m'm}\left[\mathcal{D}^{(\ell)}_{m''m}\right]^*; \tag{19.53d}$$

$$\begin{aligned}\mathcal{D}^{(\ell)}_{mm'}(\alpha,\beta,\gamma)&=\langle \ell m|\, e^{-\frac{i}{\hbar}\alpha\hat{L}_z}e^{-\frac{i}{\hbar}\beta\hat{L}_y}e^{-\frac{i}{\hbar}\gamma\hat{L}_z}\left|\ell m'\right\rangle;\\&=e^{-im\alpha}e^{-im'\gamma}r^{(\ell)}_{mm'}(\beta),\end{aligned} \tag{19.53e}$$

where

$$r^{(\ell)}_{mm'}(\beta)=\langle \ell m|\, e^{-\frac{i}{\hbar}\beta\hat{L}_y}\left|\ell m'\right\rangle. \tag{19.53f}$$

The derivation of the explicit expressions for the matrix elements is not trivial, but can be obtained using recursion relations.[1] A Mathematica program to evaluate the $\mathcal{D}^{(\ell)}_{m'm}(\alpha,\beta,\gamma)$ is given on the book's web site. Since

$$\mathcal{D}^{(\ell)}_{m'm}(\alpha,\beta,\gamma)=\left\langle \ell m'\right|\hat{R}(\alpha,\beta,\gamma)\left|\ell m\right\rangle=\langle \ell m|\,\hat{R}^{\dagger}(\alpha,\beta,\gamma)\left|\ell m'\right\rangle^*, \tag{19.54}$$

it follows from Eq. (19.53a) that

$$\hat{R}^{\dagger}(\alpha,\beta,\gamma)=\hat{R}(-\gamma,-\beta,-\alpha)=\hat{R}^{-1}(\alpha,\beta,\gamma), \tag{19.55}$$

consistent with the fact that $\hat{R}(\alpha,\beta,\gamma)$ is a unitary operator.

The rotation matrices are important. You will learn more about them if you take a graduate course in quantum mechanics. Just as I defined a scalar function under translation, I can define a scalar function under rotation. Even more important is that vector and tensor *operators* can be defined under rotation by their transformation properties. Vector and tensor operators, as well as *irreducible tensor operators*, are described briefly in Chap. 20.

The way in which the spherical harmonics transform under rotation can also be calculated. Recall that

$$Y^m_\ell(\theta,\phi)=\langle \mathbf{u}_r(\theta,\phi)\,|\ell m\rangle. \tag{19.56}$$

The transformed spherical harmonics are defined by

[1]For example, see U. Fano and G. Racah, *Irreducible Tensorial Sets* (Academic Press, Inc, New York, 1959), appendices D and E.

$$Y_\ell^m(\theta,\phi)_R \equiv \langle \mathbf{u}_r(\theta,\phi)\,|\ell m\rangle' = \sum_{m'} \mathcal{D}^{(\ell)}_{m'm}(\alpha,\beta,\gamma)\,\langle \mathbf{u}_r(\theta,\phi)\,\big|\ell m'\big\rangle$$
$$= \sum_{m'} \mathcal{D}^{(\ell)}_{m'm}(\alpha,\beta,\gamma)\,Y_\ell^{m'}(\theta,\phi)\,. \tag{19.57}$$

It is seen that the spherical harmonics transform in the same manner as the kets.

19.2.3 Extension to Include Spin Angular Momentum

The angular momentum operator is a unitary operator that preserves the length of state vectors. In other words,

$${}_R\langle \ell m\,|\ell m\rangle_R = \langle \ell m|\,\hat{R}^\dagger\hat{R}\,|\ell m\rangle = \langle \ell m\,|\ell m\rangle\,. \tag{19.58}$$

Although I derived an expression for the rotation operator using spatial wave functions, I could equally well have derived it by considering the most general unitary operator that satisfies Eq. (19.58). If I follow the same procedure for the eigenkets $|sm_s\rangle = \left|\frac{1}{2}, m_s\right\rangle$ of the spin operator $\hat{\mathbf{S}}$, I find a spin rotation operator

$$\hat{R}_{\text{spin}}(\mathbf{u_n},\omega) = e^{-\frac{i}{\hbar}\omega\mathbf{u_n}\cdot\hat{\mathbf{S}}} = e^{-i\omega\mathbf{u_n}\cdot\boldsymbol{\sigma}/2}, \tag{19.59}$$

or

$$\hat{R}_{\text{spin}}(\alpha,\beta,\gamma) = e^{-i\alpha\sigma_z/2}e^{-i\alpha\beta\sigma_y/2}e^{-i\gamma\sigma_z/2}, \tag{19.60}$$

where

$$\boldsymbol{\sigma} = \sigma_x\mathbf{u}_x + \sigma_y\mathbf{u}_y + \sigma_z\mathbf{u}_z, \tag{19.61}$$

σ_α $\{\alpha = x, y, z\}$ is a Pauli spin matrix, and $\{\alpha, \beta, \gamma\}$ are the Euler angles. The operator given by Eq. (19.59) or (19.60) is the most general operator that preserves the length of a two-component state vector. The effect of a rotation on the spin eigenkets is then given by

$$|sm_s\rangle_R = \hat{R}_{\text{spin}}\,|\ell m_s\rangle = \sum_{m_s'} \mathcal{D}^{(1/2)}_{m_s'm_s}(\alpha,\beta,\gamma)\,\big|sm_s'\big\rangle, \tag{19.62}$$

where $|sm_s\rangle_R$ is the transformed basis ket and

$$\mathcal{D}^{(1/2)}_{m_s'm_s}(\mathbf{u_n},\omega) = \big\langle sm_s'\big|\,e^{-i\omega\mathbf{u_n}\cdot\boldsymbol{\sigma}/2}\,|sm_s\rangle\,; \tag{19.63a}$$

$$\mathcal{D}^{(1/2)}_{m_s'm_s}(\alpha,\beta,\gamma) = \big\langle sm_s'\big|\,e^{-i\alpha\sigma_z/2}e^{-i\alpha\beta\sigma_y/2}e^{-i\gamma\sigma_z/2}\,|sm_s\rangle\,, \tag{19.63b}$$

with $s = 1/2$. Even though the spin wave functions are defined in an abstract space and *not* in coordinate space, their components are changed under rotation of coordinates. I have already emphasized this property in Chap. 12.

Somewhat surprisingly, Eqs. (19.63) can help in evaluating matrix elements of the rotation operator $\exp\left(-i\omega\mathbf{u_n}\cdot\hat{\mathbf{L}}/\hbar\right)$ given in Eq. (19.44). To understand why, recall that there are closed form expressions available for the $\mathcal{D}^{(\ell)}_{m'm}(\alpha,\beta,\gamma)$, which are matrix elements of the operator $e^{-\frac{i}{\hbar}\alpha\hat{L}_z}e^{-\frac{i}{\hbar}\beta\hat{L}_y}e^{-\frac{i}{\hbar}\gamma\hat{L}_z}$, but none for the matrix elements of $\exp\left(-i\omega\mathbf{u_n}\cdot\hat{\mathbf{L}}/\hbar\right)$. The most straightforward way to evaluate such matrix elements is to calculate the Euler angles $\{\alpha,\beta,\gamma\}$ that correspond to the rotation specified by $\{\mathbf{u_n},\omega\}$, and then calculate the corresponding $\mathcal{D}^{(\ell)}_{m'm}(\alpha,\beta,\gamma)$. But up to this point, we have no simple prescription for relating the $\{\mathbf{u_n},\omega\}$ to the $\{\alpha,\beta,\gamma\}$. Equations (19.63) afford us this possibility.

To relate the $\{\mathbf{u_n},\omega\}$ to the $\{\alpha,\beta,\gamma\}$, I write

$$e^{-i\omega\mathbf{u_n}\cdot\boldsymbol{\sigma}/2} = \cos\left(\omega\mathbf{u_n}\cdot\boldsymbol{\sigma}/2\right) - i\sin\left(\omega\mathbf{u_n}\cdot\boldsymbol{\sigma}/2\right), \tag{19.64}$$

expand the sines and cosines, and use the identity

$$\left(\mathbf{u_n}\cdot\boldsymbol{\sigma}\right)\left(\mathbf{u_n}\cdot\boldsymbol{\sigma}\right) = \mathbf{1} \tag{19.65}$$

to show that

$$e^{-\frac{i}{\hbar}\omega\mathbf{u_n}\cdot\boldsymbol{\sigma}/2} = \mathbf{1}\cos\left(\omega/2\right) - i\,\mathbf{u_n}\cdot\boldsymbol{\sigma}\,\sin\left(\omega/2\right), \tag{19.66}$$

where $\mathbf{1}$ is the 2×2 unit matrix. I then insert sums over complete sets of spin eigenkets between the exponential in Eq. (19.63b) to calculate

$$\mathcal{D}^{(1)}_{m_s'm_s}(\alpha,\beta,\gamma) = e^{-im_s'\alpha}e^{-im_s\gamma}\left\langle sm_s'\right| e^{-i\beta\sigma_y/2}\left|sm_s\right\rangle \tag{19.67}$$

and use Eq. (19.66) with $\mathbf{u_n} = \mathbf{u}_y$ and $\omega = \beta$ to arrive at

$$\mathcal{D}^{(1/2)}(\alpha,\beta,\gamma) = \begin{pmatrix} \cos\left(\beta/2\right)e^{-i(\alpha+\gamma)/2} & -\sin\left(\beta/2\right)e^{-i(\alpha-\gamma)/2} \\ \sin\left(\beta/2\right)e^{i(\alpha-\gamma)/2} & \cos\left(\beta/2\right)e^{i(\alpha+\gamma)/2} \end{pmatrix}, \tag{19.68}$$

where, by convention, the matrix indices go from highest to lowest values of m. By equating Eqs. (19.66) and (19.68), you can calculate the Euler angles $\{\alpha,\beta,\gamma\}$ in terms of $\{\mathbf{u_n},\omega\}$.

The eigenkets of hydrogen consist of both orbital and spin angular momentum components. To determine how an eigenket of the form

$$\left|\ell,s;m_\ell,m_s\right\rangle = \left|\ell m_\ell\right\rangle\left|sm_s\right\rangle \tag{19.69}$$

transforms under rotation, you need only remember that the orbital and spin operators act in different Hilbert spaces. It then follows that the effect of a rotation on the ket $|\ell, s; m_\ell, m_s\rangle$ is to produce a new ket

$$|\ell, s; m_\ell, m_s\rangle_R = e^{-\frac{i}{\hbar}\omega \mathbf{u_n}\cdot\hat{\mathbf{L}}} |\ell m_\ell\rangle \, e^{-\frac{i}{\hbar}\omega \mathbf{u_n}\cdot\hat{\mathbf{S}}} |s m_s\rangle = e^{-\frac{i}{\hbar}\omega \mathbf{u_n}\cdot\hat{\mathbf{J}}} |\ell m_\ell\rangle \, |s m_s\rangle\,, \tag{19.70}$$

where

$$\hat{\mathbf{J}} = \hat{\mathbf{L}} + \hat{\mathbf{S}} \tag{19.71}$$

is the total angular momentum operator and I have used the fact that $\hat{\mathbf{L}}$ and $\hat{\mathbf{S}}$ commute. As we have seen in Chap. 12, it is sometimes more convenient to use the eigenkets $|\ell, s; j, m\rangle$ instead of $|\ell, s; m_\ell, m_s\rangle$. The $|\ell, s; j, m\rangle$ basis is the natural one for evaluating matrix elements of the rotation operator,

$$\begin{aligned}
\mathcal{D}^{(j)}_{m'm}(\alpha, \beta, \gamma) &= \langle jm' | \hat{R}(\alpha, \beta, \gamma) | jm\rangle \\
&= \langle jm' | e^{-\frac{i}{\hbar}\alpha \hat{J}_z} e^{-\frac{i}{\hbar}\beta \hat{J}_y} e^{-\frac{i}{\hbar}\gamma \hat{J}_z} | jm\rangle \\
&= e^{-im'\alpha} e^{-im\gamma} \langle jm' | e^{-\frac{i}{\hbar}\beta \hat{J}_y} | jm\rangle\,,
\end{aligned} \tag{19.72}$$

for reasons to be discussed in Chap. 20. Closed form expressions for the $\mathcal{D}^{(j)}_{m'm}(\alpha, \beta, \gamma)$ are given on the book's web site. Equations (19.53) remain valid when ℓ is replaced by j.

19.2.4 Transformation of Operators

Imagine there is an operator $\hat{A}$ such that

$$\psi_2(\mathbf{r}) = \hat{A}\psi_1(\mathbf{r}). \tag{19.73}$$

Under a unitary transformation $\hat{U}$ that transforms the wave function as

$$\psi_u(\mathbf{r}) = \hat{U}\psi(\mathbf{r}), \tag{19.74}$$

Eq. (19.73) is transformed into

$$\psi_{2u}(\mathbf{r}) = \hat{U}\psi_2(\mathbf{r}) = \hat{U}\hat{A}\psi_1(\mathbf{r}) = \hat{U}\hat{A}\hat{U}^\dagger\psi_2(\mathbf{r}) = \hat{A}_u\psi_2(\mathbf{r}), \tag{19.75}$$

where $\hat{A}_u$ is the transformed operator and I have used the fact that $\hat{U}^\dagger\hat{U} = \hat{1}$. Thus, under translation, an operator $\hat{A}$ is transformed into

$$\hat{A}_T = \hat{T}\hat{A}\hat{T}^\dagger, \tag{19.76}$$

while under rotation it is transformed into

$$\hat{A}_R = \hat{R}\hat{A}\hat{R}^\dagger. \tag{19.77}$$

It can be a bit confusing when one encounters a Hamiltonian of the form $\hat{H} = \alpha\hat{\mathbf{r}}\cdot\mathbf{E}$, where α is a constant and $\mathbf{E}$ is a (classical) electric field. Although conventional vectors are transformed under rotation, in quantum mechanics the only quantities that are transformed under rotation (and any other symmetry operation, for that matter) are quantum-mechanical *operators*. That is, under rotation, $\hat{H}_R = \alpha\hat{\mathbf{r}}_R \cdot \mathbf{E}$, where $\hat{\mathbf{r}}_R = \hat{R}\hat{\mathbf{r}}\hat{R}^\dagger$, only the operator is transformed. The electric field appearing in the Hamiltonian is taken to be an external field (*not* an operator) and is unaffected by the transformation. You can expand your quantum system to include the electric field (it is no longer an external field in this case), but to do so you must quantize the field. Once the field is quantized, the electric field becomes an operator that is also transformed under rotation.

19.3 Rotations: Examples

In dealing with rotations, I distinguish between the rotation matrix (all operations are for active transformations)

$$\underset{\sim}{\mathrm{R}}(\alpha,\beta,\gamma) = \begin{pmatrix} \cos\alpha\cos\beta\cos\gamma-\sin\gamma\sin\alpha & -\cos\alpha\cos\beta\sin\gamma-\sin\alpha\cos\gamma & \cos\alpha\sin\beta \\ \sin\alpha\cos\beta\cos\gamma+\cos\alpha\sin\gamma & -\sin\alpha\cos\beta\sin\gamma+\cos\alpha\cos\gamma & \sin\alpha\sin\beta \\ -\sin\beta\cos\gamma & \sin\beta\sin\gamma & \cos\beta \end{pmatrix}, \tag{19.78}$$

the rotation operator

$$\hat{R}\left(\mathbf{u}_{\mathbf{n}},\omega\right) = e^{-\frac{i}{\hbar}\omega\mathbf{u}_{\mathbf{n}}\cdot\hat{\mathbf{L}}}; \tag{19.79a}$$

$$\hat{R}\left(\alpha,\beta,\gamma\right) = e^{-\frac{i}{\hbar}\alpha\hat{L}_z}e^{-\frac{i}{\hbar}\beta\hat{L}_y}e^{-\frac{i}{\hbar}\gamma\hat{L}_z}, \tag{19.79b}$$

and the rotation matrices

$$\mathcal{D}^{(\ell)}_{mm'}\left(\mathbf{u}_{\mathbf{n}},\omega\right) = \langle \ell m|\, e^{-\frac{i}{\hbar}\omega\mathbf{u}_{\mathbf{n}}\cdot\hat{\mathbf{L}}}\left|\ell m'\right\rangle \tag{19.80a}$$

$$\mathcal{D}^{(\ell)}_{mm'}\left(\alpha,\beta,\gamma\right) = \langle \ell m|\, e^{-\frac{i}{\hbar}\alpha\hat{L}_z}e^{-\frac{i}{\hbar}\beta\hat{L}_y}e^{-\frac{i}{\hbar}\gamma\hat{L}_z}\left|\ell m'\right\rangle. \tag{19.80b}$$

Each has a specific function which I now review.

19.3.1 Rotation Matrix

I start with the rotation matrix, Eq. (19.78). The rotation matrix tells you how the *coordinates* of a vector change under an active rotation of the vector. Remember in an active rotation the axes remain fixed and the vector is rotated, first by γ about the z axis, then by β around the y axis, and finally by α around the z axis. For example, the rotation $(\alpha = 0, \beta = -\pi/2, \gamma = 0)$ rotates a unit vector $\mathbf{f} = \mathbf{u}_x$ along the x axis into one along the z axis and I find accordingly

$$\begin{pmatrix} f_{Rx} \\ f_{Ry} \\ f_{Rz} \end{pmatrix} = \underline{\mathrm{R}}(0, \pi/2, 0) \begin{pmatrix} f_x \\ f_y \\ f_z \end{pmatrix} = \begin{pmatrix} 0 & 0 & -1 \\ 0 & 1 & 0 \\ 1 & 0 & 0 \end{pmatrix} \begin{pmatrix} 1 \\ 0 \\ 0 \end{pmatrix} = \begin{pmatrix} 0 \\ 0 \\ 1 \end{pmatrix}. \tag{19.81}$$

A somewhat more complicated example is to rotate the vector

$$\mathbf{f} = \frac{1}{\sqrt{3}} \begin{pmatrix} 1 \\ 1 \\ 1 \end{pmatrix} \tag{19.82}$$

into the "negative" of itself

$$\mathbf{f}_R = -\frac{1}{\sqrt{3}} \begin{pmatrix} 1 \\ 1 \\ 1 \end{pmatrix}. \tag{19.83}$$

You can convince yourself that this is accomplished by the rotation

$$\begin{aligned} &\underline{\mathrm{R}}(\alpha = -3\pi/4, \beta = \pi - 2\cos^{-1}(1/\sqrt{3}), \gamma = -\pi/4) \\ &= \frac{1}{3} \begin{pmatrix} -2 & 1 & -2 \\ 1 & -2 & -2 \\ -2 & -2 & 1 \end{pmatrix}. \end{aligned} \tag{19.84}$$

On the other hand, the *same* transformation is given simply by the matrix

$$\underline{\mathrm{M}} = \begin{pmatrix} -1 & 0 & 0 \\ 0 & -1 & 0 \\ 0 & 0 & -1 \end{pmatrix}, \tag{19.85}$$

but this matrix *cannot* be obtained from the rotation matrix $\underline{\mathrm{R}}(\alpha, \beta, \gamma)$. The reason for this is that the rotation matrices I am using are called *proper rotations* (having determinant $+1$), since they can be generated continuously from the unit matrix. The *improper* rotation matrices form another whole set of rotation matrices having determinant -1, that cannot be generated from the identity. Since there are two ways of generating rotations in this fashion, the rotation group (groups are discussed briefly in the following chapter) is not *simply connected.*

19.3.2 Rotation Operator

Now let me move on to the rotation operator. The rotation operator acts on *state vectors or wave functions*. In general you would expect the result to be very complicated since

$$\hat{R}(\alpha,\beta,\gamma)\,\psi(\mathbf{r}) = e^{-\frac{i}{\hbar}\alpha\hat{L}_z}e^{-\frac{i}{\hbar}\beta\hat{L}_y}e^{-\frac{i}{\hbar}\gamma\hat{L}_z}\psi(\mathbf{r}) \tag{19.86}$$

appears very difficult to evaluate. However, if you remember that the wave function is a scalar, then

$$\psi(\mathbf{r})_R = \hat{R}(\alpha,\beta,\gamma)\,\psi(\mathbf{r}) = \psi[\underline{\mathrm{R}}(\alpha,\beta,\gamma)^{-1}\,\mathbf{r}], \tag{19.87}$$

where $\underline{\mathrm{R}}(\alpha,\beta,\gamma)^{-1}$ is just the transpose of the (active) rotation matrix.

As an example, let us see how the spherical harmonic $Y_1^0(\theta,\phi)$ is transformed under a rotation of $-\pi/2$ about the y-axis, for which the Euler angles are $(\alpha,\beta,\gamma) = (0,-\pi/2,0)$. The transformed function is

$$Y_1^0(\theta,\phi)_R = Y_1^0(x,y,z)_R = \underline{\mathrm{R}}(0,-\pi/2,0)\,Y_1^0(\theta,\phi) = e^{\frac{i}{2\hbar}\pi\hat{L}_y}Y_1^0(\theta,\phi)\,. \tag{19.88}$$

Recall that the spherical coordinates can be considered to be functions of x, y, and z. This looks a bit complicated, but using Eq. (19.87), I find

$$\begin{aligned} Y_1^0(x,y,z)_R &= e^{\frac{i}{2\hbar}\pi\hat{L}_y}Y_1^0(\theta,\phi) = \psi[\underline{\mathrm{R}}(0,-\pi/2,0)^{-1}\,\mathbf{r}] \\ &= \psi\left[\begin{pmatrix} 0 & 0 & 1 \\ 0 & 1 & 0 \\ -1 & 0 & 0 \end{pmatrix}\begin{pmatrix} x \\ y \\ z \end{pmatrix}\right] = Y_1^0(z,y,-x) \end{aligned} \tag{19.89}$$

The problem reduces to expressing Y_1^0 as a function of (x,y,z) instead of (θ,ϕ). To do this I write

$$Y_1^0(\theta,\phi) = \sqrt{\frac{3}{4\pi}}\cos\theta = \sqrt{\frac{3}{4\pi}}\frac{z}{r} = Y_1^0(x,y,z)\,, \tag{19.90}$$

such that

$$\begin{aligned} Y_1^0(z,y,-x) &= -\sqrt{\frac{3}{4\pi}}\frac{x}{r} = -\sqrt{\frac{3}{4\pi}}\sin\theta\cos\phi \\ &= -\sqrt{\frac{3}{4\pi}}\sin\theta\frac{e^{i\phi}+e^{-i\phi}}{2} = \frac{1}{\sqrt{2}}\left[Y_1^1(\theta,\phi) - Y_1^{-1}(\theta,\phi)\right]. \end{aligned} \tag{19.91}$$

Thus

$$\psi(\mathbf{r})_R = e^{\frac{i}{2\hbar}\pi\hat{L}_y} Y_1^0(\theta,\phi) = \frac{1}{\sqrt{2}}\left[Y_1^1(\theta,\phi) - Y_1^{-1}(\theta,\phi)\right], \tag{19.92}$$

a result I will derive now using the rotation matrices.

19.3.3 Rotation Matrices

The rotation matrices are defined by

$$\mathcal{D}_{mm'}^{(\ell)}(\alpha,\beta,\gamma) = \left\langle \ell m'\right| e^{-\frac{i}{\hbar}\alpha\hat{L}_z} e^{-\frac{i}{\hbar}\beta\hat{L}_y} e^{-\frac{i}{\hbar}\gamma\hat{L}_z} \left|\ell m\right\rangle. \tag{19.93}$$

The state vector transforms as

$$\left|\ell m\right\rangle_R = \sum_{m'} \mathcal{D}_{m'm}^{(\ell)}(\alpha,\beta,\gamma)\left|\ell m'\right\rangle, \tag{19.94}$$

as do the spherical harmonics,

$$Y_\ell^m(\theta,\phi)_R = \sum_{m'} \mathcal{D}_{m'm}^{(\ell)}(\alpha,\beta,\gamma)\, Y_\ell^{m'}(\theta,\phi). \tag{19.95}$$

Equations (19.93) and (19.94) remain valid when ℓ is replaced by the total angular momentum quantum number j, which can be integral or half-integral.

For $\ell = 1$,

$$\mathcal{D}^{(1)}(\alpha,\beta,\gamma) = \begin{pmatrix} \cos^2\left(\frac{\beta}{2}\right)e^{-i(\alpha+\gamma)} & -\frac{1}{\sqrt{2}}\sin\beta e^{-i\alpha} & \sin^2\left(\frac{\beta}{2}\right)e^{-i(\alpha-\gamma)} \\ \frac{1}{\sqrt{2}}\sin\beta e^{-i\gamma} & \cos\beta & -\frac{1}{\sqrt{2}}\sin\beta e^{i\gamma} \\ \sin^2\left(\frac{\beta}{2}\right)e^{i(\alpha-\gamma)} & \frac{1}{\sqrt{2}}\sin\beta e^{i\alpha} & \cos^2\left(\frac{\beta}{2}\right)e^{i(\alpha+\gamma)} \end{pmatrix} \tag{19.96}$$

and, for $j = 1/2$,

$$\mathcal{D}^{(1/2)}(\alpha,\beta,\gamma) = \begin{pmatrix} \cos(\beta/2)\,e^{-i(\alpha+\gamma)/2} & -\sin(\beta/2)\,e^{-i(\alpha-\gamma)/2} \\ \sin(\beta/2)\,e^{i(\alpha-\gamma)/2} & \cos(\beta/2)\,e^{i(\alpha+\gamma)/2} \end{pmatrix}. \tag{19.97}$$

I can check to see if Eq. (19.95) agrees with Eq. (19.92) when $(\alpha = 0, \beta = -\pi/2, \gamma = 0)$. In that case

$$\mathcal{D}^{(1)}(0,-\pi/2,0) = \begin{pmatrix} \frac{1}{2} & \frac{1}{\sqrt{2}} & \frac{1}{2} \\ -\frac{1}{\sqrt{2}} & 0 & \frac{1}{\sqrt{2}} \\ \frac{1}{2} & -\frac{1}{\sqrt{2}} & \frac{1}{2} \end{pmatrix} \tag{19.98}$$

and

$$
\begin{aligned}
Y_1^0(\theta,\phi)_R &= \sum_{q'=-k}^{k} \mathcal{D}_{q'0}^{(k)}(\alpha,\beta,\gamma)\, Y_1^{q'}(\theta,\phi) \\
&= \frac{1}{\sqrt{2}}\left[Y_1^1(\theta,\phi) - Y_1^{-1}(\theta,\phi)\right]
\end{aligned}
\tag{19.99}
$$

in agreement with Eq. (19.92).

19.4 Summary

In this chapter, a brief introduction was given of the role played by transformations and symmetry in quantum mechanics. The interplay between degeneracy, symmetry, and operators that commute with the Hamiltonian was elaborated. It was shown that the operators associated with conserved physical quantities can be used to generate transformations on the eigenfunctions that leave the Hamiltonian invariant. The transformations of the eigenfunctions and eigenkets under translation and rotation were derived. It was shown that the transformations of the eigenkets under rotation matrices could be expressed in terms of the rotation matrices, whose elements, $\mathcal{D}_{m'm}^{(\ell)}(\alpha,\beta,\gamma)$, are simply matrix elements of the rotation operator in the $|\ell m\rangle$ basis.

19.5 Appendix: Finite Translations and Rotations

19.5.1 Translation Operator

I can use a slightly different definition of the translation operator (in one-dimension), writing it as

$$
\hat{T}(a) = \lim_{N\to\infty}\left(1 - e^{-\frac{i}{\hbar}\frac{a}{N}\hat{p}}\right)^N = \widehat{e^{-ipa/\hbar}},
\tag{19.100}
$$

with the operator $\widehat{e^{-ipa/\hbar}}$ defined such that

$$
\widehat{e^{-ipa/\hbar}}\left|p'\right\rangle = e^{-ip'a/\hbar}\left|p'\right\rangle.
\tag{19.101}
$$

Equation (19.101) is consistent with Eq. (5.64).

It is then straightforward to show that this operator produces the correct translations of kets. That is, with

$$|x\rangle = \int_{-\infty}^{\infty} dx \langle p' \, |x\rangle \, |p'\rangle = \frac{1}{\sqrt{2\pi\hbar}} \int_{-\infty}^{\infty} dp' e^{-ip'x/\hbar} \, |p'\rangle, \tag{19.102}$$

I find that

$$\begin{aligned} |x\rangle_T &= \widehat{e^{-ipa/\hbar}} \, |x\rangle = \frac{1}{\sqrt{2\pi\hbar}} \int_{-\infty}^{\infty} dp e^{-ipx/\hbar} \widehat{e^{-ipa/\hbar}} \, |p\rangle \\ &= \frac{1}{2\pi\hbar} \int_{-\infty}^{\infty} dx' \int_{-\infty}^{\infty} dp e^{-ipx/\hbar} e^{-ipa/\hbar} \, |x'\rangle \langle x' \, |p\rangle \\ &= \frac{1}{2\pi\hbar} \int_{-\infty}^{\infty} dx' \int_{-\infty}^{\infty} dp' e^{-ipx/\hbar} e^{-ipa/\hbar} e^{ipx'/\hbar} \, |x'\rangle \\ &= \int_{-\infty}^{\infty} dx' \delta\left(x' - x - a\right) |x'\rangle = |x + a\rangle, \end{aligned} \tag{19.103}$$

Thus, the operator $\widehat{e^{-ipa/\hbar}}$ translate kets to the *left* by a.

To see the effect of this operator on functions, I use Eq. (5.66),

$$\widehat{e^{-ipa/\hbar}} f(x) = \frac{1}{(2\pi\hbar)^{3/2}} \int_{-\infty}^{\infty} dx' \, \tilde{B}(x - x') f(x'). \tag{19.104}$$

where

$$\tilde{B}(x) = \frac{1}{(2\pi\hbar)^{3/2}} \int_{-\infty}^{\infty} dp \, e^{-ipa/\hbar} e^{ipx/\hbar} = (2\pi\hbar)^{3/2} \delta\left(x - a\right) \tag{19.105}$$

is the Fourier transform of $e^{-ipa/\hbar}$. Combining Eqs. (19.104) and (19.105),

$$\widehat{e^{-ipa/\hbar}} f(x) = f(x - a); \tag{19.106}$$

the operator $\widehat{e^{-ipa/\hbar}}$ translates *functions* to the right by a.

In other words, the assumption that,

$$\widehat{e^{-ipa/\hbar}} g(p) = e^{-ipa/\hbar} g(p), \tag{19.107}$$

which led to Eq. (5.66) guarantees that the operator $\widehat{e^{-ipa/\hbar}}$ translates functions. On the other hand, when interpreted in terms of its series expansion,

$$e^{-i\hat{p}a/\hbar} f(x) = f(x - a) \tag{19.108}$$

only if the series expansion of $f(x - a)$ about $a = 0$ converges for all x.

19.5.2 Rotation Operator

Similar considerations hold for the rotation operator. I can define the rotation operator via

$$\hat{R}(\mathbf{u_n},\omega) = \lim_{N\to\infty}\left(1 - \frac{i}{\hbar}\mathbf{u_n}\cdot\hat{\mathbf{L}}\frac{\omega}{N}\right)^N = \widehat{e^{-\frac{i}{\hbar}\omega\mathbf{u_n}\cdot\hat{\mathbf{L}}}}. \tag{19.109}$$

In this case, however, I am not able to prove that

$$\widehat{e^{-\frac{i}{\hbar}\omega\mathbf{u_n}\cdot\hat{\mathbf{L}}}}\psi(\mathbf{r}) = \psi(\underline{\mathrm{R}}(\mathbf{u_n},\omega)^{-1}\,\mathbf{r}), \tag{19.110}$$

since the angular momentum operator is a function of both momentum and position and I have no simple prescription for the action of such an operator on functions such as $f(\mathbf{r})$ or $g(\mathbf{p})$. I can resort to setting

$$\widehat{e^{-\frac{i}{\hbar}\omega\mathbf{u_n}\cdot\hat{\mathbf{L}}}} = e^{-\frac{i}{\hbar}\omega\mathbf{u_n}\cdot\hat{\mathbf{L}}} \tag{19.111}$$

and define $e^{-\frac{i}{\hbar}\omega\mathbf{u_n}\cdot\hat{\mathbf{L}}}$ in terms of its Taylor expansion. Fortunately, when defined in this manner, the rotation operator acting on a spherical harmonic produces a series that is always convergent. Thus, for all cases of practical interest, I can take the rotation operator to be given by Eq. (19.44) or (19.46).

19.6 Problems

1. Explain in general terms why symmetry, energy degeneracy, and conserved dynamic variables are related concepts? Give several reasons why it is useful to identify operators that commute with the Hamiltonian.

2. The parity operator is defined such that

$$\hat{P}\psi(\mathbf{r}) = \psi(-\mathbf{r}).$$

Prove that $\hat{P}$ is a Hermitian operator having eigenvalues ± 1. Under what conditions does the parity operator commute with the Hamiltonian for a one-dimensional potential? For potentials $V(r)$, prove the parity operator commutes with the Hamiltonian, which implies that simultaneous eigenfunctions of the Hamiltonian and the parity operator can be found, namely

$$\psi_{E\ell m}(\mathbf{r}) = R_{E\ell}(r)Y_\ell^m(\theta,\phi).$$

Prove that the parity of these eigenfunctions is $(-1)^\ell$.

3. Starting from the operator

$$\hat{T}(\epsilon) = e^{-\frac{i}{\hbar}\epsilon\hat{p}} \approx 1 - \frac{i}{\hbar}\epsilon\hat{p}$$

which is the generator of infinitesimal translations ϵ, prove that that $\hat{T}(a) = e^{-ia\hat{p}/\hbar}$. To derive this result expand $\hat{T}(a+\epsilon)$ to obtain a differential equation for $\hat{T}(a)$. It is often stated that $\hat{T}(a)$ is the translation operator for a finite displacement a. As shown in the Appendix, however, this assignment is valid only when $\hat{T}(a)$ acts on functions $f(x)$ for which the series expansion of $f(x-a)$ about $a = 0$ converges for all x.

4. Calculate

$$\left(1 - \epsilon\frac{d}{dx}\right)f(x)$$

and show that it corresponds to the function $f(x)$ translated ϵ units to the right in the limit that $\epsilon \to 0$, provided $f(x)$ is an analytic function. This implies that $\hat{T}(\epsilon)$ is the generator of infinitesimal translations for such functions.

Show, however, that the operator $\hat{T}(a) = e^{-\frac{i}{\hbar}a\hat{p}}$ acting on the function $f(x) = 1/\left(1+x^2\right)$ does *not* translate the function a units to the right.

5. Why does

$$e^{-\frac{i}{\hbar}\theta\mathbf{n}\cdot\hat{\mathbf{L}}}e^{-r^2} = e^{-r^2},$$

where $\mathbf{n}$ is a unit vector?

6. Evaluate

$$Y_1^0(\theta,\phi)_R = e^{-\frac{i}{\hbar}\frac{\pi}{4}\hat{L}_x}Y_1^0(\theta,\phi),$$

using the fact that

$$Y_1^0(\theta,\phi)_R = Y_1^0\left(\underline{\mathrm{R}}^{-1}\mathbf{r}\right),$$

where $\underline{\mathrm{R}}$ is the rotation matrix. (Hint: What Euler angles give you a rotation of $\pi/4$ about the x axis?).

7. Evaluate

$$e^{-\frac{i}{\hbar}\frac{\pi}{4}\hat{L}_x}Y_1^0(\theta,\phi),$$

using the fact that the $Y_\ell^m(\theta,\phi)$ transform under rotation as

$$\left[Y_\ell^m(\theta,\phi)\right]_R = \sum_{m'} \mathcal{D}^{(1)}_{m'm}(\alpha,\beta,\gamma) Y_\ell^{m'}(\theta,\phi)$$

Compare your answer with that of the previous problem.

8. If you apply the rotation operator to a state $|n, \ell, m\rangle$ of the hydrogen atom, how many distinct new eigenkets can you produce using different values of the rotation angles. Use four different sets of rotation angles to calculate the effect of a rotation on the state vector $|n = 1, \ell = 1, m = 0\rangle$. Show that the four new eigenkets that are generated are not linearly independent. Why must this be the case? Do not take any angles equal to zero or integral or half integral values of π in choosing your angles.

9. Suppose that you are given a state vector

$$|\psi\rangle = |\uparrow\rangle$$

for a spin 1/2 quantum system. Find the transformed state vector

$$|\psi\rangle_R = |x\rangle$$

for a rotation that takes a vector along the z-axis to the x-axis and

$$|\psi\rangle_R = |y\rangle$$

for a rotation that takes a vector along the z-axis to the y-axis. Prove that

$$\langle x|\,\hat{S}_x\,|x\rangle = \langle y|\,\hat{S}_y\,|y\rangle = 1;$$
$$\langle y|\,\hat{S}_x\,|y\rangle = \langle x|\,\hat{S}_y\,|x\rangle = 0,$$

In other words, if the quantum system is in the state $|x\rangle$ ($|y\rangle$), a measurement of the spin along the $x(y)$-direction always yields a value of $\hbar/2$.

10. You might expect that, in the absence of any external interactions, the Schrödinger equation would be invariant under time reversal. Show, however, that the solution of the time-dependent Schrödinger equation *changes* under the substitution $t \to -t$, that is

$$i\hbar \frac{\partial \psi(\mathbf{r},-t)}{\partial t} \neq \hat{H}\psi(\mathbf{r},-t),$$

but, instead,

$$i\hbar \frac{\partial\left[\psi(\mathbf{r},-t)\right]^*}{\partial t} = \hat{H}\left[\psi(\mathbf{r},-t)\right]^*. \tag{19.112}$$

The *time-independent, time-reversal operator* $\hat{T}$ (not to be confused with the translation operator) is defined such that

$$\hat{T}\psi(\mathbf{r}) = \psi_t(\mathbf{r}) = [\psi(\mathbf{r})]^*$$

and

$$\hat{T}\sum_n a_n\psi(\mathbf{r}) = \sum_n a_n^*\hat{T}\psi(\mathbf{r}) = \sum_n a_n^*\left[\psi(\mathbf{r})\right]^*.$$

Prove that $\hat{T}^{-1} = \hat{T} = \hat{T}^{\dagger}$. Show that by applying the operator $\hat{T}$ to the time-dependent Schrödinger equation, you reproduce Eq. (19.112), provided the Hamiltonian is invariant under time reversal, $\hat{T}\hat{H}\hat{T}^{-1} = \hat{H}$.

11. Under the time reversal operation, $\hat{T}\hat{\mathbf{r}}\hat{T}^{-1} = \hat{\mathbf{r}}$, $\hat{T}\hat{\mathbf{p}}\hat{T}^{-1} = -\hat{\mathbf{p}}$, and $\hat{T}\hat{\mathbf{L}}\hat{T}^{-1} = -\hat{\mathbf{L}}$. Prove these transformation properties by considering the actions of the operators on a wave function $\psi(\mathbf{r})$, using $\hat{T}\psi(\mathbf{r}) = \hat{T}^{-1}\psi(\mathbf{r}) = [\psi(\mathbf{r})]^*$.

12. The transformation $\hat{T}\psi(\mathbf{r}) = [\psi(\mathbf{r})]^*$ equation holds only for wave functions corresponding to particles without spin. Prove that, in order to have $\hat{T}\hat{\mathbf{S}}\hat{T}^{-1} = -\hat{\mathbf{S}}$ for a spin 1/2 particle, you can take $\hat{T}\psi(\mathbf{r}) = \exp\left(-i\pi\hat{S}_y/\hbar\right)[\psi(\mathbf{r})]^*$. This definition also extends to the total angular momentum operator $\hat{\mathbf{J}}$. The operator corresponds to a rotation of π about the y-axis.

13. For a time-reversal invariant Hamiltonian $\hat{H}$, prove that, if $|\alpha\rangle$ is an eigenket of $\hat{H}$, then $\hat{T}|\alpha\rangle$ is also an eigenket with the same energy eigenvalue. As a consequence if $\hat{T}|\alpha\rangle \neq |\alpha\rangle$, then there is a degeneracy related to time-reversal invariance that is referred to as *Kramers degeneracy* (after Hans Kramers). Show that for a spinless particle, there is no Kramers degeneracy but, for a particle having spin 1/2, $\hat{T}|\uparrow\rangle = |\downarrow\rangle$; there is a two-fold Kramers degeneracy.

14. For a single-particle system, the charge conjugation operator $\hat{C}$ changes the particle into its antiparticle—it effectively changes the sign of the particle's charge without affecting its other properties. The equations of physics are believed to be invariant under the combined action of the parity, time-reversal, and charge conjugation operators. Explain why the interaction potentials $-\hat{\mathbf{p}}_e\cdot\mathbf{E}$ and $-\hat{\mu}\cdot\mathbf{B}$ are invariant under $\hat{C}\hat{P}\hat{T}$ by considering the action of each of these operators separately on the atomic dipole moment $\mathbf{p}_e = -e\hat{\mathbf{r}}$ and the magnetic dipole moment operator $\hat{\mu} = -e\hat{\mathbf{L}}/2m_e$. The electric and magnetic fields appearing in these equations are taken to be external fields (not operators) and are unaffected by the transformations. Also determine if these interactions preserve parity and if they are time-reversal invariant.

15. Prove that the Bell state $|\Psi_-\rangle = (|\uparrow\downarrow\rangle - |\downarrow\uparrow\rangle)/\sqrt{2}$ of two electrons is invariant under an arbitrary rotation. Prove that the Bell state $|\Phi_+\rangle = (|\uparrow\uparrow\rangle + |\downarrow\downarrow\rangle)/\sqrt{2}$ is unchanged under a rotation that takes the z-axis into the x-axis, but that it is

transformed (to within a phase) into the Bell state $|\Phi_-\rangle = (|\uparrow\uparrow\rangle - |\downarrow\downarrow\rangle)/\sqrt{2}$ under a rotation that takes the z-axis into the y-axis.

16. Prove that

$$(\mathbf{u_n} \cdot \boldsymbol{\sigma})(\mathbf{u_n} \cdot \boldsymbol{\sigma}) = \mathbf{1},$$

where $\mathbf{u_n}$ is a unit vector, and use this identity to prove that

$$\cos(\omega \mathbf{u_n} \cdot \boldsymbol{\sigma}/2) - i \sin(\omega \mathbf{u_n} \cdot \boldsymbol{\sigma}/2) = \mathbf{1} \cos(\omega/2) - i\mathbf{u_n} \cdot \boldsymbol{\sigma} \sin(\omega/2).$$

In turn, use this expression to show that

$$e^{-i\beta\sigma_y/2} = \begin{pmatrix} \cos(\beta/2) & -\sin(\beta/2) \\ \sin(\beta/2) & \cos(\beta/2) \end{pmatrix}.$$

17–18. Use Eqs. (19.66) and (19.68) to express the Euler angles in terms of $\mathbf{u_n}$ and ω. Verify that your solution is correct for $\mathbf{u_n} = \mathbf{u}_x$, $\mathbf{u_n} = \mathbf{u}_y$, and $\mathbf{u_n} = \mathbf{u}_z$ by calculating the rotation matrix $\underline{\mathrm{R}}(\alpha, \beta, \gamma)$ in each case.

19–21. In Problem 12.7–8 you were asked to obtain the electron spin eigenkets for a quantization axis in the direction of the unit vector

$$\mathbf{u}(\theta, \phi, \psi) = \cos\psi\, \mathbf{u}_\theta + \sin\psi\, \mathbf{u}_\varphi.$$

The angles θ and ϕ are the polar and azimuthal angles of a spherical coordinate system, while ψ is the angle of the quantization axis relative to the $\mathbf{u}_\theta$ direction in a plane perpendicular to $\mathbf{u}_r$. To obtain the eigenkets using the rotation matrices, find the Euler angles that correspond to a rotation that takes the unit vector $\mathbf{v} = \mathbf{u}_z$ to the direction $\mathbf{u}(\theta, \phi, \psi)$ (since this takes the original quantization axis into the new one). Then use Eqs. (19.48) and (19.97) to calculate the eigenkets. Show that, to within a phase factor, they agree with the ones given in Problem 12.7–8. [Hint: Since $\mathbf{v} = \mathbf{u}_z$, the first rotation does not affect the vector $\mathbf{v} = \mathbf{u}_z$; that is, $\gamma = 0$. To get the remaining Euler angles, set $\mathbf{u}(\theta, \phi, \psi) = \underline{\mathrm{R}}(\alpha, \beta, \gamma)\mathbf{u}_z$.]

22. Using Eq. (5.86),

$$e^{\hat{A}}\hat{O}e^{-\hat{A}} = \hat{O} + \left[\hat{A}, \hat{O}\right] + \frac{1}{2!}\left[\hat{A}, \left[\hat{A}, \hat{O}\right]\right] + \cdots,$$

prove that, under an active translation produced by the operator $\hat{T}(a) = e^{-ia\hat{p}_x/\hbar}$, the operators $\hat{x}$ and $\hat{x}^2$ are transformed as

$$\hat{x}_T = \hat{x} - a; \qquad \hat{x}_T^2 = (\hat{x} - a)^2,$$

and that under an active *momentum boost* produced by the *momentum translation* operator $\hat{B}(q) = e^{iq\hat{x}/\hbar}$, the operators $\hat{p}_x$ and $\hat{p}_x$ are transformed as

$$(\hat{p}_x)_T = \hat{p}_x - q; \qquad \left(\hat{p}_x^2\right)_T = (\hat{p}_x - q)^2 .$$

23. Consider one-dimensional motion with $\hat{p} = \hat{p}_x$. The combined operation of a translation and a momentum boost of the previous problem,

$$\hat{B}(q)\hat{T}(a) = e^{iq\hat{x}/\hbar} e^{-ia\hat{p}_x/\hbar}$$

translates both the momentum and position wave functions. If $a = v_b t$ and $q = mv_b$ for a particle having mass m, this combined operation would seem to constitute a Galilean transformation. Of course, we are free to multiply this operator by any phase factor that does not contain any operators. Show that the free particle Hamiltonian is changed under the transformation produced by the operator

$$\hat{G}(v_b, t) = e^{-imv_b^2 t/2} \hat{B}(mv_b)\hat{T}(v_b t)$$

but the wave function

$$\psi(x,t)_G = \hat{G}(v_b, t)\psi(x,t)$$

still satisfies the Schrödinger equation. This result proves that the free particle Schrödinger equation is invariant under a Galilean transformation. It can be generalized to many-particle systems whose interaction energies depend only on the relative position vectors of the particles. In that case, the transformation is made with respect to center-of-mass variables. In solving this problem, it may prove helpful to use Eq. (5.85),

$$e^{\hat{A}+\hat{B}} = e^{\hat{A}} e^{\hat{B}} e^{-[\hat{A},\hat{B}]/2}.$$

24. Prove that under the Galilean transformation,

$$\hat{G}(v_b, t) = e^{-imv_b^2 t/2} e^{imv_b\hat{x}/\hbar} e^{-iv_b t\hat{p}_x/\hbar},$$

the free particle spatial wave function,

$$\psi(x,t) = \exp(ip_0/\hbar)\exp\left(-ip_0^2 t/2m\hbar\right),$$

and free particle momentum wave function,

$$\Phi(p,t) = \delta(p - p_0)\exp\left(-ip^2 t/2m\hbar\right),$$

transform as you would expect.

Chapter 20
Addition of Angular Momenta, Clebsch-Gordan Coefficients, Vector and Tensor Operators, Wigner-Eckart Theorem

20.1 Addition of Angular Momenta and Clebsch-Gordan Coefficients

Now that we have seen how wave functions and state vectors are changed under symmetry operations, it is natural to ask how *operators* are transformed under the same operations. To do so, however, requires a slight digression. I first need to discuss the way in which angular momenta are coupled in quantum mechanics. I have already introduced the concept of coupling of orbital and spin angular momentum in Chap. 12, but want to generalize this to the coupling of any two angular momenta. For the moment let us consider two, *non-interacting* quantum systems having total angular momentum operator $\hat{J}_1$ associated with system 1 and total angular momentum operator $\hat{J}_2$ associated with system 2. Since the systems are non-interacting, the operators $\hat{J}_1$ and $\hat{J}_2$ commute and an eigenket for the combined system can be written as

$$|j_1, j_2; m_1, m_2\rangle = |j_1, m_1\rangle\, |j_2, m_2\rangle \tag{20.1}$$

which is a simultaneous eigenstate of the operators $\hat{J}_1^2, \hat{J}_2^2, \hat{J}_{1z}, \hat{J}_{2z}$.

If I form the operator

$$\hat{\mathbf{J}} = \hat{\mathbf{J}}_1 + \hat{\mathbf{J}}_2 \tag{20.2}$$

it is easy to prove that

$$\left[\hat{J}_x, \hat{J}_y\right] = i\hbar\hat{J}_z; \tag{20.3}$$

P.R. Berman, *Introductory Quantum Mechanics*, UNITEXT for Physics,
https://doi.org/10.1007/978-3-319-68598-4_20

the components of $\hat{J}$ satisfy the usual commutation laws for angular momenta. As a consequence, a simultaneous eigenket of $\hat{J}_1^2$, $\hat{J}_2^2$, $\hat{J}^2$, and $\hat{J}_z$ can also be written as $|j_1, j_2; j, m\rangle$,where j and m are the quantum numbers associated with $\hat{J}^2$ and $\hat{J}_z$. I want to relate the $|j_1, j_2; j, m\rangle$ eigenkets to the $|j_1, m_1\rangle\, |j_2, m_2\rangle$ eigenkets.

In effect, I need to see how the matrix elements of the sum operator, $\hat{\mathbf{J}} = \hat{\mathbf{J}}_1 + \hat{\mathbf{J}}_2$, is related to those of the individual operators $\hat{\mathbf{J}}_1$ and $\hat{\mathbf{J}}_2$. *Classically*, you know how to add two vectors $\mathbf{J}_1$ and $\mathbf{J}_2$ to get the sum vector $\mathbf{J} = \mathbf{J}_1 + \mathbf{J}_2$. You can prove easily that

$$|J_1 - J_2| \leq |\mathbf{J}_1 + \mathbf{J}_2| \leq J_1 + J_2. \tag{20.4}$$

Quantum-mechanically, we will find a similar condition. Classically, you can ask, "How many ways can we combine $\mathbf{J}_1 + \mathbf{J}_2$ to a given total value $\mathbf{J}$?" Clearly the z components must add, $J_{1z} + J_{2z} = J_z$. But even with this restriction there is an infinite number of ways to add the vectors together, provided they satisfy condition (20.4) with the $<$ rather than $\leq$ signs. [To see this, draw a triangle $\mathbf{J}_1 + \mathbf{J}_2 = \mathbf{J}$ in a plane. You can now rotate $\mathbf{J}_2$ around $\mathbf{J}_1$ keeping the angle between $\mathbf{J}_1$ and $\mathbf{J}_2$ constant. For each angle of rotation, $\mathbf{J}_1 + \mathbf{J}_2 = \mathbf{J}$.] At the extreme values, there is only one way to sum the vectors since they must be aligned. In quantum mechanics, there is also only one way to sum the vectors at the extreme values; however, for other than extreme values of J, there is always a *finite* rather than infinite number of ways to add the vectors to get the final vector, since angular momentum is quantized.

To relate the two bases I write

$$|j_1, j_2; j, m\rangle = \sum_{m_1, m_2} \langle j_1, j_2; m_1, m_2\, |j_1, j_2; j, m\rangle\, |j_1, j_2; m_1, m_2\rangle\, . \tag{20.5}$$

The coupling coefficients are written as

$$\langle j_1, j_2; m_1, m_2\, |j_1, j_2; j, m\rangle \equiv \langle j_1, j_2; m_1, m_2\, |j, m\rangle = \begin{bmatrix} j_1 & j_2 & j \\ m_1 & m_2 & m \end{bmatrix} \tag{20.6}$$

and are *Clebsch-Gordan coefficients.* Unconventionally, I use square brackets for the Clebsch-Gordan coefficients. The Clebsch-Gordan coefficients are given by a Mathematica function

$$\begin{bmatrix} j_1 & j_2 & j \\ m_1 & m_2 & m \end{bmatrix} = \text{ClebschGordan}\left[\{j_1, m_1\}, \{j_2, m_2\}, \{j_3, m_3\}\right]. \tag{20.7}$$

They are usually evaluated by noting first that

$$\begin{bmatrix} j_1 \; j_2 \; j_1 + j_2 \\ j_1 \; j_2 \; j_1 + j_2 \end{bmatrix} = 1 \tag{20.8}$$

(there is only one way to couple the vectors if they are aligned). One then operates with the ladder operator

$$\hat{J}_- = \hat{J}_x - i\hat{J}_y = \hat{J}_{1x} + \hat{J}_{2x} - i\left(\hat{J}_{1y} + \hat{J}_{2y}\right) = \hat{J}_{1-} + \hat{J}_{2-} \tag{20.9}$$

on Eq. (20.5) with $m = j$ to obtain a set of algebraic equations for the Clebsch-Gordan coefficients. For other values of $j \neq j_1 + j_2$ the ladder operators can be used to determine all the Clebsch-Gordan coefficients in terms of $\langle j_1 j_2, m_1 = j_1, m_2 = j - j_1 \,| j_1 j_2; j, j\rangle$, whose value is then fixed by normalization. The phase is chosen such that all the Clebsch-Gordan coefficients are real, implying that

$$\langle j_1, j_2; m_1, m_2 \,| j, m\rangle = \langle j, m \,| j_1, j_2; m_1, m_2\rangle \,. \tag{20.10}$$

In this way it is possible to get a closed form expression for all the Clebsch-Gordan coefficients in terms of a sum, which is the way Mathematica calculates these coefficients.[1] The Clebsch-Gordan coefficients are related to *3-J symbols* defined by

$$\begin{pmatrix} j_1 & j_2 & j \\ m_1 & m_2 & -m \end{pmatrix} = \frac{(-1)^{j_1 - j_2 + m}}{\sqrt{2j+1}} \begin{bmatrix} j_1 & j_2 & j \\ m_1 & m_2 & m \end{bmatrix}. \tag{20.11}$$

The 3-J symbols are, in some sense, symmetrized forms of the Clebsch-Gordan coefficients. In the "old days" one resorted to published tables of Clebsch-Gordan coefficients and 3-J symbols; now most symbolic mathematical programs have them as built-in functions. Mathematica subroutines for evaluating these functions are also listed on the book's web site.

I list some properties of Clebsch-Gordan coefficients and 3-J symbols:

$$\begin{bmatrix} j_1 & j_2 & j \\ m_1 & m_2 & m \end{bmatrix} \text{ is real;} \tag{20.12}$$

$$\begin{bmatrix} j_1 & j_2 & j \\ m_1 & m_2 & m \end{bmatrix} = 0 \quad \text{unless } m_1 + m_2 = m \text{ and } |j_2 - j_1| \le j \le j_1 + j_2; \tag{20.13}$$

$$\sum_{m_1=-j_1}^{j_1} \sum_{m_2=-j_2}^{j_2} \begin{bmatrix} j_1 & j_2 & j \\ m_1 & m_2 & m \end{bmatrix} \begin{bmatrix} j_1 & j_2 & j' \\ m_1 & m_2 & m' \end{bmatrix} = \delta_{j,j'}\delta_{m,m'}; \tag{20.14}$$

[1]See, for example, A. R. Edmonds, *Angular Momentum in Quantum Mechanics* (Princeton University Press, Princeton, N. J.,1960), Chap. 3.

$$\sum_{j=|j_2-j_1|}^{j_1+j_2} \sum_{m=-j}^{j} \begin{bmatrix} j_1 & j_2 & j \\ m_1 & m_2 & m \end{bmatrix} \begin{bmatrix} j_1 & j_2 & j \\ m_1' & m_2' & m \end{bmatrix} = \delta_{m_1,m_1'} \delta_{m_2,m_2'}; \tag{20.15}$$

$$\begin{bmatrix} j_1 & j_2 & j \\ m_1 & m_2 & m \end{bmatrix} = (-1)^{j_1+j_2-j} \begin{bmatrix} j_2 & j_1 & j \\ m_2 & m_1 & m \end{bmatrix} \tag{20.16a}$$

$$= (-1)^{j_1+j_2-j} \begin{bmatrix} j_1 & j_2 & j \\ -m_1 & -m_2 & -m \end{bmatrix} \tag{20.16b}$$

$$= (-1)^{j_1-j+m_2} \sqrt{\frac{2j+1}{2j_1+1}} \begin{bmatrix} j & j_2 & j_1 \\ m & -m_2 & m_1 \end{bmatrix} \tag{20.16c}$$

$$= (-1)^{j_2-j-m_1} \sqrt{\frac{2j+1}{2j_2+1}} \begin{bmatrix} j_1 & j & j_2 \\ -m_1 & m & m_2 \end{bmatrix}. \tag{20.16d}$$

Equation (20.13) is the quantum analogue of the restrictions encountered in the classical addition of angular momentum vectors.

The 3-J symbol,

$$\begin{pmatrix} j_1 & j_2 & j \\ m_1 & m_2 & m \end{pmatrix},$$

vanishes unless $m_1 + m_2 + m = 0$ and $|j_2 - j_1| \leq j \leq j_1 + j_2$, it is invariant under a circular permutation of all columns, and it is multiplied by $(-1)^{j_1+j_2+j}$ under a permutation of any two columns or when the signs of all the m's are changed. Also

$$\begin{pmatrix} j_1 & j_2 & j \\ 0 & 0 & 0 \end{pmatrix} = 0 \quad \text{if } j_1 + j_2 + j \text{ is odd}, \tag{20.17}$$

as is the corresponding Clebsch-Gordan coefficient.

As an example, I can write

$$|j_1 = 1, j_2 = 1; j = 1, m = 0\rangle = \sum_{m_1,m_2} \begin{bmatrix} 1 & 1 & 1 \\ m_1 & m_2 & 0 \end{bmatrix} |j_1, j_2; m_1, m_2\rangle$$

$$= \begin{bmatrix} 1 & 1 & 1 \\ 1 & -1 & 0 \end{bmatrix} |1,1;1,-1\rangle + \begin{bmatrix} 1 & 1 & 1 \\ 0 & 0 & 0 \end{bmatrix} |1,1;0,0\rangle$$

$$+ \begin{bmatrix} 1 & 1 & 1 \\ 1 & -1 & 0 \end{bmatrix} |1,1;-1,1\rangle$$

$$= \frac{1}{\sqrt{2}} |1,1;1,-1\rangle + 0\,|1,1;0,0\rangle - \frac{1}{\sqrt{2}} |1,1;-1,1\rangle. \tag{20.18}$$

The Clebsch-Gordan coefficients are very important in calculating transition rates. Often we look at transitions between two manifolds of atomic levels, each containing a number of magnetic sublevels (corresponding to different values of m). The ratio of the various transition rates between levels having different m's is equal to the square of the Clebsch-Gordan coefficients associated with the transition rates.

20.2 Vector and Tensor Operators

Some of you may have learned about *groups* in your mathematics courses. A group consists of a number of elements and some group operation. For example, all the real numbers form a group under addition since the addition of any two real numbers produces another real number which is a member of the group, there is an *identity element* zero, which when added to any real number produces the same number, and an *inverse* (the negative of a number) which, when added to an element gives the identity element $[x + (-x) = 0]$. This is not the course to go into elements of group theory as applied to quantum mechanics, but it is a powerful method. In fact I really only want to get to the Wigner-Eckart theorem, which offers a very useful method for evaluating matrix elements of operators or ratios of matrix elements of operators. To do this, I need to introduce the concept of an *irreducible tensor operator.*

Rotations also form a group, as do the rotation operators and the rotation matrices $\underset{\sim}{R}(\alpha, \beta, \gamma)$.[2] Any two successive rotations are equivalent to a single rotation. The identity element is no rotation at all (or a rotation of $2n\pi$ about an axis) and the inverse of a rotation is just the reverse rotation. However, you can convince yourself that if you perform rotations R_1 and then R_2 about different axes in three dimensions, it does not give the same result as if you reverse the order of the rotations. Rotations are said to form a *nonabelian group.* We have already seen that angular momentum is the generator of infinitesimal rotations. *The nonabelian nature of the rotation group can be linked to the fact that the different components of the angular momentum operator do not commute with one another.* In fact the commutation relations of the angular momentum operators are said to form an *algebra* (algebras have two operations) that determines the properties of the rotation group in the vicinity of the identity.

Why am I introducing these concepts? The reason is that the group structure of rotations is determined totally by the angular momentum operators. Thus we can define scalar, vector, and tensor operators under rotation in terms of their commutation relations with the angular momentum operators, as well as in the way they transform under rotation.

---

[2]The rotation matrices $\underset{\sim}{R}(\alpha, \beta, \gamma)$ form a group of orthogonal 3×3 matrices having determinant equal to $+1$, a group that is refered to as the special orthogonal group in three dimensions, SO(3). The group of unitary 2×2 matrices having determinant equal to $+1$, such as the $\mathcal{D}^{(1/2)}(\alpha, \beta, \gamma)$ matrices, is refered to as the special unitary group in two dimensions, SU(2).

A *scalar operator* $\hat{A}$ under rotation is one for which

$$\hat{A}_R = \hat{R}\hat{A}\hat{R}^\dagger = \hat{A}. \tag{20.19}$$

Consider an infinitesimal rotation $\delta\boldsymbol{\phi}$. When looking at the effects of rotation on kets that involve both orbital and spin angular momentum, rotation operators for *both* the spin and orbital angular momenta must be used. An appropriate rotation operator is

$$\hat{R}(\mathbf{u_n}, \omega) = e^{-\frac{i}{\hbar}\omega\mathbf{u_n}\cdot\hat{\mathbf{L}}} e^{-\frac{i}{\hbar}\omega\mathbf{u_n}\cdot\hat{\mathbf{S}}} = e^{-\frac{i}{\hbar}\omega\mathbf{u_n}\cdot\hat{\mathbf{J}}}. \tag{20.20}$$

For an infinitesimal rotation $\delta\boldsymbol{\phi} = \mathbf{u_n}\delta\omega$,

$$\hat{R}(\delta\boldsymbol{\phi}) = e^{-i\delta\boldsymbol{\phi}\cdot\hat{\mathbf{J}}/\hbar} \approx \left(1 - \frac{i}{\hbar}\delta\boldsymbol{\phi}\cdot\hat{\mathbf{J}}\right). \tag{20.21}$$

Under this rotation, Eq. (20.19) becomes

$$\begin{aligned}
\hat{A}_R &= \left(1 - \frac{i}{\hbar}\delta\boldsymbol{\phi}\cdot\hat{\mathbf{J}}\right)\hat{A}\left(1 + \frac{i}{\hbar}\delta\boldsymbol{\phi}\cdot\hat{\mathbf{J}}\right) \\
&\approx \hat{A} - \frac{i}{\hbar}\begin{bmatrix} \left(\delta\phi_x\hat{J}_x + \delta\phi_y\hat{J}_y + \delta\phi_z\hat{J}_z\right)\hat{A} \\ -\hat{A}\left(\delta\phi_x\hat{J}_x + \delta\phi_y\hat{J}_y + \delta\phi_z\hat{J}_z\right) \end{bmatrix} \\
&= \hat{A} - \frac{i}{\hbar}\left\{\delta\phi_x\left[\hat{J}_x, \hat{A}\right] + \delta\phi_y\left[\hat{J}_y, \hat{A}\right] + \delta\phi_z\left[\hat{J}_z, \hat{A}\right]\right\} = \hat{A},
\end{aligned} \tag{20.22}$$

where terms of order $(\delta\phi)^2$ have been neglected. For arbitrary $\delta\boldsymbol{\phi}$, the only way Eq. (20.22) can be satisfied is if

$$\left[\hat{A}, \hat{\mathbf{J}}\right] = 0, \tag{20.23}$$

which is an alternative definition of a scalar operator under rotation. That is, either Eq. (20.19) or Eq. (20.23) can be used to define a scalar operator under rotation.

A *vector operator* or *tensor operator of rank 1* is defined as a set of *three* operators that transform as a Cartesian vector under a rotation as

$$\left(\hat{A}_R\right)_i = \hat{R}\hat{A}_i\hat{R}^\dagger = \sum_{j=1}^{3} \underline{R}_{ji}\hat{A}_j. \quad i = 1 - 3 \tag{20.24a}$$

$$\hat{\mathbf{A}}_R = \underline{R}^{-1}\hat{\mathbf{A}} = \underline{R}^T\hat{\mathbf{A}}, \tag{20.24b}$$

where $\underline{R}_{ji}$ is a matrix element of the (active) rotation matrix given by Eq. (19.78). Note the order ji of the indices—in other words, although constant vectors $\mathbf{f}$ transform under rotation as $\mathbf{f}' = \underline{\mathrm{R}}\mathbf{f}$, the vector operator $\hat{\mathbf{A}} = \left(\hat{A}_1, \hat{A}_2, \hat{A}_2\right)$ transforms as $\hat{\mathbf{A}}_R = \underline{\mathrm{R}}^T\hat{\mathbf{A}}$, where $\underline{\mathrm{R}}^T$ is the transpose of the rotation matrix.

To see that this works, let us rotate the *operator* $\hat{\mathbf{r}}$ by $\pi/2$ about the y axis. Under such a rotation we would expect the $\hat{x}$ component to transform into $-\hat{z}$, the $\hat{y}$ component to remain unchanged, and the $\hat{z}$ component to transform into $\hat{x}$. The rotation matrix in this case has Euler angles $\alpha = \gamma = 0$ and $\beta = \pi/2$, giving

$$\underline{\mathrm{R}}\left(0, \frac{\pi}{2}, 0\right) = \begin{pmatrix} 0 & 0 & 1 \\ 0 & 1 & 0 \\ -1 & 0 & 0 \end{pmatrix}; \quad \underline{\mathrm{R}}^T\left(0, \frac{\pi}{2}, 0\right) = \begin{pmatrix} 0 & 0 & -1 \\ 0 & 1 & 0 \\ 1 & 0 & 0 \end{pmatrix}. \tag{20.25}$$

Then

$$\hat{\mathbf{r}}' = \begin{pmatrix} \hat{x}' \\ \hat{y}' \\ \hat{z}' \end{pmatrix} = \underline{\mathrm{R}}^T \begin{pmatrix} \hat{x} \\ \hat{y} \\ \hat{z} \end{pmatrix} = \begin{pmatrix} -\hat{z} \\ \hat{y} \\ \hat{x} \end{pmatrix}, \tag{20.26}$$

as expected.

Equations (20.24) can be used to derive commutation relations of a vector operator with the angular momentum operator. To see this, I write the rotation matrix for an infinitesimal rotation. I consider a rotation of ϵ_x about the x axis, followed by a rotation of ϵ_y about the y axis and a rotation of ϵ_z about the z axis. Normally I would have to worry about the order of rotations since the rotation group is nonabelian. However, for *infinitesimal* rotations, the errors introduced by ignoring the order of the rotations are of second order in ϵ and can be neglected. The rotation matrix to first order in $\boldsymbol{\epsilon}$ is given by

$$\underline{\mathrm{R}}(\boldsymbol{\epsilon}) = \begin{pmatrix} 1 & -\epsilon_z & \epsilon_y \\ \epsilon_z & 1 & -\epsilon_x \\ -\epsilon_y & \epsilon_x & 1 \end{pmatrix}, \tag{20.27}$$

independent of the order of the rotations. Under such an infinitesmal rotation,

$$\begin{aligned} \left(\hat{A}_R\right)_i &= \hat{R}(\boldsymbol{\epsilon})\hat{A}_i\hat{R}^\dagger(\boldsymbol{\epsilon}) \\ &= \left(1 - \frac{i}{\hbar}\boldsymbol{\epsilon}\cdot\hat{\mathbf{J}}\right)\hat{A}_i\left(1 + \frac{i}{\hbar}\boldsymbol{\epsilon}\cdot\hat{\mathbf{J}}\right). \end{aligned} \tag{20.28}$$

Taking $i = 1 \equiv x$, I calculate

$$\begin{aligned} \left(\hat{A}_R\right)_x &\approx \hat{A}_x - \frac{i}{\hbar}\boldsymbol{\epsilon}\cdot\hat{\mathbf{J}}\hat{A}_x + \frac{i}{\hbar}\hat{A}_x\boldsymbol{\epsilon}\cdot\hat{\mathbf{J}} = \hat{A}_x - \frac{i}{\hbar}\boldsymbol{\epsilon}\cdot\left[\hat{\mathbf{J}},\hat{A}_x\right] \\ &= \hat{A}_x - \frac{i}{\hbar}\left(\epsilon_x\left[\hat{J}_x,\hat{A}_x\right] + \epsilon_y\left[\hat{J}_y,\hat{A}_x\right] + \epsilon_z\left[\hat{J}_z,\hat{A}_x\right]\right). \end{aligned} \tag{20.29}$$

On the other hand, it follows from Eqs. (20.24a) and (20.27) that the x-component of a vector operator must transform as

$$\begin{aligned}\left(\hat{A}_R\right)_x &= R_{11}(\boldsymbol{\epsilon})\hat{A}_x + R_{21}(\boldsymbol{\epsilon})\hat{A}_y + R_{31}(\boldsymbol{\epsilon})\hat{A}_z \\ &= \hat{A}_x + \epsilon_z\hat{A}_y - \epsilon_y\hat{A}_z. \end{aligned} \tag{20.30}$$

Equating coefficients of $\epsilon_x, \epsilon_y, \epsilon_z$ in Eqs. (20.29) and (20.30), I find that, if $\hat{A}$ is a vector operator, its x-component must satisfy the commutation relations

$$\left[\hat{J}_x, \hat{A}_x\right] = 0; \quad \left[\hat{J}_y, \hat{A}_x\right] = -i\hbar\hat{A}_z \quad \left[\hat{J}_z, \hat{A}_x\right] = i\hbar\hat{A}_y. \tag{20.31}$$

By considering cyclic permutations of x, y, z, I can generate the remaining commutation relations. According to this definition, $\hat{\mathbf{r}}$, $\hat{\mathbf{p}}$, and $\hat{\mathbf{J}}$ are vector operators. Equations (20.31) and their cyclical permutations provide an alternative way to define a vector operator.

I can extend this technique to consider tensor operators of rank two and beyond by combining vector operators of rank 1. For example, given two vector operators $\hat{\mathbf{A}}$ and $\hat{\mathbf{B}}$, a *tensor operator of rank 2* is defined as the set of *nine* operators $\hat{C}_{ij} = \hat{A}_i\hat{B}_j$ $(i, j = 1, 2, 3)$ that transform under rotation as

$$\left(\hat{C}_R\right)_{ij} = \hat{R}\hat{C}_{ij}\hat{R}^\dagger = \sum_{i',j'=1}^{3} \underline{\mathrm{R}}_{i'i}\underline{\mathrm{R}}_{j'j}\hat{C}_{i'j'}. \tag{20.32}$$

On the other hand, if I try to establish commutation relations of these tensor operators with the angular momentum operator, the results are not particularly useful.

The reason for this is that the set of nine operators $\hat{C}_{ij} = \hat{A}_i\hat{B}_j$ do not constitute what is referred to as an *irreducible tensor operator*. To understand something about irreducible tensor operators, you need to know something about *representations* of groups. A matrix representation of a group is the assignment of a matrix to each group element. Clearly the rotation matrix is a three-dimensional representation of the rotation group. We say it is *isomorphic* to the rotation group since each element in the group corresponds to a single matrix. However, a representation of the group can also be the unit matrix in any number of dimensions—each group element is replaced by the unit matrix. This is referred to as a *homomorphism* since the same matrix corresponds to more than one group element.

The rotation matrices $\mathcal{D}^{(j)}_{mm'}(\alpha, \beta, \gamma)$ form a $(2j+1)$ *irreducible representation* of the rotation group. To understand what is meant by an irreducible representation, let me go back to coupling of two angular momenta, which I take for the sake of definiteness as $J_1 = J_2 = 1$. The eigenkets for these two independent angular momenta can be written in the direct product basis as

$$|j_1 = 1, j_2 = 1; m_1, m_2\rangle = |j_1, m_1\rangle\, |j_2, m_2\rangle\,. \tag{20.33}$$

Under rotation each of the eigenkets transforms separately and the resultant transformation matrix is a very complicated 9×9 matrix. However we know that we can couple these two angular momenta into states having $j = 0, 1, 2$ specified by the kets $|j_1, j_2; j, m\rangle$. Under rotation the *components of each value of j transform separately* such that the total transformation breaks down the 9×9 matrix into one that has a *block-diagonal* form with 1×1 $(j = 0)$, 3×3 $(j = 1)$, and 5×5 $(j = 2)$ sub-matrices along the diagonals. A matrix representation that is reduced to block diagonal form is called an irreducible representation.

This leads me to the definition of an *irreducible tensor of rank k* as a set of $(2k + 1)$ operators $\hat{T}_k^q$ that transform under rotations as

$$\left(\hat{T}_k^q\right)_R = \hat{R}\hat{T}_k^q\hat{R}^\dagger = \sum_{q'=-k}^{k} \mathcal{D}_{q'q}^{(k)}\hat{T}_k^{q'}, \tag{20.34}$$

where the $\mathcal{D}_{q'q}^{(k)}$ are elements of the rotation matrices defined by Eq. (19.49). I can use this equation to obtain commutation relations of the $\hat{T}_k^q$ with the angular momentum operators. For an infinitesimal rotation,

$$\left(\hat{T}_k^q\right)_R = \left(1 - \frac{i}{\hbar}\boldsymbol{\epsilon}\cdot\hat{\mathbf{J}}\right)\hat{T}_k^q\left(1 + \frac{i}{\hbar}\boldsymbol{\epsilon}\cdot\hat{\mathbf{J}}\right) = \sum_{q'=-k}^{k} \mathcal{D}_{q'q}^{(k)}(\boldsymbol{\epsilon})\hat{T}_k^{q'}. \tag{20.35}$$

It is convenient to express the scalar product as

$$\boldsymbol{\epsilon}\cdot\hat{\mathbf{J}} = \frac{\epsilon_+\hat{J}_- + \epsilon_-\hat{J}_+}{2} + \epsilon_z\hat{J}_z, \tag{20.36}$$

where

$$\epsilon_\pm = \epsilon_x \pm i\epsilon_x; \tag{20.37a}$$

$$\hat{J}_\pm = \hat{J}_x \pm i\hat{J}_y. \tag{20.37b}$$

Then, by substituting

$$\mathcal{D}_{q'q}^{(k)}(\boldsymbol{\epsilon}) = \langle kq'|\, e^{-\frac{i}{\hbar}\boldsymbol{\epsilon}\cdot\hat{\mathbf{J}}}\, |kq\rangle \approx \langle kq'|\left(1 - \frac{i}{\hbar}\boldsymbol{\epsilon}\cdot\hat{\mathbf{J}}\right)|kq\rangle\,, \tag{20.38}$$

into Eq. (20.35), using the relationships

$$\hat{J}_\pm\,|kq\rangle = \hbar\sqrt{(k \mp q)\,(k \pm q + 1)}\,|kq \pm 1\rangle\,; \tag{20.39a}$$

$$\hat{J}_z\,|kq\rangle = \hbar q\,|kq\rangle\,, \tag{20.39b}$$

derived in Chap. 11, and comparing coefficients of $\epsilon_{\pm}$ and ϵ_z in Eq. (20.35), I can obtain

$$\left[\hat{J}_z, \hat{T}_k^q\right] = \hbar q \hat{T}_k^q; \tag{20.40a}$$

$$\left[\hat{J}_{\pm}, \hat{T}_k^q\right] = \hbar\sqrt{(k \mp q)(k \pm q + 1)}\hat{T}_k^{q\pm 1}, \tag{20.40b}$$

which can serve as an alternative definition of an irreducible tensor of rank k under rotation.

What are some examples of irreducible tensor operators? Any scalar operator that commutes with $\hat{J}$ (such as $\hat{J}^2$) is an irreducible tensor of rank zero. It is not difficult to show that the components $\hat{A}_x, \hat{A}_y, \hat{A}_z$ of a vector operator do *not* form an irreducible tensor of rank 1. However it can be proven rather easily using Eq. (20.31) that the operators

$$A_1^{\pm 1} = \mp\frac{\hat{A}_x \pm i\hat{A}_y}{\sqrt{2}}; \qquad A_1^0 = \hat{A}_z \tag{20.41}$$

do form an irreducible tensor of rank 1 since they have the correct commutation relations with $\hat{J}$. Thus, if we have a vector operator, we can form an irreducible tensor of rank 1 from its components using Eq. (20.41).

Imagine there are two *commuting* vector operators $\hat{\mathbf{A}}$ and $\hat{\mathbf{B}}$. Then you can show (using the commutation relations of the components with the angular momentum operator) that the combination $\hat{\mathbf{A}} \cdot \hat{\mathbf{B}}$ is an irreducible tensor of rank 0, $\hat{\mathbf{A}} \times \hat{\mathbf{B}}$ is an irreducible tensor of rank 1, and

$$\begin{aligned}
&\frac{1}{2}\left(\hat{A}_x\hat{B}_y + \hat{A}_y\hat{B}_x\right); \; \frac{1}{2}\left(\hat{A}_y\hat{B}_z + \hat{A}_z\hat{B}_y\right); \; \frac{1}{2}\left(\hat{A}_z\hat{B}_x + \hat{A}_x\hat{B}_z\right); \\
&\left(\hat{A}_x\hat{B}_x - \hat{A}_y\hat{B}_y\right); \; \left(2\hat{A}_z\hat{B}_z - \hat{A}_x\hat{B}_x - \hat{A}_y\hat{B}_y\right)
\end{aligned} \tag{20.42}$$

form an irreducible tensor of rank 2 (quadrupole tensor).

The adjoint of an irreducible tensor operator $\left(\hat{T}^{(k)}\right)^{\dagger}$ of rank k can be defined as the set of operators $\left[\left(\hat{T}^{(k)}\right)^{\dagger}\right]_k^q$ for which

$$\left[\left(\hat{T}^{(k)}\right)^{\dagger}\right]_k^q = (-1)^q\left(\hat{T}_k^{-q}\right)^{\dagger}. \tag{20.43}$$

With this definition, any irreducible tensor of rank 1 formed from a Hermitian vector operator is also Hermitian. In addition, the $\hat{Y}_{\ell}^m(\theta, \phi)$ considered as *operators* in coordinate space form a Hermitian irreducible tensor operator of rank ℓ.

20.3 Wigner-Eckart Theorem

I now state the Wigner-Eckart theorem without proof (a proof is given in the Appendix). The matrix elements of an irreducible tensor operator can be written as

$$\begin{aligned}\langle\alpha jm|\,\hat{T}_k^q\,|\alpha' j' m'\rangle &= \frac{1}{\sqrt{2j+1}}\begin{bmatrix} j' & k & j \\ m' & q & m\end{bmatrix}\langle\alpha j\|\,T^{(k)}\,\|\alpha' j'\rangle \\ &= (1)^{j-m}\begin{pmatrix} j & k & j' \\ -m & q & m'\end{pmatrix}\langle\alpha j\|\,T^{(k)}\,\|\alpha' j'\rangle \end{aligned} \tag{20.44}$$

where $\langle\alpha j\|\,T^{(k)}\,\|\alpha' j'\rangle$ is referred to as a *reduced matrix element* and α and α' are additional quantum numbers (such as n in the hydrogen atom). The matrix element of an irreducible operator is a product of a term that is independent of q, m, and m', multiplied by a Clebsch-Gordan coefficient. This theorem is extremely useful since it lets you calculate matrix elements of different components of a vector or tensor operator in terms of one quantity which itself must be calculated explicitly. Some authors use a different form for Eq. (20.44) (e.g., they omit the $1/\sqrt{2j+1}$ factor), but the form I use is the most common.

As a first example, let me consider a matrix element of the operator $\hat{\mathbf{r}}$ in the $|\alpha\ell m\rangle$ basis. I will need to evaluate matrix elements of this type when I look at atom–field interactions that are proportional to $\hat{\mathbf{r}}\cdot\mathbf{E}$, where $\mathbf{E}$ is the electric field. From the definitions (20.41), I can write

$$\hat{x} = \frac{\hat{T}_1^{-1} - \hat{T}_1^1}{\sqrt{2}}; \tag{20.45a}$$

$$\hat{y} = -\frac{\hat{T}_1^{-1} + \hat{T}_1^1}{\sqrt{2}i}; \tag{20.45b}$$

$$\hat{z} = \hat{T}_1^0, \tag{20.45c}$$

so

$$\begin{aligned}\langle\alpha\ell m|\,\hat{x}\,|\alpha'\ell' m'\rangle &= -\langle\alpha\ell m|\left(\frac{\hat{T}_1^1 - \hat{T}_1^{-1}}{\sqrt{2}}\right)|\alpha'\ell' m'\rangle \\ &= -\frac{1}{\sqrt{2}}\frac{1}{\sqrt{2\ell+1}}\left(\begin{bmatrix}\ell' & 1 & \ell \\ m' & 1 & m\end{bmatrix} - \begin{bmatrix}\ell' & 1 & \ell \\ m' & -1 & m\end{bmatrix}\right)\langle\alpha\ell\|\,r^{(1)}\,\|\alpha'\ell'\rangle; \end{aligned} \tag{20.46a}$$

$$\begin{aligned}\langle\alpha\ell m|\,\hat{y}\,|\alpha'\ell m'\rangle &= -\langle\alpha\ell m|\left(\frac{\hat{T}_1^1 + \hat{T}_1^{-1}}{\sqrt{2}i}\right)|\alpha'\ell' m'\rangle \\ &- \frac{1}{\sqrt{2}i}\frac{1}{\sqrt{2\ell+1}}\left(\begin{bmatrix}\ell' & 1 & \ell \\ m' & 1 & m\end{bmatrix} + \begin{bmatrix}\ell' & 1 & \ell \\ m' & -1 & m\end{bmatrix}\right)\langle\alpha\ell\|\,r^{(1)}\,\|\alpha'\ell'\rangle; \end{aligned} \tag{20.46b}$$

$$\langle\alpha\ell m|\,\hat{z}\,|\alpha'\ell m'\rangle = \frac{1}{\sqrt{2\ell+1}}\begin{bmatrix}\ell' & 1 & \ell \\ m' & 0 & m\end{bmatrix}\langle\alpha\ell\|\,r^{(1}\,\|\alpha'\ell'\rangle. \tag{20.46c}$$

Note that the reduced matrix element, $\langle\alpha\ell\|\, r^{(1}\, \|\alpha'\ell'\rangle$, can be calculated from the last of these equations as

$$\langle\alpha\ell|\,|r^{(1)}|\,|\alpha'\ell'\rangle = \frac{\sqrt{2\ell+1}}{\begin{bmatrix}\ell' & 1 & \ell\\ m & 0 & m\end{bmatrix}}\int d\mathbf{r}\psi^*_{\alpha\ell m}(\mathbf{r})z\psi_{\alpha'\ell' m}(\mathbf{r}), \tag{20.47}$$

using any value of m you choose. The *ratio* of matrix elements depends solely on the Clebsch-Gordan coefficients. This is useful in calculating *branching ratios* for transitions originating on different degenerate (or nearly-degenerate) sublevels in a given energy manifold of levels.

As a second example, consider matrix elements of $\hat{J}$ itself. Following the same steps that led to Eq. (20.47), I can calculate

$$\begin{aligned}\langle\alpha j\|\, J^{(1)}\, \|\alpha' j'\rangle &= \frac{\sqrt{2j+1}}{\begin{bmatrix}j' & 1 & j\\ m & 0 & m\end{bmatrix}}\langle\alpha jm|\,\hat{J}_z\,|\alpha' j' m\rangle\\ &= \frac{\sqrt{2j+1}m\hbar}{\begin{bmatrix}j & 1 & j\\ m & 0 & m\end{bmatrix}}\delta_{j,j'}\delta_{\alpha,\alpha'}\end{aligned} \tag{20.48}$$

Taking $m = j$, I find

$$\begin{aligned}\langle\alpha j\|\, J^{(1)}\, \|\alpha' j'\rangle &= \frac{\sqrt{2j+1}j\hbar}{\begin{bmatrix}j & 1 & j\\ j & 0 & j\end{bmatrix}}\delta_{j,j'}\delta_{\alpha,\alpha'} = \frac{\sqrt{2j+1}j\hbar}{\sqrt{j}/\sqrt{j+1}}\delta_{j,j'}\delta_{\alpha,\alpha'}\\ &= \hbar\sqrt{j(2j+1)(j+1)}\delta_{j,j'}\delta_{\alpha,\alpha'}.\end{aligned} \tag{20.49}$$

Further examples will be given when I discuss the Zeeman effect in the next chapter.

Let me return briefly to the reduced matrix elements, which can be calculated using Eq. (20.44). You might ask about the relationship between $\langle\alpha j\|\, T^{(k)}\, \|\alpha' j'\rangle$ and $\langle\alpha' j'\|\, T^{(k)}\, \|\alpha j\rangle$. To examine this relationship, I use Eqs. (20.43) and (20.44) to write

$$\begin{aligned}&\langle\alpha' j' m'|\left[\left(\hat{T}^{(k)}\right)^\dagger\right]_k^{-q}|\alpha jm\rangle = \frac{1}{\sqrt{2j'+1}}\begin{bmatrix}j & k & j'\\ m & -q & m'\end{bmatrix}\langle\alpha' j'\|\left(\hat{T}^{(k)}\right)^\dagger\|\alpha j\rangle\\ &= (-1)^q\langle\alpha' j' m'|\left(\hat{T}_k^q\right)^\dagger|\alpha jm\rangle = (-1)^q\langle\alpha jm|\,\hat{T}_k^q\,|\alpha' j' m'\rangle^*.\end{aligned} \tag{20.50}$$

By combining this equation with Eqs. (20.44) and (20.16d), I obtain

$$\langle\alpha j\|\, T^{(k)}\, \|\alpha' j'\rangle = (-1)^{j-j'}\langle\alpha' j'\|\left(\hat{T}^{(k)}\right)^\dagger\|\alpha j\rangle^*. \tag{20.51}$$

For any Hermitian vector operator $\hat{\mathbf{V}}$, it then follows that

$$\left\langle \alpha j \right\| V^{(1)} \left\| \alpha' j' \right\rangle = (-1)^{j-j'} \left\langle \alpha' j' \right\| V^{(1)} \left\| \alpha j \right\rangle^* . \tag{20.52}$$

As a simple example, consider matrix elements of the position operator between eigenkets $|n\ell m\rangle$ of the hydrogen atom. In that case, the matrix elements vanish unless $\ell = \ell' \pm 1$ and

$$\left\langle n\ell \right\| r^{(1)} \left\| n', \ell' \pm 1 \right\rangle = -\left\langle n', \ell' \pm 1 \right\| r^{(1)} \left\| n\ell \right\rangle , \tag{20.53}$$

where $r^{(1)}$ is an irreducible tensor having components given by Eqs. (20.45). The fact that the reduced matrix element is real can be deduced from Eq. (20.44) using $q = 0$.

20.4 Summary

Several topics were covered in this chapter. Coupling of angular momentum in quantum mechanics can be formulated in terms of the Clebsch-Gordan or 3-J symbols. In contrast to classical coupling of two vectors, there are only discrete ways in which angular momentum can be coupled in quantum mechanics. The definition of vector and tensor operators under rotation was introduced and related to the commutation relations of the operators with the angular momentum operators. It turned out to be useful to introduce a new class of operators, irreducible tensor operators, that transformed under rotation in terms of the irreducible representations of the rotation group, that is, the $\mathcal{D}^{(k)}_{q'q}$. A matrix element of an irreducible tensor operators could be expressed as a product of a Clebsch-Gordan coefficients multiplied by a reduced matrix element that is independent of the magnetic quantum numbers, a result embodied in the Wigner-Eckart theorem.

20.5 Appendix: Proof of the Wigner-Eckart Theorem

Consider

$$\widetilde{|jm\rangle} = \sum_{m',q} \hat{T}^q_k \left|\alpha' j' m'\right\rangle \begin{bmatrix} k & j' & j \\ q & m' & m \end{bmatrix}, \tag{20.54}$$

where $\hat{T}^q_k$ is an irreducible tensor operator. The ket $\widetilde{|jm\rangle}$ is an implicit function of j', k, and α' (α' represents an additional quantum number such as energy). You can show that $\widetilde{|jm\rangle}$ is an eigenket of $\hat{J}^2$ and $\hat{J}_z$ by using the commutation relations between the irreducible tensor operator $\hat{T}_k^q$ and $\hat{J}$. For example,

$$\begin{aligned}
\hat{J}_z \widetilde{|jm\rangle} &= \hat{J}_z \sum_{m',q} \hat{T}_k^q \left|\alpha' j' m'\right\rangle \begin{bmatrix} k & j' & j \\ q & m' & m \end{bmatrix} \\
&= \sum_{m',q} \begin{bmatrix} k & j' & j \\ q & m' & m \end{bmatrix} \left(\left[\hat{J}_z, \hat{T}_k^q\right] + \hat{T}_k^q \hat{J}_z \right) \left|\alpha' j' m'\right\rangle \\
&= \hbar \sum_{m',q} \begin{bmatrix} k & j' & j \\ q & m' & m \end{bmatrix} (q+m') \, \hat{T}_k^q \left|\alpha' j' m'\right\rangle \\
&= m\hbar \sum_{m',q} \begin{bmatrix} k & j' & j \\ q & m' & m \end{bmatrix} \hat{T}_k^q \left|\alpha' j' m'\right\rangle = m\hbar \widetilde{|jm\rangle},
\end{aligned} \tag{20.55}$$

since the Clebsch-Gordan coefficients vanish unless $(q+m') = m$. However, although different $\widetilde{|jm\rangle}$ are orthogonal ($\widetilde{\langle jm} \widetilde{|j'm'\rangle}$ vanishes unless $j = j'$ and $m = m'$), the $\widetilde{\langle jm} \widetilde{|jm\rangle}$ are not normalized, $\widetilde{\langle jm} \widetilde{|jm\rangle} \neq 1$.

I expand $\widetilde{|jm\rangle}$ in the $|\alpha j' m'\rangle$ basis as

$$\widetilde{|jm\rangle} = \sum_{\alpha, j', m'} \left\langle \alpha j' m' \middle| \widetilde{jm\rangle} \right. \left| \alpha j' m' \right\rangle. \tag{20.56}$$

Since both $\widetilde{|jm\rangle}$ and $|\alpha j' m'\rangle$ are simultaneous eigenkets of $\hat{J}^2$ and $\hat{J}_z$, the coupling coefficients $\langle \alpha j' m' | \widetilde{jm\rangle}$ vanish unless $j = j'$, $m = m'$;

$$\widetilde{|jm\rangle} = \sum_{\alpha} \langle \alpha jm | \widetilde{jm\rangle} \, |\alpha jm\rangle \,. \tag{20.57}$$

If you act on both sides of this equation with $\hat{J}_+$ you will find that

$$\widetilde{|jm+1\rangle} = \sum_{\alpha} \langle \alpha jm | \widetilde{jm\rangle} \, |\alpha j, m+1\rangle \,; \tag{20.58}$$

however, from Eq. (20.56) it follows that

$$\widetilde{|j, m+1\rangle} = \sum_{\alpha} \langle \alpha j, m+1 | \widetilde{j, m+1\rangle} \, |\alpha j, m+1\rangle \,. \tag{20.59}$$

By comparing Eqs. (20.58) and (20.59), you see that the coupling coefficients $\langle \alpha jm | \widetilde{jm\rangle}$ *must be independent of* m.

I now invert Eq. (20.54),

$$\hat{T}_k^q \left|\alpha' j' m'\right\rangle = \sum_{j'',m''} \begin{bmatrix} k & j' & j'' \\ q & m' & m'' \end{bmatrix} \left|\widetilde{j'' m''}\right\rangle$$
$$= \sum_{\alpha'',j'',m''} \begin{bmatrix} k & j' & j'' \\ q & m' & m'' \end{bmatrix} \langle \alpha'' j'' m'' | \widetilde{j'' m''} \rangle |\alpha'' j'' m''\rangle, \tag{20.60}$$

multiply on the left by $\langle \alpha j m|$, and use Eq. (20.16a) to obtain

$$\langle \alpha j m| \hat{T}_k^q \left|\alpha' j' m'\right\rangle = \frac{1}{\sqrt{2j+1}}(-1)^{k-j'+j} \begin{bmatrix} k & j' & j \\ q & m' & m \end{bmatrix} \langle \alpha j \| T^{(k)} \|\alpha' j'\rangle$$
$$= \frac{1}{\sqrt{2j+1}} \begin{bmatrix} j' & k & j \\ m' & q & m \end{bmatrix} \langle \alpha j \| T^{(k)} \|\alpha' j'\rangle, \tag{20.61}$$

where

$$\frac{1}{\sqrt{2j+1}}(-1)^{k-j'+j} \langle \alpha j \| T^{(k)} \|\alpha' j'\rangle = \langle \alpha j m | \widetilde{jm}\rangle \tag{20.62}$$

is independent of m, but still depends on α' and j' and the properties of $T^{(k)}$ [recall that $\left|\widetilde{j' m'}\right\rangle$, as defined by Eq. (20.54) is an implicit function of j', k and α']. The choice of writing this is somewhat arbitrary, but the result states that the matrix element of an irreducible tensor operator is equal to the product of a Clebsch-Gordan coefficient and a term that is independent of m, m', q.

20.6 Problems

1. In qualitative terms, to what do the Clebsch-Gordan coefficients correspond? What does it mean to say that an operator is a scalar or vector operator under rotation? What does it mean to say that a set of operators is an irreducible tensor operator under rotation? Under a rotation about the z axis by 2π, what happens to the rotation operator?, to the spin-rotation operator?

2. Write a subroutine that will let you calculate $|j_1 j_2; jm\rangle$ in terms of $|j_1 m_1; j_2 m_2\rangle$ and the Clebsch-Gordan coefficients. (Use the Clebsch-Gordan function in Mathematica or some equivalent function). Obtain the solution for $|j_1 = 1, j_2 = 3; j = 2, m = 1\rangle$.

3. Prove that

$$\begin{bmatrix} j_1 & j_2 & j_1 + j_2 \\ j_1 & j_2 & j_1 + j_2 \end{bmatrix} = 1.$$

Derive the orthogonality relation for the Clebsch-Gordan coefficients,

$$\sum_{m_1=-j_1}^{j_1} \sum_{m_2=-j_2}^{j_2} \begin{bmatrix} j_1 & j_2 & j \\ m_1 & m_2 & m \end{bmatrix} \begin{bmatrix} j_1 & j_2 & j' \\ m_1 & m_2 & m' \end{bmatrix} = \delta_{j,j'}\delta_{m,m'}.$$

Prove that

$$|j_1 m_1; j_2 m_2\rangle = \sum_{j=|j_2-j_1|}^{j_1+j_2} \sum_{m=-j}^{j} \begin{bmatrix} j_1 & j_2 & j \\ m_1 & m_2 & m \end{bmatrix} |j_1 j_2; jm\rangle$$

and derive the orthogonality relation

$$\sum_{j=|j_2-j_1|}^{j_1+j_2} \sum_{m=-j}^{j} \begin{bmatrix} j_1 & j_2 & j \\ m_1 & m_2 & m \end{bmatrix} \begin{bmatrix} j_1 & j_2 & j \\ m_1' & m_2' & m \end{bmatrix} = \delta_{m_1,m_1'}\delta_{m_2,m_2'}.$$

4. Evaluate

$$\frac{\langle 3,2,2|\,\hat{p}_x\,|2,1,1\rangle}{\langle 3,2,0|\,\hat{p}_y\,|2,1,-1\rangle},$$

where the states $|n\ell m\rangle$ are eigenkets of the hydrogen atom. Note: You do not have to evaluate any integrals in this problem.

5. Sodium atoms have a single valence electron. In the ground state, $L = 0$ (writing $L = 0$ is actually a convention corresponding to $\ell = 0$), while in the first excited state, $L = 1$. What are the possible values for J in the ground state and first excited state? Sodium has a *nuclear* spin quantum number $I = 3/2$. The nuclear spin couples with the angular momentum $\mathbf{J}$ to give a total angular momentum $\mathbf{F} = \mathbf{J} + \mathbf{I}$. For each of the J states in the ground and first excited states, calculate the possible values for the total angular momentum quantum number F. States having different F are split in energy by the *hyperfine* interaction of the spin of the electron with the spin of the nucleus. What are the various fine and hyperfine frequency separations in the $3S$ (ground state) and $3P$ (first excited state) states of sodium?

6–7. Calculate the effect of a rotation with Euler angles ($\alpha = 0, \beta = \pi/2, \gamma = 0$) on the hydrogen eigenket $\left|n = 2, \ell = 1, j = 1/2, m_j = 1/2\right\rangle$ and obtain an expression for the wave function of the rotated state.

8. Suppose that angular momentum operators are defined such that $\hat{\mathbf{K}} = \hat{\mathbf{J}}_1 + \hat{\mathbf{J}}_2$. By operating with the operators $\hat{K}_\pm = \hat{J}_{1\pm} + \hat{J}_{2\pm}$ on the state

$$|j_1 j_2; KQ\rangle = \sum_{m_1,m_2} \begin{bmatrix} j_1 & j_2 & K \\ m_1 & m_2 & Q \end{bmatrix} |j_1 m_1\rangle\,|j_2 m_2\rangle$$

prove the recursion relation

$$\sqrt{(K \mp Q)(K \pm Q + 1)} \begin{bmatrix} j_1 & j_2 & K \\ m_1 & m_2 & Q \pm 1 \end{bmatrix}$$

$$= \sqrt{(j_1 \mp (m_1 \mp 1))(j_1 \pm (m_1 \mp 1) + 1)} \begin{bmatrix} j_1 & j_2 & K \\ m_1 \mp 1 & m_2 & Q \end{bmatrix}$$

$$+ \sqrt{(j_2 \mp (m_2 \mp 1))(j_2 \pm (m_2 \mp 1) + 1)} \begin{bmatrix} j_1 & j_2 & K \\ m_1 & m_2 \mp 1 & Q \end{bmatrix}.$$

9–10. In Chap. 11, I showed that any operator in Dirac notation could be expanded in terms of a complete set of basis operators $|\alpha\rangle\langle\beta|$. If the eigenkets are specified as $|jm\rangle$, then a complete set of basis operators can be specified by $\underline{u}(j_1m_1; j_2m_2) = |j_1m_1\rangle\langle j_2m_2|$. The $\underline{u}(j_1m_1; j_2m_2)$ do *not* constitute a set of irreducible tensor basis operators.

(a) Prove that the set of basis operators defined by

$$\underline{u}_Q^K(j_1, j_2) = \sum_{m_1, m_2} (-1)^{j_2 - m_2} \begin{bmatrix} j_1 & j_2 & K \\ m_1 & -m_2 & Q \end{bmatrix} |j_1m_1\rangle\langle j_2m_2|$$

do form an irreducible tensor of rank K by showing they obey Eqs. (20.39).

(b) Prove that these operators are orthogonal in the sense that

$$\mathrm{Tr}\left(\underline{u}_Q^K(j_1, j_2) \left[\underline{u}_{Q'}^{K'}(j_1', j_2') \right]^\dagger \right) = \delta_{j_1, j_1'} \delta_{j_2, j_2'} \delta_{K, K'} \delta_{Q, Q'}.$$

(c) As a consequence, an arbitrary operator $\underline{A}$ can be expanded in terms of its irreducible tensor components as

$$\underline{A} = \sum_{j_1 j_2, K, Q} A_Q^K(j_1, j_2)\, \underline{u}_Q^K(j_1, j_2).$$

Show that the expansion coefficients $A_Q^K(j_1, j_2)$ are given by

$$A_Q^K(j_1, j_2) = \mathrm{Tr}\left(\underline{A} \left[\underline{u}_Q^K(j_1, j_2) \right]^\dagger \right)$$

$$= \sum_{m_1, m_2} (-1)^{j_2 - m_2} \begin{bmatrix} j_1 & j_2 & K \\ m_1 & -m_2 & Q \end{bmatrix} \langle j_1m_1| \hat{A} |j_2m_2\rangle.$$

The use of an irreducible tensor basis for operators is useful since the components transform in a simple way under rotation.

11. Suppose that angular momentum operators are defined such that $\hat{\mathbf{J}} = \hat{\mathbf{J}}_1 + \hat{\mathbf{J}}_2$. By operating with the rotation operator on the state

$$|j_1 j_2; jm\rangle = \sum_{m_1, m_2} \begin{bmatrix} j_1 & j_2 & j \\ m_1 & m_2 & m \end{bmatrix} |j_1 m_1\rangle \, |j_2 m_2\rangle$$

prove the *decomposition relation*

$$\mathcal{D}^{(j)}_{mm'}(\alpha, \beta, \gamma) = \sum_{\substack{m_1, m_2 \\ m_1', m_2'}} \begin{bmatrix} j_1 & j_2 & j \\ m_1 & m_2 & m \end{bmatrix} \begin{bmatrix} j_1 & j_2 & j \\ m_1' & m_2' & m' \end{bmatrix} \times \mathcal{D}^{(j_1)}_{m_1 m_1'}(\alpha, \beta, \gamma) \, \mathcal{D}^{(j_2)}_{m_2 m_2'}(\alpha, \beta, \gamma) \, .$$

12. For a rotation specified by the rotation operator $\hat{R}(\alpha, \beta, \gamma)$, find the transformed angular momentum operator

$$\hat{\mathbf{L}}_R(\alpha, \beta, \gamma) = \hat{R}(\alpha, \beta, \gamma) \, \hat{\mathbf{L}} \hat{R}^{-1}(\alpha, \beta, \gamma) \, .$$

If

$$\begin{aligned} \mathbf{B} &= B_x \mathbf{u}_x + B_y \mathbf{u}_y + B_z \mathbf{u}_z \\ &= B \left(\sin\theta_B \cos\phi_B \mathbf{u}_x + \sin\theta_B \sin\phi_B \mathbf{u}_y + \cos\theta_B \mathbf{u}_z \right) , \end{aligned}$$

is a constant vector having polar angles (θ_B, ϕ_B), prove that an interaction Hamiltonian of the form $\hat{H} = \alpha \hat{\mathbf{L}} \cdot \mathbf{B}$, where α is a constant, is transformed into

$$\hat{H}_R = \alpha \hat{\mathbf{L}}_R(0, -\theta_B, -\phi_B) \cdot \mathbf{B} = \hat{L}_z B$$

under the action of the rotation operator $\hat{R}(0, -\theta_B, -\phi_B)$; in other words, in the transformed Hamiltonian, it is as if the vector $\mathbf{B}$ lies along the z-axis, even though this constant vector is unaffected by the transformation (recall that only *operators* are transformed under quantum transformations).

13–14. If a constant magnetic field induction $\mathbf{B}$ is applied to a hydrogen atom, there is an extra interaction term in the Hamiltonian of the form

$$\hat{H}' = \frac{e}{2m_e} \hat{\mathbf{L}} \cdot \mathbf{B},$$

where m_e is the electron mass and spin has been neglected. I have already shown that, if the quantization axis is taken along the field direction, the eigenkets are unchanged and each ℓ level is split into $(2\ell + 1)$ equally spaced levels separated in frequency by $\omega = qB/2m_e$. Sometimes it is more convenient to take the quantization axis along a different direction (e.g., along the polarization vector of an optical field that is also present). Evaluate the matrix elements $\langle n\ell m|\hat{H}'|n'\ell' m'\rangle$ for a magnetic field induction

$$\begin{aligned}\mathbf{B} &= B_x + B_y\mathbf{u}_y + B_z\mathbf{u}_z \\ &= B\left(\sin\theta\cos\phi\mathbf{u}_x + \sin\theta\sin\phi\mathbf{u}_y + \cos\theta\mathbf{u}_z\right),\end{aligned}$$

where θ and ϕ are the polar angles of the field. Explicitly diagonalize the $\ell = 1$ submatrix to obtain the changes in the energy produced by the field and the new eigenkets.

The $\mathcal{D}^{(\ell)}(\alpha, \beta, \gamma)$ matrices can be used to get a *general* expression for the new eigenkets. Under a rotation $(0, -\theta_B, -\phi_B)$ the eigenkets transform as

$$|\ell m\rangle_R' = \hat{R}(0, -\theta_B, -\phi_B)\,|\ell m\rangle'.$$

But we have seen in the previous problem that such a transformation produces an effective interaction in which the field is along the z-direction; that is, the transformed kets are simply the *standard* kets when the quantization axis is taken along the z-axis, $|\ell m\rangle_R' = |\ell m\rangle$, which implies that

$$|\ell m\rangle = \hat{R}(0, -\theta_B, -\phi_B)\,|\ell m\rangle'.$$

Invert this equation to prove that

$$|\ell m\rangle' = \sum_{m'} \mathcal{D}^{(\ell)}_{m'm}(\phi_B, \theta_B, 0)\,|\ell m'\rangle,$$

the eigenkets for which the quantization axis is along the field can be obtained from the standard kets by rotation that takes the z-axis into the field direction. Check to see if your result for $\ell = 1$ agrees with that obtained by direct diagonalization. The result derived is quite general for any Euler angles (α, β, γ); that is, the eigenkets relative to a quantization axis defined by the Euler angles (α, β, γ) are given by

$$|\ell m\rangle' = \sum_{m'} \mathcal{D}^{(\ell)}_{m'm}(\alpha, \beta, \gamma)\,|\ell m'\rangle.$$

15–16. For two irreducible tensor operators, $\hat{T}_k^q(1)$ and $\hat{T}_k^q(2)$, associated with two *separate* atoms, you can form a set of irreducible tensor operators as

$$\hat{T}_K^Q(k_1, k_2) = \sum_{q_1, q_2} \begin{bmatrix} k_1 & k_2 & K \\ q_1 & q_2 & Q \end{bmatrix} \hat{T}_{k_1}^{q_1}(1)\hat{T}_{k_2}^{q_2}(2).$$

An interaction potential $\hat{V}$ connecting the two atoms can then be expanded as

$$\hat{V} = \sum_{q_1,q_2} A(k_1q_1, k_2q_2)\hat{T}^{q_1}_{k_1}(1)\hat{T}^{q_2}_{k_2}(2)$$
$$= \sum_{K,Q} A^Q_K(k_1, k_2)\hat{T}^Q_K(k_1, k_2).$$

As a specific example, take $\hat{V}$ as the dipole–dipole interaction between the atoms,

$$\hat{V} = \frac{1}{4\pi\epsilon_0}\frac{\hat{\mathbf{p}}_e(1)\cdot\hat{\mathbf{p}}_e(2) - 3\,[\hat{\mathbf{p}}_e(1)\cdot\mathbf{u}_R]\,[\hat{\mathbf{p}}_e(2)\cdot\mathbf{u}_R]}{R^3},$$

where $\hat{\mathbf{p}}_e(j)$ is the electric dipole moment operator of atom j and $\mathbf{u}_R$ is a unit vector in the direction of the interatomic separation $\mathbf{R}$ from atom 1 to atom 2. Choose a coordinate system in which atom 1 is located at the origin and θ and ϕ are the polar angles of $\mathbf{R}$. For this interaction, prove that

$$A^Q_K(k_1, k_2) = -3\frac{(4\pi)^{1/2}}{\epsilon_0 R^3}\left(\frac{2}{15}\right)^{1/2}\left[Y^Q_2(\theta, \phi)\right]^* \delta_{k_1,1}\delta_{k_2,1}\delta_{K,2},$$

provided $\hat{T}^{q_1}_{k_1}(j)$ is taken to be an irreducible tensor component of the electric dipole operator of atom j. In this form, both the expansion coefficients $A^Q_K(k_1, k_2)$ and the irreducible tensor operators $\hat{T}^Q_K(k_1, k_2)$ have simple transformation properties under rotation. Explain on physical grounds why the $K = 0$ component (which must be invariant under rotation) and the $K = 1$ component (which is odd under parity) must vanish.

Chapter 21
Hydrogen Atom with Spin in External Fields

Having established the properties of irreducible tensor operators, I am now in a position to use perturbation theory to study how the energy levels of the hydrogen atom are modified in external magnetic and electric fields. Before doing so, however, I will use perturbation theory to obtain the relativistic corrections to the energy levels of hydrogen in the *absence* of any external fields. This will lead naturally to a discussion of the *fine-structure* and *hyperfine structure* of hydrogen.

In the absence of any external fields, the energy levels of hydrogen (including electron spin) are given by the solution of the Dirac equation. The Dirac equation consistently treats all relativistic effects, but does not include hyperfine structure (arising from the magnetic moment of the proton), nor the Lamb shift (to be mentioned below). Explicitly, in the absence of any external fields, the energy levels are given by

$$E_{nj} = \frac{m_e c^2}{\left\{1 + \frac{\alpha_{FS}^2}{\left[n-j-\frac{1}{2}+\sqrt{\left(j+\frac{1}{2}\right)^2-\alpha_{FS}^2}\right]^2}\right\}^{1/2}}; \quad \left\{ \begin{array}{c} n = 1, 2, \ldots; \\ j = \frac{1}{2}, \frac{3}{2} \ldots, n - \frac{1}{2} \end{array} \right., \tag{21.1}$$

where

$$\alpha_{FS} = \frac{e^2}{4\pi\epsilon_0 \hbar c} \approx \frac{1}{137} \tag{21.2}$$

is the fine structure constant and m_e should be thought of as the reduced mass of the electron. Note that ℓ does not appear explicitly in this equation; j is the quantum number corresponding to the total angular momentum operator $\hat{\mathbf{J}} = \hat{\mathbf{L}} + \hat{\mathbf{S}}$ of the electron.

P.R. Berman, *Introductory Quantum Mechanics*, UNITEXT for Physics,
https://doi.org/10.1007/978-3-319-68598-4_21

Expanding this equation in a power series in α_{FS}^2, you can show that

$$E_{nj} - m_e c^2 \approx -\frac{E_R}{n^2} - \alpha_{FS}^2 \frac{E_R}{n^4}\left(\frac{n}{j+\frac{1}{2}} - \frac{3}{4}\right) + \cdots, \tag{21.3}$$

where

$$E_R = \frac{1}{2} m_e c^2 \alpha_{FS}^2 \approx 13.6 \text{ eV} \tag{21.4}$$

is the Rydberg energy.

Even though the Schrödinger equation is a non-relativistic equation, it is possible to use perturbation theory to calculate the relativistic corrections to the hydrogenic energy levels, corrections that can be attributed to relativistic mass effects, spin–orbit interactions, and the so-called Darwin term. In this manner, the α_{FS}^2 term appearing in Eq. (21.3) can be obtained within the context of non-relativistic quantum mechanics. I have already discussed spin and the spin–orbit interaction in Chap. 12, so you should review those results.

In perturbation theory, the Hamiltonian is written as $\hat{H} = \hat{H}_0 + \hat{H}'$, where $\hat{H}'$ includes relativistic corrections, hyperfine interactions, and any atom–external field interactions, while $\hat{H}_0$ is the Hamiltonian of hydrogen in the absence of such effects. The exact eigenkets are expanded in terms of eigenkets of $\hat{H}_0$ and approximate expressions are obtained for the eigenenergies and eigenkets. If there is degeneracy and if $\hat{H}'$ couples degenerate states, I must first exactly diagonalize any degenerate sub-blocks in which there are off-diagonal matrix elements.

In dealing with any stationary state problem in quantum mechanics, *the first step is to identify the constants of the motion.* Operators corresponding to dynamic variables that are constants of the motion commute with the Hamiltonian. I can find simultaneous eigenkets of the Hamiltonian and these operators. In problems involving perturbation theory, there are *two* Hamiltonians, the unperturbed Hamiltonian $\hat{H}_0$ and the total Hamiltonian $\hat{H} = \hat{H}_0 + \hat{H}'$. If an operator $\hat{A}$ having eigenvalues labeled by a commutes with *both* $\hat{H}_0$ and $\hat{H}'$, it corresponds to an *exact* constant of the motion. As a consequence, the eigenkets of both $\hat{H}_0$ and $\hat{H}$ can be chosen as simultaneous eigenkets of $\hat{A}$. *If the eigenkets are chosen in this manner, only states having the same value of the eigenvalue a can be coupled by* $\hat{H}'$, since, under these circumstances, $\hat{H}'$ is diagonal in the $|a\rangle$ basis.

Let me be more specific. Suppose that both $\hat{L}^2$ and $\hat{L}_z$ commute with both $\hat{H}_0$ and $\hat{H}'$, and that $\hat{H}_0$ is the Hamiltonian operator for hydrogen without spin. This means that the *exact* eigenkets of $\hat{H} = \hat{H}_0 + \hat{H}'$ can be written as $|E, \ell, m\rangle$, whereas the eigenkets of $\hat{H}_0$ are given by $|E, \ell, m\rangle^{(0)} \equiv |n, \ell, m\rangle$. In carrying out perturbation theory using the $|n, \ell, m\rangle$ basis for a given n, only states having the *same* ℓ and the *same* m can be coupled by $\hat{H}'$; in other words, *no* degenerate states of $\hat{H}_0$ are coupled by the perturbation and I can use nondegenerate perturbation theory. By identifying the constants of the motion, I have dramatically simplified the problem. Note that since n does not correspond to the exact energy, states having different n (but the same ℓ and the same m) *can* be coupled by $\hat{H}'$; however, contributions from these terms are usually relatively small since they involve large energy denominators.

I first consider the hydrogen atom without spin and calculate the relativistic mass correction. Spin is then included and the modifications to the energy levels from the spin–orbit interaction are obtained. The role played by the Darwin term is discussed. Finally I allow for interactions with constant external magnetic or electric fields. The hyperfine interaction is discussed in detail in Appendix 2.

21.1 Energy Levels of Hydrogen

21.1.1 Unperturbed Hamiltonian

The Hamiltonian for the hydrogen atom without any relativistic corrections is

$$\hat{H}_0 = \frac{\hat{p}^2}{2m_e} - \frac{e^2}{4\pi\epsilon_0 r} = \frac{\hat{p}^2}{2m_e} - \frac{K_e}{r}, \tag{21.5}$$

where

$$K_e = \frac{e^2}{4\pi\epsilon_0}. \tag{21.6}$$

The eigenenergies of $\hat{H}_0$ are given by

$$E_n^{(0)} = -\frac{1}{2n^2}\alpha_{FS}^2 m_e c^2 = -\frac{1}{2n^2}\frac{e^2}{4\pi\epsilon_0 a_0} = -\frac{E_R}{n^2} = -\frac{13.6\text{ eV}}{n^2}, \tag{21.7}$$

where a_0 is the Bohr radius. There is an n^2 degeneracy for states having a given n; for a given n, $\ell = 0, 1, \ldots (n-1)$, and for a given ℓ, m_ℓ varies from $-\ell$ to ℓ in integer steps.

21.1.2 Relativistic Mass Correction

Relativistic corrections are best treated by the exact solution of the Dirac equation for hydrogen. However I can get a perturbative solution for the relativistic mass correction (actually a relativistic energy correction, since mass is an invariant) if I replace the classical kinetic energy term by its relativistic equivalent,

$$\begin{aligned}\frac{p^2}{2m_e} &\rightarrow \sqrt{m_e^2c^4 + p^2c^2} - m_ec^2 = m_ec^2\left(\sqrt{1+\frac{p^2}{2m_e^2c^2}} - 1\right)\\ &\approx \frac{p^2}{2m_e} - \frac{1}{8}\frac{p^4}{m_e^3c^2},\end{aligned} \tag{21.8}$$

implying that, to lowest order, the relativistic mass correction adds a term to the Hamiltonian,

$$\hat{H}'_{\text{rel}} = -\frac{1}{8}\frac{\hat{p}^4}{m_e^3 c^2}, \tag{21.9}$$

and the total Hamiltonian is

$$\hat{H} \approx \hat{H}_0 - \frac{1}{8}\frac{\hat{p}^4}{m_e^3 c^2}. \tag{21.10}$$

The operators $\hat{H}, \hat{L}^2, \hat{L}_z$ still form a set of commuting operators allowing me to specify the states by the quantum numbers E, ℓ, m_ℓ. However, *there is no longer a single quantum number n that uniquely determines the energy.* The energy of a given level now depends on *both n and ℓ*, since the relativistic term has broken the dynamic symmetry (the classical orbit precesses—it is no longer closed and the Lenz vector is not a constant of the motion). To get the energies exactly I would need to solve the Dirac equation. To get *approximate* energies and eigenfunctions I can use the Schrödinger equation and perturbation theory.

Since the perturbation matrix is diagonal in *all* the quantum numbers except *n*, I do *not* need to use degenerate perturbation theory. The first order change in the energy is given simply by

$$\Delta E_{n\ell}^{(1)} = -\frac{1}{8m_e^3 c^2} \langle n, \ell, m | \hat{p}^4 | n, \ell, m \rangle \tag{21.11}$$

It is possible to evaluate this term as

$$\begin{aligned}
\langle n, \ell, m | \hat{p}^4 | n, \ell, m \rangle &= \int d\mathbf{r} \psi_{nlm}^*(\mathbf{r}) \hat{p}^4 \psi_{nlm}(\mathbf{r}) \\
&= \int d\mathbf{r} \left[\hat{p}^2 \psi_{nlm}(\mathbf{r})\right]^* \hat{p}^2 \psi_{nlm}(\mathbf{r}) \\
&\approx (2m_e)^2 \int d\mathbf{r} \left[\left\{\hat{H}_0 - V(r)\right\} \psi_{nlm}(\mathbf{r})\right]^* \left\{\hat{H}_0 - V(r)\right\} \psi_{nlm}(\mathbf{r}) \\
&= (2m_e)^2 \int d\mathbf{r}\, \psi_{nlm}(\mathbf{r})^* \left\{ \begin{array}{c} \hat{H}_0^2 - V(r)\hat{H}_0 \\ -\hat{H}_0 V(r) + [V(r)]^2 \end{array} \right\} \psi_{nlm}(\mathbf{r}) \\
&= (2m_e)^2 \left[\begin{array}{c} E_n^2 + 2\frac{e^2}{4\pi\epsilon_0} E_n \langle n, \ell, m | \frac{1}{r} | n, \ell, m \rangle \\ + \left(\frac{e^2}{4\pi\epsilon_0}\right)^2 \langle n, \ell, m | \frac{1}{r^2} | n, \ell, m \rangle \end{array} \right].
\end{aligned} \tag{21.12}$$

The needed matrix elements can be calculated as (see Appendix 1),

$$\int d\mathbf{r}\psi^*_{nlm}(\mathbf{r})\frac{1}{r}\psi_{nlm}(\mathbf{r}) = \frac{1}{n^2 a_0}; \tag{21.13a}$$

$$\int d\mathbf{r}\psi^*_{nlm}(\mathbf{r})\frac{1}{r^2}\psi_{nlm}(\mathbf{r}) = \frac{1}{\left(\ell+\frac{1}{2}\right)n^3 a_0^2}, \tag{21.13b}$$

where a_0 is the Bohr radius. The final energy shift is

$$\Delta E_{n\ell} = -\left|E_n^{(0)}\right| \frac{\alpha_{FS}^2}{n^2}\left[\frac{n}{\ell+1/2} - \frac{3}{4}\right]. \tag{21.14}$$

You see that the energy now *now depends on both n and ℓ*; the "accidental" degeneracy is lifted by the relativistic term. Since

$$\frac{n}{\ell+1/2} \geq \frac{(\ell+1)}{\ell+1/2} > 1, \tag{21.15}$$

it follows that $\Delta E_{n\ell} < 0$ and the energy of each state is lowered, which is not surprising since H'_{rel} is negative. The energies of the low angular momentum states for a given n are lowered the most, since the electron's average speed decreases with increasing angular momentum for a fixed energy, giving a smaller relativistic correction.

The relativistic mass corrections in frequency units are given by

$$\Delta f_{n\ell}^{\text{mass}} = -\frac{175\text{ GHz}}{n^4}\left(\frac{n}{\ell+\frac{1}{2}} - \frac{3}{4}\right). \tag{21.16}$$

For the two lowest energy states of hydrogen,

n	ℓ	$\Delta f_{n\ell}^{\text{mass}}$
1	0	− 219 GHZ
2	0	− 35.6 GHZ
2	1	− 6.38 GHZ

To *zeroth* order in the perturbation, the *eigenkets are unchanged.* What about higher order relativistic mass corrections to the eigenkets and eigenenergies? You can use nondegenerate perturbation theory to get higher order corrections. For example,

$$|E,\ell,m_\ell\rangle \approx |n,\ell,m_\ell\rangle + \sum_{n'\neq n}\frac{\langle n',\ell,m_\ell|\, H'_{\text{rel}}\,|n,\ell,m_\ell\rangle\,|n',\ell,m_\ell\rangle}{E_n^{(0)} - E_{n'}^{(0)}} \tag{21.17}$$

and

$$E_{n\ell} \approx E_n^{(0)} + \Delta E_{n\ell}^{(1)} + \sum_{n' \neq n} \frac{\left| \langle n', \ell, m_\ell | H'_{\text{rel}} | n, \ell, m_\ell \rangle \right|^2}{E_n^{(0)} - E_{n'}^{(0)}}, \tag{21.18}$$

where I used the fact that H'_{rel} is diagonal in both ℓ and m_ℓ, since both $\hat{L}^2$ and $\hat{L}_z$ are constants of the motion. The second order energy corrections are α_{FS}^2 times smaller than the first order corrections.[1] I neglect any such corrections throughout this chapter.

21.1.3 Spin Included

If spin is included, but spin-dependent interactions are neglected, the degeneracy of each state is doubled since the electron can have spin quantum numbers $m_s = \pm 1/2$. The total (orbital plus spin) angular momentum operator is

$$\hat{\mathbf{J}} = \hat{\mathbf{L}} + \hat{\mathbf{S}}. \tag{21.19}$$

Since $\hat{J}^2$ and $\hat{J}_z$ and are constants of the motion, states can be labeled by *either* $|n, \ell, m_\ell, s, m_s\rangle$ *or* $|n, \ell, s, j, m_j\rangle$. You cannot use *both* sets of labels since $\hat{J}^2$ does not commute with $\hat{L}_z$ and $\hat{S}_z$ *separately*, only with their sum, $\hat{J}_z = \hat{L}_z + \hat{S}_z$.

The results obtained in Sect. 21.1.2 are unchanged since the relativistic mass correction to the Hamiltonian given in Eq. (21.9) is spin-independent. States having a given ℓ and different j are still degenerate. Thus, the $n = 2, \ell = 1$ state is six-fold degenerate and can be labeled *either* by $n = 2, \ell = 1, m_\ell = \pm 1, 0; m_s = \pm 1/2$ *or* by $n = 2, \ell = 1, j = 1/2, 3/2$ with $m_j = -3/2, -1/2, 1/2, 3/2$ for $j = 3/2$ and $m_j = -1/2, 1/2$ for $j = 1/2$.

21.1.4 Spin–Orbit Interaction

The spin–orbit interaction leads to an additional term in the Hamiltonian given by Eq. (12.33),

$$\hat{H}'_{\text{so}} = \frac{1}{8\pi\epsilon_0} \frac{e^2 \mathbf{S} \cdot \mathbf{L}}{m_e^2 r^3 c^2}. \tag{21.20}$$

---

[1] For consistency, if I were to include such corrections, I would need to take the next term in the expansion of Eq. (21.8) and calculate its effect in first order perturbation theory.

(From this point onwards, I drop the carets on the angular momentum and spin operators, unless there is some possible cause for confusion.) For angular momentum states having $\ell = 0$ there is no spin–orbit interaction. *All equations in this subsection are restricted to values* $\ell \neq 0$.

Since the states having the same n are degenerate, it is best to use a basis in which the levels *within* a given n manifold are not coupled. In that way I do not have to use degenerate perturbation theory. To find such a basis, I write

$$\mathbf{S}\cdot\mathbf{L} = S_xL_x + S_yL_y + S_zL_z = \frac{J^2 - L^2 - S^2}{2}. \tag{21.21}$$

Clearly, $\mathbf{S}\cdot\mathbf{L}$ is *not* diagonal in the $|n, \ell, s; m_\ell, m_s\rangle$ basis; however in the $|n, \ell, s, ; j, m\rangle$ basis

$$\begin{aligned}
&\langle n, \ell, s, ; j, m| \frac{\mathbf{S}\cdot\mathbf{L}}{r^3} |n', \ell', s, ; j', m'\rangle \\
&= \langle n, \ell, s, ; j, m| \frac{J^2 - L^2 - S^2}{2r^3} |n', \ell', s, ; j', m'\rangle \\
&= \hbar^2 \frac{j(j+1) - \ell(\ell+1) - 3/4}{2} \\
&\quad \times \langle n, \ell, s, ; j, m| \frac{1}{r^3} |n', \ell, s, ; j, m\rangle \delta_{\ell,\ell'}\delta_{j,j'}\delta_{m,m'}.
\end{aligned} \tag{21.22}$$

In the $|n, \ell, s, ; j, m\rangle$ basis, there *is* coupling between *different* electronic state manifolds (which I already said I will neglect), but *within* a given n state manifold, there is no coupling! The quantum numbers ℓ, j, m are said to be *good* quantum numbers since they correspond to exact constants of the motion. In effect I am exploiting the fact that the total angular momentum is conserved, even if $\mathbf{L}$ and $\mathbf{S}$ are not conserved separately.

Thus, to lowest order perturbation theory (that is, neglecting coupling between different electronic state manifolds),

$$\begin{aligned}
E_{n\ell j} &= E_n^{(0)} + \frac{1}{8\pi\epsilon_0} \frac{e^2\hbar^2}{m_e^2c^2} \frac{j(j+1) - \ell(\ell+1) - 3/4}{2} \\
&\quad \times \langle n, \ell, s, ; j, m| \frac{1}{r^3} |n, \ell, s, ; j, m\rangle
\end{aligned} \tag{21.23}$$

and $j = \ell \pm 1/2$. The matrix elements of r^{-3} can be evaluated analytically (see Appendix 1) as

$$\begin{aligned}
\langle n, \ell, s, ; j, m| \frac{1}{r^3} |n, \ell, s, ; j, m\rangle &= \int d\mathbf{r}\psi^*_{nlm}(\mathbf{r})\frac{1}{r^3}\psi_{nlm}(\mathbf{r}) \\
&= \frac{1}{\ell\left(\ell + \frac{1}{2}\right)(\ell+1)\, n^3a_0^3},
\end{aligned} \tag{21.24}$$

leading to

$$E_{n\ell\, j=\ell-\frac{1}{2}} = E_n^{(0)} - \frac{\alpha_{FS}^2}{2n} \left|E_n^{(0)}\right| \frac{1}{\ell\left(\ell+\frac{1}{2}\right)}; \tag{21.25a}$$

$$E_{n\ell\, j=\ell+\frac{1}{2}} = E_n^{(0)} + \frac{\alpha_{FS}^2}{2n} \left|E_n^{(0)}\right| \frac{1}{\left(\ell+\frac{1}{2}\right)(\ell+1)}. \tag{21.25b}$$

The *fine structure splitting* for a state having $\ell \neq 0$ is equal to

$$\delta E_{n\ell}^{\text{so}} = E_{n\ell j=\ell+\frac{1}{2}} - E_{n\ell j=\ell-\frac{1}{2}} = \frac{\alpha_{FS}^2}{n} \left|E_n^{(0)}\right| \frac{1}{\ell(\ell+1)}. \tag{21.26}$$

For the $2P$ state of hydrogen, this splitting is equal to 4.53×10^{-5} eV which is about 11 GHz in frequency units. For the alkali atoms the fine structure splitting is much larger—it is about 515 GHz for the $3P$ state of sodium (in sodium the ground state is a $3S$ state and the first excited states are $3P$ states).

The spin orbit corrections in frequency units are given by

$$\Delta f_{n\ell\, j=\ell-\frac{1}{2}}^{\text{so}} = -\frac{175\ \text{GHz}}{2n^3} \frac{1}{\ell\left(\ell+\frac{1}{2}\right)}; \tag{21.27}$$

$$\Delta f_{n\ell\, j=\ell+\frac{1}{2}}^{\text{so}} = \frac{175\ \text{GHz}}{2n^3} \frac{1}{(\ell+1)\left(\ell+\frac{1}{2}\right)}; \tag{21.28}$$

$$\delta f_{n\ell}^{\text{so}} = \Delta f_{n\ell\, j=\ell+\frac{1}{2}}^{\text{so}} - \Delta f_{n\ell\, j=\ell-\frac{1}{2}}^{\text{so}} = \frac{175\ \text{GHz}}{n^3} \frac{1}{\ell(\ell+1)}. \tag{21.29}$$

For the $n = 2$ state

n	ℓ	j	$\Delta f_{n\ell j}^{\text{so}}$
2	1	1/2	– 7.30 GHZ;
2	1	3/2	3.65 GHZ

n	ℓ	$\delta f_{n\ell}^{\text{so}}$
2	1	10.94 GHZ.

The prediction for the fine structure splitting obtained using the Dirac equation is 10.943 GHz, when the reduced mass is used in the calculation, while the experimental value is 10.969 GHz. There are corrections to the Dirac equation arising from the finite size of the nucleus and radiative corrections, involving fluctuations of the vacuum field. When these are included, experiment and theory are in excellent agreement.

21.1.5 Darwin Term

Although $\ell = 0$ states are not shifted by the spin–orbit interaction, they *are* shifted by what is called the *Darwin term* (named for Charles Galton Darwin). The Darwin term shifts only states having $\ell = 0$. In some rough sense this term arises owing to the fact that the electron cannot have a speed greater than c. As a consequence of the uncertainty principle, the electron wave packet must then always be constrained to a sphere of radius of order of the Compton wavelength to insure that $\Delta v < c$. The potential energy is averaged over this sphere. That is, the average potential seen by the electron is

$$\begin{aligned} V(\mathbf{r}) &= \frac{1}{\frac{4}{3}\pi\bar{\lambda}_c^3}\int_{\text{sphere}} d\boldsymbol{\epsilon}\, V(\mathbf{r}+\boldsymbol{\epsilon}) \\ &\approx V(\mathbf{r}) + \frac{1}{\frac{4}{3}\pi\bar{\lambda}_c^3}\int_{\text{sphere}} d\boldsymbol{\epsilon}\left[\boldsymbol{\epsilon}\cdot\nabla V(\mathbf{r}) + \frac{1}{2}\sum_{i,j=1}^{3}\epsilon_i\epsilon_j\frac{\partial^2 V(\mathbf{r})}{\partial x_i\partial x_j}\right], \end{aligned} \tag{21.30}$$

where the integral is over a sphere having radius $\bar{\lambda}_c = \hbar/m_e c$ centered at position $\mathbf{r}$ and $V(\mathbf{r})$ is the Coulomb potential. The first term in the integral vanishes on integrating over $\boldsymbol{\epsilon}$ and the second contributes only for $i = j$. In this manner, you can obtain

$$H'^{\text{Darwin}}(\mathbf{r}) = \frac{1}{\frac{4}{3}\pi\bar{\lambda}_c^3}\frac{1}{2}\int_{\text{sphere}} d\boldsymbol{\epsilon}\sum_i^3 \epsilon_i^2\frac{\partial^2 V(\mathbf{r})}{\partial x_i^2} \tag{21.31}$$

But

$$\int_{\text{sphere}} d\boldsymbol{\epsilon}\,\epsilon_i^2 = \frac{1}{3}\int_{\text{sphere}} d\boldsymbol{\epsilon}\,\epsilon^2 = 4\pi\frac{\bar{\lambda}_c^5}{15} \tag{21.32}$$

since each component gives the same result. As a consequence,

$$H'^{\text{Darwin}}(\mathbf{r}) \approx \frac{\bar{\lambda}_c^2}{10}\sum_i^3\frac{\partial^2 V(\mathbf{r})}{\partial x_i^2} = \frac{\bar{\lambda}_c^2}{10}\nabla^2 V(\mathbf{r}) = \frac{e^2\bar{\lambda}_c^2}{10\epsilon_0}\delta(\mathbf{r}), \tag{21.33}$$

which is of the same order of magnitude as the result that can be derived from the Dirac equation,

$$H'^{\text{Darwin}}(\mathbf{r}) = \frac{e^2\bar{\lambda}_c^2\delta(\mathbf{r})}{8\epsilon_0}. \tag{21.34}$$

The energy shift vanishes for all but $\ell = 0, j = 1/2$ states. The (correct) Darwin energy shift is given by

$$\Delta E_n^{\text{Darwin}} = \frac{e^2 \bar{\lambda}_c^2}{8\epsilon_0} \langle n00 | \delta(\mathbf{r}) | n00 \rangle , \tag{21.35}$$

where $|n\ell m_\ell\rangle$ is an eigenket of the hydrogen atom neglecting spin. The matrix element is evaluated easily [see Eq. (21.108) in Appendix 2] and one finds

$$\Delta E_n^{\text{Darwin}} = \frac{E_R \alpha_{FS}^2}{n^3}; \tag{21.36a}$$

$$\Delta f_n^{\text{Darwin}} = \frac{175 \text{ GHz}}{n^3}. \tag{21.36b}$$

The frequency shifts for $n = 1, 2$ states are given by

n	ℓ	j	$\Delta f_n^{\text{Darwin}}$
1	0	1/2	175 GHZ .
2	0	1/2	21.9 GHZ

21.1.6 Total Fine Structure

The total fine structure (mass+spin-orbit+Darwin terms) is obtained by combining Eqs. (21.14), (21.25), and (21.36). The result is

$$\Delta E_{nj}^{\text{tot}} = -\alpha_{FS}^2 \frac{E_R}{n^4} \left(\frac{n}{j+\frac{1}{2}} - \frac{3}{4} \right); \quad j = \frac{1}{2}, \frac{3}{2}, \ldots, n - \frac{1}{2}; \tag{21.37a}$$

$$\Delta f_{nj}^{\text{tot}} = -\frac{175 \text{ GHz}}{n^4} \left(\frac{n}{j+\frac{1}{2}} - \frac{3}{4} \right); \quad j = \frac{1}{2}, \frac{3}{2}, \ldots, n - \frac{1}{2}. \tag{21.37b}$$

These results agree with the solutions of the Dirac equation [see Eq. (21.3)], if corrections of order $\alpha_{FS}^4 E_R$ are neglected. Note that the ℓ-dependence is now implicit, with $j = \ell \pm \frac{1}{2}$. For $n = 1, 2$

n	ℓ	j	$\Delta f_{n\ell}^{\text{tot}}$
1	0	1/2	– 43.8 GHZ
2	0	1/2	– 13.7 GHZ .
2	1	1/2	– 13.7 GHZ
2	1	3/2	– 2.74 GHZ

All energies are lowered, a result attributable mainly to the relativistic mass correction. States having the same values of n and j are still degenerate.

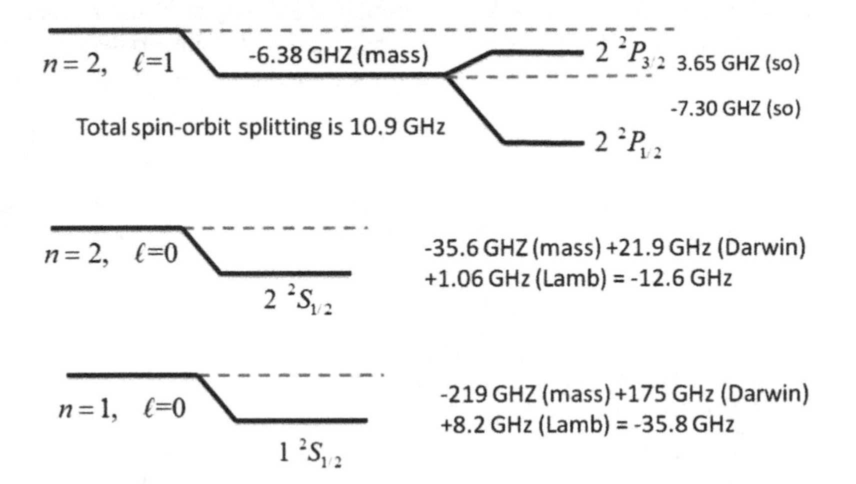

Fig. 21.1 Hydrogen fine structure for the $n = 1$ and $n = 2$ states (not to scale), including relativistic mass, spin-orbit and Darwin contributions, as well as the Lamb shift. The values shown are those obtained using the results of perturbation theory with m_e equal to the reduced mass and differ slightly from the experimental values

21.1.7 Lamb Shift

The Lamb shift [named for Willis E. Lamb, Jr. (my thesis advisor)] arises from quantum electrodynamic corrections, that is, effects related to changes in atomic energy levels resulting from fluctuations of the vacuum radiation field. The Lamb shift affects mainly $\ell = 0$ states. The $n = 2, \ell = 0, j = 1/2$ level is shifted *up* by about 1.06 GHz and the $n = 1, \ell = 0, j = 1/2$ level is shifted *up* by about 8.17 GHz. The Lamb shift breaks the degeneracy of states having the same n but different j.

In Fig. 21.1,the relativistic corrections to the $n = 1, 2$ states of hydrogen are illustrated. In the remainder of this chapter, I neglect the Lamb shift.

21.1.8 Hyperfine Structure

As does the electron, the proton in hydrogen possesses an intrinsic spin angular momentum having quantum number $I = 1/2$ and an intrinsic magnetic moment. In hydrogen, this results in an interaction of the magnetic moment of the proton with the magnetic field produced by the electron. The magnetic moment of the proton is

$$\mathbf{m}_p = \frac{eg_p\mathbf{I}}{2m_p}, \tag{21.38}$$

where m_p is the proton mass and g_p is the proton *g-factor*, equal to 5.586. The hyperfine interaction strength is roughly $m_e/m_p = 1/1836$ times smaller than the fine structure interaction strength. A calculation of the energy level shifts produced by the hyperfine interaction is given in Appendix 2. It is possible to define a total (orbital plus electronic spin plus nuclear spin) angular momentum operator by

$$\mathbf{F} = \mathbf{J} + \mathbf{I}, \tag{21.39}$$

having associated quantum number $f = j \pm 1/2$. In frequency units, the hyperfine level shifts, $\Delta f_{n\ell jf}^{\mathrm{hf}}$, and splittings, $\delta f_{n\ell j}^{\mathrm{hf}}$, for the $n = 1, 2$ states of hydrogen are:

n	ℓ	j	f	$\Delta f_{n\ell jf}^{\mathrm{hf}}$
1	0	1/2	0	– 1.06 GHZ
1	0	1/2	1	0.355 GHZ
2	0	1/2	0	– 133 MHZ
2	0	1/2	1	44.4 MHZ
2	1	1/2	0	– 44.4 MHZ
2	1	1/2	1	14.8 MHZ
2	1	3/2	1	– 14.8 GHZ
2	1	3/2	2	8.88 MHZ

n	ℓ	j	$\delta f_{n\ell j}^{\mathrm{hf}}$
1	0	1/2	1.42 GHZ
2	0	1/2	177 MHZ
2	1	1/2	59.2 MHZ
2	1	3/2	23.7 MHZ

These results are indicated schematically in Fig. 21.2.

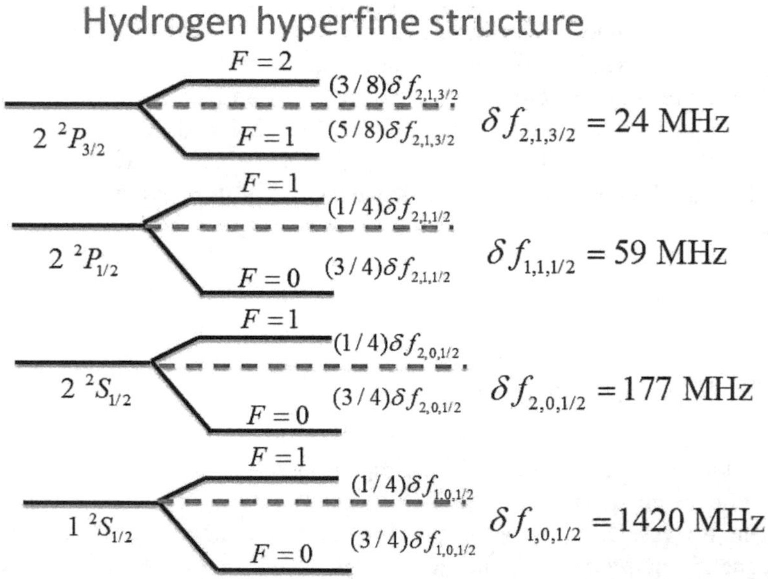

Fig. 21.2 Hydrogen hyperfine structure for the $n = 1$ and $n = 2$ states (not to scale) with splittings $\delta f_{n,\ell,j}$

21.1.9 Addition of a Magnetic Field

I now want to consider the influence of external magnetic and electric fields on the energy levels of hydrogen. Although external magnetic fields couple different hyperfine levels much more strongly than different fine structure levels (see Appendix 2), I neglect hyperfine structure in discussing magnetic and electric field effects. This is not because they are unimportant, but simply because I want to illustrate some physical concepts that are more easily understood if the hyperfine interaction is neglected.

The modifications of the energy levels of atoms by an external magnetic field are often referred to as the *Zeeman effect* (after Pieter Zeeman). If a magnetic field is present along the z direction, a term

$$\hat{H}'_B = \frac{\beta_0 B}{\hbar}(L_z + 2S_z) = \frac{\beta_0 B}{\hbar}(J_z + S_z) \tag{21.40}$$

must be added to the Hamiltonian. Recall that $\beta_0 = e\hbar/2m_e$ is the Bohr magneton and that $\beta_0/h \approx 14$ GHz/T. The total Hamiltonian then consists of the unperturbed Hamiltonian of hydrogen plus contributions from both the relativistic corrections (relativistic mass, spin-orbit, and Darwin terms) and the Zeeman effect. The only constants of the motion for the total Hamiltonian are E, L^2, S^2, J_z. The magnitude of the total angular momentum is not a constant of the motion since $\hat{J}^2$ does not commute with $\hat{S}_z$, and the z-components of the orbital and spin angular momenta are not constants of the motion since $\hat{L}_z$ and $\hat{S}_z$ do not commute with $\hat{\mathbf{L}} \cdot \hat{\mathbf{S}}$, which appears as a factor in the spin-orbit term of the Hamiltonian. Thus the eigenkets can be labeled by $|E, \ell, s = 1/2, m_j\rangle$. I will drop the s label since $s = 1/2$. In a given n manifold, the energy E of a state having a given ℓ, m_j now depends on some complicated combination of the spin–orbit and Zeeman interactions.

To apply degenerate perturbation theory, I can use either the $|n, \ell, j, m_j\rangle$ or the $|n, \ell, m_\ell, m_s\rangle$ basis kets, remembering that only states having the same ℓ and the same $m_j = m_\ell + m_s$ can be coupled since L^2 and J_z are constants of the motion. In other words, ℓ and $m_j = m_\ell + m_s$ remain good quantum numbers in the presence of an external magnetic field. It is relatively easy to calculate the matrix elements in either basis in a state of given n. Once the matrix elements are calculated, you need to diagonalize each degenerate sub-block. If the spin-orbit coupling is much larger than the magnetic field interaction strength, then j is an *approximate* good quantum number since J^2 is an approximate constant of the motion. In this limit the $|n, \ell, j, m_j\rangle$ are *approximate* eigenkets that can be used to carry out *nondegenerate perturbation theory* for the magnetic-field interaction. On the other hand, if the magnetic field interaction strength is much larger than the spin-orbit coupling strength, then m_ℓ and m_s are *approximate* good quantum number since L_z and S_z are an approximate constants of the motion. In this case, the $|n, \ell, m_\ell, m_s\rangle$ are *approximate* eigenkets that can be used to carry out *nondegenerate perturbation theory* for the spin–orbit and other relativistic interactions.

In the $\left|n,\ell,j,m_j\right\rangle$ basis, the only matrix elements that are a little tricky to evaluate are those involving $\left\langle n,\ell,j,m_j\right|\hat{S}_z\left|n,\ell,j',m_j\right\rangle$. You must convert the $\left|n,\ell,j,m_j\right\rangle$ kets back to the $\left|n,\ell,m_\ell,m_s\right\rangle$ basis using

$$\left|n,\ell,j,m_j\right\rangle = \sum_{m_\ell,m_s}\begin{bmatrix}\ell & s=1/2 & j\\ m_\ell & m_s & m_j\end{bmatrix}\left|n,\ell,m_\ell,m_s\right\rangle \tag{21.41}$$

to arrive at

$$\left\langle n,\ell,j,m_j\right|\hat{S}_z\left|n,\ell,j',m_j\right\rangle = \hbar\sum_{m_\ell,m_s} m_s\begin{bmatrix}\ell & \frac{1}{2} & j\\ m_\ell & m_s & m_j\end{bmatrix}\begin{bmatrix}\ell & \frac{1}{2} & j'\\ m_\ell & m_s & m_j\end{bmatrix}. \tag{21.42}$$

In this way, using tabulated values of the Clebsch-Gordan coefficients, you can obtain

$$\begin{aligned}
\left\langle n,\ell,j,m_j\right|\hat{H}\left|n,\ell,j',m_j\right\rangle &= E_n^{(0)}\delta_{j,j'}\\
&-\frac{\alpha_{FS}^2}{4n^2}\left(\frac{4n}{\ell+1/2}-3\right)\left|E_n^{(0)}\right|\delta_{j,j'}\\
&+\frac{\alpha_{FS}^2}{2n}\left(\frac{\delta_{j,\ell+1/2}}{(\ell+1/2)(\ell+1)}-\frac{\delta_{j,\ell-1/2}}{\ell(\ell+1/2)}\right)\left|E_n^{(0)}\right|\delta_{j,j'}\\
&+\beta_0 Bm_j\left[\left(1+\frac{1}{2\ell+1}\right)\delta_{j,\ell+1/2}+\left(1-\frac{1}{2\ell+1}\right)\delta_{j,\ell-1/2}\right]\delta_{j,j'}\\
&-\beta_0 B\frac{\sqrt{(\ell+1/2)^2-m_j^2}}{2\ell+1}\left(\delta_{j,\ell+1/2}\delta_{j',\ell-1/2}+\delta_{j,\ell-1/2}\delta_{j',\ell+1/2}\right),
\end{aligned} \tag{21.43}$$

or

$$\begin{aligned}
\left\langle n,\ell,j,m_j\right|\hat{H}\left|n,\ell,j',m_j\right\rangle &= E_n^{(0)}\delta_{j,j'}\\
&-\frac{\alpha_{FS}^2}{n^2}\left(\frac{n}{j+1/2}-\frac{3}{4}\right)\left|E_n^{(0)}\right|\delta_{j,j'}\\
&+\beta_0 Bm_j\left[\left(1+\frac{1}{2\ell+1}\right)\delta_{j,\ell+1/2}+\left(1-\frac{1}{2\ell+1}\right)\delta_{j,\ell-1/2}\right]\delta_{j,j'}\\
&-\beta_0 B\frac{\sqrt{(\ell+1/2)^2-m_j^2}}{2\ell+1}\left(\delta_{j,\ell+1/2}\delta_{j',\ell-1/2}+\delta_{j,\ell-1/2}\delta_{j',\ell+1/2}\right).
\end{aligned} \tag{21.44}$$

On the other hand, in the $\left|n,\ell,m_\ell,m_s\right\rangle$ basis it is simple to evaluate matrix elements of the magnetic field interaction, but the $\mathbf{S}\cdot\mathbf{L}$ factor in the spin–orbit interaction has to be written as $(L_+S_- + L_-S_+)/2 + L_zS_z$ to evaluate its matrix elements. In this manner you can find

$$\begin{aligned}
\langle n,\ell,m_\ell,m_s|\,\hat{H}\,|n,\ell,m'_\ell,m'_s\rangle &= E_n^{(0)}\delta_{m_\ell,m'_\ell}\delta_{m_s,m'_s} \\
&-\frac{\alpha_{FS}^2}{n^2}\left(\frac{n}{\ell+1/2}-\frac{3}{4}\right)\left|E_n^{(0)}\right|\delta_{m_\ell,m'_\ell}\delta_{m_s,m'_s} \\
&+\frac{\alpha_{FS}^2\left|E_n^{(0)}\right|(1-\delta_{\ell,0})}{n\ell(\ell+1/2)(\ell+1)}\left\{m_\ell m_s\delta_{m_\ell,m'_\ell}\delta_{m_s,m'_s}\right. \\
&+\frac{1}{2}\sqrt{(\ell-m'_\ell)(\ell+m'_\ell+1)}\delta_{m_\ell,m'_\ell+1}\delta_{m_s,-1/2}\delta_{m'_s,1/2} \\
&\left.+\frac{1}{2}\sqrt{(\ell+m'_\ell)(\ell-m'_\ell+1)}\delta_{m_\ell,m'_\ell-1}\delta_{m_s,1/2}\delta_{m'_s,-1/2}\right\} \\
&+\frac{\alpha_{FS}^2\left|E_n^{(0)}\right|\delta_{\ell,0}\delta_{m_s,m'_s}}{n}+\beta_0 B\left(m_\ell+2m_s\right)\delta_{m_\ell,m'_\ell}\delta_{m_s,m'_s}.
\end{aligned} \tag{21.45}$$

To obtain the exact (exact only in the sense that I neglect coupling between different electronic state manifolds) eigenkets and eigenenergies, I must diagonalize these matrices for a given value of n and ℓ. Before doing so, let me look at the weak and strong field limits.

Consider the $n=2,\ell=1$ manifold which consists of six degenerate levels in the absence of any magnetic fields and any relativistic corrections. The fine structure splitting is about 11 GHz and, since $\beta_0/h = 14.0$ GHz/T, the magnetic and fine structure interactions are comparable for field strengths of 1 T or above. In weak fields of much less than 1 T, the $|n,\ell,j,m_j\rangle$ are *approximate* eigenkets. That is, to lowest order I can neglect the magnetic field interaction entirely. The corrections arising from the magnetic field interaction can then be calculated using nondegenerate perturbation theory in the $|n,\ell,j,m_j\rangle$ basis. In this limit, the change in the energy levels produced by the magnetic field is given approximately by the diagonal matrix elements in Eq. (21.44), namely

$$\begin{aligned}
\Delta E_{n=2,\ell=1,j,m_j} &= \langle 2,1,j,m_j|\,\hat{H}-\hat{H}_0\,|2,1,j,m_j\rangle \\
&= -\frac{\alpha_{FS}^2}{4}\left(\frac{2}{j+1/2}-\frac{3}{4}\right)\left|E_2^{(0)}\right| \\
&+\beta_0 B m_j\left[\frac{4}{3}\delta_{j,3/2}+\frac{2}{3}\delta_{j,1/2}\right],
\end{aligned} \tag{21.46}$$

which implies that

$$\frac{\Delta E_{n=2,\ell=1,3/2,m_j}}{h} = \left[-2.74 + 18.7 m_j B(\mathrm{T})\right] \mathrm{GHz}; \tag{21.47a}$$

$$\frac{\Delta E^{(0)}_{n=2,\ell=1,1/2,m_j}}{h} = \left[-13.7 + 9.33 m_j B(\mathrm{T})\right] \mathrm{GHz}, \tag{21.47b}$$

where $B(\mathrm{T})$ is the magnetic induction in units of Tesla.

On the other hand, in fields much greater than 1 T, the spin–orbit interaction can be neglected to lowest order compared with the magnetic field interaction. In that limit, the $|n,\ell,m_\ell,m_s\rangle$ are approximate eigenkets and the magnetic field coupling strength is sufficiently large to allow me to use nondegenerate perturbation theory in this basis to approximate the changes in the energy levels produced by the spin–orbit interaction. These changes are given approximately by the diagonal matrix elements in Eq. (21.45), namely

$$\begin{aligned}\Delta E_{n=2,\ell=1,m_\ell,m_s} &= \langle 2,1,m_\ell,m_s|\,\hat{H}-\hat{H}_0\,|2,1,m_\ell,m_s\rangle \\ &= -\alpha_{FS}^2\frac{7}{48}\left|E_2^{(0)}\right| + \beta_0 B\left(m_\ell + 2m_s\right) + \frac{\alpha_{FS}^2\left|E_2^{(0)}\right|}{6} m_\ell m_s,\end{aligned} \tag{21.48}$$

which implies that

$$\frac{\Delta E_{n=2,\ell=1,m_\ell,m_s}}{h} = \left[-6.38 + 14.0\left(m_\ell + 2m_s\right)B(\mathrm{T}) + 7.30 m_\ell m_s\right] \mathrm{GHz}. \tag{21.49}$$

Note that the energy of the $m_\ell = 1, m_s = 1/2$ level in Eq. (21.49) agrees with that of the $j = 3/2, m_j = 3/2$ level in Eq. (21.47a), as does the energy of the $m_\ell = -1, m_s = -1/2$ level in Eq. (21.49) with that of the $j = 3/2, m_j = -3/2$ level in Eq. (21.47a). These pairs of states are identical in both bases.

For arbitrary field strengths I can use either basis. I will use $|n = 2, \ell = 1, m_\ell, m_s\rangle$ basis kets, for which the perturbation Hamiltonian in frequency units, obtained using Eqs. (21.45), is

$$\frac{\Delta \underline{\mathrm{H}}}{h} = \begin{pmatrix} -2.73+28B & 0 & 0 & 0 & 0 & 0 \\ 0 & -10.0 & 5.16 & 0 & 0 & 0 \\ 0 & 5.16 & -6.38+14B & 0 & 0 & 0 \\ 0 & 0 & 0 & -6.38-14B & 5.16 & 0 \\ 0 & 0 & 0 & 5.16 & -10.0 & 0 \\ 0 & 0 & 0 & 0 & 0 & -2.73-28B \end{pmatrix} \mathrm{GHz} \tag{21.50}$$

where the order for $|m_\ell, m_s\rangle$ is

$$\left|1,\frac{1}{2}\right\rangle, \left|1,-\frac{1}{2}\right\rangle, \left|0,\frac{1}{2}\right\rangle, \left|0,-\frac{1}{2}\right\rangle, \left|-1,\frac{1}{2}\right\rangle, \left|-1,-\frac{1}{2}\right\rangle. \tag{21.51}$$

The eigenenergies and eigenkets are obtained easily by diagonalizing the 2×2 submatrices of the Hamiltonian. The eigenkets $\left|E,\ell = 1, m_j\right\rangle$ are of the form

$$|E_1,1,3/2\rangle = |2,1,1,1/2\rangle\,; \tag{21.52a}$$

$$|E_2,1,1/2\rangle = \alpha\,|2,1,1,-1/2\rangle + \beta\,|2,1,0,1/2\rangle\,; \tag{21.52b}$$

$$|E_3,1,1/2\rangle = \gamma\,|2,1,1,-1/2\rangle + \delta\,|2,1,0,1/2\rangle\,; \tag{21.52c}$$

$$|E_4,1,-1/2\rangle = \alpha'\,|2,1,0,-1/2\rangle + \beta'\,|2,1,-1,1/2\rangle\,; \tag{21.52d}$$

$$|E_5,1,-1/2\rangle = \gamma'\,|2,1,0,-1/2\rangle + \delta'\,|2,1,-1,1/2\rangle\,; \tag{21.52e}$$

$$|E_6,1,-3/2\rangle = |2,1,-1,-1/2\rangle\,, \tag{21.52f}$$

where the coefficients $\alpha,\beta,\gamma,\delta,\alpha',\beta',\gamma',\delta'$ emerge in the diagonalization procedure, but are not written explicitly. The corresponding eigenfrequencies in GHZ are

$$\Delta f_1 = -2.74 + 28B; \tag{21.53a}$$

$$\Delta f_2 = -8.21 + 7B + 7\sqrt{0.611 + 0.521B + B^2}; \tag{21.53b}$$

$$\Delta f_3 = -8.21 + 7B - 7\sqrt{0.611 + 0.521B + B^2}; \tag{21.53c}$$

$$\Delta f_4 = -8.21 - 7B + 7\sqrt{0.611 - 0.521B + B^2}; \tag{21.53d}$$

$$\Delta f_5 = -8.21 - 7B - 7\sqrt{0.611 - 0.521B + B^2}; \tag{21.53e}$$

$$\Delta f_6 = -2.74 - 28B, \tag{21.53f}$$

where B is in Tesla. The eigenvalues (change in frequency, Δf) in GHz are plotted in Fig. 21.3 as a function of magnetic field strength in Tesla (T).

For weak fields ($B \ll 1$ T), the frequencies are given by Eq. (21.47), and for strong fields ($B \approx 1$ T), the frequencies begin to approach those given by Eq. (21.49). Note that in very strong fields the states having $m_\ell = 1, m_s = -1/2$ and $m_\ell = -1, m_s = 1/2$ approach degeneracy, a result predicted in Eq. (21.49). The strong field region is referred to as the *Paschen-Back* region.

21.1.10 Addition of an Electric Field

If, instead of a magnetic field, there is a constant electric field having amplitude $\mathcal{E}_0$ along the z axis, then an additional term,

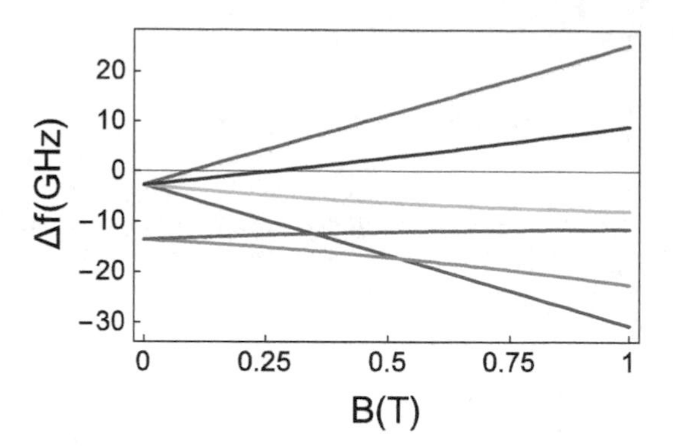

Fig. 21.3 Changes in the eigenfrequencies of the $n = 2$, $\ell = 1$ states of hydrogen produced by a constant external magnetic field

$$\hat{H}'_E = e\mathcal{E}_0\hat{z}, \tag{21.54}$$

must be added to the Hamiltonian.[2] The splitting of atomic energy levels by an electric field is commonly referred to as the Stark effect (Johannes Stark discovered this effect in 1913). I have already discussed the linear Stark effect in hydrogen in the context of degenerate perturbation theory, but relativistic corrections and spin were neglected in that discussion. In the presence of a constant external electric field, L^2, **L**, and J^2 are no longer constants of the motion for the total Hamiltonian. The only constants of the motion are E, S^2, and J_z. For example, in the $n = 2$ manifold there are eight states that can be characterized by their energy and m_j values. The states having $m_j = 3/2$ $[m_\ell = 1, m_s = 1/2]$ and $m_j = -3/2$ $[m_\ell = -1, m_s = -1/2]$ are not coupled to any other states by the field and remain a degenerate doublet. The other states break up into two *triplets* of states coupled by both the spin–orbit and electric field interactions. It turns out that the energies of the levels do not depend on the sign of m_j.

In the $|n, \ell, m_\ell, m_s\rangle$ basis it is simple to evaluate matrix elements of the electric field interaction since

$$\langle n, \ell, m_\ell, m_s | \hat{z} | n', \ell', m'_\ell, m'_s \rangle = \frac{\delta_{m_\ell, m'_\ell} \delta_{m_s, m'_s}}{\sqrt{2\ell + 1}} \begin{bmatrix} \ell' & 1 & \ell \\ m_\ell & 0 & m_\ell \end{bmatrix} \langle n, \ell \| z \| n', \ell' \rangle . \tag{21.55}$$

The Clebsch-Gordan coefficients restrict $\Delta\ell = \ell - \ell'$ to values $\pm 1, 0$, but $\Delta\ell = 0$ is ruled out owing to parity considerations; in other words, the reduced matrix elements vanish if $\Delta\ell = 0$. A general formula for the reduced matrix elements can be given in terms of hypergeometric functions, but I do not give it here. The matrix

[2]Technically there are no longer any bound states of hydrogen in an external electric field. However, the lifetimes of the low-lying energy states are effectively infinite for static fields and the shifts of these levels can be calculated using perturbation theory.

elements for the entire Hamiltonian in a state of fixed n are

$$\begin{aligned}
\langle n,\ell,m_\ell,m_s|\,\hat{H}\,\left|n,\ell',m'_\ell,m'_s\right\rangle &= E_n^{(0)}\delta_{m_\ell,m'_\ell}\delta_{m_s,m'_s}\delta_{\ell,\ell'} \\
&-\frac{\alpha_{FS}^2}{4n^2}\left(\frac{4n}{\ell+1/2}-3\right)\left|E_n^{(0)}\right|\delta_{m_\ell,m'_\ell}\delta_{m_s,m'_s}\delta_{\ell,\ell'} \\
&+\frac{\alpha_{FS}^2\left|E_n^{(0)}\right|(1-\delta_{\ell,0})}{n\ell\left(\ell+1/2\right)\left(\ell+1\right)}\left\{m_\ell m_s\delta_{m_\ell,m'_\ell}\delta_{m_s,m'_s}\right. \\
&+\frac{1}{2}\sqrt{\left(\ell-m'_\ell\right)\left(\ell+m'_\ell+1\right)}\delta_{m_\ell,m'_\ell+1}\delta_{m_s,-1/2}\delta_{m'_s,1/2} \\
&\left.+\frac{1}{2}\sqrt{\left(\ell+m'_\ell\right)\left(\ell-m'_\ell+1\right)}\delta_{m_\ell,m'_\ell-1}\delta_{m_s,1/2}\delta_{m'_s,-1/2}\right\}\delta_{\ell,\ell'} \\
&+\frac{\alpha_{FS}^2\left|E_n^{(0)}\right|\delta_{\ell,0}\delta_{m_s,m'_s}}{n} \\
&+e\mathcal{E}_0\left\langle n,\ell,m_\ell\right|\hat{z}\left|n,\ell',m_\ell\right\rangle\delta_{m_\ell,m'_\ell}\delta_{m_s,m'_s}. \qquad (21.56)
\end{aligned}$$

For the $n=2$ manifold, $\langle 2,1,0,|\,\hat{z}\,|2,0,0\rangle=-3a_0$, and, with the states denoted by $|n,\ell;m_\ell,m_s\rangle$, there are eight unperturbed eigenkets that can be written as

$$|1\rangle=\left|2,0;0,\frac{1}{2}\right\rangle;\quad m_j=1/2, \qquad (21.57a)$$

$$|2\rangle=\left|2,0;0,-\frac{1}{2}\right\rangle;\quad m_j=-1/2, \qquad (21.57b)$$

$$|3\rangle=\left|2,1;1,\frac{1}{2}\right\rangle;\quad m_j=3/2, \qquad (21.57c)$$

$$|4\rangle=\left|2,1;1,-\frac{1}{2}\right\rangle;\quad m_j=1/2, \qquad (21.57d)$$

$$|5\rangle=\left|2,1;0,\frac{1}{2}\right\rangle;\quad m_j=1/2, \qquad (21.57e)$$

$$|6\rangle=\left|2,1;0,-\frac{1}{2}\right\rangle;\quad m_j=-1/2, \qquad (21.57f)$$

$$|7\rangle=\left|2,1;-1,\frac{1}{2}\right\rangle;\quad m_j=-1/2, \qquad (21.57g)$$

$$|8\rangle=\left|2,1;-1,-\frac{1}{2}\right\rangle;\quad m_j=-3/2. \qquad (21.57h)$$

If I set

$$A = -3.84 \times 10^{-5}\mathcal{E}_0(\text{V/m}) \tag{21.58}$$

and use Eq. (21.56), I can write the perturbation Hamiltonian sub-matrix for the $n = 2$ manifold as

$$\frac{\Delta \underline{\mathrm{H}}}{h} = \begin{pmatrix} -13.7 & 0 & 0 & 0 & A & 0 & 0 & 0 \\ 0 & -13.7 & 0 & 0 & 0 & A & 0 & 0 \\ 0 & 0 & -2.74 & 0 & 0 & 0 & 0 & 0 \\ 0 & 0 & 0 & -10.0 & 5.16 & 0 & 0 & 0 \\ A & 0 & 0 & 5.16 & -6.38 & 0 & 0 & 0 \\ 0 & A & 0 & 0 & 0 & -6.38 & 5.16 & 0 \\ 0 & 0 & 0 & 0 & 0 & 5.16 & -10.0 & 0 \\ 0 & 0 & 0 & 0 & 0 & 0 & 0 & -2.74 \end{pmatrix} \text{GHz} \tag{21.59}$$

States $|3\rangle$ and $|8\rangle$ having $m_j = \pm 3/2$ are degenerate and are uncoupled to other states by the field. States $|1\rangle$,$|4\rangle$,$|5\rangle$ having $m_j = 1/2$ are coupled to each other as are states $|2\rangle$,$|6\rangle$,$|7\rangle$ having $m_j = -1/2$. However, *the eigenvalues do not depend on the sign of* m_j. Thus, these six states break up into three degenerate pairs of levels, with one state in each doublet having $m_j = 1/2$ and the other $m_j = -1/2$. These six states contain an admixture of both $\ell = 1$ and $\ell = 0$ states.

The eigenfrequencies (change in frequency) in GHz, obtained by numerically diagonalizing Eq. (21.59), are plotted in Fig. 21.4 as a function of electric field strength in units of 10^5 V/m. Note that there are four distinct energies. The horizontal straight line in the figure corresponds to the $m_j = \pm 3/2$ states, while each other line corresponds to a doublet having $m_j = \pm 1/2$. Explicit expressions for eigenenergies and eigenkets are not given, but can be obtained easily by numerically diagonalizing the Hamiltonian. The energy degeneracy is not totally lifted by the field, which implies there is still some remaining symmetry in the problem. The Hamiltonian for a hydrogen atom in an external electric field remains invariant under time reversal. Since the time-reversal operator effectively reverses the sign of m_j in the eigenkets, the energy must be independent of the sign of m_j.[3] In the case of hydrogen in an external magnetic field, time reversal symmetry is broken since the magnetic moment operator changes sign under time reversal; as a consequence, the magnetic field can totally lift the degeneracy of the different magnetic sublevels.

[3]In other words, the time-reversal operator $\hat{T}$ (see Problem 19.12) acting on a ket $|Em_j\rangle$ produces $|E, -m_j\rangle$, but $\hat{T}\,|Em_j\rangle$ must be an eigenket of the Hamiltonian having the same energy as that associated with $|Em_j\rangle$ if the Hamiltonian is invariant under time-reversal, implying that there is a two-fold degeneracy for states having $m_j \neq 0$.

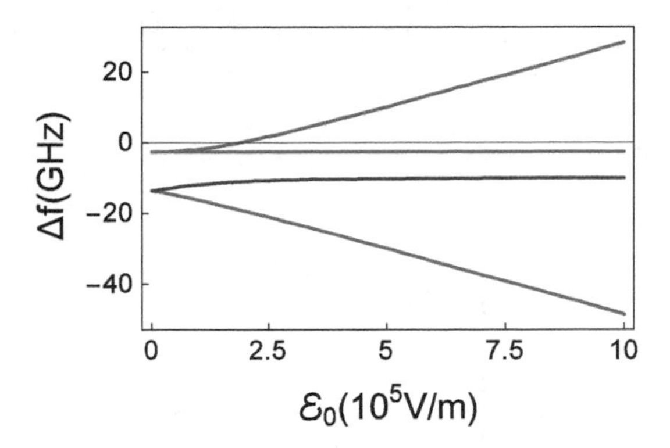

Fig. 21.4 Frequency shift Δf in GHz of the $n = 2$, $\ell = 1$ states of hydrogen as a function of electric field amplitude in units of 10^5 V/m

21.2 Multi-Electron Atoms in a Magnetic Field: Vector Model

In multi-electron atoms the coupling of the orbital and spin angular momentum of all the electrons constitutes a very difficult problem. For atoms having atomic number $Z \lesssim 30$, however, it is often a good approximation to couple the spin angular momentum of the valence band electrons into a total spin angular momentum **S** and the orbital angular momentum of the valence band electrons into a total orbital angular momentum **L**. The total angular momentum of all the electrons is then $\mathbf{J} = \mathbf{L} + \mathbf{S}$ (*Russel-Saunders* or L-S coupling scheme).[4] If a multi-electron atom whose structure can be approximated by an L-S coupling scheme is placed in a magnetic field, it is relatively easy to calculate the level splitting *approximately* in the limits that the spin-orbit coupling is much less than or much greater than the Zeeman splitting. Moreover, there is a *classical vector model* that can be used in each of these cases. In all cases, I take the magnetic field along the z-axis. In an external magnetic field, L, S, and $m_J = m_L + m_s$ remain good quantum numbers.

21.2.1 Zeeman Splitting Much Greater Than Spin-Orbit Splitting

As in the case of hydrogen, I can simply ignore the spin-orbit coupling to lowest order. In this limit the magnetic field splitting in a state of given S and L is

[4]In the opposite limit of high Z atoms, a $j{-}j$ coupling scheme can be used in which the total angular momentum of the atom is the sum of the total angular momenta of the individual electrons.

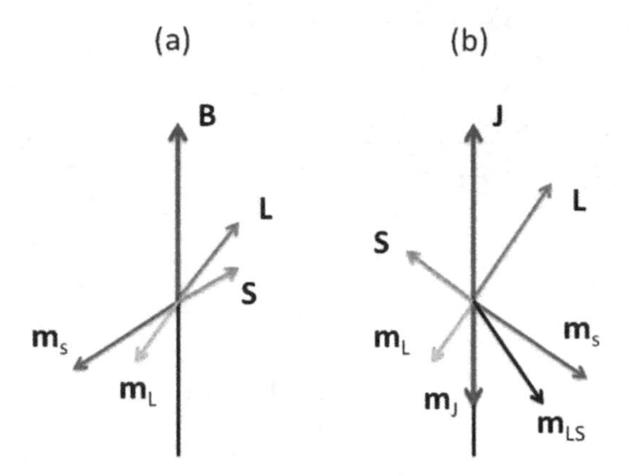

Fig. 21.5 Vector coupling schemes for calculating the Zeeman splitting. (**a**) Strong field limit (**b**) Weak field limit

$$\Delta E_{LS} \approx \beta_0 B\,(m_L + 2m_S)\,. \tag{21.60}$$

In a vector model [see Fig. 21.5 (a)], the magnetic field is sufficiently strong to cause the total spin and total orbital angular momentum to precess *separately* around the field vector. All components of the spin and orbital magnetic moments average to zero owing to the precession, *except* their components along the z-axis. These projections account for ΔE_{LS}.

21.2.2 Spin-Orbit Splitting Much Greater Than Zeeman Splitting

In analogy with hydrogen, J is now an approximate good quantum number and the magnetic field splitting in a state of given J is

$$\begin{aligned}\Delta E_{Jm_J} &= \frac{\beta_0 B}{\hbar}\left\langle L,S,J,m_J\right|\hat{J}_z + \hat{S}_z\left|L,S,J,m_J\right\rangle \\ &= \beta_0 B m_J + \frac{\beta_0 B}{\hbar}\left\langle L,S,J,m_J\right|\hat{S}_z\left|L,S,J,m_J\right\rangle .\end{aligned} \tag{21.61}$$

To calculate $\left\langle L,S,J,m_J\right|\hat{S}_z\left|L,S,J,m_J\right\rangle$, I can use the Wigner-Eckart theorem. First, I prove the following lemma: For any vector operator $\hat{A}$,

$$\left\langle \alpha',J,m_J'\right|\hat{A}_1^q\left|\alpha,J,m_J\right\rangle = \frac{\left\langle \alpha',J,m_J\right|\hat{\mathbf{J}}\cdot\hat{\mathbf{A}}\left|\alpha,J,m_J\right\rangle}{\hbar^2 J\,(J+1)}\left\langle \alpha,J,m_J'\right|\hat{J}_1^q\left|\alpha,J,m_J\right\rangle , \tag{21.62}$$

where the $\hat{A}_1^q$ are the components of an irreducible tensor operator of rank 1 given by Eq. (20.41) and α, α' represent additional quantum numbers. The proof is relatively simple. It follows from the Wigner-Eckart theorem that

$$\begin{aligned}\left\langle \alpha', J, m_J' \right| \hat{A}_1^q \left| \alpha, J, m_J \right\rangle &= \frac{1}{\sqrt{2J+1}} \begin{bmatrix} J & 1 & J \\ m_J & q & m_J' \end{bmatrix} \left\langle \alpha', J \right\| A^{(1)} \left\| \alpha, J \right\rangle \\ &= \frac{\left\langle \alpha', J \right\| A^{(1)} \left\| \alpha, J \right\rangle}{\left\langle \alpha, J \right\| J^{(1)} \left\| \alpha, J \right\rangle} \left\langle \alpha, J, m_J' \right| J_1^q \left| \alpha, J, m_J \right\rangle . \quad (21.63)\end{aligned}$$

Since $\hat{\mathbf{J}} \cdot \hat{\mathbf{A}}$ is a scalar or irreducible tensor of rank zero, it also follows from the Wigner-Eckart theorem that

$$\left\langle \alpha', J, m_J \right| \hat{\mathbf{J}} \cdot \hat{\mathbf{A}} \left| \alpha, J, m_J \right\rangle = \frac{1}{\sqrt{2J+1}} \left\langle \alpha', J \right\| \mathbf{J} \cdot \mathbf{A} \left\| \alpha, J \right\rangle . \quad (21.64)$$

In other words, these matrix elements are independent of m_J. Moreover, if you expand

$$\hat{\mathbf{J}} \cdot \hat{\mathbf{A}} = -\hat{J}_1^1 \hat{A}_1^{-1} - \hat{J}_1^{-1} \hat{A}_1^1 + \hat{J}_1^0 \hat{A}_1^0 = -\frac{\left(\hat{J}_+ \hat{A}_1^{-1} + \hat{J}_- \hat{A}_1^1 \right)}{\sqrt{2}} + \hat{J}_1^0 \hat{A}_1^0, \quad (21.65)$$

use Eqs. (22.123) and the Wigner-Eckart theorem, you will find that the left-hand side of Eq. (21.64) is proportional $\left\langle \alpha', J \right\| \mathbf{J} \cdot \mathbf{A} \left\| \alpha, J \right\rangle$, with a proportionality constant that appears to depend on J and m_J, but *not* on $\hat{\mathbf{A}}$, α, and α'. However, I know from Eq. (21.64) that the proportionality constant is independent of m_J, so

$$\left\langle \alpha', J, m_J \right| \hat{\mathbf{J}} \cdot \hat{\mathbf{A}} \left| \alpha, J, m_J \right\rangle = C(J) \left\langle \alpha', J \right\| A^{(1)} \left\| \alpha, J \right\rangle , \quad (21.66)$$

where the constant $C(J)$ is independent of α, α', and the properties of the operator $\hat{A}$. As a consequence,

$$\begin{aligned}\frac{\left\langle \alpha', J \right\| A^{(1)} \left\| \alpha, J \right\rangle}{\left\langle \alpha, J \right\| J^{(1)} \left\| \alpha, J \right\rangle} &= \frac{\left\langle \alpha', J, m_J \right| \hat{\mathbf{J}} \cdot \hat{\mathbf{A}} \left| \alpha, J, m_J \right\rangle}{\left\langle \alpha, J, m_J \right| \hat{\mathbf{J}} \cdot \hat{\mathbf{J}} \left| \alpha, J, m_J \right\rangle} \\ &= \frac{\left\langle \alpha', J, m_J \right| \hat{\mathbf{J}} \cdot \hat{\mathbf{A}} \left| \alpha, J, m_J \right\rangle}{\hbar^2 J (J+1)} . \quad (21.67)\end{aligned}$$

By combining Eqs. (21.67), and (21.63), I arrive at Eq. (21.62).

Using Eq. (21.62), I find that the diagonal matrix elements of $\hat{S}_z$ are

$$\begin{aligned}\langle L,S,J,m_J|\,\hat{S}_z\,|L,S,J,m_J\rangle &= \frac{\langle L,S,J,m_J|\,\hat{\mathbf{J}}\cdot\hat{\mathbf{S}}\,|L,S,J,m_J\rangle}{\hbar^2 J\,(J+1)}\langle J,m_J|\,\hat{J}_z\,|J,m_J\rangle \\ &= \hbar m_J\frac{\langle L,S,J,m_J|\,\hat{J}^2+\hat{S}^2-\hat{L}^2\,|L,S,J,m_J\rangle}{2\hbar^2 J\,(J+1)} \\ &= \hbar m_J\frac{J\,(J+1)+S\,(S+1)-L\,(L+1)}{2J\,(J+1)},\end{aligned} \tag{21.68}$$

having used the relationship

$$\mathbf{J}\cdot\mathbf{S}=(\mathbf{L}+\mathbf{S})\cdot\mathbf{S}=\mathbf{L}\cdot\mathbf{S}+\mathbf{S}^2=\frac{J^2-S^2-L^2}{2}+S^2. \tag{21.69}$$

When this result is substituted into Eq. (21.61), I obtain the Zeeman splitting

$$\Delta E_{Jm_J}=\beta_0 B g_J m_J, \tag{21.70}$$

where

$$g_J=1+\frac{J\,(J+1)+S\,(S+1)-L\,(L+1)}{2J\,(J+1)} \tag{21.71}$$

is referred to as the Landé g factor (after Alfred Landé).

In the vector model, the spin-orbit coupling couples $\mathbf{L}$ and $\mathbf{S}$ into $\mathbf{J}$. Both $\mathbf{L}$ and $\mathbf{S}$ precess rapidly about $\mathbf{J}$, as do the magnetic moments. Only the projection of the total magnetic moment $\mathbf{m}_{LS}$ onto $\mathbf{J}$ does not average to zero. In other words, in terms of Fig. 21.5 (b),

$$\begin{aligned}\mathbf{m}_J &\approx -\frac{e}{2m_e}\frac{(\mathbf{L}+2\mathbf{S})\cdot\mathbf{J}}{J}\frac{\mathbf{J}}{J} \\ &= -\frac{e}{2m_e}\left[L\cos\left(\mathbf{L}\cdot\mathbf{J}\right)+2S\cos\left(\mathbf{S}\cdot\mathbf{J}\right)\right]\frac{\mathbf{J}}{J} \\ &= -\frac{e\mathbf{J}}{2m_e}\left[\frac{-S^2+L^2+J^2}{2J^2}+\frac{-L^2+S^2+J^2}{J^2}\right] \\ &= -\frac{e\mathbf{J}}{2m_e}\left[1+\frac{J^2+S^2-L^2}{2J^2}\right] \\ &\Rightarrow -\frac{e\mathbf{J}}{2m_e}\left[1+\frac{J\,(J+1)+S\,(S+1)-L\,(L+1)}{2J\,(J+1)}\right],\end{aligned} \tag{21.72}$$

where I replaced J^2, S^2, L^2 by their quantum equivalents. The interaction energy $\Delta E = -\mathbf{m}_J \cdot \mathbf{B} = -(\mathbf{m}_J)_z B$ agrees with Eq. (21.70) if J_z in Eq. (21.72) is replaced by $\hbar m_J$.

21.3 Summary

I have presented a rather detailed account of the relativistic corrections to the energy levels of hydrogen. In addition, the modifications of the energy levels produced by external magnetic and electric fields were calculated. Special emphasis was placed on the role played by approximate constants of the motion. The vector model of coupling of angular momentum was introduced and used to calculate the Zeeman splitting of atomic energy levels characterized by some total angular momentum quantum number. A calculation of the hyperfine splitting in hydrogen is given in Appendix 2, where the Zeeman splitting, including the hyperfine interaction, is also discussed.

21.4 Appendix A: Radial Integrals for Hydrogen

In this Appendix, I calculate matrix elements of the form

$$\left\langle r^\beta \right\rangle = \int d\mathbf{r}\psi^*_{nlm}(\mathbf{r}) r^\beta \psi_{nlm}(\mathbf{r}) = \int_0^\infty dr \left[u_{nl}(r)\right]^2 r^\beta, \tag{21.73}$$

where the $\psi_{nlm}(\mathbf{r})$ are hydrogenic wave functions and the $u_{nl}(r) = rR_{n\ell}(r)$ are radial wave functions. These matrix elements can be calculated using the generating function for the associated Laguerre polynomials. Each integral is simply the integral of a power with an exponential, so the integrals are fairly simple to carry out. However there is a simpler method for obtaining the results that involves a few "tricks." To start I derive a recursion relation for $\left\langle r^\beta \right\rangle$.

The equation for the radial wave function $u_{n\ell}$ for hydrogen is

$$\frac{d^2 u_{n\ell}}{dr^2} + \left(-\frac{1}{n^2 a_0^2} + \frac{2}{a_0 r} - \frac{\ell(\ell+1)}{r^2} \right) u_{n\ell} = 0, \tag{21.74}$$

where $a_0 = 4\pi\epsilon_0\hbar^2/m_e e^2$ is the Bohr radius. I start from

$$\int_0^\infty dr \frac{d^2 u_{n\ell}(r)}{dr^2} r^\beta u_{nl}(r) = \int_0^\infty dr \left(\frac{1}{n^2 a_0^2} - \frac{2}{a_0 r} + \frac{\ell(\ell+1)}{r^2} \right) r^\beta \left[u_{nl}(r)\right]^2$$

$$= \frac{1}{n^2 a_0^2} \left\langle r^\beta \right\rangle - \frac{2}{a_0} \left\langle r^{\beta-1} \right\rangle + \ell(\ell+1) \left\langle r^{\beta-2} \right\rangle. \tag{21.75}$$

As $r \to 0$, $u_{n\ell}(r) d^2 u_{n\ell}(r)/dr^2 \sim r^{2\ell}$ for $\ell \neq 0$ and as $\sim r$ for $\ell = 0$; as a consequence, the integral on the left-hand side of Eq. (21.75) is convergent only if

$$\begin{cases} \beta > -2 & \ell = 0 \\ \beta > -2\ell - 1 & \ell > 0 \end{cases}. \tag{21.76}$$

Integrating the left-hand side of this equation by parts and using the fact that the endpoint contributions vanish if condition (21.76) is satisfied, I find

$$\int_0^\infty dr \frac{d^2 u_{n\ell}(r)}{dr^2} r^\beta u_{nl}(r) = -\int_0^\infty dr \frac{du_{n\ell}(r)}{dr} \left(\beta r^{\beta-1} u_{nl}(r) + r^\beta \frac{du_{n\ell}(r)}{dr} \right). \tag{21.77}$$

Integration by parts can be used on the first term on the right-hand side of this equation to transform it into

$$\begin{aligned} &-\int_0^\infty dr \frac{du_{n\ell}(r)}{dr} \beta r^{\beta-1} u_{nl}(r) \\ &= \int_0^\infty dr u_{n\ell}(r) \left(\beta (\beta - 1) r^{\beta-2} u_{nl}(r) + \beta r^{\beta-1} \frac{du_{n\ell}(r)}{dr} \right), \end{aligned} \tag{21.78}$$

which implies that

$$-\int_0^\infty dr \frac{du_{n\ell}(r)}{dr} \beta r^{\beta-1} u_{nl}(r) = \frac{1}{2} \beta (\beta - 1) \left\langle r^{\beta-2} \right\rangle. \tag{21.79}$$

I integrate the second term in Eq. (21.77) by parts and use Eq. (21.74) to obtain

$$\begin{aligned} -\int_0^\infty dr\, r^\beta \left[\frac{du_{n\ell}(r)}{dr} \right]^2 &= 2 \int_0^\infty dr \frac{r^{\beta+1}}{\beta+1} \frac{du_{n\ell}(r)}{dr} \frac{d^2 u_{n\ell}(r)}{dr^2} \\ &= 2 \int_0^\infty dr \frac{r^{\beta+1}}{\beta+1} \frac{du_{n\ell}(r)}{dr} \left(\frac{1}{n^2 a_0^2} - \frac{2}{a_0 r} + \frac{\ell(\ell+1)}{r^2} \right) u_{n\ell}(r). \end{aligned} \tag{21.80}$$

I can now use Eq. (21.79) to rewrite this equation as

$$-\int_0^\infty dr\, r^\beta \left[\frac{du_{n\ell}(r)}{dr} \right]^2 = \frac{1}{\beta+1} \left[\begin{array}{c} -\frac{(\beta+1)\left\langle r^\beta \right\rangle}{n^2 a_0^2} + \frac{2\beta \left\langle r^{\beta-1} \right\rangle}{a_0} \\ -\ell(\ell+1)(\beta-1)\left\langle r^{\beta-2} \right\rangle \end{array} \right]. \tag{21.81}$$

Combining Eqs. (21.75)–(21.81), I arrive at the recursion relation

$$\frac{1}{n^2 a_0^2} \left\langle r^\beta \right\rangle - \frac{(2\beta+1)}{(\beta+1) a_0} \left\langle r^{\beta-1} \right\rangle + \frac{\beta}{4} \frac{\left[(2\ell+1)^2 - \beta^2 \right]}{(\beta+1)} \left\langle r^{\beta-2} \right\rangle = 0. \tag{21.82}$$

Equation (21.82) is valid only when conditions (21.76) hold, consistent with the fact that $\langle r^{\beta-2}\rangle$ is non-divergent only for $\beta > -2\ell - 1$ (a result that is also derived easily using the fact that $u_{n\ell}(r) \sim r^{\ell+1}$ as $r \to 0$).

Since the wave functions are normalized, it follows immediately that $\langle r^0\rangle = 1$. By choosing $\beta = 0$ in Eq. (21.82), I then find

$$\langle r^{-1}\rangle = \frac{1}{n^2 a_0}, \tag{21.83}$$

a result that also follows from the Virial theorem. Using this result and the fact that $\langle r^0\rangle = 1$, you can take $\beta = 1$ in Eq. (21.82) to obtain

$$\frac{1}{n^2 a_0^2}\langle r\rangle - \frac{3}{2a_0} + \frac{1}{4}\frac{\left[(2\ell+1)^2 - 1\right]}{2}\frac{1}{n^2 a_0} = 0, \tag{21.84}$$

or

$$\langle r\rangle = \frac{a_0}{2}\left[3n^2 - \ell(\ell+1)\right]. \tag{21.85}$$

All higher powers of $\langle r^q\rangle$ can be generated in the same manner by taking $\beta = q$ in Eq. (21.82); for example, with $q = 2$,

$$\begin{aligned}\langle r^2\rangle &= \frac{5n^2 a_0}{3}\langle r\rangle - \frac{n^2 a_0^2}{2}\frac{\left[(2\ell+1)^2 - 4\right]}{3}\langle r^0\rangle \\ &= \frac{n^2 a_0^2}{2}\left[5n^2 + 1 - 3\ell(\ell+1)\right].\end{aligned} \tag{21.86}$$

To get values of $\langle r^{-q}\rangle$ for $q \geq 2$, I first set $\beta = -1$ in Eq. (21.82) and find

$$\langle r^{-2}\rangle = \ell(\ell+1)\, a_0 \langle r^{-3}\rangle. \tag{21.87}$$

By setting $\beta = -q$ in Eq. (21.82) with $q \geq 2$, you can get expressions for $\langle r^{-q-2}\rangle$ in terms of $\langle r^{-q-1}\rangle$ and $\langle r^{-q}\rangle$. Thus, to obtain *all* values of $\langle r^{-q}\rangle$ for $q \geq 2$, it is sufficient to calculate since $\langle r^{-2}\rangle$ and $\langle r^{-3}\rangle$, since all the others can be expressed in terms of these quantities; moreover I need only calculate $\langle r^{-2}\rangle$, since $\langle r^{-3}\rangle$ can then be determined from Eq. (21.87).

The value of $\langle r^{-2}\rangle$ can be found using a "trick." Suppose the hydrogen atom is subjected to a perturbation $H' = \epsilon/r^2$. The first order energy change produced by this perturbation is

$$\delta E_{n\ell} = \int d\mathbf{r}\psi^*_{nlm}(\mathbf{r})\frac{\epsilon}{r^2}\psi_{nlm}(\mathbf{r}) = \epsilon\langle r^{-2}\rangle. \tag{21.88}$$

However, the total Hamiltonian is exactly solvable in this case since the $\hbar^2\ell(\ell+1)/2m_er^2$ term in the effective potential is replaced by

$$\frac{\hbar^2}{2m_er^2}\left[\ell(\ell+1)-\frac{2m_e\epsilon}{\hbar^2}\right]\equiv\frac{\hbar^2}{2m_e}\ell'(\ell'+1)/r^2, \tag{21.89}$$

where

$$\ell'(\ell'+1)=\ell(\ell+1)-\frac{2m_e\epsilon}{\hbar^2} \tag{21.90}$$

and ℓ' is *not* an integer. To have an acceptable solution of the Schrödinger equation, it is necessary that $q=n-\ell'-1$ be an integer (note that n is no longer an integer, except in the limit that $\epsilon\to 0$). In that case the energy is

$$E_{n\ell}=-\frac{e^2}{4\pi\epsilon_0}\frac{1}{2a_0n^2}=-\frac{e^2}{4\pi\epsilon_0}\frac{1}{2a_0(q+\ell'+1)^2}. \tag{21.91}$$

Expanding this result to first order in ϵ, I find

$$\begin{aligned}\delta E_{n\ell}&\approx\left.\frac{\partial E_{n\ell}}{\partial\ell'}\right|_{\epsilon=0}\left.\frac{\partial\ell'}{\partial\epsilon}\right|_{\epsilon=0}\epsilon\\&=-\frac{e^2}{4\pi\epsilon_0}\left(\frac{-1}{a_0n^3}\right)\frac{m_e}{\hbar^2(\ell+1/2)}\epsilon.\end{aligned} \tag{21.92}$$

Comparing this result with Eq. (21.88), I can identify

$$\left\langle r^{-2}\right\rangle=\frac{1}{\left(\ell+\frac{1}{2}\right)n^3a_0^2} \tag{21.93}$$

and, using Eq. (21.87), that

$$\left\langle r^{-3}\right\rangle=\frac{1}{\ell\left(\ell+\frac{1}{2}\right)(\ell+1)n^3a_0^3};\quad \ell\neq 0. \tag{21.94}$$

Remember that $\left\langle r^{\beta}\right\rangle$ is defined only if $\beta>-2\ell-3$ or $\ell>-(\beta+3)/2$.

21.5 Appendix B: Hyperfine Interactions

All baryons have half-integral nuclear spin. Both the neutron and proton have nuclear spin $I=1/2$. They also have associated magnetic moments given by

$$\hat{\mathbf{m}}_p = \frac{eg_p\mathbf{I}_p}{2m_p}; \tag{21.95a}$$

$$\hat{\mathbf{m}}_n = \frac{eg_n\mathbf{I}_n}{2m_n}, \tag{21.95b}$$

where m_p is the proton mass, m_n is the neutron mass, and $g_{p,n}$ is a *nuclear g-factor*. For the proton $g_p \approx 5.586$ and for the neutron $g_n \approx -3.826$. Note that the neutron has a magnetic moment even though it is neutral. The reason for this is that the neutron, like the proton, is a composite particle, composed of three quarks bound by gluons. Whereas the magnetic moment of the electron arises naturally from the Dirac equation, there is no analogous equation that gives the nuclear magnetic moments. Theories based on *quantum chromodynamics* attempt to predict the values of nuclear magnetic moments by considering the baryons to be composed of quarks and gluons.

For hydrogen, the nucleus consists of a single proton, whose magnetic moment interacts with the magnetic field produced by the electron. There are two contributions to the magnetic field produced by the electron, one from its orbital motion and one from its intrinsic spin. There is no Thomas precession in calculating the hyperfine interaction as there was for the fine structure calculation since the proton is in an inertial frame whereas the electron was in an accelerating frame. Classically, the electron can be modeled as a point magnetic dipole having mass m_e, magnetic moment $\mathbf{m}_s$, and charge $-e$. If this point dipole is located at position $\mathbf{r}$ and has orbital angular momentum $\mathbf{L}$ about the origin, it produces a field at the origin given by

$$\mathbf{B}(0) = \frac{\mu_0}{4\pi}\left[\frac{8\pi}{3}\mathbf{m}_s\delta(\mathbf{r}) + \frac{3\mathbf{u}_r(\mathbf{u}_r \cdot \mathbf{m}_s) - \mathbf{m}_s}{r^3} + \frac{2\mathbf{m}_\ell}{r^3}\right], \tag{21.96}$$

where $\mathbf{u}_r$ is a unit vector in the radial direction,

$$\mathbf{m}_\ell = -\frac{e\mathbf{L}}{2m_e} \tag{21.97}$$

is the orbital magnetic moment, and μ_0 is the vacuum permeability. The first two terms in Eq. (21.96) give the field associated with the point magnetic moment of the electron—the delta function term is necessary for the consistency of Maxwell's equations and is related to the fact that the field at the center of a "point" dipole diverges. The third term is the field arising from the orbital motion of the electron.

The proton is modeled classically as a point dipole at the origin having magnetic moment $\mathbf{m}_p$. The interaction energy of the proton's magnetic moment with the magnetic field of the electron is

$$H = -\mathbf{m}_p \cdot \mathbf{B}(0). \tag{21.98}$$

To go over to a quantum description, I replace $\mathbf{m}_p$ by Eq. (21.95a), $\mathbf{m}_s$ by

$$\hat{\mathbf{m}}_s = -\frac{e\mathbf{S}}{m_e}, \tag{21.99}$$

and all angular momenta by operators (although I leave off the carets) to arrive at the *hyperfine interaction* contribution to the Hamiltonian,

$$\begin{aligned}\hat{H}_{hf} &= -\hat{\mathbf{m}}_p \cdot \mathbf{B}(0) \\ &= \frac{\mu_0 e^2 g_p}{8\pi m_e m_p}\left[\frac{8\pi\delta(\mathbf{r})}{3}\mathbf{S}\cdot\mathbf{I} + \frac{3\,(\mathbf{u}_r\cdot\mathbf{S})\,(\mathbf{u}_r\cdot\mathbf{I}) - \mathbf{S}\cdot\mathbf{I}}{r^3} + \frac{\mathbf{L}\cdot\mathbf{I}}{r^3}\right].\end{aligned} \tag{21.100}$$

If I define the total angular momentum as

$$\mathbf{F} = \mathbf{J} + \mathbf{I} = \mathbf{L} + \mathbf{S} + \mathbf{I}, \tag{21.101}$$

then it is convenient to use basis kets labeled by $|n\ell jfm_f\rangle$ where the electron spin quantum number $S = 1/2$ and the proton spin quantum number $I = 1/2$ have been suppressed in the label. In perturbation theory, the change in the energy levels produced by the hyperfine interaction is given by

$$\Delta E_{n\ell jf} = \langle n\ell jfm_f | \hat{H}_{hf} | n\ell jfm_f \rangle . \tag{21.102}$$

The ket $|n\ell jfm_f\rangle$ is the eigenket in the absence of the hyperfine interaction. The constants of the motion are $L^2, S^2, J^2, I^2, F^2, F_z$.

21.5.1 Hyperfine Splitting

I first derive the contribution to $\Delta E_{n\ell jf}$ from the first term in Eq. (21.100), the so-called *contact interaction*. I will then outline a method for calculating the other terms. Only $\ell = 0$ states contribute to the contact interaction since they are the only states for which the eigenfunctions do not vanish at the origin. If $\ell = 0$, then $j = 1/2$, and f can be equal to 0 or 1 Thus, the contact term gives rise to a hyperfine splitting of S states. To evaluate this term I need to calculate

$$\begin{aligned}\Delta E_{nf}^{\text{contact}} &= \langle n, \ell = 0, j = 1/2, f, m_f | \hat{H}_{hf} | n, \ell = 0, j = 1/2, f, m_f \rangle \\ &= \frac{\mu_0 e^2 g_p}{3 m_e m_p} \langle n, \ell = 0, j = 1/2, f, m_f | \delta(\mathbf{r})\mathbf{S}\cdot\mathbf{I} | n, \ell = 0, j = 1/2, f, m_f \rangle .\end{aligned} \tag{21.103}$$

However, since $\ell = 0$, it follows that $\mathbf{F} = \mathbf{S} + \mathbf{I}$, implying that

$$\mathbf{S} \cdot \mathbf{I} = \frac{F^2 - I^2 - S^2}{2};$$

$$\langle n,f,m_f | \delta(\mathbf{r}) \mathbf{S} \cdot \mathbf{I} | n,f,m_f \rangle = \hbar^2 \frac{f(f+1) - 3/2}{2} \times \langle n,f,m_f | \delta(\mathbf{r}) | n,f,m_f \rangle. \tag{21.104}$$

The hyperfine energy shifts of the $f = 0, 1$ levels are given by

$$\Delta E_{f=1}^{\text{contact}} = \beta \hbar^2 \frac{f(f+1) - 3/2}{2} = \frac{\beta \hbar^2}{4}; \tag{21.105a}$$

$$\Delta E_{f=0}^{\text{contact}} = -\frac{3\beta \hbar^2}{4}, \tag{21.105b}$$

where

$$\beta = \frac{e^2 \mu_0 g_p}{3 m_p m_e} \langle n,f,m_f | \delta(\mathbf{r}) | n,f,m_f \rangle. \tag{21.106}$$

It is straightforward to show that

$$\langle n,\ell,j,f,m_f | g(\mathbf{r}) | n,\ell,j,f,m_f \rangle = \langle n,\ell,m_\ell | g(\mathbf{r}) | n,\ell,m_\ell \rangle \tag{21.107}$$

for any function $g(\mathbf{r})$ that does not depend on electron or nuclear spin. Therefore,

$$\begin{aligned}
\beta &= \frac{e^2 \mu_0 g_p}{3 m_p m_e} \langle n00 | \delta(\mathbf{r}) | n00 \rangle \\
&= \frac{e^2 \mu_0 g_p}{3 m_p m_e} \int d\mathbf{r} \left| \psi_{n00}(r) \right|^2 \delta(\mathbf{r}) \\
&= \frac{e^2 \mu_0 g_p}{3 m_p m_e} \int d\mathbf{r} \left| R_{n0}(r) \right|^2 \left| Y_{00}(\theta,\phi) \right|^2 \delta(\mathbf{r}) \\
&= \frac{e^2 \mu_0 g_p}{3 m_p m_e} \frac{\left| R_{n0}(0) \right|^2}{4\pi},
\end{aligned} \tag{21.108}$$

where $R_{n0}(0)$ is the value of the radial wave function at the origin.[5] The hyperfine splitting of any S ($\ell = 0$) state of hydrogen is then

---

[5] In spherical coordinates

$$\int d\mathbf{r} \delta(\mathbf{r}) = \int_0^{2\pi} d\phi \int_{-1}^{1} d(\cos\theta) \int_0^{\infty} dr r^2 \frac{\delta(r)\delta(\cos\theta)\delta(\phi)}{r^2}$$

$$\delta E^{\text{contact}}_{n,\ell=0} = \Delta E^{\text{contact}}_{n,\ell=0,f=1} - \Delta E^{\text{contact}}_{n,\ell=0,f=0} = \frac{\mu_0 e^2 g_p \hbar^2}{3m_e m_p} \frac{|R_{n0}(0)|^2}{4\pi}. \tag{21.109}$$

The value of $|R_{n0}(0)|^2$ can be obtained from Eq. (10.114) as

$$|R_{n0}(0)|^2 = \frac{1}{a_0^3}\left(\frac{4}{n^5}\right)\left[L^1_{n-1}(0)\right]^2 = \frac{1}{a_0^3}\left(\frac{4}{n^3}\right), \tag{21.110}$$

where $a_0 = \hbar/(\alpha_{FS} m_e c)$ is the Bohr radius and $L^1_{n-1}(0) = n$ is a Laguerre polynomial. Combining Eqs. (21.109) and (21.110) and using the fact that $\mu_0\epsilon_0 = 1/c^2$, I find for the ground state of the hydrogen atom that

$$\delta E_{1,\ell=0} = \frac{\mu_0 e^2 g_p}{3m_p m_e} \frac{\hbar^2}{\pi a_0^3}$$

$$= \left|E^{(0)}_1\right| \frac{8 g_p \alpha^2_{FS} m_e}{3m_p} = 5.88 \times 10^{-6} \text{ eV}; \tag{21.111a}$$

$$\frac{\delta E_{1,\ell=0}}{h} = 1.42 \text{ GHz}; \tag{21.111b}$$

$$\lambda = c/f = 21.1 \text{ cm}. \tag{21.111c}$$

This corresponds to the famous 21 cm astrophysical line of radio astronomy.

It turns out that the spin–spin interaction [second term in Eq. (21.100)] averages to zero for $\ell = 0$ states; moreover, the contribution to the hyperfine interaction from the orbital motion of the electron also vanishes for $\ell = 0$ states. As a consequence, the *only* contribution to the hyperfine splitting of $\ell = 0$ states arises from the contact interaction term. This is why the "contact" label was suppressed in Eqs. (21.111).

For $\ell \neq 0$, the calculation is more complicated and is given at the end of this Appendix. The hyperfine shift for $\ell \neq 0$ is

$$\Delta E_{n\ell jf} = \left|E^{(0)}_n\right| \frac{g_p \alpha^2_{FS} m_e}{n m_p} \frac{f(f+1) - j(j+1) - 3/4}{j(j+1)(2\ell+1)} \tag{21.112}$$

and the hyperfine splitting is

is somewhat ambiguous owing to the endpoint contributions to the integrals. I use the convention that, for any spherically symmetric function $f(r)$,

$$\int d\mathbf{r}\delta(\mathbf{r}) f(r) = f(0).$$

This will insure that $\int d\mathbf{r}\delta(\mathbf{r}) = 1$.

$$\delta E_{n\ell j} = \Delta E_{n\ell jf=j+1/2} - \Delta E_{n\ell jf=j-1/2} = \left|E_n^{(0)}\right| \frac{g_p \alpha_{FS}^2 m_e}{n m_p} \frac{2j+1}{j(j+1)(2\ell+1)}. \tag{21.113}$$

Equations (21.112) and (21.113) are also valid for $\ell = 0$ states [see Eq. (21.111a)], even though they are derived assuming $\ell \neq 0$.

For a spin 1/2 atom such as hydrogen, the only contribution to the hyperfine interaction is associated with the magnetic dipole moment of the proton. Nuclei having spin $I \geq 1$ possess an electric quadrupole moment as well, which also gives rise to a hyperfine interaction. In this case, the additional hyperfine interaction results from the interaction of the nuclear quadrupole with the electric field gradient at the nucleus produced by the atomic electrons.

21.5.2 Zeeman Splitting

In the presence of an external magnetic field, there is an additional interaction Hamiltonian that is still given approximately by Eq. (21.40) (the interaction of the *nuclear* magnetic moment with the field can be neglected since it is about m_e/m_p times smaller than the electronic contribution). Within a given fine structure manifold j there are two hyperfine levels having $f = j \pm 1/2$. The calculation of the effect of a magnetic field on these levels is similar to that of a magnetic field on the two fine structure levels $j = \ell \pm 1/2$ within a given ℓ manifold that was considered in Sect. 21.1.9. In other words, in weak magnetic fields, m_f is approximately a good quantum number and each hyperfine level splits into $(2f+1)$ components. With increasing field strength there is a transition to the Paschen-Back region in which m_j and m_I are approximately good quantum numbers. This region actually corresponds to the *weak field regime* of the Zeeman splitting of a state characterized by quantum number j—that is, the components coalesce into a set of $(2j+1)$ distinct energy levels, each of which is doubly-degenerate, owing to the nuclear spin. The $(2j+1)$ energy levels are those of the weak field Zeeman effect for this $j = 1/2$ state. I concentrate mainly on the weak field regime, since the strong-field regime reproduces the weak field results that were obtained neglecting hyperfine structure.

I first consider S states, for which the hyperfine and Zeeman interaction Hamiltonian can be written as

$$\hat{H}' = \frac{\mu_0 e^2 g_p}{3 m_e m_p} \delta(\mathbf{r}) \mathbf{S} \cdot \mathbf{I} + \frac{\beta_0 B}{\hbar} (J_z + S_z). \tag{21.114}$$

Using the $\left|m_I m_j\right\rangle$ basis and writing

$$\mathbf{S} \cdot \mathbf{I} = \mathbf{J} \cdot \mathbf{I} = \frac{J_+ I_- + J_- I_+}{2} + J_z I_z \tag{21.115}$$

($\mathbf{S} = \mathbf{J}$ since $\ell = 0$), I calculate matrix elements of $\hat{H}'$ for $n, \ell = 0$ states as

$$\frac{\underline{\mathrm{H}}_n'}{h} = \begin{pmatrix} \frac{A_n}{4} - bB(T) & 0 & 0 & 0 \\ 0 & -\frac{A_n}{4} - bB(T) & \frac{A_n}{2} & 0 \\ 0 & \frac{A_n}{2} & -\frac{A_n}{4} + bB(T) & 0 \\ 0 & 0 & 0 & \frac{A_n}{4} + bB(T) \end{pmatrix}, \tag{21.116}$$

where the order is

$$\left|m_I, m_j\right\rangle = \left|-\frac{1}{2}, -\frac{1}{2}\right\rangle, \left|-\frac{1}{2}, \frac{1}{2}\right\rangle, \left|\frac{1}{2}, -\frac{1}{2}\right\rangle, \left|\frac{1}{2}, \frac{1}{2}\right\rangle, \tag{21.117}$$

$A_n = \left(1.42/n^3\right)$ GHZ, $b = 14$ GHZ, and $B\,(T)$ is the magnetic induction in units of Tesla. The eigenkets and eigenergies are

$$|E_1\rangle = \left|-\frac{1}{2}, -\frac{1}{2}\right\rangle; \quad E_1 = A_n/4 - bB(T); \tag{21.118a}$$

$$|E_2\rangle = a_{22}\left|-\frac{1}{2}, \frac{1}{2}\right\rangle + a_{23}\left|\frac{1}{2}, -\frac{1}{2}\right\rangle;$$

$$E_2 = -A_n/4 - \sqrt{A_n^2 + 4b^2B(T)^2}/2; \tag{21.118b}$$

$$|E_3\rangle = a_{32}\left|-\frac{1}{2}, \frac{1}{2}\right\rangle + a_{33}\left|\frac{1}{2}, -\frac{1}{2}\right\rangle;$$

$$E_3 = -A_n/4 + \sqrt{A_n^2 + 4b^2B(T)^2}/2; \tag{21.118c}$$

$$|E_4\rangle = \left|\frac{1}{2}, \frac{1}{2}\right\rangle; \quad E_4 = A_n/4 + bB(T), \tag{21.118d}$$

where

$$a_{22} = NA_n; \tag{21.119a}$$

$$a_{23} = -N\left[-2bB(T) + \sqrt{A_n^2 + 4\left[bB(T)\right]^2}\right]; \tag{21.119b}$$

$$a_{32} = -N\left[2bB(T) - \sqrt{A_n^2 + 4\left[bB(T)\right]^2}\right]; \tag{21.119c}$$

$$a_{33} = NA_n; \tag{21.119d}$$

$$N = 1/\left\{\left(2bB(T) - \sqrt{A_n^2 + 4\left[bB(T)\right]^2}\right)^2 + A_n^2\right\}^{1/2}. \tag{21.119e}$$

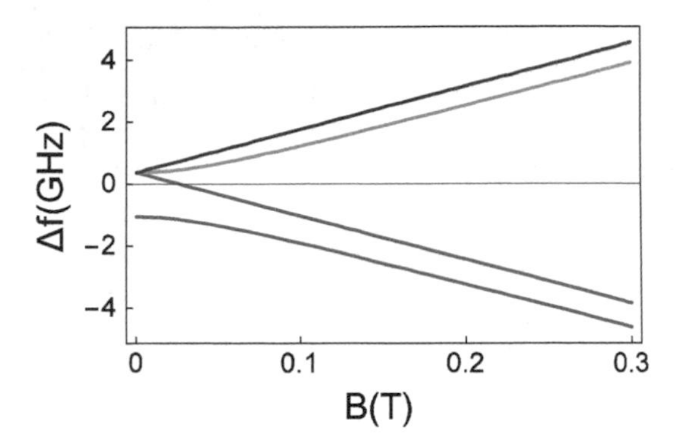

Fig. 21.6 Frequency shifts of the ground state of hydrogen resulting from the hyperfine interaction and the interaction with an external magnetic field

Each S-state manifold consists of four non-degenerate levels (one $f = 0$ level and three $f = 1$ levels) in weak magnetic fields. With increasing field strength, these levels approximately coalesce into two pairs of levels (the $m_j = \pm 1/2$ levels of the $j = 1/2$ state). For $B(T) \ll A_n/b$, the eigenkets and eigenenenergies are those associated with the $|fm_f\rangle$ basis. For $B(T) \gg A_n/b$, the eigenkets and eigenvalues are those associated with the $|jm_j\rangle$ basis, with an (almost) two-fold degeneracy for each energy. The eigenfrequencies for the ground state sublevels as a function of magnetic induction are plotted in Fig. 21.6.

There is an important difference from the Zeeman effect that we encountered for fine structure levels. Since the hyperfine splitting is 10 to a 1000 times smaller than the fine structure splitting, the crossover to the Paschen-Back region occurs at much lower field strengths for hyperfine levels. In fact, the ability to tune the energy of the hyperfine levels using magnetic field strengths that are easily accessible in the laboratory plays an important role in modern atomic spectroscopy.

In weak fields for other than S states, the Zeeman shift of the energy levels of a hydrogenic state characterized by quantum numbers $n\ell jfm_f$ is given approximately by

$$\begin{aligned}\Delta E_{\ell jfm_f} &= (\beta_0 B/\hbar)\left\langle n\ell jfm_f\right|\hat{L}_z + 2\hat{S}_z\left|n\ell jfm_f\right\rangle/\hbar \\ &= (\beta_0 B/\hbar)\sum_{m_j,m_j',m_I,m_I'}\begin{bmatrix} j & 1/2 & f \\ m_j & m_I & m_f\end{bmatrix}\begin{bmatrix} j & 1/2 & f \\ m_j' & m_I' & m_f\end{bmatrix} \\ &\quad\times\left\langle n\ell jm_j\right|\hat{L}_z + 2\hat{S}_z\left|n\ell jm_j\right\rangle\delta_{m_I,m_I'}\delta'_{m_j,m_j}.\end{aligned} \tag{21.120}$$

The matrix element $\left\langle n\ell jm_j\right|\hat{L}_z+2\hat{S}_z\left|n\ell jm_j\right\rangle$ is the one we encountered in the Zeeman effect without hyperfine structure—it is equal to $\hbar g_j m_j$; moreover, m_j must equal $(m_f - m_I)$, owing to the properties of the Clebsch-Gordan coefficients. Therefore

$$\Delta E_{\ell jfm_f} = \beta_0 B g_j \sum_{m_I=-1/2}^{1/2} \left(m_f - m_I\right) \begin{bmatrix} j & 1/2 & f \\ m_f - m_I & m_I & m_f \end{bmatrix}^2 . \tag{21.121}$$

The sum can be carried out to give

$$\Delta E_{\ell jfm_f} = \beta_0 B g_f m_f$$

where the Landé g_f factor is

$$g_f = g_j \frac{f(f+1) + j(j+1) - 3/4}{2f(f+1)}, \tag{21.122}$$

with g_j given by Eq. (21.71) with $J = j$, $L = \ell$, and $S = 1/2$. For a multi-electron, multi-nucleon atom characterized by angular momentum quantum numbers L, S, J, I, and F, the corresponding result is

$$g_f = g_J \frac{F(F+1) + J(J+1) - I(I+1)}{2F(F+1)}, \tag{21.123}$$

with g_J given by Eq. (21.71). The same result can be obtained using a vector model in which the vector $\mathbf{F} = \mathbf{J} + \mathbf{I}$ precesses about the magnetic field direction.

In the strong field vector model, $\mathbf{J}$ and $\mathbf{I}$ and their magnetic moments precess separately about the magnetic field direction; however, the contribution to level shifts resulting from the nuclear spin magnetic moment is m_e/m_p times smaller than that of the electronic magnetic moment. This is the reason the strong field result approximately reproduces the weak field results neglecting hyperfine interactions.

21.5.3 Contribution to the Hyperfine Splitting from Non-contact Interaction Terms

I now give some details of the calculation of the hyperfine level shifts from the second and third terms in Eq. (21.100). The hyperfine splitting from these non-contact interaction terms can be written as

$$\Delta E_{n\ell jf} = \frac{\mu_0 e^2 g_p}{8\pi m_e m_p} \left(A_{n\ell jf} + 3B_{n\ell jf}\right), \tag{21.124}$$

where

$$A_{n\ell jf} = \langle n, \ell, j, f, 0| \frac{\left(\hat{\mathbf{L}} - \hat{\mathbf{S}}\right) \cdot \hat{\mathbf{I}}}{r^3} |n, \ell, j, f, 0\rangle \tag{21.125}$$

and

$$B_{n\ell jf} = \langle n, \ell, j, f, 0| \frac{\left(\mathbf{u}_r \cdot \hat{\mathbf{S}}\right) \left(\mathbf{u}_r \cdot \hat{\mathbf{I}}\right)}{r^3} |n, \ell, j, f, 0\rangle . \tag{21.126}$$

I have set $m_f = 0$, since the result is independent of m_f. I now calculate $A_{n\ell jf}$ and $B_{n\ell jf}$ separately.

21.5.3.1 $\mathbf{A}_{n\ell jf}$

If Eq. (21.62) is written as a vector equation involving $\hat{\mathbf{A}}$ and $\hat{\mathbf{J}}$ with $\hat{\mathbf{A}} = \hat{\mathbf{L}}$, it follows that[6]

$$\begin{aligned}\langle n, \ell, j, f, 0| \hat{\mathbf{L}} \cdot \hat{\mathbf{I}} |n, \ell, j, f, 0\rangle &= \frac{\langle n, \ell, j, f, 0| \hat{\mathbf{L}} \cdot \hat{\mathbf{J}} |n, \ell, j, f, 0\rangle}{\hbar^2 j\,(j+1)} \\ &\times \langle n, \ell, j, f, 0| \hat{\mathbf{I}} \cdot \hat{\mathbf{J}} |n, \ell, j, f, 0\rangle \\ &= \hbar^2 \left[\frac{j(j+1) + \ell(\ell+1) - 3/4}{2j\,(j+1)}\right] \left[\frac{f(f+1) - j(j+1) - 3/4}{2}\right],\end{aligned} \tag{21.127}$$

where I used the relations $\mathbf{S} = (\mathbf{J} - \mathbf{L})$ and $\mathbf{F} = (\mathbf{J} + \mathbf{I})$. Similarly,

$$\begin{aligned}\langle n, \ell, j, f, 0| \hat{\mathbf{S}} \cdot \hat{\mathbf{I}} |n, \ell, j, f, 0\rangle &= \frac{\langle n, \ell, j, f, 0| \hat{\mathbf{S}} \cdot \hat{\mathbf{J}} |n, \ell, j, f, 0\rangle}{\hbar^2 j\,(j+1)} \\ &\times \langle n, \ell, j, f, 0| \hat{\mathbf{I}} \cdot \hat{\mathbf{J}} |n, \ell, j, f, 0\rangle \\ &= \hbar^2 \left[\frac{j(j+1) - \ell(\ell+1) + 3/4}{2j\,(j+1)}\right] \left[\frac{f(f+1) - j(j+1) - 3/4}{2}\right].\end{aligned} \tag{21.128}$$

Combining Eqs. (21.125), (21.127), and (21.128), I obtain

$$A_{n\ell jf} = \hbar^2 \left[\frac{\ell(\ell+1) - 3/4}{j\,(j+1)}\right] \left[\frac{f(f+1) - j(j+1) - 3/4}{2}\right] \left\langle r^{-3} \right\rangle, \tag{21.129}$$

where $\left\langle r^{-3} \right\rangle$ is given by Eq. (21.94).

[6] Equations (21.127) and (21.128) can also be obtained by expanding the kets in the uncoupled basis and directly evaluating the matrix elements (see the *Supplementary Material* at the book's web site).

21.5.3.2 $B_{n\ell jf}$

To evaluate Eq. (21.126), I use the fact that

$$\mathbf{u}_r = \sin\theta\cos\phi\mathbf{u}_x + \sin\theta\sin\phi\mathbf{u}_y + \cos\theta\mathbf{u}_z \tag{21.130}$$

to write

$$\frac{(\mathbf{u}_r\cdot\mathbf{S})\ (\mathbf{u}_r\cdot\mathbf{I})}{r^3} = \sum_{q,q'=-1}^{1}(-1)^{q+q'}\frac{u_q I_{-q} u_{-q'} S_{q'}}{r^3}, \tag{21.131}$$

where

$$u_{\pm 1} = \mp\frac{(u_r)_x \pm i\,(u_r)_y}{\sqrt{2}} = \mp\frac{\sin\theta\, e^{\pm i\phi}}{\sqrt{2}}; \qquad u_0 = \cos\theta. \tag{21.132}$$

Then, to evaluate the matrix elements needed in Eq. (21.126), I expand the basis kets as

$$\begin{aligned}|n,\ell,j,f,0\rangle = &\sum_{m_I,m_s=-1/2}^{1/2}\begin{bmatrix}\ell & 1/2 & j\\ -m_I-m_s & m_s & -m_I\end{bmatrix}\begin{bmatrix}j & 1/2 & f\\ -m_I & m_I & 0\end{bmatrix}\\ &\times|n,\ell,m_\ell=-m_I-m_s\rangle\,|Im_I\rangle\,|Sm_s\rangle\end{aligned} \tag{21.133}$$

with a corresponding equation for the bra, and use the relationships

$$\hat{G}_{\pm 1}\left|g,m_g\right\rangle = \hbar\sqrt{\left(g\mp m_g\right)\left(g\pm m_g+1\right)}\left|g,m_g\pm 1\right\rangle; \tag{21.134a}$$

$$\hat{G}_0\left|g,m_g\right\rangle = m_g\hbar\left|g,m_g\right\rangle, \tag{21.134b}$$

where the $\hat{G}_{\pm 1}$,$\hat{G}_0$ ($\hat{G} = \hat{S},\hat{I}$) are components of an irreducible tensor operator of rank 1. It then follows that

$$\begin{aligned}B_{n\ell jf} = \left\langle r^{-3}\right\rangle &\sum_{m_I,m_s=-1/2}^{1/2}\sum_{q,q'=-1}^{1}(-1)^{q+q'}\begin{bmatrix}\ell & 1/2 & j\\ -m_I-m_s & m_s & -m_I\end{bmatrix}\\ &\times\begin{bmatrix}j & 1/2 & f\\ -m_I & m_I & 0\end{bmatrix}\begin{bmatrix}\ell & 1/2 & j\\ -m_I'-m_s' & m_s' & -m_I'\end{bmatrix}\begin{bmatrix}j & 1/2 & f\\ -m_I' & m_I' & 0\end{bmatrix}\\ &\times S_{q'm_s'm_s}I_{qm_I'm_I}B_\ell(q,q',-m_I'-m_s',-m_I-m_s),\end{aligned} \tag{21.135}$$

where

$$B_\ell(q,q',m_\ell',m_\ell) = \left\langle\ell,m_\ell'\right|u_q u_{-q'}\left|\ell,m_\ell\right\rangle \tag{21.136}$$

and

$$S_{qm_s'm_s} = \langle Sm_s' | \hat{S}_1^q | Sm_s \rangle = \hbar \left[\begin{array}{c} m_s \delta_{q,0} \delta_{m_s,m_s'} \\ +\frac{1}{\sqrt{2}} \left(\begin{array}{c} -\delta_{q,1} \delta_{m_s,-1/2} \delta_{m_s',1/2} \\ +\delta_{q,-1} \delta_{m_s,1/2} \delta_{m_s',-1/2} \end{array} \right) \end{array} \right]; \quad (21.137a)$$

$$I_{qm_s'm_s} = \langle Im_I' | \hat{I}_1^q | Im_I \rangle = \hbar \left[\begin{array}{c} m_I \delta_{q',0} \delta_{m_I,m_I'} \\ +\frac{1}{\sqrt{2}} \left(\begin{array}{c} -\delta_{q,1} \delta_{m_I,-1/2} \delta_{m_I',1/2} \\ +\delta_{q,-1} \delta_{m_I,1/2} \delta_{m_I',-1/2} \end{array} \right) \end{array} \right]. \quad (21.137b)$$

To evaluate the $B_\ell(q, q', m_\ell', m_\ell)$, I use the identities

$$(u_0)^2 = \cos^2\theta = \frac{\sqrt{4\pi} Y_0^0(\theta,\phi) + \sqrt{\frac{16\pi}{5}} Y_2^0(\theta,\phi)}{3}, \quad (21.138a)$$

$$u_1 u_{-1} = -\frac{\sin^2\theta}{2} = \frac{\sqrt{\frac{16\pi}{5}} Y_2^0(\theta,\phi) - 2\sqrt{4\pi} Y_0^0(\theta,\phi)}{6}, \quad (21.138b)$$

$$u_0 u_{\pm 1} = \mp \frac{\cos\theta \sin\theta e^{\pm i\phi}}{\sqrt{2}} = \sqrt{\frac{4\pi}{15}} Y_2^{\pm 1}(\theta,\phi), \quad (21.138c)$$

$$(u_{\pm 1})^2 = \frac{\sin^2\theta e^{\pm 2i\phi}}{2} = \sqrt{\frac{8\pi}{15}} Y_2^{\pm 2}(\theta,\phi), \quad (21.138d)$$

and

$$\int_0^{2\pi} d\phi \int_0^{\pi} \sin\theta d\theta \left[Y_{\ell_3}^{m_3}(\theta,\phi) \right]^* Y_{\ell_2}^{m_2}(\theta,\phi) Y_{\ell_1}^{m_1}(\theta,\phi)$$
$$= \sqrt{\frac{(2\ell_1+1)(2\ell_2+1)}{4\pi(2\ell_1+1)}} \begin{bmatrix} \ell_1 & \ell_2 & \ell_3 \\ 0 & 0 & 0 \end{bmatrix} \begin{bmatrix} \ell_1 & \ell_2 & \ell_3 \\ m_1 & m_2 & m_3 \end{bmatrix} \quad (21.139)$$

to obtain

$$B_\ell(1,1,m_\ell',m_\ell) = B_\ell(-1,-1,m_\ell',m_\ell) = \frac{1-\ell(\ell+1)-m_\ell^2}{(2\ell+3)(2\ell-1)} \delta_{m_\ell,m_\ell'}; \quad (21.140a)$$

$$B_\ell(1,0,m_\ell',m_\ell) = B_\ell(0,-1,m_\ell',m_\ell)$$
$$= \frac{(2m_\ell'-1)\sqrt{\ell(\ell+1)-m_\ell'(m_\ell'-1)}}{\sqrt{2}(2\ell+3)(2\ell-1)} \delta_{m_\ell,m_\ell'-1}; \quad (21.140b)$$

$$B_\ell(1,-1,m'_\ell,m_\ell) = -\frac{\sqrt{(1+\ell-m'_\ell)\,(2+\ell-m'_\ell)\,(\ell+m'_\ell-1)\,(\ell+m'_\ell)}}{(2\ell+3)\,(2\ell-1)} \times\delta_{m_\ell,m'_\ell-2}; \tag{21.140c}$$

$$B_\ell(0,1,m'_\ell,m_\ell) = B_\ell(-1,0,m'_\ell,m_\ell) = -\frac{(2m'_\ell+1)\,\sqrt{\ell(\ell+1)-m'_\ell\,(m'_\ell+1)}}{\sqrt{2}\,(2\ell+3)\,(2\ell-1)}\delta_{m_\ell,m'_\ell+1};$$

$$B_\ell(0,0,m'_\ell,m_\ell) = \frac{2\ell(\ell+1)-2m_\ell^2-1}{(2\ell+3)\,(2\ell-1)}\delta_{m_\ell,m'_\ell}; \tag{21.140d}$$

$$B_\ell(-1,1,m'_\ell,m_\ell) = -\frac{\sqrt{(\ell-m'_\ell-1)\,(2+\ell+m'_\ell)\,(\ell+m'_\ell+1)\,(\ell-m'_\ell)}}{(2\ell+3)\,(2\ell-1)} \times\delta_{m_\ell,m'_\ell+2}. \tag{21.140e}$$

The sum in Eq. (21.135) can then be carried out to arrive at

$$B_{n\ell jf} = \hbar^2\frac{4\,[f(f+1)-j\,(j+1)]-3}{32j\,(j+1)}\left\langle r^{-3}\right\rangle. \tag{21.141}$$

Using Eqs. (21.129) and (21.141), I find

$$A_{n\ell jf}+3B_{n\ell jf} = \hbar^2\frac{\ell(\ell+1)}{j\,(j+1)}\left[\frac{f(f+1)-j(j+1)-3/4}{2}\right]\left\langle r^{-3}\right\rangle, \tag{21.142}$$

which, when combined with Eqs. (21.124) and (21.94), leads to Eq. (21.112). Note that the right-hand side of Eq. (21.142) vanishes if a small volume about the origin is excluded in evaluating $\left\langle r^{-3}\right\rangle$, implying that the only contributing term to the hyperfine level shifts from $\ell = 0$ states arises from the contact interaction. The hyperfine splitting associated with the non-contact, spin–spin interaction alone can be obtained from Eqs. (21.124)–(21.126), (21.128), and (21.141) as

$$\begin{aligned}\Delta E_{n\ell jf}^{\text{spin-spin}} &= -\frac{\mu_0 e^2 g_p\hbar^2}{8\pi m_e m_p}\left[\frac{j(j+1)+\ell(\ell+1)+3/4}{2j\,(j+1)}\right] \\ &\quad\times\left[\frac{f(f+1)-j(j+1)-3/4}{2}\right]\left\langle r^{-3}\right\rangle \\ &\quad+3\frac{\mu_0 e^2 g_p\hbar^2}{8\pi m_e m_p}\frac{4\,[f(f+1)-j\,(j+1)]-3}{32j\,(j+1)}\left\langle r^{-3}\right\rangle\end{aligned}$$

$$= -\frac{\mu_0 e^2 g_p \hbar^2}{8\pi m_e m_p}\left[\frac{j(j+1)-\ell(\ell+1)-3/4}{4j(j+1)}\right]$$
$$\times\left[f(f+1)-j(j+1)-3/4\right]\left\langle r^{-3}\right\rangle, \quad (21.143)$$

which vanishes for $\ell = 0, j = 1/2$ states.

21.6 Problems

1. What is meant by a "good" quantum number. If the spin-orbit coupling in hydrogen is much greater than the Zeeman splitting produced by an external magnetic field, what are the *approximate* good quantum numbers? In the opposite limit, what are the *approximate* good quantum numbers? For hydrogen in an external magnetic field, what are the only good eigenvalue labels? Explain. For hydrogen in an external electric field, what are the only good eigenvalue labels? Explain. For hydrogen in both external electric and magnetic fields, what are the only good eigenvalue labels? Explain.

2. Given a Hamiltonian of the form

$$\hat{H} = \frac{\hat{p}^2}{2m_e} - \frac{e^2}{4\pi\epsilon_0 r} - \frac{1}{8}\frac{\hat{p}^4}{m_e^3 c^2},$$

use the facts that $\hat{p}^4$ commutes with $\hat{L}^2$ and $\hat{L}_z$ to prove that

$$\langle n,\ell,m|\,\hat{p}^4\,|n',\ell',m'\rangle \propto \delta_{\ell,\ell'}\delta_{m,m'}.$$

Why isn't the matrix element also diagonal in the n quantum number?

3. Calculate the change in the 2P-1S transition frequencies in GHz (a) including only the relativistic mass corrections and (b) including both the relativistic mass, spin-orbit, and Darwin corrections. For case (b) compare your result with the exact result of the Dirac equation given in Eq. (21.1).

4. For an electron in a spherically symmetric potential *plus* a constant magnetic field, prove that L^2 is a constant of the motion, but **L** is not a constant of the motion. For an electron in a spherically symmetric potential *plus* a constant electric field, prove that L^2 and **L** are not constants of the motion. Neglect any spin–orbit interactions. Give a simple physical explanation of the results based on classical considerations.

5–7. Consider the hydrogen atom in the $n = 2, \ell = 1$ manifold of levels. Calculate the eigenvalues and eigenkets when the atom is subjected to an external, constant magnetic field along the z axis. Use the $|n\ell jm_j\rangle$ basis. Plot the energy eigenvalues divided by h in units of GHZ as a function of magnetic field strength for fields between 0 and 10 T. Plot this in two graphs, one for 0 T$\leq B \leq 1$ T and one for 1 T$\leq B \leq 10$ T. Show that the slopes of your energy graphs agree with perturbation theory predictions in the limits of weak and strong fields. Include all relativistic corrections.

8. For a hydrogen atom in a constant electric field $\mathbf{E} = \mathcal{E}_0 \mathbf{u}_z$ aligned along the z axis, the perturbation Hamiltonian is

$$\hat{H}' = \frac{1}{8\pi\epsilon_0}\frac{e^2 \mathbf{S}\cdot\mathbf{L}}{m_e^2 r^3 c^2} + e\mathcal{E}_0\hat{z},$$

where the relativistic mass and Darwin correction terms have been neglected. Write matrix elements of H' in both the $|n\ell s m_\ell m_s\rangle$ and $|n\ell s j m_j\rangle$ bases, allowing for all values of the quantum numbers. You need not evaluate the radial integrals, but determine the states that are coupled by the perturbation in each basis. What are the only exact constants of the motion in this problem?

9–10. For a hydrogen atom in a constant electric field aligned along the x axis, prove that $\hat{L}_x$ is a constant of the motion. Now consider the $n = 2$ electronic state manifold of hydrogen, neglecting all relativistic corrections. Find the eigenkets of the $n = 2$ manifold in this field. Show that the *nondegenerate* eigenkets of the Hamiltonian are eigenkets of $\hat{L}_x$ (as they must be), and that a linear combination of the degenerate eigenkets of the Hamiltonian can be chosen such that they are eigenkets of $\hat{L}_x$. Conversely show that the *nondegenerate* eigenkets of $\hat{L}_x$ are eigenkets of the Hamiltonian (as they must be), and that a linear combination of the degenerate eigenkets of $\hat{L}_x$ can be chosen such that they are eigenkets of the Hamiltonian.

11. Consider the hyperfine plus Zeeman interaction Hamiltonian for the *ground* state of hydrogen in the $|m_I m_j\rangle$ basis. Take the interaction Hamiltonian in frequency units as

$$\hat{H}'/h = 1.42 \text{ GHz}\,\frac{\mathbf{S}\cdot\mathbf{I}}{\hbar^2} + \left(\frac{2S_z}{\hbar}\right) 14 \text{ GHz/T } B(\text{T})$$

Use the fact that $\mathbf{S} = \mathbf{J}$ and obtain an explicit form for the interaction Hamiltonian matrix. Diagonalize the Hamiltonian to obtain the eigenkets and eigenfrequencies. Show that for $B(T) = 0$, the eigenkets and eigenvalues are those associated with states characterized by $f = 0, 1$, and, for $B(T) \gg 0.1$ T, the eigenkets and eigenvalues are those associated with a (doubly-degenerate) $j = 1/2$ state whose magnetic sublevels are split by the Zeeman effect.

12–13. The ground state of ^{87}Rb is a $5^2S_{1/2}$ state having $L = 0, S = 1/2, J = 1/2$ while the first excited states are a $5^2P_{1/2}$ state having $L = 1, S = 1/2, J = 1/2$ and a $5^2P_{3/2}$ state having $L = 1, S = 1/2, J = 3/2$. The ^{87}Rb has nuclear spin $I = 3/2$. What are the possible total angular momentum states for each of these levels? When a pair of off-resonant, optical fields that are counter-propagating in the Z-direction drive a transition between the $F = 1$ hyperfine levels of the $5^2S_{1/2}$ state and the $F = 2$ hyperfine levels of the $5^2P_{1/2}$ state, they produce an effective perturbation Hamiltonian for the $F = 1$ ground state hyperfine levels whose matrix elements are given by[7]

$$H'_{m_F,m'_F} = A \sum_{j,j'=1}^{2} \sum_{K,Q} e^{i\mathbf{k}_{jj'}\cdot\mathbf{R}} (-1)^{K+Q+1} \times \epsilon_Q^K(j,j')(-1)^{m'_F} \begin{bmatrix} 1 & 1 & K \\ m_F & -m'_F & Q \end{bmatrix} \begin{Bmatrix} 1 & 1 & K \\ 1 & 1 & 2 \end{Bmatrix},$$

where

$$\mathbf{k}_{jj'} = \mathbf{k}_j - \mathbf{k}_{j'}; \quad \mathbf{k}_1 = -\mathbf{k}_2 = k\mathbf{u}_z;$$

$$\epsilon_Q^K(1,1) = -\frac{1}{\sqrt{3}}\delta_{K,0}\delta_{Q,0} + \delta_{K,2}\left[-\frac{1}{\sqrt{6}}\delta_{Q,0} + \frac{1}{2}\left(\delta_{Q,2} + \delta_{Q,-2}\right)\right];$$

$$\epsilon_Q^K(2,2) = -\frac{1}{\sqrt{3}}\delta_{K,0}\delta_{Q,0} - \delta_{K,2}\left[\frac{1}{\sqrt{6}}\delta_{Q,0} + \frac{1}{2}\left(\delta_{Q,2} + \delta_{Q,-2}\right)\right];$$

$$\epsilon_Q^K(1,2) = \left[\epsilon_{-Q}^K(2,1)\right]^* = \frac{i}{\sqrt{2}}\left[\delta_{K,1}\delta_{Q,0} + \frac{1}{\sqrt{2}}\delta_{K,2}\left(\delta_{Q,2} - \delta_{Q,-2}\right)\right],$$

A is a constant proportional to the field intensity and inversely proportional to the frequency difference between the applied field and the atomic transition, and the curly bracket term is a 6-J symbol [Mathematica symbol, SixJSymbol[{1, 1, K}, {1, 1, 2}]. Diagonalize the perturbation matrix and plot the eigenenergies E/A as a function of kZ to obtain the *optical lattice potentials.* Counter-propagating optical fields can be used to cool atoms to microKelvin temperatures and the optical potentials associated with these fields can be used to trap the atoms.

[7]See, for example, P. R. Berman and V. Malinovsky, *Principles of Laser Spectroscopy and Quantum Optics* (Princeton University Press, Princeton, N.J., 2011) pp. 414–418. The atoms are pre-cooled using *Doppler cooling* (see Sect. 5.5.3 in the cited work) before these fields produce *sub-Doppler cooling* (see Chap. 18 in the cited work).

Chapter 22
Time-Dependent Problems

The topics I will cover in the last three chapters of this book relate to time-dependent problems. Generally speaking, these will involve problems in which some classical, *time-dependent* interaction, such as that produced by an applied electric or magnetic field, induces transitions between states of a quantum system. I will first consider some very general features of time-dependent problems and then look in detail at a spin 1/2 system in a magnetic field and a two-level atom in an optical field. The density matrix of a single quantum system will be defined and the Bloch and optical Bloch equations will be derived. After studying these "exact" problems, I will look at approximation techniques involving time-dependent problems in Chap. 23, including both the sudden and adiabatic limits. Finally, I will discuss the transitions between a discrete state and a continuum of states in Chap. 24, including Fermi's golden rule.

22.1 Time-Dependent Problems

There are classes of problems in quantum mechanics in which an isolated quantum system interacts with some externally applied fields. The isolated quantum system is characterized by a Hamiltonian $\hat{H}_0$ and its interaction with the external fields by a time-dependent contribution $\hat{V}(t)$ to the total Hamiltonian $\hat{H}$. In Dirac notation the total Hamiltonian is written in matrix form as

$$\underline{\mathrm{H}}(t) = \underline{\mathrm{H}}_0 + \underline{\mathrm{V}}(t). \tag{22.1}$$

For example, $\underline{\mathrm{H}}_0$ can be the Hamiltonian associated with an isolated atom and $\underline{\mathrm{V}}(t)$ can represent the interaction energy of the atom with a classical optical field. If $\underline{\mathrm{H}}(t)$ depends on time, the energy is no longer a constant of the motion. Let the eigenkets of $\underline{\mathrm{H}}_0$ be denoted by $|n\rangle$, such that

P.R. Berman, *Introductory Quantum Mechanics*, UNITEXT for Physics,
https://doi.org/10.1007/978-3-319-68598-4_22

$$\underline{\mathrm{H}}_0|n\rangle = E_n|n\rangle, \tag{22.2}$$

where E_n is the eigenenergy associated with the eigenket $|n\rangle$.

Since the eigenkets are complete, I can expand the state vector as

$$|\psi(t)\rangle = \sum_n a_n(t)|n\rangle \tag{22.3}$$

and substitute this expansion into Schrödinger's equation,

$$i\hbar\frac{\partial|\psi(t)\rangle}{\partial t} = [\underline{\mathrm{H}}_0 + \underline{\mathrm{V}}(t)]\,|\psi(t)\rangle, \tag{22.4}$$

to obtain the set of differential equations

$$i\hbar\sum_n \dot{a}_n(t)|n\rangle = \underline{\mathrm{H}}_0\sum_n a_n(t)|n\rangle + \underline{\mathrm{V}}(t)\sum_n a_n(t)|n\rangle. \tag{22.5}$$

Using the orthogonality of the eigenkets [that is, multiplying Eq. (22.5) by $\langle n'|$], I find that the state amplitudes evolve as

$$i\hbar\dot{a}_n(t) = E_n a_n(t) + \sum_n V_{nm}(t)a_n(t)\,, \tag{22.6}$$

where the matrix element $V_{nm}(t)$ is defined as

$$\begin{aligned} V_{nm}(t) &= \langle n|\hat{V}(t)|m\rangle \\ &= \int \psi_n^*(\mathbf{r})\hat{V}(t)\psi_m(\mathbf{r})d\mathbf{r}\,. \end{aligned} \tag{22.7}$$

I can write Eq. (22.6) as the matrix equation

$$i\hbar\dot{\mathbf{a}}(t) = \underline{\mathrm{E}}\mathbf{a}(t) + \underline{\mathrm{V}}(t)\mathbf{a}(t)\,, \tag{22.8}$$

in which $\mathbf{a}(t)$ is a vector [or, equivalently, a column matrix $\underline{\mathrm{a}}(t)$], $\underline{\mathrm{E}} = \underline{\mathrm{H}}_0$ is a diagonal matrix whose elements are the eigenvalues of $\hat{H}_0$ ($\underline{\mathrm{E}}$ is simply equal to $\underline{\mathrm{H}}_0$ written in the energy representation) and $\underline{\mathrm{V}}(t)$ is a matrix having elements $V_{nm}(t)$. The fact that $\underline{\mathrm{V}}(t)$ is not diagonal, in general, implies that there are *transitions* between the eigenstates of $\underline{\mathrm{H}}_0$.

To obtain the dynamics, I must solve Eq. (22.8) for the state amplitudes. If $\underline{\mathrm{V}}(t)$ is a finite matrix, the coupled equations can be solved numerically. As with any differential equation, you can obtain an *analytic* solution only if you already *know* the solution. You can guess a solution based upon what others have learned in the past. For example, based on the solution of the *scalar* equation

$$i\hbar\dot{x}(t) = f(t)x(t), \tag{22.9}$$

which is

$$x(t) = \exp\left[-\frac{i}{\hbar}\int_0^t f(t')dt'\right]x(0), \tag{22.10}$$

you might think (and you would have some company) that a solution to Eq. (22.8) is

$$\mathbf{a}(t) = \exp\left[-\frac{i}{\hbar}\left(\underline{\text{E}}t + \int_0^t \underline{\text{V}}(t')dt'\right)\right]\mathbf{a}(0), \tag{22.11}$$

but you would be wrong. That is, if you substitute the trial solution given by Eq. (22.11) into Eq. (22.8), it does not work if $\underline{\text{V}}(t)$ and $\underline{\text{V}}(t')$ do not commute. If $\underline{\text{V}}$ is independent of time, you *can* write the solution as

$$\mathbf{a}(t) = \exp\left[-\frac{i}{\hbar}(\underline{\text{E}} + \underline{\text{V}})t\right]\mathbf{a}(0); \tag{22.12}$$

however, in general, it is impossible to obtain analytic solutions to Eq. (22.8) when $\underline{\text{V}}$ is a function of time.

22.1.1 Interaction Representation

In some sense, I am done. Either I can solve Eq. (22.8) or I cannot. That does not prevent me from modifying the equation into what may be a more convenient form. Remember, however, that modifying the equation does not make it solvable, but it may reveal a structure where the solution is more apparent. The first such modification that I use, applicable to any time-dependent quantum problem, involves an *interaction representation*. The idea behind the interaction representation is to have the state amplitudes be *constant* in the absence of the interaction $\underline{\text{V}}(t)$. To accomplish this, I must remove the rapidly varying phase factor $\exp(-iE_nt/\hbar)$ from the state amplitudes by writing

$$|\psi(t)\rangle = \sum_n c_n(t)\exp\left(-iE_nt/\hbar\right)|n\rangle. \tag{22.13}$$

It then follows from Schrödinger's equation that the amplitudes $c_n(t)$ of the interaction representation obey the differential equation

$$i\hbar\dot{c}_n(t) = \sum_m V_{nm}(t)c_m(t)\exp\left(i\omega_{nm}t\right) \equiv \sum_m V_{nm}^I(t)c_m(t), \tag{22.14}$$

where

$$\omega_{nm} = (E_n - E_m)/\hbar \tag{22.15}$$

is a transition frequency and

$$V_{nm}^{I}(t) = \exp(i\omega_{nm}t)\, V_{nm}(t) \tag{22.16}$$

is a matrix element in the interaction representation. Note that

$$c_n(t) = a_n(t) \exp(iE_n t/\hbar), \tag{22.17}$$

such that $|c_n(t)|^2 = |a_n(t)|^2$ is equal to the probability for the quantum system to be in state n.

In the interaction representation, the state vector can be written as

$$|\psi(t)\rangle = \sum_n c_n(t) \exp(-iE_n t/\hbar)\, |n\rangle \equiv \sum_n c_n(t)|n^I(t)\rangle, \tag{22.18}$$

where "eigenkets" $|n^I(t)\rangle = \exp(-iE_n t/\hbar)\, |n\rangle$ are time-dependent. It is important not to forget this time dependence when calculating expectation values of operators. In general, the interaction representation is used often in numerical solutions rather than the Schrödinger representation; in this manner, you need not start the integration until the interaction is turned on. In the Schrödinger representation, the phases of the state amplitudes evolve even in the absence of the interaction.

22.2 Spin 1/2 Quantum System in a Magnetic Field

As a first example, I consider the interaction of the spin 1/2 quantum system with an external magnetic field induction

$$\mathbf{B}(t) = B_0\mathbf{u}_z + |B_x(t)| \cos[\omega t - \phi(t)]\, \mathbf{u}_x, \tag{22.19}$$

In other words, the z-component of the field is constant and the x-component is an oscillatory field having carrier frequency ω, amplitude $|B_x(t)|$, and phase $\phi(t)$. The interaction of a spin 1/2 quantum system with a magnetic field of the type given in Eq. (22.19) is not only of theoretical interest. For example, when a strong constant magnetic field is applied in the z-direction, and pulsed oscillating fields are applied in the x-direction that are in resonance with the transition between the spin states, it is possible to control the population difference between the spin up and spin down states, as well as the relative phase between the state amplitudes. The ability to control the dynamics of a spin 1/2 quantum system with a series of radio frequency pulses is the basis for *NMR* (nuclear magnetic resonance) and

MRI (magnetic resonance imaging). In MRI, the protons in hydrogen in the water molecule serve as the spin 1/2 quantum system.

For the sake of definiteness, I take the spin 1/2 quantum system to be an electron having $s = j = 1/2$ that is bound in the ground state of an atom having no hyperfine structure. The Hamiltonian characterizing the electron spin—magnetic field system is

$$\underline{H}_B(t) = \beta_0 \boldsymbol{\sigma} \cdot \mathbf{B}(t), \tag{22.20}$$

where

$$\boldsymbol{\sigma} = \sigma_x \mathbf{u}_x + \sigma_y \mathbf{u}_y + \sigma_z \mathbf{u}_z, \tag{22.21}$$

$$\sigma_x = \begin{pmatrix} 0 & 1 \\ 1 & 0 \end{pmatrix}; \quad \sigma_y = \begin{pmatrix} 0 & -i \\ i & 0 \end{pmatrix}; \quad \sigma_z = \begin{pmatrix} 1 & 0 \\ 0 & -1 \end{pmatrix}, \tag{22.22}$$

m_e is the electron mass, and $\beta_0 = e\hbar/2m_e$ is the Bohr magneton.

With $\mathbf{B}(t)$ given by Eq. (22.19), the Hamiltonian (22.20) can be written in terms of the Pauli matrices as

$$\underline{H}_B(t) = \frac{\hbar}{2} \left\{ \omega_0 \sigma_z + 2\omega_x(t) \cos\left[\omega t - \phi(t)\right] \sigma_x \right\}, \tag{22.23}$$

where

$$\omega_0 = \frac{2\beta_0 B_0}{\hbar} = \frac{eB_0}{m_e} = 1.76 \times 10^{11}\, B_0(\mathrm{T})\, s^{-1} = 1.76 \times 10^{7}\, B_0(\mathrm{Gauss})\, s^{-1} \tag{22.24}$$

is the frequency spacing of the two spin states in the absence of the oscillating component of the field and

$$\omega_x(t) = \frac{\beta_0 \left|B_x(t)\right|}{\hbar}. \tag{22.25}$$

is a measure of the coupling field strength in frequency units. In frequency units, $\omega_0/2\pi = 28$ GHz/T$=$ 2.8 MHz/Gauss. The physical system is represented schematically in Fig. 22.1a.

From Eqs. (22.6) and (22.23), it follows that the equation for the state amplitudes in the Schrödinger representation is

$$\dot{\mathbf{a}}_B(t) = -i \begin{pmatrix} \omega_0/2 & \omega_x(t) \cos[\omega t - \phi(t)] e^{i\omega_0 t} \\ \omega_x(t) \cos[\omega t - \phi(t)] e^{-i\omega_0 t} & -\omega_0/2 \end{pmatrix} \mathbf{a}_B(t), \tag{22.26}$$

with $\mathbf{a}_B = \left(a_\uparrow, a_\downarrow\right)$, and the up (down) arrow refers to the state having magnetic quantum number $1/2$ $(-1/2)$.

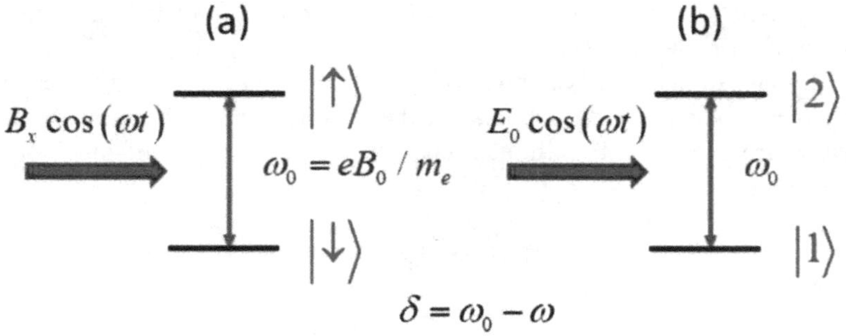

(a) (b)

$B_x \cos(\omega t)$ $|\uparrow\rangle$ $\omega_0 = eB_0/m_e$ $|\downarrow\rangle$

$E_0 \cos(\omega t)$ $|2\rangle$ ω_0 $|1\rangle$

$\delta = \omega_0 - \omega$

Fig. 22.1 Two-level quantum systems interacting with an external field. (**a**) A two-level spin system corresponding to an electron in an atom. An external magnetic field having a constant component B_0 in the z direction splits the spin up and spin down energy eigenkets, while an oscillating component $B_x \cos(\omega t)$ in the x direction drives transitions between the two spin states. (**b**) The analogous situation for a "two-level" atom in which an optical field [having electric field $E_0 \cos(\omega t)$] drives transitions between two electronic states. In both cases the two-level quantum system is assumed to be fixed at the origin. Generalizations to allow for time-dependent amplitudes and phases of the fields are included in the text

In general, these equations must be solved numerically (see problems). To get some insight into the nature of the response of the spin to the external magnetic field, I consider two cases where it is possible to get analytic solutions of the equations.

22.2.1 Analytic Solutions

22.2.1.1 $\omega_x(t) = \omega_x =$ Constant; $\omega = 0$; $\phi(t) = \phi =$ Constant

This limit corresponds to a constant field

$$\mathbf{B} = B_0 \mathbf{u}_z + B_x \mathbf{u}_x , \tag{22.27}$$

such that

$$\underline{\mathrm{H}}_B = \hbar \left(\frac{\omega_0 \sigma_z}{2} + \omega_x \cos\phi \, \sigma_x \right) = \frac{\hbar}{2} \begin{pmatrix} \omega_0 & 2\omega_x \cos\phi \\ 2\omega_x \cos\phi & -\omega_0 \end{pmatrix} . \tag{22.28}$$

This Hamiltonian is representative of a generic two-level problem in which two levels are coupled by a *constant* interaction. The solution of Eq. (22.26) is

$$\begin{aligned} \mathbf{a}_B(t) &= \exp\left[-\frac{i}{2} \begin{pmatrix} \omega_0 & 2\omega_x \cos\phi \\ 2\omega_x \cos\phi & -\omega_0 \end{pmatrix} t \right] \mathbf{a}_B(0) \\ &= \exp\left[-i \left(\frac{\omega_0 \sigma_z}{2} + \omega_x \cos\phi \, \sigma_x \right) t \right] \mathbf{a}_B(0). \end{aligned} \tag{22.29}$$

The exponential is of the form

$$e^{-i\theta\mathbf{n}\cdot\boldsymbol{\sigma}} = \mathbf{1}\cos\theta - i\mathbf{n}\cdot\boldsymbol{\sigma}\sin\theta, \tag{22.30}$$

with

$$\mathbf{n} = \frac{2\omega_x\cos\phi\mathbf{u}_x + \omega_0\mathbf{u}_z}{\sqrt{4\left(\omega_x\cos\phi\right)^2 + \omega_0^2}}; \; \theta = \frac{\sqrt{4\left(\omega_x\cos\phi\right)^2 + \omega_0^2}}{2}t, \tag{22.31}$$

leading to the solution

$$a_\uparrow(t) = -i\frac{2\omega_x\cos\phi}{X}\sin\frac{Xt}{2}a_\downarrow(0) + \left[\cos\frac{Xt}{2} - i\frac{\omega_0}{X}\sin\frac{Xt}{2}\right]a_\uparrow(0), \tag{22.32a}$$

$$a_\downarrow(t) = \left[\cos\frac{Xt}{2} + i\frac{\omega_0}{X}\sin\frac{Xt}{2}\right]a_\downarrow(0) - i\frac{2\omega_x\cos\phi}{X}\sin\frac{Xt}{2}a_\uparrow(0), \tag{22.32b}$$

where

$$X = \sqrt{\omega_0^2 + 4\left(\omega_x\cos\phi\right)^2}. \tag{22.33}$$

If the initial condition is $a_\downarrow(0) = 1$, $a_\uparrow(0) = 0$, then

$$\left|a_\uparrow(t)\right|^2 = \frac{4\omega_x^2\cos^2\phi}{\omega_0^2 + 4\omega_x^2\cos^2\phi}\sin^2\left(Xt/2\right); \tag{22.34a}$$

$$\left|a_\downarrow(t)\right|^2 = \frac{\omega_0^2 + 4\omega_x^2\cos^2\phi\cos^2\left(Xt/2\right)}{\omega_0^2 + 4\omega_x^2\cos^2\phi}. \tag{22.34b}$$

Both upper and lower state populations oscillate as a function of time. These oscillations are referred to as *Rabi oscillations* (after I. I. Rabi). If $\omega_0 = 0$, the maximum value of the upper state population is equal to unity—in this limit, complete *inversion* of the population is possible whenever $\sin^2\left(Xt/2\right) = 1$.

22.2.1.2 $\omega_0 = 0$

If there is no longitudinal field $[B_z = 0]$, the energy levels are degenerate and are coupled by the field oscillating in the x direction. Equations for the probability amplitudes obtained from Eq. (22.26) have the form

$$\dot{a}_\uparrow(t) = -if(t)a_\downarrow(t); \tag{22.35a}$$

$$\dot{a}_\downarrow(t) = -if(t)a_\uparrow(t), \tag{22.35b}$$

where

$$f(t) = \omega_x(t)\cos\left[\omega t - \phi(t)\right] . \tag{22.36}$$

By adding and subtracting the equations, I obtain

$$\dot{a}_\uparrow(t) \pm \dot{a}_\downarrow(t) = \mp i f(t)\left[a_\uparrow(t) \pm a_\downarrow(t)\right] , \tag{22.37}$$

for which the solution is

$$a_\downarrow(t) \pm a_\downarrow(t) = \left[a_\downarrow(0) \pm a_\downarrow(0)\right] e^{\mp i\theta(t)}, \tag{22.38}$$

where

$$\theta(t) = \int_0^t f(t')dt'. \tag{22.39}$$

Using this equation you can deduce easily that the state amplitudes evolve as

$$a_\uparrow(t) = \cos\left[\theta(t)\right] a_\uparrow(0) - i\sin\left[\theta(t)\right] a_\downarrow(0); \tag{22.40a}$$

$$a_\downarrow(t) = -i\sin\left[\theta(t)\right] a_\uparrow(0) + \cos\left[\theta(t)\right] a_\downarrow(0) , \tag{22.40b}$$

where

$$\theta(t) = \frac{1}{\hbar}\int_0^t V_{12}(t')dt' = \int_0^t \omega_x(t')\cos\left[\omega t' - \phi(t')\right] dt'. \tag{22.41}$$

Note that this solution is quite general; it remains valid for *any* degenerate, two-level quantum system whose degenerate energy levels are coupled by a time-dependent interaction having real matrix elements $V_{12}(t)$.

Although this is a simple solution, it can be used to illustrate some interesting physical concepts. If I take $\phi(t) = 0$ and $\omega_x(t) = \omega_x =$ constant, then

$$\theta(t) = \frac{\omega_x}{\omega}\sin\omega t, \tag{22.42}$$

which implies that the probability amplitudes contain *all* harmonics of the field. To see this more clearly I can expand $\cos\left[(\omega_x/\omega)\sin\omega t\right]$ in terms of a series of Bessel functions J_n using

$$\cos\left(z\sin\alpha\right) = J_0(z) + 2\sum_{n=1}^{\infty} J_{2n}(z)\cos(2n\alpha) . \tag{22.43}$$

For example, if the initial state has $a_{\downarrow}(0) = 1$, $a_{\uparrow}(0) = 0$, then

$$\begin{aligned}\left|a_{\uparrow}(t)\right|^2 &= \sin^2\left(\frac{\omega_x}{\omega}\sin\omega t\right) = \frac{1-\cos\left(\frac{2\omega_x}{\omega}\sin\omega t\right)}{2} \\ &= \frac{1-J_0\left(\frac{2\omega_x}{\omega}\right)}{2} - \sum_{n=1}^{\infty} J_{2n}\left(\frac{2\omega_x}{\omega}\right)\cos(2n\omega t).\end{aligned} \tag{22.44}$$

In contrast to a harmonic oscillator, which is intrinsically a *linear* device, a two-level quantum system acts as a nonlinear device—the response does not depend linearly on the applied field and contains all harmonics of the driving field frequency. For $\omega_x/\omega \geq \pi/2$, there are still times for which complete inversion can occur, $\left|a_{\uparrow}(t)\right|^2 = 1$.

The time-averaged, spin-up population is

$$\overline{\left|a_{\uparrow}(t)\right|^2} = \frac{1-J_0\left(\frac{2\omega_x}{\omega}\right)}{2}. \tag{22.45}$$

The larger the applied frequency, the smaller is the time-averaged value of the spin-up state population, provided $\omega > 0.52\omega_x$. This is not surprising since the degenerate levels are resonant with a *static* field; the more rapid the oscillation of the field, the less effective it is in driving the transition. Interestingly, there are values of ω_x/ω for which $\overline{\left|a_{\uparrow}(t)\right|^2} > 1/2$. The maximum value of $\overline{\left|a_{\uparrow}(t)\right|^2} = 0.70$ occurs for $2\omega_x/\omega = 3.83$.

22.3 Two-Level Atom

The problem of a spin 1/2 particle in a magnetic field is isomorphic to the problem of a two-level atom interacting with an optical field. It is not difficult to imagine a situation where such a two-level approximation is valid. For example, if an optical field is nearly resonant with the ground to first excited state transition frequency of an atom whose ground and excited states have angular momentum quantum numbers $J = 0$ and $J = 1$, respectively, and if the field is z-polarized, then the field interacts effectively with only the ground state and the $m = 0$ sublevel of the excited state. To make matters simple, you can think of the atom as a one electron atom whose nucleus is located at position $\mathbf{R}$. The position of the electron relative to the nucleus is denoted by $\mathbf{r}$.

In dipole approximation, the interaction Hamiltonian is given by

$$\hat{V}(\mathbf{R},t) = \hat{V}_{AF}(\mathbf{R},t) \approx -\hat{\mathbf{p}}_e\cdot\mathbf{E}(\mathbf{R},t) = e\hat{\mathbf{r}}\cdot\mathbf{E}(\mathbf{R},t)\,, \tag{22.46}$$

where $\hat{\mathbf{p}}_e = -e\hat{\mathbf{r}}$ is the atomic dipole moment operator (a matrix in the Dirac picture) and $\mathbf{E}(\mathbf{R}, t)$ is the applied electric field, evaluated at the nuclear position. Recall that the charge of the electron is $-e$ in my notation. If atomic motion is neglected as I assume, I can set $\mathbf{R} = \mathbf{0}$.

The applied electric field at the nucleus of the atom is assumed to vary as

$$\mathbf{E}(t) = \mathbf{u}_z \left|E_0(t)\right| \cos\left[\omega t - \varphi(t)\right] = \frac{1}{2}\mathbf{u}_z \left|E_0(t)\right| \left[e^{i\varphi(t)}e^{-i\omega t} + e^{-i\varphi(t)}e^{i\omega t}\right], \tag{22.47}$$

where

$$\frac{1}{2}E_0(t)e^{-i\omega t} = \frac{1}{2}\left|E_0(t)\right| e^{i\varphi(t)}e^{-i\omega t} \tag{22.48}$$

is the *positive frequency component* of the field,

$$E_0(t) = \left|E_0(t)\right| e^{i\varphi(t)} \tag{22.49}$$

is the complex field amplitude, ω is the carrier frequency of the field, $|E_0(t)|$ is the field amplitude, and $\varphi(t)$ is the field phase. Both the amplitude and phase can be functions of time. A time-varying amplitude could correspond to a pulse envelope, while a time-varying phase gives rise to a frequency "chirp" (a frequency that varies in time). With this choice of field, the interaction Hamiltonian becomes

$$\hat{V}(t) = e\hat{z}\left|E_0(t)\right| \cos\left[\omega t - \varphi(t)\right], \tag{22.50}$$

where $\hat{z}$ is the z-component of the position operator.

For the two-level atom, I take the energy of the ground state as $-\hbar\omega_0/2$ and that of the excited state as $\hbar\omega_0/2$. Denoting the ground state eigenket as $|1\rangle$ and the excited state eigenket by $|2\rangle$, I write the probability amplitudes as

$$\mathbf{a}(t) = \begin{pmatrix} a_1(t) \\ a_2(t) \end{pmatrix} \tag{22.51}$$

and matrix elements of the interaction Hamiltonian as

$$V_{12}(t) = ez_{12}\left|E_0(t)\right| \cos\left[\omega t - \varphi(t)\right]; \tag{22.52a}$$

$$V_{21}(t) = ez_{21}\left|E_0(t)\right| \cos\left[\omega t - \varphi(t)\right]; \tag{22.52b}$$

$$V_{11} = V_{22} = 0, \tag{22.52c}$$

where

$$z_{12} = \langle 1|\hat{z}|2\rangle = \langle 2|\hat{z}|1\rangle^* = z_{21}^*. \tag{22.53}$$

The diagonal elements of the interaction Hamiltonian vanish since the operator $\hat{z}$ has odd parity. In general the matrix element z_{12} is complex, but any *single* transition matrix element can be taken as real by an appropriate choice of phase in the wave function (however, if z_{12} is taken to be real, then one is not at liberty to take x_{12} real since the phase of the electronic part of the wave function has been fixed—the matrix element of *one* component only of $\mathbf{r}_{12}$ can be taken as real and this choice determines whether or not the other components are real or complex). Therefore I set

$$ez_{12} = ez_{21} = -(p_{ez})_{12} \text{ (real)}, \tag{22.54}$$

$$V_{12}(t) = V_{21}(t) = -(p_{ez})_{12}|E_0(t)| \cos[\omega t - \phi(t)], \tag{22.55}$$

and write the Hamiltonian as

$$\begin{aligned}\underline{H}(t) = \underline{H}_0 + \underline{V}(t) &= \frac{\hbar}{2}\begin{pmatrix} -\omega_0 & 0 \\ 0 & \omega_0 \end{pmatrix} \\ &+ \hbar\begin{pmatrix} 0 & |\Omega_0(t)| \cos[\omega t - \phi(t)] \\ |\Omega_0(t)| \cos[\omega t - \phi(t)] & 0 \end{pmatrix}, \end{aligned}\tag{22.56}$$

where

$$\Omega_0(t) = \frac{-(p_{ez})_{12}E_0(t)}{\hbar} = \frac{ez_{12}E_0(t)}{\hbar} = |\Omega_0(t)|\, e^{i\varphi(t)} \tag{22.57}$$

is referred to as the *Rabi frequency* and is a measure of the atom–field interaction strength in frequency units. The Rabi frequency is defined such that it is positive for positive $E_0(t)$ and z_{12}. For the Hamiltonian given in Eq. (22.56), Eq. (22.8) for the probability amplitudes $\mathbf{a}(t)$ can be written as

$$\dot{\mathbf{a}}(t) = -\frac{i}{2}\begin{pmatrix} -\omega_0 & 2\,|\Omega_0(t)| \cos[\omega t - \phi(t)] \\ 2\,|\Omega_0(t)| \cos[\omega t - \phi(t)] & \omega_0 \end{pmatrix}\mathbf{a}(t). \tag{22.58}$$

This equation can be solved numerically.

The Hamiltonian given in Eq. (22.56) can be recast as

$$\underline{H}(t) = -\frac{\hbar\omega_0}{2}\sigma_z + \hbar\,|\Omega_0(t)| \cos[\omega t - \phi(t)]\,\sigma_x. \tag{22.59}$$

This is the same type of Hamiltonian that we encountered for the interaction of the electron spin with a magnetic-field. The sign of the lead terms in the Hamiltonians (22.59) and (22.23) differ, however, since, for the optical case, I have chosen the basis $\mathbf{a} = (a_1, a_2)$, while, for the magnetic case, the standard convention for the Pauli matrices requires that I take $\mathbf{a}_B = \left(a_\uparrow, a_\downarrow\right)$. The connection between the two cases is given by the substitutions:

$$(a_1, a_2) \Leftrightarrow \left(a_\downarrow, a_\uparrow\right); \tag{22.60a}$$

$$\omega_0 \Leftrightarrow \frac{2\beta_0 B_0}{\hbar}; \tag{22.60b}$$

$$|\Omega_0(t)| \cos\left[\omega t - \phi(t)\right] \Leftrightarrow \omega_x(t) \cos\left[\omega t - \phi(t)\right]. \tag{22.60c}$$

Although the equations for the two-level atom interacting with an electric field and the spin system interacting with a magnetic field look the same, the physical values of the parameters for the two systems differ markedly. That is, the frequency separation of the spin up and spin down states in a constant external magnetic field can range between 0 Hz and 10 GHz, while radio-frequency (rf) coupling strengths $\omega_x(t)$ are typically less than 1.0 MHz. In the optical case, electronic transition frequencies ω_0 are of order 10^{14}–10^{16} Hz and coupling strengths vary, but are typically much less than the frequency separation of the levels. Only for intense laser pulses having intensities greater than 10^{17} W/cm^2 can the coupling strength be comparable to the optical frequency separations. For a typical cw (continuous-wave) lasers having a few mW of power, coupling strengths are typically in the MHz to GHz range. Given this qualitative difference in the magnetic and electric cases, it is not surprising that different approximation schemes are used in the two cases. You will see that a *rotating wave approximation* is usually a good approximation for atom–optical field interactions, based on the assumption that $\omega_0 \gg |\Omega_0(t)|$, but this is not necessarily so for the magnetic case.

I will restrict the discussion to the optical case, but you should appreciate the fact that the discussion applies equally well to the magnetic field case if $|\omega_x(t)| \ll \omega_0$ and $|\omega - \omega_0| / |\omega + \omega_0| \ll 1$, conditions under which many NMR experiments are carried out.

22.3.1 Rotating-Wave or Resonance Approximation

Although Eq. (22.58) can be solved numerically, it is best to gain some physical insight into this equation before launching into any solutions. You shouldn't be deceived by the apparent simplicity of these coupled equations. There are books devoted to these equations and even numerical solutions can be difficult to obtain in certain limits.

Without solving the problem, you can ask under what conditions the field is effective in driving transitions between levels 1 and 2. I assume that the amplitude $|\Omega_0(t)|$ and phase $\phi(t)$ of the field are slowly varying on a time scale of order ω^{-1}. In this limit, the field can be considered to be *quasi-monochromatic*. Moreover, I assume that $|(\omega_0 - \omega)/(\omega_0 + \omega)| \ll 1$ and $|\Omega_0(t)/(\omega_0 + \omega)| \ll 1$. Under these assumptions, the field is effective in driving the 1-2 transition provided $|\omega_0 - \omega|$ is small compared with $|\Omega_0(t)|$.

The equation for $\dot{\mathbf{a}}(t)$ can be written as

$$\dot{\mathbf{a}}(t) = -\frac{i}{2}\begin{pmatrix} -\omega_0 & \Omega_0(t)e^{-i\omega t} + \Omega_0^*(t)e^{i\omega t} \\ \Omega_0(t)e^{-i\omega t} + \Omega_0^*(t)e^{i\omega t} & \omega_0 \end{pmatrix}\mathbf{a}(t)\,. \tag{22.61}$$

In the interaction representation, the corresponding equation for $\dot{\mathbf{c}}(t)$ is

$$\dot{\mathbf{c}}(t) = -\frac{i}{2}\begin{pmatrix} 0 & \Omega_0(t)e^{-i(\omega_0+\omega)t} + \Omega_0^*(t)e^{-i\delta t} \\ \Omega_0(t)e^{i\delta t} + \Omega_0^*(t)e^{i(\omega_0+\omega)t} & 0 \end{pmatrix}\mathbf{c}(t)\,, \tag{22.62}$$

where

$$\mathbf{c}(t) = \begin{pmatrix} c_1(t) \\ c_2(t) \end{pmatrix}, \tag{22.63}$$

and

$$\delta = \omega_0 - \omega \tag{22.64}$$

is the *atom-field detuning.*

In the interaction representation we see that there are terms that oscillate with frequency $\omega_0+\omega$ and those that oscillate at frequency δ. Moreover there can also be oscillation at frequency $|\Omega_0(t)|$. As long as $|\Omega_0(t)/(\omega_0+\omega)| \ll 1$, $|\delta/(\omega_0+\omega)| \ll 1$, as is assumed, the rapidly oscillating terms do not contribute much since they average to zero in a very short time. In other words, if I take a *coarse-grain time average* over a time interval much greater than $1/(\omega_0+\omega)$, the contribution from these rapidly varying terms is negligibly small compared with the slowly varying terms. The neglect of such terms is called the *rotating-wave approximation* (RWA) or *resonance approximation.* The reason for the nomenclature "rotating-wave" will soon become apparent. In the RWA, Eqs. (22.61) and (22.62) reduce to

$$\dot{\mathbf{a}}(t) = -\frac{i}{2}\begin{pmatrix} -\omega_0 & \Omega_0^*(t)e^{i\omega t} \\ \Omega_0(t)e^{-i\omega t} & \omega_0 \end{pmatrix}\mathbf{a}(t)\,; \tag{22.65}$$

$$\dot{\mathbf{c}}(t) = -\frac{i}{2}\begin{pmatrix} 0 & \Omega_0^*(t)e^{-i\delta t} \\ \Omega_0(t)e^{i\delta t} & 0 \end{pmatrix}\mathbf{c}(t)\,. \tag{22.66}$$

In component form, Eq. (22.66) is

$$\dot{c}_1(t) = -i\chi^*(t)e^{-i\delta t}c_2(t) \tag{22.67a}$$

$$\dot{c}_2(t) = -i\chi(t)e^{i\delta t}c_1(t)\,. \tag{22.67b}$$

where

$$\chi(t) = \Omega_0(t)/2, \tag{22.68}$$

Equations (22.67) also look deceptively simple. For a wide range of parameters, they are easy to solve numerically; however, if the envelope $\chi(t)$ corresponds to a pulse having duration T and if $|\delta| T \gg 1$, the numerical solutions can become extremely challenging. The reason for this is that the transition amplitudes are exponentially small in $|\delta| T$ requiring very small round-off errors, while the step size required for the calculations varies inversely with $|\delta| T$. The effective two-level atom-optical field system is depicted schematically in Fig. 22.1b. To gain some insight into the atom-field dynamics, I look at some limits in which an analytic solution of Eqs. (22.67) can be obtained.

22.3.2 Analytic Solutions

When

$$\chi(t) = |\chi(t)| e^{i\phi(t)} \tag{22.69}$$

is a function of time, there are very few analytic solutions of Eqs. (22.67), although there are certain combinations of $|\chi(t)|$ and $\phi(t)$ for which such solutions can be found. If $\phi(t)$ is constant and $\delta \neq 0$ the only smooth symmetric pulse shape for which an analytic solution is possible is the hyperbolic secant pulse shape.[1] In that case the amplitudes can be expressed as hypergeometric functions. Analytic solutions are also possible for [$\chi(t)$ real] and $\delta = 0$, or for $\chi(t) =$ constant, limiting cases that I now consider.

22.3.2.1 $\chi(t)$ real, $\delta = 0$

In this case, the amplitude equations (22.67) become

$$\dot{c}_1(t) = -i\chi(t)c_2(t); \tag{22.70a}$$

$$\dot{c}_2(t) = -i\chi(t)c_1(t). \tag{22.70b}$$

These equations have the same form as Eqs. (22.35), so the solution is

$$c_1(t) = \cos[\theta(t)]\, c_1(0) - i \sin[\theta(t)]\, c_2(0); \tag{22.71a}$$

$$c_2(t) = -i \sin[\theta(t)]\, c_1(0) + \cos[\theta(t)]\, c_2(0), \tag{22.71b}$$

where

$$\theta(t) = \int_0^t \chi(t')dt'. \tag{22.72}$$

[1] N. Rosen and C. Zener, *Double Stern-Gerlach experiment and related collision phenomena*, Physical review **40**, 502–507 (1932).

In the case of an applied field pulse which "turns on" at $t = -\infty$, the initial conditions should be taken at $t = -\infty$, in which case Eqs. (22.71) are replaced by

$$c_1(t) = \cos[\theta(t)]\, c_1(-\infty) - i \sin[\theta(t)]\, c_2(-\infty) \tag{22.73a}$$

$$c_2(t) = -i \sin[\theta(t)]\, c_1(-\infty) + \cos[\theta(t)]\, c_2(-\infty), \tag{22.73b}$$

with

$$\theta(t) = \int_{-\infty}^{t} \chi(t')dt'. \tag{22.74}$$

Note that, if at $t = -\infty$, $c_1(-\infty) = 1$ and $c_2(-\infty) = 0$, then

$$|c_2(\infty)|^2 = \sin^2(A/2), \tag{22.75}$$

where

$$A \equiv 2\theta(\infty) = \int_{-\infty}^{\infty} \Omega_0(t')dt' = 2\int_{-\infty}^{\infty} \chi(t')dt' \tag{22.76}$$

is referred to as the *pulse area*. The pulse area determines how much population is transferred from the initial to final state. For reasons to be discussed in connection with the Bloch vector, A is defined such that a pulse area of π corresponds to a complete inversion, $|c_1(\infty)| = 0$, $|c_2(\infty)| = 1$, while $A = \pi/2$, results in an equal superposition of ground and excited states, $|c_1(\infty)| = |c_2(\infty)| = 1/\sqrt{2}$.

I have arrived at a pretty interesting result. You can control the degree of excitation that is achieved by a proper choice of pulse area, I might note, however, that the use of a π pulse for level inversion is not a "robust" method (when I was young, "robust" was used only to describe coffee). One must insure that the pulse intensity is uniform over the entire sample and that the pulse area is exactly equal to π to insure that all the atoms are inverted. There are other methods for achieving level inversion that are more robust. One such method is discussed in the next chapter.

22.3.2.2 $\chi(t) = \Omega_0/2 = $ Constant

In this case, Eqs. (22.67) reduce to

$$\dot{c}_1(t) = -\frac{i}{2}\Omega_0^* e^{-i\delta t} c_2(t); \tag{22.77a}$$

$$\dot{c}_2(t) = -\frac{i}{2}\Omega_0 e^{i\delta t} c_1(t). \tag{22.77b}$$

Taking the derivative of Eq. (22.77b) and using Eq. (22.77a), I obtain

$$\ddot{c}_2(t) - i\delta\dot{c}_2(t) + \frac{|\Omega_0|^2}{4}c_2(t) = 0. \tag{22.78}$$

The solution of Eq. (22.78) is

$$c_2(t) = e^{i\delta t/2}\left[A\cos\left(\frac{\Omega t}{2}\right) + B\sin\left(\frac{\Omega t}{2}\right)\right]. \tag{22.79}$$

where

$$\Omega = \sqrt{\delta^2 + |\Omega_0|^2} \tag{22.80}$$

is known as the *generalized Rabi frequency*. In a similar manner you can obtain the solution

$$c_1(t) = e^{-i\delta t/2}\left[D\cos\left(\frac{\Omega t}{2}\right) + E\sin\left(\frac{\Omega t}{2}\right)\right]. \tag{22.81}$$

Only two of the integration constants A, B, D, E can be independent since I started with two, first order coupled differential equations—the constants are related through the differential equations. It is convenient to take A and D as independent since, clearly, $A = c_2(0)$ and $D = c_1(0)$. Using Eqs. (22.77), (22.78) and (22.79), you can then show that

$$E = i\frac{\delta}{\Omega}c_1(0) - i\frac{\Omega_0^*}{\Omega}c_2(0) \tag{22.82}$$

and

$$B = -i\frac{\Omega_0}{\Omega}c_1(0) - i\frac{\delta}{\Omega}c_2(0). \tag{22.83}$$

Finally, the solution of Eqs. (22.77) is

$$c_1(t) = e^{-i\delta t/2}\left\{\begin{array}{c}\left[\cos\left(\frac{\Omega t}{2}\right) + i\frac{\delta}{\Omega}\sin\left(\frac{\Omega t}{2}\right)\right]c_1(0)\\ -i\frac{\Omega_0^*}{\Omega}\sin\left(\frac{\Omega t}{2}\right)c_2(0)\end{array}\right\}; \tag{22.84a}$$

$$c_2(t) = e^{i\delta t/2}\left\{\begin{array}{c}-i\frac{\Omega_0}{\Omega}\sin\left(\frac{\Omega t}{2}\right)c_1(0)\\ +\left[\cos\left(\frac{\Omega t}{2}\right) - i\frac{\delta}{\Omega}\sin\left(\frac{\Omega t}{2}\right)\right]c_2(0)\end{array}\right\}. \tag{22.84b}$$

This solution is of *fundamental* importance since it gives the response of a two-level atom to a monochromatic field. Note that the amplitudes depend *nonlinearly* on the applied amplitude—an atom acts as a nonlinear device, in contrast to an harmonic oscillator.

If $c_1(0) = 1; c_2(0) = 0$, then

$$|c_2(t)|^2 = \frac{|\Omega_0|^2}{\Omega^2} \sin^2\left(\frac{\Omega}{2}t\right) = \frac{|\Omega_0|^2}{\delta^2 + |\Omega_0|^2} \sin^2\left(\frac{\sqrt{\delta^2 + |\Omega_0|^2}}{2}t\right). \tag{22.85}$$

The population undergoes Rabi oscillations or *Rabi flopping* as a function of time. The time-averaged excited state population is

$$\overline{|c_2(t)|^2} = \frac{1}{2}\frac{|\Omega_0|^2}{\Omega^2} = \frac{1}{2}\frac{|\Omega_0|^2}{\delta^2 + |\Omega_0|^2}. \tag{22.86}$$

For $|\delta| \gg |\Omega_0|$, $\overline{|c_2(t)|^2} \approx |\Omega_0|^2 / (2\delta^2)$. This is a somewhat surprising result. Even though the field is off resonance, the time-averaged transition probability falls off inversely only as $1/\delta^2$. In fact, I can turn off the field at a time $t = T = \pi/\Omega$ and find that $|c_2(T)|^2 \approx |\Omega_0|^2 / \delta^2$. From the energy-time "uncertainty principle," you might expect that the transition probability would vanish at least *exponentially* with $|\delta| T$ (recall that increasing the light intensity in the photoelectric experiment does not increase the number of photoelectrons if the field frequency is not sufficiently high). What is going on?

Had the field been turned on and off *smoothly*, one would indeed find that $|c_2(t)|^2 \sim e^{-f(|\delta|T)}$, where f is some positive function. However, in the calculation that was carried out, the field is turned on *instantaneously* at $t = 0$ and turned off *instantaneously* at $t = T$. Owing to the point jump discontinuities, the Fourier components of a step function vary inversely with δT, rather than as an exponentially decaying function of $|\delta| T$.

22.3.3 Field-Interaction Representation

There is another representation that is especially useful when a single quasi-monochromatic field drives transitions between two levels or two manifolds of levels. Instead of extracting the atomic frequency, I can extract the *laser* frequency (and phase) and write

$$\begin{aligned} |\psi(t)\rangle &= \tilde{c}_1(t)e^{i[\omega t/2 - \phi(t)/2]}|1\rangle + \tilde{c}_2(t)e^{-i[\omega t/2 - \phi(t)/2]}|2\rangle \\ &\equiv \tilde{c}_1(t)|\tilde{1}(t)\rangle + \tilde{c}_2(t)|\tilde{2}(t)\rangle, \end{aligned} \tag{22.87}$$

where

$$|\tilde{1}(t)\rangle = e^{i[\omega t/2 - \phi(t)/2]}|1\rangle; \tag{22.88a}$$

$$|\tilde{2}(t)\rangle = e^{-i[\omega t/2 - \phi(t)/2]}|2\rangle \tag{22.88b}$$

are time-dependent kets, as were the standard interaction representation kets. The transformation from the Schrödinger representation to this *field interaction representation* can be written as

$$a_1(t) = \tilde{c}_1(t)e^{i[\omega t/2-\phi(t)/2]}; \tag{22.89a}$$

$$a_2(t) = \tilde{c}_2(t)e^{-i[\omega t/2-\phi(t)/2]}, \tag{22.89b}$$

and from the interaction representation to the field interaction as

$$c_1(t) = \tilde{c}_1(t)e^{-i[\delta t/2+\phi(t)/2]}; \tag{22.90a}$$

$$c_2(t) = \tilde{c}_2(t)e^{i[\delta t/2+\phi(t)/2]}. \tag{22.90b}$$

Note that

$$\left|a_j(t)\right|^2 = \left|c_j(t)\right|^2 = \left|\tilde{c}_j(t)\right|^2;$$

state population probabilities are the same in all representations. It follows from Eqs. (22.90) and (22.62) that

$$\dot{\tilde{c}}_1(t) = i\frac{\delta(t)}{2}\tilde{c}_1(t) - i\frac{|\Omega_0(t)|}{2}\tilde{c}_2(t) - i\frac{|\Omega_0(t)|}{2}e^{-2i[\omega t-\phi(t)]}\tilde{c}_2(t); \tag{22.91a}$$

$$\dot{\tilde{c}}_2(t) = -i\frac{\delta(t)}{2}\tilde{c}_2(t) - i\frac{|\Omega_0(t)|}{2}\tilde{c}_1(t) - i\frac{|\Omega_0(t)|}{2}e^{2i[\omega t-\phi(t)]}\tilde{c}_1(t), \tag{22.91b}$$

where

$$\delta(t) = \delta + \dot{\phi}(t) = \omega_0 - \left[\omega - \dot{\phi}(t)\right] \tag{22.92}$$

and $\left[\omega - \dot{\phi}(t)\right]$ can be viewed as the time-varying frequency of the optical field. In a frame rotating at the field frequency, there are rapidly varying terms oscillating at twice the field frequency. The neglect of such terms in this rotating frame is the origin of the nomenclature RWA. In the RWA, the equations reduce to

$$\dot{\tilde{c}}_1(t) = i\frac{\delta(t)}{2}\tilde{c}_1(t) - i\frac{|\Omega_0(t)|}{2}\tilde{c}_2(t); \tag{22.93a}$$

$$\dot{\tilde{c}}_2(t) = -i\frac{\delta(t)}{2}\tilde{c}_2(t) - i\frac{|\Omega_0(t)|}{2}\tilde{c}_1(t), \tag{22.93b}$$

or

$$i\hbar\dot{\tilde{\mathbf{c}}} = \widetilde{\underline{\mathrm{H}}}(t)\tilde{\mathbf{c}}(t), \tag{22.94}$$

where

$$\underset{\sim}{\widetilde{\mathrm{H}}}(t) = \frac{\hbar}{2}\begin{pmatrix} -\delta(t) & |\Omega_0(t)| \\ |\Omega_0(t)| & \delta(t) \end{pmatrix}. \tag{22.95}$$

In terms of the Pauli matrices, the RWA Hamiltonian is

$$\underset{\sim}{\widetilde{\mathrm{H}}}(t) = \frac{\hbar}{2}\left[-\delta(t)\sigma_z + |\Omega_0(t)|\,\sigma_x\right]. \tag{22.96}$$

The effective energy levels in the field interaction representation are $\mp\hbar\delta(t)/2$ and the magnitude of the Rabi frequency determines the coupling of the levels. Remember that the *eigenkets* of $\underset{\sim}{\widetilde{\mathrm{H}}}(t)$ are time-dependent [even if $\Omega_0(t)$ is constant].

You might ask which representation is the best to use. The answer is, "It depends." Rarely does one use the Schrödinger representation, except in formal manipulations of the equations. For the two-level problem and arbitrary $\Omega_0(t)$, there is not much difference between the interaction and field interaction representations. Differences between the two representations arise in problems involving more than two levels and more than a single field. Generally speaking, the field interaction representation is most useful when a single field drives transitions between two manifolds of levels, while the interaction representation should be used when fields having two or more frequencies drive transitions between two levels or two manifolds of levels.

22.4 Density Matrix for a Single Atom

In the previous section, I showed in detail how to obtain solutions of the time-dependent Schrödinger equation for the state amplitudes of a two-level spin system interacting with a magnetic field and a two-level atom interacting with an optical field. To make connection with experiment, however, you need to know how the state amplitudes are related to the possible outcomes of measurements. Any physical measurement yields real results, while the state amplitudes are complex. From the considerations of Chap. 5, you know that the expectation value of any Hermitian operator corresponding to a physical observable depends on bilinear products of the state amplitudes. For example, if you were to calculate the expectation value of the electric dipole moment of a two-level atom, you would find that it depends on the bilinear products $a_1^*(t)a_2(t)$ and $a_2^*(t)a_1(t)$ since

$$\begin{aligned}\langle\hat{\mathbf{r}}\rangle &= \langle\psi(t)|\hat{\mathbf{r}}|\psi(t)\rangle \\ &= \langle a_1(t)\psi_1 + a_2(t)\psi_2|\hat{\mathbf{r}}|a_1(t)\psi_1 + a_2(t)\psi_2\rangle \\ &= a_1^*(t)a_2(t)\langle 1|\hat{\mathbf{r}}|2\rangle + \text{c.c.},\end{aligned} \tag{22.97}$$

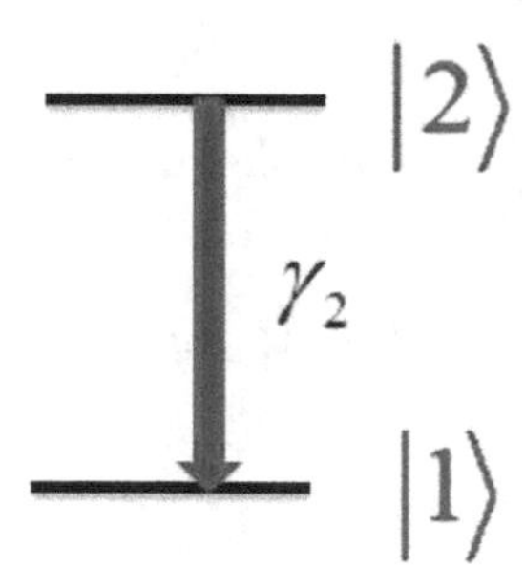

Fig. 22.2 Spontaneous emission in a two-level atom

where the fact that $\hat{\mathbf{r}}$ is an odd parity operator has been used.

A knowledge of the state amplitudes allows you to calculate the expectation values of any operators. Although all the information is contained in the state amplitudes, we are not necessarily interested in *all* the information. If we measure only *part* of the information content of a system, an amplitude approach is often no longer satisfactory. This concept can be illustrated with a simple example.

Let us look at spontaneous decay in a two-level atom, Fig. 22.2. As a result of spontaneous decay (to be discussed in Chap. 24), the upper state population $n_2(t) = |a_2(t)|^2$ decreases and the lower state population $n_1(t) = |a_1(t)|^2$ increases according to

$$\dot{n}_2(t) = -\gamma_2 n_2(t); \tag{22.98a}$$

$$\dot{n}_1(t) = \gamma_2 n_2(t), \tag{22.98b}$$

where the excited state decay rate γ_2 is real. Is it possible to account for this decay in an amplitude picture? You can try the equation

$$\dot{a}_2(t) = -\frac{\gamma_2}{2} a_2(t), \tag{22.99}$$

which implies

$$\frac{d}{dt}|a_2(t)|^2 = \frac{d}{dt}[a_2(t)a_2^*(t)] = \dot{a}_2(t)a_2^*(t) + a_2(t)\dot{a}_2^*(t) = -\gamma_2|a_2(t)|^2. \tag{22.100}$$

It works! But try to reproduce Eq. (22.98b) in a simple amplitude picture—you will find it impossible to do so.

In problems of this nature, where atoms interact with a thermal bath, such as the vacuum field for spontaneous emission, the total Hamiltonian consists of a term for the atoms, a term for the fields and a term for the atom–field interaction. The vacuum field is quantized, so all terms in the Hamiltonian are time-independent. The state amplitudes are labeled by the eigenvalues associated with *both* the atoms and the fields. If we are interested in the atomic state variables only, it is necessary to average over the field variables. A convenient method for carrying out this average involves the *density matrix* of the atom-field system. In statistical mechanics, the density matrix is used to characterize an ensemble of atoms. Even when considering a *single* atom interacting with external fields, the density matrix approach is useful—so let me start there.

The expectation value of an arbitrary Hermitian operator $\hat{A}$ in the state

$$|\psi(t)\rangle = \sum_n a_n(t)|n\rangle \tag{22.101}$$

of a single atom (possibly interacting with time-dependent external fields, but *not* with a thermal bath) is given by

$$\begin{aligned}\langle\psi(t)|\hat{A}|\psi(t)\rangle &= \sum_{n,m}\langle m|a_m^*(t)\hat{A}a_n(t)|n\rangle = \sum_{n,m} a_m^*(t)a_n(t)\langle m|\hat{A}|n\rangle \\ &= \sum_{n,m} a_m^*(t)a_n(t)A_{mn}\,,\end{aligned} \tag{22.102}$$

where the $a_n(t)$ are the atomic state amplitudes. I want to define a matrix $\underline{\rho}(t)$ whose elements in the energy basis are equal to $a_n(t)a_m^*(t)$. A matrix satisfying this criterion is the *density matrix*

$$\underline{\rho}(t) = |\psi(t)\rangle\langle\psi(t)|\,, \tag{22.103}$$

since

$$\langle n|\underline{\rho}(t)|m\rangle = \langle n|\psi(t)\rangle\langle\psi(t)|m\rangle = \sum_{p,p'} a_p(t)a_{p'}^*(t)\,\langle n\,|p\rangle\,\langle p'\,|m\rangle = a_n(t)a_m^*(t)\,. \tag{22.104}$$

Note that $\underline{\rho}(t)$ is *not* a bona fide Schrödinger operator since it is time-dependent, whereas all operators in the Schrödinger picture are taken as time independent. With this definition

$$\langle\psi(t)|\hat{A}|\psi(t)\rangle = \sum_{n,m}\rho_{nm}(t)A_{mn} = \mathrm{Tr}\left(\underline{\rho}(t)\underline{A}\right)\,. \tag{22.105}$$

where Tr stands for "trace."

In the energy basis

$$\underline{\rho}(t) = \sum_{n,m}\rho_{nm}(t)|n\rangle\langle m|\,; \tag{22.106}$$

recall that $|n\rangle\langle m|$ is a matrix with a 1 in the nm location and zeroes elsewhere. It is also possible to expand $\underline{\rho}(t)$ in other bases, such as an irreducible tensor basis. It is easy to establish some properties for $\underline{\rho}(t)$ for the single atom case. First, I note that

$$\underline{\rho}(t) = \underline{\mathrm{a}}(t)\underline{\mathrm{a}}^{\dagger}(t)\,, \tag{22.107}$$

where $\underline{\mathrm{a}}$ is interpreted as a column vector and $\underline{\mathrm{a}}^{\dagger}$ as a row vector,

$$\underline{\mathrm{a}} = \begin{pmatrix} a_1 \\ a_2 \\ \vdots \end{pmatrix}; \tag{22.108}$$

$$\underline{\mathrm{a}}^{\dagger} = \left(a_1^* \; a_2^* \; \cdots\right), \tag{22.109}$$

such that

$$\underline{\rho} = \underline{\mathrm{a}}\underline{\mathrm{a}}^{\dagger} = \begin{pmatrix} a_1 \\ a_2 \\ \vdots \end{pmatrix} \left(a_1^* \; a_2^* \; \cdots\right) = \begin{pmatrix} |a_1|^2 & a_1 a_2^* & \cdots \\ a_2 a_1^* & |a_2|^2 & \cdots \\ \vdots & \vdots & \ddots \end{pmatrix}. \tag{22.110}$$

The density matrix is an *idempotent* operator since

$$\underline{\rho}^2 = |\psi\rangle\langle\psi|\psi\rangle\langle\psi| = |\psi\rangle\langle\psi| = \underline{\rho}\,. \tag{22.111}$$

Critical to the development is the equation for the time evolution of ρ,

$$i\hbar\frac{d\underline{\rho}}{dt} = i\hbar\frac{d}{dt}(\underline{\mathrm{a}}\underline{\mathrm{a}}^{\dagger}) = [\underline{\mathrm{H}}\underline{\mathrm{a}}\underline{\mathrm{a}}^{\dagger} - \underline{\mathrm{a}}(\underline{\mathrm{H}}\underline{\mathrm{a}})^{\dagger}] = \underline{\mathrm{H}}\underline{\rho} - \underline{\rho}\underline{\mathrm{H}} = [\underline{\mathrm{H}}, \underline{\rho}]\,. \tag{22.112}$$

Note that the sign is *different* from that in the evolution equation for the expectation values of operators,

$$i\hbar\frac{\langle\psi(t)|\hat{A}|\psi(t)\rangle}{dt} = \langle\psi(t)|[\hat{A}, \hat{H}]|\psi(t)\rangle\,. \tag{22.113}$$

I have suppressed the explicit time dependence in the amplitudes and operators in Eqs. (22.108)–(22.113). For the most part, I do not indicate such time dependence explicitly from this point onwards, although I retain it in some of the equations as a reminder.

As an example, consider a two-level atom interacting with an optical field in the RWA. In this case,

$$\underline{\mathrm{H}}(t) = \frac{\hbar}{2}\begin{pmatrix} -\omega_0 & \Omega_0^*(t)e^{i\omega t} \\ \Omega_0(t)e^{-i\omega t} & \omega_0 \end{pmatrix}, \tag{22.114}$$

and

$$i\hbar\begin{pmatrix}\dot{\rho}_{11} & \dot{\rho}_{12}\\ \dot{\rho}_{21} & \dot{\rho}_{22}\end{pmatrix} = \frac{\hbar}{2}\left[\begin{pmatrix}-\omega_0 & \Omega_0^*(t)e^{i\omega t}\\ \Omega_0(t)e^{-i\omega t} & \omega_0\end{pmatrix}\begin{pmatrix}\rho_{11} & \rho_{12}\\ \rho_{21} & \rho_{22}\end{pmatrix} - \begin{pmatrix}\rho_{11} & \rho_{12}\\ \rho_{21} & \rho_{22}\end{pmatrix}\begin{pmatrix}-\omega_0 & \Omega_0^*(t)e^{i\omega t}\\ \Omega_0(t)e^{-i\omega t} & \omega_0\end{pmatrix}\right], \tag{22.115}$$

or, since $\chi(t) = \Omega_0(t)/2$,

$$\dot{\rho}_{11} = -i\chi^*(t)e^{i\omega t}\rho_{21} + i\chi(t)e^{-i\omega t}\rho_{12}; \tag{22.116a}$$

$$\dot{\rho}_{22} = i\chi^*(t)e^{i\omega t}\rho_{21} - i\chi(t)e^{-i\omega t}\rho_{12}; \tag{22.116b}$$

$$\dot{\rho}_{12} = i\omega_0\rho_{12} - i\chi^*(t)e^{i\omega t}(\rho_{22} - \rho_{11}); \tag{22.116c}$$

$$\dot{\rho}_{21} = -i\omega_0\rho_{21} + i\chi(t)e^{-i\omega t}(\rho_{22} - \rho_{11}). \tag{22.116d}$$

One can solve these equations for a given $\chi(t)$, but it is easier to solve in the amplitude picture and then simply construct $\rho_{11}(t) = |a_1(t)|^2$, $\rho_{22}(t) = |a_2(t)|^2$, $\rho_{12}(t) = a_1(t)a_2^*(t)$, $\rho_{21}(t) = a_2(t)a_1^*(t)$. It looks like I have not gained *anything*, except making the equations more difficult! In a sense that is correct. The density matrix becomes useful and essential when dealing with ensembles of particles or particles interacting with a bath.

The key point is that it is often possible to get a simple equation for atomic density matrix elements that incorporates the effects of some thermal bath acting on the atoms. A general method for doing this is developed in the Appendix. For example, spontaneous emission results from an atom interacting with the vacuum field. Although the vacuum field plays a critical role in spontaneous decay, its net effect on *atomic* state density matrix elements is given simply by

$$\dot{\rho}_{11})_{sp} = \gamma_2\rho_{22}; \tag{22.117a}$$

$$\dot{\rho}_{22})_{sp} = -\gamma_2\rho_{22}; \tag{22.117b}$$

$$\dot{\rho}_{12})_{sp} = -\frac{\gamma_2}{2}\rho_{12}; \tag{22.117c}$$

$$\dot{\rho}_{21})_{sp} = -\frac{\gamma_2}{2}\rho_{21}. \tag{22.117d}$$

These contributions can be added to the terms given in Eqs. (22.116a)–(22.116d). Equation (22.107) loses its meaning once relaxation is introduced.

Including spontaneous decay,

$$\dot{\rho}_{11} = -i\chi^*(t)e^{i\omega t}\rho_{21} + i\chi(t)e^{-i\omega t}\rho_{12} + \gamma_2\rho_{22}; \tag{22.118a}$$

$$\dot{\rho}_{22} = i\chi^*(t)e^{i\omega t}\rho_{21} - i\chi(t)e^{-i\omega t}\rho_{12} - \gamma_2\rho_{22}; \tag{22.118b}$$

$$\dot{\rho}_{12} = i\omega_0\rho_{12} - i\chi^*(t)e^{i\omega t}(\rho_{22} - \rho_{11}) - \gamma\rho_{12}\,; \tag{22.118c}$$

$$\dot{\rho}_{21} = -i\omega_0\rho_{21} + i\chi(t)e^{-i\omega t}(\rho_{22} - \rho_{11}) - \gamma\rho_{21}\,, \tag{22.118d}$$

where

$$\gamma = \gamma_2/2. \tag{22.119}$$

Note that $\dot{\rho}_{11}(t) + \dot{\rho}_{22}(t) = 0$, consistent with conservation of population. Now I *have* gotten somewhere. It is impossible to write analogous equations using state amplitudes, since Eqs. (22.118) are already averaged over a thermal bath. Equations (22.118) are the starting point for many applications involving the interaction of radiation with matter.

The corresponding equations in the field-interaction representation are

$$\dot{\rho}_{11} = -i\,|\chi(t)|\,\tilde{\rho}_{21} + i\,|\chi(t)|\,\tilde{\rho}_{12} + \gamma_2\rho_{22}\,; \tag{22.120a}$$

$$\dot{\rho}_{22} = i\,|\chi(t)|\,\tilde{\rho}_{21} - i\,|\chi(t)|\,\tilde{\rho}_{12} - \gamma_2\rho_{22}\,; \tag{22.120b}$$

$$\dot{\tilde{\rho}}_{12} = i\delta(t)\tilde{\rho}_{12} - i\,|\chi(t)|\,(\rho_{22} - \rho_{11}) - \gamma\tilde{\rho}_{12}; \tag{22.120c}$$

$$\dot{\tilde{\rho}}_{21} = -i\delta(t)\tilde{\rho}_{21} + i\,|\chi(t)|\,(\rho_{22} - \rho_{11}) - \gamma\tilde{\rho}_{21}\,, \tag{22.120d}$$

where

$$\begin{aligned} \tilde{\rho}_{12} &= e^{-i[\omega t - \phi(t)]}\rho_{12}\,, \\ \tilde{\rho}_{21} &= e^{i[\omega t - \phi(t)]}\rho_{21}\,. \end{aligned} \tag{22.121}$$

Even if $|\chi(t)|$ and $\delta(t)$ are constant, it is not easy to obtain analytic solutions of Eqs. (22.120). You can eliminate $\rho_{11}(t)$ from the equations using $\rho_{11}(t) = 1 - \rho_{22}(t)$, but you are still faced with solving an auxiliary equation for the roots r of the trial solution, $\tilde{\rho}_{ij}(t) = \tilde{\rho}_{ij}(0)e^{rt}$, that is cubic. On the other hand, for constant χ and δ, it is a simple matter to obtain the *steady-state* solutions of these equations by setting the derivatives equal to zero. For constant $\chi = \Omega_0/2$ and δ, the steady-state population of the excited state is given by

$$\rho_{22} = \frac{\frac{2\gamma}{\gamma_2}|\chi|^2}{\delta^2 + \gamma_B^2}\,, \tag{22.122}$$

where

$$\gamma_B = \gamma\sqrt{1 + \frac{4|\chi|^2}{\gamma\gamma_2}}. \tag{22.123}$$

The excitation spectrum (that is, ρ_{22} as a function of δ) is a Lorentzian centered at $\delta = \omega_0 - \omega = 0$, having width (half-width at half-maximum) equal to γ_B. The fact that γ_B increases with increasing field intensity is referred to as *power broadening*.

22.4.1 Magnetic Bloch Equations

The off-diagonal density matrix elements are complex, in general, and cannot correspond to any measurable physical observable. However, it is possible to define real variables that correspond to physically measurable quantities. I do this first for the density matrix elements associated with an electron spin in a magnetic field and then for the density matrix elements for a two-level atom interacting with an optical field. For the magnetic case, I use the Schrödinger representation and, in the optical case, I use the field interaction representation. Both approaches lead to *Bloch equations* (named after Felix Bloch), but you will see that the field-interaction representation allows one to obtain a simple geometric picture of the quantum state evolution.

The Hamiltonian for the spin-magnetic field case is

$$\underline{\mathrm{H}}_B = \frac{e}{m_e}\underline{\mathbf{S}} \cdot \mathbf{B} = \frac{e\hbar}{2m_e}\boldsymbol{\sigma} \cdot \mathbf{B} = \boldsymbol{\omega}_M \cdot \underline{\mathbf{S}}, \tag{22.124}$$

where

$$\boldsymbol{\omega}_M = \frac{e}{m_e}\mathbf{B} \tag{22.125}$$

is the *cyclotron frequency* and

$$\underline{\mathbf{S}} = \frac{\hbar}{2}\left[\begin{pmatrix} 0 & 1 \\ 1 & 0 \end{pmatrix}\mathbf{u}_x + \begin{pmatrix} 0 & -i \\ i & 0 \end{pmatrix}\mathbf{u}_y + \begin{pmatrix} 1 & 0 \\ 0 & -1 \end{pmatrix}\mathbf{u}_z\right]. \tag{22.126}$$

The density matrix associated with the two-component spin system is

$$\underline{\rho} = \begin{pmatrix} \rho_{\uparrow\uparrow} & \rho_{\uparrow\downarrow} \\ \rho_{\downarrow\uparrow} & \rho_{\downarrow\downarrow} \end{pmatrix}. \tag{22.127}$$

Since the spin operator corresponds to an angular momentum and is Hermitian, $\left\langle \hat{\mathbf{S}} \right\rangle = \hbar \left\langle \boldsymbol{\sigma} \right\rangle /2$ is real. Using

$$\left\langle \hat{\mathbf{S}} \right\rangle = \mathrm{Tr}\left(\underline{\rho}\underline{\mathbf{S}}\right) \tag{22.128}$$

with $\underline{\mathbf{S}}$ given by Eq. (22.126), you can show easily that

$$\left\langle \hat{S}_z \right\rangle = \frac{\hbar}{2} \left(\rho_{\uparrow\uparrow} - \rho_{\downarrow\downarrow} \right); \quad (22.129a)$$

$$\left\langle \hat{S}_x \right\rangle = \frac{\hbar}{2} \left(\rho_{\uparrow\downarrow} + \rho_{\downarrow\uparrow} \right); \quad (22.129b)$$

$$\left\langle \hat{S}_y \right\rangle = \frac{\hbar}{2} \left[i \left(\rho_{\uparrow\downarrow} - \rho_{\downarrow\uparrow} \right) \right]. \quad (22.129c)$$

If you can calculate the expectation value of the spin components, you can use Eqs. (22.129), along with conservation of probability, $\rho_{\uparrow\uparrow} + \rho_{\downarrow\downarrow} = 1$, to obtain the density matrix elements. Thus, specifying the expectation value of the spin components is equivalent to specifying the density matrix elements.

The time evolution of $\hat{\mathbf{S}}$ can be obtained from

$$\frac{d\left\langle \hat{\mathbf{S}} \right\rangle}{dt} = \frac{1}{i\hbar} \left\langle \left[\hat{\mathbf{S}}, \hat{H}_B \right] \right\rangle = \frac{1}{i\hbar} \left\langle \left[\hat{\mathbf{S}}, \boldsymbol{\omega}_M \cdot \hat{\mathbf{S}} \right] \right\rangle = \boldsymbol{\omega}_M \times \left\langle \hat{\mathbf{S}} \right\rangle. \quad (22.130)$$

The three equations represented by Eq. (22.130) are referred to as the magnetic Bloch equations. The vector $\left\langle \hat{\mathbf{S}} \right\rangle$ is the *Bloch vector*. Equation (22.130) is the same equation you would find for the components of a classical magnetic moment precessing in a magnetic field. Note that $\boldsymbol{\omega}_M$ can be an arbitrary function of time. For the magnetic field of Eq. (22.19), the x-component of $\boldsymbol{\omega}_M$ oscillates with frequency ω. For such a field, the axis about which the spin angular momentum is precessing is also oscillating as a function of time and it is not easy to get a simple picture of the dynamics.

22.4.2 Optical Bloch Equations

When the RWA is valid, as is often the case for fields driving optical transitions, it is possible to use the field interaction representation to remove the rapid oscillation of the precession axis associated with the frequency ω of the field. For a two-level atom, it is conventional to define real variables

$$\begin{aligned} u &= \tilde{\rho}_{12} + \tilde{\rho}_{21}; \\ v &= i \left(\tilde{\rho}_{21} - \tilde{\rho}_{12} \right); \\ w &= \rho_{22} - \rho_{11}; \\ m &= \rho_{22} + \rho_{11}, \end{aligned} \quad (22.131)$$

in terms of the density matrix elements in the field-interaction representation. The inverse transformation is

$$\begin{aligned}\tilde{\rho}_{12} &= \frac{u+iv}{2}\,; \\ \tilde{\rho}_{21} &= \frac{u-iv}{2}\,; \\ \rho_{11} &= \frac{m-w}{2}\,; \\ \rho_{22} &= \frac{m+w}{2}\,.\end{aligned} \tag{22.132}$$

Note that a matrix element such as $\rho_{12} = \tilde{\rho}_{12}e^{i[\omega t-\phi(t)]}$ can be written in terms of these variables as

$$\rho_{12} = (u+iv)e^{i[\omega t-\phi(t)]}/2. \tag{22.133}$$

It is especially important not to forget that the variables (u, v) are related to density matrix elements in the *field-interaction* representation. In calculating expectation values of operators, it is necessary to convert back to the Schrödinger representation.

In the absence of relaxation, it follows from these definitions and Eqs. (22.120) that

$$\begin{aligned}\dot{u} &= -\delta(t)v\,; \\ \dot{v} &= \delta(t)u - |\Omega_0(t)|\,w; \\ \dot{w} &= |\Omega_0(t)|\,v\,; \\ \dot{m} &= 0\,.\end{aligned} \tag{22.134}$$

Conservation of probability is expressed by the relation $m = 1$. If I construct column vectors

$$\boldsymbol{\Omega}(t) = \begin{pmatrix}|\Omega_0(t)| \\ 0 \\ \delta(t)\end{pmatrix}, \tag{22.135}$$

and

$$\mathcal{B} = \begin{pmatrix}u \\ v \\ w\end{pmatrix}, \tag{22.136}$$

Eq. (22.134) takes the vectorial form

$$d\mathcal{B}/dt = \boldsymbol{\Omega}(t) \times \mathcal{B}\,. \tag{22.137}$$

The vector $\mathcal{B}$ is referred to as the *optical Bloch vector*, the vector $\boldsymbol{\Omega}(t)$ is referred to as the *pseudofield vector*, and Eq. (22.137) is referred to as the *optical Bloch equation.*

The elements of the Bloch vector have a simple interpretation. The quantity m is the total population of the levels and w is the population difference. For the electric dipole transitions under consideration, u and v correspond to components of the atomic dipole moment that are in-phase and out-of-phase with the applied field. One often refers to u and v (as well as $\tilde{\rho}_{12}$ and $\tilde{\rho}_{21}$) as "coherence." There is a simple geometric interpretation of Eq. (22.137) as well. The Bloch vector $\mathcal{B}$ precesses about the pseudofield vector $\mathbf{\Omega}(t)$ with angular frequency

$$\Omega(t) = \sqrt{\delta(t)^2 + |\Omega_0(t)|^2}.$$

Often it is easy to picture the dynamics produced by the interaction using the Bloch vector.

Since $\mathcal{B}$ precesses about the pseudofield vector, its magnitude must remain constant. I can show this explicitly by using Eq. (22.137) to write

$$\frac{d}{dt}|\mathcal{B}|^2 = \frac{d}{dt}(\mathcal{B}\cdot\mathcal{B}) = 2\mathcal{B}\cdot\frac{d\mathcal{B}}{dt} = 2\mathcal{B}\cdot[\mathbf{\Omega}(t)\times\mathcal{B}] = 0\,. \tag{22.138}$$

Therefore

$$|\mathcal{B}|^2 = u^2 + v^2 + w^2 = \text{constant}\,. \tag{22.139}$$

With the definitions given in Eqs. (22.131), I find

$$\begin{aligned}|\mathcal{B}|^2 &= u^2 + v^2 + w^2 \\ &= \tilde{\rho}_{12}^2 + 2\tilde{\rho}_{12}\tilde{\rho}_{21} + \tilde{\rho}_{21}^2 - \tilde{\rho}_{12}^2 + 2\tilde{\rho}_{12}\tilde{\rho}_{21} - \tilde{\rho}_{21}^2 + \rho_{22}^2 - 2\rho_{22}\rho_{11} + \rho_{11}^2 \\ &= \rho_{22}^2 + 2\rho_{22}\rho_{11} + \rho_{11}^2 = (\rho_{22}+\rho_{11})^2 = 1\,, \end{aligned} \tag{22.140}$$

where the relationship $\tilde{\rho}_{12}\tilde{\rho}_{21} = |\tilde{c}_1|^2|\tilde{c}_2|^2 = \rho_{11}\rho_{22}$ was used. Note that this relationship is valid only in the *absence* of any relaxation, allowing me to set $\tilde{\rho}_{12} = \tilde{c}_1\tilde{c}_2{}^* = \tilde{\rho}_{21}^*$. Since the magnitude of $\mathcal{B}$ is unity, the Bloch vector traces out a curve on the surface of the *Bloch sphere*, a sphere having radius unity in u, v, w space.

As a simple example consider the case when $\delta(t) = 0$ and $|\Omega_0| =$ constant, with initial conditions $\rho_{11}(0) = 1$ $[\tilde{\rho}_{12}(0) = \tilde{\rho}_{21}(0) = \rho_{22}(0) = 0]$. This implies that $u(0) = v(0) = 0$ and $w(0) = -1$, see Fig. 22.3. Since $\Omega = |\Omega_0|\mathbf{u}_u$, it follows that $\mathcal{B}$ precesses about the u-axis with frequency $|\Omega_0|$, namely

$$\begin{aligned} u(t) &= 0\,; \\ v(t) &= \sin(|\Omega_0|t)\,; \\ w(t) &= -\cos(|\Omega_0|t)\,. \end{aligned} \tag{22.141}$$

If, instead, $|\Omega_0(t)|$ corresponds to a time-varying pulse envelope whose amplitude vanishes for $t < 0$, the precession phase angle at any time is given by

$$A(t) = \int_0^t |\Omega_0(t')|dt'. \tag{22.142}$$

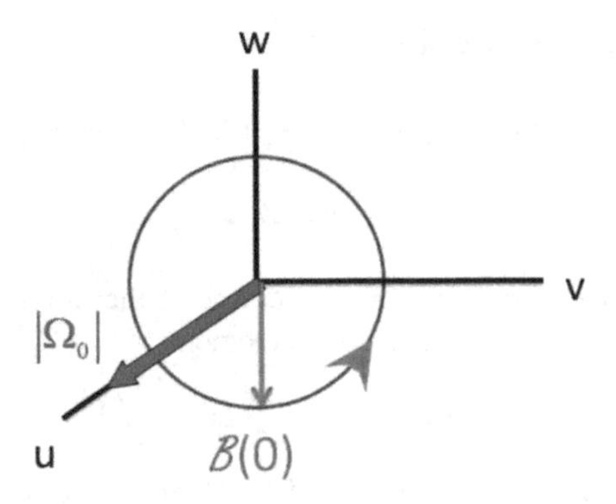

Fig. 22.3 When $\delta(t) = 0$ and $|\Omega_0|$ =constant, the Bloch vector $\mathcal{B}(t)$ rotates in the (w, v) plane with angular velocity $|\Omega_0|$

For this field, Eqs. (22.141) are replaced by

$$\begin{aligned} u(t) &= 0\,; \\ v(t) &= \sin A(t)\,; \\ w(t) &= -\cos A(t)\,, \end{aligned} \tag{22.143}$$

where $A(t)$ can be viewed as the pulse area at time t [see Eq. (22.76)]. For times when $A = \pi$, the population is completely inverted ($w = 1$), while for times when $A = \pm\pi/2$ the coherence is at a maximum ($|v| = 1$). Additional examples are given in the problems and the next chapter.

22.4.3 *Selection Rules for Electric Dipole and Magnetic Dipole Transitions*

You can use what you have learned about irreducible tensor operators and the Wigner-Eckart theorem to derive *selection rules* for electric dipole and magnetic dipole transitions. The selection rules tell you what states can be coupled by these interactions. For the sake of definiteness, I assume the eigenkets of an atom in the absence of the atom–field interactions can be written as $|nLSIJFm_F\rangle$, where n labels the electronic state manifold. In other words, I assume a Russell-Saunders coupling scheme in which the total orbital and spin angular momenta remain good quantum numbers.

22.4.3.1 Electric Dipole Transitions

The electric dipole interaction is given by Eq. (22.46), namely

$$\hat{H}'_e = -\hat{\mathbf{p}}_e \cdot \mathbf{E}(\mathbf{R}, t), \tag{22.144}$$

where $\hat{\mathbf{p}}_e$ is the electric dipole moment operator of the atom. To derive selection rules, I must determine which of the matrix elements

$$\left\langle n'L'S'IJ'F'm'_F\right| \hat{\mathbf{p}}_e \left|nLSIJFm_F\right\rangle \tag{22.145}$$

are non-vanishing. Since $\hat{\mathbf{p}}_e$ is a vector operator that can be written in terms of the components of an irreducible tensor of rank 1, it follows from the Wigner-Eckart theorem and the properties of the Clebsch-Gordan coefficients that this matrix element vanishes unless $\Delta F = F' - F = \pm 1, 0$, and that any $F = F' = 0$ matrix element also vanishes. Furthermore if I use the Clebsch-Gordan coefficients to expand $|nLSIJFm_F\rangle$ in terms of $|nLSJm_J\rangle\,|I\,m_I\rangle$ eigenkets, it also follows that a second selection rule is $\Delta J = \pm 1, 0$ and $J = 0 \to 0$ transitions are forbidden. The reason for this is that the operator $\hat{\mathbf{p}}_e$ depends only on spatial coordinates and must be diagonal in the nuclear spin quantum numbers. Continuing the development by expanding $|nLSJm_J\rangle$ in terms of $|nLm_L\rangle\,|Sm_S\rangle$ eigenkets, I arrive at the selection rules $\Delta S = 0$, $\Delta L = \pm 1, 0$. However, since $\hat{\mathbf{p}}_e$ is an odd parity operator, L must change by an odd integer—$\Delta L = 0$ is ruled out. To summarize, for the electric dipole matrix elements to be non-vanishing, the selection rules $\Delta F = \pm 1, 0$, $\Delta J = \pm 1, 0$, $\Delta S = 0$, and $\Delta L = \pm 1$ must be satisfied, with $F = 0 \to 0$ and $J = 0 \to 0$ transitions also forbidden.

It also follows from the Wigner-Eckart theorem and the properties of the Clebsch-Gordan coefficients that the selection rules for the azimuthal quantum numbers are $\Delta m_F = \pm 1, 0$. The relative contribution of the $\Delta m_F = \pm 1, 0$ matrix elements depends on the *polarization* of the electric field. I take the electric field to be of the form

$$\mathbf{E}(\mathbf{R}, t) = \frac{1}{2}E_0 \boldsymbol{\epsilon} e^{i(\mathbf{k}\cdot\mathbf{R}-\omega t)} + \frac{1}{2}E_0 \boldsymbol{\epsilon}^* e^{-i(\mathbf{k}\cdot\mathbf{R}-\omega t)}, \tag{22.146}$$

which corresponds to a classical monochromatic field propagating in the $\mathbf{k}$ direction. You can show that this field satisfies Maxwell's equations in vacuum provided $\omega = kc$ and $\mathbf{k} \cdot \boldsymbol{\epsilon} = 0$. I have taken the field amplitude E_0 to be real, but the field polarization

$$\boldsymbol{\epsilon} = \epsilon_x \mathbf{u}_x + \epsilon_y \mathbf{u}_y + \epsilon_x \mathbf{u}_z \tag{22.147}$$

can be complex. I will restrict the discussion to optical fields which resonantly couple different electronic levels, $n \neq n'$.

Let us imagine that the field couples a state $|n_1 F_1 m_{F_1}\rangle$ to a state $|n_2 F_2 m_{F_2}\rangle$ in an atom located at position $\mathbf{R}$ and that the energy of state 2 is greater than that of state

1 (all other state labels have been suppressed). In the interaction representation, the state amplitude c_2 evolves according to

$$\begin{aligned}\dot{c}_2 &= \frac{1}{i\hbar}\left\langle n_2F_2m_{F_2}\right|\hat{H}'_e\left|n_1F_1m_{F_1}\right\rangle e^{i\omega_0 t}c_1 \\ &= \frac{i}{2\hbar}\left\langle n_2F_2m_{F_2}\right|\boldsymbol{\epsilon}\cdot\hat{\mathbf{p}}_e\left|n_1F_1m_{F_1}\right\rangle E_0e^{i\mathbf{k}\cdot\mathbf{R}}e^{i\delta t}c_1,\end{aligned} \tag{22.148}$$

where I have made the RWA approximation by neglecting a term rapidly oscillating at the sum frequency $\omega+\omega_0$. This equation will allow me to determine the Δm_F selection rules for specific field polarizations.

To do so, I write

$$\boldsymbol{\epsilon}\cdot\hat{\mathbf{p}}_e = \sum_{q=-1}^{1}(-1)^q\epsilon_{-q}\hat{p}_{e,q}, \tag{22.149}$$

where

$$\hat{p}_{e,\pm1} = \mp(\hat{p}_{e,x}\pm i\hat{p}_{e,y})/\sqrt{2}; \tag{22.150a}$$

$$\hat{p}_{e,0} = \hat{p}_z. \tag{22.150b}$$

and

$$\epsilon_{\pm1} = \mp\frac{\epsilon_x\pm i\epsilon_y}{\sqrt{2}}; \tag{22.151a}$$

$$\epsilon_0 = \epsilon_z. \tag{22.151b}$$

I consider three field polarizations, *linearly polarized light* with

$$\epsilon_x^{\mathrm{lp}} = \epsilon_y^{\mathrm{lp}} = \epsilon_{\pm1}^{\mathrm{lp}} = 0;\quad \epsilon_0^{\mathrm{lp}} = 1;\ \ \mathbf{k} = k\mathbf{u}_x, \tag{22.152}$$

left-circularly polarized light (LCP) with

$$\epsilon_x^{\mathrm{lcp}} = \sqrt{\frac{1}{2}};\ \ \epsilon_y^{\mathrm{lcp}} = i\sqrt{\frac{1}{2}};\ \ \epsilon_1^{\mathrm{lcp}} = 0;\ \ \epsilon_{-1}^{\mathrm{lcp}} = 1;\ \ \epsilon_0^{\mathrm{lcp}} = 0;\ \ \mathbf{k} = k\mathbf{u}_z, \tag{22.153}$$

and *right-circularly polarized light* (RCP) with

$$\epsilon_x^{\mathrm{rcp}} = \sqrt{\frac{1}{2}};\ \ \epsilon_y^{\mathrm{rcp}} = -i\sqrt{\frac{1}{2}};\ \ \epsilon_1^{\mathrm{rcp}} = -1;\ \ \epsilon_{-1}^{\mathrm{rcp}} = 0;\ \ \epsilon_0^{\mathrm{rcp}} = 0;\ \ \mathbf{k} = k\mathbf{u}_z. \tag{22.154}$$

In the case of linear polarization, the field propagates in the X-direction and the electric field is

$$\mathbf{E}^{\mathrm{lp}}(\mathbf{R},t) = E_0 \cos(kX - \omega t)\,\mathbf{u}_z, \tag{22.155}$$

while, for circular polarization, the field propagates in the Z-direction and electric field is

$$\mathbf{E}^{\mathrm{lcp}}(\mathbf{R},t) = \frac{1}{\sqrt{2}} E_0 \left[\mathbf{u}_x \cos(kZ - \omega t) - \mathbf{u}_y \sin(kZ - \omega t) \right]; \tag{22.156a}$$

$$\mathbf{E}^{\mathrm{rcp}}(\mathbf{R},t) = \frac{1}{\sqrt{2}} E_0 \left[\mathbf{u}_x \cos(kZ - \omega t) + \mathbf{u}_y \sin(kZ - \omega t) \right]. \tag{22.156b}$$

Left circular polarization radiation has angular momentum directed *along* its propagation direction. When viewed "head-on" with the radiation approaching you, the polarization vector has constant amplitude and rotates in a counterclockwise direction. Right circular polarization (RCP) radiation has angular momentum directed *opposite* to its propagation direction. When viewed "head-on" with the radiation approaching you, the polarization has constant amplitude rotates in a clockwise direction. Note that the polarization given in Eq. (22.153) would correspond to LCP and that in Eq. (22.154) to RCP if the field propagates in the $-Z$ direction.

Now that I have written the interaction in terms of the irreducible components of the electric dipole operator, I can use Eqs. (22.148)–(22.154) and the Wigner-Eckart theorem to evaluate

$$\begin{aligned} &\langle n_2 F_2 m_{F_2} | \sum_{q=-1}^{1} (-1)^q \epsilon_{-q} \hat{p}_{e,q} | n_1 F_1 m_{F_1} \rangle \\ &= \sum_q (-1)^q \langle n_2 F_2 || p_e^{(1)} || n_1 F_1 \rangle \\ &\times \frac{\epsilon_{-q}}{\sqrt{2F_2+1}} \begin{bmatrix} F_1 & 1 & F_2 \\ m_{F_1} & q & m_{F_2} \end{bmatrix}, \end{aligned} \tag{22.157}$$

from which it follows that the selection rules on absorption are:

$$\Delta m_F = 0 \quad \text{linear polarization}; \tag{22.158a}$$

$$\Delta m_F = 1 \quad \text{LCP with } \mathbf{k} = k\mathbf{u}_z;\ \text{RCP with } \mathbf{k} = -k\mathbf{u}_z; \tag{22.158b}$$

$$\Delta m_F = -1 \quad \text{RCP with } \mathbf{k} = k\mathbf{u}_z;\ \text{LCP with } \mathbf{k} = -k\mathbf{u}. \tag{22.158c}$$

The signs are reversed for emission, that is, for transitions from a higher to lower energy state in the RWA.[2]

[2]Had I *not* made the RWA, the selection rules for LCP and RCP resulting from the rapidly varying term would be reversed from those of the resonant term.

22.4.3.2 Magnetic Dipole Transitions

The magnetic dipole interaction is given by

$$\hat{H}'_m = -\left(\hat{\mathbf{m}}_L + \hat{\mathbf{m}}_s\right) \cdot \mathbf{B}(\mathbf{R}, t) = \frac{\beta_0 B}{\hbar}\left(\hat{\mathbf{L}} + 2\hat{\mathbf{S}}\right) \cdot \mathbf{B}(\mathbf{R}, t), \tag{22.159}$$

Since $\hat{\mathbf{L}}$ and $\hat{\mathbf{S}}$ are vector operators, the selection rules for ΔF and ΔJ are the same as those for electric dipole transitions, namely $\Delta F = \pm 1, 0$, $\Delta J = \pm 1, 0$, with $F = 0 \rightarrow 0$ and $J = 0 \rightarrow 0$ transitions also forbidden. The selection rule on spin angular momentum is still $\Delta S = 0$ since the interaction cannot change the magnitude of the spin; however, the selection rule for ΔL is now $\Delta L = 0$, since the interaction (22.159) couples only those states having the *same* parity. Moreover, since the interaction does not depend on spatial operators such as $\hat{\mathbf{p}}_e$, there is now a selection rule on n, $\Delta n = 0$. Magnetic fields couple only those states *within* a given electronic state manifold. The selection rules for Δm_F are unchanged from the electric dipole case, except that it may no longer be a good approximation to make the RWA for radio-frequency fields that drive transitions between different hyperfine states.

22.5 Summary

In this chapter, I presented an introduction to problems that can be categorized as *semi-classical*, in which a classical time-dependent field interacts with a quantum system. Some general results were derived and specific examples were given related to the interaction of a magnetic field with a spin 1/2 quantum system and an optical field with a two-level atom. The concept of the density matrix was introduced and the Bloch equations were derived.

22.6 Appendix: Interaction of an Atom with a Thermal Bath

Consider what happens when a single atom interacts with a thermal bath. The total Hamiltonian is

$$\hat{H} = \hat{H}_1 + \hat{H}_2 + \hat{V}, \tag{22.160}$$

where $\hat{H}_1$ is the Hamiltonian of the atomic system (possibly including interactions with time-dependent external fields), $\hat{H}_2$ is the time-independent Hamiltonian of the bath, and $\hat{V}$ is the atom–bath interaction energy. The eigenkets of the time-independent part of $\hat{H}_1$ are denoted by $|n_1\rangle$ and the eigenkets of $\hat{H}_2$ are denoted by

$|n_2\rangle$. Suppose we have an operator $\hat{A}_1$ that acts *only* in the space of $\hat{H}_1$; that is, $\hat{A}_1$ acts only on atomic state variables. For a wave function

$$|\psi(t)\rangle = \sum_{n_1,n_2} a_{n_1n_2}(t)|n_1\rangle|n_2\rangle\,, \tag{22.161}$$

I calculate

$$\begin{aligned}\langle\hat{A}_1\rangle = \langle\psi|\hat{A}_1|\psi\rangle &= \sum_{n_1,n_2}\sum_{n_1',n_2'} a^*_{n_1n_2}a_{n_1'n_2'}\langle n_1|\langle n_2|\hat{A}_1|n_2'\rangle|n_1'\rangle \\ &= \sum_{n_1,n_2}\sum_{n_1',n_2'} a^*_{n_1n_2}a_{n_1'n_2'}\delta_{n_2,n_2'}\langle n_1|\hat{A}_1|n_1'\rangle\end{aligned} \tag{22.162}$$

or

$$\langle\psi|\hat{A}_1|\psi\rangle = \sum_{n_1,n_1'} (A_1)_{n_1n_1'}\left(\sum_{n_2} a^*_{n_1n_2}a_{n_1'n_2}\right). \tag{22.163}$$

I define the *reduced density matrix* for the atom as the total density matrix, traced over the states of the bath,

$$\underline{\rho}^{(1)} = \mathrm{Tr}_2\underline{\rho}; \tag{22.164}$$

that is,

$$\rho^{(1)}_{nn'} = \sum_{n_2}\rho_{nn_2;n'n_2} = \sum_{n_2} a_{nn_2}a^*_{n'n_2}\,. \tag{22.165}$$

It is clear from Eqs. (22.163) and (22.165) that

$$\langle\hat{A}_1\rangle = \sum_{n_1,n_1'} (A_1)_{n_1n_1'}\,\rho^{(1)}_{n_1'n_1} = \mathrm{Tr}(\underline{A}_1\underline{\rho}^{(1)})\,. \tag{22.166}$$

The key point is that it is often possible to get a simple equation for $\underline{\rho}^{(1)}$ that incorporates the effects of the bath. For example, I show explicitly in Chap. 24 that the result of spontaneous emission is to introduce terms in the equations of motion for the reduced density matrix elements for the atom given by Eqs. (22.120). One can also consider the bath as an ensemble of perturber atoms that collide with the atom of interest. The collisions cause sudden changes in energy of the various levels, but are not energetically able to cause transitions between the two levels. In this model,

$$\dot{\rho}_{11}(t)_{\mathrm{coll}} = 0\,; \tag{22.167a}$$

$$\dot{\rho}_{22}(t)_{\text{coll}} = 0\,; \tag{22.167b}$$

$$\dot{\rho}_{12}(t)_{\text{coll}} = -\left(\Gamma + iS\right)\rho_{12}(t)\,; \tag{22.167c}$$

$$\dot{\rho}_{21}(t)_{\text{coll}} = -\left(\Gamma - iS\right)\rho_{21}(t)\,, \tag{22.167d}$$

where Γ is a decay rate and S is a shift, both of which are proportional to the perturber density. The collision effects can be incorporated into Eqs. (22.118) and (22.120) by replacing γ with $\gamma_2/2 + (\Gamma + iS)$ in Eqs. (22.118c) and (22.120c) and γ with $\gamma_2/2+(\Gamma - iS)$ in Eqs. (22.118d) and (22.120d) and can be incorporated into Eq. (22.122) by replacing δ with $\delta - S$ and γ with $\gamma_2/2 + \Gamma$.

22.7 Problems

1. In time-dependent problems, the Hamiltonian is often taken to be of the form $\hat{H}(t) = \hat{H}_0 + \hat{V}(t)$. On average, is energy conserved for this Hamiltonian? Explain. Even if $\hat{H}_0$ consists of only two states (such as for a spin 1/2 system), why is it impossible, in general, to solve for the state amplitudes analytically given some initial conditions. What are some general conditions on $\hat{V}(t)$ if it is to effectively cause transitions between two states of $\hat{H}_0$.

2. Two *degenerate* states, 1 and 2, are coupled by a *constant* interaction potential, $\langle 1|\,\hat{V}\,|2\rangle = \langle 2|\,\hat{V}\,|1\rangle = V_{12}$ =constant. If at $t = 0$, the system is in state 1, find the probability amplitudes for the system to be in state 1 and in state 2 for all $t > 0$. Assume that $V_{11} = V_{22} = 0$.

3–4. In dimensionless units, the equations for the probability amplitudes for a spin 1/2 system in an oscillating magnetic field are given by

$$d\mathbf{a}_B(\tau)/d\tau = -i\begin{pmatrix} x/2 & z\cos(y\tau) \\ z\cos(y\tau) & -x/2 \end{pmatrix}\mathbf{a}_B(\tau)$$

where $\mathbf{a}_B = (a_\uparrow, a_\downarrow)$, $x = \omega_0 T; y = \omega T; z = \omega_x T$, and $\tau = t/T$. Use a computer program such as NDSolve in Mathematica to obtain and plot solutions for $P_{up}(\tau) = \left|a_\uparrow(\tau)\right|^2$ as a function of τ for

i. $\omega T = 0$; $\omega_x T = 0.5, 1, 2, 10$; $\omega_0 T = 5$;

ii. $\omega T = 2$; $\omega_x T = 0.5, 1, 2, 10$; $\omega_0 T = 0$;

iii. $\omega T = 10$; $\omega_x T = 0.5, 1, 2, 10$; $\omega_0 T = 10$,

given $a_\uparrow(0) = 0; a_\downarrow(0) = 1$. In each case, are there single or multiple frequencies present in $P_{up}(\tau)$?

5–6. Consider a *classical* spin having magnetic moment $\mathbf{m}_s = -e\mathbf{S}/m_e$ where $\mathbf{S}$ is a spin angular momentum. In an external magnetic field $\mathbf{B}$ the spin experiences a torque

$$\boldsymbol{\tau} = \frac{d\mathbf{S}}{dt} = \mathbf{m}_s \times \mathbf{B}.$$

Show that the equation for the spin angular momentum is

$$\frac{d\mathbf{S}}{dt} = \boldsymbol{\omega}_B \times \mathbf{S},$$

where

$$\boldsymbol{\omega}_B = \frac{e\mathbf{B}}{m_e}.$$

Write the differential equation for the spin in component form.

Now take

$$\boldsymbol{\omega}_B = \omega_0 \mathbf{u}_z + \omega_s \cos(\omega t)\, \mathbf{u}_x$$

and solve numerically and plot S_z as a function of time with initial condition $S_z(0) = -1$ (in arbitrary units). Take time to have dimensionless units, $\tau = t/T$, and consider two cases (i) $\omega_0 T = 10$, $\omega_s T = 1$, $\omega T = 6$ (off-resonant) and (ii) $\omega_0 T = 10$, $\omega_s T = 1$, $\omega T = 10$ (resonant). What is the difference between the two cases? Also plot $S_x(t)$ for the resonant case with the same initial conditions; in this case you will see that there are rapid oscillations in $S_x(t)$ even though the rotating-wave approximation is valid—you need to go to a rotating frame (field interaction representation) to remove these rapid oscillations.

7–8. The optical Bloch equations with decay for a monochromatic field are

$$\begin{aligned}
\dot{u} &= -\delta v - \gamma u\,; \\
\dot{v} &= \delta u - |\Omega_0| w - \gamma v\,; \\
\dot{w} &= |\Omega_0| v - \gamma_2 (w + 1); \\
\dot{m} &= 0\,,
\end{aligned}$$

where $\gamma = \gamma_2/2$. Derive these equations starting from Eqs. (22.131) and (22.120). Solve these equations in steady-state. Look at the results in the limit that $\Omega_0 \gg \gamma_2$ and interpret your result. In particular show that the expression for the steady-state, upper state population, $\rho_{22} = (1 + w)/2$, is a Lorentzian having HWHM $\sqrt{\gamma^2 + \frac{\Omega_0^2}{2}}$; since the width increases with increasing field intensity, this is known as *power broadening*.

9. Now solve the Bloch equations of Problem 22.7–8 numerically for $\gamma T = \gamma_2 T/2 = 1/2$ and ($\delta T = 0.1$, $\Omega_0 T = 0.2$) and ($\delta T = 0.1$, $\Omega_0 T = 3$) and plot w as a function of dimensionless time $\tau = t/T$, assuming the atom is initially in its ground state. In which case does the Bloch vector approach its steady-state value monotonically? Why would you expect this from the Bloch vector picture?

10–11. Assuming a constant field amplitude, solve Eqs. (22.120) analytically for $\gamma = \gamma_2/2$, $\delta = 0$, and $\rho_{11}(0) = 1$ to show that the upper state population is given by

$$\rho_{22}(t) = \frac{|\Omega_0|^2/2}{2\gamma^2 + |\Omega_0|^2}\{1 - [\cos(\lambda t) + \frac{3\gamma}{2\lambda}\sin(\lambda t)]e^{-3\gamma t/2}\},$$

where $\lambda = (|\Omega_0|^2 - \gamma^2/4)^{1/2}$. Evaluate ρ_{22} for $|\Omega_0| \gg \gamma$ and give an interpretation in terms of the Bloch vector. Show that, as $t \to \infty$, result is consistent with Problem 22.7–8.

12. In the *field interaction representation*, neglecting relaxation, the state vector can be written quite generally as

$$|\psi(t)\rangle = \sin(\theta/2)\left|\tilde{1}(t)\right\rangle + \cos(\theta/2)e^{-i\phi}\left|\tilde{2}(t)\right\rangle,$$

where θ and ϕ are arbitrary real functions of time with $0 \leq \theta \leq \pi$ and $0 \leq \phi \leq 2\pi$. Show that the angles θ and ϕ correspond to the spherical angles of the Bloch vector on the Bloch sphere.

13–15. Consider the differential equation $\dot{\mathbf{y}}(t) = \underline{\mathrm{A}}\mathbf{y}(t)$, where $\mathbf{y}(t)$ is a column vector and $\underline{\mathrm{A}}$ is a **constant** matrix.

(a) Show, by direct substitution, that a solution to this equation is $\mathbf{y}(t)=e^{\underline{\mathrm{A}}t}\mathbf{y}(0)$.
In the following parts, take

$$\underline{\mathrm{A}} = -i\begin{pmatrix} -a & b \\ b & a \end{pmatrix}; \quad \mathbf{y}(0) = \begin{pmatrix} y_1(0) \\ y_2(0) \end{pmatrix}; \quad \mathbf{y}(t) = \begin{pmatrix} y_1(t) \\ y_2(t) \end{pmatrix},$$

where a and b are real.

(b) Solve the differential equation directly by assuming a solution of the form $\mathbf{y}(t) = \mathbf{y}e^{\lambda t}$.

(c) Solve the equations using the identity

$$e^{-i\theta\hat{\mathbf{n}}\cdot\boldsymbol{\sigma}} = \mathbf{1}\cos\theta - i\,\hat{\mathbf{n}}\cdot\boldsymbol{\sigma}\sin\theta,$$

where $\hat{\mathbf{n}}$ is a unit vector and $\boldsymbol{\sigma}$ is a vector having matrix components

$$\boldsymbol{\sigma}_x = \begin{pmatrix} 0 & 1 \\ 1 & 0 \end{pmatrix}; \quad \boldsymbol{\sigma}_y = \begin{pmatrix} 0 & -i \\ i & 0 \end{pmatrix}; \quad \boldsymbol{\sigma}_z = \begin{pmatrix} 1 & 0 \\ 0 & -1 \end{pmatrix}.$$

(d) Solve the equation using the MatrixExp[{{*I a*,–*I b*},{–*I b*,–*I a*}}] function of Mathematica or some equivalent program.

(e) Find a matrix $\underline{\mathrm{T}}$ such that $\underline{\mathrm{T}}\underline{\mathrm{A}}\underline{\mathrm{T}}^{\dagger} = \underline{\Lambda}$, where $\underline{\Lambda} = \begin{pmatrix} \Lambda_1 & 0 \\ 0 & \Lambda_2 \end{pmatrix}$ is a diagonal matrix. Prove that

$$\mathbf{y}(t) = \underline{\mathrm{T}}^{\dagger} e^{\underline{\Lambda} t} \underline{\mathrm{T}} \mathbf{y}(0) = \underline{\mathrm{T}}^{\dagger} \begin{pmatrix} e^{\Lambda_1 t} & 0 \\ 0 & e^{\Lambda_2 t} \end{pmatrix} \underline{\mathrm{T}} \mathbf{y}(0),$$

and evaluate this explicitly.

(f) Show that all your results give the same solution. Note that the last method can be used for matrices of any dimension.

16. Imagine that a circularly polarized field drives a $J = 0$ to $J = 1$ transition in an atom. At the atomic position, $Z = 0$, take the field to be of the form

$$\mathbf{E}(t) = E_0 \left[\mathbf{u}_x \cos(\omega t) + \mathbf{u}_y \sin(\omega t) \right],$$

where E_0 is constant. Prove that, if one considers transitions between the $J = 0$ and $J = 1, m_J = 1$ levels only, it is *not* necessary to make any RWA to arrive at equations of the form in Eq. (22.65). On the other hand, show that the "counter-rotating" (rapidly oscillating) terms drive transitions between the $J = 0$ and $J = 1, m_J = -1$ levels. If the field is far off-resonance, both transitions contribute comparable amounts to the atomic response.

Chapter 23
Approximation Techniques in Time-Dependent Problems

In this chapter, I discuss some general techniques that can be used to obtain approximate solutions to time-dependent problems in quantum mechanics.

23.1 Time-Dependent Perturbation Theory

I already have derived the equations of motion for the state amplitudes when the Hamiltonian is of the form $\hat{H} = \hat{H}_0 + \hat{V}(t)$. In the Schrödinger representation they are

$$i\hbar\dot{a}_n(t) = E_n a_n(t) + \sum_n V_{nm}(t)a_n(t)\,, \tag{23.1}$$

when the wave function is expanded as $|\psi(t)\rangle = \sum_n a_n(t)|n\rangle$. In the interaction representation they are

$$i\hbar\dot{c}_n(t) = \sum_m V_{nm}(t)c_m(t)\exp(i\omega_{nm}t)\,, \tag{23.2}$$

when the wave function is expanded as $|\psi(t)\rangle = \sum_n c_n(t)|n\rangle \exp(-iE_nt/\hbar)$. Recall that

$$\omega_{nm} = (E_n - E_m)/\hbar \tag{23.3}$$

is the transition frequency between levels n and m.

Generally speaking, these equations correspond to an infinite number of coupled equations. Imagine, however, that the system starts in its ground state, denoted by $n = 0$, at $t = -\infty$ and that the perturbation is weak. Then I can use *time-dependent perturbation theory* to estimate the amplitude of the other states ($n \neq 0$) as

P.R. Berman, *Introductory Quantum Mechanics*, UNITEXT for Physics,
https://doi.org/10.1007/978-3-319-68598-4_23

$$c_n(t) \approx \frac{1}{i\hbar}\int_{-\infty}^{t} dt' V_{n0}(t')c_0(t') \exp\left(i\omega_{n0}t'\right) \approx \frac{1}{i\hbar}\int_{-\infty}^{t} dt' V_{n0}(t') \exp\left(i\omega_{n0}t'\right), \tag{23.4}$$

where I have assumed that the initial state is not affected very much by the perturbation, allowing me to replace $c_0(t')$ by unity in the integrand. Clearly this approach is valid only if

$$\sum_{n\neq 0} |c_n(t)|^2 \ll 1. \tag{23.5}$$

It turns out that this condition is necessary, but not sufficient, for perturbation theory to be valid. Since state amplitudes have phases, it is possible for these phases to become greater than unity even when condition (23.5) holds. In that case there is also a breakdown of perturbation theory.

As an example of the use of time-dependent perturbation theory, consider a pulse of laser radiation having electric field amplitude at the position of an atom that varies as

$$\mathbf{E}(t) = \boldsymbol{\epsilon} E_0(t) \cos(\omega t), \tag{23.6}$$

where $\boldsymbol{\epsilon}$ is the field polarization and $E_0(t)$ is the field amplitude envelope function. The atom–field interaction is taken as

$$\hat{V}(t) \approx e\hat{\mathbf{r}} \cdot \mathbf{E}(t) = e\hat{\mathbf{r}}\cdot\boldsymbol{\epsilon} E_0(t) \cos(\omega t), \tag{23.7}$$

where $\hat{\mathbf{r}}$ is the position operator for the electron. As a consequence

$$c_n(t) \approx \frac{e\mathbf{r}_{n0} \cdot \boldsymbol{\epsilon}}{i\hbar} \int_{-\infty}^{t} dt' E_0(t') \cos\left(\omega t'\right) \exp\left(i\omega_{n0}t'\right), \tag{23.8}$$

where $\mathbf{r}_{n0}$ is a matrix element of the position operator.

For the sake of definiteness I assume that

$$E_0(t) = E_0 e^{-t^2/T^2}. \tag{23.9}$$

The excited state amplitudes $c_n(\infty)$ after the pulse has passed are given by

$$\begin{aligned} c_n(\infty) &= \frac{e\mathbf{r}_{n0} \cdot \boldsymbol{\epsilon} E_0}{i\hbar} \int_{-\infty}^{\infty} dt' e^{-t^2/T^2} \cos(\omega t) \exp(i\omega_{n0}t) \\ &= \frac{e\mathbf{r}_{n0} \cdot \boldsymbol{\epsilon} E_0 \sqrt{\pi} T}{2\hbar} \left[e^{-(\omega-\omega_{n0})^2 T^2/4} + e^{-(\omega+\omega_{n0})^2 T^2/4} \right] \end{aligned} \tag{23.10}$$

and the final state probabilities, P_n, by

$$\begin{aligned} P_n &= |c_n(\infty)|^2 \\ &= \frac{A^2}{4}\left[e^{-(\omega-\omega_{n0})^2T^2/2} + e^{-(\omega+\omega_{n0})^2T^2/2} + 2e^{-\left(\omega^2+\omega_{n0}^2\right)T^2/2}\right], \end{aligned} \tag{23.11}$$

where

$$A = \sqrt{\pi}\frac{e\mathbf{r}_{n0} \cdot \boldsymbol{\epsilon} E_0 T}{\hbar} \tag{23.12}$$

is the pulse area defined by Eq. (22.76). For perturbation theory to be valid, the pulse area must be much less than unity.[1] We see here a manifestation of the "time-energy" uncertainty principle. To significantly excite a transition with a pulse that is detuned by $\omega - \omega_{n0}$, the pulse duration must be less than or comparable to $|\omega - \omega_{n0}|^{-1}$. In general, significant excitation occurs only if the perturbation has frequency components at the transition frequency.

As a second example, I use perturbation theory to calculate the induced dipole moment of a 3-D harmonic oscillator having mass m produced by an external electric field

$$\mathbf{E}(t) = \begin{cases} 0 & t < 0 \\ \mathbf{u}_x E_0 \cos(\omega t) & t > 0 \end{cases}. \tag{23.13}$$

For $t > 0$, the Hamiltonian is

$$\hat{H}(t) = \hbar\omega_0 \left(a^\dagger a + 3/2\right) + \hat{V}(t),$$

where

$$\hat{V}(t) \approx -q\hat{\mathbf{r}} \cdot \mathbf{E}(t) = -q\hat{x}E_0 \cos(\omega t), \tag{23.14}$$

q is the charge of the oscillating mass, ω_0 its natural frequency, and a and $a^\dagger$ are lowering and raising operators. At $t = 0$, the oscillator is in its ground state.

The average induced dipole moment $\mathbf{p}_d(t)$ is given by

$$\mathbf{p}_d(t) = q\langle x(t)\rangle \mathbf{u}_x. \tag{23.15}$$

I use number state eigenkets for the oscillator, such that the initial condition is $|\psi(0)\rangle = |0\rangle$. To lowest order in the applied field amplitude, only the $n = 1$ state is excited so I can set

[1] In this example, condition (23.5) is necessary but not sufficient for perturbation theory to be valid. It turns out that a *necessary and sufficient* condition is that the pulse area be much less than unity.

$$|\psi(t)\rangle \approx |0\rangle + c_1(t)e^{-i\omega_0 t}|1\rangle . \tag{23.16}$$

It then follows that

$$\langle x(t)\rangle = \sqrt{\frac{\hbar}{2m\omega_0}}\left[x_{01}c_1(t)e^{-i\omega_0 t} + x_{10}c_1^*(t)e^{i\omega_0 t}\right], \tag{23.17}$$

where

$$\begin{aligned} x_{nn'} &= \langle n|\hat{x}|n'\rangle = \sqrt{\frac{\hbar}{2m\omega_0}}\langle n|a + a^\dagger|n'\rangle \\ &= \sqrt{\frac{\hbar}{2m\omega_0}}\left(\sqrt{n'}\delta_{n,n'-1} + \sqrt{n'+1}\delta_{n,n'+1}\right), \end{aligned} \tag{23.18}$$

such that

$$x_{01} = x_{10} = \sqrt{\frac{\hbar}{2m\omega_0}}. \tag{23.19}$$

In lowest order perturbation theory, $c_0(t) = 1$, and Eqs. (23.8), (23.14), and (23.19) can be used to calculate

$$\begin{aligned} c_1(t) &= -(i/\hbar)\int_0^t dt' V_{10}(t')e^{i\omega_0 t'} = iqE_0\sqrt{\frac{1}{2\hbar m\omega_0}}\int_0^t dt' \cos(\omega t')\, e^{i\omega_0 t'} \\ &= \frac{qE_0}{2}\sqrt{\frac{1}{2\hbar m\omega_0}}\left[\frac{e^{i(\omega_0+\omega)t}-1}{\omega_0+\omega} + \frac{e^{i(\omega_0-\omega)t}-1}{\omega_0-\omega}\right], \end{aligned} \tag{23.20}$$

which, together with Eqs. (23.17) and (23.19), implies that

$$\begin{aligned} \langle x(t)\rangle &= \frac{qE_0}{4m\omega_0}\left[\frac{e^{i\omega t}-e^{-i\omega_0 t}}{\omega_0+\omega} + \frac{e^{-i\omega t}-e^{-i\omega_0 t}}{\omega_0-\omega}\right] + \text{c.c.} \\ &= \frac{qE_0}{m}\frac{\cos(\omega_0 t) - \cos(\omega t)}{\omega^2-\omega_0^2}. \end{aligned} \tag{23.21}$$

In this case, I can get the *exact* solution for $\langle x(t)\rangle$ from the equation of motion

$$\frac{d^2\langle x(t)\rangle}{dt^2} + \omega_0^2\langle x(t)\rangle = \frac{qE_0}{m}\cos(\omega t). \tag{23.22}$$

With initial conditions $\langle x(0)\rangle = 0$, $d\langle x(0)\rangle/dt = 0$, corresponding to the oscillator in its ground state, the solution of this equation is

$$\langle x(t)\rangle = \frac{qE_0}{m}\frac{\cos(\omega_0 t) - \cos(\omega t)}{\omega^2 - \omega_0^2}. \tag{23.23}$$

This is an amazing result—the *exact* solution agrees with the solution to *first order* in perturbation theory. The reason for this is clear. The response of a simple harmonic oscillator to an applied electric field is always linear in the field. Since higher order perturbation theory would produce terms that are *nonlinear* in the field amplitude, any such contributions to $\langle x(t)\rangle$ must vanish! This does not mean that successive terms in higher order perturbation theory vanish. On the contrary, such terms drive the quantum oscillator up and down the ladder of n states; however, they do so in such a fashion that these higher order terms do not contribute to $\langle x(t)\rangle$.

23.2 Adiabatic Approximation

The adiabatic approximation seems mysterious and even magical, but it is really based on a simple premise. Consider the equation of motion

$$\dot{\mathbf{a}}(t) = -i\mathbf{W}(t)\mathbf{a}(t) \tag{23.24}$$

where

$$\mathbf{W}(t) = \mathbf{H}(t)/\hbar, \tag{23.25}$$

and I have changed the notation somewhat by representing matrices by boldface variables, rather than with underscores. You can think of $\mathbf{W}(t)$ as an effective Hamiltonian in frequency units. Imagine that $\mathbf{W}$ is *constant*. Then I could simply obtain the eigenvalues and eigenkets of $\mathbf{W}$ and expand the state vector in terms of these eigenkets. The state populations remain constant and their amplitudes acquire a phase factor, $e^{-i\omega_n t}$, where the ω_n are eigenvalues of $\mathbf{W}$. If $\mathbf{W}$ is *not* constant, it no longer possesses time-independent eigenkets; however, if $\mathbf{W}(t)$ does not contain Fourier components equal or nearly equal to the frequency separation of the *instantaneous* eigenvalues of $\mathbf{W}(t)$ (that is, the eigenvalues at time t), then transitions between the instantaneous eigenkets of $\mathbf{W}(t)$ can be neglected. In that limit it is relatively easy to solve for the state amplitudes.

To see this more formally, I set

$$\mathbf{x}(t) = \mathbf{T}(t)\mathbf{a}(t), \tag{23.26}$$

where the unitary matrix $\mathbf{T}(t)$ is chosen such that

$$\boldsymbol{\Lambda}(t) = \mathbf{T}(t)\mathbf{W}(t)\mathbf{T}^{\dagger}(t) \tag{23.27}$$

is diagonal. Then

$$\begin{aligned}
\dot{\mathbf{x}}(t) &= \mathbf{T}(t)\dot{\mathbf{a}}(t) + \dot{\mathbf{T}}(t)\mathbf{a}(t) \\
&= -i\mathbf{T}(t)\mathbf{W}(t)\mathbf{a}(t) + \dot{\mathbf{T}}(t)\mathbf{a}(t) \\
&= -i\mathbf{T}(t)\mathbf{W}(t)\mathbf{T}^{\dagger}(t)\mathbf{x}(t) + \dot{\mathbf{T}}(t)\mathbf{T}^{\dagger}(t)\mathbf{x}(t) \\
&= -i\boldsymbol{\Lambda}(t)\mathbf{x}(t) + \dot{\mathbf{T}}(t)\mathbf{T}^{\dagger}(t)\mathbf{x}(t) .
\end{aligned} \tag{23.28}$$

The adiabatic approach is useful only if the $\dot{\mathbf{T}}(t)\mathbf{T}^{\dagger}(t)$ term can be neglected, which is generally the case if the frequency difference between the diagonal elements of $\boldsymbol{\Lambda}(t)$ is much larger than the inverse of the time interval in which $\mathbf{T}(t)$ changes significantly. In that limit,

$$x_n(t) \approx x_n(0) \exp\left(-i\int_0^t \Lambda_{nn}\left(t'\right) dt'\right) \tag{23.29}$$

and

$$\mathbf{a}(t) = \mathbf{T}^{\dagger}(t)\mathbf{x}(t). \tag{23.30}$$

23.2.1 A Simple Example

Let me consider the interaction of the spin magnetic moment of an electron with the magnetic field

$$\mathbf{B} = B_0 \left\{\mathbf{u_z} \cos\left[\theta(t)\right] + \mathbf{u}_x \sin\left[\theta(t)\right]\right\}. \tag{23.31}$$

where $\theta(t)$ is some arbitrary function of time. The Hamiltonian in frequency units is

$$\begin{aligned}
\mathbf{W}(t) &= -\frac{\boldsymbol{\mu}_s\cdot\mathbf{B}}{\hbar} = \frac{e\mathbf{S}\cdot\mathbf{B}}{m_e\hbar} = \frac{eB_0}{m_e\hbar}\left\{\mathbf{S}_x \sin\left[\theta(t)\right] + \mathbf{S}_z \cos\left[\theta(t)\right]\right\} \\
&= \frac{\omega_0}{2}\begin{pmatrix} \cos\left[\theta(t)\right] & \sin\left[\theta(t)\right] \\ \sin\left[\theta(t)\right] & -\cos\left[\theta(t)\right] \end{pmatrix},
\end{aligned} \tag{23.32}$$

where

$$\omega_0 = \frac{eB_0}{m_e}. \tag{23.33}$$

It is a simple matter to diagonalize this matrix. The eigenvalues are $w_{1,2} = \pm\omega_0/2$ (which are time-independent in this problem) and the eigenkets are

$$|\uparrow\rangle' = \cos\left(\frac{\theta(t)}{2}\right)|\uparrow\rangle + \sin\left(\frac{\theta(t)}{2}\right)|\downarrow\rangle ; \tag{23.34a}$$

$$|\downarrow\rangle' = \cos\left(\frac{\theta(t)}{2}\right)|\downarrow\rangle - \sin\left(\frac{\theta(t)}{2}\right)|\uparrow\rangle . \tag{23.34b}$$

The matrix $\mathbf{T}(t)$ that diagonalizes $\mathbf{W}(t)$ is

$$\mathbf{T}(t) = \begin{pmatrix} \cos\left(\frac{\theta(t)}{2}\right) & -\sin\left(\frac{\theta(t)}{2}\right) \\ \sin\left(\frac{\theta(t)}{2}\right) & \cos\left(\frac{\theta(t)}{2}\right) \end{pmatrix} \tag{23.35}$$

and

$$\dot{\mathbf{T}}(\mathbf{t})\mathbf{T}^{\dagger}(\mathbf{t}) = \frac{\dot{\theta}}{2}\begin{pmatrix} 0 & -1 \\ 1 & 0 \end{pmatrix}. \tag{23.36}$$

Therefore, if $\left|\dot{\theta}\right| \ll \omega_0$, the adiabatic approximation should be good. In that limit, if $|\psi(0)\rangle = |\downarrow\rangle$, the spin *stays adiabatically in the instantaneous spin down state.* In other words, the state vector is approximately

$$|\psi(t)\rangle = |\downarrow(t)\rangle' = \cos\left(\frac{\theta(t)}{2}\right)|\downarrow\rangle - \sin\left(\frac{\theta(t)}{2}\right)|\uparrow\rangle , \tag{23.37}$$

such that the probability to find the spin down in a basis quantized along the z-axis is simply

$$P_{\downarrow}(t) = \cos^2\left(\frac{\theta(t)}{2}\right). \tag{23.38}$$

The solution given in Eq. (23.38) can be compared with the exact solution, $P_{\downarrow}(t) = \left|a_{\downarrow}(t)\right|^2$, where $a_{\downarrow}(t)$ is obtained as a numerical solution of the differential equations,

$$\begin{pmatrix} \dot{a}_{\uparrow}(t) \\ \dot{a}_{\downarrow}(t) \end{pmatrix} = -i\frac{\omega_0}{2}\begin{pmatrix} \cos[\theta(t)] & \sin[\theta(t)] \\ \sin[\theta(t)] & -\cos[\theta(t)] \end{pmatrix}\begin{pmatrix} a_{\uparrow}(t) \\ a_{\downarrow}(t) \end{pmatrix}, \tag{23.39}$$

subject to the initial condition $a_{\downarrow}(0) = 1$. As an example, I choose

$$\theta(t) = \frac{\pi}{2}\left(1 - e^{-t^2/T^2}\right)\Theta(t), \tag{23.40}$$

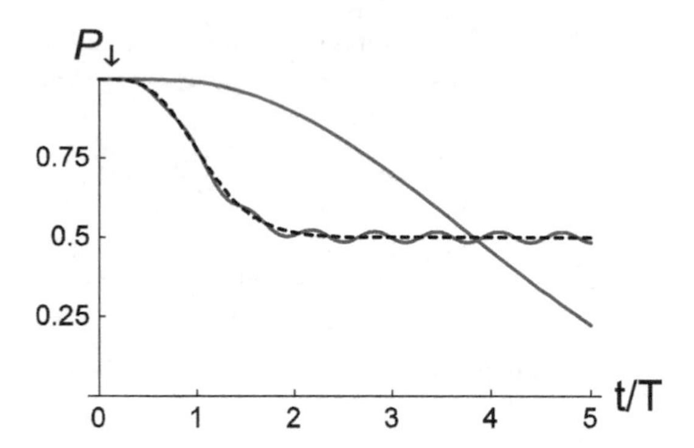

Fig. 23.1 A graph of $P_\downarrow(t)$ vs t/T. The upper solid blue curve is for $\omega_0 T = 0.5$ and the lower solid red curve for $\omega_0 T = 10$. The dashed black curve is the adiabatic solution

where $\Theta(t)$ is a Heaviside function, such that θ varies from 0 to $\pi/2$ in a time of order T. I expect the adiabatic approximation to be good if $\omega_0 T \gg 1$. In Fig. 23.1, the numerical solution for $P_\downarrow(t)$ is plotted as a function of t/T for $\omega_0 T = 0.5$ (upper solid curve) and $\omega_0 T = 10$ (lower solid curve), along with the adiabatic solution (dashed curve). You can see that the adiabatic solution is very good for $\omega_0 T = 10$.

23.3 Sudden Approximation

The "opposite" of the adiabatic approximation is the *sudden approximation*. In the sudden approximation, the Hamiltonian is changed suddenly from one Hamiltonian, say $\hat{H}_1$, to another, say $\hat{H}_2$, at time $t = t_0$. If this is the case the result is simple. You can expand the wave function at $t = t_0$ in terms of the *new* eigenkets. The state vector then evolves under the new Hamiltonian with this initial condition. The sudden approximation is valid if the Hamiltonian is changed in a time that is much shorter than *all* the inverse transition frequencies of the original Hamiltonian.

For example, consider the interaction of the spin magnetic moment of the electron with a magnetic field

$$\mathbf{B} = B_0 \left[\mathbf{u_z} \cos[\theta(t)] + \mathbf{u}_x \sin[\theta(t)] \right]. \tag{23.41}$$

Suppose that at $t = 0$, the magnetic field direction is switched suddenly from the z direction ($\theta = 0$) to some angle θ_f. The initial state before switching is spin down (in the original basis). I write the initial state in terms of the *new* eigenkets as

$$|\downarrow\rangle = |\psi(0)\rangle = \cos\left(\theta_f/2\right) |\downarrow\rangle' + \sin\left(\theta_f/2\right) |\uparrow\rangle', \tag{23.42}$$

where the new eigenkets $\left(|\downarrow\rangle', |\uparrow\rangle'\right)$ are given by Eq. (23.34). The eigenfrequency associated with the $|\uparrow\rangle'$ ket is $\omega_0/2$ and that associated with the $|\downarrow\rangle'$ ket is $(-\omega_0/2)$, with ω_0 given by Eq. (23.33). As a consequence, the state vector for $t > 0$ is

$$|\psi(t)\rangle = e^{i\omega_0 t/2} \cos\left(\theta_f/2\right)|\downarrow\rangle' + \sin\left(\theta_f/2\right) e^{-i\omega_0 t/2}|\uparrow\rangle'. \tag{23.43}$$

Using Eq. (23.34) to re-express the new kets in terms of the original basis eigenkets, I find

$$\begin{aligned}
|\psi(t)\rangle &= \begin{bmatrix} \cos\left(\theta_f/2\right) e^{i\omega_0 t/2}\left(\cos\left(\theta_f/2\right)|\downarrow\rangle - \sin\left(\theta_f/2\right)|\uparrow\rangle\right) \\ +e^{-i\omega_0 t/2} \sin\left(\theta_f/2\right)\left(\cos\left(\theta_f/2\right)|\uparrow\rangle + \sin\left(\theta_f/2\right)|\downarrow\rangle\right) \end{bmatrix} \\
&= \left[\cos^2\left(\theta_f/2\right) e^{i\omega_0 t/2} + \sin^2\left(\theta_f/2\right) e^{-i\omega_0 t/2}\right]|\downarrow\rangle \\
&\quad - \sin\left(\theta_f/2\right)\cos\left(\theta_f/2\right)\left[e^{i\omega_0 t/2} - e^{-i\omega_0 t/2}\right]|\uparrow\rangle \\
&= \left[\cos\left(\omega_0 t/2\right) + i\sin\left(\omega_0 t/2\right)\cos\theta_f\right]|\downarrow\rangle - i\sin\theta_f \sin\left(\omega_0 t/2\right)|\uparrow\rangle,
\end{aligned} \tag{23.44}$$

such that

$$P_\downarrow(t) = \cos^2\left(\frac{\omega_0 t}{2}\right) + \sin^2\left(\frac{\omega_0 t}{2}\right)\cos^2\theta_f. \tag{23.45}$$

Note that if $\theta_f = \pi$, $P_\downarrow(t) = 1$ since the initial state is an eigenstate of the new Hamiltonian.

I can check the validity of the sudden approximation by again taking $\theta(t)$ given by Eq. (23.40). For $\theta_f = \pi/2$

$$P_\downarrow(t) = \cos^2\left(\frac{\omega_0 t}{2}\right) = \cos^2\left[\frac{\omega_0 T}{2}(t/T)\right]. \tag{23.46}$$

The sudden approximation should be valid if $\omega_0 T \ll 1$, since this constitutes a rapid change in the field direction. In Fig. 23.2, the numerical solution for $P_\downarrow(t)$ is plotted as a function of t/T for $\omega_0 T = 0.2$ and in Fig. 23.3 for $\omega_0 T = 2$, along with the sudden solution (dashed curve). You can see that the sudden solution is almost exact for $\omega_0 T = 0.2$.

23.4 Summary

In this chapter, I have explored several techniques for obtaining approximate solutions to the time-dependent Schrödinger equation when a time-dependent interaction is present. For weak interaction potentials, a perturbative approach can be used. Under certain conditions it is also possible to use adiabatic or sudden approximations. In general the adiabatic approximation is valid if the matrix elements coupling any instantaneous eigenstates of the system does not contain the

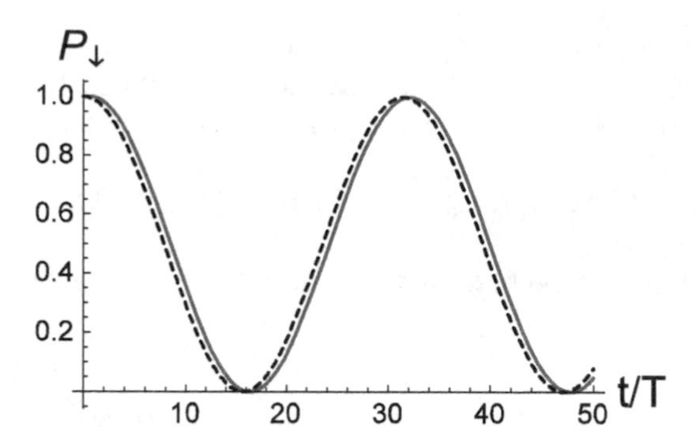

Fig. 23.2 A graph of $P_\downarrow(t)$ vs t/T. The solid red curve is for $\omega_0 T = 0.2$. The dashed black curve is the sudden approximation solution

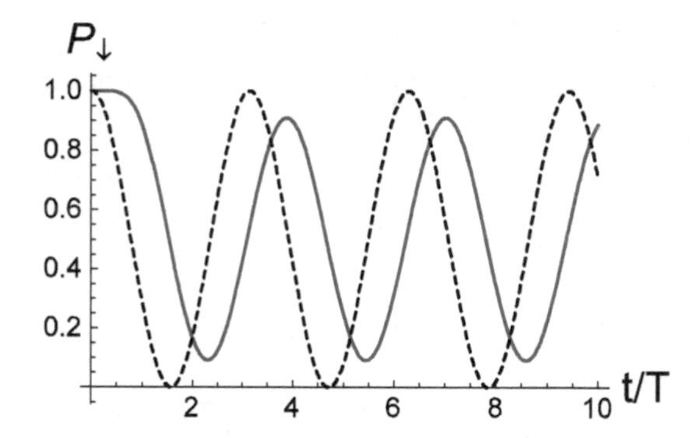

Fig. 23.3 A graph of $P_\downarrow(t)$ vs t/T. The solid red curve is for $\omega_0 T = 2$. The dashed black curve is the sudden approximation solution

Fourier components needed to drive transitions between those states. The sudden approximation is valid when the change in the interaction potential occurs on a time scale that is sufficiently fast to contain Fourier components that cover all the relevant states that can be coupled by the potential.

23.5 Problems

1. Under what conditions is the adiabatic approximation valid? Under what conditions is the sudden approximation valid? Explain.

Problems 2–11 involve the interaction of a two-level atom with a classical optical field. In the interaction representation and the RWA, the equations for the state amplitudes are given by Eqs. (22.77), namely

$$\dot{c}_1(t) = -\frac{i}{2}\Omega_0^* e^{-i\delta t} c_2(t);$$

$$\dot{c}_2(t) = -\frac{i}{2}\Omega_0 e^{i\delta t} c_1(t) ,$$

and, in the field interaction representation, by Eqs. (22.93), namely

$$\dot{\tilde{c}}_1(t) = i\frac{\delta(t)}{2}\tilde{c}_1(t) - i\frac{|\Omega_0(t)|}{2}\tilde{c}_2(t) ;$$

$$\dot{\tilde{c}}_2(t) = -i\frac{\delta(t)}{2}\tilde{c}_2(t) - i\frac{|\Omega_0(t)|}{2}\tilde{c}_1(t) ,$$

where $\Omega_0(t)$ is the Rabi frequency,

$$\delta(t) = \delta + \dot{\phi}(t) = \omega_0 - \left[\omega - \dot{\phi}(t)\right] ,$$

$\delta = \omega_0 - \omega$ is the atom-field detuning, and $\phi(t)$ if the phase of the applied field. In all these problems, assume that $c_1(-\infty) = \tilde{c}_1(-\infty) = 1$, $c_2(-\infty) = \tilde{c}_2(-\infty) = 0$, and $\delta > 0$.

2. Solve for $P_2(\infty) = |c_2(\infty)|^2$ using first-order time-dependent perturbation theory with $\phi(t) = 0$ and

$$(1) \quad \Omega_0(t) = \Omega_0 \exp[-(t/T)^2];$$

$$(2) \quad \Omega_0(t) = \Omega_0 \operatorname{sech}(\pi t/T) ;$$

$$(3) \quad \Omega_0(t) = \begin{cases} \Omega_0 & 0 \le t \le T \\ 0 & \text{otherwise} \end{cases} .$$

In cases (1) and (2) show that $P_2(\infty)$ is consistent with what you might have expected from the time-energy uncertainty principle, that is, the excitation probability falls off exponentially with some power of δT, but that in case (3), the fall off is as an inverse power law. Why is this so?

3. You might think that the criterion for applying perturbation theory in the previous problem is that $|\tilde{c}_2(t)|^2 \ll 1$. It turns out that this condition is necessary, but not sufficient. To see this, you can use the *exact* solution for the hyperbolic secant pulse. Rosen and Zener solved this problem in 1932 and found that, if

$$\Omega_0(t) = (A/T) \operatorname{sech}(\pi t/T) ,$$

where

$$A = \int_{-\infty}^{\infty} \Omega_0(t) dt$$

is the pulse area, then

$$P_2(\infty) = |\tilde{c}_2(\infty)|^2 = \sin^2(A/2)\text{sech}^2\,(\delta T/2)\,.$$

Show that this result is consistent with the perturbation theory result only if $A \ll 1$. Moreover, show that if $\delta T \gg 1$, it is possible that the perturbation theory result is wrong even though $|c_2(\infty)|^2 \ll 1$. Explain why the fact that $P_2(\infty)$ saturates with increasing A is consistent with what you learned from the photoelectric effect.

4–5. Prove that, in the adiabatic limit [that is, if $\delta(t)$ and $\Omega_0(t)$ are slowly varying over the duration of the field pulse],

$$|c_2(t)|^2 = |\tilde{c}_2(t)|^2 \approx \frac{1}{2}\left[1 - \frac{\delta(t)}{\Omega(t)}\right],$$

where

$$\Omega(t) = \sqrt{\delta^2 + |\Omega_0(t)|^2}$$

is the generalized Rabi frequency. To solve this problem you need to instantaneously diagonalize the matrix

$$\mathbf{A}(t) = \frac{1}{2}\begin{pmatrix} -\delta(t) & |\Omega_0(t)| \\ |\Omega_0(t)| & \delta(t) \end{pmatrix}$$

that appears in the field interaction representation equations. This can be accomplished with a matrix of the form

$$\mathbf{T}(t) = \begin{pmatrix} \cos\left[\theta(t)\right] & -\sin\left[\theta(t)\right] \\ \sin\left[\theta(t)\right] & \cos\left[\theta(t)\right] \end{pmatrix}.$$

Find the value of $\theta(t)$ for which

$$\mathbf{T}(t)\mathbf{A}(t)\mathbf{T}^{\dagger}(t) = \frac{1}{2}\begin{pmatrix} -\Omega(t) & 0 \\ 0 & \Omega(t) \end{pmatrix}.$$

6. Obtain an integral expression for $c_2(t)$ using first-order time-dependent perturbation theory, assuming that $\phi(t) = 0$. Assume that $\delta \gg \left|\dot{\Omega}_0(t)\right|$, and use integration by parts to show that

$$P_2(t) = |c_2(t)|^2 \approx \frac{[\Omega_0(t)]^2}{4\delta^2}.$$

Prove that this result is consistent with the previous problem. This would seem to imply that $P_2(\infty) \sim 0$, whereas you found this *not* to be the case in Problem 23.2.

This result illustrates the fact that the adiabatic approximation (as do many asymptotic expansions) misses terms that are exponentially small in the adiabaticity parameter (which is $\left|\dot{\Omega}_0(t)\right|/\delta$ in this example).

7–8. Numerically calculate the transition probability $P_2(t)$ as a function of time for a pulse envelope,

$$|\Omega_0(t)| = |\Omega_0| \exp[-(t/T)^2]$$

and detunings,

$$\begin{aligned} &(1) \quad \delta(t) = \delta_0 ; \\ &(2) \quad \delta(t) = -\delta_0 \tan^{-1}(t/T) , \end{aligned}$$

where $T > 0$ is the pulse duration. Use dimensionless variables with $\tau = t/T$ and take (i) $\delta_0 T = 3$ and $|\Omega_0| T = 3$ and (ii) $\delta_0 T = 10$ and $|\Omega_0| T = 10$. Plot $P_2(\tau T) = |\tilde{c}_2(\tau T)|^2$ as a function of τ for cases (1)(i), (1)(ii), (2)(i), and (2)(ii). Show that in case (1)(ii), the population returns to state 1 at the end of the pulse while in case (2)(ii) (*adiabatic switching*) the population is transferred to state 2 by the pulse. Compare your numerical solutions with the predictions of adiabatic following obtained in Problem 23.4–5.

9. Give a simple argument to explain why the Bloch vector stays aligned (or anti-aligned) with the pseudofield vector if $\left|\dot{\Omega}(t)/\Omega(t)\right| \ll 1, \Omega_0(\pm\infty) = 0, \delta(\pm\infty) \neq 0$, and $\tilde{c}_1(-\infty) = 1$. Use this result to explain why the atom returns to its ground state in case *i* of the previous problem, but is transferred to its excited state in case *ii* of the previous problem.

10. Prove that, in the sudden approximation limit [that is, if the Rabi frequency is changed "instantaneously" from 0 to Ω_0 at $t = 0$], for $t > 0$,

$$|\tilde{c}_2(t)|^2 \approx \frac{|\Omega_0|^2}{\Omega^2} \sin^2\left(\frac{\Omega t}{2}\right),$$

where

$$\Omega = \sqrt{\delta^2 + |\Omega_0|^2}$$

is the generalized Rabi frequency.

11. Numerically calculate the transition probability $P_2(t)$ as a function of time for

$$|\Omega_0(t)| = |\Omega_0| \left(1 - \exp[-(t/T)^2]\right)$$

and $\phi(t) = 0$. Use dimensionless variables with $\tau = t/T$ and take (i) $\delta T = 0$ and $|\Omega_0| T = 0.1$, (ii) $\delta T = 0$ and $|\Omega_0| T = 3$, and (iii) $\delta T = 3$ and $|\Omega_0| T = 0.1$.

Numerically calculate and plot $P_2(\tau) = |\tilde{c}_2(\tau)|^2$ as a function of τ for these cases. Compare your numerical solutions with the predictions of the sudden approximation solution obtained in the previous problem. Under what conditions do you expect the sudden approximation to give results that are in good agreement with the exact solutions?

12–13. Consider a one-dimensional harmonic oscillator (modeled as a mass on a spring having mass m and charge q) driven by an external electric field $E(t)$. The Hamiltonian is

$$\hat{H}(t) = \frac{\hat{p}^2}{2m} + \frac{1}{2}m\omega_0^2\hat{x}^2 - q\hat{x}E(t).$$

Assume that the oscillator is in its ground state at $t = 0$. Treating the last term in the Hamiltonian as a perturbation, find the state vector as a function of time. Using this state vector, calculate $\langle x(t)\rangle$. Now calculate $\langle x(t)\rangle$ *exactly*, using $d\langle x(t)\rangle/dt = (i\hbar)^{-1}\left\langle\left[\hat{x}, \hat{H}(t)\right]\right\rangle$ and show that it agrees with the result of first order perturbation theory. How can this be?

14. Consider a one-dimensional harmonic oscillator that is in its ground state at $t = 0$. Suppose that the natural frequency ω of the oscillator is changed as a function of time, that is $\omega \to \omega(t)$ for $t > 0$. If the frequency is changed adiabatically, find the wave function and average energy as a function of time. What is the condition that must be satisfied for the adiabatic approximation to hold?

15. Repeat the calculation of the previous problem assume the frequency is changed suddenly from ω to 2ω at $t = 0$. Your answer will involve a sum of terms involving integrals that are tabulated. The sum converges very rapidly (why?) so that only a few terms are needed. What is the condition that must be satisfied for the sudden approximation to hold?

Chapter 24
Decay of a Discrete State into a Continuum of States: Fermi's Golden Rule

The last topics I will cover are Fermi's Golden Rule and irreversible decay. These are pretty interesting topics that involve situations in which a discrete state of a quantum system is coupled to an infinite *continuum* of quantum states. Although one starts with a Hermitian Hamiltonian, there is *irreversible* behavior in the system. This is approximately true in many cases of practical interest. For example, an atom that is prepared in an excited state decays as a result of its interaction with the continuum of vacuum field modes. A hydrogen atom placed in a high frequency optical field is photo-ionized into a continuum of (Coulomb modified) free-particle states. The reason we can get irreversible behavior in these cases is connected with the fact that the quantization volume goes to infinity. This feature in already seen in the quantum revivals of a wave packet in the infinite square square well discussed in Chap. 6. There are always quantum revivals, but the revival time goes to infinity as the well size approaches infinity. In other words, an outgoing wave packet can never be reflected back by the boundary of the quantization volume. In problems of this nature, I start by writing equations in which *all* states are discrete and then take the limit in which *some* of the states form a continuum.

24.1 Discrete State Coupled to a Continuum

The generic problem that I am interested in is illustrated in Fig. 24.1. A discrete state, labeled by 0 and having energy E_0, is coupled to a large number of states labeled by n. Eventually, I take the limit of the number of states going to infinity and the energy spacing ϵ between the states going to zero. In doing so I will replace any sums over n by an integral over energy using the prescription

$$\sum_n \rightarrow \int \rho(E)dE, \tag{24.1}$$

P.R. Berman, *Introductory Quantum Mechanics*, UNITEXT for Physics,
https://doi.org/10.1007/978-3-319-68598-4_24

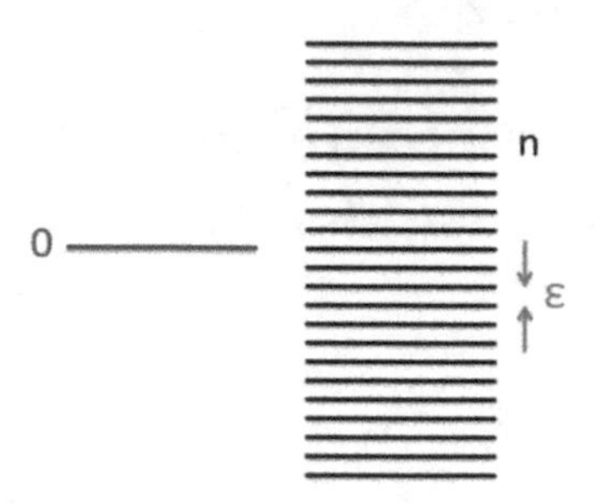

Fig. 24.1 Discrete state 0 coupled to a discrete continuum of states labeled by n. The energy difference between successive levels is ϵ

where $\rho(E)$ is referred to as the (energy) *density of states*. I have already calculated the density of states for the radiation field in Chap. 1 and for matter waves in Chap. 5.

There are two general classes of problems in which levels schemes such as those shown in Fig. 24.1 are encountered. In processes such as spontaneous emission, the Hamiltonian is of the form

$$\hat{H} = \hat{H}_{\text{atom}} + \hat{H}_{\text{field}} + \hat{V}_{AF}, \tag{24.2}$$

where $\hat{H}_{\text{atom}}$ is the atomic Hamiltonian, $\hat{H}_{\text{field}}$ is the (quantized) Hamiltonian for the vacuum radiation field, and $\hat{V}_{AF}$ is the atom–field interaction potential. The eigenstates of $\hat{H}_{\text{field}}$ constitute the continuum levels of the problem. The atom is prepared initially in some excited state and decays to a lower energy state as a result of the interaction with the vacuum field. In this problem, all operators are bona-fide, time-independent Schrödinger operators.

In the second class of problems such as photoionization, a time-dependent classical field drives an electron in an atom from some initial bound state to a continuum of unbound energy states. Although the ionized electron still sees a residual field from the ion that is left behind, it is sometimes a good approximation to consider the final states of the electron as free-particle states. The external field is assumed to have a constant amplitude and to oscillate at frequency ω.

In addressing both types of problems, I assume that at $t = 0$ the atom is prepared in eigenket $|0\rangle$ and begins to make transitions to states $|n\rangle$. In the interaction representation, the state amplitudes evolve according to

$$\dot{c}_0(t) = -\frac{i}{\hbar}\sum_n V_{0n}(t)e^{-i\omega_{n0}t}c_n(t) \tag{24.3}$$

$$\dot{c}_n(t) = -\frac{i}{\hbar}V_{n0}(t)e^{i\omega_{n0}t}c_0(t), \tag{24.4}$$

where

$$\omega_{n0} = (E_n - E_0)/\hbar. \tag{24.5}$$

and $V_{n0}(t)$ are matrix elements of the interaction that couple the initial state to the continuum states. For the problems I consider, either $V_{n0}(t)$ is constant or it oscillates at frequency ω_L. In the latter case, I will make the RWA and replace $V_{n0}(t)$ by $V_{n0}e^{-i\omega_L t}/2$; the net effect is that of a constant perturbation with

$$\omega_0 \to \omega_0' = \omega_0 + \omega_L. \tag{24.6}$$

Thus without loss of generality, I can take the equations for the state amplitudes to be

$$\dot{c}_0(t) = -\frac{i}{\hbar}\sum_n V_{0n}e^{-i\omega_{n0}t}c_n(t) \tag{24.7a}$$

$$\dot{c}_n(t) = -\frac{i}{\hbar}V_{n0}e^{i\omega_{n0}t}c_0(t), \tag{24.7b}$$

where the coupling matrix elements are *constants* and any time dependence of an applied oscillatory field is incorporated by a redefinition of the initial state energy

$$E_0 \to E_0' = \hbar\omega_0' = \hbar\left(\omega_0 + \omega_L\right). \tag{24.8}$$

I integrate Eq. (24.7b) and substitute the result into Eq. (24.7a) to obtain

$$\dot{c}_0(t) = -\frac{1}{\hbar^2}\sum_n |V_{n0}|^2 \int_0^t dt' e^{-i\omega_{n0}(t-t')}c_0(t'). \tag{24.9}$$

The prescription (24.1) is used to convert the sum to an integral,

$$\dot{c}_0(t) = -\frac{1}{\hbar^2}\int dE\rho(E)\left|V(E)\right|^2 \int_0^t dt' e^{-i(E-E_0)(t-t')/\hbar}c_0(t'), \tag{24.10}$$

where I have set $V_{n0} \equiv V(E)$ to allow for the possibility that the matrix elements depend on energy. To proceed further I need to know something about the nature of $\rho(E)$ and $V(E)$. For the moment I assume that these quantities are slowly varying functions of energy compared with the exponential factor in the integrand. Since the exponential factor makes it maximum contribution at $E = E_0$, I evaluate both $\rho(E)$ and $V(E)$ at $E = E_0$ and remove them from the integral. Moreover I assume that the range of allowed energies in the integration extends from $-\infty$ to ∞. If this is the case, then

$$\begin{aligned}\dot{c}_0(t) &= -\frac{\left|V(E_0)\right|^2}{\hbar^2}\rho(E_0)\int_0^t dt' \int_{-\infty}^{\infty} dE e^{-i(E-E_0)(t-t')/\hbar}c_0(t') \\ &= -\frac{2\pi\left|V(E_0)\right|^2}{\hbar}\rho(E_0)\int_0^t dt'\delta\left(t-t'\right)c_0(t') = -\frac{\Gamma}{2}c_0(t),\end{aligned} \tag{24.11}$$

where

$$\Gamma = \frac{2\pi}{\hbar} |V(E_0)|^2 \rho(E_0) \tag{24.12}$$

is the decay rate of the initial state population. (In deriving this result, I used the relationship $\int_0^t dt' \delta(t - t') = 1/2$, based on the fact that $\delta(t - t')$ is a symmetric function about $t' = t$.) Equation (24.12) is known as Fermi's *Golden Rule*, although it is usually derived using a somewhat different approach.

From Eq. (24.11), we see that

$$c_0(t) = e^{-\Gamma t/2}; \qquad |c_0(t)|^2 = e^{-\Gamma t}. \tag{24.13}$$

Both the initial state amplitude and initial state probability undergo *exponential decay*. The irreversible behavior occurs because I have taken an infinite quantization volume—the emitted ionized electron or emitted photon cannot return to re-excite the quantum system undergoing decay.

Although the behavior is irreversible, the fact that the decay is *exponential* depends on the assumptions that were made concerning the density of states and $V(E)$. If both $\rho(E)$ and $V(E)$ are constant for $-\infty < E < \infty$, then Eq. (24.11) is *exact*. This results in what is known is a *Markov process* since the effective correlation time of the continuum of states giving rise to the decay is equal to zero. The delta function appearing in the integrand of Eq. (24.11) is a signature of the fact that the process is Markovian—it has no temporal memory. At any instant of time, the decay is independent of past events—it is *always* exponential. Of course, both $\rho(E)$ and $V(E)$ usually depend on energy. Moreover the energy spectrum is usually bounded from below since negative kinetic energies of particles and negative energies of photons are not allowed. As a consequence, the decay of a discrete state into the continuum can never be *exactly* exponential. However, to a very good approximation, processes such as spontaneous decay and particle decay can be represented as Markovian in nature. I will return to this point in discussing the Zeno effect later in this chapter. I will also discuss how the decay process is modified if the continuum is bounded from above and below. For the present, however, I consider two examples, photoionization and spontaneous decay.

24.1.1 Photoionization

As a first example of Fermi's Golden Rule, I consider the photoionization of hydrogen from its ground state produced by a high frequency (X-ray) radiation field. It is assumed that the frequency of the field is much greater than the ionization energy of hydrogen divided by h. In this limit, the electron that emerges can be treated in first approximation as a free particle. Moreover, since the matrix element involves a coupling from the ground state, the integral that determines the value of the matrix element is restricted to radii on the order of the Bohr radius. As a consequence, I use the dipole approximation and assume an interaction potential of the form

$$\hat{V}(t) = -\hat{\mathbf{p}}_e \cdot \mathbf{E}_0 \cos(\omega_L t) = e\mathbf{E}_0 \cdot \hat{\mathbf{r}} \cos(\omega_L t), \tag{24.14}$$

where $\mathbf{E}_0$ is the field amplitude, $\hat{\mathbf{p}}_e$ is the dipole moment operator of the atom, and $\hat{\mathbf{r}}$ is the position operator of the atom. In the rotating wave approximation, the equations of motion for the state amplitudes are

$$\dot{c}_0(t) = -\frac{i}{\hbar}\sum_n V_{0n} e^{-i(E_n - E_0 - \hbar\omega_L)t/\hbar} c_n(t) \tag{24.15a}$$

$$\dot{c}_n(t) = -\frac{i}{\hbar} V_{n0} e^{i(E_n - E_0 - \hbar\omega_L)t/\hbar} c_0(t), \tag{24.15b}$$

where

$$V_{0n} = -\langle n|\hat{\mathbf{p}}_e \cdot \mathbf{E}_0|0\rangle/2 \tag{24.16}$$

is one-half the matrix element of the interaction operator between the ground state $|0\rangle = |n = 1, \ell = 0, m = 0\rangle$ of the electron in hydrogen and the final state $|n\rangle$ of the electron, which is approximated as a free-particle state. For the ionization problem, I have to evaluate matrix elements of $e\mathbf{E}_0 \cdot \hat{\mathbf{r}}$ between the final free particle states (quantized using periodic boundary conditions having period L in all directions),

$$(\psi_f)_n = \frac{e^{i\mathbf{k}_n \cdot \mathbf{r}}}{\sqrt{L^3}} \to \frac{e^{i\mathbf{k}\cdot\mathbf{r}}}{\sqrt{L^3}}, \tag{24.17}$$

and the initial state

$$\psi_0 = \frac{e^{-r/a_0}}{\sqrt{\pi a_0^3}}, \tag{24.18}$$

where a_0 is the Bohr radius. In the final state wave function

$$k_{nx} = \frac{2\pi n_y}{L}; \quad k_{ny} = \frac{2\pi n_y}{L}; \quad k_{nz} = \frac{2\pi n_z}{L}, \tag{24.19}$$

and the n_x, n_y, n_z are integers (positive, negative, or zero), as discussed in the Appendix of Chap. 5. To go over to continuum states I use the prescription given by Eq. (5.161), namely

$$\sum_{n_x,n_y,n_z} \to \left(\frac{L}{2\pi}\right)^3 \int d\mathbf{k} = \left(\frac{L}{2\pi}\right)^3 \int k^2 dk d\Omega_k. \tag{24.20}$$

I can then talk about a *density of states per unit solid angle* $\rho(E, \Omega_k)$ by setting

$$\left(\frac{L}{2\pi}\right)^3 \int k^2 dk = \int \rho(E, \Omega_k) dE. \tag{24.21}$$

With $k_E = \sqrt{2mE/\hbar^2}$ and m the electron mass, I find that $\rho(E, \Omega_k)$ is then given by

$$\rho(E, \Omega_k) = \frac{mL^3}{8\pi^3\hbar^2} k_E. \tag{24.22}$$

According to Eq. (24.6), I must evaluate k_E at

$$k_E = k_f = \sqrt{\frac{2m}{\hbar}(\omega_L - \omega_0)} \tag{24.23}$$

where

$$\hbar\omega_0 = 13.6 \text{ eV} = -E_0 \tag{24.24}$$

is the ionization energy of hydrogen. It is assumed that $\omega_L \gg \omega_0$, but that $(\omega_L - \omega_0) / (\omega_L + \omega_0) \ll 1$ to insure the validity of the RWA.

If I take the field along the z axis, then

$$\hat{V}(t) = e\mathcal{E}_0 \hat{z} \cos(\omega_L t) = e\mathcal{E}_0 \widehat{r \cos\theta} \cos(\omega_L t), \tag{24.25}$$

where $\mathcal{E}_0$ is the magnitude of the field amplitude. From Eq. (24.12), I can calculate the *photoionization rate per unit solid angle* as

$$\Gamma(\Omega_k) = \frac{d\Gamma}{d\Omega_k} = \frac{2\pi}{\hbar} \left|V(E_f, \Omega_k)\right|^2 \rho(E_f = \hbar^2 k_f^2/2m, \Omega_k) \tag{24.26}$$

where

$$V(E_f, \Omega_k) = \frac{1}{2} \int d\mathbf{r} \frac{e^{i\mathbf{k}\cdot\mathbf{r}}}{\sqrt{L^3}} r\cos\theta \frac{e^{-r/a}}{\sqrt{\pi a^3}}\bigg|_{k=k_f}. \tag{24.27}$$

As you can see, the matrix element $V(E_f)$ depends on the direction of emission $\mathbf{k}$ of the electron.

To evaluate the matrix element, I use the spherical wave expansion

$$e^{i\mathbf{k}\cdot\mathbf{r}} = 4\pi \sum_{\ell=0}^{\infty} \sum_{m=-\ell}^{\ell} i^\ell j_\ell(kr) \left[Y_\ell^m(\theta, \phi)\right]^* Y_\ell^m(\theta_k, \phi_k), \tag{24.28}$$

where (θ, ϕ) are the spherical angles of $\mathbf{r}$ and (θ_k, ϕ_k) are the spherical angles of $\mathbf{k}$. Since $\cos\theta = \sqrt{\frac{4\pi}{3}} Y_1^0(\theta, \phi)$, the integral over angles (θ, ϕ) in Eq. (24.27) is nonvanishing only if $\ell = 1$ and $m = 0$, yielding

$$\int d\mathbf{r} e^{i\mathbf{k}\cdot\mathbf{r}} r\cos\theta e^{-r/a} = 4\pi i\sqrt{\frac{3}{4\pi}} Y_1^0(\theta_k, \phi_k) \int_0^\infty dr j_1(kr) r^3 e^{-r/a}$$
$$= 4\pi i \frac{8a^5 k\cos\theta_k}{(1+k^2a^2)^3}. \tag{24.29}$$

By combining Eqs. (24.26), (24.22), (24.27), and (24.29) I arrive at

$$\Gamma(\Omega_k) = \frac{64mk_f^3 a_0^7 e^2 \mathcal{E}_0^2 \cos^2\theta_k}{\pi\hbar^3\left(1+k_f^2a_0^2\right)^6}. \tag{24.30}$$

The total ionization rate is

$$\Gamma = \int d\Omega_k \Gamma(\Omega_k) = \frac{4\pi}{3}\frac{64mk_f^3 a_0^7 e^2 \mathcal{E}_0^2}{\pi\hbar^3\left(1+k_f^2a_0^2\right)^6}. \tag{24.31}$$

Note that

$$k_f a_0 = \sqrt{\frac{2mca_0^2}{\hbar}\frac{(\omega_L - \omega_0)}{c}} \approx \sqrt{\frac{2mca_0}{\hbar}\frac{\omega_L a_0}{c}}$$
$$= \sqrt{\frac{2mca_0}{\hbar} k_L a_0} \approx 17\sqrt{k_L a_0}. \tag{24.32}$$

For the dipole approximation to be valid, it is necessary that $k_L a_0 \ll 1$. On the other hand, the theory is probably valid only in the limit that $k_f a_0 \gg 1$. Thus the result should not be viewed as an accurate description, except for a limited range of parameter space.

24.1.2 Spontaneous Decay

As a second example, I calculate the spontaneous emission rate at which an atom decays from an excited state to its ground state. To solve this problem it is necessary to quantize the radiation field since the continuum states that lead to decay are those of the vacuum field. You can find discussions of field quantization in most any book on quantum optics. I will not present any formal derivation of field quantization. A standard treatment involves the use of periodic boundary conditions for the field.

Once the normal modes of the field are found (in this case, plane waves subject to periodic boundary conditions), creation and annihilation operators are assigned to each field mode. The net effect is that the Hamiltonian for the quantized field can be written as

$$\hat{H}_{\text{field}} = \sum_{\mathbf{k}_n} \hbar\omega_{k_n} a^{\dagger}_{\mathbf{k}_n} a_{\mathbf{k}_n} \tag{24.33}$$

where a_{k_n} ($a^{\dagger}_{k_n}$) is a destruction (creation) operator for mode n having frequency $\omega_{k_n} = ck_n$. The summation index $\mathbf{k}_n$ is meant to imply a summation over $\{n_x, n_y, n_z\}$ with

$$\mathbf{k}_n = \frac{2\pi n_x}{L}\mathbf{u}_x + \frac{2\pi n_y}{L}\mathbf{u}_y + \frac{2\pi n_z}{L}\mathbf{u}_z. \tag{24.34}$$

The corresponding electric field is

$$\mathbf{E}(\mathbf{R}) = i\sum_{\mathbf{k}_n} \left(\frac{\hbar\omega_{k_n}}{2\epsilon_0\mathcal{V}}\right)^{1/2} \boldsymbol{\epsilon}_{\mathbf{k}_n} \left(a_{\mathbf{k}_n} e^{i\mathbf{k}_n\cdot\mathbf{R}} - a^{\dagger}_{\mathbf{k}_n} e^{-i\mathbf{k}_n\cdot\mathbf{R}}\right), \tag{24.35}$$

where $\mathcal{V} = L^3$ is the quantization volume and $\boldsymbol{\epsilon}_{\mathbf{k}_n}$ is the field polarization unit vector. There are two, independent polarizations for each $\mathbf{k}$. The creation and destruction operators for the field are similar to those for the harmonic oscillator, having the same commutation relations. Instead of creating and destroying number states of the oscillator, they create and destroy *photon states* of the field. That is,

$$a^{\dagger}_{\mathbf{k}_n} |0\rangle = |1_{\mathbf{k}_n}\rangle, \tag{24.36}$$

where $|0\rangle$ is the *vacuum state* and $|1_{\mathbf{k}_n}\rangle$ is a state with one photon in mode $\mathbf{k}_n$ of the field. These are the only states I need to discuss spontaneous decay.

Rather than use Fermi's Golden Rule to derive an expression for the transition rate, I will calculate the state vector of the system as a function of time directly. The total Hamiltonian is given by

$$\hat{H} = \hat{H}_{\text{atom}} + \hat{H}_{\text{field}} + \hat{V}_{AF},$$

where $\hat{H}_{\text{atom}}$ is the atomic Hamiltonian, $\hat{H}_{\text{field}}$ is the free-field Hamiltonian given by Eq. (24.33), and $\hat{V}_{AF}$ is the atom–field interaction potential given by

$$\hat{V}_{AF} = -\hat{\mathbf{p}}_e \cdot \mathbf{E}(\mathbf{R} = \mathbf{0}), \tag{24.37}$$

where $\hat{\mathbf{p}}_e$ is the dipole moment operator of the atom. The nucleus of the atom is assumed to be fixed at the origin.

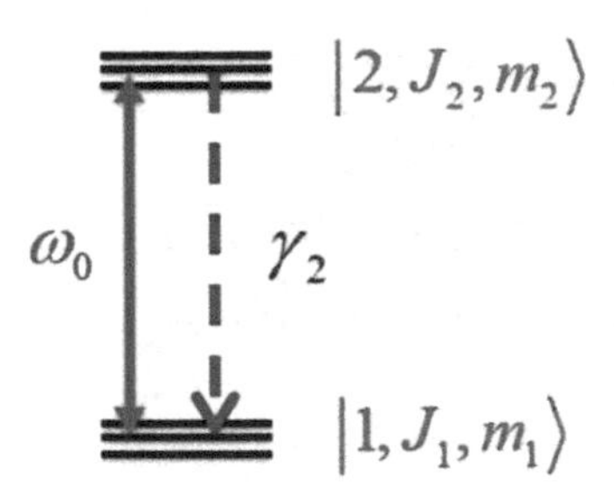

Fig. 24.2 Spontaneous emission between an excited and ground state manifold of levels

The eigenkets of $\hat{H}_{\text{atom}}$ are written as $|n, J_n, m_n\rangle$, where J_n is the total angular momentum of a manifold of levels in electronic state n and m_n a magnetic quantum number associated with J_n. In practice I consider emission from a state $|2, J_2, m_2\rangle$ which has been excited at time $t = 0$ to the manifold of ground states denoted by $|1, J_1, m_1\rangle$ (see Fig. 24.2). The transition is driven by the atom-vacuum field interaction. Of course, it is impossible to *instantaneously* excite the atom at a given time, so what I really mean is that the atom is excited in a time τ that is much less than the decay rate of the system, but much longer than the inverse of the transition frequency. It is important not to forget that it is simply a vacuum field induced decay process with which we are dealing, since the algebra can get a little messy.

Initially, the system is in the state $|2, J_2, m_2; 0\rangle$ where the 0 labels the vacuum state of the field. In the interaction representation, it follows from Eq. (24.7a) that the excited state amplitude evolves as

$$\dot{c}_{2,J_2,m_2;0}(t) = \frac{1}{i\hbar} \sum_{\mathbf{k},\boldsymbol{\epsilon}_{\mathbf{k}}} \sum_{m_1} e^{-i(\omega_k - \omega_0)t} \times \langle 2, J_2, m_2; 0 | \hat{V}_{AF} | 1, J_1, m_1; \mathbf{k}, \boldsymbol{\epsilon}_{\mathbf{k}} \rangle c_{1,J_1,m_1;\mathbf{k},\boldsymbol{\epsilon}_{\mathbf{k}}}(t) . \quad (24.38)$$

where $\omega_0 = \omega_{21} = (E_2 - E_1)/\hbar$ is the transition frequency, $\{\mathbf{k}, \boldsymbol{\epsilon}_{\mathbf{k}}\}$ labels a one-photon state of the field having propagation vector $\mathbf{k}$ and polarization $\boldsymbol{\epsilon}_{\mathbf{k}}$. The sum over $\boldsymbol{\epsilon}_{\mathbf{k}}$ refers to a sum over the two independent field polarizations for each $\mathbf{k}$ and the sum over $\mathbf{k}$ is a shorthand notation for summing over all $\mathbf{k}_n$. A formal solution for the ground state amplitude, obtained using Eq. (24.7b), is

$$c_{1,J_1,m_1;\mathbf{k},\boldsymbol{\epsilon}_{\mathbf{k}}}(t) = \frac{1}{i\hbar} \int_0^t e^{i(\omega_k - \omega_0)t'} \sum_{J_2', m_2'} \times \langle 1, J_1, m_1; \mathbf{k}, \boldsymbol{\epsilon}_{\mathbf{k}} | \hat{V}_{AF} | 2, J_2', m_2'; 0 \rangle c_{2,J_2',m_2';0}(t') , \quad (24.39)$$

which, when substituted back into Eq. (24.38), yields

$$\dot{c}_{2,J_2,m_2;0}(t) = -\frac{1}{\hbar^2}\sum_{\mathbf{k},\boldsymbol{\epsilon}_\mathbf{k}}\sum_{J_2',m_2'}\int_0^t e^{-i(\omega_k-\omega_0)(t-t')}c_{2,J_2',m_2';0}(t')$$
$$\times\langle 2,J_2,m_2;0|\hat{V}_{AF}|1,J_1,m_1;\mathbf{k},\boldsymbol{\epsilon}_\mathbf{k}\rangle$$
$$\langle 1,J_1,m_1;\mathbf{k},\boldsymbol{\epsilon}_\mathbf{k}|\hat{V}_{AF}|2,J_2',m_2';0\rangle\,, \tag{24.40}$$

where I have allowed for transitions back to another level J_2' in the same state 2 electronic state manifold.

The sum over **k** is converted to an integral using

$$\sum_{\mathbf{k}_n}\to\frac{\mathcal{V}}{(2\pi)^3}\int d^3k = \frac{\mathcal{V}}{(2\pi)^3}\int_0^\infty k^2dk\int d\Omega_k$$
$$= \frac{\mathcal{V}}{(2\pi)^3}\int_0^\infty\frac{\omega_k^2}{c^3}d\omega_k\int d\Omega_k. \tag{24.41}$$

In this continuum limit, Eq. (24.35) is replaced by

$$\mathbf{E}(\mathbf{R}=\mathbf{0}) = i\frac{1}{(2\pi)^3}\int_0^\infty\frac{\omega_k^2}{c^3}d\omega_k\int d\Omega_k\left(\frac{\hbar\omega_k\mathcal{V}}{2\epsilon_0}\right)^{1/2}\boldsymbol{\epsilon}_\mathbf{k}\left(a_\mathbf{k}-a_\mathbf{k}^\dagger\right), \tag{24.42}$$

By combining Eq. (24.37) with Eqs. (24.40)–(24.42), I find that the quantization volume cancels (as it must, if the results are to make any sense) and that I am faced with an integral of the form

$$\int_0^t dt'\int_0^\infty\omega_k^3 d\omega_k e^{-i(\omega_k-\omega_0)(t-t')}c_{2,J_2',m_2';0}(t'). \tag{24.43}$$

Although the integral over ω_k diverges, it is not unreasonable to cut off the integral at some value of $(\omega_k-\omega_0)$ of order ω_0.[1] With such a cutoff, the major contribution to the ω_k integral comes from a region $|\omega_k-\omega_0|\sim\gamma\ll\omega_0$. This allows me to replace ω_k^3 by ω_0^3, remove it from the integral and extend the lower integration limit of the ω_k integral to $-\infty$. These two approximations constitute the so-called *Weisskopf-Wigner approximation*,[2] and are equivalent to the Markov

[1]Based on theoretical considerations, the cutoff frequency could be determined by the range of validity of the dipole approximation, $ka_0=\omega_k a_0/c\approx 1$, which gives an ω_k(cutoff) of order 10^{18} s^{-1}. Based on *experimental* considerations, the cutoff can be taken at $\omega_k(\text{cutoff})=\omega_0+(1/T)$, where $\gamma\ll(1/T)\ll\omega_0$ and T is the time it takes to excite the atom to its ground to excited state [see P. R. Berman, *Wigner-Weisskopf approximation under typical experimental conditions*, Physical Review A **72**, 025804 (2005)].

[2]V. Weisskopf and E. Wigner, *Berechnung der natürlichen Linienbreite auf Grund der Diracschen Lichttheorie*, Zeitschrift für Physik 63, pp. 54–73 (1930); a translation is available, *Calculation of the Natural Line Width on the Basis of Dirac's Theory of Light,* in W. R. Hindmarsh, *Atomic*

approximation used in arriving at Fermi's Golden Rule. In the Weisskopf–Wigner approximation,

$$\begin{aligned}
&\int_0^t dt' \int_0^\infty \omega_k^3 d\omega_k e^{-i(\omega_k-\omega_0)(t-t')} c_{2,J_2',m_2';0}(t') \\
&\approx \omega_0^3 \int_0^t dt' \int_{-\infty}^\infty d\omega_k e^{-i(\omega_k-\omega_0)(t-t')} c_{2,J_2',m_2';0}(t') \\
&= 2\pi\omega_0^3 \int_0^t \delta\left(t-t'\right) c_{2,J_2',m_2';0}(t')dt' = \pi\omega_0^3 c_{2,J_2',m_2';0}(t).
\end{aligned} \tag{24.44}$$

Combining this result with Eq. (24.37) with Eqs. (24.40)–(24.42), I find

$$\begin{aligned}
\dot{c}_{2,J_2,m_2;0}(t) = &-\frac{1}{\hbar}\frac{\pi\omega_0^3}{2\epsilon_0(2\pi)^3c^3}\sum_{J_2',m_2'}\int d\Omega_k \\
&\times\sum_{\epsilon_{\mathbf{k}}}\langle 2,J_2,m_2|\hat{\mathbf{p}}_e\cdot\boldsymbol{\epsilon}_{\mathbf{k}}|1,J_1,m_1\rangle \\
&\times\langle 1,J_1 m_1|\hat{\mathbf{p}}_e\cdot\boldsymbol{\epsilon}_{\mathbf{k}}|2,J_2',m_2'\rangle c_{2,J_2',m_2';0}(t)\,.
\end{aligned} \tag{24.45}$$

I now need to carry out the angular integration. To do so I must write explicit expressions for the unit polarization vectors. The unit vectors $\boldsymbol{\epsilon}_{\mathbf{k}}^{(1)}$, $\boldsymbol{\epsilon}_{\mathbf{k}}^{(2)}$, and $\hat{\mathbf{k}}$ make up a right-handed system with

$$\hat{\mathbf{k}} = \sin\theta_k\cos\phi_k\mathbf{u}_x + \sin\theta_k\sin\phi_k\mathbf{u}_y + \cos\theta_k\mathbf{u}_z\,; \tag{24.46a}$$

$$\boldsymbol{\epsilon}_{\mathbf{k}}^{(1)} = \hat{\boldsymbol{\theta}}_k = \cos\theta_k\cos\phi_k\mathbf{u}_x + \cos\theta_k\sin\phi_k\mathbf{u}_y - \sin\theta_k\mathbf{u}_z\,; \tag{24.46b}$$

$$\boldsymbol{\epsilon}_{\mathbf{k}}^{(2)} = \hat{\boldsymbol{\phi}}_k = -\sin\phi_k\mathbf{u}_x + \cos\phi_k\mathbf{u}_y\,. \tag{24.46c}$$

It is then straightforward to carry out the summations and integrations in Eq. (24.45) to obtain

$$\begin{aligned}
\sum_{\lambda=1}^{2}\sum_{i,j=1}^{3} d_i\bar{d}_j \int d\Omega_k \left(\boldsymbol{\epsilon}_{\mathbf{k}}^{(\lambda)}\right)_i \left(\boldsymbol{\epsilon}_{\mathbf{k}}^{(\lambda)}\right)_j &= \frac{8\pi}{3}\sum_{i,j}\delta_{ij}d_i\bar{d}_j \\
&= \frac{8\pi}{3}\mathbf{d}\cdot\bar{\mathbf{d}},
\end{aligned} \tag{24.47}$$

Spectra (Pergamon Press, London, 1967), pp. 304–327. For an extensive discussion of the validity of the Weisskofp-Wigner approximation, see the article by Paul R. Berman and George W. Ford, *Spontaneous Decay, Unitarity, and the Weisskopf-Wigner Approximation*, in *Advances in Atomic, Molecular, and Optical Physics*, edited by E. Arimondo, P. R. Berman, and C. C. Lin (Elsevier-Academic Press, New York, 2010), volume **59**, pp. 175–221.

where

$$\mathbf{d} = \langle 2, J_2, m_2|\hat{\mathbf{p}}_e|1, J_1, m_1\rangle; \tag{24.48a}$$

$$\bar{\mathbf{d}} = \langle 1, J_1 m_1|\hat{\mathbf{p}}_e|2, J_2', m_2'\rangle. \tag{24.48b}$$

I have denoted the first and second matrix elements in Eq. (24.45) by $\mathbf{d}$ and $\bar{\mathbf{d}}$, respectively.

To proceed further, I write the components of the matrix elements $\mathbf{d}$ and $\bar{\mathbf{d}}$, as well as those of the dipole moment operator, in spherical form as

$$\hat{p}_{e1} = -\frac{\hat{p}_{ex} + i\hat{p}_{ey}}{\sqrt{2}}, \quad \hat{p}_{e,-1} = \frac{\hat{p}_{ex} - i\hat{p}_{ey}}{\sqrt{2}}, \quad \hat{p}_{e0} = \hat{p}_{ez}, \tag{24.49a}$$

$$d_1 = -\frac{d_x + id_y}{\sqrt{2}}, \quad d_{-1} = \frac{d_x - id_y}{\sqrt{2}}, \quad d_0 = d_z, \tag{24.49b}$$

such that

$$\mathbf{d} \cdot \bar{\mathbf{d}} = \sum_q (-1)^q d_q \bar{d}_{-q} \,. \tag{24.50}$$

I can now use the Wigner-Eckart theorem and the relationship

$$\left(\hat{p}_{e,q}\right)^\dagger = (-1)^q \hat{p}_{e,-q} \tag{24.51}$$

to evaluate

$$\begin{aligned} d_q &= \langle 2, J_2, m_2|\hat{p}_{e,q}|1, J_1, m_1\rangle \\ &= \frac{1}{\sqrt{2J_2+1}} \begin{bmatrix} J_1 & 1 & J_2 \\ m_1 & q & m_2 \end{bmatrix} \langle 2, J_2\|p^{(1)}\|1, J_1\rangle \,, \end{aligned} \tag{24.52}$$

and

$$\begin{aligned} \bar{d}_{-q} &= \langle 1, J_1 m_1|\hat{p}_{e,-q}|2, J_2', m_2'\rangle \\ &= \left(\langle 2, J_2', m_2'|\left(\hat{p}_{e,-q}\right)^\dagger|1, J_1, m_1\rangle\right)^* \\ &= (-1)^q \left(\langle 2, J_2' m_2'|\hat{p}_{e,q}|1, J_1, m_1\rangle\right)^* \\ &= \frac{1}{\sqrt{2J_2+1}} (-1)^q \begin{bmatrix} J_1 & 1 & J_2 \\ m_1 & q & m_2' \end{bmatrix} \left(\langle 2, J_2\|p_e^{(1)}\|1, J_1\rangle\right)^* . \end{aligned} \tag{24.53}$$

It then follows that

$$\sum_{q,m_1}(-1)^q d_q \bar{d}_{-q} = \frac{1}{2J_2+1}|\langle 2, J_2\|p_e^{(1)}\|1, J_1\rangle|^2$$
$$\times \sum_{q,m_1}\begin{bmatrix} J_1 & 1 & J_2 \\ m_1 & q & m_2 \end{bmatrix}\begin{bmatrix} J_1 & 1 & J_2' \\ m_1 & q & m_2' \end{bmatrix}$$
$$= \frac{1}{2J_2+1}|\langle 2, J_2\|p_e^{(1)}\|1, J_1\rangle|^2 \delta_{J_2,J_2'}\delta_{m_2,m_2'}, \tag{24.54}$$

where the orthogonality property of the Clebsch-Gordan coefficients has been used. The sum is proportional to $\delta_{J_2,J_2'}\delta_{m_2,m_2'}$. Incorporating Eqs. (24.47) and (24.54) into Eq. (24.45), I find the upper state amplitude decays as

$$\dot{c}_{2,J_2,m_2;0}(t) = -\gamma c_{2,J_2,m_2;0}(t), \tag{24.55}$$

which has as solution [given $c_{2,J_2,m_2;0}(0) = 1$],

$$c_{2,J_2,m_2;0}(t) = e^{-\gamma t}, \tag{24.56}$$

where

$$\gamma = \gamma_{2,J_2;1,J_1}/2 = \frac{2}{3}\frac{\alpha_{FS}}{2J_2+1}|\langle 2, J_2\|r^{(1)}\|1, J_1\rangle|^2 \frac{\omega_{21}^3}{c^2}, \tag{24.57}$$

α_{FS} is the fine structure constant, $\langle 2, J_2\|r^{(1)}\|1, J_1\rangle$ is the reduced matrix element for the position operator, and $\gamma_{2,J_2;1,J_1}$ is the *decay rate* from state $|2, J_2\rangle$ to $|1, J_1\rangle$. For dipole allowed optical transitions from an excited state to the ground state, $\gamma_{2,J_2;1,J_1}/\omega_{21}$ is of order 10^{-7}; that is, decay rates from the first excited states of atoms back to the ground state are typically in the nanosecond to tens of nanoseconds range.

Using the quantized vacuum field, I have derived a fundamental and important equation for the spontaneous emission rate. The fact that $\dot{c}_{2,J_2,m_2;0}$ is not coupled to $c_{2,J_2',m_2';0}$ for $J_2' \neq J_2$ or $m_2' \neq m_2$ can be understood simply in terms of conservation of angular momentum. States in the excited state manifold differing in total angular momentum or the z-component of angular momentum cannot be coupled by "emitting" and "absorbing" the *same* photon. Equation (24.57) also contains the important result that the decay rate is proportional to the cube of the transition frequency.[3] I had to invoke the Weisskopf-Wigner approximation to carry out the calculation. When used consistently, the Weisskopf-Wigner approximation

[3]This assumes that the reduced matrix element is independent of frequency as it is for atoms. For an oscillator, however, the square of the reduced matrix element varies inversely with the transition frequency such that the decay rate depends on the square of the transition frequency.

results in overall conservation of probability for the atom-field system. You can use Eqs. (24.39) and (24.56) to evaluate $\sum_{m_1} \left|c_{1,J_1,m_1;\mathbf{k},\epsilon_\mathbf{k}}(\infty)\right|^2$, which provides a measure of the spectral and directional properties of the emitted radiation (see problems).

24.2 Bounded Continuum

In many situations, the continuum is *bounded*. For example, in spontaneous decay, you cannot have emission at negative frequencies—the continuum is bounded from below at $E = 0$. When the continuum is bounded, there can never be purely exponential decay and the exact solutions are very complicated. However, there are two limiting cases that can be solved approximately. If the energy of the continuum encompasses the discrete state (as it does in spontaneous emission since the discrete state has energy $\hbar\omega_0$), and the width of the continuum is much larger than the decay rate (as in spontaneous emission), then the decay is exponential to a very good approximation. In this limit, the infinite continuum model provides a good approximation to the exact result.

The second case is very different. Imagine in the photoelectric effect that you send in a field having frequency ω_L for which $\hbar\omega_L$ is *less* than the work function $E_a = \hbar\omega_a$. In that case, you might think that the initial state amplitude remains equal to unity and this is *approximately* true. But some of the initial state amplitude is lost. The situation is similar to that in off-resonant Rayleigh scattering, where there is a small, but non-vanishing, excited state probability amplitude. To model this effect, I take the zero of energy such that $E_0 = 0$ and assume that the continuum extends from energy $E_a = \hbar\omega_a$ to $E_b = \hbar\omega_b$, with $E_b > E_a > \hbar\omega_L$. In other words, the field frequency is not sufficiently high to effectively drive transitions from the initial state into the continuum.

The coupling matrix elements are taken as $V_{E0}(t) = V_{E0}\cos(\omega_L t)$. I assume the effect of the field is sufficiently weak to allow me to solve Eq. (24.7b) with $c_0(t) \approx 1$. In that case, I find

$$c_E(t) \approx -\frac{i}{\hbar}V_{E0}\int_0^t dt' e^{i\omega_E t'}\cos\left(\omega_L t'\right) \approx \frac{V_{E0}}{2\hbar}\frac{\left(1 - e^{i(\omega_E-\omega_L)t}\right)}{\omega_E - \omega_L}, \tag{24.58}$$

where $\omega_E = E/\hbar$ and I have assumed that $\omega_E - \omega_L \ll \omega_E + \omega_L$. The initial state probability is then given by

$$|c_0(t)|^2 \approx 1 - \int_{-\infty}^{\infty} dE \left|\frac{V_{E0}}{\hbar}\right|^2 \rho(E)\frac{\sin^2\left(\frac{(\omega_E-\omega_L)t}{2}\right)}{(\omega_E - \omega_L)^2}$$

$$= 1 - \left|\frac{V}{\hbar}\right|^2 \rho \int_{E_a}^{E_b} dE \frac{\sin^2\left(\frac{(\omega_E - \omega_L)t}{2}\right)}{(\omega_E - \omega_L)^2}, \tag{24.59}$$

where, for simplicity, I assumed that both the matrix elements and density of states are constant over the energy range of the continuum. This can be rewritten as

$$\begin{aligned} |c_0(t)|^2 &\approx 1 - \frac{|V|^2 \rho}{\hbar} \int_{\omega_{aL}}^{\omega_{bL}} d\omega \frac{\sin^2\left(\frac{\omega t}{2}\right)}{\omega^2} \\ &= 1 - \frac{|V|^2 \rho}{\hbar \omega_{aL}} \int_1^{\omega_{bL}/\omega_{aL}} dx \frac{\sin^2\left(\frac{\omega_{aL} t}{2} x\right)}{x^2}, \end{aligned} \tag{24.60}$$

where

$$\omega_{aL} = \omega_a - \omega_L; \qquad \omega_{bL} = \omega_b - \omega_L. \tag{24.61}$$

This whole treatment is valid only if $|c_0(t)|^2 \approx 1$, so the correction term must be small. It *is* small, provided

$$\beta = \frac{|V|^2 \rho}{2\hbar \omega_{aL}} \ll 1, \tag{24.62}$$

a condition that is obtained by replacing $\sin^2(\omega t/2)$ by $1/2$ in the integrand, but β is *not* equal to zero. Let me take $\omega_b = \infty$. The integral is a tabulated function, but not particularly simple. The behavior of $|c_0(t)|^2$ *is* simple, however. It starts from a value of unity, oscillates and decays to a final value equal to

$$P_0(\infty) = |c_0(\infty)|^2 \approx 1 - \frac{1}{2\hbar \omega_{aL}} |V|^2 \rho \int_1^{\infty} dx \frac{1}{x^2} = 1 - \beta. \tag{24.63}$$

A graph of $P_0(t) = |c_0(t)|^2$, with $|c_0(t)|^2$ given by Eq. (24.60), is shown in Fig. 24.3 for $\beta = 0.1$ as a function of $\omega_{aL} t$. The oscillations appear because the field is turned on suddenly at $t = 0$. If, instead, the matrix coupling element is of the form $V_{E0}(t) = V(t) \cos(\omega_L t)$, where $V(t)$ increases from an initial value of zero to a final value of V in a time that is long compared with ω_{aL}^{-1}, $P_0(t)$ would not be oscillatory; it would decay smoothly as $1 - |V(t)|^2 \rho / (4\hbar \omega_{aL})$, assuming a final value of $1 - \beta/2$, instead of $1 - \beta$ (see problems).

If we apply this idea to photoionization with a field frequency below the ionization frequency, it looks like we violate energy conservation if an electron is ionized. Actually, the "free" electron amplitude oscillates rapidly between zero and a small value for *each* possible final energy state of the ionized electron. Thus the "free" electron cannot get very far away from the atom since each excited state returns to the initial state periodically. To measure the free electron you would have to make a measurement on a time scale less than the inverse of the oscillation

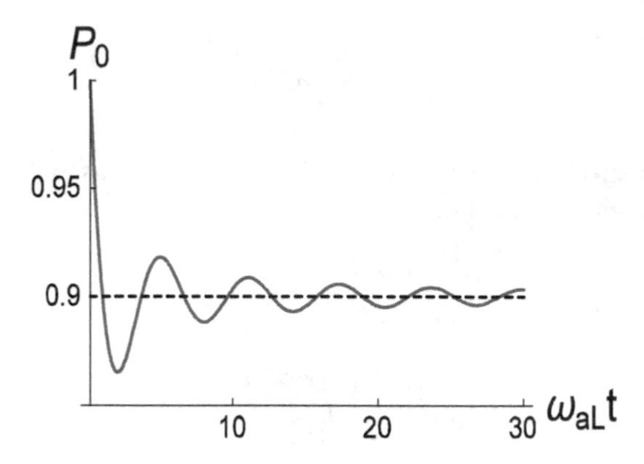

Fig. 24.3 Initial state probability P_0 as a function of $\omega_{aL}t$ for $\beta = 0.1$. The asymptotic value $P_0 \sim 0.9$ is indicated by the dashed line

frequency which would introduce Fourier components to compensate for the energy mismatch. The situation is the same for the anti-resonant component of the vacuum field interacting with a ground state atom. The "emitted" field never gets very far from the atom. On the other hand, such fields, which are of quantum origin, can be viewed as being responsible for the van der Waals interaction. Two ground state atoms separated by a distance less than the wavelength associated with a ground to excited state transition in the atoms "emit" fields that lead to an interaction energy between the atoms.

24.3 Zeno Paradox and Zeno Effect

It has become fashionable for introductory and graduate quantum mechanics texts to mention the *quantum Zeno paradox*.[4] The original Zeno paradoxes consist of a number of scenarios in which it appears that motion is impossible. For example, in the stadium paradox, the question is raised as to how long it will take to cross a stadium. To cross the stadium, you must first reach the midway point. But to reach the midway point, you must first reach the quarter way point, and so on. Since you have to reach an infinite number of points to cross the stadium, it will take you an infinite time to cross the stadium. The "paradox" is that you know it takes a finite time to cross the stadium. Although it is obvious to us that we can traverse an infinite number of points in a finite time, Zeno's paradoxes were troubling to the ancients—Aristotle spent considerable time trying to refute them.

[4]See, for example, David Griffiths, *Introduction to Quantum Mechanics*, Second Edition (Pearson Prentiss Hall, Upper Saddle River, N.J., 2005) (Sect. 12.5) and Eugen Merzbacher, *Quantum Mechanics,* Third Edition (John Wiley and Sons, New York, 1998), Sect. 19.8.

Just like the ancients were confused by Zeno's paradoxes, it appears that the quantum Zeno paradox is confusing to contemporary scientists. Part of the reason for this is related to semantics. The quantum Zeno paradox was originally formulated by Misra and Sudarshan in terms of a bubble chamber experiment.[5] Imagine that an unstable particle enters a bubble chamber and leaves a track of bubbles. Each bubble is an indication that the particle has not yet decayed. In some sense, therefore, the bubbles constitute measurements in which the particle is projected into its original state. By increasing the bubble density, it would seem that the particle is continuously projected into its initial state—it does not decay at all! The paradox is that, in the experiment, the particle decays with its normal decay rate, independent of any bubbles. What is going on?

I have shown that a Markovian decay process has no memory—decay is always exponential. For a Markovian process, you cannot change the decay rate, period. There is no Zeno paradox in this limit. The reason is simple. To modify the decay rate, you must act on the quantum system many times within the *correlation time* of the bath. For a Markovian process the correlation time of the bath giving rise to the relaxation is zero, so it is impossible to inhibit decay. Of course, no decay process is strictly Markovian. However, processes such as spontaneous emission and particle decay are close enough to Markovian to render any attempts to inhibit decay useless.

Why, then, is there all this interest in the Zeno effect. I like to distinguish the *Zeno effect* from the *Zeno paradox*. I have just resolved the Zeno paradox—there can be no change in spontaneous decay rates produced by measurements on the system owing to the Markovian nature of the decay. On the other hand, there can be a Zeno *effect* for coherently driven transitions. I like to illustrate this by asking someone to explain something to me. As soon as she starts talking, I begin to yell gibberish so she must stop and begin the explanation anew. While not really an example of the quantum Zeno effect, it provides the central idea behind the effect. To cause a transition from one state to another, an interaction must build up a phase. We have already seen this in the optical Bloch equations where the application of a pulse having area equal to π leads to an inversion of a two level quantum system. If, during the pulse, you apply some *other* pulses, it is possible to inhibit the coherent build-up of the phase responsible for level inversion.[6] There is nothing magical about this—in fact you often try to *avoid* processes that lead to phase decoherence. In certain quantum information protocols, you can reverse the effects of phase decoherence using the quantum Zeno effect, but this is not phase

[5]B. Misra and E. C. G. Sudarshan, *The Zeno's paradox in quantum theory*, Journal of Mathematical Physics **18**, 756–763 (1977).

[6]In fact, a revival of interest in the Zeno effect was generated by an article by W. M. Itano, D. J. Heinzen, J. J. Bollinger, and D. J. Wineland, *Quantum Zeno effect*, Physical Review **41**, 2295–2300 (1990). In that article, they showed that excitation of a long-lived state from the ground state using by a π pulse could be totally inhibited if an auxiliary field is used to drive a coupled ground to dipole-allowed optical transition. The observation of the spontaneously emitted radiation (or lack therof) on the dipole-allowed transition constituted the "measurements" on the atom.

decoherence caused by Markovian processes. Rather it is phase decoherence caused by effects such as fluctuating magnetic or electric fields whose coherence time is finite.

Misra and Sudarshan proved that, *under certain conditions*, any quantum system can have its decay inhibited by continuously measuring the system. To understand their argument and to see why it fails for Markovian processes, imagine that there is a Hamiltonian $\hat{H}$ that characterizes a quantum system in which some initial discrete state $|\psi_0\rangle$ is coupled to a continuum. We can ask for the probability that we measure the quantum system in the state $|\psi_0\rangle$ at time t, predicated on the fact that we measured in state $|\psi_0\rangle$ at times $t_1, t_2 \dots t_n$. This probability is given by

$$\begin{aligned} P(t) = & \left|\langle\psi_0| e^{-i\hat{H}t_1/\hbar} |\psi_0\rangle\right|^2 \left|\langle\psi_0| e^{-i\hat{H}(t_2-t_1)/\hbar} |\psi_0\rangle\right|^2 \cdots \\ & \times \cdots \left|\langle\psi_0| e^{-i\hat{H}(t-t_n)/\hbar} |\psi_0\rangle\right|^2 . \end{aligned} \tag{24.64}$$

Misra and Sudarshan showed rigorously that $P \sim 1$ as $n \sim \infty$, provided $\hat{H}$ is a Hermitian and semi-bounded operator (e.g., $\hat{H}$ is semi-bounded if all its eigenenergies are greater than or equal to some energy E_0). In other words, in the limit of *continuous* measurement on the system, the initial state never decays! They then went on to say that a bubble chamber does not really constitute *continuous* measurement of the undecayed particle, which explains why the decay is not inhibited.

For early times when $\omega_{n0}t \ll 1$, it follows from Eqs. (24.7) that

$$c_n(t) \approx -\frac{i}{\hbar} V_{n0} t; \tag{24.65a}$$

$$c_0(t) \approx 1 - \sum_n \frac{|V_{0n}|^2 t^2}{2\hbar^2}. \tag{24.65b}$$

If the sum is replaced by an integral over continuum states, it follows that

$$|c_0(t)|^2 \approx 1 - t^2/t_c^2, \tag{24.66}$$

where

$$t_c = \frac{\sqrt{2}}{\left[\int_{-\infty}^{\infty} dE \left|\frac{V(E)}{\hbar}\right|^2 \rho(E)\right]^{1/2}} \tag{24.67}$$

can be viewed as the correlation time of the bath. If t_c is *finite*, Eq. (24.64) reduces to

$$P(t) = \left|1 - t_1^2/t_c^2\right| \left|1 - (t_2 - t_1)^2/t_c^2\right| \cdots \left|1 - (t - t_n)^2/t_c^2\right| . \tag{24.68}$$

Taking equal time intervals $(t_{j+1} - t_j) = t/n$ and letting $n \sim \infty$, I find

$$P(t) = \lim_{n \sim \infty} \left|1 - t^2/n^2 t_c^2\right|^n = \lim_{n \sim \infty} e^{-t^2/nt_c^2} = 1. \tag{24.69}$$

As long as t_c is finite, continuous measurement produces a quantum Zeno effect. For a Markovian process, however, $\rho(E)$ and $V(E)$ are constant for $-\infty < E < \infty$, implying that $t_c = 0$. In that limit, I found that, for $\Gamma t \ll 1$

$$|c_0(t)|^2 \approx 1 - \Gamma t, \tag{24.70}$$

which implies that, for a Markovian process,

$$P(t) = \lim_{n \sim \infty} |1 - \Gamma t/n|^n = e^{-\Gamma t}. \tag{24.71}$$

There is no quantum Zeno effect for a Markovian process. Spontaneous emission is very close to a Markovian process, for which t_c is of order of an inverse optical frequency, making it all but impossible to make "continuous" measurements on such a system.

24.4 Summary

The important problem of transitions from a discrete state to a continuum of states has been studied. In such cases, irreversible behavior can occur, such as that observed in photoemission and spontaneous decay. Fermi's Golden Rule can be used to evaluate the decay rate, although a better picture of the decay processes can be obtained by solving the equations of motion for the state amplitudes and forming the appropriate probability distributions for the final and initial states. Although it might appear to be possible to inhibit such decay by constant measurements on a quantum system, I have shown that this is not possible for Markovian processes.

24.5 Problems

1–2. Estimate the $2P$–$1S$ and the $2S$–$2P$ decay rates in hydrogen using Eq. (24.57). You will need to use the Wigner-Eckart theorem and look up the wave functions and frequency spacings of the levels.

3. For two fine structure levels within the same electronic state manifold, the ratio of the decay rates to some lower level is approximately equal to the ratio of cube of the transition frequencies of the two transitions. Estimate the ratio of the decay

rates for the D2 ($\lambda = 588.995$ nm) and D1 ($\lambda = 589.522$ nm) transitions in Na. The experimental ratio is approximately equal to 1.0028.[7]

4. Prove that

$$\sum_{\lambda=1}^{2} \int d\Omega_k \sum_i \left(\boldsymbol{\epsilon}_{\mathbf{k}}^{(\lambda)}\right)_i d_i \sum_j \left(\boldsymbol{\epsilon}_{\mathbf{k}}^{(\lambda)}\right)_j \bar{d}_j = \frac{8\pi}{3} \sum_i d_i \bar{d}_i .$$

5–6. The frequency spectrum $P(\omega_k)\, d\omega_k$ of the radiation emitted in spontaneous decay is given by the probability that a photon is emitted into a mode of the radiation field having that frequency. In terms of continuous frequency variables,

$$P(\omega_k)\, d\omega_k = \frac{\mathcal{V}}{(2\pi)^3} \frac{\omega_k^2}{c^3} \sum_{m_1=-J_1}^{J_1} \sum_{\lambda=1}^{2} \int d\Omega_k \left| c_{1,J_1,m_1;\mathbf{k},\boldsymbol{\epsilon}_{\mathbf{k}}^{(\lambda)}}(\infty) \right|^2 d\omega_k,$$

where $\mathcal{V}$ is the quantization volume. Show that $P(\omega_k)$ is proportional to ω_k^3 times a Lorentzian function of ω_k. In the Weisskopf-Wigner approximation prove that all the initial energy in the atom is converted into the energy of the radiated field,

$$\int_0^\infty \hbar\omega_k P(\omega_k)\, d\omega_k = \hbar\omega_0 n_2(0),$$

where $n_2(0)$ is the initial excited state probability in the atom. Thus, when applied in a consistent manner, the Weisskopf-Wigner approximation leads to conservation of energy, although there was no guarantee that this would be the case.

7–8. Complementary to the frequency spectrum of spontaneous decay is the radiation pattern $I_{\mathbf{k},\boldsymbol{\epsilon}_{\mathbf{k}}^{(\lambda)}} d\Omega_k$ (direction and polarization of radiation emitted into an element of solid angle) given by

$$I_{\mathbf{k},\boldsymbol{\epsilon}_{\mathbf{k}}^{(\lambda)}} d\Omega_k = \frac{\mathcal{V}}{(2\pi)^3} \frac{1}{c^3} \sum_{m_1=-J_1}^{J_1} \int_0^\infty \omega_k^2 d\omega_k \left| c_{1,J_1,m_1;\mathbf{k},\boldsymbol{\epsilon}_{\mathbf{k}}^{(\lambda)}}(\infty) \right|^2 d\Omega_k.$$

In the Weisskopf-Wigner approximation, show that if an atom is prepared in the $m = 0$ sublevel of a $J_2 = 1$ excited state and decays to a $J_1 = 0$ ground state, the radiation pattern is the same as that emitted by a classical dipole oscillator whose dipole moment is in the z-direction.

---

[7]U. Volz, M. Majerus, H. Liebel, A. Schmitt, and H. Schmoranzer, *Precision Lifetime Measurements on NaI $3p^2P_{1/2}$ and $3p^2P_{3/2}$ by Beam-Gas-Laser Spectroscopy,* Physical Review Letters **76**, 2862 (1996)].

9. An atom undergoes spontaneous decay with a central wavelength of 600 nm and a lifetime of 100 ns. Estimate the Rabi frequency if the field from this atom interacts with a similar atom one cm away. If, instead, the single photon pulse can be focused to an area of λ^2 when it strikes the atom, show that the Rabi frequency is of the order of the decay rate and the pulse area is of order unity; that is, a single photon pulse focused to a wavelength can fully excite an atom.

10–11. A particle having mass m moves in a one-dimensional potential

$$V(x) = \begin{cases} -V_0 & x \leq |a|/2 \\ 0 & \text{otherwise} \end{cases},$$

where V_0 is a positive constant.

Assume that the particle is in the ground state of the potential and, in addition, there is a perturbative contribution to the Hamiltonian of the form

$$\hat{V}(x,t) = \alpha \hat{x} \cos(\omega t),$$

where $\hbar\omega$ is much greater than the magnitude of the ground state energy and α is a constant. Calculate the rate at which the particle is "ionized" (escapes from the potential into unbound states). You can leave your answer in terms of the ground state energy E_1 and an integral involving the ground state eigenfunction $\psi_{E_1}(x)$. You need not evaluate the integral. [Hint: To solve this problem, assume that the final states can be approximated as free particle states. You need to calculate the density of states for these free particle states in one-dimension.]

12–13. In the calculation of Sect. 24.3 for a bounded continuum, I assumed that the field was turned on instantaneously at $t = 0$. Suppose, instead, that the coupling matrix element is $V_{E0}(t) = V(t)\cos(\omega_L t)$, where $V(t)$ is a smooth function that increases from an initial value of zero at $t = 0$ to a final value of V in some characteristic time T. Repeat the calculation of Sect. 24.3 with $\omega_b = \infty$ and show that

$$P_0(t) \approx 1 - \frac{|V(t)|^2 \rho}{4\hbar\omega_{aL}}, \tag{24.72}$$

provided $\omega_{aL} T \gg 1$. In other words, the approach to the steady-state value occurs without oscillations if $V(t)$ is turned on adiabatically with respect to the frequency ω_{aL}.

Now derive an expression for $P_0(t)$ by solving Eq. (24.7b) with $c_0(t) \approx 1$ and

$$V(t) = \begin{cases} 0 & t < 0 \\ V\left(1 - e^{-t/T}\right) & t \geq 0 \end{cases}.$$

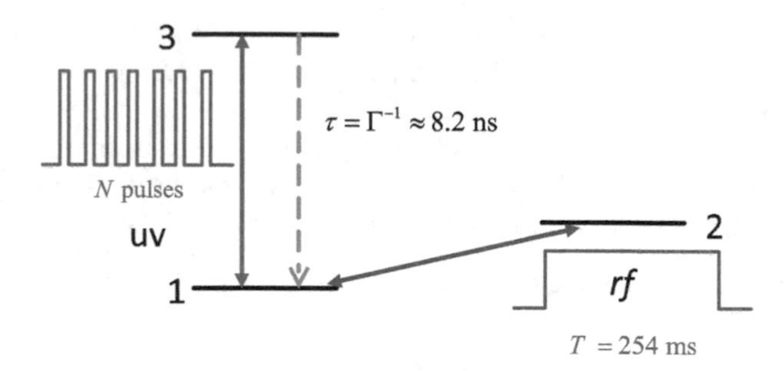

Fig. 24.4 Level scheme for Problem 24.14. Level 3 is the $2^2P_{3/2}$ that decays back to level 1. The parameters shown refer to the article of Itano et al. In their experiment the uv pulse width was 2.4 ms and up to $N = 64$ pulses were used

Plot P_0 as a function of $\omega_{aL}t$ for $\beta = |V|^2 \rho/(2\hbar\omega_{aL}) = 0.1$ and $1/\omega_{aL}T = 0.1$ and compare the result with the prediction of Eq. (24.72). There are still some very small oscillations owing to the fact that the derivative of $V(t)$ is not continuous at $t = 0$. Also plot P_0 as a function of $\omega_{aL}t$ for $\beta = 0.1$ and $1/\omega_{aL}T = 40$ (sudden turn-on of field) and compare the result with that shown in Fig. 24.3. Note that the asymptotic value for $P_0(\infty)$ is $1 - \beta$ for a sudden turn-on of the field and $1 - \beta/2$ for an adiabatic turn on of the field. The parameter

14. In the experiment of Itano et al.,[8] a transition between two, $2^2S_{1/2}$, ground state hyperfine levels of a Be ion is driven by a radio-frequency (rf) field. Let the initial state be denoted by 1 and the final state by 2. A π pulse having duration $T = 256$ ms transfers the population from state 1 to 2, assuming no other interactions played a role. To demonstrate a Zeno effect, Itano et al. added a number of ultraviolet pulses that could excite state 1 to a $2^2P_{3/2}$ level that undergoes rapid spontaneous decay back to level 1 on a time scale of order 10 ns. (see Fig. 24.4). The optical pulses are sufficiently long (but still have duration $\ll T$) and intense to insure that several spontaneously emitted photons can be observed, provided the $2^2P_{3/2}$ level is excited during the pulse. The emission of the spontaneous radiation (or lack thereof) is said to effect a "measurement" on the atom. The ion "collapses" into state 1 if the emission occurs and into state 2 if no emission occurs (no emission implies the ion must have been in state 2 when the uv pulse was applied). If N such equally spaced pulses are applied, calculate the population of the initial $2^2S_{1/2}$ at time $t = T$. Show that in the limit $N \to \infty$, this population approaches zero—there is a Zeno effect produced by the measurement pulses. To carry out this calculation, you can assume that at $t = 0$ the Bloch vector is $(u(0), v(0), w(0)) = (0, 0, -1)$. Following the first measurement pulse at time t_1, the Bloch vector "collapses" into $(u(t_1), v(t_1), w(t_1)) = (0, 0, w(t_1))$, since there is a probability $\rho_{11}(t_1)$ that the

[8] W. M. Itano, D. J. Heinzen, J. J. Bollinger, and D. J. Wineland, *Quantum Zeno effect*, Physical Review **41**, 2295–2300 (1990).

ion is in state 1 and a probability $\rho_{22}(t_1)$ that the ion is in state 2 at the time of the measurement. The ion then evolves freely under the influence of the rf pulse until the second measurement pulse at time t_2 "collapses" the Bloch vector into $(u(t_2), v(t_2), w(t_2)) = (0, 0, w(t_2))$; and so on.[9]

[9] As you know I am not a fan of the "collapse" picture. It is not necessary to invoke the collapse picture to arrive at the final result, a simple density matrix calculation gives the same result. See Ellen Block and P. R. Berman, *Quantum Zeno effect and quantum Zeno paradox in atomic physics*, Physical Review A **44**, 1466–1472 (1991).

Index

P.R. Berman, *Introductory Quantum Mechanics*, UNITEXT for Physics,
https://doi.org/10.1007/978-3-319-68598-4

S

CPSIA information can be obtained
at www.ICGtesting.com
Printed in the USA
LVHW061355090619
620638LV00005B/452/P

9 783319 886282